MEMBRANE
MIMETIC CHEMISTRY

MEMBRANE MIMETIC CHEMISTRY

CHARACTERIZATIONS AND APPLICATIONS OF MICELLES, MICROEMULSIONS, MONOLAYERS, BILAYERS, VESICLES, HOST-GUEST SYSTEMS, AND POLYIONS

JANOS H. FENDLER
Department of Chemistry
Clarkson College of Technology

A Wiley-Interscience Publication

JOHN WILEY & SONS

New York • Chichester • Brisbane • Toronto • Singapore

Library of Congress Cataloging in Publication Data:

Fendler, Janos H.
Membrane mimetic chemistry.

"A Wiley-Interscience publication."
Includes bibliographies and index.
1. Membranes (Technology) I. Title. [DNLM:
1. Chemistry, Physical. 2. Surface-Active agents.
QD 506 F331m]
TP159.M4F46 660.2′8423 82-2583

ISBN 0-471-07918-9 AACR2

Printed in the United States of America

10 9 8 7 6 5 4 3 2

PREFACE

This book heralds the appearance of a new discipline: membrane mimetic chemistry. This explosively growing research area is directed to the practical exploitation of membrane-mediated processes in relatively simple chemical systems. The catch-phrase has been coined to describe the philosophy of approach. Although the label is recognized as imperfect, no apology is offered for its introduction. The acute need for bringing together all aspects of organized assemblies—membrane mimetic agents—has prompted the writing of this book. As indicated in the subtitle, micelles, microemulsions, monolayers, bilayers, vesicles, host-guest systems, and polyions have been used as membrane mimetic agents.

Increasing numbers of scientists with different training are entering into membrane mimetic chemistry. They employ a variety of approaches and frequently use different terminologies. This book is primarily addressed to them. It is sincerely hoped that synthetic, physical, colloid, polymer, and biological chemists, as well as physicists, biophysicists, pharmacologists, and engineers, will find the subject stimulating enough to consider making their own contributions. My desire to reach such a broad audience has undoubtedly led to uneven treatment of topics. To the expert, sections relating to his or her own competence may appear to belabor obvious points, while to the neophyte the same sections may not appear to supply sufficient detail.

By and large, the depth of topic coverage in this book is inversely proportional to the availability of recent reviews. A serious attempt has been made to present all points of view and to provide all pertinent references to original publications and review articles. The literature survey is current through the first half of 1981. Typically, the vast majority of papers referred to have been published within the last five years. Exhaustive data compilations are also provided in tables in the book to facilitate quick and critical grasp of published works. Naturally, the explosion of activity would result in exponential growth of such tables if they were to be updated in a few years time. I think this is a healthy situation. The interested researcher will find it relatively easy to fit new data into the conceptual framework provided in the book.

A large number of people have critically read earlier versions of the book. Particularly, Drs. M. Almgren, E. W. Anacker, H. F. Eicke, J. B. F. N. Engberts, P. Fromherz, E. Goddard, D. Jager, L. Magid, G. S. Manning, H. Morawetz, C. J. O'Connor, L. S. Romsted, J. Sagiv, Z. A. Schelly, H. T. Tien, and D. G. Whitten have provided helpful criticisms and pointed out omissions.

My sincere thanks to all these good people for giving their time so unselfishly. I alone accept responsibility, of course, for any imperfections or misrepresentations. I subscribe to the hackneyed but very true statement, "If one would wait until producing the perfect piece of work, one would never produce anything."

An enormous amount of typing, retyping, editing and proofreading are involved in a book of this size. Ms. Rhonda Wichman most competently translated my illegible scribbling into a very respectable manuscript. She, along with Nelson Prieto, repeatedly proofread the manuscript. Their contributions have been invaluable.

JANOS H. FENDLER

Potsdam, New York
January, 1982

CONTENTS

PART 2. APPLICATION OF MEMBRANE MIMETIC AGENTS

MEMBRANE MIMETIC CHEMISTRY

PART 1

CHARACTERIZATION OF MEMBRANE MIMETIC AGENTS

CHAPTER 1 INTRODUCTION

Biological membranes provide compartments of defined sizes, shapes, and microenvironments. They organize living matter in the cell, create a fluid two-dimensional matrix, and allow for the controlled transport of solutes. Such cellular functions as recognition, fusion, endocytosis, exocytosis, intercellular interaction, excitability, translocation, transport, and osmosis are all membrane mediated processes. Higher organisms contain intracellular membranes and perform specialized functions (Jain and Wagner, 1980; Bittar, 1980).

Membranes of plant and animal cells are typically composed of 40–50% lipids and 50–60% proteins. There are wide variations in the types of lipids and proteins as well as in their ratios. Arrangements of lipids and proteins in membranes are best considered in terms of the fluid-mosaic model, proposed by Singer and Nicolson (1972). According to this model, the matrix of the membrane, a lipid bilayer composed of phospholipids and glycolipids, incorporates proteins, either on the surface or in the interior, and acts as a permeability barrier (Figure 1.1). Molecules are free to diffuse laterally in the plane of the membrane. This often produces selective ordering and segregated *domain* formation. There is a variety of molecular motions within the membrane matrix. Lipids undergo rotation and segmental motions, kink formation, and transverse motion from one interface to the other (flip-flop). Lipids contain a hydrophobic moiety, generally an aliphatic double chain, phosphate or carboxylate ester polar headgroups, and intermediate regions where hydrogen bonding can occur (Figure 1.1).

Much of our chemical understanding of membrane structures has been obtained through the investigation of models (Khorana, 1980). Surfactant monolayers (Gaines, 1966; Goddard, 1975; Gershfeld, 1976) and bilayers (Tien, 1974), as well as phospholipid vesicles (Bangham, 1968; Tyrrell et al., 1976; Chapman, 1980; Quinn and Chapman, 1980), have been used most extensively as membrane models. It should, however, be realized that no model is perfect or is able to mimic faithfully all aspects of complex membrane assemblies. Important differences have been noted, for example, between liposomes and biological membranes (Conrad and Singer, 1979). Using hygroscopic desorption, these authors found neligible binding of small amphiphatic molecules to various cell membranes. Conversely, appreciable binding constants have been found for the interaction of the same molecules with phospholipid vesicles. Differences have

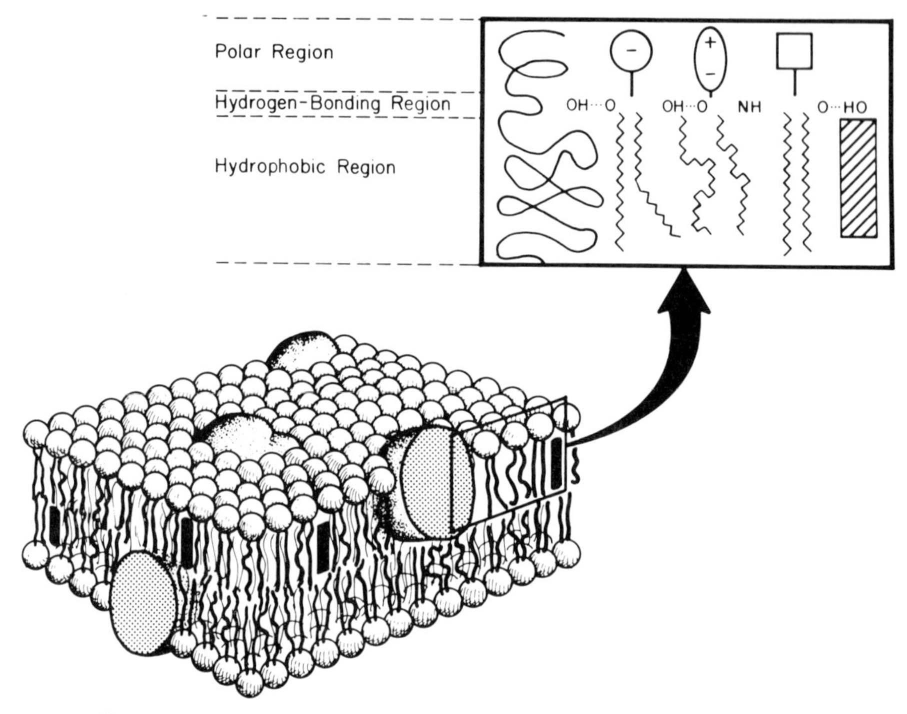

Figure 1.1. Fluid-mosaic model of a cell membrane (taken from Fendler, 1980).

been ascribed to the presence in membranes of a large internal pressure that precludes the uptake of amphiphatic molecules. Integral proteins, present in membranes, are, at least partially, responsible for generating the internal pressure. In the absence of proteins in lipid vesicles there is no internal pressure to prevent the incorporation of surfactants (Conrad and Singer, 1979). Consequently, many biophysical properties determined on membrane models may have little relevance to intact membranes.

There is, however, a more important objective in studying membrane models. It is to develop novel chemistry of practical utility based on mimicking membrane mediated processes in relatively simple systems. Mother nature need not be slavishly reproduced. Advantage can be taken of micelles, monolayers, bilayers, vesicles, host-guest systems, and polyions, referred to collectively as membrane mimetic agents, to organize substrates, to alter microenvironments and reactivities, as well as to act as carriers. Micelles, monolayers, bilayers, and vesicles are typically formed from readily available, easily purifiable, and well-defined surfactants. Membrane mimetic agents have been utilized in reactivity control, photochemical solar energy conversion and storage, molecular recognition and transport, drug encapsulation, and in providing unique environments for substrates and enzymes. Emphasis is placed on these aspects of membrane mimetic chemistry in Chapters 10–14. Chapters 2–9 summarize properties of the different membrane mimetic agents.

REFERENCES

Bangham, A. D. (1968). *Prog. Biophys. Mol. Biol.* **18**, 29–95. Membrane Models with Phospholipids.

Bittar, E. E. (1980). *Membrane Structure and Function*, Wiley-Interscience, New York.

Chapman, D. (1980). In *Membrane Structure and Function* (E. E. Bittar, Ed.), pp. 103–152, John Wiley, New York. Studies Using Model Biomembrane Systems.

Conrad, M. J. and Singer, S. J. (1979). *Proc. Natl. Acad. Sci. USA* **76**, 5202–5206. Evidence for a Large Internal Pressure in Biological Membranes.

Fendler, J. H. (1980). *Acc. Chem. Res.* **13**, 7–13. Surfactant Vesicles as Membrane Mimetic Agents: Characterization and Utilization.

Gaines, G. L., Jr. (1966). *Insoluble Monolayers at Liquid-Gas Interfaces*, Interscience, New York.

Gershfeld, N. L. (1976). *Annu. Rev. Phys. Chem.* **27**, 349–368. Physical Chemistry of Lipid Films at Fluid Interfaces.

Goddard, E. D. (1975). *Monolayers, Advances in Chemistry Series*, Vol. 144, American Chemical Society, Washington, D.C.

Jain, M. H. and Wagner, R. C. (1980). *Introduction to Biological Membranes*, Wiley-Interscience, New York.

Khorana, H. G. (1980). *Bioorg. Chem.* **9**, 363–405. Chemical Studies of Biological Membranes.

Quinn, P. J. and Chapman, D. (1980). *CRC Crit. Rev. Biochem.* **8**, 1–117. The Dynamics of Membrane Structure.

Singer, S. J. and Nicolson, G. L. (1972). *Science* **175**, 720–731. Fluid Mosaic Model of the Structure of Cell Membranes.

Tien, H. T. (1974). *Bilayer Lipid Membranes (BLM) Theory and Practice*, Marcel Dekker, New York.

Tyrrell, D. A., Heath, T. D., Colley, C. M., and Ryman, B. E. (1976). *Biochim. Biophys. Acta* **457**, 259–302. New Aspects of Liposomes.

CHAPTER

2

SURFACTANTS IN WATER

Properties of aqueous micelles, formed in the dynamic association of large numbers of surfactant molecules in water, have been treated comprehensively in recent books and reviews (Schick, 1967; Shinoda, 1967; Elworthy et al., 1968; Jungermann, 1970; Mukerjee and Mysels, 1971; Berezin et al., 1973; Bunton, 1973, 1979; Tanford, 1973; Fendler and Fendler, 1975; Frank, 1975; Linfield, 1976; Fisher and Oakenfull, 1977; Menger, 1977, 1979a; Mittal, 1977, 1979; Kalyanasundaram, 1978; Goodman and Walker, 1979; Brown, 1979; Sudhölter et al., 1979; Wennerström and Lindman, 1979a; Lindman and Wennerström, 1980). Only a brief summary is provided, therefore, in this chapter. Attention is focused on principles that govern the use of micelles as membrane mimetic agents. Recent developments and future potentials are emphasized.

1. MICELLIZATION AND CRITICAL MICELLE CONCENTRATION

Surfactants, commonly called detergents, are amphiphatic molecules having distinct hydrophobic and hydrophilic regions. Depending on the chemical structure of their polar headgroups, surfactants can be neutral, cationic, anionic, or zwitterionic. The apolar moiety can be of differing lengths, contain unsaturated bond(s), and/or consist of two or more chains. Functional groups can also be incorporated into surfactants. Figure 2.1 shows structures of typical surfactants.

Over a narrow range of concentration there is a sudden transition in the physical properties of aqueous surfactants. This transition corresponds to the formation of aggregates and is used to define the critical micelle concentration or CMC. An IUPAC (International Union of Pure and Applied Chemists) commission has defined this and other terms used in miceller chemistry (IUPAC, 1972). A large number of methods exist for the determination of CMC (Fendler and Fendler, 1975). CMC values depend somewhat on the methods used for their determination. Positron annihilation has recently been shown to be an extremely sensitive method for detecting phase changes and determining CMC values (Jean and Ache, 1978; Ache, 1979; Jean et al., 1979). Neutral positrons, formed in the combination of positrons and electrons, exist in a singlet and triplet state (Ache, 1979). The lifetime and intensity of emission from triplet

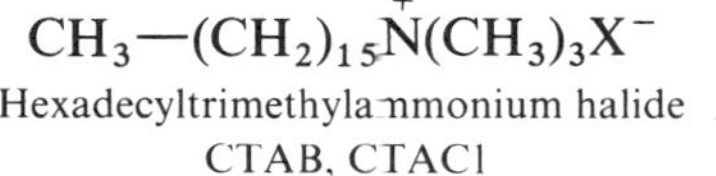

$CH_3-(CH_2)_{15}\overset{+}{N}(CH_3)_3X^-$

Hexadecyltrimethylammonium halide
CTAB, CTACl

$[CH_3-(CH_2)_{11}]_xSO_4^-M^+$

$x = 1$; M = Na	Sodium dodecyl sulfate, SDS
$x = 1$; M = Ag	Silver dodecyl sulfate, AgDS
$x = 2$; M = Cu	Copper dodecyl sulfate, $Cu(DS)_2$
$x = 2$; M = Co	Cobalt dodecyl sulfate, $Co(DS)_2$
$x = 2$; M = Ni	Nickel dodecyl sulfate, $Ni(DS)_2$

$CH_3(CH_2)_{15}SO_4^-Na^+$

Sodium hexadecyl sulfate, SHS

$CH_3(CH_2)_7CH{=}CH(CH_2)_7COO^-Na^+$

Sodium oleate

$CH_3(CH_2)_7(OCH_2CH_2)_6OH$

Polyoxyethylene(6) alcohol

$CH_3(CH_2)_3-CH(C_2H_5)CH_2-O-C(=O)-CH(SO_3^-Na^+)-C(=O)-O-CH_2CH(C_2H_5)-(CH_2)_3CH_3$

Sodium bis(2-ethylhexyl)sulfosuccinate,
aerosol-OT

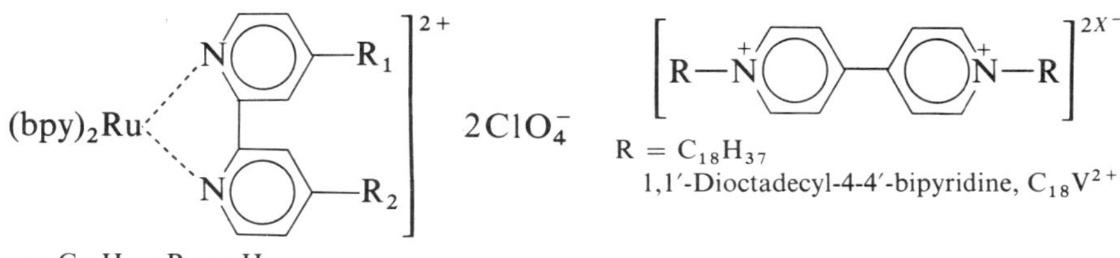

$[(bpy)_2Ru(R_1,R_2\text{-bpy})]^{2+}\ 2ClO_4^-$

$R_1 = C_{18}H_{37}$; $R_2 = H$
(*N*-octadecyl-2,2′-bipyridine)-bis(2,2′-bipyridine)ruthenium(II)$^{2+}$, $RuC_{18}(bpy)_3^{2+}$

$R_1 = R_2 = CO_2C_{18}H_{37}$
(*N*,*N*′-Dioctadecyl-2-2′-bipyridine)-bis(2,2′-bipyridine)ruthenium(II)$^{2+}$, $RIIC_{18}(bpy)_3^{2+}$

$R_1-C(=O)-O-CH_2-CH(-O-C(=O)-R_2)-CH_2-O-P(=O)(=O)-OX$

$R_1 = R_2 = C_{15}H_{31}$; $X = -CH_2CH(OH)CH_2OH$
3-*sn*-Dipalmitoylphosphatidyl-1′-*sn*-glycerol

$R_1 = R_2 = C_{15}H_{31}$; $X = -CH_2CH_2-\overset{+}{N}(CH_3)_3$
3-*sn*-Dipalmitoylphosphatidyl-1′-*sn*-cholin(lecithin)

$R_1 = R_2 = C_{15}H_{31}$; $X = -CH_2-OH_2-NH_2$
3-*sn*-Dipalmitoylphosphatidyl-1′-*sn*-ethanolamine

$[R-\overset{+}{N}C_5H_4-C_5H_4\overset{+}{N}-R]\ 2X^-$

$R = C_{18}H_{37}$
1,1′-Dioctadecyl-4-4′-bipyridine, $C_{18}V^{2+}$

$[CH_3-(CH_2)_{17}]_2\overset{+}{N}(CH_3)_2\ Cl^-$

Dioctadecyldimethylammonium chloride,
DODAC

$[CH_3-(CH_2)_{15}-O]_2P(O)O^-$

Dihexadecylphosphate, DHP

Figure 2.1. Structures of typical surfactants.

positrons are sensitive to phase changes. Since these species are neutral and since they are present only in negligible amounts, the system remains virtually unperturbed. Figure 2.2 illustrates typical plots of positronium triplet emission intensities as functions of surfactant concentrations. Mukerjee and Mysels (1971) have compiled CMC values of surfactants covering the literature until December 1966, and have evaluated the different methods used for their determinations. Table 2.1 lists CMC values for the most commonly used surfactants.

CMC values depend on the hydrophobicity of the hydrocarbon chain, on the net charge of the surfactant, on the nature of the polar head and counterion, and on the type and concentrations of added electrolytes. CMCs are also affected by the temperature, the pressure, and added solubilizates. This latter factor needs to be considered in experiments involving micelles. Addition of solubilizates, in general, lowers the CMC (Mukerjee and Mysels, 1971). As a first approximation ideal mixing between the solubilizates and the aggregated surfactant may be assumed. Accordingly, the CMC value is expected to be reduced by a factor proportional to the mole fraction of the solubilizate in the micelle. There are, however, pronounced deviations from this ideality for solubilizates that are present at high concentrations and/or that decrease the hydrophobic associations between monomeric surfactants. Reduction of hydrophobic interactions would result, of course, in decreased micellar stabilities and hence in increased CMCs. The situation is even more complex for ionic micelles. On the one hand, solubilizates at higher concentrations would tend to decrease the ionic strength and hence increase the CMC. On the other hand, they also increase the volume of

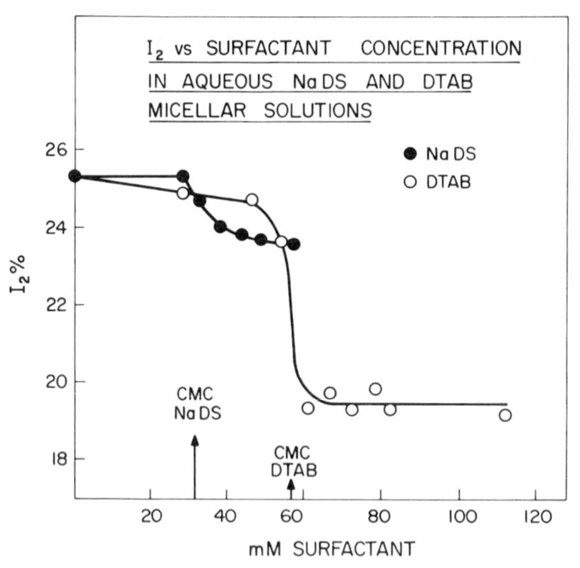

Figure 2.2. Positronium triplet emission intensities as functions of sodium decylsulfate (NaDS) and decyltrimethylammonium bromide (DTAB) concentrations (taken from Jean and Ache, 1978).

Table 2.1. CMC Values of Surfactants at 25°C

Surfactant	CMC
Cationic	
Decylammonium bromide	5.0×10^{-3} *M*
Dodecylammonium chloride	1.5×10^{-2} *M*
Dodecyltrimethylammonium bromide	1.5×10^{-2} *M*
Dodecylpyrodinium bromide	1.1×10^{-2} *M*
Hexadecyltrimethylammonium bromide (CTAB)	9.2×10^{-4} *M*
Anionic	
Sodium decanoate	9.4×10^{-2} *M*
Sodium dodecanoate	2.4×10^{-2} *M*
Sodium dodecyl sulfate (SDS)	8.1×10^{-3} *M*
Copper(II) dodecyl sulfate	1.2×10^{-3} *M*
Zwitterionic	
3-(Dimethyldodecylammonio)-propane-1-sulfonate (DDAPs)	1.2×10^{-3} *M*
N-Dodecyl-*N*,*N*-dimethylglycine	1.8×10^{-3} *M*
Nonionic	
Polyoxyethylene(6) octanol	9.9×10^{-3} *M*
Polyoxyethylene(9.5) octylphenol (Triton X-100)	3.0×10^{-4} *M*
Polyoxyethylene(15) nonylphenol (Igepal CO-730)	2.0×10^{-4} *M*
Polyoxyethylene(20–24) hexadecanol (Cetomacrogol)	7.7×10^{-5} *M*

the micelle, thereby decreasing the charge densities of the headgroups. Since micellization is driven by hydrophobic attractions between the hydrocarbon chains of the surfactants but opposed by repulsions between the headgroups (Tanford, 1973), decreasing the charge densities by substrate solubilization also decreases the tendency to oppose micellization.

Until recently (Fendler and Fendler, 1975), most applications of aqueous micelles were restricted to regions extending slightly below and slightly above the CMC. Interesting effects are being discovered at surfactant concentrations well below and well above the critical point. Background to these systems is provided in Sections 2.6 and 2.7.

2. STRUCTURES OF MICELLES

Although aqueous micelles have been investigated for more than six decades, detailed understanding of their structures only began to emerge recently (Menger, 1979a; Fromherz, 1980; Dill and Flory, 1981). The classical micelle (Hartley, 1935, 1948) is pictured as a roughly spherical aggregate containing

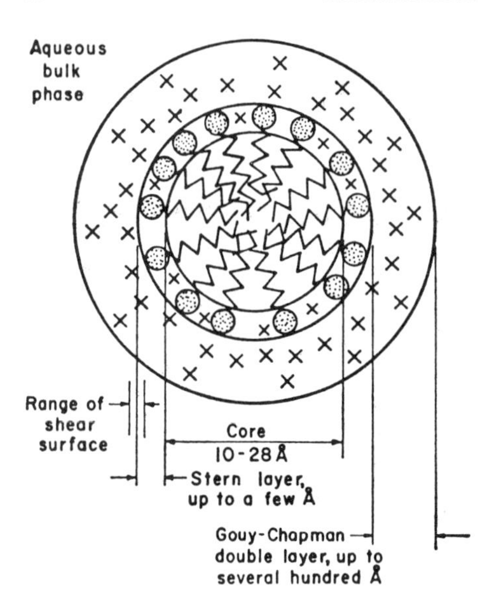

Figure 2.3. An oversimplified two-dimensional representation of a spherical ionic micelle. The counterions (×), the headgroups (☺), and the hydrocarbon chains (~) are schematically indicated to denote their relative locations but not their numbers, distribution, or configuration (taken from Fendler and Fendler, 1975).

50–200 monomers (Figure 2.3; Wennerström and Lindman, 1979a). The surface of the micelle, as pointed out by Hartley (1948), is much rougher than Figure 2.3 suggests. The radius of the sphere approximately corresponds to the extended length of the hydrocarbon chain of the surfactant. The micelle has a hydrocarbon core and a polar surface; it resembles an oil drop with a polar coat (Mukerjee and Cardinal, 1978). The headgroups and associated counterions of ionic micelles are found in the compact Stern layer. Some of the counterions are bound within the shear surface, but most are located in the Gouy–Chapman electrical double layer where they are dissociated from the micelle and are free to exchange with ions distributed in the bulk aqueous phase (Figure 2.3). The amount of free counterions in bulk water, expressed as the fractional charge, typically varies between 0.1–0.4.

The validity of the classical Hartley model (Figure 2.3) has been seriously questioned by Menger (1979a). He summarized the conflicting experimental data and interpretations concerning microviscosities, solubilization sites, and water penetration in micelles. Uncertainties in the determinations of miceller shapes, aggregation numbers, and polydispersities have also been pointed out. As an alternative to the Hartley model, Menger (1979a) has deduced micellar structures from CPK models. The model was constructed about a point of symmetry and empty space within the micellar core was held to a minimum. Figure 2.4 illustrates cationic dodecyltrimethylammonium halide micelles having aggregation numbers of 58 with fully extended and with somewhat compacted hydrocarbon chains. There are several important differences between the Hartley (1935, 1948) and the Menger (1979a) models. The latter model has a relatively small interior and it allows for considerable water penetration into the micellar interior. Interestingly, the extent of water-to-hydrocarbon contact does not appreciably

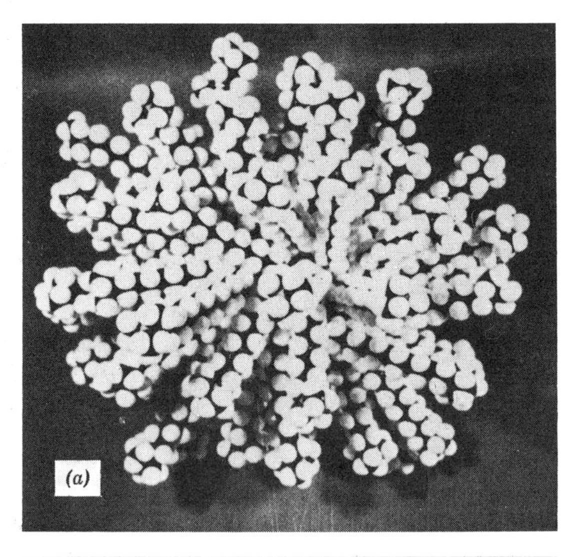

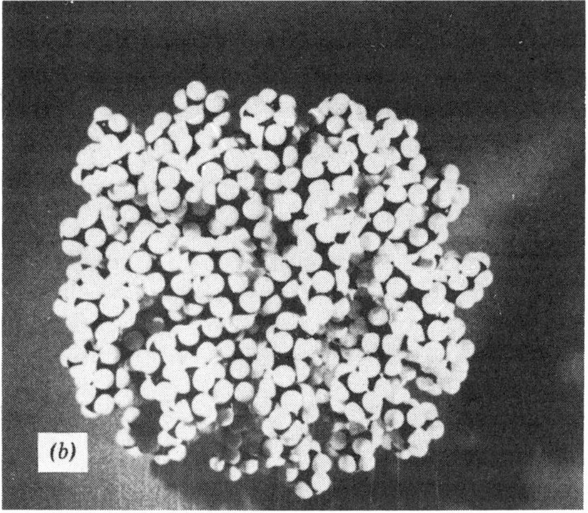

Figure 2.4. (*a*) A dodecyltrimethylammonium ion micelle with an aggregation number of 58. All chains are fully extended. A pyrene molecule is situated among the chains close to the micelle surface. (*b*) A dodecyltrimethylammonium ion micelle with an aggregation number of 58. Chains consist mainly of trans conformations but possess one or more "kinks" that shorten the chains (taken from Menger, 1979a).

decrease upon compacting of the micelle (Figure 2.4). Chain folding results in deeper inward penetration of the headgroups and in a concomitant exposure of larger portions of the hydrocarbon chains at the water surface. Headgroup protrusion has also been deduced from the kinetic treatment (see Chapter 2.4) of micellization (Aniansson, 1978). In the Menger model the micellar surface is much more rugged and the Stern layer is more poorly defined than in the Hartley model. These characteristics are in accord with the experimentally observed

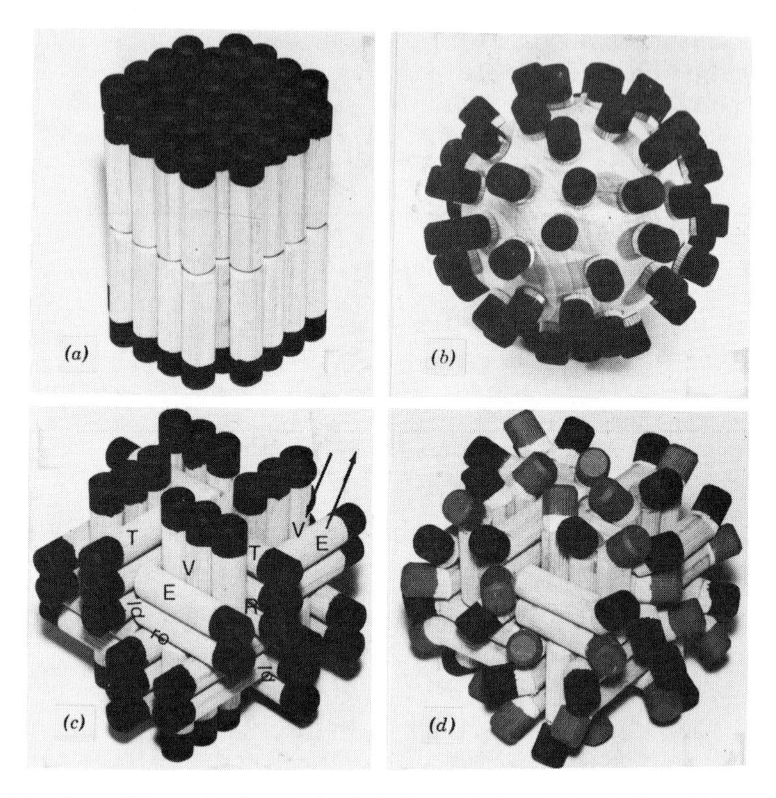

Figure 2.5. Space filling molecular models of micelles made from dodecylsulfate with aggregation number 64. The surfactant molecules are represented by cylindrical sticks. The relative length of hydrocarbon and headgroup (black) is taken from Pauling–Corey models. The ratio of diameter and length of hydrocarbon is chosen such that the ratio of volume to length in a square lattice equals the experimental ratio. This construction implies an average contribution of some gauge conformations to the fairly extended hydrocarbon chain. A square lattice is chosen because of the nature of packing in the final surfactant-block micelle. Evaluation in a triagonal lattice would lead to slightly thicker sticks. (*a*) Bilayer-fragment model: The molecules are assembled in a square lattice as closely as possible to a cylindrical shape. The hydrocarbon core is isometric. The hydrocarbon-water surface corresponds on average to about two wetted methylenes per surfactant molecule. (*b*) Compact droplet: The model is constructed on the same scale as the bilayer by coalescing the volume of the hydrocarbon chains up to the α-carbon, which protrudes from the core together with the headgroup. The radius of the core equals 1.2 times the length of the hydrocarbon chain up to the α-carbon. The hydrocarbon-water surface of the sphere between the headgroups corresponds to about two wetted methylenes per surfactant monomer. (*c*) The assembly obtained after application of rules 1 and 2. Saturating cube stoichiometry is 72. Some vacancies (V), edge sites (E), radial (R), and tangential (T) sites are indicated. Growth beyond the cube, to rods and plates, proceeds by insertion of blocks at the perimeters indicated (ro/pl). Dissociation and association reactions are marked by arrows. (*d*) The assembly obtained after application of rule 3. Bending of the models behind the headgroup corresponds to one gauge conformation. Headgroup contact is released without affecting the core. Counterion binding is indicated by the grey tone of half of the headgroups, which are neutralized on average. Shape, diameter, and headgroup distribution of this block-assembly correspond on average closely to the ideal droplet as in upper right (taken from Fromherz, 1980).

predominant solubilization sites of hydrophobic molecules in relatively aqueous environments within the micelle. Even large hydrocarbons fit with apparent ease into the polar grooves without disturbing the micellar structure. Viscosity and solvent polarity values, determined in different regions of the micelle, can also be rationalized in terms of the proposed model (Menger, 1979a). The kinetic behavior of micellar catalyzed reactions (Fendler and Fendler, 1975) is likewise explicable using the Menger model. Taking into consideration effective reagent concentrations in the micellar environments, second order rate constants are not appreciably altered with respect to those in water (Martinek et al., 1977; Romsted, 1977; Bunton, 1979). These results imply, of course, aqueous like solubilization sites for reactants that are provided by the extensive micelle-water interface.

Structures of micelles have been rationalized in terms of a surfactant-block model (Fromherz, 1980). Figure 2.5 shows space filling models of dodecyl sulfate micelles. The following rules were used to define micellar structures:

1. The orientation of the surfactant molecules in a micelle is parallel correlated. The width of a correlated surfactant block equals the length of the hydrocarbon chain up to the hydrated α-carbon, the chain being in an extended conformation perturbed by a few gauche-trans-gauche kinks.
2. The surfactant blocks are assembled with headgroups separated as far as possible, preferentially at right angles.
3. Steric contact of headgroups within blocks is released by a gauche conformation near the headgroup. Electrostatic repulsion within blocks is released by counterion binding.

The features of the surfactant block model were delineated as follows (Fromherz, 1980):

1. Surfactant molecules in a micellar assembly are assigned to sites in an isometric aperiodic lattice. A molecular parameter, the hydrocarbon chain length measured in units of the hydrocarbon chain diameter (chain valency v), governs the stoichiometry. The saturating aggregation number m_s of an isometric micelle is $m_s = 2(2v)^2$. Headgroup repulsion leads to vacancies in the cube spanned by the chains such that the actual aggregation number m is smaller than m_s.
2. The micelle structure consists of a large set of configurations with similar state of the surfactant components and with similar energy. Fluctuations beyond this set of compact configurations towards more open ones is low because dislocation energy by one methylene is distinctly higher than the thermal energy kT for all bound surfactant molecules.
3. The spherical average corresponds to the classical droplet shape. The cuboid deformation ensures appropriate packing within an average radius exceeding the length of a hydrocarbon chain.
4. Low surface energy and low headgroup repulsion are combined with low conformation energy and perfect packing. The thermodynamic description as a droplet is valid as a first approximation.

5. Conformation, mobility and interactions of the hydrocarbon chains resemble those in lamellar lyotropic liquid-crystalline phases and lipid membranes. Similar microbulk properties are a consequence.

6. Non-identical sites of molecules exist, as central (radial) sites with hydrocarbon completely buried in the core and peripheral (tangential) sites with high lateral water contact. An identification of chain coordinate with micellar radius is not possible.

7. Occupied and vacant sites are in a steady equilibrium by internal rearrangement and by the dissociation/association reaction. The aggregation reaction proceeds preferentially through surfactants and vacancies at the edges.

8. The entire hydrocarbon chain of all molecules is wettable in time average. The average number of wetted methylenes per chain is given by the wetted α-carbon, by the smooth surface of saturating cube configuration—about two methylenes—and by roughness.

9. Roughness of the hydrocarbon core is created by discrete vacancies and by headgroup bending. The latter increases with headgroup size. The headgroups are protruding beyond the rough core. Hydrated counterions are condensed in the grooves.

10. The whole sequence of aggregation from monomer to cuboid micelle follows block-assembly rules—with reduced block size at intermediate aggregation numbers (and with antiparallel correlation at small aggregation). Also growth beyond saturating cube stoichiometry follows block-assembly rules: Blocks are inserted along one (for rods) or along two perpendicular (for plates) perimeters of the cube [Figure 2.5]. Due to locally enhanced headgroup density the formation of those structures requires small headgroups.

Molecular organization of surfactants has also been discussed in terms of a statistical theory using a lattice model (Dill and Flory, 1980, 1981). The micelle is considered to comprise J_1 chains, each with $n + 1$ segments connected by n flexible bonds. The segments were assigned to locations on a three-dimensional curved lattice such that each lattice site accommodates one chain segment. The lattice was constructed with constant radial interlayer spacing and with equal volumes of lattice sites (Figure 2.6; Dill and Flory, 1981). It should be realized that the lattice representation of molecular configurations is artificial. The function of the lattice is to apportion the volume equitably to all radii. Configurational freedom of the alkyl chains is greatest at the outer surface of the micelle regardless of the chain length. This results in ordered center and disordered surface of the micelle. The lattice model supports Menger's postulate of high degree of water penetration into the micellar core; for a 14-carbon chain in a spherical micelle there is a substantial probability that a methylene group even halfway down the chain may occur in the outermost layer of the lattice (Dill and Flory, 1981).

Models, even sophisticated ones, cannot truly represent all the functions of that which they were intended to mimic. They serve as mental crutches to help our understanding. Hartley's model can be considered to be a time-averaged representation of several dynamic structures. The spherical average of all surfactant block configurations can be considered, for example, to describe the

Figure 2.6. Lattice model representation. Because the diagrams are two-dimensional, they most nearly resemble the cross-section of a cylindrical micelle (Dill and Flory, 1981).

classical droplet micelle. On the whole, the classical model underestimates the extent of water penetration, The CPK, the surfactant block, and the lattice models, on the other hand, overestimate somewhat the extent of water penetration. Micellar models stimulate experimentation and provide frameworks for the rationalization of the data obtained. Within the last couple of years, fluorescence and light scattering spectroscopy have contributed most significantly toward the experimental verification of micellar models. Methods and results pertaining to the elucidation of structural parameters are discussed here.

Application of luminescence techniques to micelles has been reviewed (Grätzel and Thomas, 1976; Thomas, 1977; Kalyanasundaram, 1978; Turro et al., 1980a). Attention is focused here on the determination of mean aggregation numbers, $\bar{n}$ (Turro and Yekta, 1978; Infelta, 1979; Yekta et al., 1979; Lianos and Zana, 1980). The method is based on quenching of the fluorescence intensity of a probe, P, by a quencher, Q. Conditions need to be carefully adjusted. Both P and Q must be exclusively localized in the micellar pseudophase. The concentration of P and the excitation light intensity must be kept as low as possible. The excited state of P, P^*, must be rapidly and completely quenched by Q (static quenching, manifested in a single exponential decay of fluorescence intensity, which is independent of [Q]). Under these conditions, emission occurs only from that fraction of P^* containing micelles that are free of Q. If the distribution of quenchers in micelles obeys Poisson statistics (Almgren et al., 1979a; Infelta and

Grätzel, 1979), the measured luminescence intensity of P^* in the absence ($I0$) and in the presence (I) of Q is given by

$$\frac{I}{I0} = \exp\left\{-\frac{[Q]}{[M]}\right\} \tag{2.1}$$

where the micelle concentration, [M], is related to the total (stoichiometric) detergent concentration, [DET], and to the concentration of free monomers, $[S_1]$, by

$$[M] = \frac{[DET] - [S_1]}{\bar{n}} \tag{2.2}$$

Combination of equations 2.1 and 2.2 leads to

$$\ln\frac{I0}{I} = \frac{[Q]\bar{n}}{[DET] - [S_1]} \tag{2.3}$$

which allows the determination of $\bar{n}$ and $[S_1]$. Using ruthenium-2,2′-bipyridine cations ($Ru(bpy)_3^{2+}$) as P and 9-methylanthracene as Q, Turro and Yekta (1978) have determined $\bar{n}$ and $[S_1]$ for aqueous micellar SDS to be 60 ± 2 and 7.5×10^{-3} M. These values are in good agreement with those given in the literature (Fendler and Fendler, 1975). Since either [Q] or [DET] can be varied, the method is not limited to measurements of $\bar{n}$ close to the CMC value of the surfactant. It is also useful for determining the dependency of $[S_1]$ on [DET]. Using $Ru(bpy)_3^{2+}$ and 9-methylanthracene as P and Q, S_1 was shown to be equal to the CMC and to remain constant over a reasonably large range of SDS concentrations (Turro and Yekta, 1978). Attention has, however, been drawn to solubilizates whose distributions do not obey Poisson's distribution (Hunter, 1980).

Low-angle laser light scattering and photon correlation spectroscopy (Chu, 1974; Cummins and Pike, 1974, 1977) are increasingly utilized as versatile and sensitive techniques for determining weight averaged molecular weights, $\overline{Mw}$, mean translational diffusion coefficients, $\bar{D}$, and polydispersities of micelles. Classical techniques of measuring scattered light intensities of micelles relative to that of pure solvent provide a convenient method for determining $\overline{Mw}$ (Kratohvil and Dellicolli, 1970). Due to the Brownian motion of macromolecules, the intensity of scattered light fluctuates, however, around a mean value. The time dependency of this fluctuation, τ_c, determined by quasi-elastic (low-angle) light scattering, is related to the translational diffusion coefficient, D, by

$$\tau_c = \frac{1}{DK^2} \tag{2.4}$$

where K, the scattering vector, is calculated from

$$K = \frac{4\pi\chi \sin(\frac{1}{2}\theta)}{\lambda} \tag{2.5}$$

where θ is the scattering angle, λ is the wavelength of the incident light *in vacuo*, and χ is the index of refraction of the scattering medium. Substitution of appropriate constants into equations 2.4 and 2.5 and determination of τ_c yields values for D. For a monodisperse solution of noninteracting aggregates, the time dependent term is a single exponential with a time constant given by equation 2.4. For a polydisperse solution the decay is, however, nonexponential. Deviation from single exponential decay contains information about the size distribution. Calculations indicate that the width of the size distribution is narrow for small micelles and that it broadens with increasing aggregation number (Mukerjee, 1977). The ratio of the weight, N_w, and number, N_n, average degree of aggregation defines the polydispersity index, N_w/N_n. For completely monodisperse micelles $N_w/N_n = 1$, while for those with wide size distributions N_w/N_n is in the order of 2. Assuming the micelles to be spherical particles, the measured mean translation diffusion coefficient, $\bar{D}$, in a medium of viscosity, η, is related to the mean hydrodynamic radius, $\bar{R}_H$, by the Stokes–Einstein equation:

$$\bar{D} = \frac{kT}{6\pi\eta\bar{R}_H} \tag{2.6}$$

where k is Boltzmann's constant and T is the absolute temperature. Since $\bar{D}$ is an average translational diffusion coefficient, $\bar{R}_H$ represents the average hydrodynamic radius of all aggregates present in the solution. $\bar{R}_H$ also implies the averaging of hydrodynamic frictional factors over all orientations of the aggregates with respect to their translational motion. Micellar shapes can be assessed by assuming negligible hydration and by comparing the determined form factor (obtained by dividing the measured $\bar{R}_H$ by the radius of a sphere calculated for the micelle from its determined $\bar{M}w$) with that calculated by means of the Perrin equation (Chu, 1974) for prolate or oblate ellipsoids. Alternatively, advantage can be taken of temperature dependent changes of $\bar{R}_H$. In this method the experimental dependency of scattered light intensity ratios, $I/I_{\min}$ values, on $\bar{R}_H$ is compared with theoretical values calculated for spherical, prolate, or oblate ellipsoidal micelles (Mazer and Benedek, 1976; Mazer et al., 1977a, 1979). For ellipsoids the semiaxis is assumed to be equal to the radius of the minimum spherical micelle. Figure 2.7 compares experimental points with theoretical curves for 6.9×10^{-2} M SDS in the presence of 0.6 M NaCl (Mazer and Benedek, 1976). Provided that the temperature range is sufficiently wide and that it covers regions well below the critical micellar temperature, the treatment appears to be sufficiently sensitive to allow differentiation between the different micellar shapes. Data shown in Figure 2.7 are clearly compatible only with prolate ellipsoidal dodecyl sulfate micelles.

Once the micellar shape is deduced it becomes possible to estimate the mean aggregation number, $\bar{n}$, from

$$\bar{n} = \frac{n_{\min} a'}{a_0} \tag{2.7}$$

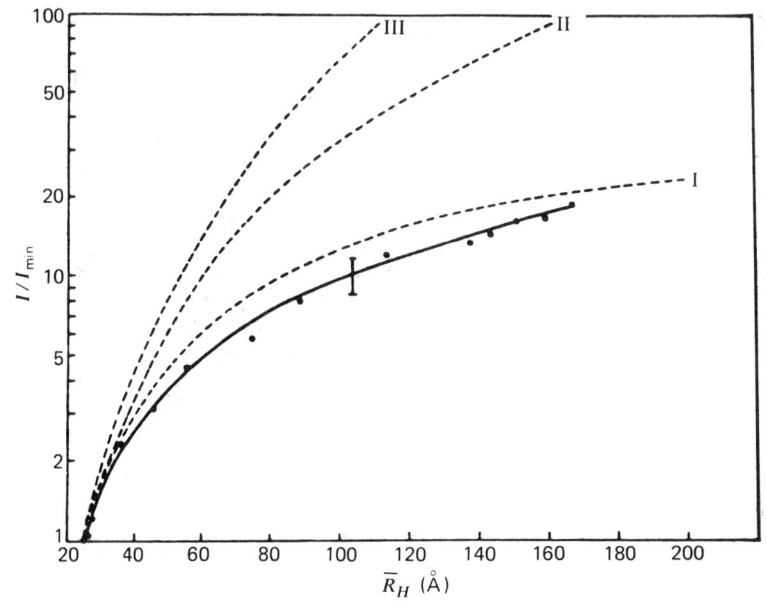

Figure 2.7. Scattered light intensity ratio as a function of $\overline{R}_H$. Dots represent experimental dedendence obtained from $\overline{R}_H$ values and intensity ratios. Error bar indicates the maximum uncertainty ($\pm$ 15%) in all data points. Dashed curves represent model calculations for the following shapes: I, prolate ellipsoid; II, oblate ellipsoid; III, sphere (taken from Mazer et al., 1979).

where $n_{\min}$ is the aggregation number of the minimum spherical micelle with a hydrodynamic radius $\overline{R}_H$, a' is the semiaxis of a prolate or oblate micelle, while a_0 is the radius of the minimal spherical micelle. Determinations of aggregation number by sedimentation equilibrium–isopiestic distillation, luminescence quenching, and light scattering methods have been critically evaluated (Kratohvil, 1980).

Laser light scattering and photon correlation spectroscopy have provided structural information on aqueous micellar SDS (Mazer and Benedek, 1976; Mazer et al., 1977a; Young et al., 1978, Corti and Degiorgio, 1978, 1981; Briggs et al., 1980; Missel et al., 1980), CTAB (Corti and Degiorgio, 1978), and bile salts (Mazer et al., 1977b, 1979; Holzbach et al., 1977). These techniques have also been utilized to assess the dependence of micellar sizes on pressure (Nicoli et al., 1979). Salient points emerging from these studies are that, close to their CMC, micelles are ellipsoids and that changes of temperature, pressure, surfactant, and added electrolyte concentrations substantially affect sizes and size distributions. The aggregation number determined for SDS by fluorescence quenching (Turro and Yekta, 1978) agrees well with that obtained by quasi-elastic light scattering and photon correlation spectroscopy (Mazer and Benedek, 1976). At high ionic strengths distribution of the quencher no longer obeys Poisson statistics and hence the determined aggregation numbers are unreliable (Almgren, 1980). Fluorescence quenching methods for determining aggregation numbers (Turro and Yekta, 1978) have been critically reexamined both by steady state and nanosecond time resolved fluorescence techniques (Almgrem and Löfroth, 1980).

An important function of micelles is to provide microenvironments that differ substantially from that of bulk water. Experimental assessment of "effective" polarities and viscosities at different regions of the micelle is important (Fendler and Fendler, 1975; Mukerjee et al., 1977). Most of the spectroscopic techniques necessarily rely on the use of probes. Interpretation of the results requires care since the exact location of the probe is not always known and the structure of the micelle may well be perturbed (Rodgers et al., 1976). Additionally, the diffusion coefficient of the probe in a restricted volume may be much smaller than that in the bulk even if the microscopic and bulk viscosities are the same. Be that as it may, microviscosities of aqueous micelles range between 15 and 30 cP (Kalyanasundaram, 1978; Turro et al., 1980a). These values indicate viscosities higher than those observed for neat hydrocarbon solvents or water (1–2 cP) but they are considerably lower than those determined for bilayer lipid vesicles ($\sim$200 cP). Thus microviscosity measurements substantiate the proposed fluidities of micellar interiors. Nuclear magnetic relaxation (Williams et al., 1973; Roberts and Chachaty, 1973) and laser Raman scattering studies (Okabayashi et al., 1975; Kalyanasundaram and Thomas, 1976) provided evidence for a gradient in the segmental motion of the surfactant hydrocarbon chains in micelles. Terminal methyl groups behaved as if they were in neat hydrocarbons. Fluorescence polarizations and excimer fluorescence intensities have recently been determined for alkylbenzene sulfonate micelles in which the position of the intrinsic fluorophore (the benzene sulfonate moiety) was varied along the hydrocarbon chain (Aoudia and Rodgers, 1979). Information on rotational and translational motion was deduced from fluorescence polarizations and extents of excimer formation, respectively. Interestingly, both motions changed in a similar manner, suggesting micellar structures in which the benzene sulfonate moieties are located close to the micelle-water interface.

Absorption spectroscopic studies using alkyl pyridinium iodies indicated that the effective polarities of the Stern region resemble that of methanol (Mukerjee et al., 1977). Using micellar 1-methyl-4-dodecylpyridinium iodide as an intrinsic microscopic medium polarity reporter provided information on the innermost part of the electrical double layer (Sudhölter and Engberts, 1979). The micropolarity reported corresponded to that of ethanol ($Z = 80.6 \pm 0.2$) and indicated a fair degree of homogeneity of the Stern layer. The presence of ions at high local concentrations, the closeness of hydrophobic hydrocarbon chains, and the dielectric saturation, or a combination of these factors contribute to the observed apparent dielectric constant at the micelle-water interface. Experiments have been carried out to assess the relative contribution of these different parameters. The lack of alteration of the absorption spectrum of 1-ethyl-4-carbomethoxy pyridinium iodide (Kosower, 1968) in the presence of various concentrated salt solutions was taken to imply a negligible role of high local ion concentrations in affecting micellar micropolarities (Mukerjee et al., 1977). Addition of electrolytes to 1-methyl-4-dodecylpyridinium iodide micelles resulted, however, in substantial deviations from Beer–Lambert's law. These results were interpreted to imply electrolyte induced increases in the Stern layer thickness and decreased standard chemical potentials of the micelle (Sudhölter and Engberts, 1979). The observed identical water activities at micellar interfaces and in bulk aqueous solutions (Perez de Albrizzio and Cordes, 1979) support this postulate. Structural and saturation effects were evaluated by determining effective dielectric constants, using dodecyl pyridinium iodide in different micellar systems (Mukerjee et al., 1977). Table 2.2 shows the results. The increased effective dielectric constant in nonionic octyl glucoside underlines the importance of the proximity of hydrocarbons to the micellar surface. The lower effective dielectric constant of Brij-35 compared with that of octyl glucoside is the consequence of high surface concentrations of polyoxyethylene units in the former micelle causing exclusion of water. Charge separation zwitterionic dodecyl *N*-betaine leads to a strong electric field. The lower effective dielectric constant in this surfactant compared with that of octyl glucoside indicates, therefore, an appreciable contribution from dielectric saturation. Microscopic polarities at micelle-water interfaces can be rationalized, at least for the system investigated (Mukerjee et al., 1977), both in terms of alkyl chain exposure and dielectric saturation. Variations in ΔpK_a values (Table 2.2) indicate the complexity of problems related to micellar microenvironments. The ΔpK_a value obtained for the zwitterionic micelle does not follow the trend observed for the nonionic micelles with respect to their effective dielectric constants. Changes of dissociation constants of indicators at micellar surfaces are used to assess surface potentials (Fernandez and Fromherz, 1977).

Water penetration into micelles, or stated more precisely, the extent of water-to-hydrocarbon contact, is an important, much debated, yet unsettled problem. At one time or another, water was believed to penetrate the micelle completely (Svens and Rosenholm, 1973), not at all (Stigter, 1974a, 1974b; Lindman et al., 1977), or indeed to reach to any intermediate depth (Muller and Birkhahn, 1967;

Table 2.2. Effective Dielectric Constants and Δp*Ka* Values of Bromothymol Blue in Different Micelles[a]

Micelle	Effective Dielectric Constant[b]	ΔpK_a[c]
Dodecyl pyridinium iodide	36	
Dodecyl *N*-betaine ($C_{12}H_{25}N^+(CH_3)_2CH_2CO_2^-$	37	0.85
Brij-35 ($C_{12}H_{25}(OCH_2CH_2)_{23}OH$)	36	1.81
β-D-Octyl glucoside ($C_8H_{17}OCHC_5H_{10}O_5$)	46	1.32

[a] *Source:* Drawn from Mukerjee et al., 1977.

[b] From the spectrum of solubilized dodecyl pyridinium iodide using alkanols and alkanol-water mixtures as reference solvents.

[c] Change in p*Ka* values upon solubilization of bromothymol blue.

Podo et al., 1973; Mukerjee et al., 1977). It is interesting to note that even the Hartley model assumes "wet" micelles. Based on a spherical SDS micelle with a hydrocarbon chain length of 16.6 Å, a headgroup diameter of 4.6 Å, and a micellar radius of 21.2 Å, one calculates the total micellar volume to be 3.99×10^4 Å^3 ($\frac{4}{3}\pi \cdot (21$ Å$)^3$), the micellar core volume to be 1.92×10^4 Å^3 ($\frac{4}{3}\pi \cdot (16.6$ Å$)^3$), and hence the Stern layer volume to be 2.07×10^4 Å^3 (Romsted, 1975). Thus the Stern layer is about 50% of the total micellar volume. If water penetrates only two methylene groups toward the center ($\sim$2.77 Å), then the wet portion of the micelle is about 75% of the total micellar volume (Romsted, 1981). Neither of the two extremes, picturesquely described as the "fjord" and the "reef" models (Figure 2.8; Menger et al., 1978), is likely to represent the real situation. Water was concluded not to reach the very center of the micellar core but to penetrate as far as seven carbon atoms away from the headgroup. This conclusion was based on a comparison of the ^{13}C nuclear magnetic resonance (nmr) shifts of carbonyl carbons of nonaggregated aldehydes in different solvents with those observed for aqueous micellar hexanal and 8-ketohexadecyltrimethylammonium bromide, $CH_3(CH_2)_7CO(CH_2)_7N^+(CH_3)_3Br^-$ (Menger et al., 1978) on examining optical rotary dispersion of (+)-*trans*-2-chloro-5-methylcyclohexanone (Menger and Boyer, 1980) and following the solvolyses of bis(4-nitrophenyl)carbonate and *p*-chlorobenzhydryl chloride, whose rates are highly sensitive to solvent composition (Menger et al., 1980). Although great pains were taken to justify this method, due caution is needed, as the authors themselves point out, in the interpretation of data based on the use of probes that may perturb the system. A distinction should be made between wetted and wettable methylenes. The former is a time averaged thermodynamic quantity while the latter relates to the observable time domain during nmr and fluorescence measurements. Menger's interpretation (Menger et al., 1978) has been seriously questioned (Wennerström and Lindman, 1979b). Protrusion of a micellar headgroup to bulk water (Aniansson, 1978) is an alternative way of rationalizing these results. The final word has not been said on the extent of water penetration into aqueous micelles.

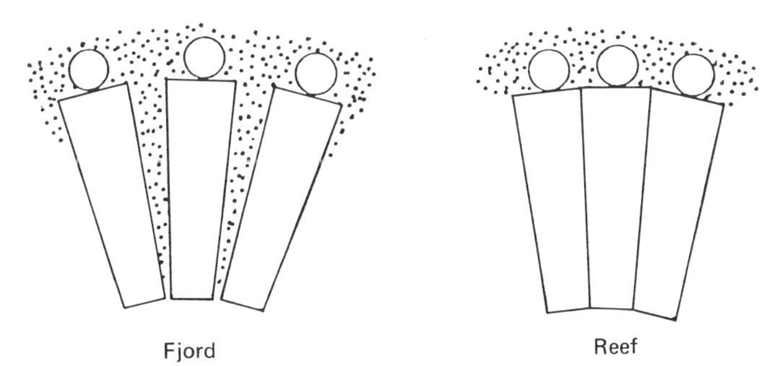

Figure 2.8. The "fjord" and the "reef" micellar models illustrating extremes of water penetration (taken from Menger et al., 1978).

For ionic micelles information on the net charge, on the charge density, and on factors influencing these properties are of significance and are related to their potential application in membrane mimetic chemistry. Most recent data have been deduced from nmr and electron paramagnetic resonance (epr) investigations (Oakes, 1973; Robb and Smith, 1974; Gustavsson and Lindman, 1975; Nakagawa and Tokiwa, 1976; Lindman et al., 1977). Although the results have been mostly interpreted in terms of a two-state model involving bound and free counterions, a distinction between these states is not always clear. Bound counterions are hydrated approximately to the same extent as free ions and they are relatively mobile. Bromide ion binds in the Stern region more strongly than chloride ion (Patterson and Vieil, 1973). This difference is also manifested in the shape of aggregates. Hexadecyltrimethylammonium chloride (CTACl) micelles appear to be spherical at all concentrations, while the corresponding bromide aggregates change from spherical to rodlike structures at high surfactant concentrations (Ulmius et al., 1978). Transformation from spherical to rodlike micelles increases the degree of counterion association. Differences have also been noted for anionic surfactants having different counterions (Moroi et al., 1979a, 1979b). Formation of giant ($n \simeq 1000$) micelles has been reported in aqueous solutions of magnesium and dimethylammonium salts of perfluorononanoate (Hoffmann et al., 1978). Effects of systematic changes in the structure of the surfactant headgroup on the aggregation number and dye solubilization efficiency have been summarized by Anacker (1979).

Counterion binding has been assessed in terms of the Gouy–Chapman electrical double-layer theory (Stigter, 1974a, 1974b, 1975a, 1975b). In the classical form this theory assumes a uniform and continuous surface charge on the micelle, which is balanced by a diffuse atmosphere of point charges distributed in the bulk solution. Attempts are being made to improve this approach by taking into consideration the discrete nature and size of headgroups and counterions as well as the effects of ion fluctuations (Stigter, 1975a, 1975b). Taking advantage of surface potential measurements, properties of the micellar electrical double layer have been calculated by means of a modified nonlinearized Poisson–Boltzmann equation (Frahm and Diekmann, 1979; Frahm et al., 1980; Gunnarsson et al., 1980) in a manner analogous to that used for polyelectrolytes (see Section 8.3). Pulse radiolytic techniques have been used to establish the direct exchange of Ag_2^+ between two anionic micelles (Henglein and Proske, 1978a). Once again, additional experimental data and improved theoretical calculations are needed for more precise understanding of counterion specificities.

When the surfactant concentration is increased above the CMC, the spherical micelle elongates to assume cylindrical and lamellar structures. At high surfactant concentrations liquid crystals predominate (see Section 2.7).

Electron microscopy provides the most direct determination of micellar structures. Micelles with relatively low CMCs have been visualized in a reproducible manner by electron microscopy of spray-frozen solutions (Kutter et al., 1976). Electron spin resonance and spin echo studies of photoproduced tetramethylbenzidine cation radical in SDS have also indicated the maintenance of the integrity of the micelle upon freezing to 77 K (Narayana et al., 1981). These studies have also provided evidence for micellar models that require extensive exposure of the surfactant hydrocarbons to water.

3. THERMODYNAMICS OF MICELLIZATION

Several models have been used for treating the thermodynamics of micellization. The mass-action and phase-separation models are the simplest and most frequently used treatments. The former model applies the law of mass-action between monomeric surfactant, $S_1^{-\text{ or }+}$, with counterions, $B^{+\text{ or }-}$, and micelles, M, formed in a single step:

$$nS_1^{-\text{ or }+} + (n-m)B^{+\text{ or }-} \rightleftharpoons M^{m-\text{ or }+} \tag{2.8}$$

where m represents the concentration of free counterions (the degree of ionization is $\alpha = m/n$). The equilibrium constant for micellization, K_M, is given by

$$K_M = \frac{[M]}{[S_1^{-\text{ or }+}]^n[B^{+\text{ or }-}]^{n-m}} \tag{2.9}$$

Concentrations are usually expressed in mole fractions and activity coefficients are neglected, even though nonideality prevails. The standard free energy of micellization per S is given by

$$\Delta G_M = \frac{RT}{n}\{n \ln [S_1^{-\text{ or }+}] + (n-m)[B^{+\text{ or }-}] - \ln [M]\} \tag{2.10}$$

At the CMC $[S_1] \simeq [B] \simeq$ CMC and hence equation 2.10 approximates to

$$\Delta G_M \sim RT\left(1 + \frac{m}{n}\right) \ln \text{CMC} \tag{2.11}$$

The alternative phase-separation model considers micelles, together with their counterions, as a separate phase that appears at the CMC. The standard free energy of micellization is given by

$$\Delta G_M = 2RT \ln \text{CMC} \tag{2.12}$$

Both the mass-action and the phase-separation models are gross oversimplifications. In both derivations micelles are assumed to be monodisperse and noninteractive. Definition of CMC, using the mass-action approximation, is somewhat arbitrary. In spite of the approximations involved, both the mass-action and the phase-separation models predict properties of large micelles ($\bar{n} \geq 100$) fairly well.

Small-system thermodynamics and statistical thermodynamics (Hall, 1970a, 1970b, 1972) have also been applied to micelles. Original publications should be consulted for details.

Tanford (1972, 1977; Reynolds et al., 1974) has treated the thermodynamics of micellization in terms of a relatively simple geometric model. He expressed an idealized equilibrium constant for micelle formation by

$$K_M = \frac{X_n}{nX_1^n} \tag{2.13}$$

where X_1 is the mole fraction of surfactant monomers and X_n/n represents the surfactant concentration present in micelles of a given size. A plot of X_n against n defines the distribution function of the aggregation numbers of the micelles. The optimum aggregation number, n^*, is defined as that value of n for which X_n is maximum:

$$\left(\frac{d \ln X_n}{dn}\right)_{n=n^*} = 0 \tag{2.14}$$

The overall concentration of the surfactant in micellar form, X_{mic}, is related to X_1 by

$$X_{\text{mic}} = \sum_{n=2}^{\infty} X_n = \sum_{n=2} nK_M X_1^n \tag{2.15}$$

and the total stoichiometric concentration of surfactant in the solution, X_{tot}, is expressed by

$$X_{\text{tot}} = \sum_{n=1}^{\infty} X_n = X_1 + X_{\text{mic}} \tag{2.16}$$

The advantage of using mole fraction units in equations 2.15 and 2.16 is that free energy transfer of a monomeric surfactant from the bulk solution to a micelle, ΔG_n^0, having an aggregation number of n, is automatically given in unitary units (Tanford, 1973):

$$n\Delta G_n^0 = -RT \ln K_M \tag{2.17}$$

The use of unitary units has been criticized (Ben-Naim, 1978), however, Calculations can be carried out by splitting ΔG^0 into terms relating to contributions from the hydrophobic associations of the apolar hydrocarbon chains, ΔU_n^0, and to those from the repulsions between the polar headgroups, W_n (Tanford, 1977):

$$\Delta G_n^0 = \Delta U_n^0 + W_n \tag{2.18}$$

Values for ΔU_n^0 have been evaluated for all possible n values by assuming disklike aggregates to minimize the hydrocarbon water contacts.

Numerical assignments have been based on available thermodynamic data for the free energy transfer of the hydrophobic part of the surfactant from water to the pure anhydrous state. Allowances have been made for the ordered nature of the micellar core and for the incomplete removal of the surfactant tail from the aqueous phase. ΔU_n^0 can, therefore, be calculated from equation 2.19 (Tanford, 1977):

$$\Delta U_n^0 = -1400 - 700n_c + C' + 25(A_{Hm} - 21) \tag{2.19}$$

where n_c is the number of hydrophobic carbon atoms per chain (taken to be one less than the actual chain length), C' is an empirical parameter related to the effects of alkyl chain length on CMCs (approximately equal to 100 cal mole^{-1}) and A_{Hm} is the surface area per headgroup in contact with water. Only this latter parameter depends on the micelle size. Electrostatic repulsion factors, W_n values, have been estimated from monolayer compression data. Availability of calculated ΔG_n^0 values (equation 2.18) allows thermodynamic predictions of micellar size distribution (equations 2.14–2.16), and CMCs as well as the dependencies of these parameters on alkyl chain lengths and on other structural parameters. A thermodynamic model of ionic surfactant systems has recently been developed in terms of free energy contributions from hydrophobic energy, surface free energy of the aggregates, electrostatic free energy, and entropy of mixing (Jönsson and Wennerström, 1981).

4. DYNAMICS OF MICELLIZATION

Micelles are dynamic species. They rapidly break up and reform. Kinetics of these processes have been investigated by stopped flow, temperature jump, pressure jump, ultrasonic absorption, nmr and epr spectroscopy (Fendler and Fendler, 1975). The early results were confusing and often irreproducible. By and large, relaxation times, depending on the methods used, fell into two time domains: one occurring in the microsecond (τ_1) and the other in the millisecond (τ_2) range. It is now realized that these relaxation times represent two different processes (Muller, 1972, 1979), Loosely stated, the faster process, represented by τ_1, is dissociation of the micelles and it corresponds to the release of a single surfactant molecule and to its subsequent reincorporation. The slower process, represented by τ_2, is the stepwise dissolution of the micelles to monomers and their subsequent reassociation. Quantitative treatment of these processes has been derived recently in terms of the multiple association of monomeric uncharged surfactants using heat conduction and diffusion as an analogy (Aniansson and Wall, 1974, 1975, 1977; Aniansson, 1975; Aniansson et al., 1976; Almgren et al., 1977a; 1977b):

$$
\begin{aligned}
S_1 &+ S_1 \underset{k_{-1}}{\overset{k_1}{\rightleftharpoons}} S_2 \\
&\vdots \\
S_i &+ S_1 \underset{k_{-i}}{\overset{k_i}{\rightleftharpoons}} S_{i+1} \\
&\vdots \\
S_{n-1} &+ S_1 \underset{k_{-n}}{\overset{k_n}{\rightleftharpoons}} S_n
\end{aligned}
\tag{2.20}
$$

In the fast relaxation the population of the micelles is redistributed but their net numbers remain unaltered (Figure 2.9). The reciprocal relaxation time for this process is given by

$$
\tau_1^{-1} = \frac{k_{-n}}{\sigma^2 + (k_{-n}/n)(S_{eq}/S_1)} \tag{2.21}
$$

where k_{-n} is the dissociation rate constant for a monomer leaving the micelle of aggregation number n, σ is the variance of population distribution, and S_{eq} and S_1 are the equilibrium concentrations of the surfactant in the micelle and in the monomer, respectively. If the aggregation number remains constant and is

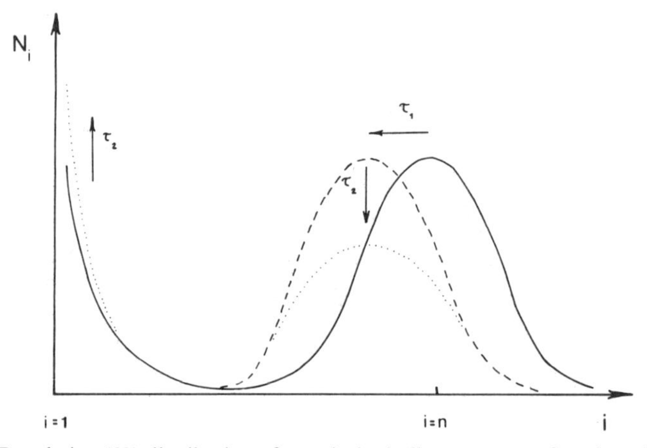

Figure 2.9. Population (N_i) distribution of a typical micelle system as a function of aggregation number i. The dashed curve indicates the distribution that arises after a small perturbation of the system that leads to dissociation and decrease of the mean aggregation number of micelles. The dotted line is the final equilibrium distribution. τ_1 and τ_2 are the characteristic relaxation times of the processes indicated by the arrows (taken from Schelly et al., 1979).

known independently, k_{-n} and σ can be evaluated from the experimental dependence of τ_1^{-1} on the stoichiometric surfactant concentration (equation 2.21). Strictly speaking, Aniansson's treatment is only valid for systems that are at least partially heterodisperse (i.e., $\sigma > 1$). Under this condition $[S_{n-1}] \simeq [S_n]$ and, therefore, k_n can be evaluated from the relationship

$$\frac{k_n}{k_{-n}} = \frac{1}{\mathrm{CMC}} \tag{2.22}$$

In the slower relaxation the dissolution of micelles (Figure 2.9) was shown to obey equation 2.23 (Aniansson and Wall, 1975, 1977):

$$\tau_2^{-1} \simeq \frac{(1/RC_3)(S_1 + n^2C_3)}{S_1 + \sigma^2 C_3} \tag{2.23}$$

where $R = \Sigma(k_{-i}S_i)^{-1}$, with k_{-i} and S_i denoting the dissociation rate constant and equilibrium concentration in the domain of the distribution minimum, and C_3 standing for ΣS_i in the domain of the micellar aggregation space. Validities of equations 2.21 and 2.22 have also been tested at different temperatures, thus allowing the experimental verification of thermodynamic predictions for micellization (Pakusch and Strey, 1980). Aniansson's theory for relaxation of monomers in aqueous micelles has been found to apply to ultrasonic absorption (Teubner, 1979), to pressure-jump (Baumgardt et al., 1979), and to conductance stopped-flow spectroscopy (Okubo et al., 1979). Pulse radiolytic techniques have also been applied to the determination of entry and exit rates of monomeric surfactant units into micelles (Henglein and Proske, 1978b; Almgren et al., 1979b).

Aniansson's treatment has recently been extended to allow the evaluation of relaxation parameters of charged aggregates (Chan and Kahlweit, 1977; Chan et al., 1977a, 1977b, 1978). This method allows the detailed investigation of dynamic parameters affecting micellar surface potentials and charge densities. Such information is of great significance in the meaningful design of electrochemical experiments in micellar media.

Micellar relaxations have also been treated in terms of a shell model distribution function (Kegeles, 1980). This approach has been, however, criticized (Aniansson and Wall, 1980).

5. SUBSTRATE SOLUBILIZATION AND ORGANIZATION

Many applications of micelles are related to their ability to dissolve or solubilize water-insoluble substances. Solubilization is the consequence either of the mutal association of the solubilized molecules, the solubilizates, with monomers, or of comicellization. Hydrophobic substances do not associate appreciably with monomeric surfactants. A sudden increase in solubility occurs in the region of the

CMC. Substrate solubilization has, in fact, often been used for CMC determinations. In contrast to dissolution in an organic solvent, micellar solubilization is a highly cooperative phenomenon. Solubilization is the basis of many industrial processes and is of fundamental importance in rationalizing micellar effects on reactivities. Methods for determining solubilities (Mader and Grady, 1970) and the colloid chemistry of solubilization (McBain and Hutchinson, 1955; Shinoda, 1967; Elworthy et al., 1968) are treated in standard books and are not reiterated here. Likewise, the 1975 complilation (Fendler and Fendler, 1975) should be consulted for the available methods of investigating micellar solubilization sites and for data obtained prior to that date. This complilation listed the determined solubilization sites for benzene, naphthalene, xylene, pyrene, nitrobenzene(s), short chain aliphatic alcohols, acetone, dioxane, phenol(s), benzoic acid(s), aryl sulfates, aryl sulfonates, fluorobenzene(s), benzophenone, acetophenone, steroids, and stable free radicals in uncharged and charged micelles (Fendler and Fendler, 1975). Solubilization was established to be a dynamic process involving both hydrophobic and electrostatic interactions. Based on the "like dissolves like" adage polar molecules were considered to associate with the surfactant headgroups while apolar ones were predominantly localized in the micellar core. What was meant by micellar core or interior was not always clear, however. Subsequent evidence has negated the idea of deep substrate penetration into micelles. Lack of appreciable micellar effects on the reduction rates of rigid steroids having different polarities is only compatible with their binding close to the micelle-water interface (Menger, 1979a). Deep penetration of the steroids into the micellar interior would have resulted in a substantial decrease in the rates of their reduction. Microenvironments of micelle solubilized substrates in no case resemble those of hydrocarbons. This is, of course, in accord with the loose micellar model proposed by Menger (1979a).

Substrate solubilization in micelles has been treated recently in terms of a two-phase process (Mukerjee and Cardinal, 1978; Mukerjee, 1979). Solubilizate distribution between two extreme sites is considered in this model. The two sites correspond to a hydrophobic "dissolved state," in the micellar interior (presumably operationally defined), and a more polar "adsorbed state," at the micelle-water interface. Adsorption is largely the consequence of substrate surface activity and it is greatly amplified by the high micellar surface to volume ratios. This factor is responsible for the fact that some polar molecules are more soluble in micelles than in either water or a pure hydrocarbon solvent (McBain and Hutchinson, 1955). The predominant solubilization site of benzene near the interface of cationic micellar CTAB (Eriksson and Gillberg, 1966; Fendler and Patterson, 1971) is also explicable in terms of the surface activity of benzene, present in low concentrations at the micelle-water interface. At high benzene concentration, however, the solubilizate is redistributed in favor of the dissolved state (Eriksson and Gillberg, 1966). Solubilization of high mole fractions of solutes thus results in the swelling of the micelle. Micellar solubilization of other aromatic molecules follows a pattern similar to that of benzene (Mukerjee and Cardinal, 1978). Substrate solubilization is best understood in terms of a nonclassical micelle model (Menger, 1979a; Fromherz, 1980; Dill and Flory, 1981).

Solubilizates not exhibiting any surface activity, aliphatic hydrocarbons for example, can be considered to be present exclusively in the dissolved state (Mukerjee, 1979). The extent of their solubilization depends upon the size and shape of the micelles, as well as on the surface charge and potential of the head-groups. Entry of hydrophobic molecules into the dissolved micellar state is opposed by the Laplace pressure, resulting from the curvature of micelles. The observed decrease of micellar solubilization of hydrocarbon homologs with increasing chain length (McBain and Hutchinson, 1955) supports this postulate. Assuming the interactions of the hydrocarbons with the micelle are ideal except for the Laplace pressure effect, Mukerjee (1979) has estimated the Laplace pressure, P, by means of

$$PV = -RT \ln X \tag{2.24}$$

where V is the partial molal volume and X is the mole fraction of the solubilizate in the micelle. A knowledge of the Laplace pressure allows, in turn, the calculation of the interfacial tension, ξ:

$$P = \frac{2\xi}{r} \tag{2.25}$$

where r is the radius of the spherical micellar core. Satisfactory agreement has been obtained between calculated and independently determined interfacial tensions (Mukerjee, 1979). The lesser solubility of hydrophobic molecules in micelles compared with that in hydrocarbons is also due, at least in part, to the Laplace pressure in micelles. The necessity of evoking Laplace pressure to rationalize micellar properties has, however, been questioned (Menger, 1979b).

Solubilization thermodynamics has considered micelles to be spherical droplets providing free energy wells for solubilizates (Almgren et al., 1979a). Standard free energies of a nonpolar solubilizate in water, μ^0_{aq}, in a hydrocarbon solvent, μ^0_{HC}, and in the micelle, μ^0_M, are shown in Figure 2.10a. The Laplace pressure in the micelle and the entropy restrictions were proposed to be responsible for increasing μ^0_M with respect to μ^0_{HC} (Almgren et al., 1979a). If the solubilizate distribution is uniform in the micelle, the probability of finding the solubilizate, $\rho(r)$, between r and $r + dr$ depends parabolically on r and the mean distance from the center of the sphere, $\bar{r}$, is equal to $3R_C/4$ and $r_m = 0.797R_C$ (Figure 2.10b). Restriction in the mobilities of the hydrocarbon chains in the micelle reduces further substrate solubilization.

Substrate adsorption at the micelle-water interface perturbs the micellar charge densities. This, in turn, may facilitate the transition from spherical to rodlike micelles. As expected, solubilization induced transitions are more prevalent for polar than for apolar solubilizates (Mukerjee, 1979).

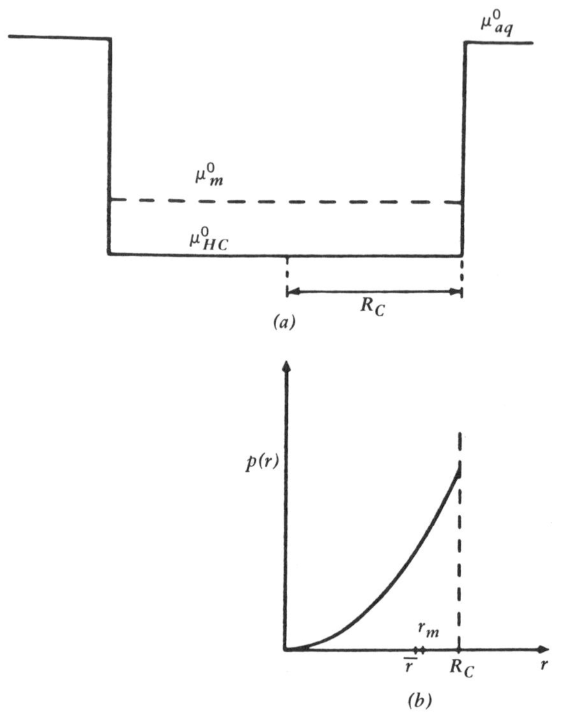

Figure 2.10. (*a*) Free energy well for a solubilized molecule in a micelle, regarded as a spherical oil droplet. The standard free energy μ°_{HC} of the molecule in the hydrocarbon solvent is lower than that in the spherical micelle owing to the Laplace pressure in the micelle and to entropy factors resulting from restrictions on the freedom of motion of the probes. (*b*) The probability $\rho(r)\,dr$ of finding the probe between r and $r + dr$ in a spherical well as in (*a*) depends parabolically on r. The mean distance from the center of the sphere is $\bar{r} = 3R_C/4$, and the median is $r_m = 0.797R_C$ (taken from Almgren et al., 1979a).

Although the two-stage solubilization model represents an advance over the one-stage model (substrate distribution is considered between the bulk solvent and a micellar phase) in providing satisfactory rationalizations for experimental observations, its limitations must be realized. Microenvironments available for solubilization change continuously and solubilizates dynamically distribute among the many available sites.

Stopped-flow and concentration-jump techniques (Schelly and Chao, 1979) have been utilized for the elucidation of solubilization dynamics. In an early study interaction of pinacyanyl chloride, a cationic dye, with anionic micellar SDS was rationalized in a two-step process (Takeda et al., 1974). The first step was considered to be a fast equilibrium formation of a dye-surfactant complex that reacted in a rate determining step with the micelle. Based on stopped-flow studies, Robinson et al. (1975) proposed that acridine dyes initially adsorb onto the surface and subsequently enter into the interior of *n*-alkyl sulfate micelles. Timescales for these events are in the micro- and millisecond range, respectively.

Effects of surfactant headgroups, counterions, and charges on the dynamics of substrate-micelle interactions have also been investigated (James et al., 1977). Solubilization of acridine orange by nonionic Triton-305 micelles has been described by nucleation of surfactant on the dye adsorbed on the surface of the micelle (Schelly et al., 1979). Although two component relaxation profiles have been obtained in these studies for solubilizate-micelle interactions (Takeda et al., 1974; Schelly et al., 1979), interpretation of the results is not unambiguous. Relaxations may well involve the dissociation of the micelle (see Section 2.4). Rotational diffusion rates of micelle solubilized dyes have been determined by subnanosecond laser spectroscopy (Lessing and Von Jena, 1979). Useful information has been obtained by measuring fluorescence lifetimes (Rodgers, 1981) and rotational correlation times (Reed et al., 1981) of rose bengal—a rigid, negatively charged molecule—in aqueous micelles. This probe binds both to anionic and cationic micelles. Significantly, the observed rotational correlation times (2–3 nsec) could only be rationalized by assuming extensive exposure of the surfactant hydrocarbons to water (Reed et al., 1981).

Alternative treatments for solubilization dynamics have recently been derived through the use of luminescence quenching and Poisson statistics (Quina and Toscano, 1977; Atik and Singer, 1978; Infelta and Grätzel, 1979; Almgren et al., 1979a; Yekta et al., 1979; Atik et al., 1979; Aikawa et al., 1979, 1980; Turro et al., 1980a; 1980b; Ziemiecki et al., 1980; Waka et al., 1980). Equation 2.26 describes the general scheme for treatment of fluorescence quenching in micelles (Yekta et al., 1979):

$$
\begin{array}{ccccc}
(P^*) + Q_{aq} & \underset{k_-}{\overset{k_+}{\rightleftharpoons}} & (P^* \cdot Q) + Q_{aq} & \underset{2k_-}{\overset{k_+}{\rightleftharpoons}} & (P^* \cdot 2Q) \\
h\nu \uparrow\downarrow & & h\nu \uparrow\downarrow \; \tau_0^{-1} + k_r & & h\nu \uparrow\downarrow \; \tau_0^{-1} + 2k_r \\
(P) + Q_{aq} & \underset{k_-}{\overset{k_+}{\rightleftharpoons}} & (P \cdot Q) + Q_{aq} & \underset{2k_-}{\overset{k_+}{\rightleftharpoons}} & (P \cdot 2Q)
\end{array}
\qquad (2.26)
$$

where the probe P and its excited state P* are completely micellized, the quencher Q may be in the aqueous, Q_{aq}, or in the micellar phase, P · Q, the circles symbolize the micelles, M, without implying any preferred sites for P, P*, P* · Q_{aq}, and so on (i.e., it is a one-stage model), τ_0 is the luminescence lifetime of the donor in the absence of quenchers, Q_{aq} is the water solubilized quencher, k_+ and k_- are rate constants for the entry and exit of the quencher into the micelle, $K = k_+/k_-$, and k_r is the quenching rate within the micelle (i.e., the rate constant for intramicellar quenching). Four limiting cases have been recognized.

The first case (case 1) is static quenching for systems in which both the probe and the quencher are exclusively localized in the micelles and, therefore, emission occurs only from that fraction of P* containing micelles that are free of quencher molecules. Case 1 is characterized by a single exponential decay of the fluorescence lifetime, which is independent of [Q] (see Figure 2.11*a*). The steady state luminescence intensity, under these conditions, is given by equation 2.1. As discussed previously (Section 2.2), treatment of the data allows the determination of aggregation numbers.

The second possibility (case 2) is that while P resides exclusively in the micellar phase, Q is distributed in the micellar and bulk aqueous phases, and that the quenching is static. Steady state emission still occurs only from the fraction of micellized P* that is free of Q and for steady state illumination

$$\frac{I0}{I} = \frac{(1 + k_+ \tau_0[Q_w]) \exp K[Q]}{(1 + K[M])} \tag{2.27}$$

where the first and second terms are contributions to intensity quenching from dynamic and static quenching, respectively. Equation 2.27 reduces to equation 2.1 when Q is completely in the micellar phase ($K[M] \gg 1$). The lifetime of

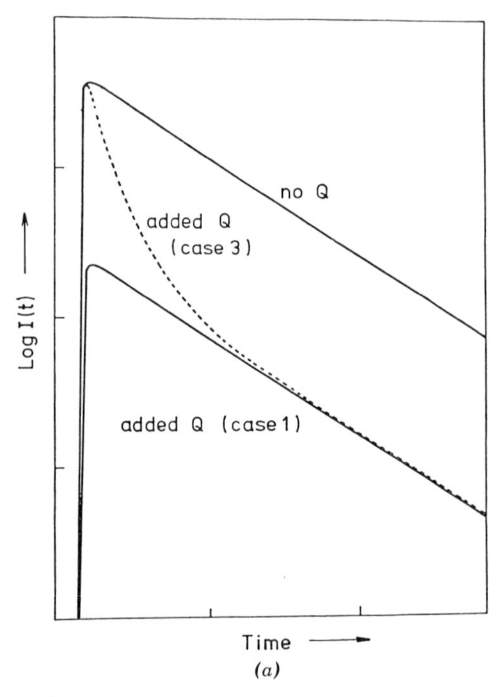

(a)

Figure 2.11. (*a*) Schematic representations of case 1 (limit of complete static quenching of P^* when a micelle contains at least one Q and no quenching of P* when a micelle contains no Q), and case 3 (limit of nonstatic quenching). Decay of P* is strictly exponential for case 1 and nonexponential for case 3. In both cases Q is completely micellized.

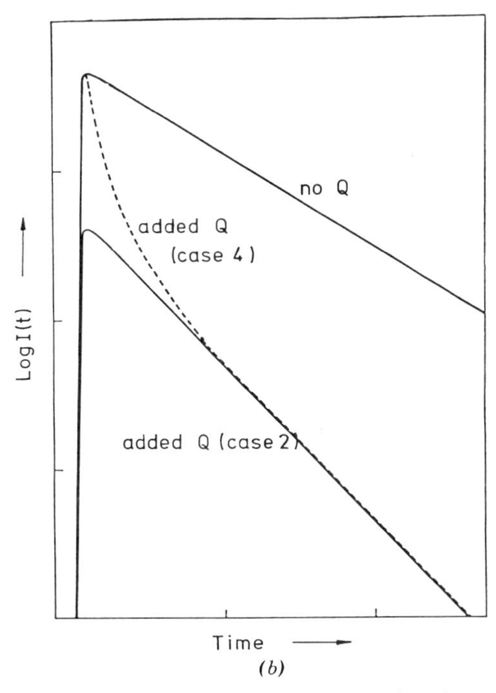

Figure 2.11. (*b*) Schematic representations of case 2 (limit of static quenching with partially micellized quencher) and case 4 (limit of nonstatic quenching with quencher partially micellized) (taken from Yekta et al., 1979).

emission, τ, is, however, reduced by dynamic diffusional quenching by the water solubilized quenchers, Q_{aq} (Figure 2.11*b*):

$$\tau^{-1} = \frac{\tau_0^{-1} + k_+[Q]}{(1 + K[M])} \tag{2.28}$$

Since increasing concentrations of surfactants decrease Q_{aq}, case 2 is characterized by increased luminescence lifetimes with increasing concentrations of micelles. Steady state and time resolved data for case 2 afford values for K, k_+, the mean residence time (k_-), and $\bar{n}$.

Dynamic quenching with complete micellization of Q represents the third mode of quenching (case 3). For a given micelle the quenching probability is proportional to the number of quenchers in the micelle with the resultant nonexponential decay of the fluorescence life (Figure 2.11*a*). Fluorescence intensity under steady state irradiation is given by

$$\frac{I}{I0} = e^{-\langle q \rangle} \sum_{q=0} \frac{\langle q \rangle^q}{[1 + q(k_r \tau_0)]q!} \tag{2.29}$$

where $q = [Q]/[M]$ and k_r is the intramicellar quenching constant. The luminescence decay function is expressed as

$$\frac{I(t)}{I0} = \exp\left\{-\left[\frac{t}{\tau_0} + \langle q\rangle(1 - e^{-k_r t})\right]\right\} \tag{2.30}$$

Equation 2.30 reduces to equation 2.1 when $k_r\tau_0 \rightarrow 0$ (i.e., when quenching becomes static). Treatment of data under conditions corresponding to case 3 provides information on the intramicellar quenching constant, k_r (Rodgers and Wheeler, 1978).

Dynamic quenching with partially micellized quencher represents the typical situation (case 4). Fluorescence decay for case 4 is also nonexponential (Figure 2.11*b*). Fluorescence intensity under steady state irradiation is given by (Tachiya, 1975)

$$\frac{I}{I0} = \frac{\tau}{\tau_0} e^{-\langle q\rangle\alpha_r^2} \sum_{q=0} \frac{(\langle q\rangle\alpha_r^2)^q}{[1 + q(k_b + k_r)_0]q!} \tag{2.31}$$

and the decay function is expressed by

$$\frac{I(t)}{I0} = \exp\left\{-\left[\frac{t}{\tau} + \langle q\rangle\alpha_r^2(1 - e^{-(k_r+k_b)^t})\right]\right\} \tag{2.32}$$

In principle, steady state and time dependent analysis of fluorescence quenching in terms of equations 2.31 and 2.32 should yield all the parameters for the interaction of Q with the micellized probe (equation 2.26). Quenching of pyrene fluorescence by iodide ion and by nitromethane was treated by equations similar to 2.32 (Yekta et al., 1979).

An alternative situation exists when P is distributed between the micellar phase and water and Q may or may not reside exclusively in the micelles. If all quenching occurs in the aqueous phase, then the rate is determined by the escape of the probe from the micelle (Turro and Yekta, 1978). A treatment has been described to allow for the exchange of quenchers by micellar collisions (Dederen et al., 1979). Kinetic treatments of systems in which both P and Q are distributed between the aqueous and micellar phases, and in which quenching occurs in both of these phases, are complex and have not yet been examined.

Phosphorescence quenching studies have provided direct information on entry and exit rates of substituted naphthalene triplets in aqueous cationic and anionic micelles (Almgren et al., 1979a). Low concentrations of hydrophobic probes and relatively high concentrations of ionic quenchers having charges identical to those on the micelles were used. Under these conditions all quenching occurred in the aqueous bulk phase, and for long lived phosphorescence k_- could be determined from the quenching data (Almgren et al., 1979a). Satisfactory agreement has been obtained between experimentally determined and calculated solubilizate entry rates.

Table 2.3. Solubilize Binding Constants and Entry and Exit Rate Constants in Micellar SDS[a]

Solubilizate	K_M, M^{-1}	k_+, M^{-1} sec^{-1}	k_{-1}, sec$^{-1\,b}$
Oxygen	3×10^2	1×10^{10}	5×10^7
Benzene	2×10^3		4×10^6 (4.4×10^6)
Naphthalene	2×10^4		2×10^5 (2.5×10^5)
Anthracene	4×10^5		2×10^4 (1.7×10^4)
Pyrene	2×10^6		4×10^3 (4.1×10^3)
1-Bromonaphthalene	2×10^5		3×10^4 (3.3×10^4)
Perylene			(4.1×10^2)

Source: Compiled from Turro et al., 1980a, unless otherwise stated.

[a] See equation 2.26 for definitions.

[b] Values in parentheses are taken from Almgren et al., 1979a.

Substrate binding constants and residence times in aqueous micelles, determined by the above outlined fluorescence and phosphorescence quenching methods, are collected in Table 2.3.

It should be borne in mind that luminescence quenching does not provide a mechanistic insight into the dynamic substrate-micelle interactions. Further, no allowances can be made for solubilizate induced micellization or alteration of micellar structures or behavior.

6. PREMICELLAR AGGREGATES, HYDROPHOBIC ION PAIRS, AND HYDROPHOBIC ASSOCIATIONS

Manifestation of micellar behavior has often been noted in surfactant solutions below their critical micelle concentrations (Fendler and Fendler, 1975). In some instances this effect could be attributed to solubilizate induced micellization (Mukerjee and Mysels, 1971). In other cases substantial effects, particularly rate enhancements, have been observed in aqueous surfactant solutions at concentrations orders of magnitude lower than the CMC. Submicellar aggregates are particularly evident in aqueous solutions of bulky hydrophobic molecules such as tetraalkylammonium halides. These molecules are not prone to form micelles and are often used as phase transfer agents (Weber and Gokel, 1977; Starks and Liotta, 1979; Makosza, 1979; Dehmlow and Dehmlow, 1980). Kunitake and his co-workers reported extremely large rate enhancements for deacylation of carboxylic esters by hydrophobic hydroxamate ions in the presence of nonmicellar quaternary ammonium salts (Shinkai and Kunitake, 1976a, 1976b; Kunitake et al., 1976; Okahata et al., 1977; Kunitake and Sakamoto, 1979; Kunitake et al., 1979). In many cases rate constants in the presence of these hydrophobic ion pairs have been found to be larger than those observed for the same process in aqueous micellar solutions (Kunitake et al., 1979). The larger rate enhancements in hydrophobic ion pairs are readily rationalized in terms of the kinetic treatments derived for second order reactions in aqueous micelles (Martinek et al., 1977; Romsted, 1977; Bunton, 1979). According to this treatment, concentrations of both reactants *present in the aggregates* determine the rate. In the smaller hydrophobic ion pairs the number of reactants per aggregate is higher than those in larger micelles at the same reagent concentrations. More efficient reagent concentration in the smaller aggregates results in higher second order rate constants for reactions carried out in hydrophobic ion pairs than for those observed in micelles.

Formation of small ion pair aggregates of **2.1** and **2.2** in solutions well below the CMC of **2.1** has been demonstrated recently by epr and fluorescence spectroscopic techniques (Atik and Singer, 1979).

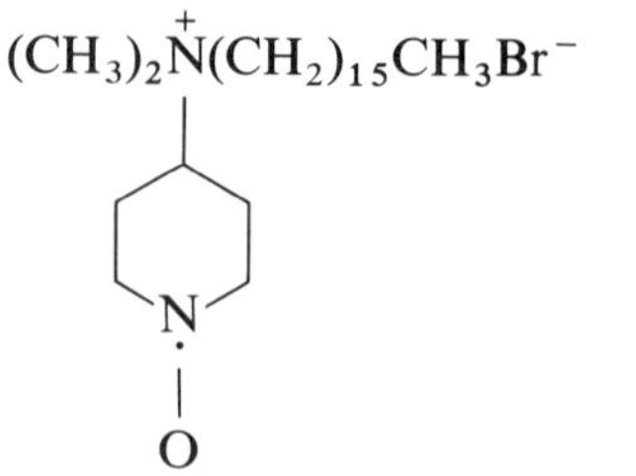

$(CH_3)_2\overset{+}{N}(CH_2)_{15}CH_3Br^-$ $(CH_2)_4CO_2^-Na^+$

N

O

2.1 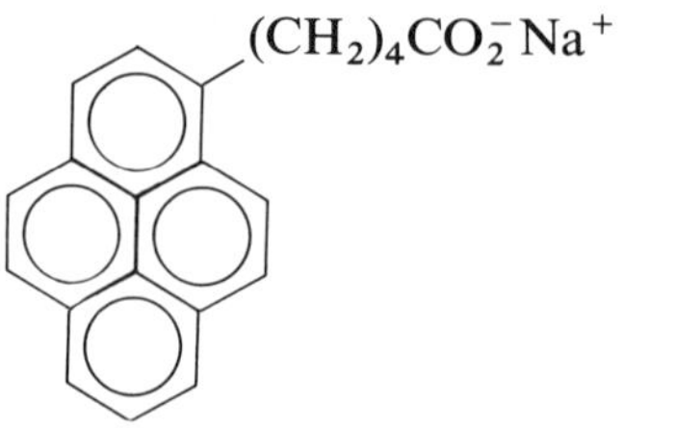 **2.2**

Spectroscopic studies of this type will contribute toward our detailed understanding of the properties of small hydrophobic aggregates.

Hydrophobic ion pairs in general, and hydrophobic interactions in particular, are, of course, of vital importance in several areas of biochemistry (Tanford, 1973; Weber, 1975; Frank, 1975; Chan et al., 1979). The driving force for the formation of hydrophobic ion pairs is similar to that of micellization. Free energies are decreased for both types of aggregates by water displacements (hydrophobic interaction) and by charge neutralization (electrostatic effect). Hydrophobic interactions are best considered in terms of partial reversal of the solution process (Nemethy, 1967; Jencks, 1969; Tanford, 1973; Roseman and Jencks, 1975). The free energy of solution, ΔG_s^0, is given by (Ben-Naim, 1974)

$$\Delta G_s^0 = \Delta G_{cav}^0 + \Delta G_{rgt}^0 \tag{2.32}$$

where ΔG_{cav}^0 is the free energy expended to form a cavity in the solvent to accommodate the solute and ΔG_{rgt}^0 is the free energy released upon the rearrangement of water molecules subsequent to insertion of the solute. The free energy of hydrophobic interaction, ΔG_H^0, resulting from bringing together two apolar molecules, is given by (Pierotti, 1963, 1965)

$$\Delta G_H^0 = f(\Delta G_{rgt}^0) \tag{2.33}$$

Hydrophobic interaction can, therefore, be considered to be the consequence of an extension or partial reversal of that part of the solution that produced ΔG^0_{rgt} (Oakenfull and Fenwick, 1977). Based on the flickering three-dimensional continuously changing water structure, hydrogen bond breaking and making is defined by (Symons, 1978)

$$(H_2O)_{bound} \rightleftharpoons (OH)_{free} + (LP)_{free} \qquad (2.34)$$

where $(OH)_{free}$ and $(LP)_{free}$ represent free hydroxide groups and free lone pairs. Substrate dissolution is expected to shift equilibrium 2.34 to the right while hydrophobic interaction between two apolar solutes has the opposite effect. Taking advantage of the $2\nu_{OH}$ vibration of HOD molecules in D_2O, Symons (1978) determined the concentrations of $(OH)_{free}$ in aprotic solvents and electrolytes. As expected, addition of tetraalkylammonium salts decreased the $(OH)_{free}$ concentration. It would be interesting to extend these infrared determinations to hydrophobic ion pairs that elicit large rate enhancements in deacylation reactions (Kunitake et al., 1979).

7. POSTMICELLAR ASSOCIATION—LYOTROPIC LIQUID CRYSTALS

Increasing the surfactant concentration above the CMC results in the formation of different types of aggregates. Initially, there is a transition from more or less spherical to rodlike or cylindrical micelles. Interestingly, in concentrated aqueous solutions of CTAB and CTACl both spherical and rodlike micelles coexist; the chloride counterions tend to associate with the former, while the bromide counterions predominate around the rodlike micelles (Almgren et al., 1979c). At higher concentrations the liquid crystalline state predominates (Gray, 1962; de

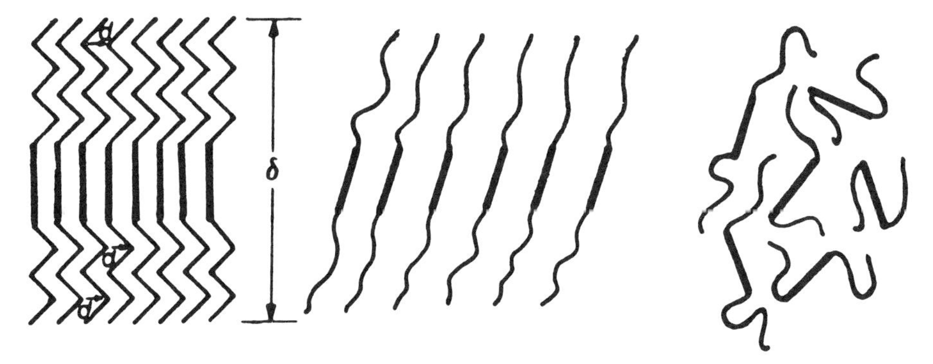

Figure 2.12. Schematic illustrations of the crystalline (left), liquid crystalline (middle), and liquid (right) states. Both short range (d) and long distance (δ) order prevails for the crystalline state. The liquid state is disordered, while the liquid crystalline state has short range disorder and long range order (taken from Friberg, 1977).

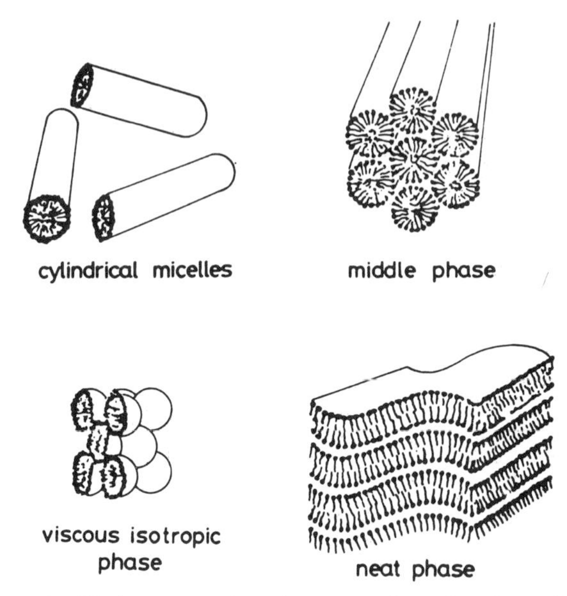

Figure 2.13. Oversimplified representation of structures formed as the surfactant concentration increases in a surfactant-water two-component system.

Gennes, 1974; Brown, 1974; Friberg, 1977; Saeva, 1979). This state is characterized by short range disorder and long range order (Figure 2.12). Several concentration dependent liquid crystalline phases have been recognized. Figure 2.13 shows oversimplified structures of the so-called middle, viscous, and neat liquid crystalline phases. Transitions among these different phases are generally observed by low angle X-ray diffraction techniques (Brown et al., 1971). The presence of a third component, such as alcohol, gives rise to even larger varieties of structures. A lecithin–ethylene glycol system provides the first example of a nonaqueous lyotropic liquid crystalline system (Moucharafich and Friberg, 1979). Polymerization of lyotropic liquid crystals opens the door to interesting technical applications (Thundathil et al., 1980; Friberg et al., 1980).

REFERENCES

Ache, H. J. (1979). In *Positron and Muonium Chemistry*, Advances in Chemistry Series, Vol. **175**, American Chemical Society, Washington, D.C. Positron Chemistry: Present and Future Directions.

Aikawa, M., Yekta, A., and Turro, N. J. (1979). *Chem. Phys. Lett.* **68**, 285–290. Photoluminescence Probes of Micelle Systems. Cyclic Azoalkanes as Quenchers of 1,5-Dimethylnaphthalene Fluorescence.

Aikawa, M., Yekta, A., Liu, J.-M., and Turro, N. J. (1980). *Photochem. Photobiol.* **32**, 297–303. Useful Photoluminescence Probes of Micellar Systems—Cyclic Azoalkanes as Fluorescence Acceptors and 1,5-Dimethylnaphthalene as a Fluorescence Donor.

Almgren, M. (1980). Private communication.

Almgren, M. and Löfroth, J.-E. (1980). *J. Colloid Interface Sci.* Determination of Micelle Aggregation Numbers and Micelle Fluidities from Time Resolved Fluorescence Quenching (in press).

Almgren, M., Aniansson, E. A. G., Wall, S. N., and Holmåker, K. (1977a). In *Micellization, Solubilization and Microemulsions* (K. L. Mittal, Ed.), Plenum Press, New York, pp. 329–345. On the Kinetics of Redistribution of Micellar Sizes.

Almgren, M., Aniansson, E. A. G., and Holmåker, K. (1977b). *Chem. Phys.* **19**, 1–16. The Kinetics of Redistribution of Micellar Sizes. Systems with Exponential Monomer Relaxation.

Almgren, M., Grieser, F., and Thomas, J. K. (1979a). *J. Am. Chem. Soc.* **101**, 279–291. Dynamic and Static Aspects of Solubilization of Neutral Arenes in Ionic Micellar Solutions.

Almgren, M., Grieser, F., and Thomas, J. K. (1979b). *J. Chem. Soc. Faraday Trans. I* **75**, 1674–1687. Rate of Exchange of Surfactant Monomer Radicals and Long Chain Alcohols Between Micelles and Aqueous Solutions. A Pulse Radiolysis Study.

Almgren, M., Löfroth, J.-E., and Rydholm, R. (1979c). *Chem. Phys. Lett.* **63**, 265–268. Co-existence of Rod-like and Globular Micelles in the CTAB-CTAC-H_2O System. Evidence from the Fluorescence of Solubilized Pyrene.

Anacker, E. W. (1979). In *Solution Chemistry of Surfactants* (K. L. Mittal, Ed.), Plenum Press, New York, pp. 247–265. Electrolyte Effect on Micellization.

Aniansson, E. A. G. (1975). In *Chemical and Biological Applications of Relaxation Spectrometry* (E. Wyn-Jones, Ed.), D. Reidel Co., Dordrecht, Holland, 245–253. On the theory of Many Step Processes.

Aniansson, E. A. G. (1978). *J. Phys. Chem.* **82**, 2805–2808. Dynamics and Structure of Micelles and Other Amphiphile Structures.

Aniansson, E. A. G. and Wall, S. N. (1974). *J. Phys. Chem.* **78**, 1024–1030. On the Kinetics of Step-wise Micelle Association.

Aniansson, E. A. G. and Wall, S. N. (1975). *J. Phys. Chem.* **79**, 857–858. A Correction and Improvement of "On the Kinetics of Step-wise Micelle Association" by E. A. G. Aniansson and S. N. Wall.

Aniansson, E. A. G. and Wall, S. N. (1977). *Ber. Bunsenges. Phys. Chem.* **81**, 1293–1294. Comments on "The Kinetics of the Formation of Ionic Micelles. II. Analysis of the Time Constants" by S. K. Chan, U. Herrmann, W. Ostner and M. Kahlweit.

Aniansson, E. A. G. and Wall, S. N. (1980). *J. Colloid Interface Sci.* **78**, 567–568. Comments on "The Dissolution of Micelles in Relaxation Kinetics" by G. Kegeles.

Aniansson, E. A. G., Wall, S. N., Almgren, M., Hoffmann, H., Keilman, I., Ulbricht, W., Zana, R., Lang, J., and Tondre, C. (1976). *J. Phys. Chem.* **80**, 905–922. Theory of the Kinetics of Micellar Equilibria and Quantitative Interpretation of Chemical Relaxation Studies of Micellar Solutions of Ionic Surfactants.

Aoudia, M. and Rodgers, M. A. J. (1979). *J. Am. Chem. Soc.* **101**, 6777–6779. An Investigation of the Fluidity of Alkylbenzene Sulfonate Aqueous Micelles by Fluorescence Spectroscopy.

Atik, S. and Singer, L. A. (1978). *J. Am. Chem. Soc.* **100**, 3234–3235. Fluorescence Quenching by Nitroxyl Radicals in Micellar Environments. A Useful Probe for Studying Micelle-Substrate Interactions.

Atik, S. S. and Singer, L. A. (1979). *J. Am. Chem. Soc.* **101**, 6759–6761. Spectroscopic Studies on Small Aggregates of Amphiphatic Molecules in Aqueous Solution.

Atik, S. S., Kwan, C. L., and Singer, L. A. (1979). *J. Am. Chem. Soc.* **101**, 5696–5702. Multiphase Fluorescence Quenching by a Surfactant Nitroxyl Radical.

Baumgardt, K., Klar, G., and Strey, R. (1979). *Ber. Bunsenges. Phys. Chem.* **83**, 1222–1229. On the Kinetics of Micellization, Measured with Pressure-Jump and Stopped-Flow.

Ben-Naim, A. (1974). *Water and Aqueous Solutions*, Plenum Press, New York.

Ben-Naim, A. (1978). *J. Phys. Chem.* **82**, 792–803. Standard Thermodynamics of Transfer. Uses and Misuses.

Berezin, I. V., Martinek, K., and Yatsimirskii, A. K. (1973). *Russ. Chem. Rev. Eng. Transl.* **42**, 787–802. Physicochemical Foundations at Micellar Catalysis.

Briggs, J., Nicoli, D. F., and Ciccolello, R. (1980). *Chem. Phys. Lett.* **73**, 149–152. Light Scattering from Polydisperse SDS Micellar Solutions.

Brown, G. H. (1974). *Advances in Liquid Crystals*, Academic Press, New York.

Brown, G. H., Doane, J. W., and Neff, U. D. (1971). *A Review of the Structure and Physical Properties of Liquid Crystals*, CRC Press, Cleveland, Ohio.

Brown, J. M. (1979). In *Colloid Science, A Specialist Periodical Report* (D. H. Everett, Senior Reporter), Vol. 3, The Chemical Society, London, pp. 253–292. Structure and Reactivity in Micellar Aggregates.

Bunton, C. A. (1973). *Prog. Solid State Chem.* **8**, 239–281. Micellar Catalysis and Inhibition.

Bunton, C. A. (1979). *Catal. Rev. Sci. Eng.* **20**, 1–56. Reaction Kinetics in Aqueous Surfactant Solutions.

Chan, S. K. and Kahlweit, M. (1977). *Ber. Bunsenges. Phys. Chem.* **81**, 1294–1296. Reply to "Comments on: On the Kinetics of the Formation of Ionic Micelles. II. Analysis of the Time Constants" by E. A. G. Aniansson and S. N. Wall.

Chan, S. K., Herrmann, U., Ostner, W., and Kahlweit, M. (1977a). *Ber. Bunsenges. Phys. Chem.* **81**, 396–402. On the Kinetics of the Formation of Ionic Micelles. II. Analysis of the Time Constants.

Chan, S. K., Herrmann, U., Ostner, W., and Kahlweit, M. (1977b). *Ber. Bunsenges. Phys. Chem.* **81**, 60–66. On the Kinetics of the Formation of Ionic Micelles. I. Analysis of Amplitudes.

Chan, S. K., Herrmann, U., Ostner, W., and Kahlweit, M. (1978). *Ber. Bunsenges. Phys. Chem.* **82**, 380–384. On the Kinetics of Formation of Micelles in Aqueous Solutions. III.

Chan, D. Y. C., Mitchell, D. J., Ninham, B. W., and Pailthorpe, B. A. (1979). In *Water. A Comprehensive Treatise* (F. Franks, Ed.), Plenum Press, New York, pp. 239–278. Solvent Structure and Hydrophobic Solutions.

Chu, B. (1974). *Laser Light Scattering*, Academic Press, New York.

Corti, M., and Degiorgio, V. (1978). *Chem. Phys. Lett.* **53**, 237–240. Investigation of Micelle Formation in Aqueous Solution by Laser-Light Scattering.

Corti, M. and Degiorgio, V. (1981). *J. Phys. Chem.* **85**, 711–717. Quasi-Elastic Light Scattering Study of Intermicellar Interactions in Aqueous Sodium Dodecyl Sulfate.

Cummins, H. Z. and Pike, E. R. (1974). *Photon Correlation and Light Beating Spectroscopy*, Plenum Press, New York.

Cummins, H. Z. and Pike, E. R. (1977). *Photon Correlation Spectroscopy and Velocimetry*, Plenum Press, New York.

Dederen, J. C., Van der Avweraer, M., and Schryver, F. C. (1979). *Chem. Phys. Lett.* **68**, 451–454. Quenching of 1-Methylpyrene by Cu^{2+} in Sodium Dodecylsulfate. A More General Kinetic Model.

de Gennes, P. G. (1974). *The Physics of Liquid Crystals*, Oxford University Press, Oxford, England.

Dehmlow, E. V. and Dehmlow, S. S. (1980). *Phase Transfer Catalysis*, Verlag Chemie, New York.

Dill, K. A. and Flory, P. J. (1980). *Proc. Natl. Acad. Sci. USA* **77**, 3115–3119. Interphases of Chain Molecules: Monolayers and Lipid Bilayer Membranes.

Dill, K. A. and Flory, P. J. (1981). *Proc. Natl. Acad. Sci. USA* **78**, 676–680. Molecular Organization in Micelles and Vesicles.

Elworthy, P. H., Florence, A. T., and Macfarlane, C. B. (1968). *Solubilization of Surface Active Agents*, Chapman and Hall, London.

Eriksson, J. C. and Gillberg, G. (1966). *Acta Chem. Scand.* **20**, 2019–2027. NMR-Studies of the Solubilization of Aromatic Compounds in Cetyltrimethylammonium Bromide Solution.

Fendler, J. H. and Fendler, E. J. (1975). *Catalysis in Micellar and Macromolecular Systems*, Academic Press, New York.

Fendler, J. H. and Patterson, L. K. (1971). *J. Phys. Chem.* **75**, 3907. Comment on "Solubilization of Benzene in Aqueous Cetyltrimethylammonium Bromide Measured by Differential Spectroscopy."

Fernandez, M. S. and Fromherz, P. (1977). *J. Phys. Chem.* **81**, 1755–1761. Lipoid pH Indicators as Probes of Electrical Potential and Polarity of Micelles.

Fisher, L. R. and Oakenfull, D. G. (1977). *Chem. Soc. Rev.* **6**, 25–42. Micelles in Aqueous Solutions.

Frahm, J. and Diekmann, S. (1979). *J. Colloid Interface Sci.* **70**, 440–447. Numerical Calculation of Diffuse Double Layer Properties for Spherical Colloidal Particles by Means of a Modified Nonlinearized Poisson–Boltzmann Equation.

Frahm, J., Diekmann, S., and Haase, A. (1980). *Ber. Bunsenges. Phys. Chem.* **84**, 566–571. Electrostatic Properties of Ionic Micelles in Aqueous Solutions.

Frank, F. (1975). *Water, A Comprehensive Treatise*, Vol. 4, Plenum Press, London.

Friberg, S. (1977). *Naturwissenshaften* **64**, 612–618. Lyotropic Liquid Crystals.

Friberg, S. E., Thundathil, R., and Stoffer, J. O. (1980). *Science* **205**, 607–608. Changed Lyotropic Liquid Crystalline Structure Due to Polymerization of the Amphiphilic Component.

Fromherz, P. (1980). *Chem. Phys. Lett.* **77**, 460–466. Micelle Structure: A Surfactant-Block Model.

Goodman, J. F. and Walker, T. (1979). In *Colloid Science, A Specialist Periodical Report* (C. H. Everett, Senior Reporter), Vol. 3, The Chemical Society, London, pp. 230–252. Micellization in Aqueous Solution.

Grätzel, M. and Thomas, J. K. (1976). In *Modern Fluorescence Spectroscopy* (E. L. Wehry, Ed.), Vol. 2, Plenum Press, New York, pp. 169–216. The Application of Fluorescence Techniques to the Study of Micellar Systems.

Gray, G. W. (1962). *Molecular Structure and the Properties of Liquid Crystals*, Academic Press, New York.

Gunnarsson, G., Jönsson, B., and Wennerström, H. (1980). *J. Phys. Chem.* **84**, 3114–3121. Surfactant Association into Micelles. An Electrostatic Approach.

Gustavsson, H. and Lindmann, B. (1975). *J. Am. Chem. Soc.* **97**, 3923–3930. Nuclear Magnetic Resonance Studies of the Interaction Between Alkali Ions and Micellar Aggregates.

Hall, D. G. (1970a). *Trans. Faraday Soc.* **66**, 1351–1358. Thermodynamics of Solutions of Ideal Multi-component Micelles, Part 1.

Hall, D. G. (1970b). *Trans. Faraday Soc.* **66**, 1359–1368. Thermodynamics of Solutions of Ideal Multi-component Micelles, Part 2.

Hall, D. G. (1972). *Kolloid-Z.* **250**, 895–899. The Application of the Thermodynamic Theory of Ideal Multi-component Micelles to Ionic Micelles.

Hartley, G. S. (1935). *Trans. Faraday Soc.* **31**, 31–50. The Application of the Debye–Hückel Theory of Colloidal Electrolytes.

Hartley, G. S. (1948). *Quart. Rev. Chem. Soc.* **2**, 152–183. State of Solution of Colloidal Electrolytes.

Henglein, A. and Proske, T. (1978a). *Ber. Bunsenges. Phys. Chem.* **82**, 471–476. Two and Three Dimensional Reactions of Ag^{+} Radical Cations in Aqueous Solutions of Anionic Tensides.

Henglein, A. and Proske, T. (1978b). *J. Am. Chem. Soc.* **100**, 3706–3709. Formation and Disappearance of Free Radicals, and the Micellar Equilibrium in the Detergent Sodium 4-(6′-Dodecyl)benzene-sulfonate.

Hoffmann, H., Ulbricht, W., and Tagesson, B. (1978). *Z. Physik. Chem. Neve Folge* **113**, 17–36. Investigations on Micellar Systems of Perfluorodetergents. Evidence for Emulsion-Droplet-like Giant Micelles.

Holzbach, R. T., Oh, S. Y., McDonnell, M. E., and Jamieson, A. M. (1977). In *Micellization, Solubilization and Microemulsions* (K. L. Mittal, Ed.), Plenum Press, New York, pp. 403–418.

Quasielastic Laser Spectrometry Studies of Pure Bile Salt and Bile Salt-Derived Lipid Micellar Systems.

Hunter, T. F. (1980). *Chem. Phys. Lett.* **75**, 152–155. The Distribution of Solubilizate Molecules in Micellar Assemblies.

Infelta, P. P. (1979). *Chem. Phys. Lett.* **61**, 88–91. Fluorescence Quenching in Micellar Solutions and Its Application to the Determination of Aggregation Numbers.

Infelta, P. P. and Grätzel, M. (1979). *J. Chem. Phys.* **70**, 179–186. Statistics of Solubilizate Distribution and Its Application to Pyrene Fluorescence in Micellar Systems. A Concise Kinetic Model.

IUPAC (1972). *Pure Appl. Chem.* **31**, 577–638. IUPAC: Division of Physical Chemistry. Manual of Symbols and Terminology for Physico-chemical Quantities and Units. Appendix II.

James, A. D., Robinson, B. H., and White, N. C. (1977). *J. Colloid Interface Sci.* **59**, 328–336. Dynamics of Small Molecule-Micelle Interactions: Charge and pH Effects on the Kinetics of Interaction of Dyes with Micelles.

Jean, Y. C. and Ache, H. J. (1978). *J. Am. Chem. Soc.* **100**, 984–985. Determination of Critical Micelle Concentrations in Micellar and Reversed Micellar Systems by Positron Annihilation Techniques.

Jean, Y. C., Djermouni, B., and Ache, H. J. (1979). In *Solution Chemistry of Surfactants* (K. L. Mittal, Ed.), Plenum Press, New York, p. 129. Micellar Systems Studied by Positron Annihilation Techniques.

Jencks, W. P. (1969). *Catalysis in Chemistry and Enzymology*, McGraw-Hill, New York.

Jönsson, B. and Wennerström, H. (1981). *J. Colloid Interface Sci.* **80**, 482–496. Thermodynamics of Ionic Amphiphile-Water Systems.

Jungermann, E. (1970). *Cationic Surfactants*, Marcel Dekker, New York.

Kalyanasundaram, K. (1978). *Chem. Soc. Rev.* **7**, 453–472. Photophysics of Molecules in Micelle Forming Surfactant Solutions.

Kalyanasundaram, K. and Thomas, J. K. (1976). *J. Phys. Chem.* **80**, 1462–1473. On the Conformational State of Surfactants in the Solid State and in Micellar Form. A Laser Excited Raman Scattering Study.

Kegeles, G. (1980). *Arch. Biochem. Biophys.* **200**, 279–287. Relaxation Times and Size Distribution Readjustments in Micellar Kinetics.

Kosower, E. M. (1968). *An Introduction to Physical Organic Chemistry*, John Wiley, New York.

Kratohvil, J. P. (1980). *J. Colloid Interface Sci.* **75**, 271–275. Comments on Some Novel Approaches for the Determination of Aggregation Numbers.

Kratohvil, J. P. and Dellicolli, H. T. (1970). *Fed. Proc.* **29**, 1335–1342. Measurement of the Size of Micelles; The Case of Sodium Taurodeoxylate.

Kunitake, T. and Sakamoto, T. (1979). *Bull. Chem. Soc. Jpn.* **52**, 2624–2629. Nucleophilic Ion Pairs. 8. Facile Nucleophilic Cleavage of Dinitrophenylsulfate in the Presence of Micellar Zwitterionic Hydroxymates.

Kunitake, T., Shinkai, S., and Okahata, Y. (1976). *Bull. Chem. Soc. Jpn.* **49**, 540–545. Nucleophilic Ion Pairs I. Enhanced Esterolytic Reactivity of Hydrophobic Ion Pairs in Micellar Systems.

Kunitake, T., Okahata, Y., Tanamachi, S., and Ando, R. (1979). *Bull. Chem. Soc. Jpn.* **52**, 1967–1971. Nucleophilic Ion Pairs. 6. Catalytic Hydrolysis of *p*-Nitrophenyl Acetate by Zwitterionic Hydroxamate Nucleophiles in Representative Micellar Systems.

Kutter, P. Schmitt-Fumian, W. W., and Bachmann, L. (1976). *Proc. Eur. Reg. Congr. Electron Microsc.*, **6**, 119–121. Freeze Etching of Micellar Solutions.

Lessing, H. E. and Von Jena, A. (1979). *Chem. Phys.* **41**, 395–406. Rotational Diffusion of Dyes in Micellar Media from Transient Absorption.

Lianos, P. and Zana, R. (1980). *Chem. Phys. Lett.* **76**, 62–67. Fluroescence Probing Study of the Effect of Medium Chain-Length Alcohols on the Properties of Tetradecyltrimethylammonium Bromide Aqueous Micelles.

Lindman, B. and Wennerström, H. (1980). *Top. Curr. Chem.* **87**, 1–83. Micelles. Amphiphile Aggregation in Aqueous Solution.

Lindman, B., Lindblom, G., Wennerström, H., and Gustavsson, H. (1977). In *Micellization, Solubilization and Microemulsions* (K. L. Mittal, Ed.), Plenum Press, New York, pp. 195–227. Ionic Interactions in Amphiphilic Systems Studied by NMR.

Linfield, W. M. (1976). *Anionic Surfactants*, Marcel Dekker, New York.

McBain, M. E. L. and Hutchinson, E. (1955). *Solubilization*, Academic Press, New York.

Mader, W. J. and Grady, L. T. (1970). In *Physical Methods of Chemistry* (A. Weissberger, Ed.), Part V, Vol. I, John Wiley, New York, pp. 257–308. Determination of Solubility.

Makosza, M. (1979). *Surv. Prog. Chem.* **9**, 1–50. Two Phase Reactions in Organic Chemistry.

Martinek, K., Yatsimirski, A. K., Levashov, A. V., and Berezin, I. V. (1977). In *Micellization, Solubilization and Microemulsions* (K. L. Mittal, Ed.), Plenum Press, New York, pp. 489–508. The Kinetic Theory and the Mechanism of Micellar Effects on Chemical Reactions.

Mazer, N. A. and Benedek, G. B. (1976). *J. Phys. Chem.* **80**, 1075–1085. An Investigation of the Micellar Phase of Sodium Dodecyl Sulfate in Aqueous Sodium Chloride Solutions Using Quasielastic Light Scattering Spectroscopy.

Mazer, N. A., Carey, M. C., and Benedek, G. B. (1977a). In *Micellization Solubilization and Microemulsions* (K. L. Mittal, Ed.), Plenum Press, New York, pp. 359–382. The Size, Shape and Thermodynamics of Sodium Dodecyl Sulfate (SDS) Micelles Using Quasielastic Light Scattering Spectroscopy.

Mazer, N. A., Kwasnick, R. F., Carey, M. C., and Benedek, G. B. (1977b). In *Micellization, Solubilization and Microemulsions* (K. L. Mittal, Ed.), Plenum Press, New York, pp. 383–402. Quasielastic Light Scattering Spectroscopic Studies of Aqueous Bile Salt, Bile Salt-Lecithin and Bile Salt-Lecithin-Cholesterol Solutions.

Mazer, N. A., Carey, M. C., Kwasnick, R. F., and Benedek, G. B. (1979). *Biochemistry* **18**, 3064–3075. Quasielastic Light Scattering Studies of Aqueous Biliary Lipid Systems. Size, Shape, and Thermodynamics of Bile Salt Micelles.

Menger, F. M. (1977). In *Bioorganic Chemistry III. Macro- and Multicomponent Systems* (E. E. van Tanelen, Ed.), Academic Press, New York, pp. 137–152.

Menger, F. M. (1979a). *Acc. Chem. Res.* **12**, 111–117. On the Structures of Micelles.

Menger, F. M. (1979b). *J. Phys. Chem.* **83**, 893. Laplace Pressure Inside Micelles.

Menger, F. M. and Boyer, B. J. (1980). *J. Am. Chem. Soc.* **102**, 5936–5938. Water Penetration into Micelles as Determined by Optical Rotary Dispersion.

Menger, F. M., Jerkunica, J. M., and Johnston, J. C. (1978). *J. Am. Chem. Soc.* **100**, 4676–4678. The Water Content of a Micelle Interior. The Fjord *vs.* Reef Model.

Menger, F. M., Yoshinaga, H., Venkatasubban, K. S., and Das, A. R. (1980). *J. Org. Chem.* **46**, 415–419. Solvolyses of Carbonate and a Benzhydryl Chloride Inside Micelles. Evidence for a Porous Cluster Micelle.

Missel, P. J., Mazer, N. A., Benedek, G. B., Young, C. Y., and Carey, M. C. (1980). *J. Phys. Chem.* **84**, 1044–1057. Thermodynamic Analysis of the Growth of Sodium Dodecyl Sulfate Micelles.

Mittal, K. L. (1977). *Micellization, Solubilization and Microemulsions*, Plenum Press, New York.

Mittal, K. L. (1979). *Solution Chemistry of Surfactants*, Plenum Press, New York.

Moroi, Y., Infelta, P. P., and Grätzel, M. (1979a). *J. Am. Chem. Soc.* **101**, 573–579. Light-Initiated Redox Reactions in Functional Micellar Assemblies. 2. Dynamics in Europium(III) Surfactant Solutions.

Moroi, Y., Braun, A. M., and Grätzel, M. (1979b). *J. Am. Chem. Soc.* **101**, 567–572. Light-Initiated Electron Transfer in Functional Surfactant Assemblies. 1. Micelles with Transition Metal Counterions.

Moucharafich, N. and Friberg, S. E. (1979). *Mol. Cryst. Liq. Cryst.* **49**, 231–238. A First Comparison Between Aqueous and Nonaqueous Lyotropic Liquid Crystalline Systems.

Mukerjee, P. (1977). In *Micellization, Solubilization and Microemulsions* (K. L. Mittal, Ed.), Plenum Press, New York, pp. 171–194. Size Distribution of Micelles: Monomer-Micelle Equilibria. Treatment of Experimental Molecular Weight Data, the Sphere-to-Rod Transition and a General Association Model.

Mukerjee, P. (1979). In *Solution Chemistry of Surfactants* (K. L. Mittal, Ed.), Plenum Press, New York, pp. 153–174. Solubilization in Aqueous Micellar Systems.

Mukerjee, P. and Cardinal, J. R. (1978). *J. Phys. Chem.* **82**, 1620–1627. Benzene Derivatives and Naphthalene Solubilized in Micelles. Polarity of Microenvironment, Location and Distribution in Micelles, and Correlation with Surface Activity in Hydrocarbon-Water Systems.

Mukerjee, P. and Mysels, K. (1971). "Critical Micelle Concentrations of Aqueous Surfactant Systems," National Standards Reference Data Series, Vol. 36, National Bureau of Standards (U.S.).

Mukerjee, P., Cardinal, J. R., and Desai, N. R. (1977). In *Micellization, Solubilization and Microemulsions* (K. L. Mittal, Ed.), Plenum Press, New York, pp. 241–261. The Nature of the Local Microenvironments in Aqueous Micellar Systems.

Muller, N. (1972). *J. Phys. Chem.* **76**, 3017–3020. Kinetics of Micelle Dissociation by Temperature-Jump Techniques. A Reinterpretation.

Muller, N. (1979). In *Solution Chemistry of Surfactants* (K. L. Mittal, Ed.), Plenum Press, New York, pp. 267–295. Kinetics of Micellization.

Muller, N. and Birkhahn, R. H. (1967). *J. Phys. Chem.* **71**, 957–962. Investigation of Micelle Structure by Fluorine Magnetic Resonance. I. Sodium 10,10,10-Trifluorocaprate and Related Compounds.

Nakagawa, T. and Tokiwa, T. (1976). *Surf. Colloid Sci.* **9**, 70–164. Nuclear Magnetic Resonance of Surfactant Solutions.

Narayana, P. A., Li, A. S. W., and Kevan, L. (1981). *J. Am. Chem. Soc.* **103**, 3603–3604. Electron Spin Resonance and Electron Spin Echo Studies of Photoproduced Tetramethylbenzidine Cation Radical in Frozen Aqueous Micellar Solutions: Cation Surroundings and Retention of Micellar Structure in Frozen Solutions.

Nemethy, G. (1967). *Angew. Chem. Int. Ed. Eng.* **6**, 195–206. Hydrophobic Interactions.

Nicoli, D. F., Dawson, D. R., and Offen, H. W. (1979). *Chem. Phys. Lett.* **66**, 291–294. Pressure Dependence of Micelle Size by Photon Correlation Spectroscopy.

Oakenfull, D. and Fenwick, D. E. (1977). *Aust. J. Chem.* **30**, 741–752. Thermodynamics and Mechanism of Hydrophobic Interaction.

Oakes, J. (1973). *J. Chem. Soc. Faraday Trans. II* **69**, 1321–1329. Magnetic Resonance Studies in Aqueous Systems.

Okabayashi, I., Okuyama, M., and Kitagawa, T. (1975). *Bull. Chem. Soc. Jpn.* **48**, 2264–2269. The Raman Spectra of Surfactants and the Concentration Dependence of Their Molecular Conformations in Aqueous Solutions.

Okahata, Y., Ando, R., and Kunitake, T. (1977). *J. Am. Chem. Soc.* **99**, 3067–3072. Remarkable Activation of Anionic Nucleophiles Toward *p*-Nitrophenyl Acetate by Aqueous Trioctylmethylammonium Chloride: A New Class of Hydrophobic Aggregate.

Okubo, T., Kitano, H., Ishiwatari, T., and Ise, N. (1979). *Proc. Roy. Soc. London Ser. A* **366**, 81–90. Conductance Stopped-Flow Study on the Micellar Equilibria of Ionic Surfactants.

Pakusch, A. and Strey, R. (1980). *Ber. Bunsenges. Phys. Chem.* **84**, 1163–1168. On Relaxation Amplitudes in T-Jump Experiments in Micellar Solutions.

Patterson, L. K. and Vieil, E. (1973). *J. Phys. Chem.* **77**, 1191–1192. Comparison of Micellar Effects on Singlet Excited States of Anthracene and Perylene.

Perez de Albrizzio, J. and Cordes, E. H. (1979). *J. Colloid Interface Sci.* **68**, 292–294. Water Activity at the Surface of Ionic Micelles as Measured by the Extent of Hydration of 3-Formyl-*N*-tetradecylpyridinium Bromide.

Pierotti, R. A. (1963). *J. Phys. Chem.* **67**, 1840–1845. The Solubility of Gases in Liquids.

Pierotti, R. A. (1965). *J. Phys. Chem.* **69**, 281–288. Aqueous Solutions of Nonpolar Gases.

Podo, F., Ray, A., and Nemethy, G. (1973). *J. Am. Chem. Soc.* **95**, 6164–6171. Structure and Hydration of Nonionic Detergent Micelles. A High Resolution Nuclear Magnetic Resonance Study.

Quina, F. H. and Toscano, U. G. (1977). *J. Phys. Chem.* **81**, 1750–1754. Photophenomena in Surfactant Media. Quenching of a Water-Soluble Fluorescence Probe by Iodide Ion in Micellar Solutions of Sodium Dodecyl Sulfate.

Reed, W., Politi, M. J., and Fendler, J. H. (1981). *J. Am. Chem. Soc.* **103**, 4591–4593. Rotational Diffusion of Rose Bengal in Aqueous Micelles: Evidence for Extensive Exposure of the Hydrocarbon Chains.

Reynolds, J. A., Gilbert, D. B., and Tanford, C. (1974). *Proc. Natl. Acad. Sci. USA* **71**, 2925–2927. Empirical Correlation Between Hydrophobic Free Energy and Aqueous Cavity Surface Area.

Robb, I. D. and Smith, R. (1974). *J. Chem. Soc. Faraday Trans. I* **70**, 287–292. Nuclear Spin-Lattice Relaxation Times of Alkali Metal Counterions at Micellar Interfaces.

Roberts, R. T. and Chachaty, C. (1973). *Chem. Phys. Lett.* **22**, 348–351. ^{13}C Relaxation Measurements of Molecular Motion in Micellar Solutions.

Robinson, B. H., White, N. C., and Mateo, C. (1975). *Adv. Mol. Relaxation Interact. Processes* **7**, 321–338. Dynamics of Small Molecule-Micelle Interactions. A Stopped-Flow Investigation of Kinetics of Absorption of Acridine (and Related) Dyes into Anionic Micelles and the Characterization of Surfactant-Dye Interactions.

Rodgers, M. A. J. (1981). *Chem. Phys. Lett.* **78**, 509–514. Picosecond Fluorescence Studies of Rose Bengal in Aqueous Micellar Dispersions.

Rodgers, M. A. J. and Wheeler, M. F. D. (1978). *Chem. Phys. Lett.* **53**, 165–169. Quenching of Fluorescence from Pyrene in Micellar Solutions by Cationic Quenchers.

Rodgers, M. A. J., Da Silva, E., and Wheeler, M. E. (1976). *Chem. Phys. Lett.* **43**, 587–591. Fluorescence from Pyrene Solubilized in Aqueous Micelles. A Model for Quenching by Inorganic Ions.

Romsted, L. (1975). Thesis, Indiana University, Bloomington, Indiana. Rate Enhancements in Micellar Systems.

Romsted, L. S. (1977). In *Micellization, Solubilization and Microemulsions* (K. L. Mittal, Ed.), Plenum Press, New York, pp. 509–530. A General Kinetic Theory of Rate Enhancements for Reactions Between Organic Substrates and Hydrophilic Ions in Micellar Systems.

Romsted, L. (1981). Private communication.

Roseman, M. and Jencks, W. P. (1975). *J. Am. Chem. Soc.* **97**, 631–640. Interactions of Urea and Other Polar Compounds in Water.

Saeva, F. D. (1979). *Liquid Crystals, the Fourth State of Matter*, Marcel Dekker, New York.

Schelly, Z. A. and Chao, D. Y. (1979). *Adv. Mol. Relaxation Interact. Processes* **14**, 191–202. Thermodynamic Theory of Solvent-Jump Relaxation Method. Coupled Equilibria [1].

Schelly, Z. A., Chao, D. Y., and Sundani, G. (1979). In *Solution Chemistry of Surfactants* (K. L. Mittal, Ed.), Plenum Press, New York, pp. 323–335. Relaxation Amplitude of Non-ionic Systems Perturbed by Solvent Jump.

Schick, M. J. (1967). *Nonionic Surfactants*, Marcel Dekker, New York.

Shinkai, S. and Kunitake, T. (1976a). *Chem. Lett.* 109–112. Nucleophilic Ion Pairs. Facile Cleavage of an Amide Substrate by Hydroxamate Anions.

Shinkai, S. and Kunitake, T. (1976b). *J. Chem. Soc. Perkin Trans. II*, 980–985. Nucleophilic Ion Pairs. Part II. Micellar Catalysis of Proton Abstraction by Hydroxamate Anions.

Shinoda, K. (1967). *Solvent Properties of Surfactant Solutions*, Marcel Dekker, New York.

Starks, C. M. and Liotta, C. (1979). *Phase Transfer Catalysis: Principles and Techniques*, Academic Press, New York.

Stigter, D. (1974a). *J. Phys. Chem.* **78**, 2480–2485. Micelle Formation by Ionic Surfactants. II. Specificity at Head Groups, Micelle Structure.

Stigter, D. (1974b). *J. Colloid Interface Sci.* **47**, 473–482. Micelle Formation by Ionic Surfactants.

Stigter, D. (1975a). *J. Phys. Chem.* **79**, 1008–1114. Micelle Formation by Ionic Surfactants. III. Model of Stern Layer, Ion Distribution and Potential Fluctuations.

Stigter, D. (1975b). *J. Phys. Chem.* **79**, 1015–1022. Micelle Formation by Ionic Surfactants. IV. Electrostatic and Hydrophobic Free Energy From Stern Gouy Ionic Double Layer.

Sudhölter, E. J. R. and Engberts, J. B. F. N. (1979). *J. Phys. Chem.* **83**, 1854–1859. Salt Effects on the Critical Micellar Concentration, Iodide Counterion Binding, and Surface Micropolarity of 1-Methyl-4-dodecylpyridinium Iodide Micelles.

Sudhölter, E. J. R., Van de Langkruis, G. B., and Engberts, J. B. F. N. (1979). *Recl. Trav. Chim. Pays-Bas Belg.* **99**, 73–82. Micelles. Structure and Catalysis.

Svens, B. and Rosenholm, B. (1973). *J. Colloid Interface Sci.* **44**, 495–504. An Investigation of the Size and Structure of the Micelles in Sodium Octanoate Solutions by Small Angle X-Ray Scattering.

Symons, M. C. R. (1978). *J. Chem. Res.* 140–141. Electrolyte and Cosolvent Effects on the Rates of Water-Catalyzed Deprotonation Reactions: An Explanation Based on Water Structures.

Tachiya, M. (1975). *Chem. Phys. Lett.* **33**, 289–292. Application of a Generating Function to Reaction Kinetics in Micelles. Kinetics of Quenching of Luminescent Probes in Micelles.

Takeda, K., Tatsumoto, N., and Yasunaga, T. (1974). *J. Colloid Interface Sci.* **47**, 108–133. Kinetic Study on Solubilization of Pinacylanol Chloride into the Micelles of Sodium Dodecyl Sulfate by a Stopped-Flow Technique.

Tanford, C. (1972). *J. Phys. Chem.* **76**, 3020–3024. Micelle Shape and Size.

Tanford, C. (1973). *The Hydrophobic Effect*, John Wiley, New York.

Tanford, C. (1977). In *Micellization, Solubilization and Microemulsions* (K. L. Mittal, Ed.), Plenum Press, New York, pp. 119–131. Thermodynamics of Micellization of Simple Amphiphiles in Aqueous Media.

Teubner, M. (1979). *J. Phys. Chem.* **83**, 2917–2920. Theory of Ultrasonic Absorption in Micellar Solution.

Thomas, J. K. (1977). *Acc. Chem. Res.* **10**, 133–138. Effect of Structure and Charge on Radiation-Induced Reactions in Micellar Systems.

Thundathil, R., Stoffer, J. O., and Friberg, S. E. (1980). *J. Polym. Sci. Polym. Chem. Ed.* **18**, 2629–2640. Polymerization in Lyotropic Liquid Crystals. I. Change of Structure During Polymerization.

Turro, N. J. and Yekta, A. (1978). *J. Am. Chem. Soc.* **100**, 5951–5952. Luminescent Probes for Detergent Solutions. A Simple Procedure for Determination of the Mean Aggregation Number of Micelles.

Turro, N. J., Grätzel, M., and Braun, A. M. (1980a). *Angew. Chem. Int. Ed. Eng.* **19**, 675–696. Photophysical and Photochemical Processes in Micellar Systems.

Turro, N. J., Tanimoto, Y., and Gabor, G. (1980b). *Photochem. Photobiol.* **31**, 527–532. Functional Detergent Probes of Micelle Structure. Absorption, Fluorescence and Quenching Measurements.

Ulmius, J., Lindman, B., Lindblom, G., and Drakenberg, T. (1978). *J. Colloid Interface Sci.* **65**, 88–97. ^{1}H, ^{13}C, ^{35}Cl and ^{81}Br NMR of Aqueous Hexadecyltrimethylammonium Salt Solutions: Solubilization, Viscoelasticity, and Counterion Specificity.

Waka, Y., Hamamoto, K., and Mataga, N. (1980). *Photochem. Photobiol.* **32**, 27–35. Heteroexcimer Systems in Aqueous Micellar Solutions.

Weber, G. (1975). *Adv. Protein Chem.* **29**, 1–84. Energetics of Ligand Binding to Proteins.

Weber, W. P. and Gokel, G. W. (1977). *Phase Transfer Catalysis in Organic Synthesis*, Springer, Heidelberg.

Wennerström, H. and Lindman, B. (1979a). *Phys. Rep.* **52**, 1–86. Micelles, Physical Chemistry of Surfactant Association.

Wennerström, H. and Lindman, B. (1979b). *J. Phys. Chem.* **83**, 2931–2932. Water Penetration into Surfactant Micelles.

Williams, E., Sears, B., Allerhand, A., and Cordes, E. H. (1973). *J. Am. Chem. Soc.* **95**, 4871–4873. Segmental Motion of Amphiphatic Molecules in Aqueous Solutions and Micelles. Application of Natural Abundance Carbon-13 Partially Relaxed Fourier Transformation Nuclear Magnetic Resonance Spectroscopy.

Yekta, A., Aikawa, M., Turro, N. J. (1979). *Chem. Phys. Lett.* **63**, 543–548. Photoluminescence Methods for Evaluation of Solubilization Parameters and Dynamics of Micellar Aggregates. Limiting Cases Which Allow Estimation of Partition Coefficients, Aggregation Numbers, Entrance and Exit Rates.

Young, C. Y., Missel, P. J., Mazer, N. A., Benedek, G. B., and Carey, M. C. (1978). *J. Phys. Chem.* **82**, 1375–1378. Deduction of Micellar Shape from Angular Dissymmetry Measurements of Light Scattered from Aqueous Sodium Dodecyl Sulfate Solutions at High Sodium Chloride Concentrations.

Ziemiecki, H. W., Holland, R., and Cherry, W. R. (1980). *Chem. Phys. Lett.* **73**, 145–148. Solute-Micelle Binding Constants. A Simple Fluorescence Quenching Method Which Is Independent of the Fluorescence Lifetime.

CHAPTER

3

SURFACTANTS IN APOLAR SOLVENTS

Molecular association is not limited to aqueous solutions. Formation of ion pairs, small and large molecular weight aggregates in hydrocarbon solvents, has been recognized for some time. In spite of their importance in many industrial processes, the colloid chemical characterization of surfactant aggregation in organic solvents is rather limited (Singleterry, 1955; Pilpel, 1963; Little and Singleterry, 1964; Little, 1966; Fowkes, 1967; Becher, 1967; Elworthy et al., 1968; Kitahara, 1970, 1980; Fendler and Fendler, 1975; Fendler, 1976; Eicke, 1980). A 1976 survey (Kertes and Gutmann, 1976) reported fewer than 200 publications on the basic chemistry of surfactants in organic solvents. By comparison, the number of papers concerned with the characterization of aqueous surfactants exceeds 10,000. Confusion prevailing in the early literature of surfactant aggregates in apolar solvents, reversed or inverted micelles is, therefore, hardly surprising.

Surfactant association in apolar solvents is predominantly the consequence of dipole–dipole and ion pair interactions between the amphiphiles. This is quite different from the opposing hydrophobic attractions–electrostatic repulsions responsible for micellization in water. Concepts derived for surfactant association in water cannot be extended, therefore, to those in apolar solvents. Aggregation behavior in organic solvents is complex. It depends on the nature and concentration of the surfactant as well as on the property of the solvent. Furthermore, solubilities and aggregations of surfactants in organic solvents are dramatically affected by the presence of a deliberately added or unintentionally present third component (Kertes and Gutmann, 1976). Experimental evidence indicates the virtual impossibility of complete water removal from the surfactants (Eicke and Christen, 1978; Herrman and Schelly, 1979; Djermouni and Ache, 1979). The presence of water, at least in trace amounts, was, in fact, suggested to be a prerequisite for surfactant aggregation in organic solvents (Eicke and Christen, 1978; Rouviere et al., 1979a). Most thermodynamic treatments have been limited to two-component systems (Muller, 1975, 1978; Kertes and Gutmann, 1976; Kertes, 1977; Eicke, 1977a; Magid, 1979), although analyses of the aggregation behavior of some surfactant–organic solvent–solubilizate three-component systems have recently become available (Eicke, 1977b, 1979). Section 3.1 presents these treatments with the implied understanding that they are, at best, insufficient for describing the behavior of systems used in membrane mimetic studies. Greater emphasis is placed on the properties of surfactant entrapped water pools in organic solvents (Section 3.2) and on microemulsions (Section 3.3).

1. SURFACTANT ASSOCIATION IN APOLAR SOLVENTS

Solubilities of surfactants in apolar solvents are related to their aggregation behavior. Those with large ions and small counterions are only moderately soluble in organic solvents and are prone, therefore, to aggregate formation. Increasing the temperature increases the solubility of the surfactant. The narrow range of temperature at which there is a steep increase in the solubility, the critical solution temperature or Kraft point, corresponds to enhanced aggregation.

Both charged and uncharged surfactants undergo self-association in apolar solvents. Table 3.1 collects average aggregation numbers, $\bar{n}$ values, for selected surfactants in cyclohexane and benzene. Alkylammonium carboxylates are seen to be present predominantly as trimers. For these surfactants the ^{1}H nmr shift data in benzene and cyclohexane (Fendler et al., 1973a; 1973b; El Seoud et al., 1973) have been shown to be equally compatible with the monomer $\rightleftharpoons$ n-mer type and with the monomer $\rightleftharpoons$ dimer $\rightleftharpoons$ trimer $\rightleftharpoons \cdots \rightleftharpoons$ n-mer type indefinite association (Figure 3.1; Muller, 1975). Treatment of vapor pressure osmometry data allowed a clear distinction between these modes of surfactant aggregation (Lo et al., 1975). In this method values obtained for the number, and weight averaged molecular weights of aggregates and weight fractions of monomers have been utilized assuming different types of self-association. The data were analyzed in terms of monomer $\rightleftharpoons$ n-mer $\rightleftharpoons$ m-mer (1, 2, 4, . . .; 1, 2, 6, . . .; 1, 2, 8, . . .), and sequential types of indefinite self-associations. The sequential indefinite type self-association is described by

$$\begin{array}{llll} S_1 + S_1 & \xrightleftharpoons{K_2} & S_2 & K_2 = \dfrac{[S_2]}{[S_1]^2} \\ S_1 + S_2 & \xrightleftharpoons{K_3} & S_3 & K_3 = \dfrac{[S_3]}{[S_1][S_2]} \\ \vdots \quad \vdots & & \vdots & \vdots \qquad \vdots \\ S_1 + S_{n-1} & \xrightleftharpoons{K_n} & S_n & K_n = \dfrac{[S_n]}{[S_1][S_{n-1}]} \end{array} \tag{3.1}$$

Table 3.1. Average Aggregation Numbers of Surfactants in Organic Solvents

Surfactant	Solvent	$T°$, C	Concentration Range	$\bar{n}$	Method[a]	Reference
Dodecylammonium propionate	Benzene	25	$(3\text{–}7) \times 10^{-3}$ M	3	vp	Herrmann and Schelly, 1979
	Benzene	37	3×10^{-3} M	3+	vp	Lo et al., 1975
	Cyclohexane	20	$(7\text{–}10) \times 10^{-4}$ M	5+	vp	Lo et al., 1975
Dodecylammonium octanoate	Benzene	25	$(2\text{–}5) \times 10^{-3}$ M	3	vp	Herrmann and Schelly, 1979
Dodecylammonium benzoate	Benzene	25	$(6\text{–}10) \times 10^{-3}$ M	3.6	vp	Eicke and Denss, 1978
	Cyclohexane	20	0.8–3, wt %	3.6–9	vp	Kitahara, 1970
Di-*n*-dodecylammonium chloride	Toluene	50	0.1–1, wt %	2.0–3.1	vp	David-Auslaender et al., 1974
Aluminum didodecanoate	Benzene	20	0.001–1, wt %	6–573	vp	McBain and Working, 1947
Sodium bis-2-ethylhexylsulfosuccinate (aerosol-OT)	Cyclohexane	28	1–3, wt %	45–65	ls	Kitahara et al., 1962
	Benzene	28	1–3, wt %	23	ls	Kon-no and Kitahara, 1971
Lithium dinonylnaphthalene sulfonate	Benzene	35	0.5–6, wt %	7	vp	Little and Singleterry, 1964
	Cyclohexane	35	0.5–6, wt %	8	vp	Little and Singleterry, 1964
Sodium dinoylnaphthalene sulfonate	Cyclohexane	35	0.5–8, wt %	8	vp	Little and Singleterry, 1964
	Benzene	35	0.5–10, wt %	7	vp	Little and Singleterry, 1964
α-Monoglyceryl oleate	Benzene	20	0.01–0.06, wt %	19	ls	Robinson, 1960
Lecithin	Benzene	20	0.001–0.01, wt %	80	ls	Elworthy and McIntosh, 1964
	Benzene	25	0.07–1, wt %	73	vp	Elworthy, 1959

[a] vp—vapor pressure osmometry; ls—light scattering.

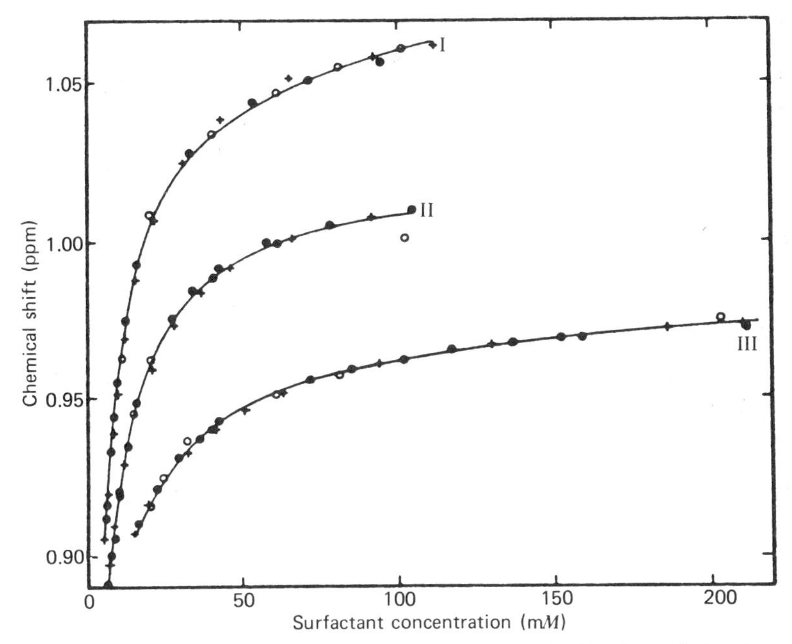

Figure 3.1. Chemical shifts as a function of surfactant concentration for the methyl protons of the propionate ions of octyl (I), hexyl (II), and butyl (III) ammonium propionates in benzene solution: (○) experimental values (Fendler et al., 1973a); (●) calculated values for a single-equilibrium model; (+) calculated values for a multiple-equilibrium model (taken form Muller, 1975).

On the molar concentration scale,

$$\begin{aligned} [S_2] &= K_2[S_1]^2 \\ [S_3] &= K_2K_3[S_1]^3 \\ \vdots\ \ & \quad \vdots \\ [S_n] &= K_2K_3\cdots K_n[S_1]^n \end{aligned} \tag{3.2}$$

and, if all equilibrium constants are assumed to be equal (i.e., $K_2 = K_3 \cdots K_n$), the weight fraction of the monomer, f_1, can be shown to be related to the molecular weight of the monomeric surfactant, M_1, and to the stoichiometric surfactant concentration, Co expressed in the g m^{-1} scale, by

$$\frac{1 - f_1^{1/2}}{f_1} = \frac{1000K_2}{\mathrm{M}_1}Co \tag{3.3}$$

Very significantly, the monomer ⇌ n-mer and the monomer ⇌ n-mer ⇌ m-mer model failed to describe the observed data, while the fit of the data with equation 3.3 is remarkably good (Lo et al., 1975).

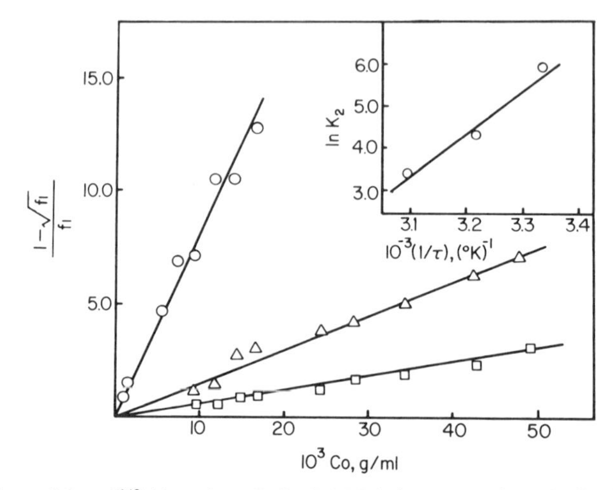

Figure 3.2. Plots of $(1 - f_1^{1/2})/f_1$ against *Co* for DAPB in benzene at 25.0°C (○), 370°C (△), and 50.0°C (□). The temperature dependence of the equilibrium constant for aggregate formation, K_2, is illustrated in the insert (taken from Tsujii et al., 1978).

Similar results were obtained for the association of dodecylammonium pyrene-1-butyrate (DAPB) in benzene (Tsujii et al., 1978). The data, obtained by vapor pressure osmometry, fitted well with equation 3.3 (Figure 3.2). Association of DAPB in benzene follows, therefore, the monomer $\rightleftharpoons$ dimer $\rightleftharpoons \cdots \rightleftharpoons$ *n*-mer indefinite-type aggregation. At each surfactant concentration there is a distribution of aggregates. Increasing the surfactant concentration leads to the formation of larger aggregates in greater percentages. Taking advantage of the fluorophore, the photophysics of DAPB excimer formation has been investigated in ethanol and in benzene (Tsujii et al., 1978). The relative quantum efficiency of excimer formation, $q_1[(\mathrm{DAPB})_2^*]/q_1[(\mathrm{DAPB})^*]$, is substantially greater in benzene (2.58) than in ethanol (0.83). Corresponding values for pyrene excimer formation in hydrocarbon solvents are around 1.1 (Birks, 1970). The ease of DAPB excimer formation efficiency in benzene is a consequence of the self-association of the surfactant, leading to an enhanced microscopic concentration of the pyrene moiety (Tsujii et al., 1978). Initial formation of triple ions has been reported for the association of 1-methyl-4-dodecylpyridinium iodide in dichloromethane (Sudhölter and Engberts, 1977).

A consequence of the sequential-type self-association model, particularly for systems whose average aggregation number is ≤ 4, is that changes in physical properties with increasing surfactant concentrations are expected to be gradual rather than abrupt. Under these circumstances it is often difficult to assign a meaningful critical micelle formation (CMC). Indeed the very concept of CMC has been questioned for surfactants in organic solvents (Kertes and Gutmann, 1976).

Eicke and Denss (1978) have recently provided a general definition of micelles. They considered the independence of aggregates above the "operational"

CMC rather than their sizes as an essential requirement for micellization. The presence of micelles can be established by fitting the experimentally obtained number of weight averaged molecular weights in a given surfactant–apolar solvent system to the monomer ⇌ n-mer or to the sequential self-association model. A CMC is expected only for systems that fit the monomer ⇌ n-mer model. For other systems, particularly those that do not fit either model, the concept of the CMC needs to be deemphasized. Neither model appears to fit the experimentally determined dependency of average aggregation numbers of dodecylammonium propionate (DAP) concentration in benzene (Figure 3.3; Eicke and Denss, 1978). Surprisingly, however, the monomer ⇌ n-mer model describes satisfactorily the concentration dependent aggregation of dodecylammonium benzoate, DABz, in the same solvent (Figure 3.3; Eicke and Denss, 1978). Clearly, subtle differences can alter the aggregation behaviour of surfactants in organic solvents.

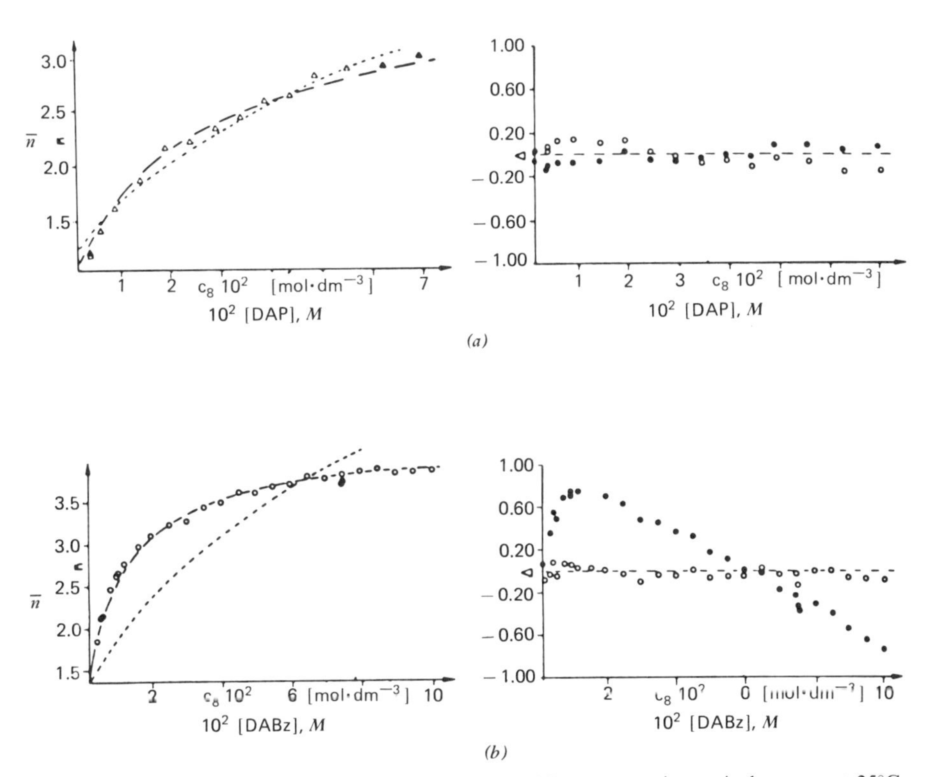

Figure 3.3. (a) Mean aggregation number $\bar{n}$ versus DAP concentration c_s in benzene at 25°C. Pseudophase model (—); multiple-equilibrium model (- - -); residuals (△) of experimental points and theoretical models of the above systems. Pseudophase model (○); multiple-equilibrium model (●). (b) Mean aggregation number, $\bar{n}$, versus DABz concentration c_s in benzene at 25°C. Pseudophase model (—); multiple-equilibrium model (- - -); residuals (△) of experimental points and theoretical models of the above systems. Pseudophase model (○); multiple-equilibrium model (●) (taken from Eicke and Denss, 1978).

Differences in the average aggregation numbers between alklammonium carboxylate and dialkylsulfosuccinate surfactants are even more striking (Table 3.1). Muller (1978) has distinguished between two types of surfactant associations in organic solvents. Type-I association (DAP in benzene, for example) is characterized by sequential indefinite self-association. The aggregation numbers are rather small ($3 \leq \bar{n} \leq 7$) and progressively increase with increasing surfactant concentrations without reaching a limiting constant value. Type-I aggregates do not have pronounced CMCs.

Surfactants that follow type-II association (aerosol-OT, for example) behave quite differently. They have relatively large average aggregation numbers ($12 \leq \bar{n} \leq 30$), which reach constant limiting values, and their CMCs are fairly well-defined (Muller, 1978). Type-II association in organic solvents is somewhat reminiscent, therefore, of aqueous micelle formation. More detailed analysis revealed that the vapor pressure osmometric results of aerosol-OT in benzene could equally well be accommodated either in terms of a monomer $\rightleftharpoons$ n-mer or by a monomer $\rightleftharpoons$ 6-mer $\rightleftharpoons$ 14-mer type association (Tamura and Schelly, 1981, 1981b).

Positronium annihilation techniques have been extensively utilized in the investigations or surfactant association in organic solvents (Jean and Ache, 1978; Djermouni and Ache, 1970 Fucugauchi et al., 1979). By including internal entropy contributions, Muller (1978) calculated the lowest free energies for strongly polar acyclic dimeric and oligomeric species for solvents that have high dielectric constants and for surfactants that have large ionic headgroups. Alkylammonium carboxylates in organic solvents correspond to these situations. Conversely, calculations predict cyclic dimers and compact clusters as favored structures in solvents that have low dielectric constants and for surfactants that have small ionic headgroups. Anionic surfactants in apolar solvents meet these requirements. Type-I and type-II associations are extremes, of course. In reality, mixed associations are likely to prevail (Muller, 1978).

Shapes and sizes of surfactant aggregates in apolar solvents depend strongly on the type and concentration of surfactant, and on the nature of the counterion and solvent. Factors controlling surfactant aggregation in nonpolar solvents have been delineated and their contribution to the standard free energy formation of the aggregates have been calculated in terms of (Ruckenstein and Nagarjan, 1980)

$$\Delta G_g^0 = \frac{\mu_g^0}{g - \mu_1} = \Delta G_{hi}^0 + \Delta G_{ltfm}^0 + \Delta G_{lrfm}^0 + \Delta G_{hbmcb}^0 \quad \text{(when applicable)} \tag{3.4}$$

where ΔG_g^0 is the difference between the standard free energies of an amphiphile within an aggregate of size g and a singly dispersed amphiphile in the solvent, ΔG_{hi}^0 is the amount of difference controlled by headgroup interactions, ΔG_{ltfm}^0 the amount of difference caused by loss of translational freedom of motion,

ΔG^0_{lrfm} the amount of difference caused by loss of rotational freedom of motion, and ΔG^0_{hbmcb} is the amount of difference caused by hydrogen bonding or metal coordination bonding.

2. SURFACTANT SOLUBILIZED WATER IN ORGANIC SOLVENTS—REVERSED MICELLES

Addition of water has dramatic effects on the aggregation of surfactants in hydrocarbons. Simple geometrical considerations allow the estimation of the effect of water on the size of DAP aggregates (Correll et al., 1978).

Average aggregation numbers in 0.08 *M* DAP in cyclohexane in the presence of different amounts of water have been approximated by calculating the radius of a given reversed micellar interior by two different approaches (Correll et al., 1978). The micellar size is taken when the agreement between the two calculations is the best. In both approaches the aggregates, as well as the water pools entrapped therein, are necessarily assumed to be spherical. In the first approach the radii of different sized micellar cavities (r_I values in Table 3.2) are calculated by assuming a surface area of 42 Å^2 for the ionic headgroups of one DAP molecule. This assumption is not unreasonable since the limiting surface area at high pressures of all surfactants with a single headgroup per hydrocarbon chain in monolayers is always found to be 21 Å^2 (Tanford, 1973). In the second approach volumes of surfactant entrapped water pools, and hence their radii (r_{II} values in Table 3.2), are calculated at each cosolubilized water concentration for different aggregation numbers. A value of 1.0 ml g^{-1} was taken for the partial specific volume of water by analogy with the available data on aerosol-OT (Mathews and Hirschorn, 1953). The data are collected in Table 3.2. Best fits, and hence the estimated average aggregation numbers, are given in italics in Table 3.2. Figure 3.4 is an idealized picture of the proposed structure of a DAP aggregate.

In the presence of the smallest amounts of cosolubilized water (0.093 *M*), the calculated radii of water pools (r_{II}) are smaller at any assumed aggregation number than those estimated from the surface areas of DAP headgroups (r_I). Since an empty space within the micellar cavity is hardly feasible, the most likely explanation is that DAP aggregates that contain little or no water are not spherical. This postulate is in accord with Muller's (1978) hypothesis. If water in excess of 0.185 *M* in 0.08 *M* DAP in cyclohexane, $r_I > r_{II}$ at low values of $\bar{n}$, but at higher values of $\bar{n}$, $r_{II} > r_I$ (Table 3.2). The crossover occurs between r_I and r_{II} when an appropriate aggregation number is reached and it is taken to be the best estimate of the size of the reversed micelle at a given cosolubilized water concentration. The estimated average aggregation number is seen to increase with increasing water concentration to 115 (Table 3.2). The observed effects of electrolytes on the solubility of water, relative viscosity, and enthalpy of hydration in aerosol-OT *n*-decane could only be rationalized by assuming spherical shapes for reversed micelles (Rouviere et al., 1979b). Conversely,

Table 3.2. Calculations of Reversed Micellar Parameters in 0.08 *M* DAP in Cyclohexane in the Presence of Different Amounts of Water and Determined Microviscosities Therein

$[H_2O]$, M^a	$\overline{N}^b$	r_I, Å^c	r_{II}, Å^d	Number of Free H_2O Molecules per Aggregate	r_f, Å^e	η, cP^f	$[H_2O]$, M^a	$\overline{N}^b$	r_I, Å^c	r_{II}, Å^d	Number of Free H_2O Molecules per Aggregate	r_f, Å^e	η, cP^f
0.093	5	4.09	3.50	None		70.2	0.417	5	4.09	5.70			
	6	4.48	3.68					10	5.80	7.19			
	7	4.87	3.87					35	10.80	10.90			
	8	5.17	4.05					*38*	*11.26*	*11.27*	46.08	6.90	23.7
	9	5.48	4.21					40	11.70	11.40			
	10	5.80	4.36					45	12.20	11.90			
	11	6.06	4.50					50	12.90	12.30			
	12	6.33	4.64					60	14.10	13.10			
0.185	5	4.09	4.35	None		59.5	0.463	5	4.09	5.91			
	6	4.48	4.63					10	5.80	7.19			
	7	*4.87*	*4.87*					40	11.70	11.82			
	8	5.17	5.09					*45*	*12.24*	*12.30*	80.43	8.31	19.0
	9	5.48	5.29					48	12.66	12.55			
	10	5.80	5.48					50	12.92	12.75			
	11	6.06	5.66					60	14.15	13.53			
	12	6.33	5.83					65	14.73	13.89			

0.277	5	4.09	4.98				0.556	5	4.09	6.28			
	10	5.80	6.28					10	5.80	7.91			
	15	7.08	7.18					50	12.92	13.50			
	18	*7.75*	*7.63*	None		49.6		60	14.15	14.40			
	20	8.18	7.91					62	14.39	14.54			
	25	9.14	8.52					*65*	*14.73*	*14.75*	191.8	11.1	16.4
	30	10.00	9.05					100	18.27	17.00			
	35	10.80	9.53					120	20.02	18.12			
0.370	5	4.09	5.48				0.739	5	4.09	6.91			
	10	5.80	6.91					10	5.80	8.70			
	28	9.67	9.75					50	12.92	14.90			
	30	*10.00*	*9.97*	18.75	5.12	44.5		100	17.00	18.60			
	32	10.30	10.20					110	19.20	19.36			
	35	10.80	10.50					*115*	*19.60*	*19.65*	602	16.26	10.3
	70	11.70	11.00					120	20.02	19.90			
	50	12.90	11.80					150	22.38	21.47			

[a] Stoichiometric concentration of added water.
[b] Assumed mean aggregation number.
[c] Radius of micellar cavities, calculated by assuming (1) surface area of 42 $Å^2$ for the headgroups of one DAP molecule, (2) spherical water pools, and (3) 1.0 mL g^{-1} for the partial specific volume of water.
[d] Radius of 0.08 *M* DAP entrapped water pools, calculated from the known volume of entrapped water assuming (1) spherical water pools and (2) 1.0 mL g^{-1} for the partial specific volume of water.
[e] Radius of *free* water molecules in micellar cavity.
[f] Microviscosity, calculated from pyranine fluorescence data.

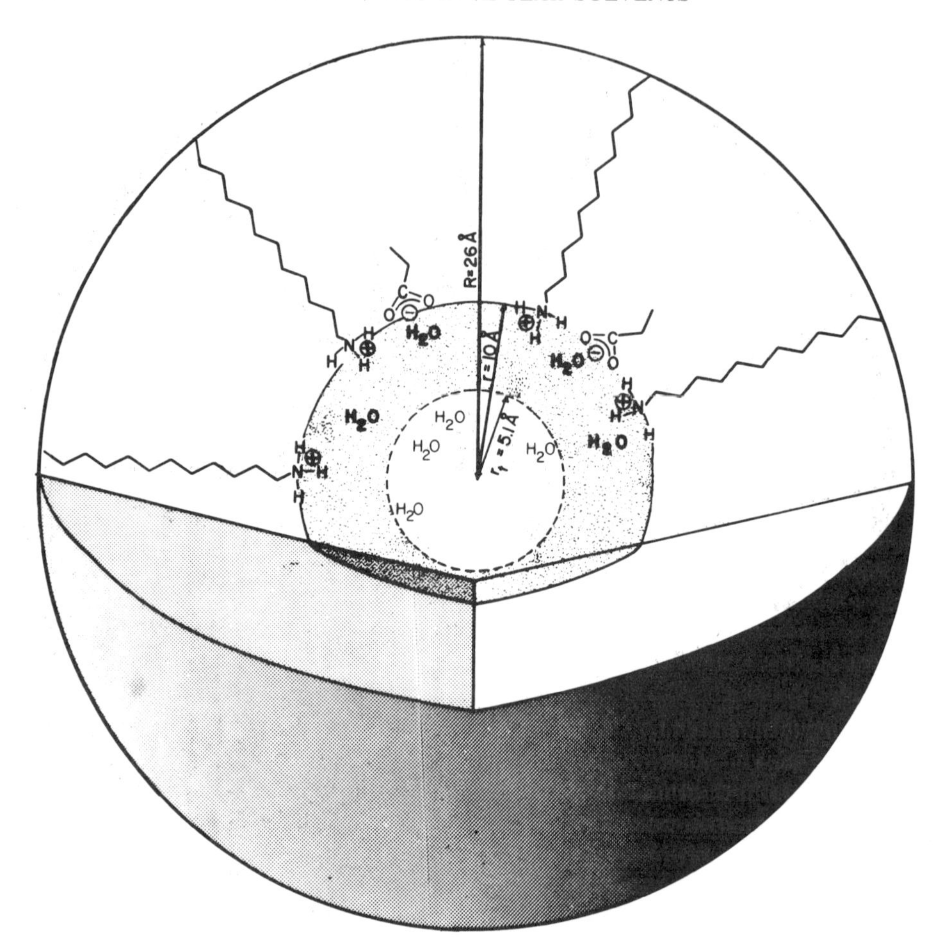

Figure 3.4. An oversimplified representation of 0.08 M DAP aggregate in cyclohexane in the presence of 0.370 M cosolubilized water. The estimated average aggregation number is 30. The radius of the aggregate, R, is 26 Å; that of the water pool, r, is 10 Å, and that of the free water pool, r_f, is 5.1 Å.

evidence has been presented for prolate ellipsoidal aerosol-OT reversed micelles (Ekwall et al., 1970).

The DAP entrapped water pool is not uniform. Addition of small amounts of water to a cyclohexane solution of the surfactant results in the hydration of the amine and the propionate headgroups. Since each DAP molecule was shown to be hydrated by four water molecules in a nonpolar solvent (Fendler et al., 1974), the radius of the free water pool, r_f, can be calculated. The data are given in Table 3.2. All the water molecules are seen to be tied up in hydrating the surfactant headgroups when the concentrations of the stoichiometric water are 0.093, 0.185, and 0.277 M. Free water molecules begin to appear at 0.370 M water concentration. With increasing amounts of cosolubilized water the enlargement of the reversed micelle is paralleled by an increase of free water molecules. Changes in the "microviscosity" of the environment of pyranine in 0.08 *M* DAP in cyclohexane in the presence of different amounts of cosolubilized water substantiate this treatment. It is seen in Table 3.2 and Figure 3.5 that when all the available water molecules are tied up hydrating pyranine, the "microviscosity" is rather high. Anionic pyranine interacts electrostatically with the cationic headgroups of DAP with the resultant partial immobilization of the probe. The appearance of free water molecules in the reversed micellar cavity, at 0.370 *M* stoichiometric water concentration, corresponds to the change in the rate of microviscosity reduction with increasing water concentration (Figure 3.5). At this point pyranine is completely surrounded by water molecules of hydration. As the concentration of free water in the DAP entrapped pool increases further, pyranine is more and more shielded from the charged surfactant headgroups. Consequently, this probe becomes increasingly more mobile and hence it will report lower and lower microviscosities.

The radii and sizes of sodium bis-2-ethylhexylsulfosuccinate (aerosol-OT) aggregates have been directly determined in cyclohexane, toluene, and chlorobenzene, as functions of added water, by viscosity and dynamic photon correlation spectroscopic techniques (Day et al., 1979) as well as by nanosecond time resolved fluorescence polarizations (Keh and Valeur, 1981). Addition of water results in a rapid increase of the average aggregation number and the

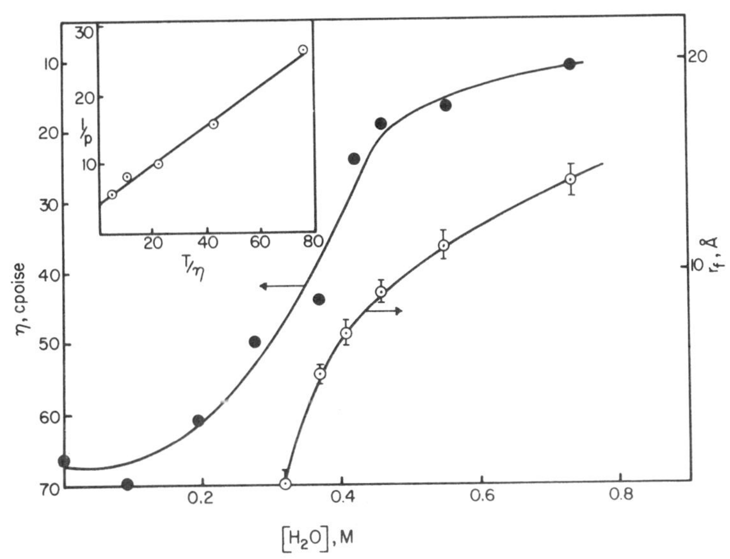

Figure 3.5. A plot of microviscosities of 0.08 *M* DAP in cyclohexane as a function of cosolubilized water concentration (left-hand scale) and a plot of the calculated radius of surfactant entrapped free water pools at different stoichiometric water concentrations (right-hand scale). The insert shows the calibration plot of pyranine fluorescence polarization in glycerol-water mixtures.

size of the surfactant entrapped water pool (Table 3.3). However, the average size of aggregates at given water to surfactant ratios is essentially independent of the concentration of aerosol-OT and of the solvent used (Day et al., 1979). The satisfactory agreement obtained between light scattering and viscosity measurements justifies the assumed spherical shape of the aggregates. Depending on the H_2O:aerosol-OT ratios, the measured diffusion coefficients varied between (1.2 and 1.8)10^{-10} m^2 sec^{-1} in toluene (Day et al., 1979). Sedimentation equilibrium (Robinson et al., 1979) and small-angle neutron scattering (Fletcher et al., 1979) techniques yielded morphological results in good agreement with those obtained by viscosity and photon correlation spectroscopic techniques (Figure 3.6; Eicke and Rehak, 1976; Day et al., 1979).

These and similar studies (Wong et al., 1975, 1977; Eicke and Zinsli, 1978; Eicke and Christen, 1978; Zulauf and Eicke, 1979) have established the formation of sizable surfactant aggregates in organic solvents in the presence of small amounts of water. Their behavior corresponds to that described for type-II aggregation by Muller (1978). It is not unreasonable, therefore, to refer to these systems of surfactant entrapped water pools in apolar solvents (Menger et al., 1973, 1975; Menger and Saito, 1978; Menger and Yamada, 1979) as reversed or inverted micelles (Fendler, 1976). Hydrogen bonding appreciably stabilizes reversed micelles (Zundel, 1969; Eicke and Christen, 1978). The stability of aerosol-OT reversed micelles in isooctance between $-85°$ and $+95°C$ was rationalized in terms of a hydrogen bond network that remained constant over this wide temperature range (Zulauf and Eicke, 1979).

Table 3.3. Average Aggregation Numbers and Radii of 0.10 *M* Aerosol-OT Entrapped Water Pools in Cyclohexane, Toluene, and Chlorobenzene

	Solvent					
	Cyclohexane[a]		Toluene[b]		Chlorobenzene[c]	
R(= $[H_2O]/[AOT]$)	$\bar{n}$	r, Å	$\bar{n}$	r, Å	$\bar{n}$	r, Å
1	27	6.4	41	9.3	25	6.7
2	36	8.1	44	9.9	34	8.7
3	47	10.0	49	10.7	46	10.9
4	59	11.9	56	11.8	59	13.0
5	72	13.7	68	13.5	73	15.0
6	86	15.4	82	15.2	85	16.6
7	101	17.1	97	17.0	99	18.2
8	114	18.6	112	18.7	108	19.3

Source: Compiled from Day et al., 1979.
[a] $\epsilon = 2.02$; temperature = 293.3 K.
[b] $\epsilon = 2.38$; temperature = 294.7 K.
[c] $\epsilon = 5.71$; temperature = 293.3 K.

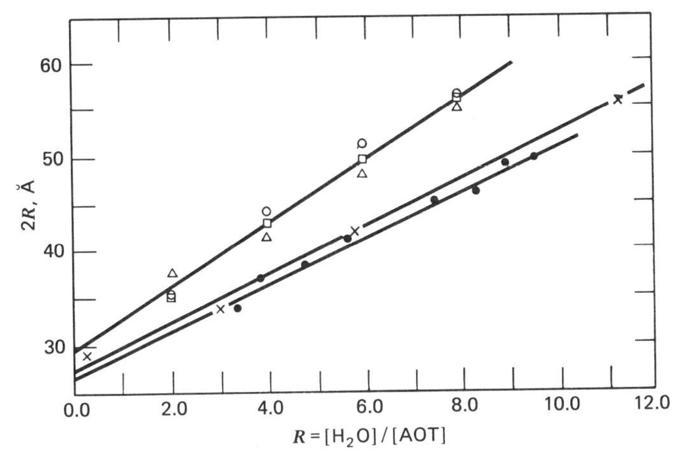

Figure 3.6. Droplet diameter, $2R$, as a function of water : aerosol-OT ratios. Open symbols are viscosity data for solvents: toleune (△); chlorobenzene (○); cyclohexane (□); (Day et al., 1979); ultracentrifuge data (+), using *n*-heptane solvent (Robinson et al., 1979); dynamic light scattering using toluene solvent (●) (taken from Day et al., 1979).

Reversed micelle entrapped water pools resemble polar pockets in enzymes and provide unique microenvironments for substrate solubilization and interactions (Fendler, 1976). Formation of unsolvated cobalt(II) complexes has been, for example, observed in phosphatidylcholine reversed micelles in ether (Sunamoto and Hamada, 1978). Water pools are formed at the polar groups of surfactant aggregates. This contention is substantiated by ^{1}H nmr experiments in DAP reversed micelles in methylene chloride (Fendler and Liu, 1975). Plots of the observed chemical shifts of the magnetically discrete surfactant protons versus the concentration of solubilized water result in a considerable upfield shift of the NH_3 protons and in a somewhat smaller downfield shift of the $CH_2NH_3^+$ and $CH_2CO_2^-$ protons, while the position of other surfactant protons is unaltered. Similar conclusions were reached for the aerosol-OT–water–heptane system (Wong et al., 1977).

Additional insight into the nature of the water of hydration has been obtained by using spin probes (Lim and Fendler, 1978). Analysis of hyperfine coupling constants and line widths of nitroxide spin labels revealed the propionate headgroup of the surfactant to be in a more polar environment than its counterion if there is a sufficiently large water pool entrapped by DAP in benzene. Consequently, there are at least two different types of bound water molecules in the reversed micellar DAP–water–benzene systems: those around the ammonium and those around the carboxylate headgroups (Lim and Fendler, 1978). Extents of bound and free water molecules have also been delineated for aerosol-OT (Wong et al., 1975, 1977) and hexadecyltrimethylammonium chloride (CTACl) reversed micelles (Kondo et al., 1980). Interactions of water with surfactant headgroups in reversed micellar hexadecyltrimethylammonium bromide (CTAB), CTACl, dimethyldioctadecylammonium chloride (DODAC), egg yolk phosphatidylcholine (EL), DAP, aerosol-OT, 3-(hexadecyldimethyl-ammonio)-1-propanesulfonate (HPS), and dodecylbenzenesulfonate (DBS), have been examined in chloroform by infrared spectroscopy (Seno et al., 1980; Sunamoto et al., 1980). Surfactant solubilized water exhibits two absorption bands: one in the 1900–1920 nm region due to free water, and one in the 1920–2020 nm region due to surfactant bound water. Each of these bands was found to shift with increasing concentrations of surfactant entrapped water. Extrapolation to zero water concentration allows the use of infrared data to estimate the tightness of water binding to the surfactant headgroups. Water is bound tighter in cationic (CTAB, CTACl, DODAC) than in DAP, zwitterionic (EL, HPS), or in anionic (Aerosol-OT, DBS) reversed micelles in chloroform (Sunamoto et al., 1980). Significantly, pulse radiolytically generated hydrated electrons could only be observed in aerosol-OT reversed micelles in heptane if sufficient amounts of water were added over and above those molecules required for hydrating the surfactant headgroups (Wong et al., 1975). The first few molecules of water per phosphatidylcholine molecule were established to be bound tightly in lecithin reversed micelles in ether or in benzene by ^{1}H (nmr), visible and fluorescence spectroscopic techniques (Klose and Stelzner, 1974; Poon and Wells, 1974). Additional water molecules occupy the core of the micelle, and their properties resemble bulk water. Bound and free water molecules coexist and exchange rapidly (Poon and Wells, 1974). In reversed micellar aerosol-OT in heptane, ^{1}H and ^{23}Na nmr indicate the initial water molecules to be highly immobilized due to strong ion–dipole interactions (Wong et al., 1977; Fujii et al., 1979). When sufficient water is present to solvate the sodium counterion (six molecules of water per Na^{+}), the rigidity in the micelle is greatly reduced. Interestingly, ^{13}C nmr spin-lattice relaxation measurements of the magnetically discrete reversed micellar aerosol-OT carbon atoms in C_6D_6 in the presence of different amounts of cosolubilized water indicated enhanced mobilities close to the polar headgroup (Eicke and Zinsli, 1978). Increasing amounts of added water enhance, of course, the mobility of the carbon atoms (Figure 3.7).

Studies of self diffusion coefficients provided evidence for distinct spatial separation of hydrophobic and hydrophilic domains in reversed micelles with well defined interfaces between them (Fabre et al., 1981). Surfactant entrapped cores contain most of the water molecules and the counterions are embedded in the environment of surfactant ions. Self diffusion data indicated that reversed micelles retained their spherical shapes even at the highest concentrations of entrapped water (Fabre et al., 1981).

Properties of surfactant entrapped water pools are intimately related to the water to surfactant ratios in the hydrocarbon solvent. Viscosities and polarities, reported by use of reversed micellar entrapped polar probes, can be altered considerably. Thus the macroscopic dielectric constant of aerosol-OT in hexane, containing up to 16.7% solubilized water, resembles that of octane (Menger et al., 1973). Effective polarities experienced by fluorescence probes in aerosol-OT reversed micelles varied as a function of added water concentration in the range

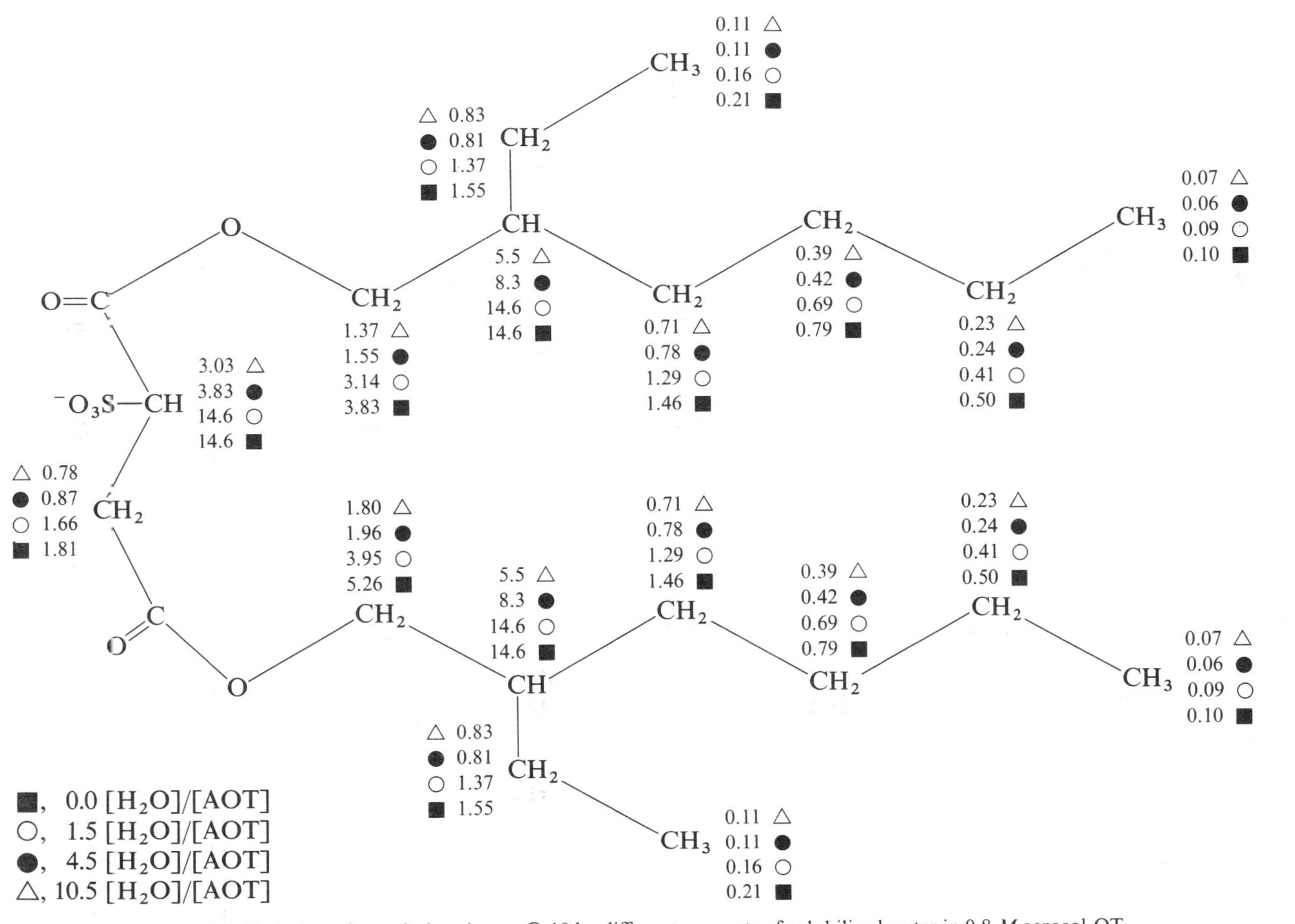

Figure 3.7. Variation of correlation time τ_c ○ 10 by different amounts of solubilized water in 0.8 *M* aerosol-OT in C_6D_6. ^{13}C-nmr measurements, $H_0 = 23.5$ kg; temperature = 25°C (Eicke and Zinsli, 1978).

that corresponded to that between methanol and water (Menger et al., 1973). Polarities of DAP and polyoxyethylene(6)-nonylphenol solubilized water in methylene chloride, benzene, and cyclohexane have been assessed by absorption and fluorescence spectroscopic techniques using 1-ethyl-4-carbomethoxy-pyridinium iodide (Fendler and Liu, 1975), vitamin B_{12} (Fendler et al., 1974), and hemin (Hinze and Fendler, 1975) as extrinsic probes. Effective polarities were related to the size of the surfactant water pools and varied from that resembling pyridine to that corresponding to water. It should be pointed out that large molecules, present even in a very low concentration, can alter the structure and size of reversed micelles. Indeed, enzymes appear to create their own water pools upon solubilizing them in reversed micelles (Menger and Yamada, 1979).

Acidities of surfactant entrapped water pools have been assessed by titrating solubilized dyes (Nome et al., 1976a, 1976b; Menger and Saito, 1978; Miyoshi and Tomita, 19080a, 1980b). Addition of increasing concentrations of $HClO_4$ alters the absorbances of the indicators in a manner analogous to that observed in aqueous titrations. From appropriate plots, pKa^{app} values for bromophenol blue, thymol, methyl orange, and malachite green have been determined and are given in Table 3.4, together with the p*Ka* values of those dyes in water. pKa^{app} values in reversed micellar entrapped water pools are up to seven units lower than the corresponding p*Ka* values in water. Preferential proton concentration in the 0.55 *M* water pools can only account for a decrease of two units in pKa^{app} values. Evidently, Igepal CO-530 restricted volume of water provides an apparently basic environment for the indicators. Increasing the size of the surfactant entrapped water pool results, as expected, in decreased differences between the dissociation constant of surfactant solubilized and bulk water distributed dyes (Nome et al., 1976b).

Apparent ionization constants of dyes depend upon their microenvironments in the reversed micelle entrapped water pools. The p*Ka* value of *p*-nitrophenol

Table 3.4. pKa^{app} Values of Indicators in Water and in Igepal CO-530 Solubilized Water in Benzene at $23 \pm 1°$[a]

Indicator	pKa^{app}	p*Ka* in Water
Bromophenol blue	−2.50	4.07
Thymol blue	−5.00	1.65
Methyl orange	−3.98	3.41
Malachite green	−5.20	2.26

Source: Compiled from Nome et al., 1976b.
[a] [Igepal CO-530] = 0.50 *M*; [H_2O] = 0.55 *M*; concentration of $HClO_4$ is calculated by assuming that all acid is localized in the "water pool," and it is expressed in the H_0 scale.

increases 4.5 units upon its adsorption at the negatively charged sulfosuccinate headgroups in reversed micellar aerosol-OT in heptane (Menger and Saito, 1978). Imidazole, added as buffer, displaces *p*-nitrophenol from its binding site into the water pool with a concomitant decrease in the observed pKa^{app} value (Menger and Saito, 1978). Due care needs to be exercised, therefore, in the use of buffers and added solubilizates.

An acidity scale for aerosol-OT entrapped water pools in isooctane, pH_{wp}, has been defined by measuring the ^{31}P chemical shifts of phosphate buffers (Smith and Luisi, 1980). Chemical shifts in bulk water were compared to those found in reversed micelles assuming the p*Ka* of phosphate ion identical in the two systems. Based on the pH_{wp} values, apparent p*Ka*'s of phenol red and 4-nitrophenol were determined in reversed micelles containing different buffers and different water contents. Since the p*Ka* values were sensitive to these changes, these dyes were considered to have limited utility for determining acidities in reversed micelles (Smith and Luisi, 1980).

Addition of acids or bases to ionic surfactants in nonpolar solvents results in charge neutralization. Infrared titrations revealed the neutralization of DAP to dodecylammonium propionic acid in benzene (Nome et al., 1976a). The rate of oxygen atom exchange between DAP solubilized $H_2^{18}O$ and propionic acid/propionate ion in benzene was found to depend on the size of the water

Table 3.5. Ground- and Excited-State Protonation of Pyrene 1-Carboxylate in Water and in DAP Entrapped Water in Benzene at 25.0°C

Parameter	Water	0.10 *M* DAP in Benzene[a]
pKa^{b}	3.1 ± 0.3	3.2 ± 0.3
pKa^{*c}	4.2 ± 0.4	5.0 ± 0.5
ΔpK^{d}	1.4 ± 0.4	2.0 ± 0.5
k_p^*, M^{-1} sec^{-1e}	$(7.7 \pm 0.4) \times 10^{10}$	$(2.0 \pm 0.2) \times 10^{12}$
k_d^*, sec^{-1}	$(5.3 \pm 0.5) \times 10^{5}$	$(2.2 \pm 0.3) \times 10^{7}$

Source: Compiled from Escabi-Perez and Fendler, 1978.

[a] Containing 0.55 *M* water.

[b] Determined from changes of the absorption spectra of PyCOOH and $PyCO_2^-$ as functions of added H^+.

[c] Determined from changes of the fluorescence lifetimes of $PyCO_2^-$ as functions of added H^+.

[d] $pKa^* - pKa$, calculated from the Forster cycle, using the absorption maxima for $PyCO_2H$ and $PyCO_2^-$ of 348 and 340 nm in water and of 355 and 346 nm in 0.10 *M* DAP in benzene as well as the emission maxima for $(PyCO_2H)^*$ and $(PyCO_2^-)^*$ of 415 and 383 nm in water and of 400 and 385 nm in 0.10 *M* DAP in benzene.

[e] Calculated from the slopes in plots of fluorescence rate parameters versus $[H^+]_{eff}$. $[H^+]_{eff}$ was estimated from infrared titrations.

pool in the presence of mineral acid but not in its absence (Lomax and O'Connor, 1978). The DAP reversed micellar system has been advantageously utilized for the demonstration of dimensionality reductions (Escabi-Perez and Fendler, 1978). Rate and equilibrium constants for the excited state protonation of pyrene 1-carboxylate, described by equation 3.5, are given in Table 3.5.

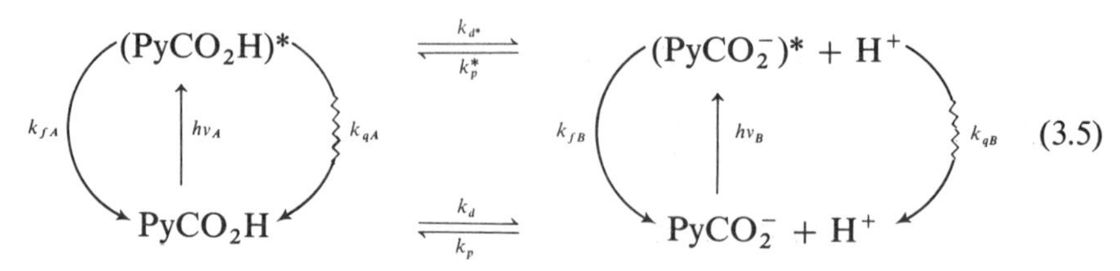

The most striking feature of the data is the extraordinarily large rate constant for the protonation of $(PyCO_2^-)^*$ in the surfactant solubilized water pool; $k_p^* = (2.0 \pm 0.2)10^{12}\ M^{-1}\ \mathrm{sec}^{-1}$! Such ultrafast proton transfer is only feasible if the donor and acceptor are in close proximity within the same hydrogen bond regime. Indeed, protonation of $(PyCO_2^-)^*$, at a given acid concentration, can be regarded as a unimolecular process within an ion pair at the hydrated inner surface of reversed micelles. The reactive partners need not diffuse together—they are there all the time. The significance of these results is that micellar surfaces are able to localize reactants to such an extent as to reduce their reaction dimensionality (Adam and Delbruck, 1968). Reduction of dimensionality is an important and recognized way to bring about fast reactions at membrane surfaces (Richter and Eigen, 1974; Eigen, 1974).

Reversed micellar entrapped water pools are in dynamic equilibrium. Appreciable imidazole catalyzed hydrolysis of *p*-nitrophenyl *p'*-guanidinobenzoate hydrochloride occurred on mixing these two reactants, present initially in separate aerosol-OT entrapped water pools in hexane (Menger et al., 1973). These results have demonstrated the ability of a substrate, entrapped in one set of water pools, to "communicate" with another substrate, located in a different set of surfactant entrapped water pools. Intermicellar communication of surfactant entrapped water pools has been investigated in detail by examining $TbCl_3$ luminescence intensities, enhanced by energy transfer from hydroxyphenylacetic acid, in reversed micelles (Eicke et al., 1976). Emission intensities of $TbCl_3$ increased exponentially with decreasing volumes of aerosol-OT entrapped water pools in osooctane. Dilution of polbe containing reversed micelles, by equal concentrations of aerosol-OT micelles containing *larger* water pools, resulted in the immediate decrease of the emission intensity. Conversely, dilution by empty micelles promptly enhanced the $TbCl_3$ emission intensity. No change in the fluorescence was observed, however, on diluting probe containing reversed micelles, with empty ones having equal concentrations of solubilized water (Eicke et al., 1976). These experimental results have been accommodated in terms of a liquid-sphere model. The model treats collisions between liquid spheres and the resultant deformations. Surfactant molecules have been considered to move outwards from the point of impact during the collision, thereby providing channels through which water and polar solubilizates diffuse from one micelle to the other (Eicke et al., 1976). Similar conclusions have been drawn from the stopped-flow investigation of the kinetics of the reaction between hydrated Ni^{2+} and murexide, contained in separate aerosol-OT solubilized water pools in heptane (Robinson et al., 1979). These studies indicated virtually no energy barrier to the communication of molecules between surfactant solubilized water pools. Water exchanged between pools, solubilized by phospholipid reversed micelles in benzene, faster than the nmr time scale (Chen and Springer, 1979). Conversely, the exchange of lanthanide ions and phospholipid molecules are slow on the nmr time scale. Figure 3.8 shows the proposed dynamics (Chen and Springer, 1979). Addition of ionophors enhanced the cation exchange between phospholipid inverse micelles (Ting et al., 1981).

Substrate solubilization in reversed micelles has been reviewed (Kitahara, 1980). Polar substrates are expected to be localized in water pools. This expectation has been amply substantiated by ^1H nmr experiments (Fendler et al., 1972, 1975a; El Seoud et al., 1974a, 1974b; Fendler and Liu, 1975; El Seoud and Fendler, 1975). Of all the chemical shifts of the magnetically discrete protons of DAP butanoate, and benzoate, only the NH_3^+, $CH_2NH_3^+$, $^-O_2CCH_2$, and $^-OCAr(H_{2,6})$ protons are affected by the addition of 2,3,4,6-tetramethyl-D-glucose (Fendler et al., 1972). The dependence of the chemical shifts on the glucose concentration was found to decrease with increasing separation from the ionic head groups. These results are compatible, of course, with the solubilization of the sugar at the polar core of the reversed micelle (Fendler et al., 1972).

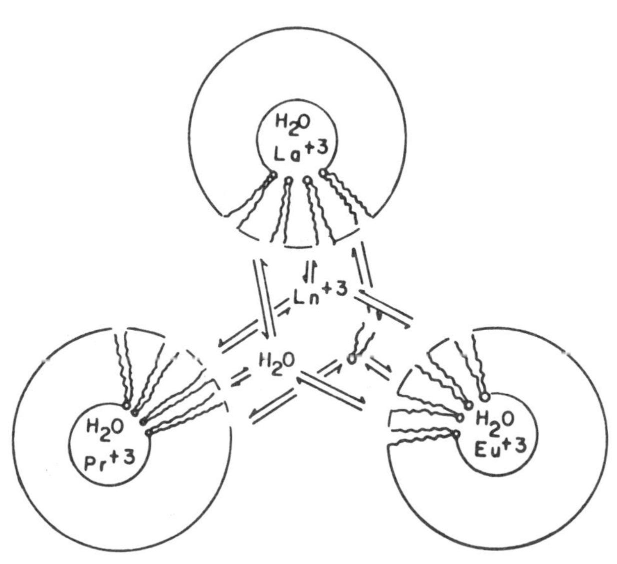

Figure 3.8. The dynamics of water and lanthanide ion exchange between phospholipid entrapped water pools in benzene (taken from Chen and Springer, 1979).

Similarly, addition of Me_2SO, methanol, pyrazole, 2-pyridone, and tetrabutylammonium perchlorate to DAP aggregates in benzene, deuteriochloroform, and dichloromethane affect the NH_3^+ protons of the surfactant most (El Seoud et al., 1974a, 1974b). The predominant interaction of ethylpyridinium bromide with micellar DAP in methylene chloride is also at the polar core (Fendler and Liu, 1975). Imidazole, methanol, and pyrazole have also been shown to be localized in the cavities of reversed micellar aerosol-OT (El Seoud and Fendler, 1975).

Positions of pyrenebutyric and pyrenesulfonic acid in DAP aggregates in cyclohexane have been assessed by studying the quenching of their fluorescence (Correll et al., 1978). Two types of quenchers were used. Potassium bromide, a polar quencher, is taken up in the surfactant solubilized water pools. Carbon tetrachloride, a nonpolar quencher, is distributed in bulk cyclohexane. The fluorescence lifetime of ionized pyrenebutyric acid, solubilized by DAP in cyclohexane, is efficiently quenched by carbon tetrachloride, but it is unaffected by potassium bromide. The excitation energy of pyrenesulfonic acid in the same system, on the other hand, is quenched by carbon tetrachloride and by potassium bromide, but rate constants for quenching by the latter depend on the size of the water pool (Correll et al., 1978). These results imply that pyrenebutyric acid is lined up along the alkyl chains of the surfactant such that its carboxylate group is close to the micellar core but its aromatic moiety is near the bulk hydrocarbon solvent. The hydrocarbon layer of the DAP surfactant is sufficiently thick to prevent energy transfer from the pyrene ring to the ionic quencher, localized in the micellar interior. Pyrenesulfonic acid, having no alkyl chain, is pulled in closer to the micellar core, presumably by dipole–dipole attractions between the micellar headgroups and the sulfonate ions; thus energy transfer occurs with both quenchers (Correll et al., 1978).

The chemical nature of the surfactant headgroups and their possible interaction with solubilizates must be kept in mind. In the solvolysis of 2,4-dinitrophenyl sulfate in polar cavities of reversed micelles in benzene, for example, constituents of the alkylammonium carboxylate surfactants caused general acid and general base catalysis (O'Connor et al., 1973). Formation of chelatelike complexes between the surfactant headgroups and metal ions, entrapped in water pools, solubilized by dodecylpyridinium chloride aggregates in chloroform has also been reported (Masui et al., 1977a, 1977b).

Free energies of transfer, ΔF_t, of amino acids from DAP trapped water in hexane to bulk water have been determined (Fendler et al., 1975b). The data can be clearly divided into two groups: hydrophobic and charged or polar amino acids. Within each group ΔF_t values increase in the order of alanine $<$ proline $<$ valine $\leq$ phenylalanine $\approx$ isoleucine $\approx$ leucine and glycine $<$ arginine $<$ glutamic acid $<$ histidine $<$ aspartic acid. More significantly, values of ΔF_t ranged from -200 to $+500$ cal mole^{-1} (Fendler et al., 1975b). Such a selectivity in the amino acid uptake substantiates the proposed micellar model for the prebiotic compartmentalization of amino acids and nucleotides (Nagyvary and Fendler, 1974).

Solubilization in reversed micelles is best discussed in terms of adsorption of the substrate onto and/or into the micellar surface (Kitahara et al., 1976; Kitahara, 1980). Concentration-jump experiments on the interaction of picric acid with reversed micellar aerosol-OT in benzene revealed a fast (millisecond range) and a slow (10^{-1} sec range) process (Tamura and Schelly, 1979, 1981a, 1981b). The fast process was attributed to adsorption of picric acid to a poor binding site of the micelle while the slow process corresponded to the transport of the solubilizate to a more polar site in the interior of the surfactant entrapped water pool (Tamura and Schelly, 1979, 1981a, 1981b).

The extent of substrate binding depends on the polarity of the substrate, the nature and concentration of the surfactant, the amount of cosolubilized water, and the polarity of the bulk nonpolar solvent. At present, there are insufficient data for establishing relationships among these parameters. Qualitatively, it appears that in a given system the more polar the solubilizate, the greater the binding constant. Conversely, for a given solubilizate and micelle the less polar the bulk solvent the greater the substrate-micelle interaction. The accumulation of considerably more data on the properties of surfactant aggregates in nonpolar solvents and on the magnitude and size of solubilizate interactions therein is clearly required in order to rationalize the kinetics and thermodynamics of substrate reversed micelle interactions.

3. MICROEMULSIONS

At present, there is not a precise, or indeed agreed upon, definition of microemulsions (Prince, 1977; Mittal, 1977, 1979; Friberg and Buraczewska, 1977; Rance and Friberg, 1977; Mackay et al., 1977; Rosoff, 1978). Discussion is limited, therefore, to systems utilized in membrane mimetic studies. In this context microemulsions are considered to be stable, optically transparent, monodisperse droplets of oil (in water) or of water (in hydrocarbon), having diameters in the range of 50–1000 Å (Fendler, 1980).

Increasing the size of surfactant entrapped water pools results in the formation of water-in-oil (w/o) microemulsions. A clear transition from aerosol-OT micelle solubilized water pools in isooctane to w/o microemulsions has been established by photon correlation spectroscopy (Zulauf and Eicke, 1979). In the absence of deliberately added water the hydrodynamic radius, $\bar{R}_H$, of aerosol-OT reversed micelles, 15.0 ± 0.3 Å, is independent of the surfactant concentration and of the temperature in the ranges of 8.0×10^{-3} to 2.0×10^{-1} M and $-20°$ to $+95°$C (Figure 3.9). At water to aerosol-OT ratios of less than 10 increasing concentrations of cosolubilized water result in increased $\bar{R}_H$. $\bar{R}_H$ remains independent of the surfactant concentration and of the temperature in the 0°–50°C range. Above 50°C $\bar{R}_H$ decreases on standing to a limiting value of 15 Å. At water to aerosol-OT ratios of greater than 10, and at $(5–8) \times 10^{-2}$ M surfactant concentrations, $\bar{R}_H$ values of the aggregates are strongly temperature dependent (Figure 3.9). These data place the appearance

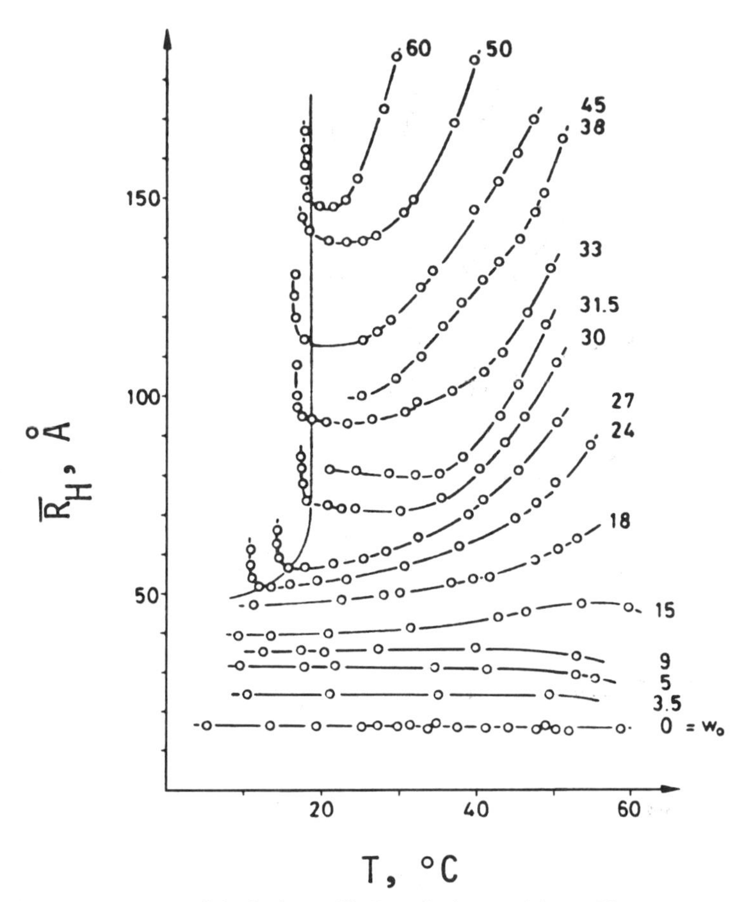

Figure 3.9. Measurements of the Stokes radii, R_H, of microemulsions of isooctane, aerosol-OT, and water as a function of the temperature, *T*. Aerosol-OT concentrations were between 5.5 and 8.1×10^{-2} *M*, the scattering angles varied between 8 and 90°. w_0 is the molar ratio of water to surfactant. The vertical line at 18°C indicates a stability boundary, to the left of which occurs spontaneous growth of the aggregates, which will ultimately lead to phase separation. The data points shown in this region indicate successive measurements in 30 sec time intervals (taken from Zulauf and Eicke, 1979).

of microemulsions in isooctane at water to aerosol-OT ratios of 10. Below this ratio most of the solubilized water molecules are assumed to be structured by hydrogen bonding to the surfactant headgroup. At higher ratios of cosolubilized water there are several pseudophases present in the w/o microemulsion whose behavior is extremely sensitive to such external parameters as temperature, pressure, and mutual solubility of the coexisting phases (Zulauf and Eicke, 1979). Electrical conductivity in w/o microemulsions has been rationalized in terms of a model that includes the dissociation of surfactants within and charge transfer between the aggregates (Eicke and Denss. 1979).

Small-angle neutron scattering has indicated w/o microemulsions to be of the hard-sphere type (Ober and Taupin, 1980).

The relationship between aqueous micelles and oil-in-water (o/w) microemulsions is not straightforward (Mackay et al., 1977). Progressive addition of a surfactant in the absence or in the presence of a cosurfactant (typically an alcohol) to water leads to microemulsions. Phase diagrams are needed for the complete characterization of microemulsions (Prince, 1977). Average sizes of sodium oleate–hexanol–hexadecane o/w microemulsions and the properties of microenvironments they provide have been assessed by photophysical investigations (Gregoritch and Thomas, 1980; Almgren et al., 1980; Tricot et al., 1981). The average dimension of the microemulsion droplets was obtained by treating the intramicellar azulane quenching of methylanthracene triplets by Poisson statistics in a manner similar to that used for aqueous micelles (see Section 2.2). The calculated average radius of a microemulsion droplet, 125 Å, agreed reasonably well with that obtained by light scattering, 158 Å (Gregoritch and Thomas, 1980; Almgren et al., 1980). Fluorescence depolarization of probes located in different parts of the microemulsions were interpreted in terms of relatively fluid environments. Microemulsions provide for pyrene, however, a somewhat more hydrophobic environment than that of aqueous micelles (Gregoritch and Thomas, 1980). This is a direct consequence of the accepted picture of o/w microemulsions. They are considered to be spherical particles having appreciably hydrocarbon interiors (the oil drop). Indicators, located at the interphase of o/w cationic CTAB, anionic hexadecyl sulfate, and nonionic B200, J300, Tween 40, and Brij-96 microemulsions were reported to have an effective dielectric constant of approximately 20 Debye for their environments (Mackay et al., 1980). The intrinsic pKa values of the indicators appeared to be independent of the nature of the microemulsion (Mackay et al., 1980). In this respect microemulsions differ substantially from aqueous micelles.

As expected, structures of o/w microemulsions depend somewhat on the type and composition of surfactants used (Bansal et al., 1979). Properties of nonionic microemulsions have been recently examined (Mackay and Agarwal, 1978; Hermansky and Mackay, 1980; Kumar and Balasubramanian, 1980). Treatment of light scattering data indicated the gradual size increases of Brij-96 and Tween 60 nonionic microemulsions with increasing oil content (Hermansky and Mackay, 1980). This behavior should be contrasted with the abrupt discontinuities observed in turbidity measurements (Hermansky and Mackay, 1980).

Microemulsions have not been as well characterized as have micelles. The emergence of readily available new techniques will surely result in improved understanding of these potentially useful systems. Positron annihilation, for example, has been found to be an extremely sensitive technique for determining phase changes in microemulsions (Boussaha et al., 1980).

REFERENCES

Adam, G. and Delbrück, M. (1968). In *Structural Chemistry and Molecular Biology* (A. Rich and N. Davidson, Eds.), Freeman & Co., San Francisco, California, pp. 198–215. Reduction of Dimensionality in Biological Diffusion Processes.

Almgren, M., Grieser, F., and Thomas, J. K. (1980). *J. Am. Chem. Soc.* **102**, 3188–3193. Photochemical and Photophysical Studies of Organized Assemblies. Interaction of Oils, Long-Chain Alcohols, and Surfactants Forming Microemulsions.

Bansal, V. K., Chinnaswamy, K., Ramachandran, C., and Shah, D. O. (1979). *J. Colloid Interface Sci.* **72**, 524–537. Structural Aspects of Microemulsions Using Dielectric Relaxation and Spin Label Techniques.

Becher, P. (1967). In *Nonionic Surfactants* (M. J. Schick, Ed.), Marcel Dekker, New York, pp. 478–515. Micelle Formation in Aqueous and Nonaqueous Solutions.

Birks, J. B. (1970). *Photophysics of Aromatic Molecules*, John Wiley, New York.

Boussaha, A., Djermouni, B., Fucugauchi, L. A., and Ache, H. J. (1980). *J. Am. Chem. Soc.* **102**, 4654–4658. Microemulsion Systems Studied by Positron Annihilation Techniques.

Chen, S.-T. and Springer, C. S. (1979). *Chem. Phys. Lipids* **23**, 23–40. Hyperfine Shift NMR Studies of Hydrated Phospholipid Inverted Micelles.

Correll, G. D., Cheser, R. N., III, Nome, F., and Fendler, J. H. (1978). *J. Am. Chem. Soc.* **100**, 1254–1265. Fluorescence Probes in Reversed Micelles. Luminescence Intensities, Lifetimes, Quenching, Energy Transfer and Depolarization of Pyrene Derivatives in Cyclohexane in the Presence of Dodecylammonium Proportionate Aggregates.

David-Auslaender, J., Gutmann, H., Kertes, A. S., and Zangen, M. (1974). *J. Solution Chem.* **3**, 251–260. Nonideal Behavior of Alkylammonium Salts in Organic Solvents at Elevated Temperatures.

Day, R. A., Robinson, B. H., Clarke, J. H. R., and Doherty, J. V. (1979). *J. Chem. Soc. Faraday Trans. I* **75**, 132–139. Characterization of Water-Containing Reversed Micelles by Viscosity and Dynamic Light Scattering Methods.

Djermouni, B. and Ache, H. J. (1979). *J. Phys. Chem.* **83**, 2476–2479. Effect of Temperature and Counterion on the Micelle Formation Process Studied by Positron Annihilation Techniques.

Eicke, H. F. (1977a). In *Micellization, Solubilization and Microemulsions* (K. L. Mittal, Ed.), Plenum Press, New York, pp. 429–444. Micelles in Apolar Media.

Eicke, H. F. (1977b). *J. Colloid Interface Sci.* **59**, 308–318. Thermodynamical and Statistical Considerations on Some Microemulsion Phenomena.

Eicke, H. F. (1979). *J. Colloid Interface Sci.* **68**, 440–450. On the Cosurfactant Concept.

Eicke, H. F. (1980). *Top. Curr. Chem.* **87**, 85–145. Surfactants in Nonpolar Solvents. Aggregation and Micellization.

Eicke, H. F. and Christen, H. (1978). *Helv. Chim. Acta* **61**, 2258–2263. Is Water Critical to the Formation of Micelles in Apolar Media?

Eicke, H. F. and Denss, A. (1978). *J. Colloid Interface Sci.* **64**, 386–388. The Definition of a Micelle Revisited.

Eicke, H.-F. and Denss, A. (1979). In *Solution Chemistry of Surfactants* (K. L. Mittal, Ed.), Plenum Press, New York, pp. 699–706. An Electrical Conductivity Model of W/O Microemulsions.

Eicke, H. F. and Rehak, J. (1976). *Helv. Chim. Acta* **59**, 2883–2891. On the Formation of Water/Oil-Microemulsions.

Eicke, H. F. and Zinsli, P. E. (1978). *J. Colloid Interface Sci.* **65**, 131–140. Nanosecond Spectroscopic Investigations of Molecular Processes in W/O Microemulsions.

Eicke, H. F., Shepherd, J. C. W., and Steinemann, A. (1976). *J. Colloid Interface Sci.* **56**, 168–176. Exchange of Solubilized Water and Aqueous Electrolyte Solutions Between Micelles in Apolar Media.

Eigen, M. (1974). In *Quantum Statistical Mechanics in the Natural Sciences* (B. Kuseneglu, S. L. Mintz, and S. Windmayer, Eds.), Plenum Press, New York, pp. 37–58. Reduction of Dimensionality.

Ekwall, P., Mandell, L., and Fontell, K. (1970). *J. Colloid Interface Sci.* **33**, 215–235. Some Observations on Binary and Ternary Aerosol-OT Systems.

El Seoud, O. A. and Fendler, J. H. (1975). *J. Chem. Soc. Faraday Trans. I* **71**, 452 460. Proton Magnetic Resonance Investigations of the Interactions of Aerosol-OT with Imidazole, Methanol and Pyrazole in Carbon Tetrachloride.

El Seoud, O. A., Fendler, E. J., Fendler, J. H., and Medary, R. T. (1973). *J. Phys. Chem.* **77**, 1876–1882. Proton Magnetic Resonance Investigations of Alkylammonium Carboxylate Micelles in Nonaqueous Solvents. III. Effects of Solvents.

El Seoud, O. A., Fendler, E. J., and Fendler, J. H. (1974a). *J. Chem. Soc. Faraday Trans. I* **70**, 450–458. Proton Magnetic Resonance Investigations of Alkylammonium Carboxylate Micelles in Nonaqueous Solvents. Part 4. Effects of Dimethylsulfoxide, Imidazole, Methanol, Pyrazole, 2-Pyridone, and Tetrabutylammonium Perchlorate on Dodecylammonium Propionate in Benzene, Deuterochloroform and Dichloromethane.

El Seoud, O. A., Fendler, E. J., and Fendler, J. H. (1974b). *J. Chem. Soc. Faraday Trans. I* **70**, 459–470. Proton Magnetic Resonance Investigations of Alkylammonium Carboxylate Micelles in Nonaqueous Solvents. Part 5. Effects of Dodecylammonium Proportionate on Solubilizates in Benzene and Deuterochloroform.

Elworthy, P. H. (1959). *J. Chem. Soc.*, 813–817. Micelle Formation by Lecithin in Benzene.

Elworthy, P. H. and McIntosh, D. S. (1964). *Kolloid-Z. Z. Polym.* **195**, 27–34. The Effect of Solvent Dielectric Constant on Micellisation by Lecithin.

Elworthy, P. H., Florence, A. T., and Macfarlane, C. B. (1968). *Solubilization of Surface Active Agents*, Chapman and Hall, London.

Escabi-Perez, J. R. and Fendler, J. H. (1978). *J. Am. Chem. Soc.* **100**, 2234–2236. Ultrafast Excited State Proton Transfer in Reversed Micelles.

Fabre, H., Kamenka, N., Lindman, B. (1981). *J. Phys. Chem.* **85**, 3493–3501. Aggregation in Three Components Surfactant Systems from Self-Diffusion Studies. Reversed Micelles, Microemulsions, and Transition to Normal Micelles.

Fendler, J. H. (1976). *Acc. Chem. Soc.* **9**, 153–161. Interactions and Reactions in Reversed Micellar Systems.

Fendler, J. H. (1980). *J. Phys. Chem.* **84**, 1485–1491. Microemulsions, Micelles and Vesicles as Media for Membrane Mimetic Photochemistry.

Fendler, J. H. and Fendler, E. J. (1975). *Catalysis in Micellar and Macromolecular Systems*, Academic Press, New York.

Fendler, J. H. and Liu, L.-J. (1975). *J. Am. Chem. Soc.* **97**, 999–1003. Charge Transfer Interactions in Nonpolar Solvents in the Presence of Surfactant Aggregates. Ethylpyridinium Bromide Charge-Transfer Complexes in Methylene Chloride Solutions of Dodecylammonium Propionate.

Fendler, J. H., Fendler, E. J., Medary, R. T., and Woods, V. A. (1972). *J. Am. Chem. Soc.* **94**, 7288–7295. Catalysis by Reversed Micelles in Nonpolar Solvents. Mutarotation of 2,3,4,6-Tetramethyl-α-D-glucose in Benzene and in Cyclohexane.

Fendler, J. H., Fendler, E. J., Medary, R. T., and El Seoud, O. A. (1973a). *J. Chem. Soc. Faraday Trans. I* **69**, 280–288. Proton Magnetic Resonance Investigations of the Formation of Alkylammonium Propionate Micelles in Benzene and in Carbon Tetrachloride.

Fendler, E. J., Fendler, J. H., Medary, R. T., and El Seoud, O. A. (1973b). *J. Phys. Chem.* **77**, 1432–1436. Proton Magnetic Resonance Investigations of Alkylammonium Carboxylate Micelles in Nonaqueous Solvents. II. Effects of Carboxylate Structure in Benzene and in Carbon Tetrachloride.

Fendler, J. H., Nome, F., and Van Woert, H. C. (1974). *J. Am. Chem. Soc.* **96**, 6745–6753. Effects of Surfactants on Ligand Exchange Reactions in Vitamin $B_{12}a$ in Water and in Benzene. Influence of Water and of Solvent Restriction.

Fendler, E. J., Chang, S. A., and Fendler, J. H. (1975a). *J. Chem. Soc. Perkin Trans. II*, 482–486. Interaction of Sodium Methoxide with 4-Nitropyridine *N*-Oxide in Benzene in the Presence of Surfactant Aggregates.

Fendler, J. H., Nome, F. J., and Nagyvary, J. (1975b). *J. Mol. Evol.* **6**, 215–232. Compartmentalization of Amino Acids in Surfactant Aggregates. Partitioning Between Water and Aqueous Micellar Sodium Dodecanoate and Between Hexane and Dodecylammonium Propionate Trapped Water in Hexane.

Fletcher, P. D. I., Robinson, B. H., Bermejo, F., Dore, J. C., and Steytler, D. C. (1979). Private communication to be published. Small-Angle Neutron Scattering from Water-Containing Reversed Micelle System.

Fowkes, F. M. (1967). In *Solvent Properties of Surfactant Solutions*, (K. Shinoda, Ed.), Marcel Dekker, New York, pp. 65–115. The Interactions of Polar Molecules, Micelles, and Polymers in Nonaqueous Media.

Friberg, S. and Buraczewska, I. (1977). In *Micellization, Solubilization and Microemulsions* (K. L. Mittal, Ed.), Plenum Press, New York, pp. 791–799. Microemulsions Containing Ionic Surfactants.

Fucugauchi, L. A., Djermouni, B., Handel, E. D., and Ache, H. J. (1979). *J. Am. Chem. Soc.* **101**, 2841–2844. Study of Micelle Formation in Solutions of Alkylammonium Carboxylates in Apolar Solvents by Positron Annihilation Techniques.

Fujii, H., Kawai, T., and Nishikawa, H. (1979). *Bull. Chem. Soc. Jpn.* **52**, 1978–1983. Hydrolytic Reactions of Para-Nitrophenyl Esters in Reversed Micellar Systems.

Gregoritch, S. J. and Thomas, J. K. (1980). *J. Phys. Chem.* **84**, 1491–1495. Photochemistry in Microemulsions: Photophysical Studies in Oleate/Hexanol/Hexadecane, Oil in Water Microemulsions.

Hermanski, C. and Mackay, R. A. (1980). *J. Colloid Interface Sci.* **73**, 324–331. Light Scattering Measurements in Nonionic Microemulsions.

Herrmann, U. and Schelly, Z. A. (1979). *J. Am. Chem. Soc.* **101**, 2665–2669. Aggregation of Alkylammonium Carboxylates and Aerosol-OT in Apolar Solvents Studied Using Absorption and Fluorescence Probes.

Hinze, W. and Fendler, J. H. (1975). *J. Chem. Soc. Dalton Trans.* 238–244. Interactions and Reactions in Restricted Polar Media. Binding of Cyanide Ion to Hemin in Surfactant-Solubilized Methanol in Benzene.

Jean, Y. C. and Ache, H. J. (1978). *J. Am. Chem. Soc.* **100**, 6320–6327. Study of the Micelle Formation and the Effect of Additives on this Process in Reversed Micellar Systems by Positron Annihilation Techniques.

Keh, E. and Valeur, B. (1981). *J. Colloid Interface Sci.* **79**, 465–478. Investigation of Water-Containing Inverted Micelles by Fluorescence Polarization. Determination of Size and Internal Fluidity.

Kertes, A. S. (1977). In *Micellization, Solubilization and Microemulsions* (K. L. Mittal, Ed.), Plenum Press, New York, pp. 445–454. Aggregation of Surfactants in Hydrocarbons. Incompatibility of the Critical Micelle Concentration Concept with Experimental Data.

Kertes, A. S. and Gutmann, H. (1976). In *Surface and Colloid Science* (E. Matijevic, Ed.), Vol. 8, John Wiley, New York, pp. 194–295. Surfactants in Organic Solvents. The Physical Chemistry of Aggregation and Micellization.

Kitahara, A. (1970). In *Cationic Surfactants* (E. Jungermann, Ed.), Marcel Dekker, New York, pp. 289–310. Micelle Formation in Nonaqueous Media.

Kitahara, A. (1980). *Adv. Colloid Interface Sci.* **12**, 109–140. Solubilization and Catalysis in Reversed Micelles.

Kitahara, A., Kabayashi, T., and Tachibana, T. (1962). *J. Phys. Chem.* **66**, 363–365. Light Scattering Study of Solvent Effect on Micelle Formation of Aerosol-OT.

Kitahara, A., Kon-no, K., and Fujiwara, M. (1976). *J. Colloid Interface Sci.* **57**, 391–392. On Solubilization Equilibrium in Reversed Micellar Systems.

Klose, G. and Stelzner, F. (1974). *Biochim. Biophys. Acta* **363**, 1–8. NMR Investigations of the Interaction of Water with Lecithin in Benzene Solutions.

Kondo, H., Hamada, T., Yamamoto, S., and Sunamoto, J. (1980). *Chem. Lett.*, 809–812. Structure and Nature of Cationic Reversed Micelles as Probed on NMR by Solubilized Cupric Chloride.

Kon-no, K. and Kitahara, A. (1971). *J. Colloid Interface Sci.* **35**, 636–642. Micelle Formation of Oil-Soluble Surfactants in Nonaqueous Solutions: Effect of Molecular Structure of Surfactants.

Kumar, C. and Balasubramanian, D. (1980). *J. Phys. Chem.* **84**, 1895–1899. Structural Features of Water-in-Oil Microemulsions.

Lim, Y. L. and Fendler, J. H. (1978). *J. Am. Chem. Soc.* **100**, 7490–7494. Spin Probes in Reversed Micelles. Electron Paramagnetic Resonance Spectra of 2,2,5,5-Tetramethylpyrrolidine-1-oxyl Derivatives in Benzene in the Presence of Dodecylammonium Propionate Aggregates.

Little, R. C. (1966). *J. Colloid Interface Sci.* **21**, 266–272. Some Physical Properties of Soap/Solvent Systems and Their Relations to the Solubility Parameter of the Solvent.

Little, R. C. and Singleterry, C. R. (1964). *J. Phys. Chem.* **68**, 3453–3465. The Solubility of Alkali Dinonylnaphthalenesulfonates in Different Solvents and a Theory for the Solubility of Oil-Soluble Soaps.

Lo, F. Y-F., Escott, B. M., Fendler, E. J., Adams, E. T., Jr., Larsen, R. D., and Smith, P. W. (1975). *J. Phys. Chem.* **79**, 2609–2621. Temperature-Dependent Self-Association of Dodecylammonium Propionate in Benzene and Cyclohexane.

Lomax, T. D. and O'Connor, C. J. (1978). *J. Am. Chem. Soc.* **100**, 5910–5914. Tracer Studies of Carboxylic Acids. 7. Oxygen-18 Exchange of Propionic Acid with Solvent Water and with Surfactant Solubilized Water in Benzene.

McBain, J. W. and Working, E. B. (1947). *J. Phys. Chem.* **51**, 974–980. Aluminium Dilaurate as Association Colloid in Benzene.

Mackay, R. A. and Agarwal, R. (1978). *J. Colloid Interface Sci.* **65**, 225–231. Conductivity Measurements in Nonionic Microemulsions.

Mackay, R. A., Letts, K., and Jones, C. (1977). In *Micellization, Solubilization and Microemulsions* (K. L. Mittal, Ed.), Plenum Press, New York, pp. 801–815. Interactions and Reactions in Microemulsions.

Mackay, R. A., Jacobson, K., and Tourian, J. (1980). *J. Colloid Interface Sci.* **76**, 515–524. Measurement of pH and p*K* in O/W Microemulsions.

Magid, L. (1979). In *Solution Chemistry of Surfactants* (K. L. Mittal, Ed.), Plenum Press, New York, pp. 427–453. Solvent Effects on Amphiphilic Aggregation.

Masui, T., Watanabe, F., and Yamagishi, A. (1977a). *J. Colloid Interface Sci.* **61**, 388–393. EPR Studies of $FeCl_3$, $MnCl_2 \cdot 4H_2O$ and $CuSO_4 \cdot 5H_2O$ Solubilized by Dodecylpyridinium Chloride in Chloroform.

Masui, T., Watanabe, F., and Yamagishi, A. (1977b). *J. Phys. Chem.* **81**, 494–496. Temperature-Jump Study on the Aquation of the Iron(III) Complex by Dodecylpyridinium Chloride Solubilized Water Pool in Chloroform.

Mathews, M. B. and Hirschorn, E. (1953). *J. Colloid Sci.* **8**, 86–96. Solubilization and Micelle Formation in a Hydrocarbon Medium.

Menger, F. M. and Saito, G. (1978). *J. Am. Chem. Soc.* **100**, 4376–4379. Adsorption, Displacement and Ionization in Water Pools.

Menger, F. M. and Yamada, K. (1979). *J. Am. Chem. Soc.* **101**, 6731–6734. Enzyme Catalysis in Water Pools.

Menger, F. M., Donohue, J. A., and Williams, R. F. (1973). *J. Am. Chem. Soc.* **95**, 286–288. Catalysis in Water Pools.

Menger, F. M., Saito, G., Sanzero, G. V., and Dodd, J. R. (1975). *J. Am. Chem. Soc.* **97**, 909–911. Motional Freedom and Polarity Within Water Pools of Different Sizes Spin Label Studies.

Mittal, K. L. (1977). *Micellization, Solubilization and Microemulsions*, Plenum Press, New York.

Mittal, K. L. (1979). *Solution Chemistry of Surfactants*, Plenum Press, New York.

Miyoshi, N. and Tomita, G. (1980a). *Z. Naturforsch.* **35b**, 741–745. Solubilization of Methylene Blue in Reversed Micelles, Effects of Water.

Miyoshi, N. and Tomita, G. (1980b). *Z. Naturforsch.* **35b**, 736–740. Buffer Action of Reversed Micelles.

Muller, N. (1975). *J. Phys. Chem.* **79**, 287–291. A Multiple-Equilibrium Model for the Micellization of Ionic Surfactants in Nonaqueous Solvents.

Muller, N. (1978). *J. Colloid Interface Sci.* **63**, 383–393. Attempt at a Unified Interpretation of the Self-Association of 1-1 Ionic Surfactants in Solvents of Low Dielectric Constant.

Nagyvary, J. and Fendler, J. H. (1974). *Origins Life* **5**, 357–362. The Origin of the Genetic Code: A Physical Chemical Model of Codon Assignments.

Nome, F., Chang, S. A., and Fendler, J. H. (1976a). *J. Chem. Soc. Faraday Trans. I* **72**, 296–302. Indicators in Benzene in the Presence of Dodecylammonium Propionate.

Nome, F., Chang, S. A., and Fendler, J. H. (1976b). *J. Colloid Interface Sci.* **56**, 146–158. Indicators in Benzene in the Presence of Polyoxyethylene(6) Nonylphenol.

Ober, R. and Taupin, C. (1980). *J. Phys. Chem.* **84**, 2418–2422. Interactions and Aggregations in Microemulsions. A Small-Angle Neutron Scattering Study.

O'Connor, C. J., Fendler, E. J., and Fendler, J. H. (1973). *J. Org. Chem.* **38**, 3371–3375. Hydrolysis of 2,4-Dinitrophenyl Sulfate in Benzene in the Presence of Alkylammonium Carboxylate Surfactants.

Pilpel, N. (1963). *Chem. Rev.* **63**, 221–234. Properties of Organic Solutions of Heavy Metal Soaps.

Poon, P. H. and Wells, M. A. (1974). *Biochemistry* **13**, 4928–4936. Physical Properties of Egg Phosphatidylcholine in Diethyl Ether–Water Solutions.

Prince, L. M. (1977). *Microemulsions, Theory and Practice*, Academic Press, New York.

Rance, D. G. and Friberg, S. (1977). *J. Colloid Interface Sci.* **60**, 207–209. Micellar Solutions Versus Microemulsions.

Richter, P. H. and Eigen, M. (1974). *Biophys. Chem.* **2**, 255–263. Diffusion Controlled Reaction Rates in Spheroidal Geometry. Application to Repressor-Operator Association and Membrane Bound Enzymes.

Robinson, N. (1960). *J. Pharm. Pharmacol.* **12**, 685–689. Micellar Size and Surface Activity of Some C_{18}-α-Monoglycerides in Benzene.

Robinson, B. H., Steytler, D. C., and Tack, R. D. (1979). *J. Chem. Soc. Faraday Trans. I* **75**, 481–496. Ion Reactivity in Reversed-Micellar Systems. Kinetics of Reaction Between Micelles Containing Hydrated Nickel(II) and Murexide-Containing Micelles in the System Aerosol-OT + Water + Heptane.

Rosoff, M. (1978). *Progress in Surface and Membrane Science, Vol. 12*, Academic Press, New York, pp. 405–477. The Nature of Microemulsions.

Rouviere, J., Couret, J.-M., Lindheimer, M., Dejardin, J.-L., and Marrony, R. (1979a). *J. Chim. Phys.* **76**, 289–296. Structure des Agregats Inverses D'AOT I. Forme et Taille des Micelles.

Rouviere, J., Couret, J.-M., Lindheimer, A., and Brun, B. (1979b). *J. Chim. Phys.* **76**, 297–301. Structure des Agregats Inverses D'AOT II. Effets de Sel sur les Micelles Inverses.

Ruckenstein, E. and Nagarjan, R. (1980). *J. Phys. Chem.* **84**, 1349–1358. Aggregation of Amphiphiles in Nonaqueous Media.

Seno, M., Sawada, K., Araki, K., Iwamoto, K., and Kise, H. (1980). *J. Colloid Interface Sci.* **78**, 57–64. Properties of Water in Hexadecyltrimethylammonium Bromide-Chloroform System.

Singleterry, C. R. (1955). *J. Am. Oil Chem. Soc.* **32**, 446–452. Micelle Formation and Solubilization in Nonaqueous Solvents.

Smith, R. E. and Luisi, P. L. (1980). *Helv. Chim. Acta* **63**, 2302–2311. Micellar Solubilization of Biopolymers in Hydrocarbon Solvents. III. Empirical Definition of Acidity Scale in Reversed Micelles.

Sudhölter, E. J. R., and Engberts, J. B. F. N. (1977). *Recl. Trav. Chim. Pays-Bas Belg.* **96**, 85–87. Aggregation Behavior of 1-Methyl-4-dodecyl and 1,4-Dimethylpyridinium Iodide in Dichloromethane. Micropolarity in the Vicinity of a Triple Ion.

Sunamoto, J. and Hamada, T. (1978). *Bull. Chem. Soc. Jpn.* **51**, 3130–3135. Solvochromism and Thermchromism of Cobalt(II) Complexes Solubilized in Reversed Micelles.

Sunamoto, J., Hamada, T., Seto, T., and Yamamoto, S. (1980). *Bull. Chem. Soc. Jpn.* **53**, 583–589. Microscopic Evaluation of Surfactant-Water Interaction in Apolar Media.

Tamura, K. and Schelly, Z. A. (1979). *J. Am. Chem. Soc.* **101**, 7643–7644. Kinetics of Penetration of a Probe into a Reversed Micelle.

Tamura, K. and Schelly, Z. A. (1981a). *J. Am. Chem. Soc.* **103**, 1013–1018. Reversed Micelles of Aerosol-OT in Benzene, 2. Equilibrium Studies Using Vapour Pressure Osmometry and Spectrometry with Picric Acid as Indicator.

Tamura, K. and Schelly, Z. A. (1981b). *J. Am. Chem. Soc.* **103**, 1018–1022. Reversed Micelles of Aerosol-OT in Benzene, 3. Dynamics of the Solubilization of Picric Acid.

Tanford, C. (1973). *The Hydrophobic Effect*, John Wiley, New York.

Ting, D. Z., Hagam, H. S., Chan, S. I., Doll, J. D., and Springer, C. S. (1981). *Biophys. J.* **34**, 189–215. Nuclear Magnetic Resonance Studies of Cation Transport Across Vesicle Bilayer Membrane.

Tricot, Y., Kiwi, J., Niederberger, W., and Grätzel, M. (1981). *J. Phys. Chem.* **85**, 862–870. Application of ^{13}C NMR, Fluorescence, and Light Scattering Techniques for Structural Studies of Oil-in-Water Microemulsions.

Tsujii, K., Sunamoto, J., Nome, F., and Fendler, J. H. (1978). *J. Phys. Chem.* **82**, 423–429. Concentration Dependent Ground and Excited State Behavior of Dodecylammonium Pyrene-1-butyrate in Ethanol and Benzene.

Wong, M., Grätzel, M., and Thomas, J. K. (1975). *Chem. Phys. Lett.* **30**, 329–333. On the Nature of Solubilized Water Clusters in Aerosol OT/Alkane Solutions. A Study of the Formation of Hydrated Electrons and 1,8-Anilinonaphthalene Sulphonate Fluorescence.

Wong, M., Thomas, J. K., and Nowak, T. (1977). *J. Am. Chem. Soc.* **99**, 4730–4736. Structure and State of H_2O in Reversed Micelles. 3.

Zulauf, M. and Eicke, H.-F. (1979). *J. Phys. Chem.* **83**, 480–486. Inverted Micelles and Microemulsions in the Ternary System H_2O/Aerosol-OT/Isooctane as Studied by Photon Correlation Spectroscopy.

Zundel, G. (1969). *Hydration and Intermolecular Interaction*. Academic Press, New York.

CHAPTER 4

MONOLAYERS AND ORGANIZED MULTILAYER ASSEMBLIES

Monolayers have been known considerably longer than have micelles or vesicles. They were investigated by Benjamin Franklin and Lord Rayleigh at the end of the last century (Gaines, 1966; Kuhn et al., 1972; Adamson, 1976; Gershfeld, 1976). Techniques for monolayer handling, developed by Pockels (1891), were standardized in the 1930s (Blodgett, 1935, 1939; Langmuir, 1939; Adam, 1941; Trurnit, 1945) and improved in the 1970s (Kuhn et al., 1972). Early investigations were intimately related to interfacial phenomena. Much momentum was gained in subsequent studies by the recognition of analogies between monolayers and biological membranes. More recent work has been prompted by the desire to use monolayers and organized multilayer assemblies for photochemical solar energy conversion and storage.

The purpose of the present chapter is to summarize the preparation and properties of monolayers and organized multilayer assemblies. Standard books and review articles (Adam, 1941; Davies and Rideal, 1963; Gaines, 1966, 1972; Kuhn and Möbius, 1971; Kuhn et al., 1972; Barnes, 1975, 1979; Goddard, 1975; Adamson, 1976; Möbius, 1981) should be consulted for additional details. In spite of their extensive use, quantitative physical chemical treatments of monolayers (Gershfeld, 1976) and information on the effects of additives (Ter-Minassian-Saraga, 1975; Blank, 1979) have only recently begun to emerge. Emphasis is placed on these recent developments.

1. PREPARATION OF MONOLAYERS

Absolute and scrupulous cleanliness of all materials and equipment is a must in monolayer studies. Impurities in parts per million may well lead to experimental artifacts and misinterpretations. Unfortunately, the need for this type of care has not always been realized. All published work should, therefore, be scrutinized for experimental details. Those that do not provide adequate descriptions or follow the required protocol should be viewed with suspicion.

All equipment is to be constructed from glass, Teflon, stainless steel, or platinum. Water should be triply distilled from a quartz container. Organic impurities introduced by the use of ion exchange resins to purify water can cause severe problems. Only the highest purity materials should be used. Special care should be taken to prevent the introduction of grease. Airborne pollutants can be avoided by placing the entire monolayer handling system into a temperature controlled box or into a laminar air flow hood. All materials should be handled by wearing inert gloves. Prior to forming the monolayer the surface of the subphase should be cleaned by sweeping it and removing the compressed interface by sucking with a glass capillary. The ultimate test of purity is performed by blank spreading experiments. No film should be formed in the absence of deliberately added film forming materials. Identical surface properties of enantiomeric *N*-α-methylbenzylstearamide monolayers can be taken as stringent criteria for the purity of these preparations (Arnett et al., 1978). Interestingly, much greater force is required to pack racemic *N*-α-methylbenzylstearamide in a given area than is required for the pure enantiomer (see Chapter 11.3; Arnett et al., 1978). Similar effects have been noted for phosphatidylcholine monolayers. At a given surface pressure *sn*-2-phosphatidylcholine is more expanded than its *sn*-3 isomer (Seelig et al., 1980).

Most monolayer forming substances are amphiphatic molecules containing alkyl chains of 12 or more carbon atoms. The polar headgroups of these surfactants are attracted to and are in contact with water, the subphase, while their hydrocarbon tails protrude above it. Figure 4.1 shows an idealized and a somewhat more realistic monolayer. Long chain carboxylic acids, alcohols, oximes, ketones, amines, ammonium salts, sulfates, and sulfonates have been used most frequently as monolayer forming substances (Gaines, 1966). Monolayers formed from phospholipids (Cadenhead, 1970), porphyrins (Bergerou et al., 1967), chlorophyll (Aghion et al., 1969), polypeptides (Malcolm, 1973), and polymers (Rosoff, 1969; Gaines, 1972) have also been investigated.

Some materials, such as a drop of oleic acid or a crystal of hexadecyl alcohol, spread spontaneously if placed on a water surface. The use of a volatile spreading solvent is more convenient, however. It allows better concentration control and provides a means of monolayer formation from substances that do not spread

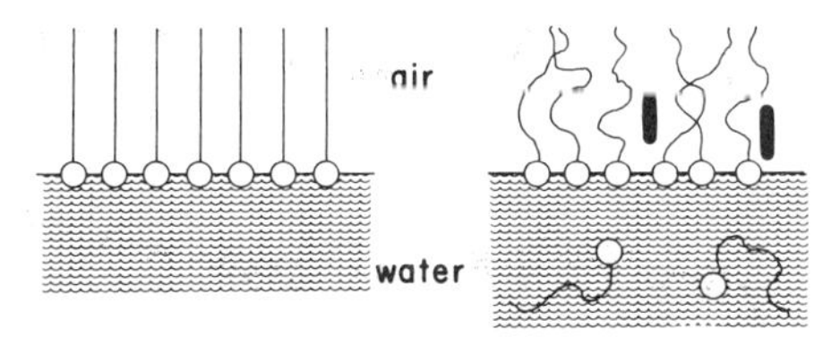

Figure 4.1. Schematic representations of idealized (left) and of somewhat more realistic (right) monolayers at atmospheric pressure. Monolayer entrapped spreading solvent and/or other impurities are illustrated by ▮. Using a film balance under high pressure, it is possible, however, to condense molecules having appropriate hydrophobic cross-sectional areas and headgroup sizes to monolayers resembling molecular crystals of paraffins in their microscopic structures.

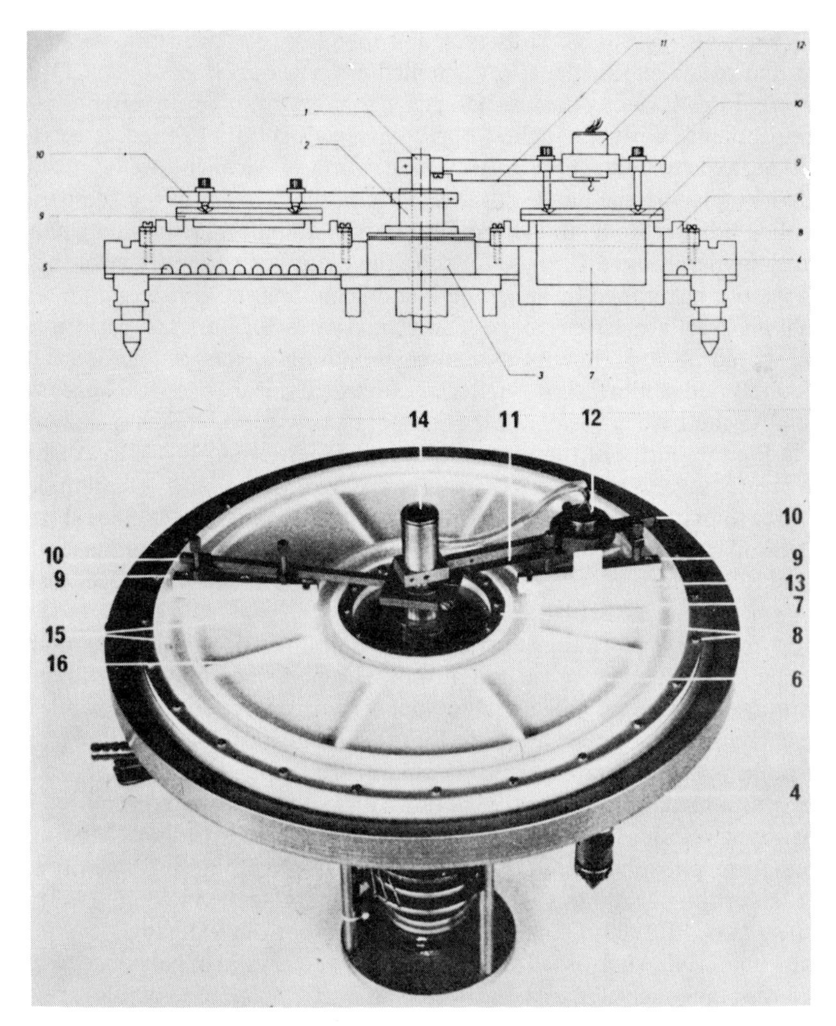

Figure 4.2. Double-axle multicompartment trough. Top—schematic; bottom—photo. 1, Inner axle; 2, outer axle; 3, frame of the driving assembly; 4, base plate; 5, grooves for thermostating; 6, trough, PTFE plate; 7, hole for multilayer deposition; 8, aluminum rings pressing the trough to the base plate; 9, barriers; 10, cams; 11, spring of Wilhelmy balance; 12, motion transducer; 13, hydrophilic place (filter paper); 14, electric connector for Wilhelmy balance; 15, channels; 16, walls (taken from Fromherz, 1975). Commercially available from Mayer-Feintechik, 34 Göttingen, Germany.

spontaneously. The spreading solvent should, of course, be pure, chemically inert, and volatile. Hexane, cyclohexane, chloroform, ethylether, benzene, dimethylformamide, dimethylsulfoxide, and their mixtures are the most frequently used spreading solvents. Film formation is accomplished by the dropwise addition of the spreading solution either by a pipette or by allowing the solution to flow down evenly on a rod, held a few millimeters above the water surface. Each drop is allowed to evaporate before the next drop is applied. Drops are distributed over the surface to be covered. Alternatively, film formation can be accomplished by immersing a stainless steel plate into the spreading solution. After draining all excess liquids the monolayer is formed by carefully pulling the plate through the solution, just below the surface of the subphase.

Monolayers are contained in trays or troughs along with various handling and measuring equipment. Although there are commercially available Langmuir troughs (Gaines, 1966), most of the recent research is being carried out on home-built equipment. Figure 4.2 illustrates a much used versatile multicompartment trough with movable barriers (Fromherz, 1975). This instrument allows the compression of monolayers and their transport from one subphase to another. It provides for isolation of monolayers as well as the sequential buildup of multilayers on any desired subphase. With the addition of appropriate accessories, absorbances, surface pressure–surface area isotherms, and surface potentials can also be determined. A classical piece of equipment (Langmuir–Adam type) is depicted in Figure 4.3.

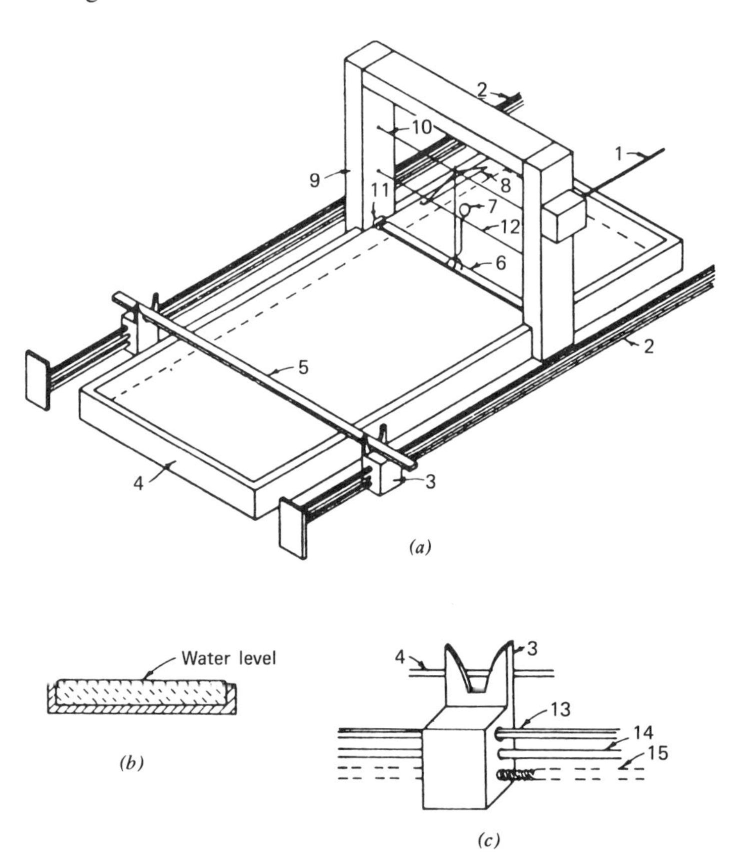

Figure 4.3. A modern film balance. (*a*) Overall view. 1, torsion wire control; 2, sweep control; 3, sweep holder; 4, trough; 5, sweep; 6, float; 7, mirror; 8, calibration arm; 9, head; 10, main torsion wire; 11, gold foil barriers; 12, wire for mirror; 13, elevation control; 14, guide; 15, traverse. (*b*) In order to sweep the surface and to contain the film in the trough, the water level is kept above the rim of the paraffin-coated trough. (*c*) The sweep holders are raised or lowered by a rack.

2. CHARACTERIZATION OF MONOLAYERS

The term monolayer implies the presence of a uniform monomolecular layer of thin film on the surface of water. Verification of this fact is essential for meaningful experimentation. It should be recognized that monomolecular arrangement is not a necessary requirement for film formation. Multilayer surface films can often be formed. The thickness of the film can, in principle, be determined by examining the behavior of transmitted and reflected light. Although potentially useful, practical problems and incomplete theories limit the applications of ellipsometry and interferometry in monolayer studies (Adamson, 1976). Grossly incomplete spreading can be visually observed by trained eyes. Surface pressure–surface area curves, surface potentials, and surface viscosities are the most frequently determined properties used for monolayer characterization (Gaines, 1966).

The surface pressure of the monolayer, the area occupied per molecule, Π, is defined by

$$\Pi = \gamma_0 - \gamma \tag{4.1}$$

as being the difference between the surface tension of the pure solvent (γ_0) and that of the film covered surface (γ). The Wilhelmy-plate or the Langmuir-float methods are generally used for determining monolayer surface pressures. In the former force due to the surface tension acting on a thin suspended plate (a sheet of filter paper for example) is determined. Figure 4.2 shows the incorporation of a Wilhelmy plate on monolayer handling equipment. The Wilhelmy method measures changes in γ, rather than that in Π. Due care is needed, therefore, to avoid any adventitious influence on the surface tension that may be misinterpreted as originating in surface pressure changes. An additional disadvantage of the Wilhelmy method is the adsorption of the film material on the plate, which may cause unknown deviations from the contact angle. The simplicity of use and the ease of construction compensate for these disadvantages.

In the Langmuir film balance (Figure 4.3) the clean subphase is separated from that covered by the monolayer by a movable float. The difference between the surface tensions on either side of the float is measured directly as a force acting on the float. With modern Langmuir balances accuracy of film pressure measurements can exceed hundredths of a dyn per centimeter.

Monitoring surface pressures of constant monolayer areas is a useful reproducibility criterion. A case in point is the observed difference between the surface pressure–surface area isotherms of long chain ruthenium bipyridine complex monolayers prepared by Valenty and Gaines (1977) and by Seefeld et al. (1977). Figure 4.4 shows differences between these two preparations as well as illustrating the significant influence of the subphase. Surface pressure–surface area isotherms are extremely sensitive to losses of monolayer materials. A decreased monolayer concentration results in a prompt decrease of surface pressure.

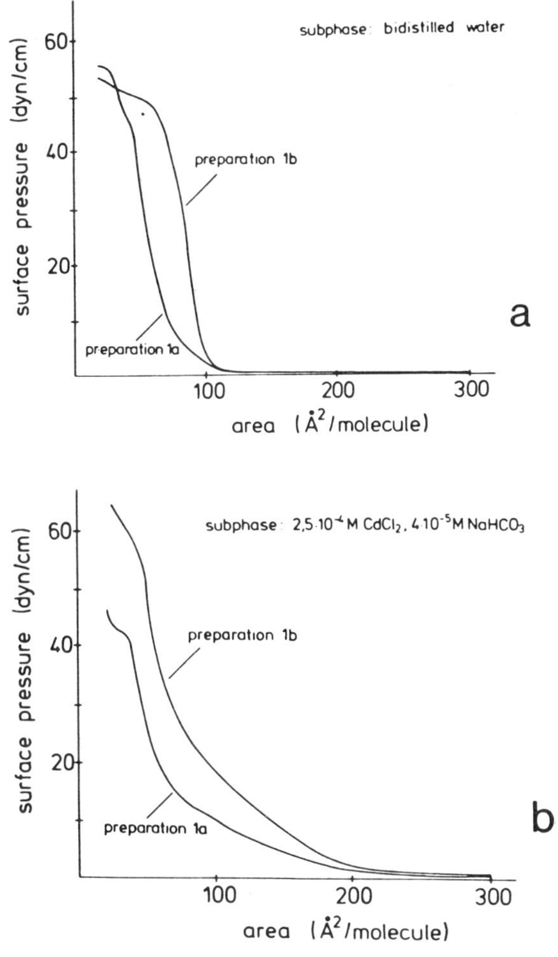

Figure 4.4. Ru(II) complex 1 monolayers, preparation 1a (Valenty and Gaines, 1977) and 1b (Seefeld et al., 1977). Surface pressure versus area per molecule at 24°C. (*a*) On bidistilled H_2O, pH 5.6. (*b*) On 2.5×10^{-4} *M* CdCl and 4×10^{-5} *M* $NaHCO_3$ at pH 6.8 (taken from Seefeld et al., 1977).

Monolayer contact or surface potential is measured as the Volta potential difference between the surface of the monolayer and that of a metal probe. The Volta potential is defined as the work required to bring a unit charge from infinity just up to, but not into, the phase. Although the interpretation of monolayer surface potentials is not unambiguous, their measurement provides an alternative means of monitoring phase changes. The standard work (Gaines, 1966) should be consulted for experimental methodology.

Surface viscosity measurements can also be used for monolayer characterizations. Quantitative interpretations of the data, like those for surface potential measurements, still await the development of exact theories.

3. PROPERTIES OF MONOLAYERS

Stability of monolayers was implicitly assumed in the early investigations. Monolayer forming materials were thought to be nonvolatile and to be completely insoluble in the subphase. These assumptions are not strictly valid. Monolayers are best characterized by metastable surface pressure–surface area isotherms, undergoing time dependent changes, and often showing hysteresis (Gershfeld, 1976).

Any thermodynamic treatment of monolayers needs to consider all the processes leading to the formation of the different states. Attention must, therefore, be paid not only to monolayer formation and to phase transitions within the confines of the monolayer but also to sublimation and evaporation into the vapor phase above the film (Gershfeld, 1976) as well as to dissolution into the subphase.

Thermodynamic treatments, so far, have necessarily been limited to phase transitions occurring within the boundaries of the monolayer (Gaines, 1966). By analogy with bulk matter, the presence of gaseous, liquid, and solid states has been visualized. Equations of states are then described by surface pressure–surface area isotherms. Figure 4.5 is an idealized representation of possible monolayer phase transitions.

In the gaseous state the available area for each molecule is large. Molecules float freely, mostly lying flat, on the surface without exerting much force on each other. Monolayers in their gaseous state may be infinitely expanded without any phase change. This state is characterized by a constant dipole moment and by an exponential decrease of Π with increasing surface area. Experimental in-

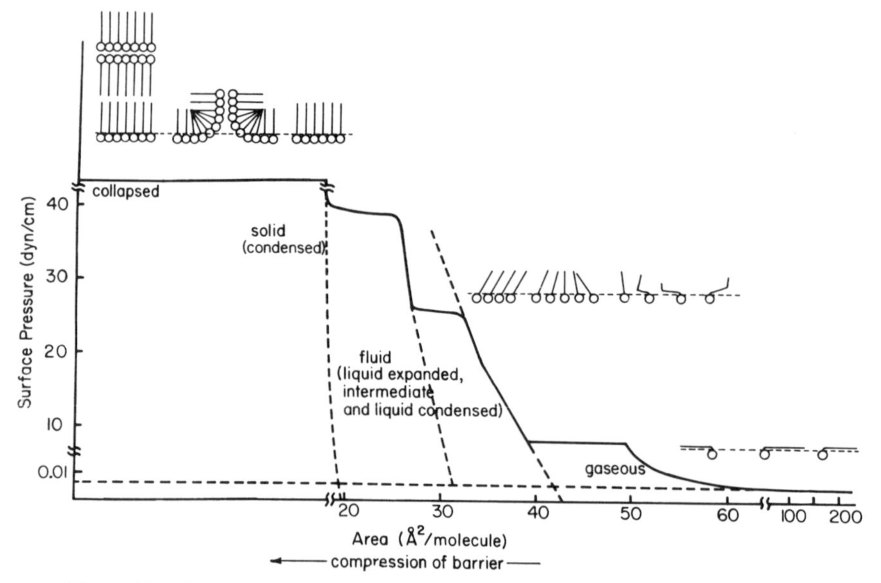

Figure 4.5. Schematic representation of a surface pressure–surface area isotherm.

vestigation of gaseous monolayers is rather difficult since the required surface pressures are generally in the ranges of 0.01–0.001 dyn/cm^{-1}.

Compressing the gaseous monolayers results in a transition to a fluid state. In this state Π decreases much more steeply with increasing surface area than is found for the gaseous state. Monolayer morphologies are least known in their fluid states. At least two subphases have been recognized. The initial transition on decreasing the surface area of gaseous monolayers results from a gradual reorganization of molecules to a position more or less perpendicular to the subphase surface. In this liquid expanded, or L_1, state the average intermolecular distances are much greater than in bulk liquids. On further compression there is a definite, albeit gradual, change in the surface pressure–surface area isotherm. This intermediate state, likely to be characterized by water separated surfactant headgroups on the monolayer surface, is followed by the liquid condensed, L_2, state. In the L_2 state monolayer forming materials are believed to be quite close to each other and tilted with respect to the subphase surface.

In the solid state the molecules in the monolayer are packed as closely as possible. This configuration requires the surfactants to be perpendicular to the subphase surface or tilted at an angle. Monolayers in their solid phase show low compressibility as indicated by the almost vertical surface pressure–surface area isotherm. At zero pressure the surface area per molecule, around 20 Å^2, approximates that of closed packed hydrocarbons.

Ultimately, compression leads to an inflection or to a break in the isotherm. At this point the surface area decreases at constant pressure and the monolayer collapses into bilayers and multilayers.

The simplest method of treating monolayer surface pressure–surface area isotherms, at least in the expanded state, is to assume the validity of the ideal two-dimensional gas equation:

$$\Pi A = kT \tag{4.2}$$

where A is the area occupied by the molecule.

Equation 4.2 is a gross approximation and is only valid at very large areas and at very low pressures. The various modifications of equation 4.2 (Gaines, 1966) have found only limited applications. More rigorous thermodynamic treatments have been derived by considering film formation in terms of the Gibbs reference surface (Eriksson, 1971; Lucassen-Reynders, 1976). Expressions for evaluating the change per mole of monolayer forming material in surface excess free energy (f^s), entropy (s^s), and energy (u^s) due to spreading to a given surface excess concentration, Γ, are given by (Gershfeld, 1976)

$$-(f_e^s - f^b) = \Pi_e A_e \tag{4.3}$$

$$(s_e^s - s^b) = \frac{A_e\, d\Pi_e}{dt} \tag{4.4}$$

$$-(u_e^s - u^b) = A_e\left(\Pi_e - \frac{T\, d\Pi_e}{dt}\right) \tag{4.5}$$

where the superscripts s and b refer to surface and bulk states, the subscript e denotes film spreading in the presence of excess monolayer forming material, Π_e is the equilibrium spreading pressure (i.e., the pressure where the monolayer is in equilibrium with the bulk phase under the experimental conditions), and $\Gamma = 1/A$ where A is the surface area occupied by a mole of monolayer forming material. Equations 4.3–4.5 are used to evaluate thermodynamic properties for forming the monolayer.

Assuming ideal behavior (i.e., the presence of monomolecular nonvolatile films), monolayers can be treated as two-dimensional partially closed systems. Under this condition equations 4.6–4.8 describe the Helmholtz free energy (Δf^s), enthalpy (ΔV_c), and entropy (ΔS_c) of film compression (Gershfeld, 1976):

$$f_c = (f^s - f_i^s) = -\int_{Ai}^{A} \Pi \, dA \tag{4.6}$$

$$-\Delta S_c = (S^s - S_i^s) = \left(\frac{\delta \Delta f_c}{\delta T}\right)_A \tag{4.7}$$

$$\Delta f_c = \Delta u_c - T\Delta S_c \tag{4.8}$$

where the subscript c refers to the bulk lipid reference state $\Pi A = RT$ denoted by subscript i.

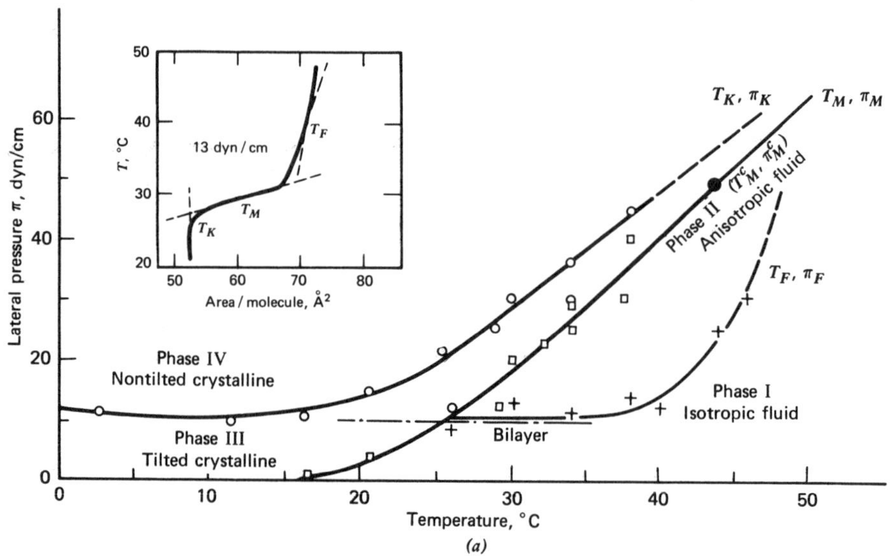

Figure 4.6. (*a*) Phase diagram of dipalmitoyl phosphatidyl choline (DPPC). The curves separate four different monolayer phases denoted by I, II, III, IV. The thick line is the first order transition, which separates the fluid and crystalline phases. The point ● most probably is a tricritical point. The line –·–·– is suggested to correspond to the main transition of the bilayer. Insert: Temperature versus area curve at a lateral pressure of 13 dyn cm^{-1}.

Phase transitions of phospholipid monolayers have been analyzed in terms of their surface pressure–surface area and surface area–temperature isobars (Albrecht et al., 1978; Sackmann, 1978). Treating the obtained data in terms of lateral pressure versus temperature plots revealed the presence of four different phases (Figure 4.6*a*). Below 15°C two different crystalline phases (Phases III and IV) predominate. Transition from the rigid tilted (Phase III) to the nontilted (Phase IV) phase takes place along the T_K, Π_K isotherm at nearly constant transition pressure in the 0°–20°C region. Above 26°C there are two fluid phases (Phases I and II). The crystalline and fluid phases are separated by a first order transition line (the thick line in Figure 4.6*a*) going over to a second order transition line at the tricritical point. The transition isotherm, T_F, Π_F, separates the isotropic and anisotropic fluid phases. Under the experimental conditions (Albrecht et al., 1978) the gaseous monolayer state could not be observed and is not included, therefore, in the phase diagram shown in Figure 4.6*a*. Phase transition temperatures are clearly seen in the surface area–temperature isobars (insert in Figure 4.6*a*). Breaks in the isobars correspond to discontinuities in the

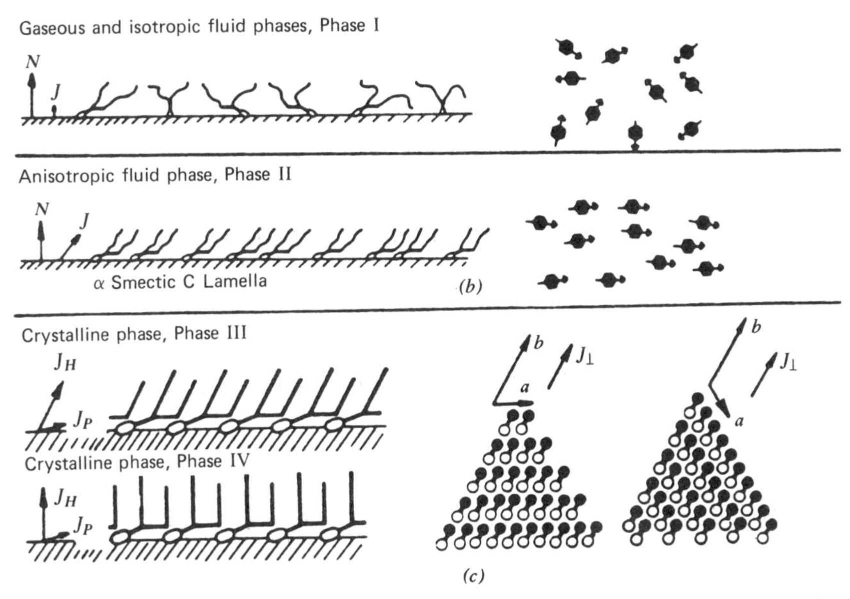

Figure 4.6. (*b*) and (*c*) Schematic representations of possible phases of phospholipid monolayers. Left side, view of monolayers parallel to water surface; right side, bird's eye view of monolayer. In (*b*) the hexagons give the centers of mass of the whole lipid molecules. The arrows indicate the direction of the molecular stretching vector projected onto the water surface. In (*c*) the empty and filled circles indicate the centers of mass of the hydrocarbon chains. The bars between these circles represent the glycerol backbone. The polar headgroup of one molecule is hidden below the chain of the adjacent molecule (cf. left side). This is indicated by the filled circle. The vectors *a* and *b* are the primitive translation vectors of the two-dimensional lattice. $J_\perp$ is the projection of the stretching vector of the whole molecule onto the water surface, J_H the stretching vector of hydrophobic part, and J_p the stretching vector of hydrophilic part of the monolayer (taken from Albrecht et al., 1978).

lateral thermal expansion coefficient. The discontinuities provide information on symmetry differences of two phases separated by second order transitions (Landau and Lifschitz, 1960). Lipid symmetry in turn is described by the stretching vector, J (Albrecht et al., 1978),

$$J = \rho(J_H + J_P) \tag{4.9}$$

where ρ is the density of the monomeric units of the lipid and J_H and J_P are orientation vectors of the hydrocarbon chain and the polar headgroup of the lipid, respectively. Figure 4.6*b* shows the proposed phospholipid monolayer phases. The cited publication (Albrecht et al., 1978) should be consulted for treatments of the phase transitions involved.

Ionic monolayers contain bound and free counterions that, like those in micelles, set up an electrical double layer (Figure 4.7). Electrical double layer theories are incomplete for monolayers and detailed experiments are scarce. Many monolayer forming materials contain functional groups that ionize under acidic and basic conditions. The extent of ionization depends on the surface pH, pH_s, which is related to the bulk pH, pH_b, by

$$pH_s = pH_b + \frac{e\Psi}{2.3kT} \tag{4.10}$$

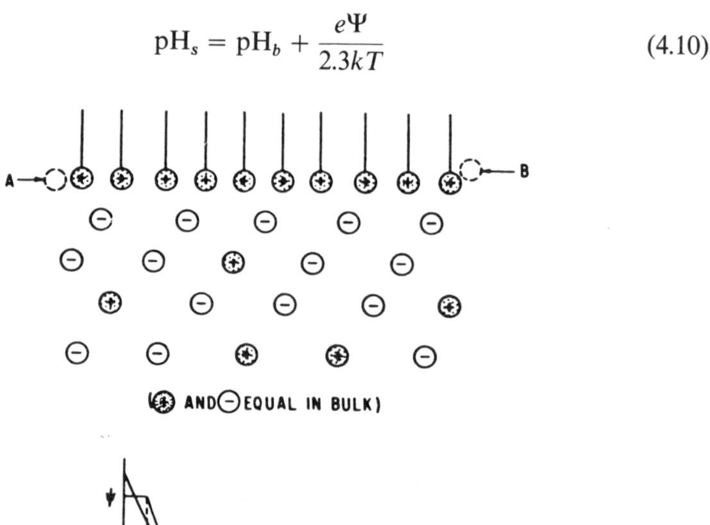

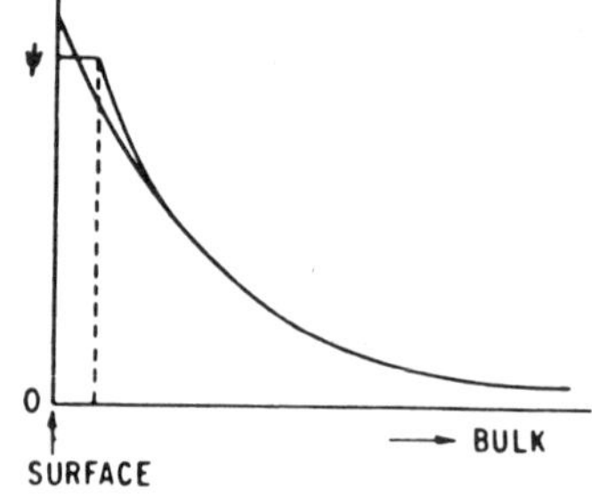

Figure 4.7. A charged monolayer sets up an ionic double layer at the surface. Since counterions might be able to penetrate between monolayer ions (position *A*) or even reach positions above them (position *B*), there is great difficulty in deciding what the effective potential distribution at and near the surface is. This problem is indicated schematically by two different ψ curves in the potential diagram (taken from Gaines, 1966).

where Ψ is the interfacial potential. Change of surface pressure due to ionization, $\Delta\Pi$, is given by (Davies and Rideal, 1963)

$$\Delta\Pi = \Pi - \Pi_0 = -\int_0^{\Psi_0} \varepsilon \, d\Psi \tag{4.11}$$

where Π and Π_0 are the surface pressures of the ionized and nonionized monolayer and ε is the surface charge density. Counterion effects on monolayer surface area–surface pressure isotherms are similar to those on micelles. The order of interaction for RCO_2^- monolayers is $Li^+ > Na^+ > K^+ > Me_4N^+ > Et_4N^+$, that for RSO_4^- monolayers is $Cs^+ = Rb^+ > K^+ > Na^+ > Li^+$, that for $R—N—(CH_3)_3^+$ monolayers is $SCN^- > I^- > NO_3^- > Br^- > Cl^- > F^-$, and those for $R—NH(CH_3)_2^+$ and RNH_3^+ monolayers are $Br^- > Cl^-$(Goddard, 1970).

Monolayers can and often do contain more than one film forming component (Cadenhead and Phillips, 1968). Two components may be completely immiscible or partially or completely miscible (Figure 4.8). For completely miscible monolayers the law of additivity applies. Molecular packing in mixed steroid + lecithin monolayers has been reexamined (Muller-Landau and Cadenhead, 1979a, 1979b; Cadenhead and Muller-Landau, 1979; Albrecht et al., 1981).

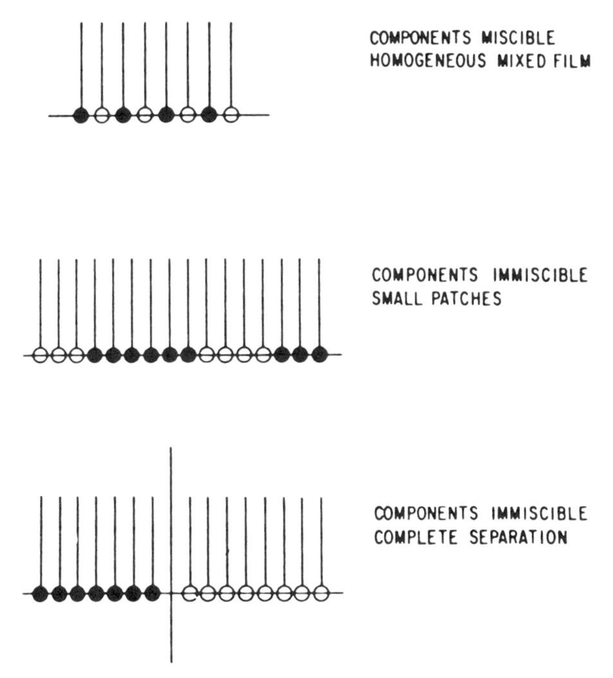

Figure 4.8. Possible molecular distributions that may result from the spreading of a "mixed" monolayer (taken from Gaines, 1966).

Surface area–surface pressure relationships have been investigated for mixed monolayers of long chain carboxylic acids and alcohols and of two different carboxylic acids as functions of chain lengths, composition, and temperature (Matuo et al., 1978, 1979a, 1979b). An equation has been derived to describe the equilibrium surface pressure for a monolayer when diluted by a surfactant (Alexander and Barnes, 1980).

Monolayers, just like micelles, are dynamic species. Their stabilities are certainly kinetic, rather than thermodynamic. Lateral diffusion dynamics of 12-(1-pyrene) dodecanoic acid (PDA) have been determined in mixed monolayers of oleic acid + PDA (Loughran et al., 1980). Ratios of PDA excimer to monomer emission intensities varied linearly with increasing mole fraction of PDA in the monolayer. The lateral diffusion coefficient of PDA in the monolayer was calculated to be 1.7×10^{-6} cm^2/sec^{-1} from the theory of random walks on lattices with traps and from Monte Carlo data simulations (Loughran et al., 1980). Molecular dynamics methods have also been used to simulate monolayers (Kox et al., 1980). In addition to possible chemical reactions in monolayers, sublimation and evaporation into the vapor phase above the monolayer, desorption and diffusion into the subphase, viscoelastic deformation, and flow of the monolayer need to be considered (Gershfeld, 1976). The general assumption that monolayers are stable during a given experiment may not be true. Dynamic effects should be investigated since, if unrecognized, they can lead to misinterpretations.

4. ORGANIZED MULTILAYER ASSEMBLIES

Multilayer assemblies provide means for organizing molecules in controlled topological arrangements (Zwick and Kuhn, 1962; Drexhage et al., 1963; Bucher et al., 1967; Kuhn, 1970, 1979; Kuhn and Möbius, 1971; Kuhn et al., 1972; Seefeld et al., 1977; Möbius, 1978a, 1978b; Whitten et al., 1978; Polymeropoulos et al., 1978; Mercer-Smith and Whitten, 1979; Janzen and Bolton, 1979; Janzen et al., 1979; Whitten, 1979).

The first requirement for assembling organized multilayers is to be able to transfer floating monolayers to a solid support. Langmuir transferred monolayers to a solid support as early as 1920 (Langmuir, 1920). Transfer, under a finite surface pressure, is accomplished by dipping a clean dry plate through the monolayer. Alternatively, in the "touching method," bringing a wet plate into contact with the edge of the monolayer results in the deposition of a monolayer on the plate. Glass slides, stainless steel, or platinum have been used as solid supports. Euqipment similar to that shown in Figure 4.2 is needed for the transfer of monolayers to a solid support. Structures of solid supported monolayers are determined by the properties of the monolayer forming material and of the support itself, the surface pressure, and the method of preparation. As an example, on the basis of electron microscopic and electrochemical investigations, it was deduced that 17-octadecenoic acid is considered to form a thin and fluid

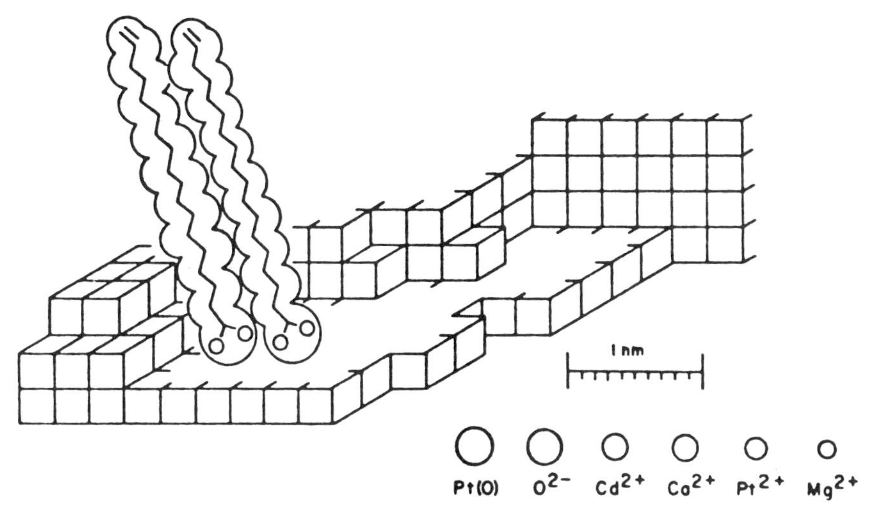

Figure 4.9. Schematic representation of a 17-octadecenoic acid monolayer on a platinum film (taken from Richard et al., 1978).

monolayer on platinum foils (Figure 4.9; Richard et al., 1978). Monolayer and mixed monolayer morphologies have been investigated using surfactants having covalently linked chromophores that are oriented perpendicular to the plane of the layer (Heesemann, 1980a, 1980b).

Mixed monolayers of controllable molecular organization have been formed on various solid supports by adsorption from organic solutions (Sagiv, 1979a, 1979b, 1980). Three different types of mixed monolayers were investigated: long chain saturated fatty acids + long chain cyanine dyes (of the types shown in the legend to Figure 4.14), *n*-octadecyltrichlorosilane ($C_{18}H_{37}SiCl_3$, OTS) + long chain substituted cyanine dyes, and fatty acids + OTS (Sagiv, 1980). Mixed monolayers were prepared both by chemisorption and by physical adsorption. Formation of covalently bound silane monolayers is illustrated in Figure 4.10. Cyanine dye molecules in mixed adsorbed monolayers on solid support are dispersed and have their chromophores oriented with the longitudinal axes parallel to the surface. Conversely, the cyanine dye in mixed monolayers on water surfaces is not dispersed. Dimers and higher aggregates are formed. Cyanine dyes could also be organized in a nonrandom fashion in mixed monolayers on thin polyvinyl alcohol (PVA) films (Sagiv, 1979a). Treatment of mixed OTS cyanine dye monolayers adsorbed on PVA films with chloroform resulted in the selective desorption of the dye (Figure 4.11). Electrical conduction has been studied at the Al/adsorbed monolayer/Al junctions (Polymeropoulos and Sagiv, 1978). The skeleton monolayer contains holes of molecular dimensions that may be used as sites for subsequent adsorption of molecules whose dimensions correspond to the vacancies (Sagiv, 1979b). The vacancies possess, therefore, a type of memory regarding the structure from which they had originated. Monolayer formation by adsorption represents an intermediate stage between

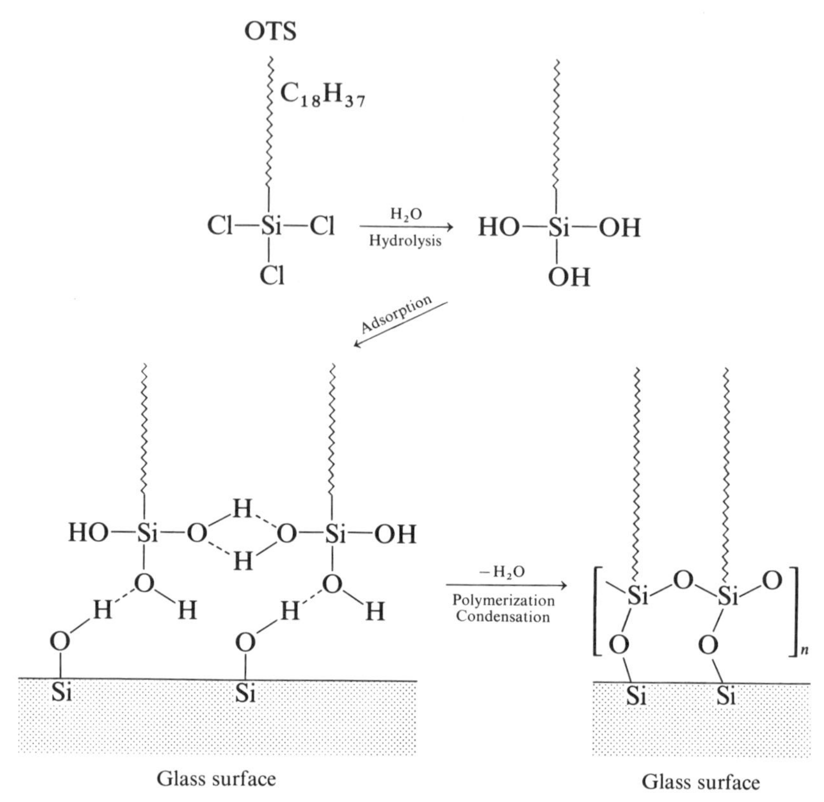

Figure 4.10. Proposed mechanism for chemisorption of *n*-octadecyltrichlorosilane (OTS) on plates (taken from Sagiv, 1980).

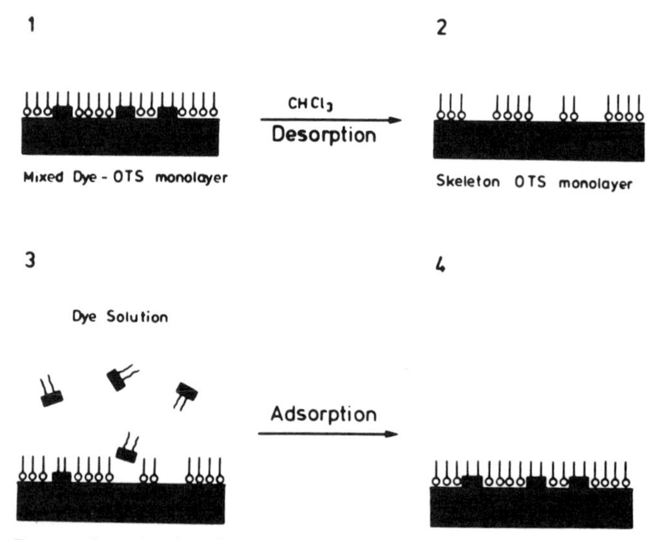

Figure 4.11. Proposed mechanism for selective desorption and adsorption of cyanine dyes from mixed cyanine-dye multilayer OTS monolayers on thin PVA films (taken from Sagiv, 1979a).

the spontaneous association occurring in aqueous micelles and the monolayers produced mechanically with a Langmuir trough or similar devices. Adsorbed monolayer may develop into an extremely powerful membrane mimetic system.

Dipping a metal plate successively in and out of a monolayer covered liquid results in the buildup of multilayers (Blodgett, 1934, 1935). Depending on the monolayer forming material and on the mode of deposition, three structurally different multilayers are recognized (Figure 4.12). X-type multilayers (plate-tail-head-tail-head, etc.) are formed by the sequential hydrophobic attachments of

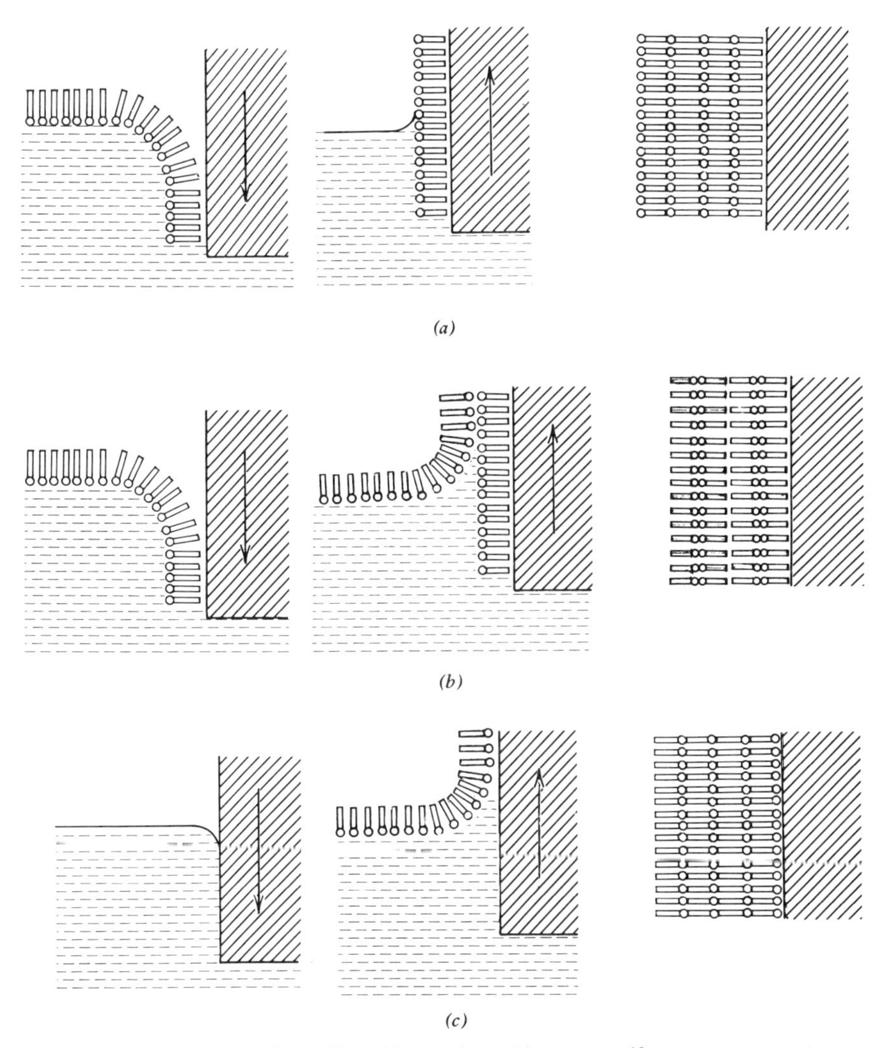

Figure 4.12. Types of monolayer deposition and resulting system if no rearrangement occurs. (*a*) X-Deposition. (*b*) Y-Deposition. (*c*) Z-Deposition. Each fatty acid molecule is represented by a circle (carboxyl group) and a bar (long hydrocarbon chain) (taken from Kuhn et al., 1972).

monolayers onto the plate upon immersion only. Conversely, *Z*-type multilayers (plate-head-tail-head-tail, etc.) are the result of sequential hydrophilic attachements of monolayers onto the plate upon withdrawal only. The *Y*-type of multilayers (plate-tail-head-head-tail, etc.) are the most common arrangements. These multilayers are built up both by dipping and by withdrawing the plate through the floating monolayer. Multilayers can be separated by depositing a thin film of PVA on the outer monolayer. Once the PVA film dries it can be removed, along with the monolayer attached to it, from the plate (Figure 4.13: Kuhn et al., 1972).

Ideally, distinct molecules are organized at controlled concentrations within given monolayers in a multilayer assembly. Kuhn and his co-workers have pioneered the investigations of these systems (Kuhn, 1970, 1979; Kuhn and Möbius, 1971; Kuhn et al., 1972; Möbius, 1978a, 1978b; Möbius and Kuhn, 1979; Polymeropoulos and Möbius, 1979). Photophysical techniques have been developed to determine intermolecular distances within the multilayer assembly. An often cited example is the Förster-type energy transfer from sensitizer *S* to

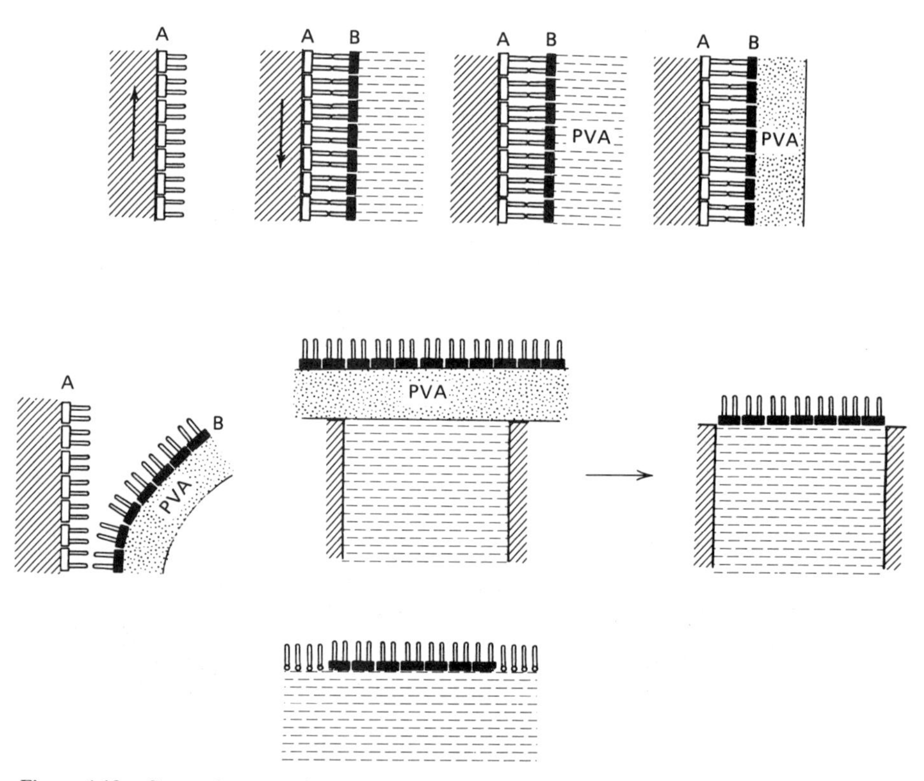

Figure 4.13. Separating monolayers. Monolayer A sticks to the glass surface, monolayer B is deposited on down trip. PVA is added to the substrate, the sample is pulled out through the clean surface, and the PVA is dried. Monolayer B is separated from A by removing the PVA foil from the slide. Monolayer B is retrieved on the water surface by dissolving the PVA (taken from Kuhn et al., 1972).

acceptor *A*, localized in monolayers different distances from each other (Figure 4.14; Kuhn et al., 1972). Energy transfer is efficient when *S* and *A* are in close proximity. This transfer manifests as a yellow fluorescence. When *S* and *A* are further apart energy transfer is less efficient and the emission is in the blue. In the absence of sensitizer there is no fluorescence.

Organized multilayer assemblies are not without unresolved problems. Foremost of these is the question of short and long term stability. On standing, multilayers rearrange (Richard et al., 1978). Small neutral molecules such as O_2, CO, N, and NO penetrate the multilayers and react at different sites (Horsey and Whitten, 1978; Mercer-Smith and Whitten, 1979). Solvents, even in trace amounts, dissolve constituents of the multilayers. The holes formed are readily refilled by impurities. Water penetration into multilayers has been examined (Windreich and Silberberg, 1980).

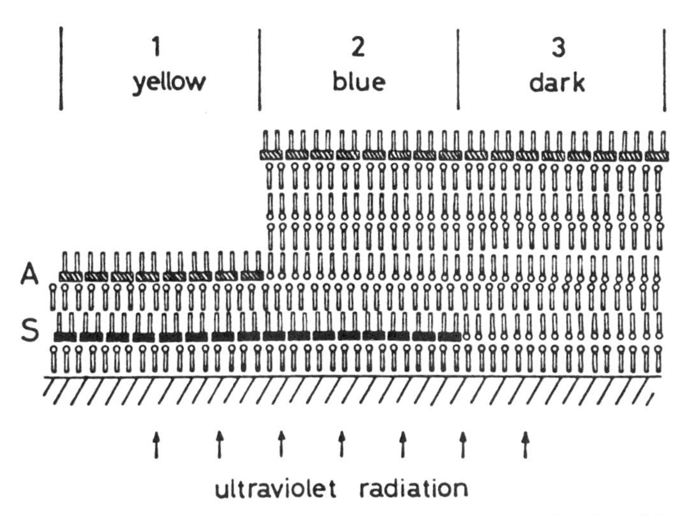

Figure 4.14. Energy transfer between dye monolayers. Cross section of a glass slide with monolayers of S (dye III and arachidate, $r = 1:20$) and A (dye II and arachidate, $r = 1:20$, illuminated with ultraviolet radiation, which is absorbed by S. Section 1 (layers of chromophores S and A at 50 Å distance): energy transfer from S to A, yellow fluorescence of A. Section 2 (S and A at 150 Å distance): no energy transfer, blue fluorescence of S. Section 3 (S absent): no fluorescence (taken from Kuhn et al., 1972).

Dye II: O, C—CH=CH—CH=C, O; $\overset{+}{N}$—$C_{18}H_{37}$, N—$C_{18}H_{37}$

Dye III: O, C—CH=C, O; $\overset{+}{N}$—$C_{18}H_{37}$, N—$C_{18}H_{37}$

An interesting development is the formation of polymerized monolayers and multilayers from diacetylenic fatty acids upon ultraviolet irradiation (Tiecke et al., 1976; Day et al., 1979; Johnston et al., 1980; Berraud et al., 1977, 1980; Hub et al., 1980, 1981; Ringsdorf and Schupp, 1980, 1981). The rigid crystalline nature of the polymerized films render them useful membrane models.

On closing this chapter, the vital need for meticulous care and scrupulous cleanliness is once again reemphasized. Close attention should be given to experimental details. Personal observation of the experimental protocol in leading laboratories is the least painful way of initiating research in monolayers and organized multilayer assemblies.

REFERENCES

Adam, N. K. (1941). *The Physics and Chemistry of Surfaces*, 3rd ed., Oxford University Press, London.

Adamson, A. W. (1976). *The Physical Chemistry of Surfaces*, 3rd ed., Wiley-Interscience, New York.

Aghion, J., Broyde, S. B., and Brody, S. S. (1969). *Biochemistry* **8**, 3120–3126. Surface Reaction of Chlorophyll-a Monolayers at a Water–Air Interface Photochemistry and Complex Formation.

Albrecht, O., Gruler, H., and Sackman, E. (1978). *J. Phys.* (Orsay, France) **39**, 301–313. Polymorphism of Phospholipid Monolayers.

Albrecht, O., Gruler, H., and Sackman, E. (1981). *J. Colloid Interface Sci.* **79**, 319–338. Pressure-Composition Phase Diagrams of Cholesterol/Lecithin, Cholesterol/Phosphatidic Acid, and Lecithin/Phosphatidic Acid Mixed Monolayers: A Langmuir Film Balance Study.

Alexander, D. M. and Barnes, G. T. (1980). *J. Chem. Soc. Faraday Trans. I* **76**, 118–125. Use of the Gibbs Equation to Calculate Adsorption into Monolayer-Covered Surfaces.

Arnett, E. M., Chao, J., Kinzig, B., Stewart, M., and Thompson, O. (1978). *J. Am. Chem. Soc.* **100**, 5575–5576. Chiral Aggregation Phenomena. I. Acid Dependent Chiral Recognition in a Monolayer.

Barnes, G. T. (1975). "Colloid Science," Specialist Periodical Reports, The Chemical Society, London, Vol. 2, pp. 173–190. Insoluble Monolayers—Equilibrium Aspects.

Barnes, G. T. (1979). In *Colloid Science, A Specialist Periodical Report* (D. H. Everett, Senior Reporter), Vol. 3, The Chemical Society, London, pp. 150–192. Insoluble Monolayers—Dynamic Aspects.

Barraud, A., Rosilo, C., Raudel-Teixier, A. (1977). *J. Colloid Interface Sci.* **62**, 509–523, Solid-State Electron-Induced Polymerization of ω-Tricosenoic Acid Multilayers.

Barraud, A., Rosilio, C., Raudel-Teixier, A. (1980). *Thin Solid Films* **68**, 91–98. Polymerized Monomolecular Layers: A New Class of Ultrathin Resins for Microlitography.

Bergerou, J. A., Gaines, G. L., Jr., and Bellamy, W. P. (1967). *J. Colloid Interface Sci.*, **25**, 97–106. Monolayers of Porphyrin Esters: Spectral Disturbances and Molecular Interactions.

Blank, M. (1979). In *Progress in Surface and Membrane Science* (D. A. Cadenhead and J. F. Danielli, Eds.), Vol. 13, Academic Press, New York, pp. 87–139. Monolayer Permeability.

Blodgett, K. B. (1934). *J. Am. Chem. Soc.* **56**, 495. Monomolecular Films of Fatty Acids on Glass.

Blodgett, K. B. (1935). *J. Am. Chem. Soc.* **57**, 1007–1022. Films Built by Depositing Successive Monomolecular Layers on a Solid Surface.

Blodgett, K. B. (1939). *Phys. Rev.* **55**, 391–404. Use of Interference to Extinguish Reflection of Light from Glass.

Bucher, H., Drexhage, K. H., Fleck, M., Kuhn, H., Möbius, D., Schäfer, F. P., Sondermann, W., Sperling, W., Tillmann, P., and Wiegand, J. (1967). *Mol. Cryst.* **2**, 199–230. Controlled Transfer of Excitation Energy Through Thin Layers.

Cadenhead, D. A. (1970). In *Recent Progress in Surface Science* (J. F. Danielli, A. C. Riddiford, and M. D. Rosenberg, Eds.), Vol. 3, Academic Press, New York. 169–192. Monolayers of Synthetic Phospholipids.

Cadenhead, D. A. and Müller-Landau, F. (1979). *Chem. Phys. Lipids* **25**, 329–343. Molecular Packing in Steroid-Lecithin Monolayers. Part III. Mixed Films of 3-Doxyl Cholestane and 3-Doxyl-17-hydroxyl Androstane with Dipalmitoylphosphatidylcholine.

Cadenhead, D. A. and Phillips, M. C. (1968). In *Advances in Chemistry* (E. D. Goddard, Ed.) Vol. 84, American Chemical Society, Washington, D.C., pp. 131–148. Molecular Interactions in Mixed Monolayers.

Davies, J. T., and Rideal, E. K. (1963). *Interfacial Phenomena*, Academic Press, New York.

Day, D., Hub, H. H., and Ringsdorf, H. (1979). *Isr. J. Chem.* **18**, 325–329. Polymerization of Mono and Bifunctional Diacetylene Derivatives in Monolayers at the Gas-Water Interface.

Drexhage, K. H., Zwick, M. M., and Kuhn, J. (1963). *Ber. Bunsenges. Phys. Chem.* **67**, 62–67. Sensibilisierte Fluoreszenz nach strahlungslosen Energieübergang durch dünne Schichten.

Eriksson, J. C. (1971). *J. Colloid Interface Sci.* **37**, 659–667. Thermodynamics of Surface-Phase Systems VI. On the Rigorous Thermodynamics of Insoluble Surface Films.

Fromherz, P. (1975). *Rev. Sci. Intrum.* **46**, 1380–1385. Instrumentation for Handling Monomolecular Films at an Air–Water Interface.

Gaines, G. L., Jr., (1966). *Insoluble Monolayers at Liquid–Gas Interfaces*, Interscience, New York.

Gaines, G. L. Jr., (1972). In *MTP International Review of Science, Vol. 7. Surface Chemistry and Colloids* (M. Kerker, Ed.), Butterworths, London, pp. 1–24. Insoluble Monolayers.

Gershfeld, N. L. (1976). *Annu. Rev. Phys. Chem.* **27**, 349–368. Physical Chemistry of Lipid Films at Fluid Interfaces.

Goddard, E. D. (1970). *Croat. Chem. Acta* **42**, 143–150. Specific Counterion Effects in Surface and Colloid Chemistry. Ionized Monolayers.

Goddard, E. D. (1975). In *Monolayers*, Advances in Chemistry Series (E. D. Goddard, Ed.), Vol. 144, American Chemical Society, Washington, D.C.

Heesemann, J. (1980a). *J. Am. Chem. Soc.* **102**, 2167–2176. Studies on Monolayers. 1. Surface Tension and Absorption Spectroscopic Measurements of Monolayers of Surface-Active Azo and Stilbene Dyes.

Heesemann, J. (1980b). *J. Am. Chem. Soc.* **102**, 2176–2181. Studies on Monolayers. 2. Designed Monolayer Assemblies of Mixed Films of Surface-Active Azo Dyes.

Horsey, B. E. and Whitten, D. G. (1978). *J. Am. Chem. Soc.* **100**, 1293–1295. Environmental Effects on Photochemical Reactions: Contrasts in the Photooxidation Behavior of Protoporphyrin IX in Solution, Monolayer Films, Organized Monolayer Assemblies and Micelles.

Hub, H. H., Hupfer, B., and Ringsdorf, H. (1980). *Am. Chem. Soc. Coat. Plast. Chem.* **42**, 2–7. Polymerization of Lipid and Lysolipid-Like Diacetylenes in Monolayers and Liposomes.

Hub, H. H., Hupfer, B., Koch, H., and Ringsdorf, H. (1981). *J. Macromol. Sci. Chem.* **A15**, 701–715. Polymerization of Lipid and Lysolipid-Like Diacetylenes in Monolayers and Liposomes.

Janzen, A. F. and Bolton, J. R. (1979). *J. Am. Chem. Soc.* **101**, 6342–6348. Photochemical Electron Transfer in Monolayer Assemblies. 2. Photoelectric Behavior in Chlorophyll-a Acceptor Systems.

Janzen, A. F., Bolton, J. R., and Stillman, M. J. (1979). *J. Am. Chem. Soc.* **101**, 6337–6341. Photochemical Electron Transfer in Monolayer Assemblies. 1. Spectroscopic Study of Radicals Produced in Chlorophyll-a/Acceptor Systems.

Johnston, D. S., Sanghera, S., Manjon-Rubio, A., and Chapman, D. (1980). *Biochim. Biophys. Acta* **602**, 213–216. The Formation of Polymeric Model Biomembranes from Diacetylenic Fatty Acids and Phospholipids.

Kox, A. J., Michels, J. P. J., and Wiegel, F. W. (1980). *Nature* **287**, 317–319. Simulation of a Lipid Monolayer Using Molecular Dynamics.

Kuhn, H. (1970). *J. Chem. Phys.* **53**, 101–108. Classical Aspects of Energy Transfer in Molecular Systems.

Kuhn, H. (1979). *J. Photochem.* **10**, 111–132. Synthetic Molecular Organizates.

Kuhn, H. and Möbius, D. (1971). *Angew. Chem. Int. Ed. Eng.* **10**, 620–637. Systems of Monomolecular Layers—Assembling and Physico-Chemical Behavior.

Kuhn, H., Möbius, D., and Bücher, H. (1972). In *Physical Methods for Chemistry*, Vol. 1, Part IIIB (A. Weissberger and B. W. Rossiter, Eds.), Wiley-Interscience, New York, pp. 577–701. Spectroscopy of Monolayer Assemblies, Part I. Principles and Applications, Part II. Experimental Procedure.

Landau, L. D. and Lifschitz, E. M. (1960). *Statistical Physics* Addison-Wesley, New York.

Langmuir, I. (1920). *Trans. Faraday. Soc.* **15**, iii, 62–74. The Mechanism of the Surface Phenomena of Floatation.

Langmuir, I. (1939). *Proc. Roy. Soc. London Ser. A* **170**, 1–39. Molecular Layers.

Loughran, T., Hatlee, M. D., Patterson, L. K., and Kozak, J. J. (1980). *J. Chem. Phys.* **72**, 5791–5797. Monomer–Excimer Dynamics in Spread Monolayers. I. Lateral Diffusion of Pyrene Dodecanoic Acid at the Air–Water Interface.

Lucassen-Reynders, E. H. (1976). In *Progress in Surface and Membrane Science, Vol.* **10**, Academic Press, New York, pp. 253–360. Adsorption of Surfactant Monolayers at Gas/Liquid and Liquid/Liquid Interfaces.

Malcolm, B. R. (1973). *Prog. Surf. Sci. Membr. Struct.* **7**, 183–229. The Structure and Properties of Monolayers of Synthetic Polypeptides at the Air–Water Interface.

Matuo, H., Hiromoto, K., Motomura, K., and Matuura, R. (1978). *Bull. Chem. Soc. Jpn.* **51**, 690–693. Mixed Monolayers of Long Normal Chain Fatty Acids with Long Normal Chain Fatty Alcohols.

Matuo, H., Yoshida, N., Motomura, K., and Matuura, R. (1979a). *Bull. Chem. Soc. Jpn.* **52**, 667–672. Mixed Monolayers of Long Normal Chain Fatty Acids with Normal Chain Esters. I. Fatty Acids–Ethyl Hexadecanoate System.

Matuo, H., Motomura, K., and Matuura, R. (1979b). *Bull. Chem. Soc. Jpn.* **52**, 673–676. Mixed Monolayers of Long Normal Chain Fatty Acids with Long Normal Chain Esters. II. Fatty Acids-Hexadecyl Acetate System.

Mercer-Smith, J. A. and Whitten, D. G. (1979). *J. Am. Chem. Soc.* **101**, 6620–6625. Photoreactions of Metalloporphyrins in Supported Monolayer Assemblies and at Assembly-Solution Interfaces. Reductive Addition of Palladium Complexes with Surfactants and Water Soluble Dialkylanilines.

Möbius, D. (1978a). In *Topics in Surface Chemistry* (E. Kay and P. S. Bagus, Eds.), Plenum Press, New York, pp. 75–101. Monolayer Assemblies.

Möbius, D. (1978b). *Ber. Bunsenges. Phys. Chem.* **82**, 848–858. Designed Monlayer Assemblies.

Möbius, D. (1981). *Acc. Chem. Res.* **14**, 63–68. Molecular Cooperation in Monolayer Organizates.

Möbius, D. and Kuhn, H. (1979). *Isr. J. Chem.* **18**, 375–384. Monolayer Assemblies of Dyes to Study the Role of Thermal Collisions in Energy Transfer.

Müller-Landau, F. and Cadenhead, D. A. (1979a). *Chem. Phys. Lipids* **25**, 299–314. Molecular Packing in Steroid-Lecithin Monolayers, Part I: Pure Films of Cholesterol, 3-Doxyl-cholestane, 3-Doxyl-17-hydroxy-andostane, Tetradecanoic Acid and Dipalmitoylphosphatidylcholine.

Müller-Landau, F. and Cadenhead, D. A. (1979b). *Chem. Phys. Lipids* **25**, 315–328. Molecular Packing in Steroid-Lecithin Monolayers, Part II. Mixed Films of Cholesterol with Dipalmitoylphosphatidylcholine and Tetradecanoic Acid.

Pockels, A. (1891). *Nature* **43**, 437–439. Surface Tension.

Polymeropoulos, E. E. and Möbius, D. (1979). *Ber. Bunsenges. Phys. Chem.* **83**, 1215–1222. Photochromism in Monolayers.

Polymeropoulos, E. E. and Sagiv, J. (1978). *J. Chem. Phys.* **69**, 1836–1847. Electrical Conduction Through Adsorbed Monolayers.

Polymeropoulos, E. E., Möbius, D., and Kuhn, H. (1978). *J. Chem. Phys.* **68**, 3918–3931. Photoconduction in Monolayer Assemblies with Functional Units of Sensitizing and Conducting Molecular Components.

Richard, M. A., Dutch, J., and Whitesides, G. M. (1978). *J. Am. Chem. Soc.* **100**, 6613–6625. Hydrogenation of Oriented Monolayers of ω-Unsaturated Fatty Acids Supported on Platinum.

Ringsdorf, H. and Shupp, H. (1980). *Am. Chem. Soc. Coat. Plast. Chem.* **42**, 379–385. Polymerization of Substituted Butadienes at the Gas-Water Interface.

Ringsdorf, H. and Schupp, H. (1981). *J. Macromol. Sci. Chem.* **A15**, 1015–1026. Polymerization of Substituted Butadienes at the Gas-Water Interface.

Rosoff, M. (1969). In *Physical Methods in Macromolecular Chemistry* (B. Carroll, Ed.), Marcel Dekker, New York, pp. 1–107. Surface Chemistry and Polymers.

Sackmann, E. (1978). *Ber. Bunsenges. Phys. Chem.* **82**, 891–909. Dynamic Molecular Organization in Vesicles and Membranes.

Sagiv, J. (1979a). *Isr. J. Chem.* **18**, 339–345. Organized Monolayers by Adsorption. I. Molecular Orientation in Mixed Dye Monolayers Built on Anisotropic Polymeric Surfaces.

Sagiv, J. (1979b). *Isr. J. Chem.* **18**, 346–353. Organized Monolayers by Adsorption. III. Irreversible Adsorption and Memory Effects in Skeletonized Silane Monolayers.

Sagiv, J. (1980). *J. Am. Chem. Soc.* **102**, 92–98. Organized Monolayers by Adsorption. IV. Formation and Structure of Oleophobic Mixed Monolayers on Solid Surfaces.

Seefeld, K.-P., Möbius, P., and Kuhn, H. (1977). *Helv. Chim. Acta* **60**, 2608–2632. Electron Transfer in Monolayer Assemblies with Incorporated Ruthenium(II) Complexes.

Seelig, J., Dijkman, R., and de Haas, G. H. (1980). *Biochemistry* **19**, 2215–2219. Thermodynamic and Conformational Studies on *sn*-2-Phosphatidylcholines in Monolayers and Bilayers.

Ter-Minassian-Saraga, L. (1975). In *Progress in Surface and Membrane Science* (D. A. Cadenhead, J. F. Danielli, and M. D. Rosenberg, Eds.), Vol. 9, Academic Press, New York, pp. 223–256. Interaction of Ions with Monolayers.

Tieke, B., Wegner, G., Naegele, D., and Ringsdorf, H. (1976). *Angew. Chem. Int. Ed. Eng.* **15**, 764–765. Polymerization of Tricosa-10,12-diynoic Acid in Multilayers.

Trurnit, H. J. (1945). *Fortschr. Chem. Org. Naturst.* **4**, 347–476. Über Monomolekulare Filme an Wassergrenzflächen und über Schichtfilme.

Valenty, S. J. and Gaines, G. L., Jr. (1977). *J. Am. Chem. Soc.* **99**, 1285–1287. Preparation and Properties of Monolayer Films of Surfactant Ester Derivatives of Tris(2,2′-bipyridine)ruthenium $(II)^{2+}$.

Whitten, D. G. (1979). *Angew. Chem. Int. Ed. Eng.* **18**, 440–450. Photochemical Reactions of Surfactant Molecules in Condensed Monolayer Assemblies. Environmental Control and Modification of Reactivity.

Whitten, D. G., Eaker, D. W., Horsey, B. E., Schmell, R. H., and Worsham, P. R. (1978). *Ber. Bunsenges. Phys. Chem.* **82**, 858–867. Photochemical and Thermal Reactions of Porphyrins and Organic Surfactants in Monolayer Assemblies. Modification of Reactivity in Condensed Hydrophobic Microenvironments.

Windreich, S. and Silberberg, A. (1980). *J. Colloid Interface Sci.* **77**, 427–434. Interaction of Lipid Multilayers with Water.

Zwick, M. M. and Kuhn, H. (1962). *Z. Naturforsch.* **17a**, 411–414. Strahlungsloser Übergang von Elektronenanregungsenergie durch dünne Schichten.

CHAPTER 5

BILAYER, BLACK LIPID MEMBRANES (BLMs)

The formation of a membraneous film on a small orifice separating aqueous compartments was described in the early 1960s (Mueller et al., 1962a, 1962b, 1963, 1964). Since the film spontaneously thins, essentially to a bilayer, it has been referred to as bimolecular or black lipid membrane, abbreviated as BLM. Characterizations and applications of BLMs have been comprehensively and repeatedly reviewed (Mueller et al., 1964; Thompson, 1964; Tien and Diana, 1968; Howard and Burton, 1968; Rothfield and Finkelstein, 1968; Haydon, 1968; Henn and Thompson, 1969; Mueller and Rudin, 1969a, 1969b; Goldup et al., 1970; Tien, 1971, 1972, 1974, 1976, 1979; Jain, 1972; Tien and Howard, 1972; Fettiplace et al., 1975; Mountz and Tien, 1978). This chapter, therefore, is limited to bare essentials.

1. FORMATION, COMPOSITION, AND STABILITY OF BLMs

The preparation of BLMs is deceptively simple. In the brush technique an organic solution of a lipid is brushed across a pinhole spearating the two aqueous phases (Figure 5.1*A*). The initially formed lipid is rather thick and reflects white light with a gray color (Figure 5.1*B*). Within a few minutes the film thins and the reflected light exhibits interference colors that ultimately turn black (hence the name "black lipid membrane"). In the dipping technique a small loop, made of stainless steel or Teflon, is pulled through a lipid solution. Alternatively, the loop is pulled through an already formed monolayer (Figure 5.2; Takagi et al., 1965). The BLM is surrounded by the Plateau–Gibbs border (the annulus), which for spherical loops takes the shape of a lens. The Plateau–Gibbs border acts as a liquid reservoir and its presence is essential for maintaining the stability and the integrity of the BLM. Properties of the Plateau–Gibbs border and that of the BLM have been discussed (Tien, 1974). Evidence has recently been presented for the dynamic nature of the BLM (Bach and Miller, 1980). The surface of the BLM was proposed not to be planar but to fluctuate ("breathe") between swollen and shrunken structures (Figure 5.3).

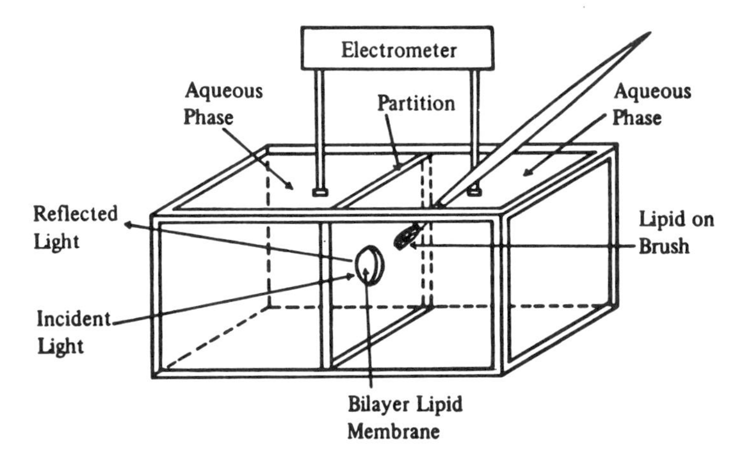

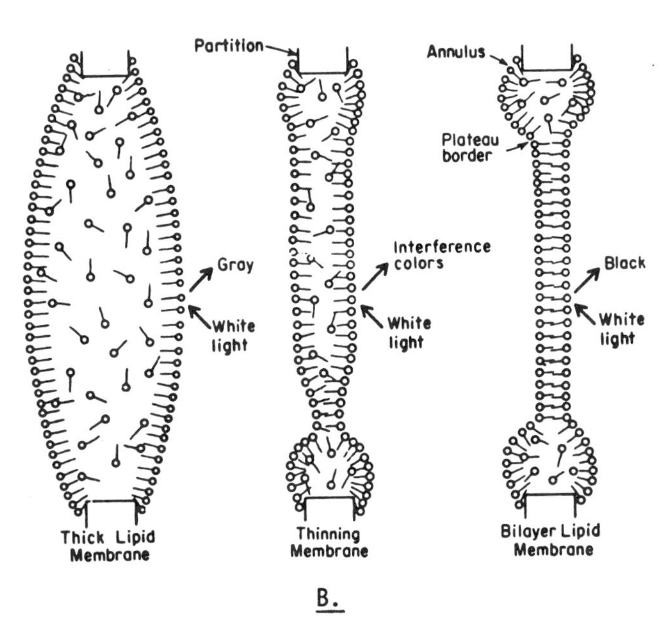

Figure 5.1. (*a*) Preparation of BLMs. A surfactant (lipid) solution is painted across a pinhole separating two aqueous compartments. Electrodes in the aqueous phases permit the measurements of bilayer conductance and other electrical properties. (*b*) Time dependent thinning of a BLM in aqueous media indicating the patterns of reflected light.

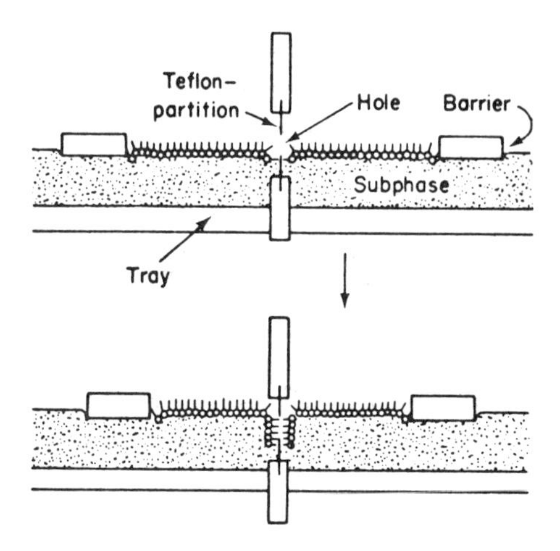

Figure 5.2. Schematic representation of BLM formation by pulling a loop through a monolayer (taken from Takagi et al., 1965).

Preparation of BLMs may sound easy. In reality they are notoriously difficult. Meticulous care and scrupulous cleanliness are needed for BLM formation. The protocol to be followed (Tien, 1974) is similar to that detailed in Chapter 4 for monolayers. Neophytes have often been frustrated by their inability to reproduce published BLMs.

Lipids of biological origins have been used in most of the early BLM studies (Tien, 1974). Subsequently, BLMs were formed from such synthetic lipids as 1,2-dipalmitoyl-*sn*-glycero-3-phosphoryl choline, 1-stearoyl-2-myristocyl-*sn*-glycero - 3 - phosphonyl choline, 1 - oleoyl - 2 - stearoyl - *sn* - glycero - 3 - phosphorylcholine, *rac* - 1 - oleoyl - 2 - O - palmitoyl - *sn* - glycero - 3 - phosphoryl choline, and *rac*-1-oleoyl-2*n*-hexadecylpropanediol-3-phosphoryl choline (Leslauer et al., 1968), as well as from such surfactants as sorbitan tristearate

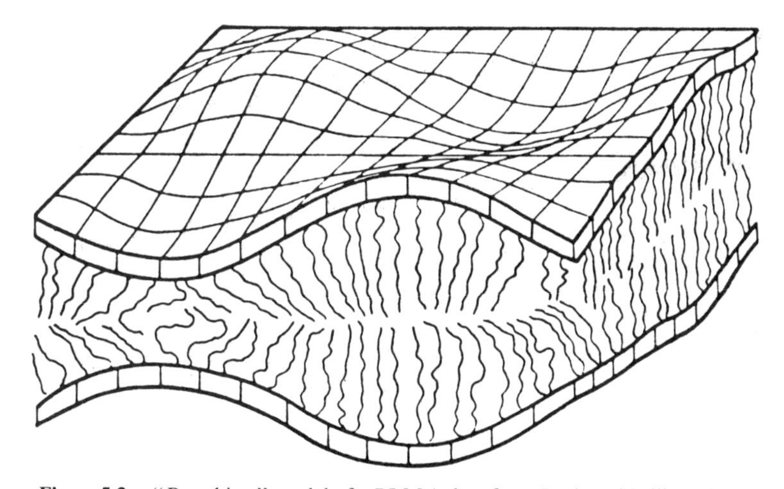

Figure 5.3. "Breathing" model of a BLM (taken from Bach and Miller, 1980).

(Tien and Dawidowicz, 1966; Bradley and Dighe, 1969), glyceryl distearate (Tien and Dawidowicz, 1966), glycerol dioleate (Tien and Diana, 1967), sorbitan monopalmitoleate (Taylor and Haydon, 1967), dioctadecylphosphite (Tien, 1967a), hexadecyltrimethylammonium bromide (CTAB) (Ter-Minassian-Saraga and Wietzerbin, 1970; Sweeney and Blank, 1973), hexadecyltrimethylammonium chloride (CTACl) (Jain, 1972), and *N*-lauryl myristoyl-β-aminopropionic acid (Tien, 1968). Interestingly, oxidized cholesterol yields a BLM while pure cholesterol does not (Tien et al., 1966). Apparently, oxidized cholesterol has higher dichotomy between polar and nonpolar regions than does pure cholesterol. The precise chemical composition, or indeed the morphology, BLMs is generally unknown. Large mole fractions of additives and/or solvents are present in the bilayer.

Although a great deal has been written on factors influencing the stability of a BLM (Tien, 1974), its actual lifetime is rarely mentioned. The simple fact is that BLMs rarely last longer than a couple of hours. Those surviving for 24 hours are hailed as major accomplishments (Jain, 1972; Tien, 1979). Sample impurities, oxidation of the lipids, impurities on the equipment, unsuitability of the spreading solvent, temperature fluctuations, vibrations, concentration, and viscosity stress are some of the factors responsible for the instability of BLMs. Attempts have been made to control these factors and to increase the longevity of BLMs by the addition of polymers (Yoshida et al., 1975; Shchipunov, 1978). It should be pointed out, however, that during their existence BLMs are quite stable. They are viscoelastic, seal themselves when punctured gently, and withstand strong electrical currents (Mueller et al., 1964; Huang et al., 1964).

2. OPTICAL AND ELECTRICAL PROPERTIES OF BLMs

Optical techniques are used to monitor BLM formation (Figure 5.1). Additionally, reflectance measurements provide a means for calculating the bilayer thickness. By analogy with soap films, either a single- or a triple-layer model is used (Figure 5.4). Using the single-layer model, the ratio of the intensity of the incident (I_i) and reflected (I_r) light (Figure 5.5) is given by the Rayleigh equation

$$\frac{I_r}{I_i} = 4\left[\frac{n_h - n_w}{n_h + n_w}\right]^2 \sin^2\left[\frac{2\pi n_n\, d \cos\theta}{\lambda}\right] \tag{5.1}$$

where n_h and n_w are refractive indices of the hydrocarbon and aqueous phases, respectively, d is the BLM thickness, θ is the internal angle of reflection, and λ is the wavelength of the light used. The single-layer model, equation 5.1, only provides a first approximation for d since it assumes BLMs to be isotropic. The triple-layer structure is a better model; it allows for separate refractive

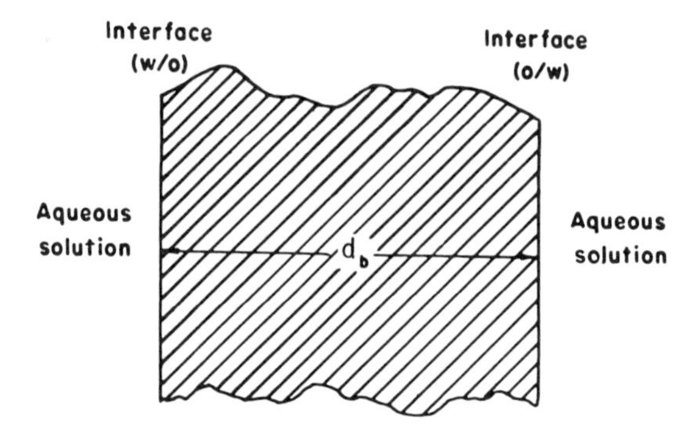

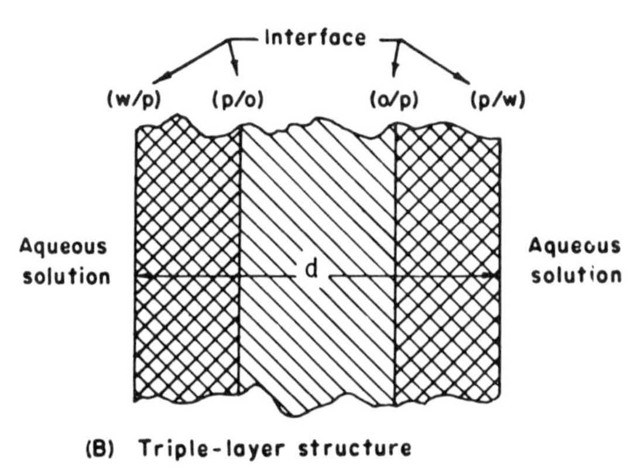

Figure 5.4. Proposed single- and triple-layer BLM models (taken from Tien, 1974).

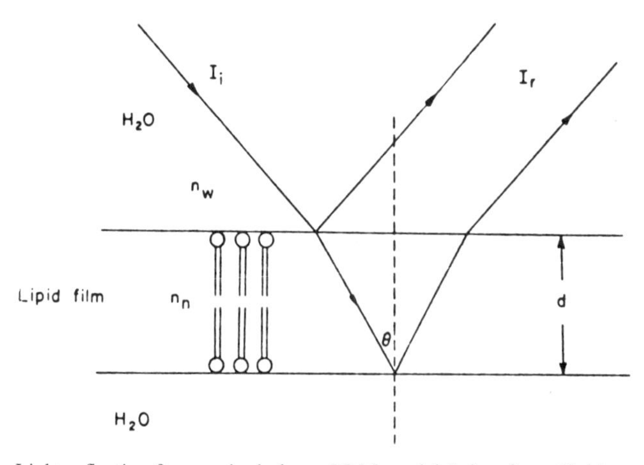

Figure 5.5. Light reflection from a single-layer BLM model (taken from Goldup et al., 1970).

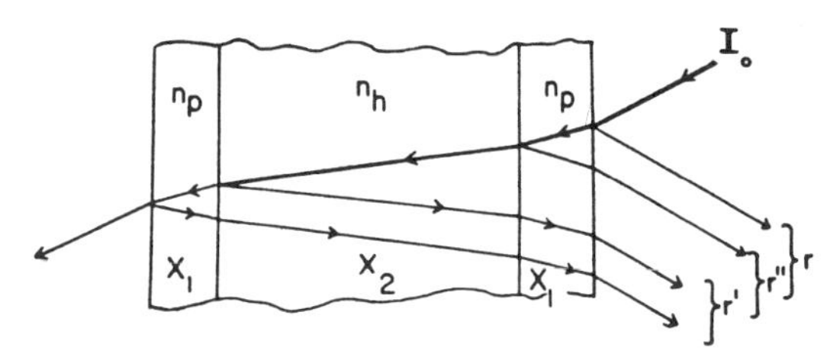

Figure 5.6. Light reflection from a triple-layer BLM model. I_0 is the incident light wave. X's refer to phase differences due to different layers. r, r', and r'' are amplitudes of the reflected light at various interfaces. n_p and n_h are reflective indices of the polar and hydrocarbon layers (taken from Tien, 1974).

indices at the middle layers (Figure 5.6). The reflectance, R, is given by (Tien, 1967b)

$$R = \frac{r_1^2 + r''^2 + 2r_1 r'' \cos(x_1 - \delta_h)}{1 + r_1^2 r''^2 + 2r_1 r'' \cos(x_1 - \delta_h)} \tag{5.2}$$

where r's represent the dimensionless Fresnel's amplitudes of the light reflected from the interfaces 1, 2, 3, and 4, designated as r_1, r_2, r_3, and r_4, respectively, and given by

$$r_1 = -r_4 = \frac{n_p - n_w}{n_p + n_w} \tag{5.3}$$

and

$$r_2 = -r_3 = \frac{n_h - n_p}{n_h + n_p} \tag{5.4}$$

where n_p is the refractive index of the polar layer. The phase differences (X_1 and X_2) and the corresponding reflectances (r' and r'') are given by

$$X_1 = X_3 = \frac{4\pi n_p\, d_p \cos\theta}{\lambda} \tag{5.5}$$

$$X_2 = \frac{4\pi n_h\, d_h \cos\theta}{\lambda} \tag{5.6}$$

and

$$r'^2 = \frac{r_3^2 + r_4^2 + 2r_3 r_3 r_4 \cos x_1}{1 + r_3^2 r_4^2 + 2r_3 r_4 \cos x_1} \tag{5.7}$$

$$r''^2 = \frac{r_2^2 + r'^2 + 2r_2 r' \cos(X_2 - \delta_p)}{1 + r_2^2 r'^2 + 2r_2 r' \cos(X_2 - \delta_p)} \tag{5.8}$$

where θ and δ are the reflected angles in the polar and hydrocarbon layers. Knowledge of the refractive indices in each layer allows the calculation of BLM thicknesses d_p and d_h. Refractive indices can, in turn, be obtained from Brewster angle measurements. Considering the experimental uncertainties and the assumptions involved, bilayer thicknesses of 70 ± 10 Å obtained from optical measurements (Tien, 1974), are reasonable.

The thicknesses of BLMs have also been determined by means of electrical capacitance measurements. The bilayer is considered as a parallel plate condenser and the thickness of the hydrophobic region, d_h, of dielectric ε_n is related to the capacitance, C_m by

$$C_m = \frac{A\varepsilon_h}{4\pi\, d_h} \tag{5.9}$$

where A is the area of the film. Assuming the dielectric constant of the bilayer to be between 2 and 5, bilayer thicknesses were determined to be around 70 Å (Tien, 1974). Asymmetric surface potentials have also been obtained from the measured dependence of bilayer capacitance on transmembrane voltage (Schoch et al., 1979). Bilayer thicknesses measured by means of electrical capacitance agreed well with those determined by optical and electron microscopic methods. Ideally, BLM thicknesses should be determined by more than one method. Staining or freeze etching, used in electron microscopy, and the presence of an electric field across the membrane, in measuring electrical capacitance, may well alter the structure of BLMs. Although reflectance measurements do not disturb the BLM, the need for evaluating refractive indices renders this method somewhat cumbersome.

The physical arrangement of BLMs in separating two aqueous compartments renders electrical measurements feasible. Thus in addition to capacitance measurements, conductivities, dielectric breakdown voltages, and membrane potentials have been determined (Tien, 1974). Conductivities are determined by applying a known voltage through the BLM and measuring the current. For most BLMs Ohm's law is followed up to about 60 mV. At higher potentials there is a gradual decrease of the slope in the voltage-current plot until the membrane breaks. This dielectric breakdown typically occurs at around 200 mV. The rise time of a BLM current on applying a dc pulse is of the order of 0.1–10 sec. While the capacitance is reproducible from BLM to BLM, that is not the case in conductivity measurements. Electrical resistances are reported, therefore, only as orders of magnitudes and are used mainly to assess the relative influence of additives (Tien, 1974). Addition of electrolytes to one side of the BLM (see Figure 5.1) often results in dramatic changes in the membrane resistance.

The decrease in membrane resistance in the presence of an electrolyte gradient has been rationalized in terms of equilibrium 5.8 occurring at the BLM surface (Tien and Diana, 1967):

$$\text{BLM}^{+}\text{—X}^{-} + \text{Y}^{-} \rightleftharpoons \text{BLM}^{+}\text{—Y}^{-} + \text{X}^{-} \tag{5.10}$$

$BLM^+—Y^-$ is likely to have different surface potential and charge densities from $BLM^+—X^-$. Additionally, ion-pair formation and activity coefficient effects may alter the membrane structure and phase transitions. A manifestation of these effects is the altered sequence of alkali cation transference numbers in BLMs ($H^+ < Li^+ < Rb^+ < Na^+ < Cs^+ < K^+$) with respect to those in bulk water ($Li^+ < Na^+ < K^+ < Rb^+ < Cs^+ < H^+$) (Miyamoto and Thompson, 1967). Conductance across BLMs has been discussed in terms of ionic and electronic mechanisms. Of these ionic conductance appears to account better for the observed data (Tien, 1974).

Effects of organic molecules and proteins on BLM resistances have also been investigated (Tien, 1974). A particularly interesting example is the early verification of Mitchell's chemiosmotic hypothesis (Mitchell, 1961, 1977). Addition of 2,4-dinitrophenol, an uncoupler, to the aqueous solution on one side of the BLM resulted in an appreciable decrease in the resistance (Bielawski et al., 1966; Hopfer et al., 1968). 2,4-Dinitrophenol acts as a hydrogen ion carrier, thereby obviating the pH gradient required for phosphorylation. Dielectric measurements have also been utilized to determine the extent of perturbation caused by spectroscopic probes on BLMs (Ashcroft et al., 1980).

Applied electric field across the BLM causes a substantial decrease in the bilayer thickness (White, 1970; Crowley, 1973; Benz et al., 1975; Requena et al., 1975). This is due to a field induced shift of solvent molecules from the bilayer to the Plateau–Gibbs border (White, 1980). Applying high voltages (~1 V) to BLMs in the 10^{-9}–10^{-6} sec timescale results in a dramatic increase of the

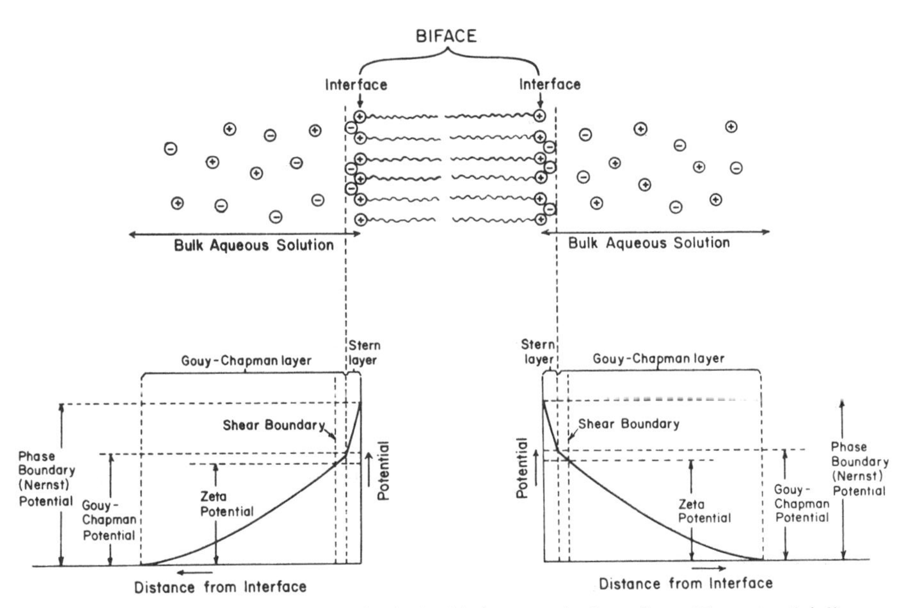

Figure 5.7. A charged bilayer sets up ionic double layers at both surfaces. The potential diagram is drawn schematically, ignoring possible penetration of the counterions into the BLM.

conductance (Benz et al., 1979). Concomitantly, there is an increase in permeability. This field induced increase in conductance has been termed electrical (dielectric) breakdown. In contrast to analogous effects in solid state physics, the electrical breakdown of cell membranes is reversible. After the removal of the field the original membrane and all of its functions are restored. Similar effects have been observed in BLMs prepared from oxidized cholesterol-*n*-decane (Benz et al., 1979).

Potential differences are generated on either side of the BLM (Figure 5.7). These are best discussed in terms of the electrical double-layer theory (MacDonald and Bangham, 1972). Close to the surfactant headgroups there are relatively immobile hydrated counterions at high local concentrations. This compact area, the Stern layer, extending a few angstroms from the interface, provides the surface potential. The shear surface is defined as the plane that separates bound and free water molecules. Beyond the shear surface the concentration of counterions and the potential gradually decrease to zero. The total potential at the BLM interface, the Nernst potential, extends to several hundred millivolts and to several hundred angstroms. The potential between the shear and diffuse layers and that between the Stern region and the shear surface are known as the Gouy–Chapman and ζ-potential, respectively. Surface and transmembrane potentials are intimately involved in membrane permeabilities.

3. PERMEABILITIES OF BLMs

Selective transport of ions and molecules is one of the most essential functions of membranes. BLMs provide relatively convenient means for the quantitative investigation of permeabilities. Extensive theoretical and experimental studies have been carried out on the transport of water, uncharged molecules, and electrolytes across BLMs in the absence and in the presence of additives (Jain, 1972; Tien, 1974; Shamoo and Goldstein, 1977; Abramson and Shamoo, 1979; Shamoo and Tivol, 1980).

Water permeabilities are determined by setting up an osmotic gradient and following the volume flux of water transported across the bilayer, either directly, or by isotopic labeling. The volume flux of water, Φ_w (in moles per second) passing across the BLM of area A (in square centimeters) at a known solute concentration gradient of ΔC_s (in moles per milliliter) is related to the permeability coefficient P (in centimeters per second) by

$$\Phi_w = -PA\phi\Delta C_s \tag{5.11}$$

where Φ is the osmotic coefficient, a constant property of the measuring apparatus. This type of water permeability determination assumes the BLM to be a perfect osmometer with respect to the solute used to set up the gradient (i.e., the solute is assumed to be completely impermeable). Typical values for water permeability coefficients in BLMs are (5–100) × 10^{-4} cm sec^{-1} (Fettiplace, 1978). Differences in permeabilities reflect changes in the composition of BLMs.

The permeation rate of polar nonelectrolytes across BLMs is appreciably lower than that of water. Permeability coefficients of such solutes as glucose, ribose, sucrose, glycerol, urea, thiourea, and indoles in BLMs range between (1 and 1000)10^{-8} cm sec^{-1} (Jain, 1972; Tien, 1974). Nonelectrolyte permeabilities and selectivities across BLMs are functions of several, as yet incompletely understood, parameters. By analogy with biomembranes (Stein, 1967; Diamond and Wright, 1969), permeation is believed to be related to the partitioning of the substrate between a hydrocarbon solvent and water. This statement implies that the rate determining permeation step is diffusion through the interior rather than through the water-membrane interface. For many systems, however, the highest potential energy barrier is at the hydrated bilayer surface. Permeability coefficients (P values) of nonelectrolytes across membranes have been related to the diffusion coefficients of the permeant from the bulk solution to the membrane, D_{sm}, and that through the membrane, D_m, by (Zwolinski et al., 1949)

$$\frac{1}{P} = \frac{2l}{D_{sm}} + \frac{d}{D_m K} \tag{5.12}$$

where l is the mean free path, d is the thickness of the membrane, and K is the partition coefficient of the permeant; $K = k_{sm}/k_{ms}$, where k_{ms} is the rate of diffusion from the membrane to the bulk solution and k_{sm} is that from the bulk solution to the membrane. Values of P can be evaluated from equation 5.12 when the rate of diffusion is membrane limiting (i.e., $k_{ms} \gg k_m$), interface limiting (i.e., $k_m \gg k_{ms}$), or membrane interface limiting. Nonelectrolyte permeabilities across membranes have been discussed in terms of hydrogen bonding, hydrophobic, and weak van der Waals interactions. Quantitative treatments of nonelectrolyte permeability in BLMs have not yet been developed.

Electrostatic interactions are additional factors that influence that transport of ions across membranes. Ion permeabilities are markedly slower in BLMs than in biomembranes (Jain, 1972). Different ion permeabilities across membranes and BLMs were shown to originate in the difference between the microscopic dielectric constants of these systems. Some bilayer components of the inner mitocondrial membrane were shown to have dielectric constants higher than 2.2 (Dilger and McLaughlin, 1979). In contrast, dielectric constants of BLMs are much lower. The observed 1000-fold increase in the permeabilities of perchlorate and thiocyanate ions in BLMs prepared from 1-chlorodecane ($\varepsilon = 4.5$), as compared with those obtained by using n-decane ($\varepsilon = 2.0$) as the solvent (Dilger et al., 1979), lends credence to the postulated importance of microscopic membrane dielectric constants.

Incorporation of macrocyclic antiobiotics and crown ethers (ionophores) into BLMs allows the modeling of carrier mediated substrate transport across biomembranes (Jain, 1972; Tien, 1974).

REFERENCES

Abramson, J. J. and Shamoo, A. E. (1979). *J. Membr. Biol.* **50**, 241–255. Anionic Detergents as Divalent Cation Ionophores Across Black Lipid Membranes.

Ashcroft, R. G., Thulborn, K. R., Smith, J. R., Coster, H. G. L., and Sawyer, W. H. (1980). *Biochim. Biophys. Acta* **602**, 299–308. Perturbations to Lipid Bilayers by Spectroscopic Probes as Determined by Dielectric Measurements.

Bach, D. and Miller, I. R. (1980). *Biophys. J.* **29**, 183–188. Glyceryl Monooleate Black Lipid Membranes Obtained from Squalene Solutions.

Benz, R., Fröhlich, O., Läuger, P., and Montal, M. (1975). *Biochim. Biophys. Acta*, **394**, 323–334. Electrical Capacity of Black Lipid Films and of Lipid Bilayers Made from Monolayers.

Benz, R., Beckers, F., and Zimmermann, V. (1979). *J. Membr. Biol.* **48**, 181–204. Reversible Electrical Breakdown of Lipid Bilayer Membranes: A Charge-Pulse Relaxation Study.

Bielawski, J., Thompson, T. E., and Lehninger, A. L. (1966). *Biochem. Biophys. Res. Commun.* **24**, 948–954. The Effect of 2,4-Dinitrophenol on the Electrical Resistance of Phospholipid Bilayer Membranes.

Bradley, J. and Dighe, A. M. (1969). *J. Colloid Interface Sci.* **29**, 157. Electrical Resistivity of Ultrathin Black Films of Sorbitan Tristearate.

Crowley, J. M. (1973). *Biophys. J.* **13**, 711–724. Electrical Breakdown of Bimolecular Lipid Membranes as an Electrochemical Instability.

Diamond, J. M. and Wright, E. M. (1969). *Annu. Rev. Physiol.* **31**, 581–646. Biological Membranes: The Basis of Ion and Nonelectrolyte Selectivity.

Dilger, J., and McLaughlin, S. (1979). *J. Membr. Biol.* **46**, 359–384. Proton Transport Through Membranes Induced by Weak Acids: A Study of Two Substituted Benzimidazoles.

Dilger, J. P., McLaughlin, S. G. A., McIntosh, T. J., and Simon, S. A. (1979). *Science* **206**, 1196–1198. The Dielectric Constant of Phospholopid Bilayers and the Permeability of Membranes to Ions.

Fettiplace, R. (1978). *Biochim. Biophys. Acta* **513**, 1–10. The Influence of the Lipid on the Water Permeability of Artificial Membranes.

Fettiplace, R., Gordon, L. G. M., Hladky, S. B., Requena, J., Zingsheim, H. P., and Haydon, D. A. (1975). *Methods Membr. Biol.* **4**, 1–75. Techniques in the formation and Examination of "Black" Lipid Bilayer Membranes.

Goldup, A., Ohki, S., and Danielli, J. F. (1970). In *Recent Progress in Surface Science*, (J. F. Danielli, A. C. Riddiford, and M. D. Rosenberg, Eds.), Vol. 3, Academic Press, New York, pp. 193–260. Black Lipid Films.

Haydon, D. A. (1968). In *Membrane Models and the Formation of Biological Membranes* (L. Bolis and B. A. Pethica, Eds.), North Holland, Amsterdam, pp. 91–97. The Electrical Properties and Permeability of Lipid Bilayers.

Henn, F. A. and Thompson, T. E. (1969). *Annu. Rev. Biochem.* **38**, 241–262. Synthetic Bilayer Membranes.

Hopfer, V., Lehninger, A. L., and Thompson, T. E. (1968). *Proc. Natl. Acad. Sci. USA* **59**, 484–490. Photonic Conductance Across Phospholipid Bilayer Membranes Induced by Uncoupling Agents for Oxidative Phosphorylation.

Howard, R. E. and Burton, R. M. (1968). *J. Am. Oil Chem. Soc.* **45**, 202–229. Thin Lipid Membranes with Aqueous Interfaces: Apparatus Design and Methods of Study.

Huang, C., Wheeldon, L., and Thompson, T. E. (1964). *J. Mol. Biol.* **8**, 148–160. The Properties of Lipid Bilayer Membranes Separating Two Aqueous Phases: Formation of a Membrane of Simple Composition.

Jain, M. H. (1972). *The Bimolecular Lipid Membrane: A System*, Van Nostrand Reinhold Co., New York.

Lesslauer, W., Slotboom, A. J., Postema, N. M., De Haas, G. H., and Van Deenen, L. L. M. (1968). *Biochim. Biophys. Acta* **150**, 306–308. Effects of Phospholipase A and Pre-phospholipase A on Black Phospholipid Membranes.

MacDonald, R. D. and Bangham, A. D. (1972). *J. Membr. Biol.* **7**, 29–53. Comparison of Double Layer Potentials in Lipid Monolayers and Lipid Bilayer Membranes.

Mitchell, P. (1961). *Nature* **191**, 144–148. Coupling of Phosphorylation to Electron and Hydrogen Transfer by a Chemi-osmotic Type of Mechanism.

Mitchell, P. (1977). *FEBS Lett.* **78**, 1–20. A Commentary on Alternative Hypotheses of Proton Coupling in the Membrane Systems Catalyzing Oxidative and Photosynthetic Phosphorylation.

Miyamoto, V. K. and Thompson, T. E. (1967). *J. Colloid Interface Sci.* **25**, 16–25. Some Electrical Properties of Lipid Bilayer Membranes.

Mountz, J. D. and Tien, H. T. (1978). *J. Bioenerg. Biomembr.* **10**, 139–151. Bilayer Lipid Membranes (BLM): Study of Antigen-Antibody Interactions.

Mueller, P. and Rudin, D. O. (1969a). In *Current Topics in Bioenergetics* (R. Sanadi, Ed.), Vol. 3, Academic Press, New York, pp. 157–249. Translocators in Bimolecular Lipid Membranes: Their Role in Dissipative and Conservative Bioenergy Transductions.

Mueller, P. and Rudin, D. O. (1969b). In *Laboratory Techniques in Membrane Biophysics* (H. Passow and R. Stampfli, Eds.), Springer, Berlin, pp. 141–156. Bimolecular Lipid Membranes: Techniques of Formation, Study of Electrical Properties, and Induction of Ion Gating Phenomena.

Mueller, P., Rudin, D. O., Tien, H. T., and Wescott, W. C. (1962a). *Nature* **194**, 979–980. Reconstitution of Cell Membrane Structure in Vitro and Its Transformation into an Excitable System.

Mueller, P., Rudin, D. O., Tien, H. T., and Wescott, W. C. (1962b). *Circulation* **26**, 1167–1170. Reconstitution of Excitable Cell Membrane Structure in Vitro.

Mueller, P., Rudin, D. O., Tien, H. T., and Wescott, W. C. (1963). *J. Phys. Chem.* **67**, 534–535. Methods for the Formation of Single Bimolecular Lipid Membranes in Aqueous Solution.

Mueller, P., Rudin, D. O., Tien, H. T., and Wescott, W. C. (1964). In *Recent Progress in Surface Science* (J. F. Danielli, K. G. A. Parkhurst, and A. C. Riddiford, Eds.), Vol. 1, Academic Press, New York, pp. 379–393. Formation and Properties of Bimolecular Lipid Membranes.

Requena, J., Haydon, D. A., and Hladky, S. B. (1975). *Biophys. J.* **15**, 77–81. Lenses and the Compression of Black Lipid Membranes by an Electric Field.

Rothfield, L. and Finkelstein, A. (1968). *Annu. Rev. Biochem.* **37**, 463–496. Membrane Biochemistry.

Schoch, P., Sargent, D. F., and Schwyzer, R. (1979). *J. Membr. Biol.* **46**, 71–89. Capacitance and Conductance as Tools for the Measurement of Asymmetric Surface Potentials and Energy Barriers of Lipid Bilayer Membranes.

Shamoo, A. E. and Goldstein, D. A. (1977). *Biochim. Biophys. Acta* **472**, 13–53. Isolation of Ionophores from Ion Transport Systems and Their Role in Energy Transduction.

Shamoo, A. E. and Tivol, W. F. (1980). *Curr. Top. Membr. Transp.* **15**, 57–126. Criteria for Reconstitution of Ion Transport Systems.

Shchipunov, Y. A. (1978). *Biophysics* **23**, 162–164. Stabilization of Bilayer Lipid Membranes by Polymer Resins.

Stein, W. D. (1967). *The Movement of Molecules Across Cell Membranes*, Academic Press, New York.

Sweeney, G. D. and Blank, M. (1973). *J. Colloid Interface Sci.* **42**, 410–417. Some Electrical Properties of Thin Lipid Films Formed from Cholesterol and Acetyltrimethylammonium Bromide.

Takagi, M., Azuma, K., and Koshimoto, U. (1965). *Annu. Rep. Biol. Works, Fac. Sci. Osaka Univ.* **13**, 107–120. A New Method for the Formation of Bilayer Membranes in Aqueous Solutions.

Taylor, J., and Haydon, D. A. (1967). *Discuss. Faraday Soc.* **51**, 42–45. Stabilization of Thin Films of Liquid Hydrocarbon by Alkyl Chain Interaction.

Ter-Minassian-Saraga, L., and Wietzerbin, J. (1970). *Biochem. Biophys. Res. Commun.* **41**, 1231–1237. Action of Hexadecyltrimethylammonium Bromide on Bilayer Lipid Membranes.

Thompson, T. E. (1964). In *Cellular Membranes in Development* (M. Locke, Ed.), Academic Press, New York, pp. 83–96. The Properties of Bimolecular Phospholipid Membranes.

Tien, H. T. (1967a). *J. Phys. Chem.* **71**, 3395–3401. Black Lipid Membranes in Aqueous Media—Interfacial Free Energy Measurements and Effect of Surfactants on Film Formation and Stability.

Tien, H. T. (1967b). *J. Theor. Biol.* **16**, 97–110. Black Lipid Membranes: Thickness Determination and Molecular Organization by Optical Methods.

Tien, H. T. (1968). *J. Phys. Chem.* **72**, 2723–2729. Thermodynamics of Bimolecular (Black) Lipid Membranes at Water-Oil-Water Biface.

Tien, T. H. (1971). In *Surface and Colloid Science* (E. Matijevic, Ed.), Vol. 4, John Wiley, New York, pp. 361–423. Bimolecular Lipid Membranes.

Tien, H. T. (1972). In *Surface Chemistry and Colloids* (M. Kerker, Ed.), MPT International Review of Science, Vol. 7, Butterworths, London, pp. 25–78. Membranes: Their Interfacial Chemistry and Biophysics.

Tien, H. T. (1974). *Bilayer Lipid Membranes (BLM) Theory and Practice*, Marcel Dekker, New York.

Tien, H. T. (1976). *Photochem. Photobiol.* **24**, 97–116. Electronic Processes and Photoelectric Aspects of Bilayer Lipid Membranes.

Tien, H. T. (1979). In *Photosynthesis in Relation to Model Systems* (J. Barber, Ed.), Elsevier, Amsterdam, pp. 115–173. Photoeffects in Pigmented Bilayer Lipid Membranes.

Tien, H. T., and Dawidowicz, E. A. (1966). *J. Colloid Interface Sci.* **22**, 438–453. Black Lipid Films in Aqueous Media: Experimental Techniques and Thickness Measurements.

Tien, H. T. and Diana, A. L. (1967). *J. Colloid Interface Sci.* **24**, 287–296. Black Lipid Membranes in Aqueous Media—The Effect of Salts on Electrical Properties.

Tien, H. T. and Diana, A. L. (1968). *Chem. Phys. Lipids* **2**, 55–101. Bimolecular Lipid Membranes: A Review and a Summary of Some Recent Studies.

Tien, H. T. and Howard, R. E. (1972). In *Techniques of Surface and Colloid Chemistry and Physics* (R. J. Good, R. R. Stromberg, and R. L. Patrick, Eds.), Vol. 1, Marcel Dekker, New York, pp. 109–211. Bimolecular Lipid Membranes.

Tien, H. T., Carbone, S., and Dawidowicz, E. A. (1966). *Kolloid-Z. Z. Polym.* **212**, 165–166. Black Lipid Films: A New Type of Interfacial Adsorption Phenomenon.

White, S. H. (1970). *Biophys. J.* **10**, 1127–1148. A Study of Lipid Bilayer Membrane Stability Using Precise Measurements of Specific Capacitance.

White, S. H. (1980). *Science* **107**, 1075–1077. How Electric Fields Modify Alkane Solubility in Lipid Bilayers.

Yoshida, T., Ogura, S., and Okuyama, M. (1975). *Bull. Chem. Soc. Jpn.* **48**, 2775–2778. Ultra-thin Layer of Lipids in Polymer Film and Its Electric Properties. New Sorption Method to Prepare a Model Biomembrane System.

Zwolinski, B. J., Eyring, H., and Reese, C. E. (1949). *J. Phys. Colloid Chem.* **53**, 1426–1453. Diffusion and Membrane Permeability I.

CHAPTER 6

VESICLES

The formation of multilamellar structures, "myelin figures," upon the swelling of dried phospholipids in water has been recognized for some time (Stoeckenius, 1959). Bangham and his co-workers subsequently showed them to be smectic mesophases of closed phospholipid bilayers capable of entrapping ions in their aqueous interiors (Bangham et al., 1965, 1967a, 1967b; Papahadjopoulos and Watkins, 1967). Liposomes, as these preparations came to be known, gained considerable popularity as membrane models (Bangham, 1968; Bangham et al., 1974; Tyrrell et al., 1976; Papahadjopoulos, 1978; Kimelberg and Mayhew, 1978; Gregoriadis and Allison, 1980; Chapman, 1980). Properties of liposomes have been investigated by electron micrography (Zingsheim and Plattner, 1976; Wagner, 1980), nuclear magnetic resonance (nmr) (Petersen and Chan, 1977; Mantsch et al., 1977; Seelig, 1977; Roberts et al., 1978; Pope and Cornell, 1978; Smith, 1979; Skarjune and Oldfield, 1979; Hauser et al., 1980a), electroparamagnetic resonance (epr) (Hubbell and McConnell, 1971; Berlinger, 1976; Likhtenschtein, 1976), fluorescence (Gaffney and Chen, 1970; Radda and Vanderkooi, 1972; Radda, 1975; Chen and Edelhoch, 1975; Wehry, 1976; Browning and Nelson, 1979), and laser Raman and infrared spectroscopy (Wallach et al., 1979; Pink et al., 1980; Bansil et al., 1980).

In the present monograph the term vesicles is used to describe spherical or ellipsoidal single- or multicompartment closed bilayer structures, regardless of their chemical compositions. Vesicles composed of naturally occurring or synthetic phospholipids are referred to as liposomes. In contrast, those formed from completely synthetic surfactants are to be designated as surfactant vesicles (Fendler, 1980a). Preparations and characterizations of liposomes and surfactant vesicles are discussed in this chapter. Since publications in these areas (already in the tens of thousands) are growing at an alarming rate, no attempt can be made at an exhaustive coverage. Our purpose is to provide an up-to-date background, sufficient to allow the novice to design meaningful experiments using liposomes and surfactant vesicles.

1. PREPARATION OF LIPOSOMES

Three types of liposomes have been investigated to date. They are the small (200–500 Å diameter), the large (0.1–10 μ diameter) single bilayer, and the multicompartment (1000–8000 Å diameter) liposomes (Pagano and Weinstein, 1978; Szoka and Papahadjopoulos, 1980). Abbreviations used in some publications are: SUV = small unilamellar vesicle, MLV = multilamellar vesicle, LUV = large unilamellar vesicle, and REV = reversed-phase evaporation vesicle (= LUV) (Szoka and Papahadjopoulos, 1980).

Hydration of dried phospholipids results in the spontaneous formation of multicompartment liposomes. The method, developed by Bangham (Bangham et al., 1965), involves the swelling of a thin lipid film. The film is deposited on the wall of a conical flask by rotary evaporating a chloroform stock solution of the phospholipid. Complete solvent removal is ensured by drying the thin film in a vacuum desiccator for several hours. Removal of the lipid film from the wall of the conical flask is accomplished by handshaking or by vortexing with an appropriate amount of distilled water. Addition of glass beads facilitates this process. Cosolution of long chain amines (e.g., stearylamine) or, alternatively, long chain phosphates (e.g., dihexadecylphosphate) with the phospholipid results in liposomes that have net positive or net negative charges. Details of multicompartment liposome preparations are provided in *Methods of Enzymology* (Kinsky, 1974).

Individual multicompartment liposome preparations are rather heterogeneous. Spherical onionlike, oblong, and tubular structures of various sizes coexist. This is clearly seen in electron micrographs (Figure 6.1). Importantly, however, liposomes are closed; their aqueous compartments are separated from each other (Figure 6.1).

Encapsulation of magnetite, Fe_3O_4, prepared by the addition of ammonium hydroxide solution (1.0 *M*) to a solution containing equimolar ratios of ferrous and ferric sulfate (0.5 *M*) until a pH of 10.0 is reached, into single-compartment phosphatidylcholine liposomes has been found to considerably aid X-ray and nmr investigations (Mann et al., 1979).

Multicompartment liposomes of defined size and homogeneity have been prepared by extrusion through polycarbonate membranes (Olson et al., 1979). Liposome solutions, prepared in the usual way, were sequentially extruded through polycarbonate membranes (Nucleopore Inc., Pleasanton, CA) with pore diameters of 1.0, 0.8, 0.6, 0.4, and 0.2 μm, under pressure. Passage through the membrane pores resulted in progressively smaller and less polydisperse multicompartment liposomes (Table 6.1). This technique provides a reproducible protocol for the formation of multicompartment liposomes of definite size distribution while maintaining high efficiency of entrapped aqueous volumes (Olson et al., 1979). Filtration has to be carried out, however, below the phase transition temperature since above the phase transition temperatures liposomes are not filterable (Brendzel and Miller, 1980).

Multicompartment liposomes have also been prepared by the introduction of an aqueous buffer into an ether solution of the phospholipid, followed by the removal of the organic solvent (Szoka and Papahadjopoulos, 1978). The proposed mechanism involves the collapse of reversed micelles into a gel-like state on the removal of the organic solvent, and this results ultimately in the

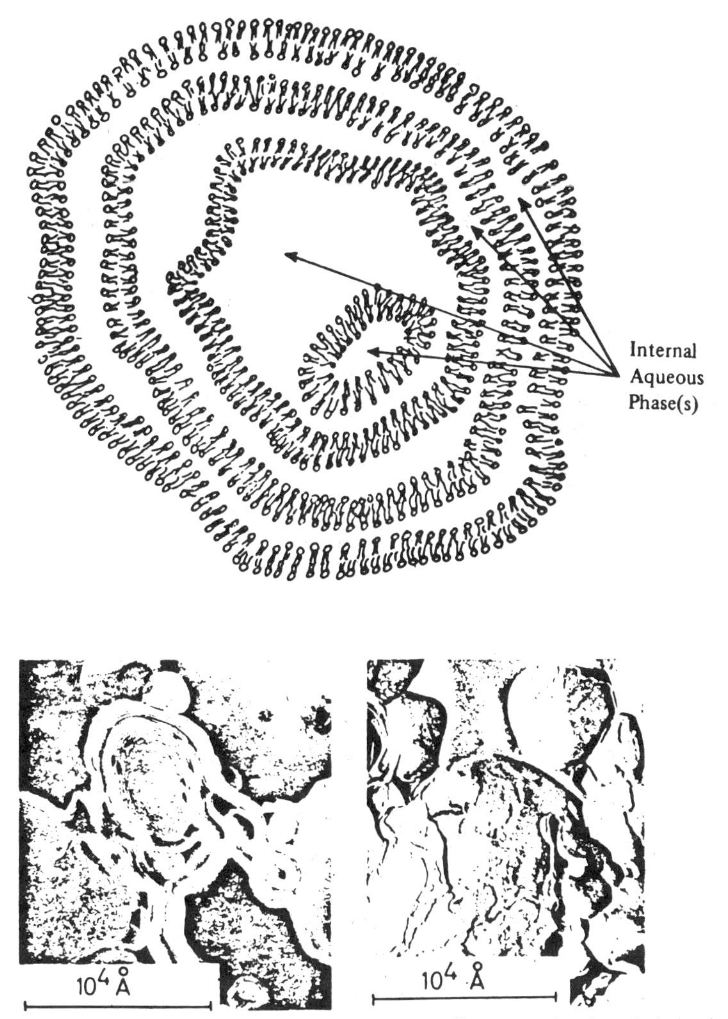

Figure 6.1. Schematic representation of multicompartment liposomes (top) and their electron microscopic appearances, using ammonium molybdate stains (bottom left) or freeze-fracture (bottom right).

assembly of bilayers. Significantly, liposomes produced by this reversed-phase evaporation have four times larger aqueous compartments than those formed by the simple swelling of phospholipids (Szoka and Papahadjopoulos, 1978).

Single-compartment liposomes of uniform size distributions are better suited for physical chemical investigations than their multicompartment analogs. The earliest and most widely used method for forming single-compartment small bilayer liposomes is sonication (Huang, 1969). Either a probe or a bath-type sonicator can be used. In the former a metal probe is directly inserted into the sample. In the latter the sample, sealed in a glass vial, is suspended in an

Table 6.1. Characteristics of Multicompartment Liposomes Sequentially Extruded Through Polycarbonate Membranes[a]

Pore Diameter of Polycarbonate Membranes Through Which the Liposomes Are Extruded, μm	Mean Liposome Diameters, μm[b]	75% Volume Limits, μm[c]	Capture Volume, 1 aqueous space $mole^{-1}$[d]
None	1.32	0.57–2.14	2.36
1.0	0.85	0.36–1.29	2.45
0.8	0.62	0.33–1.02	2.63
0.6	0.44	0.36–0.65	2.81
0.4	0.38	0.25–0.60	2.75
0.2	0.26	0.17–0.37	2.28

Source: Compiled from Olson et al., 1979.

[a] Multicompartment liposomes prepared in 300 mM glucose, $\frac{1}{10}$ phosphate buffered saline, pH = 7.4, at 37°C with 5 mM [^{14}C]glucose as the aqueous space marker and [^{3}H]phosphatidylcholine in the bilayer. The total lipid concentration extruded was 2 μmole/ml^{-1} (phosphatidylserine:phosphatidylcholine:cholesterol = 1:4:5 mole ratio).

[b] Determined by electron microscopy on negative stain preparations.

[c] Calculated diameter range, that contains 75% of the liposome volume.

[d] The glucose captured volume of total lipid.

ultrasonic cleaning bath. Inserting the probe directly into the lipid dispersion results in greater energy input and hence in sonication times shorter than those required when using the bath-type sonicator. Titanium fragments, released from the tip of the probe, need to be removed from the samples by either centrifugation or ultrafiltration. The temperature in both types of sonication is maintained slightly above the phase transition temperature of the lipid. Sonication below the phase transition temperature produces liposomes with structural defects (Lawaczek et al., 1976). Degradation of the lipid is prevented by working in a nitrogen or an argon atmosphere.

Increasing the sonication time for a given lipid dispersion results in an exponential decrease of turbidity and viscosity. At the plateau region of sonication liposomes are optically transparent and are relatively homodisperse. Liposomes of more uniform sizes can be obtained by gel filtration on Sepharose 4B (fraction II in Figure 6.2), and these have been established to be spherical single-compartment bilayers with diameters of 250 ± 10 Å (Huang, 1969).

Single-compartment liposomes have also been prepared without sonication. Several methods have been developed. In the injection method an alcoholic solution of the lipid is injected slowly by a small bore Hamilton syringe (10^{-4}–10^{-2} ml min^{-1} to 10 ml) into a well-stirred aqueous solution, thermostated above the phase transition temperature of the lipid (Batzri and Korn, 1973). Sizes of the liposomes formed are related to the concentration of the lipid in the alcohol injected (Kremer et al., 1977a).

Detergent dilutions have provided a means to form single-compartment

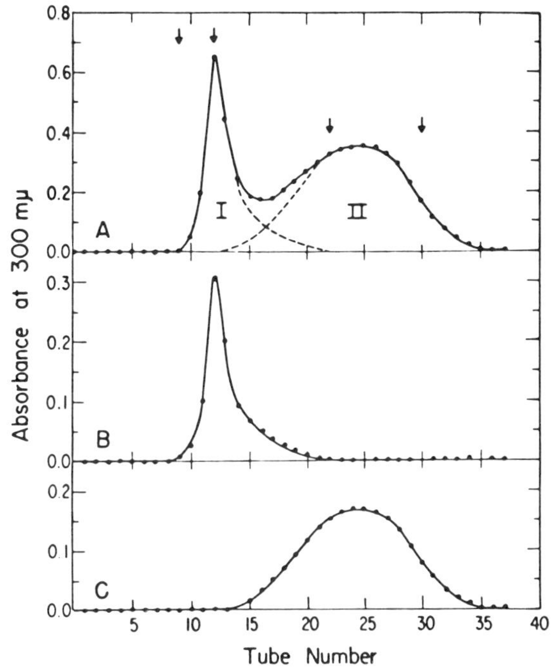

Figure 6.2. Elution patterns of phosphatidylcholine. (*A*) Dispersion obtained from ultrasonic irradiation. Phosphatidylcholine dispersion (8 ml of 3 %) was applied to a 2.5 × 50 cm lipid-treated Sepharose 4B column. (*B*) Vesicles collected and concentrated from fraction I of (*A*) as indicated by the first two arrows. (*C*) Vesicles collected and concentrated from fraction II of (*A*) as indicated by the last two arrows (taken from Huang, 1969).

liposomes. Phospholipids are cosolubilized by a detergent that is subsequently removed by gel filtration (Brunner et al., 1976) or by dialysis (Slack et al., 1973; Milsmann et al., 1978; Rhoden and Goldin, 1979). Sodium cholate is the favored detergent. Controlled dialysis of phosphatidylcholinesodium cholate comicelles typically yields liposomes with a radius of 266 ± 15 Å (Milsmann et al., 1978). Addition of differing amounts of cholesterol and a variation of the pH of the dialyzate provide means for preparing liposomes of controllable dimensions (Rhoden and Goldin, 1979). The commercially available Lipoprep^R Bilayer Liposome Preparation Device (Bachofer, Reutlinger, West Germany) is also based on detergent removal from detergent-lipid comicelles by fast and controlled flow dialysis.

Single-compartment bilayer liposomes have also been prepared by injecting lipid dispersions through the small orifice of a Power laboratory (French) press (Barenholz et al., 1979; Hamilton et al., 1980), as well as by combined reversed-phase evaporation and sequential extrusion through polycarbonate membranes (Szoka et al., 1980).

A convenient method has recently been proposed to determine the uniformity of liposomes (Barrow and Lentz, 1980). Plots of the turbidity of uniform

Table 6.2. Preparation of Liposomes

Method of Preparation	Liposomes Formed	Advantages of Method	Disadvantages of Method	Key Reference
Handshaking or vortexing thin lipid film with aqueous solution	Multicompartment heterodisperse, diameter = 1800–8000 Å	Ease of preparation, large entrapped volumes	Nonuniformity precludes basic studies	Bangham et al., 1965; Kinsky, 1974
Extrusion of liposomes through polycarbonate membranes	Multicompartment	Defined size and homogenity, large entrapped volumes	Need of special equipment	Olson et al., 1979; Szoka et al., 1980
Introduction of an aqueous buffer into an ether solution of the phospholipid	Multicompartment	Entrapped volumes are larger than vortex liposomes	Difficult to control injection rate	Szoka and Papahadjopoulos, 1978; Schieren et al., 1978
Ultrasonic dispersal of lipid	Mostly single-compartment, diameter = ca. 250 Å	Relative ease of preparation	Possible degradation, small entrapped volumes, need to remove titanium fragment, inhomogenity	Huang, 1969
Injection of an alcoholic phospholipid solution through a small bore syringe into thermostated water	Single compartment, diameters range between 150 and 500 Å, depending on lipid concentration in alcohol	No degradation, fair degree of homogenity	Difficult to control injection rate, ethanol needs to be removed	Batzri and Korn, 1973; Kremer et al., 1977a

Dilution of phospholipid-detergent micelles by gel filtration or dialysis	Single-compartment, diameter = 150–250 Å	No degradation, homogeneous liposomes, commercial equipment available	Detergents and rate of dialysis need to be controlled	Razin, 1972; Brunnert et al., 1976; Korenbrot, 1977; Milsmann et al., 1978; Rhoden and Goldin, 1979; Zumbuehl and Weder, 1981
Injection of lipid dispersion through the small orifice of a Power laboratory (French) press	Single compartment, diameter = 315–525 Å	No degradation, high degree of reproducibility, large entrapped volumes	Special equipment needed	Barenholz et al., 1977, 1979; Hamilton et al., 1980
Slow swelling of a lipid film	Large, predominantly single-compartment, diameter = 0.5–1.0 μm	Possible to entrap large molecules	Difficult to control hydration rate	Reeves and Dowben, 1969, 1970
Slow swelling of protein containing lipid film	Large, predominantly single-compartment, diameter = 0.5–1.0 μm	Suitable model for reconstituted membranes	Inhomogenity, difficult to control rate of hydration	Darszon et al., 1980
Controlled injection of an ether solution into a warm aqueous solution	Large, predominantly single-compartment, diameter = 0.1–1.0 μm	Possible to entrap large molecules	Difficult to control injection rate	Deamer and Sangham, 1976; Deamer, 1978; Schieren et al., 1978
Removal of organic phase under reduced pressure from water-in-oil (w/o) microemulsions of lipids and buffer (REV)	Large, predominantly single-compartment, diameter = 0.2–0.9 μm	Possible to entrap large macromolecules	Exposure of material to be encapsulated to organic solvents	Szoka and Papahadjopoulos, 1978
Addition of Ca^{2+}, EDTA to single-compartment liposomes	Large spiral cylinders, diameter = 2000–10,000 Å	Possible to entrap large molecules	Needs to be characterized	Papahadjopoulos et al., 1975

single-compartment liposomes between 300 and 650 nm against the reciprocal fourth power of the scattering wavelength have been shown to be linear with zero intercepts. Even minute amounts of multicompartment liposomes cause nonzero intercepts in the plots. Liposome size distributions have been determined by sedimentation field flow fractionation (Kirkland et al., 1982).

The need for entrapping macromolecules, particularly enzymes (Papahadjopoulos, 1978), prompted the development of large single-compartment bilayer liposomes. Slow swelling of a lipid film has been shown to lead to thin walled, osmotically active, 0.5–1.0 μ diameter liposomes (Reeves and Dowben, 1969, 1970). It is essential to allow the swelling to proceed completely undisturbed. Liposome formation can be monitored directly with a phase contrast microscope (Figure 6.3; Reeves and Dowben, 1969). The thin phospholipid film has the appearance of concentric domelike lamellae attached to the glass slide at their edges (Figure 6.3*A*). Blowing a gentle stream of water saturated nitrogen over the lipid results in gradual swelling and separation (Figure 6.3*B*). Addition of a drop of water causes further swelling, detachment of the lamellae from the surface (Figure 6.3*C*), and curling up into large predominantly single-walled liposomes (Figure 6.3*D*).

An important application of large single-compartment liposomes is to reconstitute membrane proteins in their functional state. Using a method similar to that developed by Reeves and Dowben (1969, 1970), bovine and squid

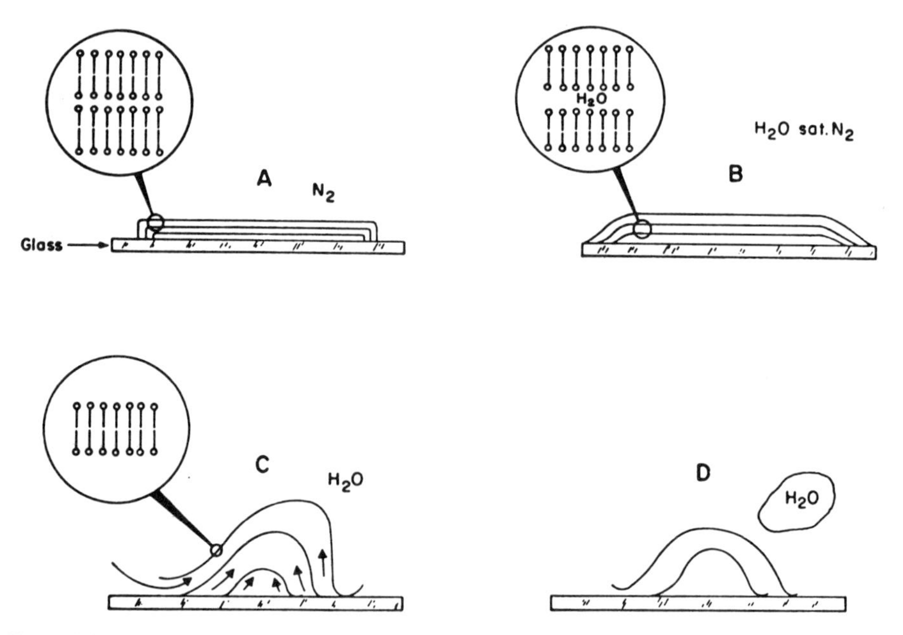

Figure 6.3. Schematic diagram illustrating the formation of vesicles. Lamellae of phospholipids (*A*) swell and become separated from one another in a moist atmosphere (*B*). When an aqueous solution is then added, it runs in between the lamellae (*C*), which finally become detached from the glass surface and round up into vesicles (*D*) (taken from Reeves and Dowben, 1969).

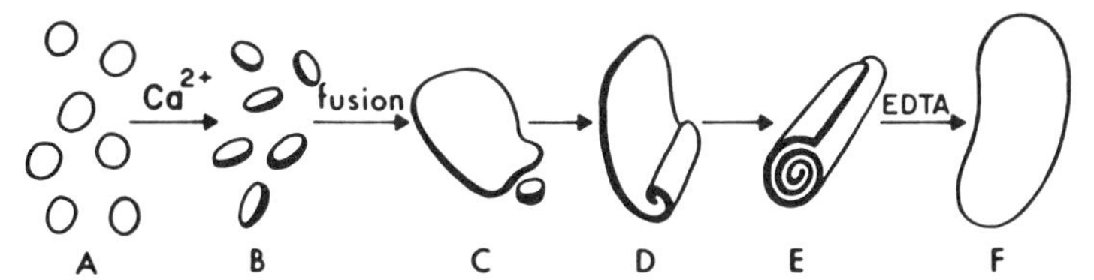

Figure 6.4. Schematic representation of the effect of Ca^{2+} on phosphatidylserine vesicles. (*A*) Small unilamellar vesicles (SUV) before the addition of Ca^{2+}. (*B*, *C*, and *D*) Hypothetical intermediates involving the aggregation and fusion of the SUV. (*E*) Cochleate cylinders as observed by freeze-fracture electron microscope. (*F*) Large unilamellar vesicles (LUV) observed after the addition of EDTA (taken from Papahadjopoulos et al., 1975).

rhodopsin, reaction centers from *Rhodopseudomonas sphaeroides*, beef heart cytochrome C oxidase, and acetylcholine receptors from *Torpedo californica* have been incorporated into large liposomes of several micrometer diameter (Darszon et al., 1980). These bilayer protein-lipid vesicles displayed biological activity.

Controlled injection of an ether solution of phospholipids into warm aqueous solutions yields directly large single-compartment liposomes (Deamer and Bangham, 1976; Deamer, 1978; Schieren et al., 1978). Uniformity is achieved by millipore and gel filtration (Deamer, 1978).

Small single-compartment phosphatidylserine liposomes fuse into multi-compartment structures in the presence of calcium ions (Papahadjopoulos et al., 1975). Addition of Ethylenediaminetetraacetic acid (EDTA) produces large cylindrical unilamellar structures, called cochleates (Figure 6.4; Papahadjopoulos et al., 1975).

The different methods of preparation and the characteristics of the liposomes formed are highlighted in Table 6.2.

2. MORPHOLOGIES OF LIPOSOMES AND THEIR CONSTITUENTS

The properties of liposomes are intimately related to the structures and conformations of their phospholipid constituents. On the supramolecular level phospholipids are amphiphatic. They contain polar headgroups and long hydrocarbon chains. As seen in Figure 6.5, phospholipid headgroups can be anionic or zwitterionic depending on the pH. The hydrocarbon chains can be of different lengths and can consist of different degrees of unsaturation. Alkyl chains of phospholipids of biological origin are quite long (16 carbon atoms at least) and contain a fair degree of unsaturation.

Liposome morphologies can be understood at the molecular level by considering the influence of alkyl chain conformations on packing (Seelig and Seelig, 1980). Much information has been obtained from X-ray investigations of bimolecular films formed between solid surfaces in layered crystals (Lagaly,

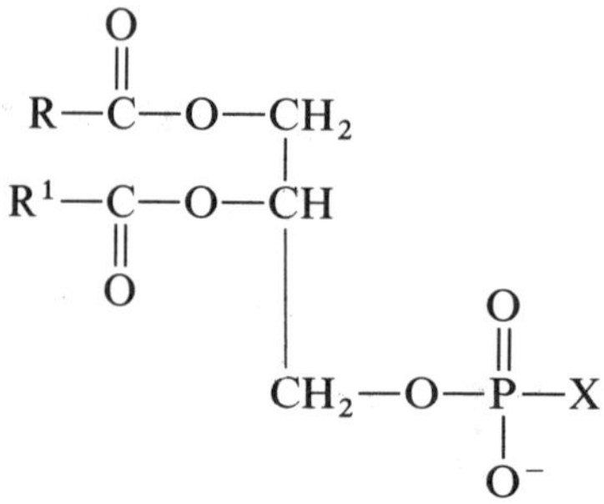

$$
\begin{array}{l}
\quad\ \ \mathrm{O} \\
\quad\ \ \| \\
\mathrm{R{-}C{-}O{-}CH_2} \\
\qquad\qquad\ | \\
\mathrm{R^1{-}C{-}O{-}CH} \\
\quad\ \ \| \qquad\quad | \\
\quad\ \ \mathrm{O} \qquad\ \ | \qquad\quad \mathrm{O} \\
\qquad\qquad\ \ | \qquad\quad \| \\
\qquad\qquad\ \mathrm{CH_2{-}O{-}P{-}X} \\
\qquad\qquad\qquad\qquad\ | \\
\qquad\qquad\qquad\qquad\ \mathrm{O^-}
\end{array}
$$

$X = -OH$ Phosphatidic acid

$X = -OCH_2CH_2\overset{+}{N}(CH_3)_3$ Phosphatidylcholine

$X = -OCH_2CH_2NH_2$ Phosphatidylethanolamine

$X = -OCH_2CH(COOH)NH_2$ Phosphatidylserine

$X = -OCH{-}CH(CH_3COOH){-}NH_2$ Phosphatidylthreonine

$X = -OCH_2CH(OH)CH_2OH$ Phosphatidylglycerol

Saturated Fatty Acids

$CH_3(CH_2)_nCOOH$

$n = 10$ lauric
$n = 12$ myristic
$n = 14$ palmitic
$n = 16$ stearic
$n = 18$ arachidic
$n = 20$ behemic

Unsaturated Fatty Acids

Oleic = 9-octadecenoic
Linoleic = 9,12-octadecenoic
Linolenic = 6,9,12-octadecenoic

Figure 6.5. Structures of typical phospholipids and fatty acids.

1976). A large number of isomers can arise from rotation about the carbon-carbon bonds in the acyl chain of phospholipids. Energetically gauche configurations are only 0.6 kcal mole^{-1} less stable than the optimal trans conformation (Figure 6.6). In an alkyl chain the all-trans conformation, in general, has the lowest energy but *not* the lowest free energy. For instance, a hexadecane has 30 conformations with one gauche bond, so that even though a trans → gauche transition has $\Delta H = +700$ cal mole^{-1}, there will be 15 times as many molecules with one gauche bond as there will be all-trans molecules at room temperature. Only those rotational isomers that essentially maintain the stretched nature of the hydrocarbon chain prevail. This is accomplished by sequential trans and gauche bonds, known as kinks. Figure 6.7 illustrates alkyl chains having all-trans, isolated gauche, and different types of kink conformations. The presence of one kink shortens the isolated chain by 1.27 Å and increase its volume by 25–50 Å^3. Two kinks shorten the alkyl chain by 2×1.27 Å and increase its volume by $2 \times (25\text{–}50\ \text{Å}^3)$, and so on. This is expressed by the *ngm* nomenclature (Pechhold and Blasenbrey, 1967, 1970) where *n* defines the number of gauche bonds and *m* refers to the number of times the alkyl chain is shortened by 1.27 Å (see Figure 6.7). Alkyl chains, aggregated into bilayers, assume trans or gauche

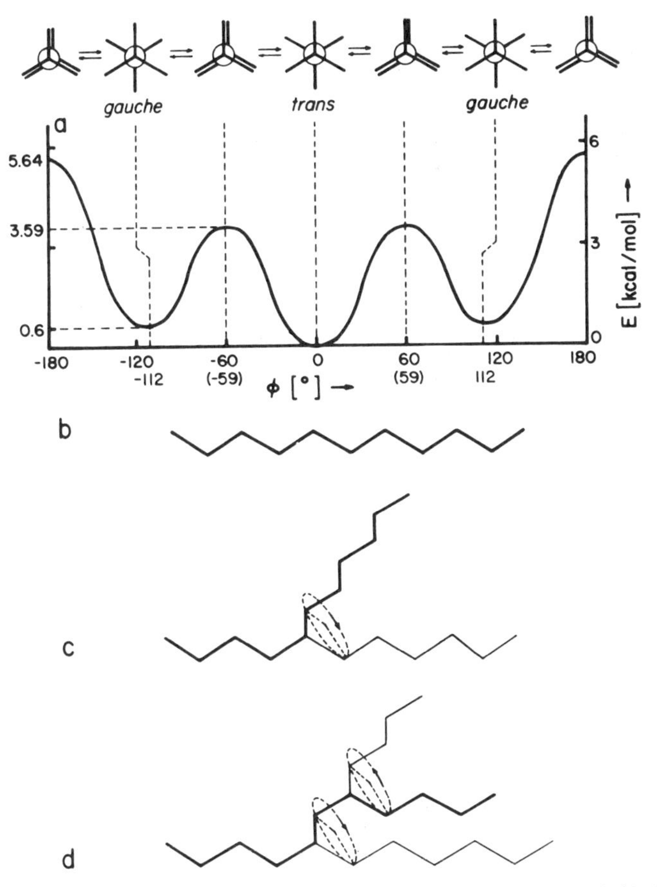

Figure 6.6. (*a*) Rotational potential (according to alkyl chains). (*b*) All-trans chain. (*c*) Chain with isolated gauche bond. (*d*) Chain with one kink (taken from Lagaly, 1976).

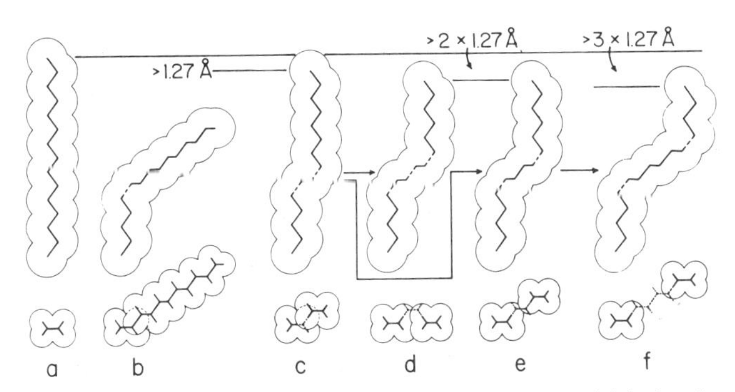

Figure 6.7. Various types of kinks (after Pechhold) and alkyl chains with isolated gauche bonds. (*a*) ttttttttttt (all-trans). (*b*) tttgttttttt (isolated gauche bond). (*c*) tttgt$\bar{g}$ttttt (2g1 kink). (*d*) tttgt$\bar{g}$tgtttt (3g2 kink). (3) tttgttt$\bar{g}$tttt (2g2 kink). (*f*) tttgtttt$\bar{g}$tt (2g3 kink) (taken from Lagaly, 1976).

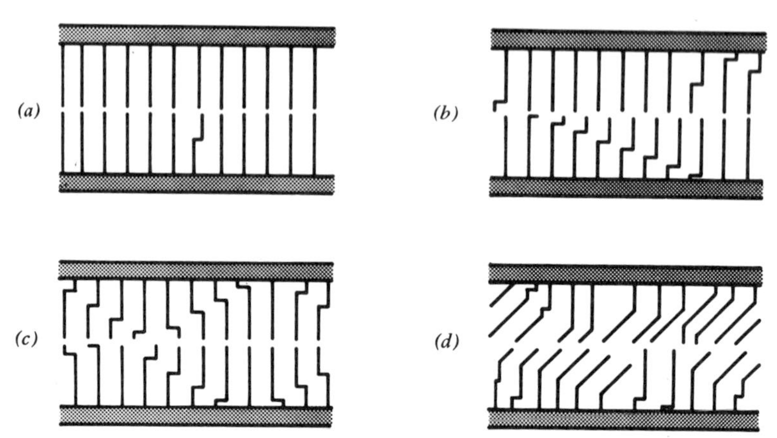

Figure 6.8. Conformation of alkyl chains in a bimolecular film. (*a*) All-trans block with a kink as an isolated defect. (*b*) Kink-block β_2. (*c*) Kink-block β_3. (*d*) gauche block (taken from Lagaly, 1976).

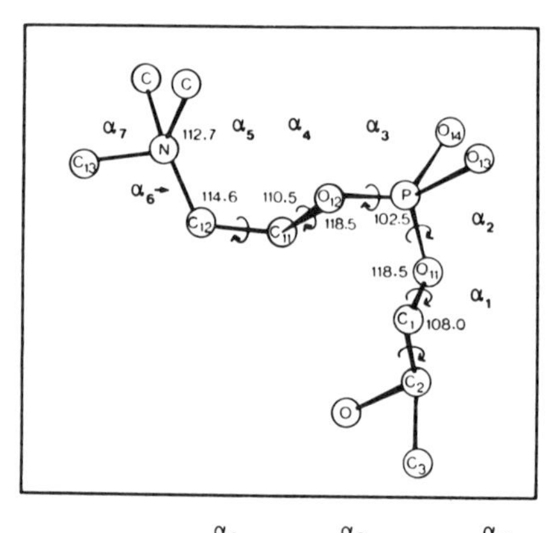

	α_1	α_2	α_3	α_4	α_5
Glycerylphosphorylcholine · $CdCl_2$	169°(t)	-69°(g^+)	-73°(g^+)	178°(t)	73°(g^-)
Glycerylphosphorylcholine - 1	165°(t)	-71°(g^+)	-59°(g^+)	-138°(t)	73°(g^-)
Glycerylphosphorylcholine - 2	-172°(t)	64°(g^-)	65°(g^-)	140°(t)	-75°(g^+)

Figure 6.9. Crystal structure of the glycerylphosphorylcholine $CdCl_2$ complex (taken from Sundaralingam, 1972).

block structures with various kinks as defects (Figure 6.8; Lagaly, 1976). Kink formation, like phase transition (see Section 6.3), is a cooperative phenomenon in the bilayer.

The nature of the headgroup also influences the conformation of phospholipids. Crystal structures, determined for the $CdCl_2 \cdot$ glycerylphosphorylcholine complex and for free glycerylphosphorylcholine, indicate the presence of two independent conformers, glycerylphosphorylcholine-1 and glycerylphosphorylcholine-2, in the unit cell (Sundaralingam, 1972). The structure of the glyceryl-

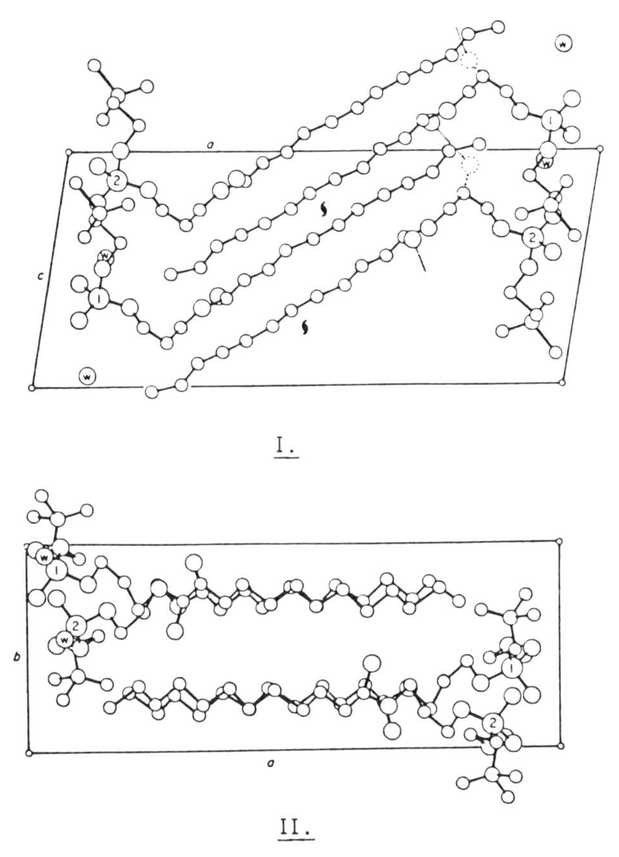

Figure 6.10. (I) Molecular arrangement of 1-laurylpropanediol-3-phosphorylcholine (LPPC) monohydrate projected onto the *a-c* plane. The four molecules constituting one unit cell are shown. Molecules 1 and 2 are mirror images. Corresponding molecules on opposite layer sides are related to each other by twofold screw axis (§). The broken circles and lines at the right side indicate the positions of the free hydroxyl groups with natural configuration and their hydrogen bonds with the carbonyl oxygens as postulated for natural lysophosphatidylcholine; w = water molecule. (II) Molecular arrangement of LPPC monohydrate projected onto the *a-b* plane (taken from Hauser et al., 1980b).

phosphorylcholine group and the corresponding torsional angles are given in Figure 6.9. All three structures are characterized by a gauche–gauche conformation of the phosphodiester linkage ($\alpha_1\alpha_2$) and by a gauche conformation of the O—C—C—N linkage (α_5). More recent X-ray data on 1-laurylpropanediol-3-phosphorylcholine are in accord with this interpretation (Hauser et al., 1980b). This compound lacks the free hydroxyl group in the 2 position of the glycerol moiety and crystallizes more readily than lysophosphatidylcholine. The unit cell contains four symmetry related molecules arranged in pairs of conformational enantiomers, packed head-to-tail, forming a common hydrocarbon chain matrix that is confined on both sides by a boundary layer of polar groups (Figure 6.10). The phosphorylcholine moieties are oriented parallel to the layer

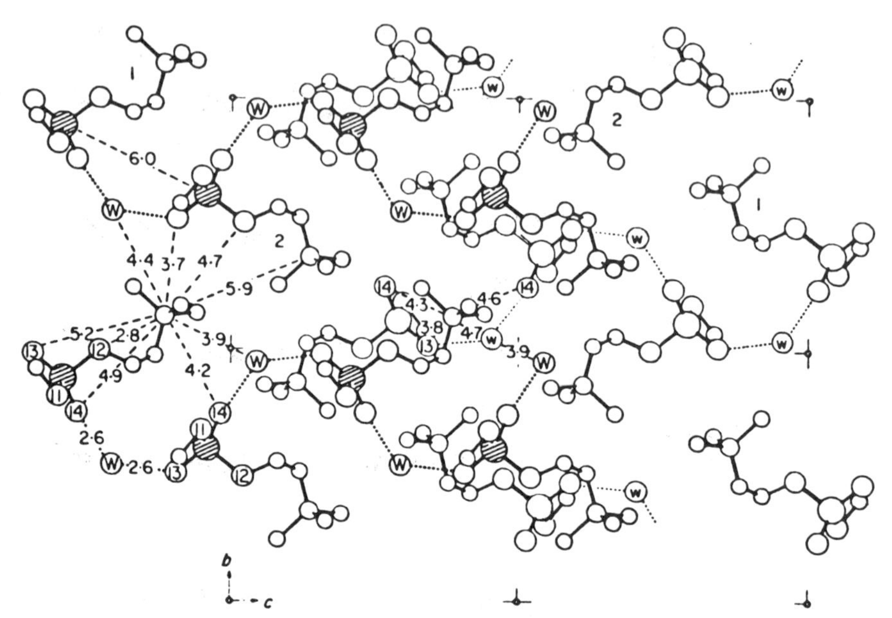

Figure 6.11. Arrangement of the polar groups of 1-laurylpropanediol-3-phosphorylcholine (LPPC) monohydrate at the contact interface (*b-c* plane) of two adjacent layers. The phosphorylcholine groups of eight molecules above (◍) and below (○) the layer interface are shown partially overlapping. The hydrogen bonds (dotted lines) and distances between phosphate oxygens and water molecules (w) are indicated. Distances between the choline nitrogen and surrounding oxygen atoms (broken lines) within one polar group layer (left) and between adjacent layers (center) are given (taken from Hauser et al., 1980b).

surface (*b–c* plane). These arrangements resemble those of phosphatidylethanolamine and glycerylphosphorylcholine in bilayers. Figure 6.11 shows the headgroup conformations perpendicular to the layer surface. Phosphate headgroups, linked by water molecules of hydration, form a zigzag pattern along the *c* axis. They are rather large and each polar group occupies an area of 52 Å^2 (Hauser et al., 1980b). Significantly, single-crystalline conformations of 1-laurylpropanediol-3-phosphorylcholine are consistant with experimental data obtained for lysophosphatidylcholine and phosphatidylcholine in their gel state. Thus X-ray crystallographic data can provide details on the organization of lipid molecules in bilayers. Figure 6.12 shows how the conformations of phosphatidylcholine molecules in the bilayer have been derived from X-ray data (Hauser et al., 1980b).

^{1}H, ^{2}H, ^{13}C, and ^{31}P nmr spectroscopic techniques have been used to obtain information on the conformation of phospholipid headgroups in liposomes (Roberts and Dennis, 1977; Seelig et al., 1977; Brown and Seelig, 1978; Smith et al., 1978; Griffin et al., 1978; Hauser et al., 1980a, 1980c). Interpretation of data in terms of static or dynamic molecular conformations is less than straightforward, however. Indeed, no evidence has been found to support unique

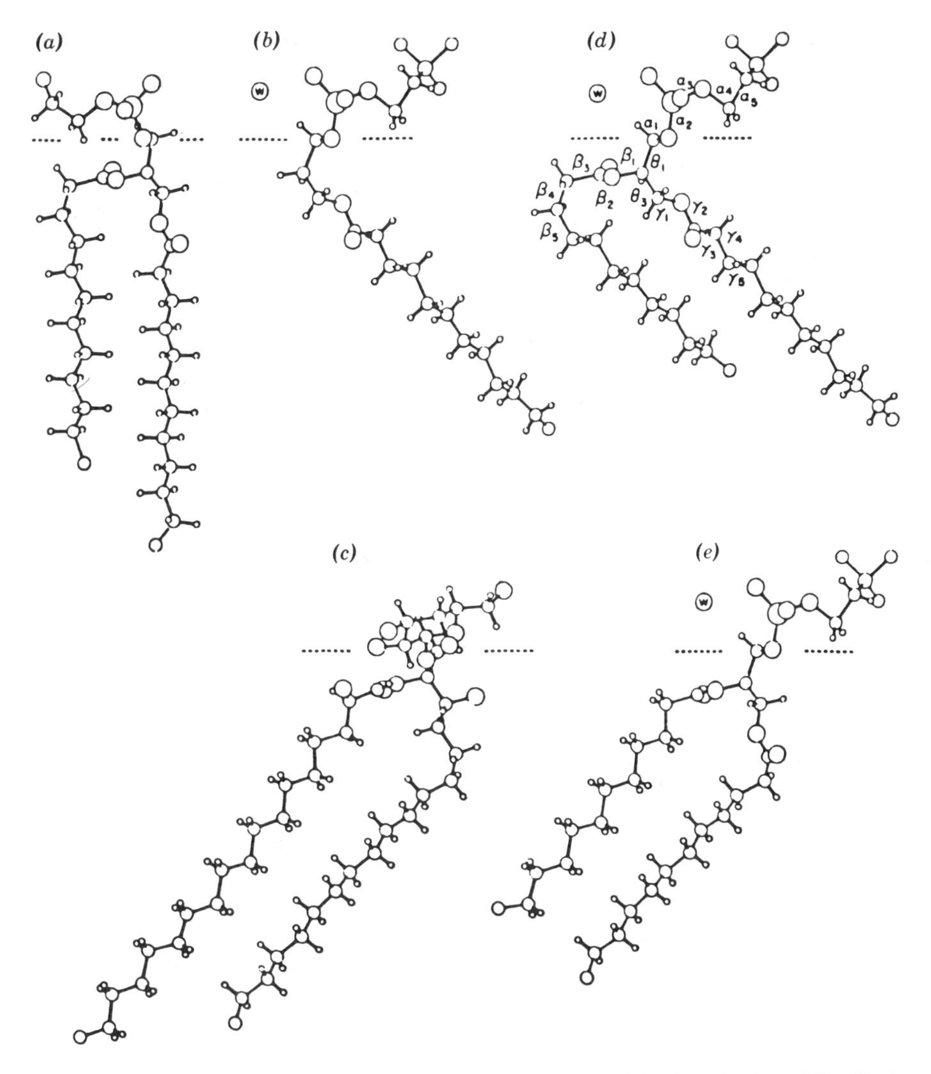

Figure 6.12. Conformation of membrane lipids. (*a*) Phosphatidylethanolamine (PE). (*b*) 1-Laurylpropanediol-3-phosphorylcholine (LPPC). (*c*) Phosphatidylcerebroside. (*a–c*) as found in crystals. (*d* and *e*) Proposed conformations of phosphatidylcholine (PC) in the gel state. Molecule (*d*) is generated from LPPC (*b*) by attachment of the β fatty acid. In molecule (*e*) the lipophilic part with a conformation similar to that of phosphatidylcerebroside (*c*) is combined with the polar group of LPPC (*b*). The carbon atoms of the glycerol part and corresponding atoms of sphingosine are arranged in a similar way. The molecules are aligned with respect to a horizontal layer plane (broken line). PE has an area per molecule of 38.6 Å^2 and untilted hydrocarbon chains. All the other lipids occupy an area per molecule of approximately 52 Å^2 and have a chain tilt of 41° with respect to the layer normal (taken from Hauser et al., 1980b).

conformational solutions to the nmr data obtained for liposome headgroups (Skarjune and Oldfield, 1979). Conversely, good agreement has been obtained between the preferred conformation of 1-laurylpropanediol-3-phosphorylcholine micelles, derived from combined ^{1}H, ^{13}C, and ^{31}P nmr data, and those obtained from X-ray analysis of the single crystal (Hauser et al., 1980c). Interestingly, the preferred conformation of the polar group of dihexanoylphosphatidyl choline was found to be essentially the same in the monomeric and in the micellar state (Hauser et al., 1980a). Headgroup conformations are likely to change as a consequence of the phase transition (see Section 6.3), added electrolytes, and substrates (see Section 6.5). Further, different headgroups have different conformations and mobilities. The phosphatidylcholine headgroup is more mobile, for example, than phosphatidylethanolamine (Seelig, 1977).

The number of water molecules that hydrate the phospholipid headgroups and the strengths of water-lipid interactions strongly influence the formation and properties of liposomes (Hauser, 1975a, 1975b). Typically, one phosphatidylcholine molecule is associated with 23 molecules of water. Of these, 11 molecules are entrapped in the interior of the lipid and 12 molecules hydrate the headgroups. Using deuteron magnetic resonance techniques, five different types of waterlipid interactions have been distinguished (Hauser, 1975a). Water molecules occupying the inner hydration shell are so tightly bound that they do not freeze even at subzero temperatures. Subsequent water molecules were assumed to occupy the main hydration shell, to be entrapped, weakly bound, or to be free (Hauser, 1975a). More recent measurements of the work of removing water from liposomes provided no evidence for discrete classes of water (Parsegian et al., 1979). The work of water removal was found to be a continuous function of water content and lattice repeat spacing. Evidence has been obtained for the presence of very large "hydration forces." Repulsive pressure between egg phosphatidylcholine liposomes was first detected at 27 Å separation. This pressure was found to grow exponentially, with a decay constant of approximately 2.6 Å, to a value of 1500 atm (1 atm $= 1.013 \times 10^{5}$ newton $m^{-2} = 1.013 \times 10^{6}$ dyn cm^{-2}) at 3 Å separation (Parsegian et al., 1979). Such a large repulsive force provides an ample kinetic barrier to prevent liposomes from approaching each other.

Hydrophobic interactions between the hydrocarbon chains and electrostatic repulsions between the headgroups—the opposing forces (Tanford, 1980)—are responsible, of course, for the self-association of lipids. Based on geometrical and theoretical arguments, vesicle formation was considered to require surfactants containing two alkyl chains; surfactants with one alkyl chain were considered to associate only to micelles (Israelachvili et al., 1976). The successful preparation of stable surfactant vesicles from single-chain amphiphiles (Hargreaves and Deamer, 1978a; Okahata and Kunitake, 1979) is not in accord with this postulate. Apparently, stable bilayer formation is a more complex and as yet incompletely understood process. The high degree of curvature in spherical vesicles requires the external surface area to be larger than the internal area. Assuming identical partial specific volumes for phosphatidylcholine molecules in both halves of the bilayers, the ratio, X, of the number of molecules in the

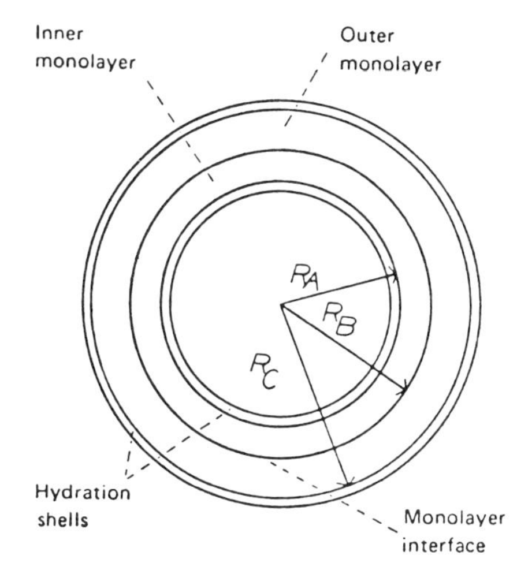

Figure 6.13. Vesicle bilayer cross section showing the definitions of the radial parameters R_A, R_B, and R_C discussed in the text (taken from Huang and Mason, 1978).

outer monolayer, n_o, to that in the inner monolayer, n_i, can be calculated by (Huang and Mason, 1978)

$$X = \frac{R_C^3 - R_B^3}{R_B^3 - R_A^3} \tag{6.1}$$

where R_A, R_B, and R_C are the radial parameters defined in Figure 6.13. Further, the anhydrous liposome weight, M, the volume available to each lipid molecule within the outer monolayer, Vo, and the inner monolayer, Vi, are given by

$$M = \frac{N[\frac{3}{4}\pi(R_C^3 - R_A^3)]}{\bar{v}} \tag{6.2}$$

$$Vo = \frac{\frac{4}{3}\pi(R_C^3 - R_B^3)}{n_o} \tag{6.3}$$

$$Vi = \frac{\frac{4}{3}\pi(R_B^3 - R_A^3)}{n_i} \tag{6.4}$$

where N is Avogadro's number and $\bar{v}$ is the partial specific volume of the liposome. Substituting values of R_A, R_B, and R_C, obtained from hydrodynamic and ^{31}P nmr measurements, leads to the calculated packing parameters for egg yolk phosphatidylcholine (EL) liposome given in Table 6.3 (Huang and Mason, 1978). There are several manifestations of asymmetry. In particular, the outer monolayer is thicker and contains more than twice as many lipid molecules as the inner monolayer. This causes a constriction for the inner bilayer, which manifests itself in a smaller headgroup area than those found for lipids in the outer monolayer (Table 6.3). In view of these packing constraints, the need to

Table 6.3. Geometric Packing Parameters for Egg Yolk Phosphatidylcholine Vesicles

Parameter	Values[a]
Radial parameters	$R_A = 62$ Å; $R_B = 78$ Å; $R_C = 99$ Å
Outer monolayer thickness	$R_C - R_B = 21$ Å
Inner monolayer thickness	$R_B - R_A = 16$ Å
Number of lipid molecules in each monolayer	$n_o = 1658$; $n_i = 790$
Volume per lipid molecule	$Vo = Vi = 1253$ Å^3
Surface area per lipid headgroup	Outer, 74 Å^2; inner, 61 Å^2
Acyl chain across section at R_B	Outer, 46 Å^2; inner, 97 Å^2
	Outer, 46 Å^2; inner, 97 Å^2
Anhydrous bilayer thickness	$R_C - R_A = 37$ Å

Source: Compiled from Huang and Mason, 1978.
[a] See Figure 6.3 for definitions.

provide thermal or ultrasonic energy for forming single-compartment liposomes is not unexpected. Indeed the relative ease of liposome formation under these conditions is quite remarkable (Tanford, 1980). A further problem is to reconcile the properties of single-compartment liposomes with Laplace's law:

$$Pi - Po = \frac{2\gamma}{Rs} \tag{6.5}$$

where $Pi - Po$ is the pressure difference across a curved surface, γ is the surface tension of each monolayer, and Rs is the radius of the curvature of the surface. Equation 6.5 requires a pressure difference at curved surfaces. However, for small water permeable vesicles in the absence of solutes the activity of water, and hence the pressure, must be the same in the inside and in the outside of the liposome. This conflict has been rationalized by assuming either that both of the monolayer surface tensions (γ) must be zero or that one must be negative and the other positive (Tanford, 1979). Alternatively, Laplace's law (equation 6.5) may not be applicable to microscopic liposomes (White, 1980).

Suitable structural models for liposomes need to account for phospholipid chain-chain, headgroup-headgroup, and chain-headgroup interactions in the bilayers. The Marčelja model takes into consideration all physical forces contributing to stability and expresses the energy of a hydrocarbon chain with configuration (i) in the bilayer $E^{(i)}$ by (Marčelja, 1974a, 1974b)

$$E^{(i)} = E^{(i)}_{\text{int}} + E^{(i)}_{\text{disp}} + \gamma A^{(i)} \tag{6.6}$$

where $E^{(i)}_{\text{int}}$ and $E^{(i)}_{\text{disp}}$ are the intramolecular energy of the chain configuration and the intermolecular van der Waal's attractions between neighboring hydrocarbon chains, respectively. The term $\gamma A^{(i)}$ takes into consideration electrostatic, hydrophobic, and steric interactions in the bilayer. $A^{(i)}$, the effective cross-sectional area of a given chain configuration (i), is approximated by

$$A^{(i)} = \frac{Aolo}{l^{(i)}} \tag{6.7}$$

where lo is the length of the hydrocarbon chain in extended all-trans configuration with a cross-sectional area of Ao (20.4 Å^2 for dipalmitoyl-3-*sn*-phosphatidylcholine) and $l^{(i)}$ is the effective length of the configuration (i) (i.e., the projected length on the bilayer normal). Values between 18 and 24 dyn cm^{-1} are taken for γ (Parsegian, 1966, 1968). Equation 6.6 allows the calculation of the order parameters, S_{mol} values, for each carbon segment of the alkyl chain constituting the liposome (Schindler and Seelig, 1975). Marčelja's model has been improved upon by many workers (Merajver et al., 1981). The order parameter, S_{mol}, relates the average orientation of an alkyl chain segment to θ, the momentary angle between the bilayer normal and direction of the chain segment, by

$$S_{mol} = \tfrac{1}{2}(3\langle \cos^2 \theta \rangle - 1) \tag{6.8}$$

where the angular brackets indicate the time average. The calculated values for S_{mol} for dipalmitoyl-3-*sn*-phosphatidylcholine (Schindler and Seelig, 1975) agreed well with those determined by deuterium magnetic resonance spectroscopy on selectively deuterated lipids (Table 6.4; Seelig and Seelig, 1974). Deuterium magnetic resonance spectroscopy has provided insight into liposome structures. In particular, the two alkyl chains were found not to be completely

Table 6.4. Order Parameters, S_{mol}, for Dipalmitoyl-3-*sn*-phosphatidylcholine Liposomes at 41°C

	S_{mol}	
Carbon Atom	Calculated[a]	Determined[b]
^{3}C	0.42	0.43
^{4}C	0.43	0.47
^{5}C	0.44	0.46
^{9}C	0.45	0.45
^{10}C	0.44	0.43
^{12}C	0.41	0.35
^{15}C	0.21	0.21

[a] From Schindler and Seelig, 1975.
[b] From Seelig and Seelig, 1974.

equivalent physically, and the order parameters appeared to be constant up to the first nine carbon segments, after which they decreased. Quantitative evaluation of deuterium magnetic resonance parameters leads to a bilayer thickness of 34–35 Å (Seelig and Seelig, 1974) in good agreement with values calculated (Table 6.4) and determined by X-ray diffraction (Wilkins et al., 1971). Topology and dynamics of lipids in membranes have been reviewed (Israelachvili et al., 1980; van Deenen, 1981; Büldt and Wohlgemuth, 1981).

3. PHASE TRANSITIONS OF LIPOSOMES

Concentration Induced Phase Transitions

Based on their interaction with water, lipids have been classified as insoluble nonswelling (Class I), insoluble swelling (Class II), and soluble (Class III) by Small (1968). Liposome forming phospholipids, belonging to Class II, undergo a multitude of structural organizations in the presence of different amounts of water. Figure 6.14 is an oversimplified presentation of a phospholipid-water phase diagram. At high lipid concentrations crystals and thermotropic liquid crystals (Brown and Wolken, 1979) prevail in a fair degree of homogeneity. Melting points of phospholipids increase with increasing length of the alkyl chains and with their degrees of unsaturation. Investigations of these thermotropic transitions of smectic mesophases of lamellar phospholipids have provided

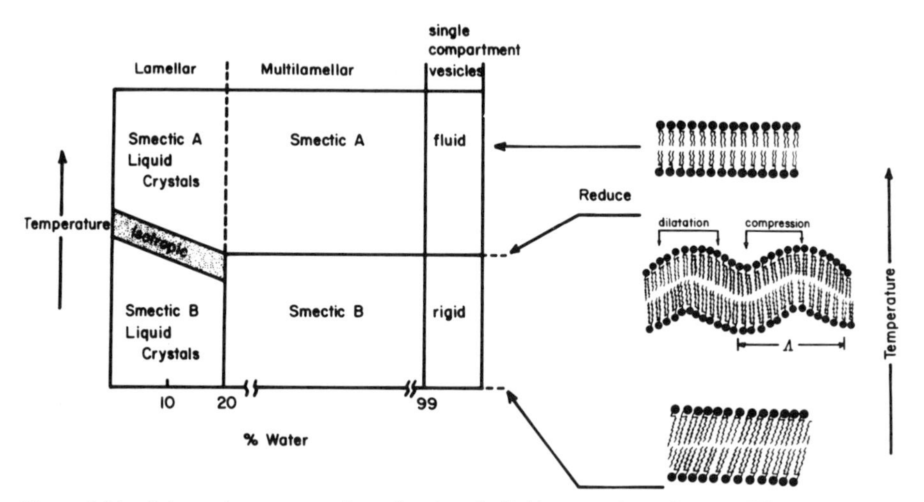

Figure 6.14. Schematic representation of a phospholipid-water phase diagram. The temperature scale is arbitrary and varies from lipid to lipid. For the sake of clarity phase separations and other complexities in the 20–99% water region are not indicated. Structures proposed for the phospholipid bilayers at different temperatures are shown on the right-hand side. At low temperature the lipids are arranged in tilted one-dimensional lattices. At the pretransition temperature two-dimensional arrangements are formed with periodic undulations. Above the main phase transitions lipids revert to one-dimensional lattice arrangements, separated somewhat from each other, and assume mobile liquidlike conformations.

much insight into liposome morphologies (Sturtevant, 1974; Lee, 1977a, 1977b). In smectic structures molecules are arranged in layers with their long axes parallel to each other. In smectic A structures liquid crystals are packed in monomolecular layers with the long axis of the molecules perpendicular to the plane. Smectic B structures have two different symmetries, $D_{\infty h}$ and C_{2v}. The former are optically uniaxial and involve hexagonal packing with the molecular axis perpendicular to the layers. In the latter lipid molecules are tilted and optically biaxial. Increasing the temperature results in phase transition from the more rigid smectic B phase to the more fluid smectic A phase (Figure 6.14). There is a narrow range of temperature (the shaded area in Figure 6.14) where the two smectic phases coexist. Increasing the amount of water incorporated between the bilayers increases the repeat distances and decreases the bilayer thickness. In the intermediate water concentration range heterodisperse multicompartment vesicles and myelin structures are formed. Single and monodisperse multicompartment liposomes (see Section 6.1 for their formation) predominate at lipid concentrations of $<1\%$. In the ensuing treatments of temperature and substrate induced phase transitions, discussion is limited to single- and multicompartment liposomes.

Temperature Induced Phase Transitions

Both multi- and single-compartment liposomes undergo distinct structural changes at a certain temperature, the phase transition temperature, when heated or cooled. These changes do not affect the gross structural features of liposomes (i.e., they remain roughly spherical closed smectic mesophases of phospholipid bilayers). Below the phase transition temperature lipids in the bilayers are in highly ordered "gel" states, with their alkyl chains in all-trans conformations. Above the phase transition temperature lipids become "gel" as the consequence of gauche rotations and kink formations. The "solid" and "fluid" states are sometimes referred to as the gel and the liquid crystalline phases. Gel-to-fluid liposome phase transitions have been observed by a large variety of intrinsic and extrinsic methods. Nmr, epr, and Raman spectroscopy, X-ray diffraction, differential calorimetry, viscosity, light scattering, and refractivity measurements have been carried out in the absence and in the presence of probes (Lee, 1977a, 1977b).

Thermodynamic parameters for liposome phase transitions are best obtained by differential scanning calorimetry (Mabrey and Sturtevant, 1978). Figure 6.15 illustrates the excess specific heat, C_{ex} (line A), and the excess enthalpy, ΔH_{cal} (curve B), determined by differential scanning calorimetry for dipalmitoylphosphatidylcholine multicompartment liposomes (Hinz and Sturtevant, 1972b). The excess enthalpy defines the ratio of molecules in the liquid crystalline phase to those in the gel phase and is related to ΔH_{cal} by

$$\Delta H_{cal} = M \int_{T_1}^{T_2} C_{ex}\, dT \tag{6.9}$$

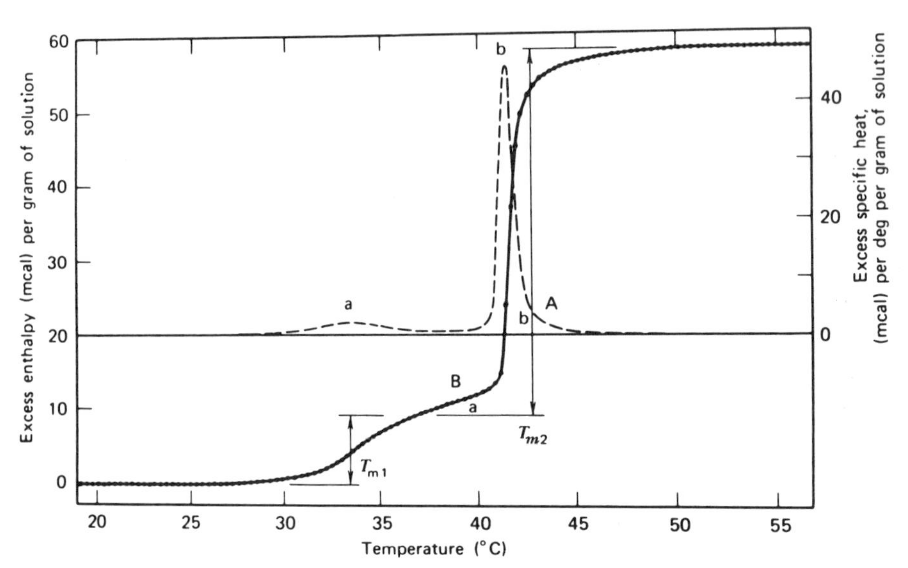

Figure 6.15. The variation with temperature of the excess specific heat (Curve *A*, right hand ordinates) and the excess enthalpy (Curve *B*, left hand ordinates) during the gel-to-liquid crystal transition of dipalmitoyl L-α-lecithin in aqueous suspension at a lipid concentration of 3.88 mg/ml^{-1}. The points on Curve B are the experimental data recorded digitally at 1 min intervals (taken from Hinz and Sturtevant, 1972b).

where M is the molecular weight of the lipids and T_1 and T_2 are temperatures prior and subsequent to the phase transition. Assuming the enthalpy change within the temperature range of the phase transition to be proportional to the transition at that temperature, and unit activity coefficients, differential calorimetric data can be used to obtain the van't Hoff enthalpy for the transition. If Θ is the fraction of the lipids in the liquid crystalline state, then assuming an equilibrium constant $\kappa = \Theta/(1 - \Theta)$, one obtains from $\partial \ln K/\partial T = \Delta H_{vH}/RT^2$ for the midpoint of the transition, $\Theta = \frac{1}{2}$, at temperature T_m,

$$\left(\frac{\partial \Theta}{\partial T}\right)_{P, T_m} = \frac{\Delta H_{vH}}{4RT_m^2} \tag{6.10}$$

where $\partial\Theta/\partial T$ is the midpoint of transition (on curve B in Figure 6.15) and T_m is the midtransition temperature. Comparison of ΔH_{vH} with ΔH_{cal} provides important information on the cooperativity of the phase transition. For non-cooperative processes $\Delta H_{vH}/\Delta H_{cal} = 1$; for cooperative processes $\Delta H_{vH}/\Delta H_{cal} \gg 1$. The number of molecules in the cooperative unit is operationally defined as the ratio of $\Delta H_{vH}/\Delta H_{cal}$. The cooperativity parameter, σ, is best expressed for two-dimensional hexagonal lattices by (Gruenewald et al., 1979)

$$\frac{\Delta H_{vH}}{\Delta H_{cal}} = \frac{1}{3\sqrt{\sigma} - 2} \tag{6.11}$$

Equation 6.11 implies strong variation of $\Delta H_{vH}/\Delta H_{cal}$ as $\sqrt{\sigma}$ approaches $\frac{2}{3}$, where $\Delta H/\Delta H_{cal}$ becomes infinity. Increasing the size (and hence decreasing the curvature) of dimyristoylphosphatidylcholine liposomes has resulted in an almost parallel increase in ΔH_{vH} and ΔH_{cal} values (Gruenewald et al., 1979). Thus bilayer curvature affects the enthalpy of transition but not the cooperativity. The calculated cooperative interaction energy for dimyristoylphosphatidylcholine was found to correspond to an infinitely steep first order transition. Thus liposomes may be considered to be ideal cooperative systems (Mitaku et al., 1978; Gruenewald et al., 1979).

Liposomes often show two phase transitions. It has been found useful to graphically resolve the excess specific heat, determined by differential scanning calorimetry, into two transitions, as illustrated in Figure 6.15 (Hinz and Sturtevant, 1972b). Using zero specific heat as the base line, an enthalpy is assigned to the lower transition, ΔH_1. The remaining enthalpy is assigned to the second transition ΔH_2. Transition temperatures corresponding to ΔH_1 and ΔH_2 are indicated by T_{m1} and T_{m2} in Figure 6.15. T_{m1}, due to the pretransition, is associated with structural transformations from a tilted one- to a tilted two-dimensional lattice of lipid bilayers, distorted by periodic undulations (Janiak et al., 1976). T_{m2} corresponds to the main or chain melting transition, in which the hydrocarbons assume liquidlike conformations and their lattices revert to one-dimensional structures. The proposed structural models associated with the pretransition and the main transition are drawn schematically on the right-hand side of Figure 6.14. A third phase transition, the "subtransition," has recently been observed for phosphatidylcholine bilayers by differential calorimetry (Chen et al., 1980).

Calorimetric measurements (e.g., Figure 6.15) indicate that liposome gel-to-fluid phase transitions are endothermic. Phase transitions occur over a narrow range of temperature and approximate well isothermal first order processes for liposomes prepared from pure phospholipid (Albon and Sturtevant, 1978). Table 6.5 collects typical thermodynamic parameters for multi- and single-compartment liposome phase transitions. Several pertinent points should be noted from consideration of liposome phase transitions. Thus enthalpy changes for single-compartment liposomes are much smaller than those for their multicompartment analogs; pretransitions are often undetectable for single-compartment liposomes. For multicompartment liposomes the pretransition is much less endothermic and occurs over a broader temperature range than the main transition. Phase transition temperatures for the main transition of multicompartment liposomes, T_{m2} values, increase linearly with increasing hydrocarbon chain length (n) of the lipid (Nagle and Scott, 1978). In the limit, $n \rightarrow \infty$, the transition temperature extrapolates to that of polyethylene. Liposome main transitions correspond, therefore, to transitions from premelted to melted states of hydrocarbons and are the consequence of hydrocarbon chain disordering (Nagle and Scott, 1978).

Fluorescence techniques have been extensively used for determining vesicle phase transitions (Georgescauld et al., 1980; Zachariasse et al., 1980).

Table 6.5. Phase Transitions of Liposomes

Liposome	Lower Transition				Upper Transition			
	T_{m1}, °C	Temperature Range of Transition, °C	ΔH_1, kcal mole^{-1}	Cooperativity Unit	T_{m1}, °C	Temperature Range of Transition, °C	ΔH_1, kcal mole^{-1}	Cooperativity Unit
Multicompartment dipalmitoylphosphatidylcholine[a]	35.0	33.3–36.5	1.83	290	41.8	40.3–44.0	8.74	260
Multicompartment dimyristoylphosphatidylcholine[a]	14.2		1.00	280	23.9		5.44	330
Multicompartment distearoylphosphatidylcholine[a]	49.1	47.5–51.2	1.85	160	54.9	52.8–56.0	10.62	130
Single-compartment dipalmitoylphosphatidylcholine[b]					36.4	29.7–40.6		
Single-compartment dimyristoylphosphatidylcholine[b]					20.9	14.3–27.4		
Single-compartment distearoylphosphatidylcholine[b]					51.3	46.5–53.8		

[a] Taken from Hinz and Sturtevant, 1972a, 1972b; Marbey and Sturtevant, 1978.
[b] Taken from Lentz et al., 1976.

Table 6.6. Morphologies of Dimyristoylphosphatidylcholine Single-Compartment Liposomes at Different Temperatures

Feature	Temperature, °C				
	10	15	20	25	30
Stokes radius, r_s, Å[a]	98.8 ± 3.9	96.2 ± 3.5	95.8 ± 2.9	113.9 ± 6.8	128.7 ± 3.1
Hydrated radius, r_h, Å[a]	94.2 ± 3.8	97.9 ± 5.0	94.5 ± 3.7	113.4 ± 6.0	124.3 ± 6.4
Internal radius, R_A, Å[b,c]	43.4 ± 4.9	44.5 ± 5.1	54.5 ± 5.1	77.2 ± 4.5	82.4 ± 4.1
External radius, R_C, Å[b,d]	92.6 ± 4.3	95.2 ± 4.6	93.2 ± 4.3	108.7 ± 5.6	112.2 ± 5.4
Thickness of external water layer, Å[e]	3.55 ± 0.72	1.14 ± 0.47	2.29 ± 0.80	4.8 ± 1.2	13.3 ± 1.0
Anhydrous bilayer thickness, Å[f]	49.22 ± 0.59	50.68 ± 0.50	38.64 ± 0.78	31.5 ± 1.0	29.8 ± 1.3
Area per lipid in bilayer, $Å^2$[g]	46.2 ± 3.5	45.3 ± 3.6	58.2 ± 4.7	70.7 ± 4.3	74.7 ± 3.7

[a] *Source:* Compiled from Watts et al., 1978a.
[b] Obtained directly from the measured diffusion coefficient and viscosity.
[b] See Figure 6.13 for definitions of R_A and R_C.
[c] Determined directly from trapped volume measurements.
[d] Calculated from anhydrous molecular weights of liposomes.
[e] Weighted mean values calculated from r_s and r_h.
[f] Anhydrous bilayer thickness = $R_C - R_A$.
[g] Area per lipid in bilayer = $4\pi(R_C^2 - R_A^2)\overline{M}w_{\text{lipid}}/\overline{M}w_{\text{liposome}}$ ($\overline{M}w$s are molecular weights).

Detailed understanding of phase transitions have been obtained from the intensities of Raman lines in multicompartment dipalmitoylphosphatidylcholine liposomes as a function of temperature (Pink et al., 1980). The observed Raman intensities fitted well with those calculated by taking into consideration the dynamics of hydrocarbon conformations in the lipid bilayers. Internal energies, chain lengths, headgroup areas, and degeneracies have been calculated for all-trans structures, for several intermediate structures involving a number of kinks (see Figures 6.6 and 6.7), and for completely melted conformations (Pink et al., 1980).

While thermotropic phase transitions do not involve the destruction of gross liposome features, a number of pronounced conformational changes occur. These temperature dependent changes have been determined for single-compartment dimyristoylphosphatidylcholine liposomes by analyzing data obtained by electron microscopy, ultracentrifugation, column chromatography, capillary-flow viscometry, and spin-label trapped volume determinations (Watts et al., 1978a). Liposomes were found to remain spherical and constant in lipid content (2850 $\pm$ 140 molecules of dimyristoylphosphatidylcholine per liposome) as a function of temperature changes, but their outer radius, internal volumes, and bound water content showed pronounced increases (Table 6.6). Indeed packing restrictions were considered to preclude fluid-to-fluid crystalline phase transitions for single-compartment liposomes (Cornell et al., 1980). Packing for multicompartment liposomes, however, allows for this transition. The use of several different techniques in the morphological characterizations of liposomes is obligatory. Quasi-elastic light scattering methods alone are not sensitive enough to detect temperature dependent changes for dimyristoylphosphatidylcholine liposomes (Aune et al., 1977).

The kinetics of liposome phase transitions have been examined. Pressure-jump experiments indicated time constants for the phase transitions of single- and multicompartment liposomes to be in the ranges of 0.1–1.0 and 1.0–10.0 msec, respectively (Gruenewald et al., 1980). Temperature-jump studies on multicompartment dimyristoylphosphatidylcholine liposomes showed the presence of two relaxations (Tsong, 1974). The slower one, in the 100 msec range, was attributed to a nucleation in which a limited number of lipid molecules initiate the phase transition. The fast relaxation, in the millisecond range, was ascribed to the propagation of the transition.

Electrolyte and Substrate Induced Phase Transitions

Liposome phase transitions are sensitive to the presence of additives. The sharp transitions observed for liposomes prepared from single pure lipids become broad, sometimes even vanish in the presence of additives or adventitious impurities. Phase transition temperatures of lipid mixtures are broad and deviate from those expected from ideal mixing (Mabrey and Sturtevant, 1976).

The mechanism for phase transitions of ionic liposomes are necessarily more complex than those for neutral ones (Lee, 1977a). Phase transitions for ionic liposomes can be triggered by external charges (Träuble and Eibl, 1972; Verkleij et al., 1974; Van Dijck et al., 1975; Watts et al., 1978b). The effect can be illustrated by the observed alteration of phase transition temperatures for multicompartment 1,2 - dihexadecyl - *sn* - glycerol - 3 - phosphoric acid, 1,3-dimyristoylglycerol - 2 - phosphoric acid, 1,2 - dimyristoyl - *sn* - glycerol - 3 - phosphoric acid, and 1,2-dipalmitoyl-*sn*-glycerol-3-phosphoric acid as functions of external pH (Figure 6.16; Eibl and Blume, 1979). At low pH phosphoric

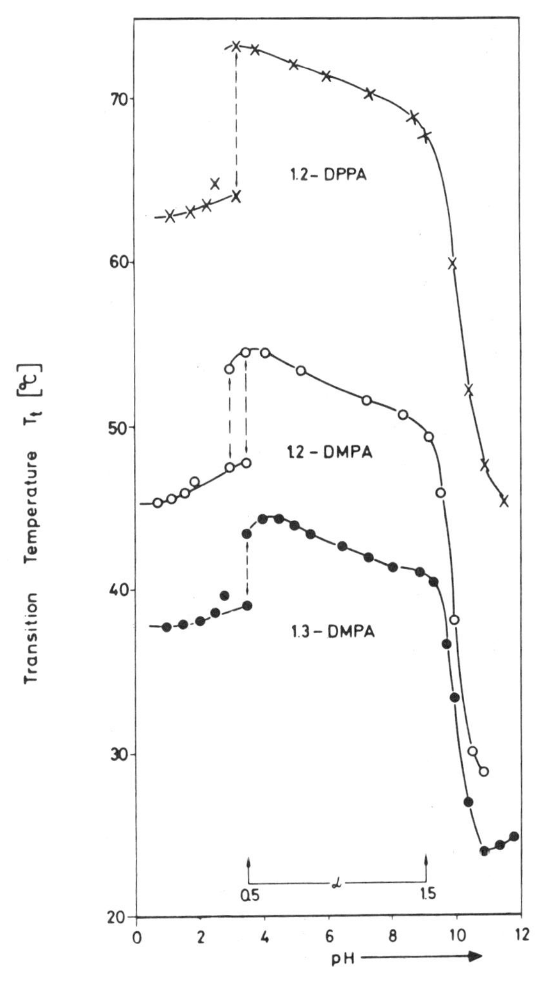

Figure 6.16. The effect of increasing ionization on the phase transition temperature of 1,2-dipalmitoyl-*sn*-glycerol-3-phosphoric acid (1,2-DPPA) (X), 1,2-dimyristoyl-*sn*-glycerol-3-phosphoric acid (1,2-DMPA) (○), and 1,3-dimyristoyl-*sn*-glycerol-3-phosphoric acid (1,3-DMPA) (●). Two distinct points of the degree of association α are shown: 0.5 and 1.5 (pK_1 and pK_2) (taken from Eibl and Blume, 1979).

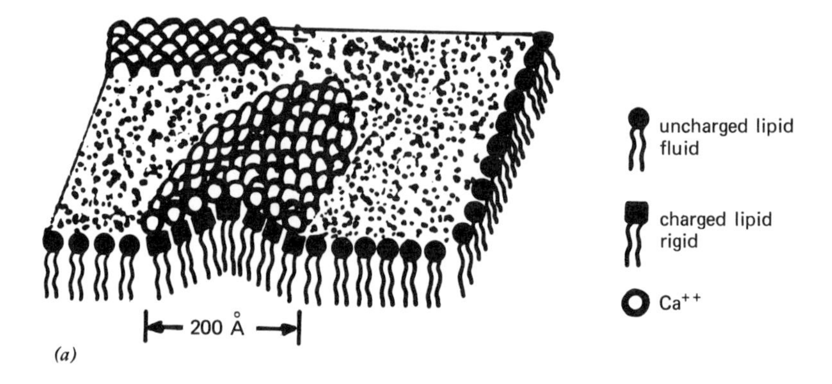

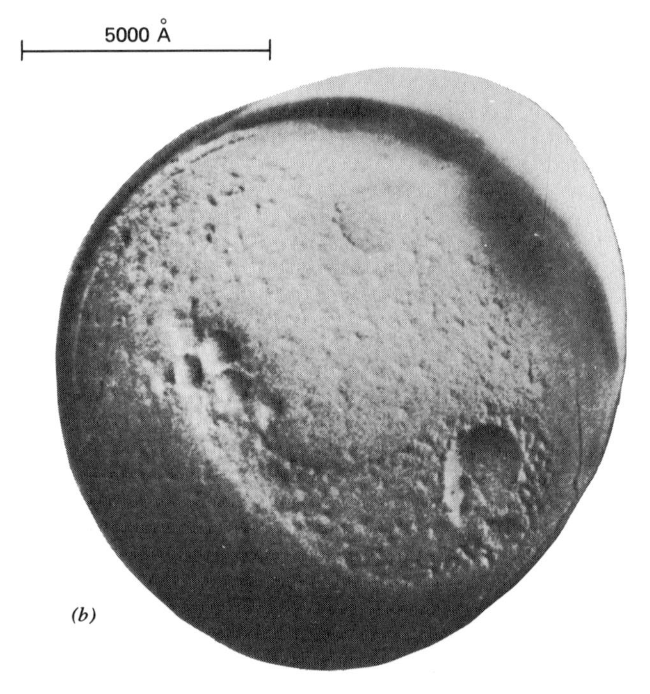

Figure 6.17. (*a*) Demonstration of charge (Ca^{2+}) induced domain formation in mixed membrane of dipalmitoylphosphatidylcholine (DPPC), and twofold charged phosphatidic acid (e.g., dipalmitoyl phosphatidic acid (DPPA)) by spin label technique. (*b*) Electron micrograph of large vesicle of 1 : 1 mixture of dioleyl phosphatidic acid and dioleyl phosphatidyl choline after addition of Ca^{2+}. Elongated protrusions are caused by domains of Ca^{2+}-bound phosphatidic acid (taken from Sackman, 1978).

acid moieties are unionized and the phase transition temperatures correspond to the melting points of the fully protonated crystals. Monoionization of phosphoric acids leads to an increase in the phase transition temperature with maxima at pH = 3.5 (pK of phosphoric acid $\simeq$ 3.5). Phase transition temperatures between pH 4 and 9, where the phosphoric acids are partially mono- and partially dianionic, are only slightly altered by pH changes (see the plateau in Figure 6.16). Negatively charged liposomes are effectively stabilized by protons on their surfaces. At higher pH values, where phosphoric acids are dianionic, liposome phase transition temperatures decrease below values of the melting point of their parent phospholipids. The decrease in transition enthalpy at high pH values is due to a change in the hydrocarbon chain interactions induced by doubly charged headgroups (Blume and Eibl, 1979). pH induced phase changes exhibit, in the case of mixed lipids, hysteresis. Hysteresis associated with phase transitions may have a bearing on information storage in biological membranes (Sackman, 1978). Addition of calcium ions to negatively charged liposomes also alters the phase transition temperatures as the result of complex formation. Calcium ions also facilitate fusion (see Sections 6.1 and 6.5).

Charge induced phase separations may occur concomittantly with lateral lipid separations on the liposome surface. These, in turn, lead to the establishment of long range order and cooperativity within such segregated domains. Figure 6.17 illustrates calcium ion induced domain formation (Sackmann, 1978). Domain formation, often triggered by charged proteins, is highly relevant to membrane functions (Gebhardt et al., 1977; Sackmann, 1978).

The presence of cholesterol in natural membranes has prompted the extensive investigations of cholesterol containing liposomes (Chapman et al., 1967; Ladbrooke et al., 1968; Oldfield and Chapman, 1972; Hinz and Sturtevant, 1972a; Lee, 1977b: Mabrey et al., 1978; De Kruijff, 1978; Cullis and De Kruijff, 1978; Hunt and Tipping, 1978). In general, addition of cholesterol increases the fluidity of liposomes below, but decreases it above, the gel-to-liquid phase transitions. Early studies indicated the decrease of the transition enthaply to be proportional to the added cholesterol (Hinz and Sturtevant, 1972a). More recently, heat absorptions of cholesterol and dipalmitoylphosphatidylcholine mixtures showed the coexistance of two immiscible solid phases (Mabrey et al., 1978). Under appropriate conditions cholesterol can, therefore, induce domain formations.

4. ELECTRICAL PROPERTIES OF LIPOSOMES

Substrate transport and energy transducing are related to the difference in electrical potential between the interior and exterior of liposomes. Although several theories have been developed for treating transmembrane potentials (Lakshminarayanaiah, 1974, 1975, 1979; Ohki, 1976; Kotyk and Janacek, 1977), complexities of the real membranes have to date precluded experimental verifications. Transmembrane potentials were investigated, therefore, on membrane models (Tien, 1974, 1975, 1979; Jain, 1972; Jain and Wagner, 1980). Due to their nature BLMs lend themselves to direct electrochemical investigations (see Section 5.2). Conversely, transliposome potentials can only be determined indirectly.

Transliposome potentials have been estimated by analyzing potential dependent optical changes of fluorescent dyes (Waggoner, 1976), or by using radioactive hydrophobic ions (Bakeeva et al., 1970) or spin probes (Kornberg et al., 1972; Cafiso and Hubbell, 1978). The method is illustrated by the use of positively charged spin labels (**6.1** and **6.2**) in estimating transliposome potentials for single-compartment phosphatidylcholine liposomes (Cafiso and Hubbell, 1978).

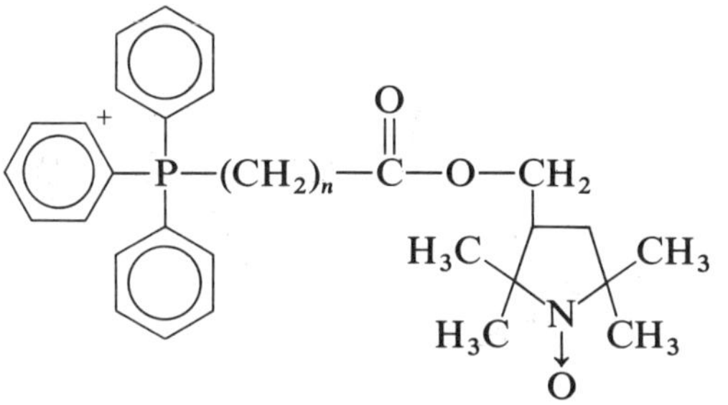

$(C_6H_5)_3P^+{-}(CH_2)_n{-}C(=O){-}O{-}CH_2$–(2,2,5,5-tetramethylpyrrolidine N→O)

6.1

$(C_6H_5)_3P^+{-}(CH_2)_2{-}C(=O){-}O{-}(CH_2)_8$–(2-$CH_3$-4,4-dimethyloxazolidine N→O)

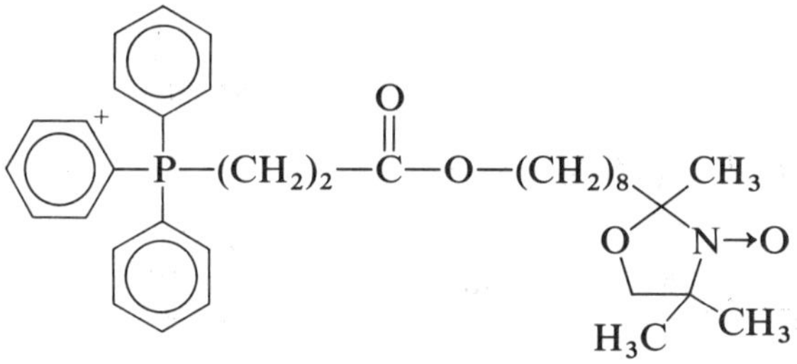

6.2

Spin labels **6.1** and **6.2** partition into the liposome phase, where they permeate freely. Their transit from the outside to the inside involves the passage through potential energy barriers on either side of the bilayers (Figure 6.18; Cafiso and Hubbell, 1978). Accordingly, there are four discrete regions the probes can occupy: the exterior of the liposome (o), the exterior interface (m_o), the interior interface (m_i), and the interior (i). Transmembrane potentials are generated by creating K^+ gradients across the liposome bilayers in the presence of valinomycin. At equilibrium the electrochemical potential, $\bar{\mu}$, of the spin label in each region $j (j = o, m_i, m_o, i)$ is given by

$$\bar{\mu}_j = \mu_j^0 + RT \ln \gamma_j \chi_j + ZF\Psi_j \qquad (6.12)$$

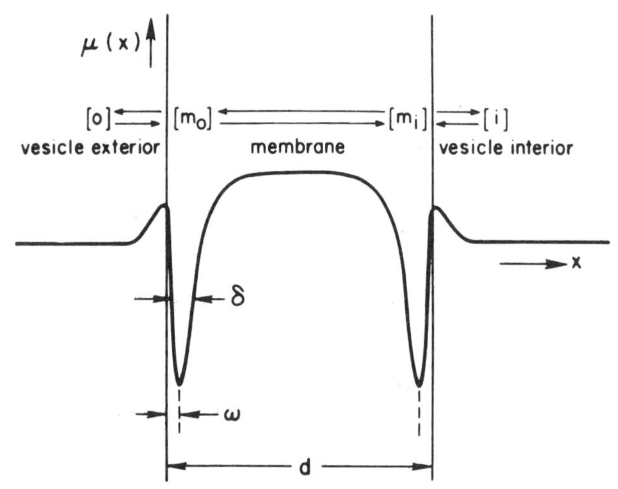

Figure 6.18. Potential energy profile across a bilayer membrane for a hydrophobic ion in the absence of a transmembrane electric potential. ω is the internal displacement of the potential minima from the membrane surface, δ is the effective width of the potential wells, d is the membrane thickness, and $\mu(\chi)$ is the potential energy of the ion. o, m_o, m_i, and i represent four discrete regions occupied by the ion: the exterior solution; the exterior interface; the interior interface; and the vesicle interior volume, respectively (taken from Cafiso and Hubbell, 1978).

where μ_j^0 is the chemical potential in the region j in the standard state, χ is the mole fraction of the charged group in the region j, γ_j is the activity coefficient of the label in region j, Ψ_j is the electric potential in region j, and Z, F, R, and T are the valence of the label, the Faraday constant, the universal gas constant, and the absolute temperature, respectively. Maintaining the ionic strengths constant, but not the gradient, activity coefficient ratios can be assumed to be unity. In dilute solution

$$\frac{\chi_j}{\chi_k} = \frac{n_j V_k}{n_k V_j} \tag{6.13}$$

where n_j and n_k are the number of moles of spin labels in regions j and k, and V_k and V_j are the corresponding volumes of the membrane phase in those regions per liposome. Taking ω as the displacement of the binding region from the plane of the ions responsible for the creation of the potential and d as the bilayer thickness, $\alpha \equiv \omega/d$ (see Figure 6.18). In the constant field approximation $\Psi_{m_o} = \alpha\Delta\Psi$ and $\Psi_{m_i} = (1 - \alpha)\Delta\Psi$. With the above approximations, definitions, and the equilibrium conditions that $\bar{\mu}_j - \bar{\mu}_k$ for all regions (j, k), the following general expression for the ratio (moles spin bound)/(moles spin free) may be readily derived (Cafiso and Hubbel, 1978):

$$\frac{n_{m_i} + n_{m_o}}{n_o + n_i} = \frac{N_b}{N_f} = \frac{V_{m_i}}{V_i} \times \left[\frac{K_{m_i} \exp \alpha ZF\Delta\Psi/RT) + K_{m_o}(V_{m_o}/V_{m_i}) \exp [ZF(1 - \alpha)\Delta\Psi/RT]}{1 + V_o/V_i \exp (ZF\Delta\Psi/RT)}\right] \tag{6.14}$$

where n_{m_o}, n_{m_i}, n_o, and n_i are the moles of spin at equilibrium in the four regions of space defined in Figure 6.18, N_b and n_f are the total numbers of bound and free spin, respectively, and K_{m_o} and K_{m_i} are the partition coefficients for the labels to the external and internal vesicle surfaces in the absence of a transmembrane potential (i.e., $K_{m_o} = \exp(-\Delta\mu^\circ_{m_o}/RT)$ and

$$K_{m_i} = \exp(-\Delta\mu^\circ_{m_i}/RT),$$

where $\Delta\mu^\circ_{m_i}$ and $\Delta\mu^\circ_{m_o}$ are the corresponding unitary free energies of transfer for the ion from the bulk aqueous phase to the membrane phase. Utilization of equation 6.14 requires knowledge of V_o/V_i, KV_{m_i}/V_i, V_{m_o}/V_{m_i}, and α. Values for V_o/V_i can be independently determined by using standard methods for determining vesicle entrapped volumes (see Section 6.1). KV_{m_i}/V_i values are obtained from binding studies and α is an adjustable parameter. The physical principle of employing equation 6.13 for transmembrane potential determinations is based on the larger surface-to-volume ratio of the liposome interior space compared with that of the external solution. When the liposome interior has a negative electric potential relative to the exterior, spin probes will migrate to this region and, since there is more membrane phase per unit volume on the interior, the net amount of bound spin label increases (Figure 6.19). Using equation 6.14, good agreement has been obtained between this spin label method and the independently determined transliposome potentials (Cafiso and Hubbell,

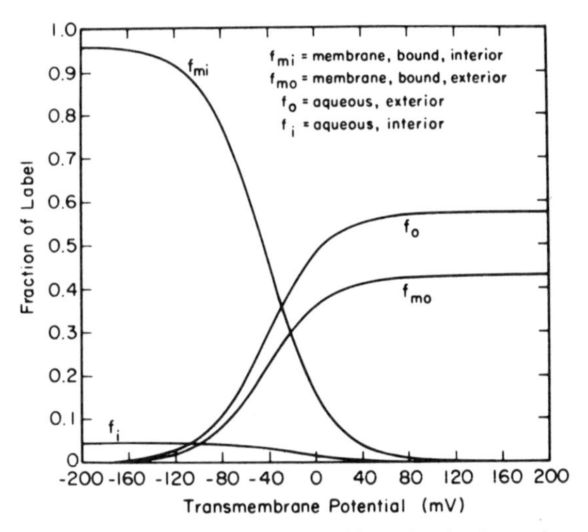

Figure 6.19. Theoretical distribution of a hydrophobic cation in the regions o, m_o, m_i, and i (see Figure 6.18) as a function of transmembrane potential. f_o, f_{m_o}, f_{m_i}, and f_i are the fractions of total label residing in each of the respective regions. This distribution has been calculated for $V_{m_o}/V_{m_i} = 2.3$, $V_o V_i = 70$, $\alpha = 70$, $\alpha = 0$, and $KV_{m_i}/V_i = 20$ (taken from Cafiso and Hubbell, 1978).

1978). The sensitivity of the method will depend on the vesicle size, geometry, and concentration. The sensitivity, S, is defined as $S \equiv \delta(N_b/N_f)/\delta\Delta\Psi$, and

$$S = K \frac{V_{m_i}}{V_i} \frac{ZF}{RT} \exp\left(+ \frac{\Delta\Psi ZF}{RT} \right) \left\{ \frac{V_{m_o}/V_{m_i} - V_o/V_i}{[1 + V_o/V_i \exp(\Delta\Psi ZF/RT)]^2} \right\} \quad (6.15)$$

Notice that when $V_{m_o}/V_{m_i} = V_o/V_i$, $S = 0$. For any arbitrary vesicle size and geometry the sensitivity increases linearly with the binding constant of the hydrophobic ion, a parameter that is readily controlled by varying the alkyl chain length of molecules of **6.2**. An important advantage of the spin label technique is that very low concentrations of probes (fewer than one probe per liposome) can be used. Further, the results are independent of the concentration of the label. Similar spin label techniques have been used for the changes in the boundary potential initiated by the absorption of a photon by membrane bound rhodopsin (Cafiso and Hubbell, 1980).

Generation of a potential gradient across the bilayer is paralleled by morphological changes of the liposome. Taking advantage of the fluorescence depolarization of 1,6-diphenyl-1,3,5-hexatriene, exponential increases in the rigidity of single-compartment phosphatidylcholine and phosphatidylserine liposomes have been observed with increasing transliposome potentials generated by Na^+/K^+ gradients (Lelkes, 1979).

Applying large potential gradients results ultimately in the dielectric breakdown of liposomes. Applying external fields to membranes and bilayers results in a reversible breakdown (Coster and Zimmermann, 1975; Zimmermann et al., 1976; Benz and Zimmermann, 1981). This opens the possibility of controlled substrate entrapments in liposomes.

5. MOLECULAR DYNAMICS OF LIPOSOMES AND THEIR CONSTITUENTS

Liposomes are dynamic entities. Configurational and conformational changes of the lipids (see Section 6.2), phase transitions (see Section 6.3), liposome-liposome fusions, osmotic shrinkages and swellings, interliposome lipid and substrate exchanges are all dynamic processes, occurring at various timescales.

Molecular motions of individual lipids within the liposome ensemble are quite varied. In addition to kink formation (see Figures 6.6 and 6.7), rotational and segmental motions, lateral diffusions, and transverse motion from one interface of the bilayer to the other (flip-flop) have been recognized (Figure 6.20). Substantial information has been obtained from nmr, epr, and Raman spectroscopic measurements, often expressed in terms of order parameters. Equation 6.8 describing the order parameter can be rewritten as

$$S_{mol} = [\tfrac{1}{2}(3\langle\cos^2\alpha\rangle - 1)][\tfrac{1}{2}(3\langle\cos^2\gamma\rangle - 1)] \quad (6.16)$$

$$S_{mol} = S_\alpha S_\gamma$$

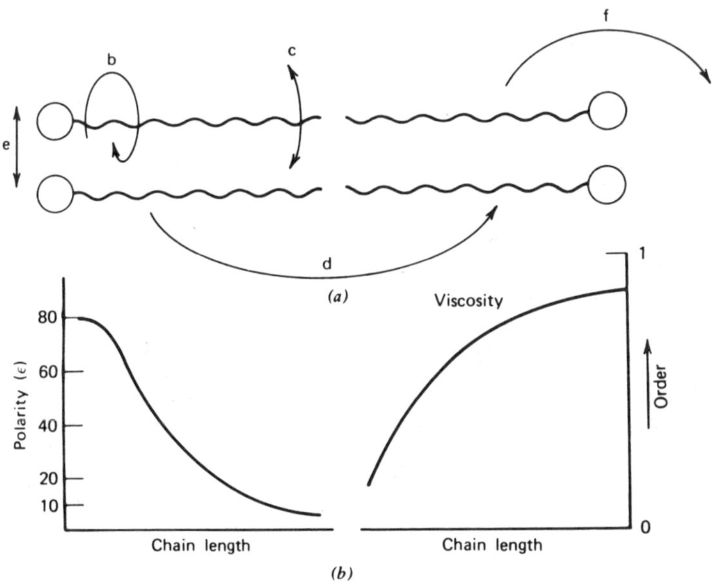

Figure 6.20. (*a*) An oversimplified representation of molecular motions in liposome bilayers. Individual lipids can rotate (b), undergo segmental motion (c), flip-flop (d), undergo lateral diffusion (e), or intervesicle exchange (f). (*b*) Approximate gradients of polarity and microviscosity (order) as a function of the acyl chain lengths (taken from Jain and Wagner, 1980).

where S_α is a chain order parameter, related to chain reorientation motion, and S_γ is an intramolecular order parameter, related to chain isomerization (trans → gauche). Nmr experiments only measure the product, S_{mol}. Conversely, Raman spectroscopic measurements allow the separate assessments of inter- and intramolecular chain orders. Epr experiments using spin labels provide information primarily on chain isomerizations and rotations.

Dynamics of chain reorientation and hindered rotation (isomerization) in liposomes have been assessed from proton and dueterium order parameters (Petersen and Chan, 1977). Spin lattice relaxation times modulated by at least two motions with correlation times $\tau_{\|}$ and $\tau_{\perp}$, can be approximated by (Kroon et al., 1976)

$$\frac{1}{T_1} \cong A\tau_{\|} + B\frac{1}{\omega_0^2\tau_{\perp}} \tag{6.17}$$

where ω_0 is the nmr frequency. The faster event, $\tau_{\|}$, occurring at a rate of $\sim 10^{10}$ sec^{-1}, is associated with rotational isomerization. The slower event, $\tau_{\perp}$, occurring at a rate of $\sim 10^7$ sec^{-1}, is attributable to chain isomerization (Petersen and Chan, 1977).

Phospholipid molecules can rotate along the axis or perpendicular to the plane of the liposomes. Only the latter rotation (as indicated in Figure 6.20) is favorable since it keeps the headgroups in the aqueous phase and the hydrocarbon chains within the bilayer. Rotational correlation times depend on the liposome phase transition. Using saturation transfer electron spin resonance spectroscopy, an upper limit of $\geq 10^{-4}$ sec was placed on the effective correlation time of lipid rotations along their long axes in multicompartment phosphatidylethanolamine and phosphatidylcholine liposome below their phase transition temperatures (Marsh, 1980). The effective correlation time decreases to $\sim 10^{-6}$ sec at a temperature corresponding to the calorimetric pretransition in phosphatidylcholine liposomes. A corresponding rapid increase in rotation occurs at the main transition temperature for phosphatidylethanolamine liposomes (Marsh, 1980).

The segmental motion (indicated by c in Figure 6.20), and hence the molecular disorder, increase on moving away from the polar lipid headgroup. At their polar ends lipids are anchored onto the liposome interface. Thus the motion of the individual chain becomes less restricted on moving toward the terminal methyl groups. This behavior is reflected in the order parameters (see Table 6.4). The timescale for segmental motion is between 10^{-8} and 10^{-4} sec.

Maintaining dissymmetry in liposomes requires the transverse motion (see Figure 6.20) of molecules, between the inner and outer monolayers, to be a slow process. Indeed, flip-flops of lipids, cholesterol, and substrates occur on the timescale of days (Jain and White, 1977; De Kruyff et al., 1977) without affecting the permeability properties of liposomes. Flip-flop rates dramatically increase, however, in the presence of proteins (De Kruijff and Wirtz, 1977; De Kruijff and Baken, 1978; Nakagawa et al., 1979). Phosphatidic acid, formed on the outside of single-compartment phosphatidylcholine liposomes by phospholipase D, transfers to the inner monolayer with a half-time of 30–40 min (De Kruijff and Baken, 1978). ^{31}P nmr spectroscopy, in the presence of Nd^{3+} in the liposome interior, allowed the simultaneous monitoring of liposome interiors and exteriors (De Kruijff and Baken, 1978). Employing similar techniques, a distinct maximum was observed in the flip-flop rate at the phase transition temperature of single-compartment dimyristoyl phosphatidylcholine liposomes (De Kruijff and Zoelen, 1978). The rates were followed by ^{13}C nmr in the presence of $DyCl_3$ using $N—^{13}CH_3$ labeled lipids, introduced into the outer monolayer of unlabeled liposomes by a phosphatidyl exchange protein. Apparently, increased lateral liposome compressibility at the phase transition facilitates transmembrane lipid movements (De Kruijff and Zoelen, 1978).

Lateral distribution of cholesterol and other molecules in phosphatidylcholine bilayers has been simulated by computer calculations (Freire and Snyder, 1980; Snyder and Freire, 1980). Lateral diffusion (see Figure 6.20) of molecules on the liposome surface is relatively unhindered. Diffusion coefficients are in the order of 10^{-7}–10^{-8} cm^2 sec^{-1} (Edinin, 1974; Jain and White, 1977). Lateral diffusion of proteins on membrane surfaces facilitates membrane fusions, cell divisions and drug-receptor interactions. Measurements of rotational and lateral diffusion of membrane proteins by flash photolysis, saturation transfer epr spectroscopy, and fluorescence recovery after photobleaching have been summarized (Cherry, 1979).

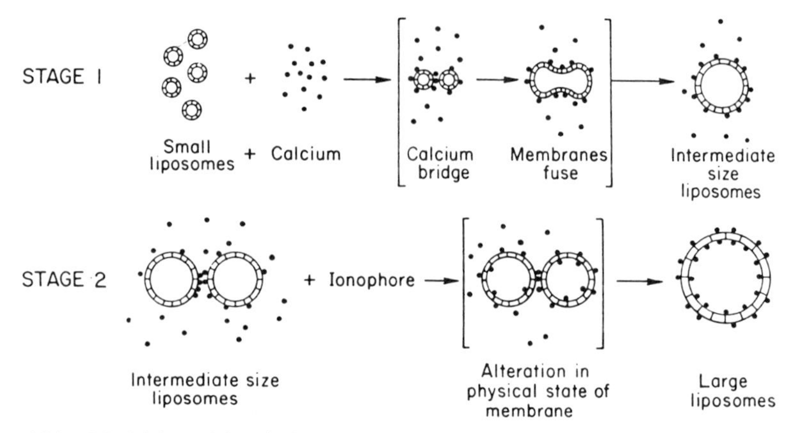

Figure 6.21. Model for calcium induced fusion. Fusion occurs in two stages. In the first stage calcium ion binds to phospholipids on the outside of small (approximately 60 nm diameter) liposomes and catalyzes fusion events that produce intermediate size liposomes (approximately 120 nm diameter). Calcium may be exerting its effects by altering the membrane properties so that fusion is facilitated and by bridging the liposomes so that the liposomes are held in close proximity. The second stage of fusion occurs when calcium ion can traverse the membrane, such as in the presence of ionophores. Calcium may catalyze this stage of fusion by binding to the phospholipids on the inside of intermediate size vesicles, thereby altering the membrane properties once again, to allow formation of large liposomes (approximately 170 nm diameter) (taken from Ingolia and Koshland, 1978).

Liposome-liposome fusions (Poste and Nicolson, 1978) are not only inherently interesting but offer opportunities to introduce materials directly into cells (see Chapter 12). Liposomes rarely fuse spontaneously. Fusion of negatively charged liposomes can be triggered by divalent metal ions (Papahadjopoulos et al., 1975, 1977; Wilschut and Papahadjopoulos, 1979; McIver, 1979) and that of neutral ones by *n*-alkyl bromides (Mason and Miller, 1979). Calcium induced fusion was seen to provide means for the preparation of large single-walled liposomes (see Section 6.1; Papahadjopoulos et al., 1975). Liposome fusions have been discussed in terms of two major mechanisms: direct and indirect. The direct mechanism involves the merger of two liposomes without any loss of entrapped substrates or labels. In the indirect mechanism at least one liposome opens up prior to or during the fusion. At present, there is no agreement as to which of these mechanisms predominates (Ginsberg, 1978; Wilschut and Papahadjopoulos, 1979).

Details of calcium induced liposome fusion have been investigated in the presence of ionophores (Papahadjopoulos et al., 1977; Ingolia and Koshland, 1978; Liao and Prestegard, 1979). Single-compartment soybean phospholipid liposomes fused in a two-stage process in the presence of calcium ion and valinomycin or calcium ionophore A23187 (Figure 6.21; Ingolia and Koshland, 1978). In the first stage calcium ion binds to the outside of the liposomes having average diameters of 600 Å. The resultant charge neutralization provides the necessary driving force for the formation of intermediate size (~1200 Å diameter) liposomes. In the second stage the ionophore mediates the permeability of calcium ions into the liposome interior, which, in turn, favors the formation of larger (~1700 Å diameter) liposomes. Fluorescence spectroscopic techniques have been used for monitoring liposome fusion (Weinstein et al., 1977; Ingolia and Koshland, 1978; Vanderwerf and Ullman, 1980; Owen, 1980). Fusion was induced between liposomes loaded with firefly extract and those containing magnesium adenosine triphosphate (MgATP). Since the bulk solution contained sufficient amounts of quenchers, luminescence was only observed if the two sets of liposomes fused without subsequently releasing the MgATP–firefly extract complex (Ingolia and Koshland, 1978). Conversely, addition of ionophores to single-compartment phosphatidic acid–phosphatidylcholine liposomes decreased the extent of calcium induced fusion and no fusion occurred in the absence of ionophores when the calcium ions had been internalized (Liao and Prestegard, 1979). Either the mechanisms are different for the different liposomes used or one or both experimental methodologies are imperfect.

Rates of liposome fusions are generally determined by following the increase of turbidities in a stopped-flow spectrometer. The obtained kinetics for calcium ion induced fusion of phosphatidic acid and phosphatidyl serine liposomes were analyzed in terms of (Lansman and Haynes, 1975; Haynes and Westine, 1980)

$$n\mathrm{Ca}^{2+} + \text{liposome} \underset{k_{-0}}{\overset{k_0}{\rightleftharpoons}} (\text{liposome} \cdot n\mathrm{Ca}^{2+}) \tag{6.18}$$

$$(\text{liposome} \cdot n\mathrm{Ca}^{2+}) + (\text{liposome} \cdot n\mathrm{Ca}^{2+}) \underset{k_{-app}}{\overset{k_{app}}{\rightleftharpoons}} (\text{liposome} \cdot n\mathrm{Ca}^{2+})_2 \tag{6.19}$$

$$(\text{liposome} \cdot n\mathrm{Ca}^{2+})_p + (\text{liposome} \cdot n\mathrm{Ca}^{2})_q \underset{k_{-pq}}{\overset{k_{pq}}{\rightleftharpoons}} (\text{liposome} \cdot n\mathrm{Ca}^{2+})_{p+q} \tag{6.20}$$

where n represents the number of calcium ions per liposome sufficient to promote aggregation and p and q are stoichiometric coefficients greater than 1. Equation 6.19 describes a composite process; dimerization involves the fast collision of calcium coated liposomes to form an encounter complex that, in a slower step, converts to a stable complex (see idealized structures in square brackets in the first stage of fusion in Figure 6.21):

$$(\text{liposome} \cdot n\mathrm{Ca}^{2+}) + (\text{liposome} \cdot n\mathrm{Ca}^{2+}) \underset{k_{-1}}{\overset{k_1}{\rightleftharpoons}} (\text{encounter complex}) \tag{6.21}$$

$$(\text{encounter complex}) \underset{k_{-2}}{\overset{k_2}{\rightleftharpoons}} (\text{stable complex}) \tag{6.22}$$

The observed second order rate constant for liposome fusion, k_{app}, is given by

$$k_{app} = \frac{k_1}{k_{-1}} k_2 \tag{6.23}$$

Values for k_{app} are obtained by fitting time dependence of turbidity increases, used to monitor the fusion, to equation 6.24 (Haynes and Westine, 1980):

$$\frac{k_{app}}{N_{pl}} = \frac{\alpha}{2(1-\alpha)[PL]t} \tag{6.24}$$

where N_{pl} is the number of phospholipids per liposome, $[PL]$ is the stoichiometric lipid concentration, α is the degree of progress of reaction, measured as a fraction of maximal absorbance change Δabs/Δabs max, and t refers to the time domain in which the fusion occurs. Table 6.7 summarizes the obtained kinetic constants for liposome dimerizations. Comparison of these rates with that calculated from the Smoluchowski equation showed that 200–700 collisions are necessary for stable liposome dimer formation (Lansman and Haynes, 1975). The rate limiting step, the formation of the stable complex, requires $(5\text{–}10) \times 10^{-4}$ sec. The extent of dimerization depends on the Ca^{2+} to phospholipid ratio. At given lipid concentrations the extent of dimerization increases with increasing Ca^{2+} concentration to a maximum, after which it decreases (Haynes and Westine, 1980). The maximum corresponds to full charge neutralization. Increasing the Ca^{2+} concentration further results in the buildup of net positive charges on the liposomes, which, in turn, electrostatically hinder dimerizations.

A kinetic model to distinguish between direct and indirect mechanisms for liposome fusion has been developed (Lawaczek, 1978). Data for phosphatidic acid liposome dimerization (*vide supra*; Lansman and Haynes, 1975) could be best accommodated in terms of a direct liposome-liposome fusion mechanism. A method has been proposed to allow a distinction between liposome aggregation and fusion (Sunamoto et al., 1980). Polysaccharide pollulan increased the turbidity of multicompartment egg lecithin liposomes. Addition of enzyme pollulanase decreased the turbidity to that due to the liposome alone. Pollulan induces, therefore, reversible aggregation of multicompartment liposomes. Conversely, the pullulan induced turbidity increase of single-compartment liposomes could not be reversed by the addition of pollulanase. Single-compartment liposomes undergo, therefore, irreversible pollulan induced fusion (Sunamoto et al., 1980).

Table 6.7. Kinetic Constants for Liposome Dimerization

Liposome	k_{app}, M^{-1} sec^{-1}	k_1/k_{-1}, M^{-1}	k_2, sec^{-1}
Phosphatidic acid	3.9×10^7 [a]	3.9×10^4 [a]	1.0×10^3 [a]
	9.9×10^6 [b]	5.1×10^3 [b]	1.9×10^3 [b]
Phosphatidylserine	6.9×10^7 [a]	3.9×10^4 [a]	1.7×10^3 [a]
	1.7×10^7 [b]	5.1×10^3 [b]	3.4×10^3 [b]

Source: Compiled from Lansman and Haynes, 1975.

[a] Calculated using a liposome radius of 250 Å and $N_{pl} = 2.66 \times 10^4$.

[b] Calculated using a liposome radius of 125 Å and $N_{pl} = 6.67 \times 10^3$.

Neutral molecules and lipids readily interchange between liposomes (Martin and MacDonald, 1976; Kremer et al., 1977b; Kremer and Wiersema, 1977; Galla et al., 1979; Kano et al., 1980; Dody et al., 1980; Almgren, 1980). Rates of interliposome exchanges depend on the system investigated as well as on its phase transition temperature. Above the phase transition temperature of dipalmitoylphosphatidylcholine-dimyristoylphosphatidylcholine liposomes, rate constants for lipid exchange are between (1–3) × 10^5 M^{-1} sec^{-1} at 50°C, while there is no exchange at 20°C (Kremer et al., 1977b). The exchange of myristic acid between dipalmitoylphosphatidylcholine liposomes is complete within a few seconds above the phase transition temperature but below it concurrent liposome-liposome fusions complicate the interpretation (Kremer and Wiersema, 1977). Pyrene and pyrene carboxylic acids exchange rapidly between single-compartment liposomes (Galla et al., 1979).

Using fluorescence detected stopped-flow spectroscopic methods, interliposome pyrene exchanges have been discussed in terms of a desolubilization-diffusion-resolubilization mechanism (Almgren, 1980) similar to that developed previously for substrate dissolution kinetics in micelles (Almgren et al., 1979). The residence time of pyrene was found to depend on the radius of the liposome, and the rate constant for its exit was determined to be 4 × 10^2 sec^{-1} at 25°C (Almgren, 1980). Time dependent energy transfer from *N*-alkylated alloxazines to isoalloxazines, located separately in single-compartment dipalmitoyl-D,L-α-phosphatidylcholine liposomes, indicated the interliposome exchange of these probes to occur on the timescale of minutes (Kano et al., 1980).

The transfer of fluorescent 9-(3-pyrenyl) nonanoic acid between single-compartment dimyristoylphosphatidylcholine and dipalmitoylphosphatidylcholine liposomes was found to be a first order process, independent of the concentrations or chemical compositions of donor and acceptor liposomes (Dody et al., 1980). The mechanism for the interliposome transfer was discussed in terms of a rate limiting dissociation from the donor vesicle, followed by a much faster diffusion to, and uptake by, a neighboring acceptor liposome. Assuming the p*Ka* for the 9-(3-pyrenyl)nonanoic acid in bulk water to be 5.0, energy diagrams have been constructed for the interliposome probe exchange (Figure 6.22). Experimental values for the free energies of transfer from the liposome to interface (indicated by A → B in Figure 6.22) are 18.8 and 19.0 kcal for the liquid crystalline and gel phases for the protonated probe (I in Figure 6.22) and 16.3 and 16.5 kcal $mole^{-1}$ for the corresponding phases of the ionized probe (VII in Figure 6.22). Free energies of transfer from the interfacial region to the aqueous phase (indicated by B → C in Figure 6.22) are much less. The experimental values for the liquid crystalline and gel phases for the protonated probe are 10.4 and 10.3 kcal $mole^{-1}$ (II in Figure 6.22) and 9.4 and 9.2 kcal $mole^{-1}$ for the corresponding phases of the ionized probe (VII in Figure 6.22; Dody et al., 1980).

Osmotic activity of liposomes is also a dynamic process. Under suitable conditions liposomes are only permeable to water. Thus placing them in hyperosmalar solutions results in osmotic shrinkage, while in hypoosmolar solutions

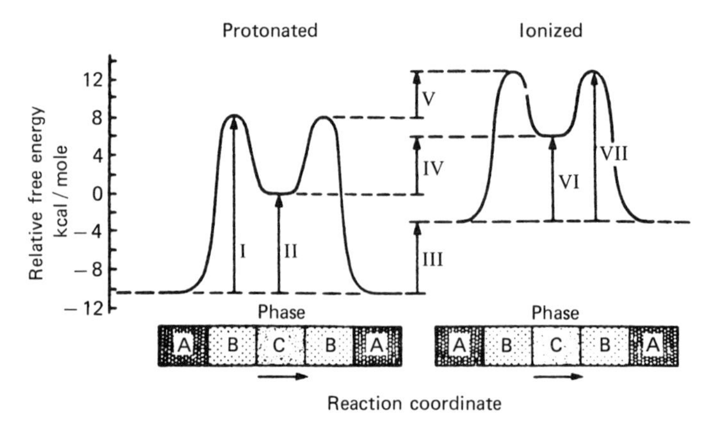

Figure 6.22. Correlation of pH and phase dependence with the thermodynamic quantities. Energy level diagrams are shown for the transfer of 9-(3-pyrenyl)-nonanoic acid between dimyristoylphosphatidylcholine vesicles in the liquid crystalline phase at 37°C and in the gel phase at 15°C. Phases A, B, and C denote the phospholipid vesicle, the interfacial region, and the aqueous phase, respectively (taken from Dody et al., 1980).

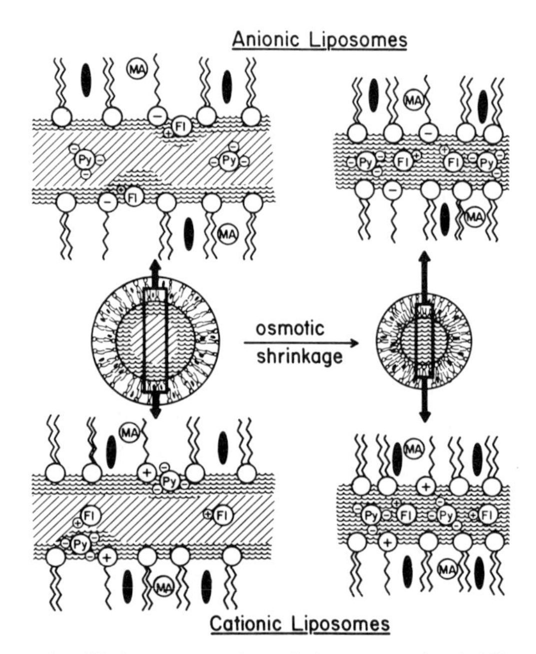

Figure 6.23. An oversimplified representation of the proposed solubilization sites of methyl anthracene (MA), pyranine (PY), and acriflavine (Fl) prior and subsequent to osmotically shrinking cationic and anionic single-compartment dipalmitoyl-D,L-α-phosphatidylcholine liposomes. Bound and free water are indicated by waved and straight lines. The charged additives are shown to be interspersed with the phospholipid and cholesterol. (Taken from Kano and Fendler, 1979).

they swell (Bangham et al., 1967b; Rendi, 1967). Taking advantage of turbidity changes that accompany the osmotic shocks, water permeabilities across the liposome bilayers were found to occur on the second to hour timescale (Bittman and Blau, 1972; Jain et al., 1973; Blok et al., 1975, 1976). Initial rates of water permeabilities depend on the nature of the liposome, on the electrolyte gradients, and on the presence of additives. Added cholesterol, for example, decreases the rates of water permeabilities above the phase transition temperature, but increase it in the solid phase (Bittman and Blau, 1972). Osmotic shrinkages of single-compartment liposomes result in increased microviscosities. Neutral methylanthracene, anionic trisodium 8-hydroxy-1,3,6-pyrenetrisulfonate (pyranine), and cationic 3,6-diamino-10-methylacridinium chloride (acriflavine) have been used as fluorescence probes to investigate effects of osmotic shrinkage on neutral, cationic, and anionic dipalmitoyl-D,L-α-phosphatidylcholine liposomes. The effect is most pronounced in the interior of liposomes, where, following the shrinkage, essentially all the water becomes bound water (Figure 6.23; Kano and Fendler, 1979).

6. SUBSTRATE INTERACTION WITH AND TRANSPORT IN LIPOSOMES

Available sites for substrate interactions and bindings in liposomes are shown schematically in Figure 6.24 (Fendler, 1980b). Highly polar and relatively small solubilizates are trapped in the aqueous compartments. Electrostatic interactions play an important part. Anionic liposomes attract cations, but repel anions. The point is illustrated by the interaction of trisodium 8-hydroxyl-1,3,6-pyrenetrisulfonate, pyranine, with charged dipalmitoyl-D,L-α-phosphatidylcholine liposomes (Kano and Fendler, 1978). Using fluorescence polarization techniques, pyranine was shown to be well shielded, in negatively charged liposome interiors, from the phospholipid headgroups by a large number of water molecules. The microviscosity of the environment of this probe, 0.96 cP, corresponds, in fact, closely to that of water ($n_{H_2O}^{25} = 0.8904$ cP). Conversely, pyranine is bound to the surface of cationic liposomes. Here the microviscosity reported by the probe is greater than 20 cP (Kano and Fendler, 1978). Due care needs to be taken in extrapolating fluorescence measurements to microviscosities (Chen et al., 1977; Hare and Lussan, 1978; Hare et al., 1979; Lakowicz and Knutson, 1980).

Addition of electrolytes to already formed liposomes has several important consequences. At low concentrations they bind to counterions, reducing the net charge of liposomes. This, particularly if divalent cations are used, results in domain formation and in liposome-liposome fusion (see Figure 6.17). Alternatively, electrolytes may permeate the liposome (see below) or create osmotic gradients.

Nonpolar molecules are intercalated between the phospholipid bilayer. Amphiphatic molecules are anchored into the vesicles by their hydrocarbon chains. The extent and the site of their binding depend on both electrostatic and hydrophobic interactions. Interactions of cholesterol with liposomes have been extensively investigated (Papahadjopoulos and Kimelberg, 1974; Demel and De Kruijff, 1976; Lentz et al., 1980). The lengths and the nature of the side chain on cholesterol influence its ordering in liposomes (Craig et al., 1978).

Addition of alcohols or detergents destroys liposomes (Zaslavsky et al., 1980). The mechanism of surfactant-liposome interaction has been investigated by examining the leakage of bromophenol blue from single-compartment liposomes containing *N*-hexadecyl-*N*-(imidazol-4-yl)-methyl-*N*,*N*-dimethylammonium ion, **6.3**, as a functional detergent (Sunamoto et al., 1978). Bromophenol blue leakage was rationalized in terms of formation of acyl derivative **6.4**, which damages the liposomal structure:

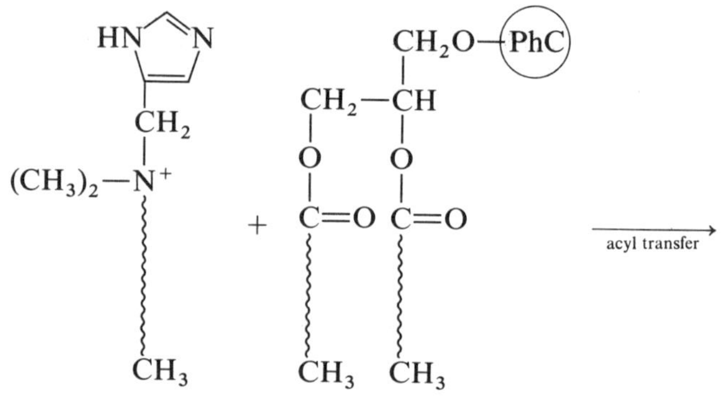

HN N — CH_2 — $(CH_3)_2$—N^+ ~~~ CH_3 (Im − I) + CH_2O—(PhC), CH_2—CH, O O, C=O C=O, ~~~ CH_3 CH_3 (DPL) $\xrightarrow{\text{acyl transfer}}$

6.3

N N—C=O (~~~ CH_3) — CH_2 — $(CH_3)_2$—N^+ ~~~ CH_3 + CH_2O—(PhC), CH_2—CH, OH O, C=O, ~~~ CH_3 or CH_2O—(PhC), CH_2—CH, O OH, C=O, ~~~ CH_3

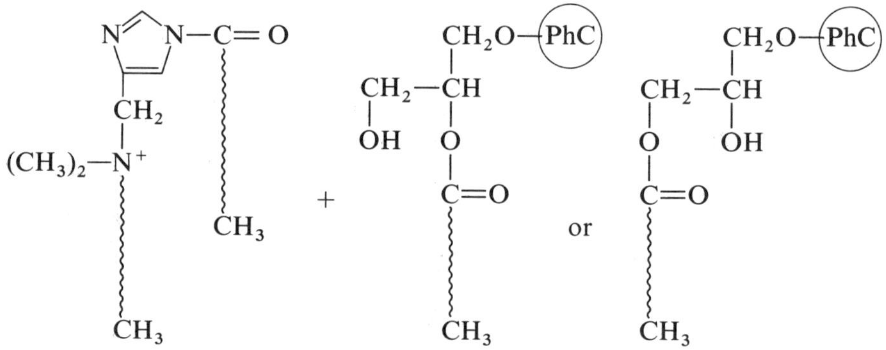

6.4

—(PhC): $-\overset{\displaystyle O}{\overset{\|}{P}}(O^-)-OCH_2CH_2\overset{+}{N}(CH_3)_2$

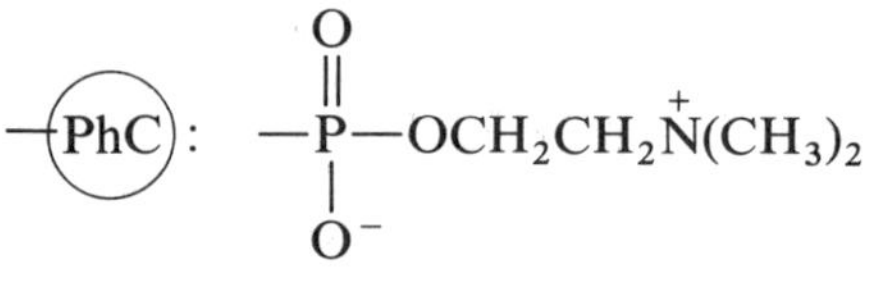

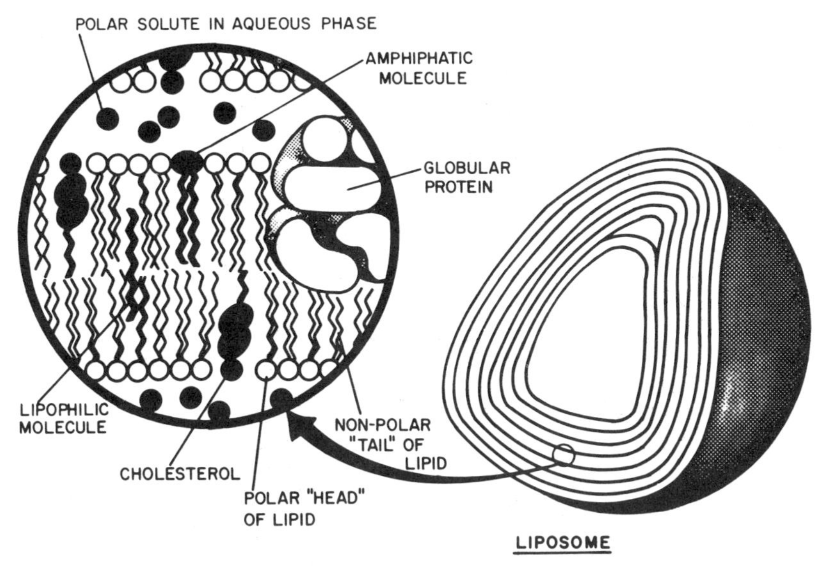

Figure 6.24. An oversimplified representation of the sites of interactions of polar, apolar, and amphiphatic molecules, cholesterol, and proteins with multicompartment liposomes (taken from Fendler, 1980b).

The damage was considered to be repaired by the rapid reaggregation of **6.4** and lysophosphatidylcholine (Sunamoto et al., 1978).

Addition of such cryoprotective agents as glycerol and dimethyl sulphoxide DMSO increased the stability of liposomes to freezing. The damage is considered to be prevented by limiting water dehydration (Strauss and Ingenito, 1980).

Macromolecules associate with liposomes in a variety of ways. They can be entrapped into the aqueous interiors, span across the bilayers, or coat the liposome surface. Evidence has been obtained from fluorescence polarization for the rapid uptake of polymyxin, an antibiotic peptide, exclusively in the outer monolayers of negatively charged phosphatidic liposomes (Sixl and Galla, 1980). Distributions of proteins are nonrandom and asymmetric (Gebhardt et al., 1977). Furthermore, proteins are fairly mobile laterally. The expression that "proteins freely swim in the fluid lipid matrix" (Edinin, 1974; Jain and Wagner, 1980) is particularly apt. The presence of substrates modulates, of course, liposome phase transitions (see discussion of electrolyte and substrate induced phase transitions in Section 6.3), morphologies (see Section 6.2), and other physical chemical properties.

Just like electrical properties (see Section 6.4), substrate permeabilities have been investigated in BLMs (see Section 5.2) to a greater extent than in liposomes. Solutes either diffuse across bilayers or their transport may be facilitated by the presence of pores or carriers (Shamoo, 1975). Energetically, substrate binding to one side of the bilayer, penetration and diffusion through the hydrocarbon regions, binding to and release from the other side of the bilayer are the steps

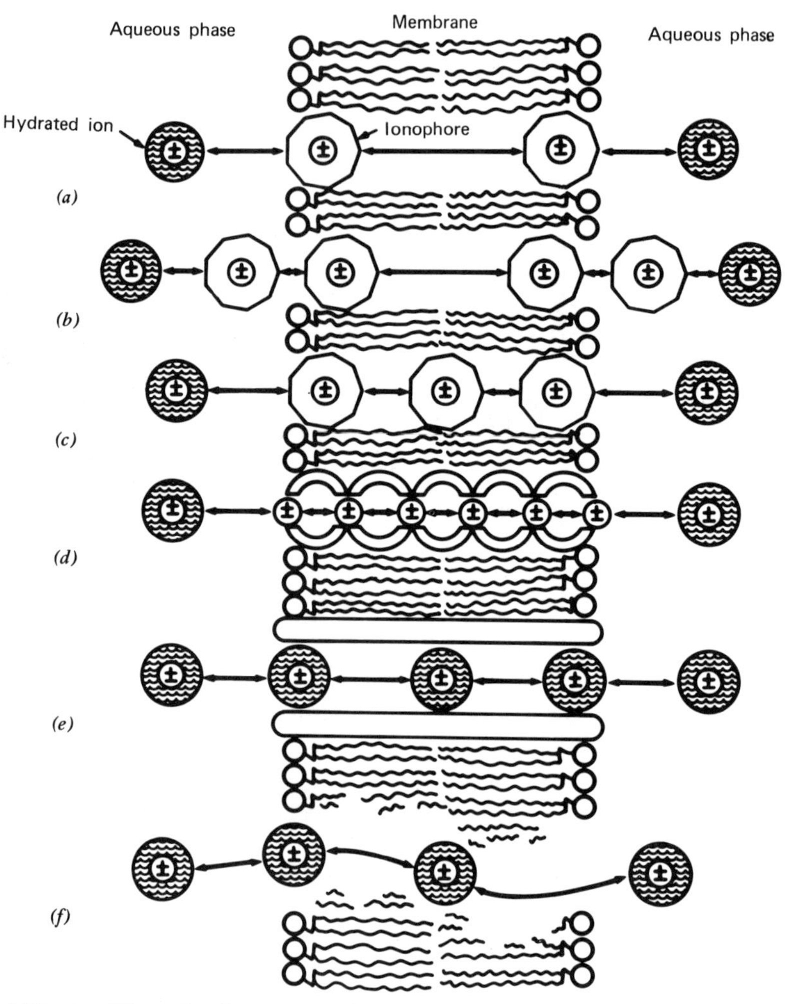

Figure 6.25. Possible modes of movement of an ion through a lipid bilayer membrane. (*a*) Shuttle (reaction at interface). (*b*) Shuttle (reaction in aqueous phase). (*c*) Relay shuttle. (*d*) Chain or channel. (*e*) Aqueous pore. (*f*) Membrane lysis.

that can be visualized in diffusion driven permeabilities. Proton–hydroxide ion permeabilities in large unilamellar liposomes are several orders of magnitude greater than those found for monovalent ions (Nichols and Deamer, 1980; Nichols et al., 1980). The permeabilities of protons 1.44×10^{-4} cm sec^{-1} (Nichols et al., 1980), are greater than those of cations, 10^{-13}–10^{-14} cm sec^{-1}, or of anions, 10^{-11}–10^{-12} cm sec^{-1}; this has been rationalized in terms of transport via hydrogen bond exchange with water molecules in the bilayer (Nichols and Deamer, 1980). Water permeabilities across single-compartment liposomes are in the 10^{-3}–10^{-6} cm sec^{-1} range (Lawaczek, 1979). The presence of proteins or additives in distinct domains alters, of course, substrate diffusions.

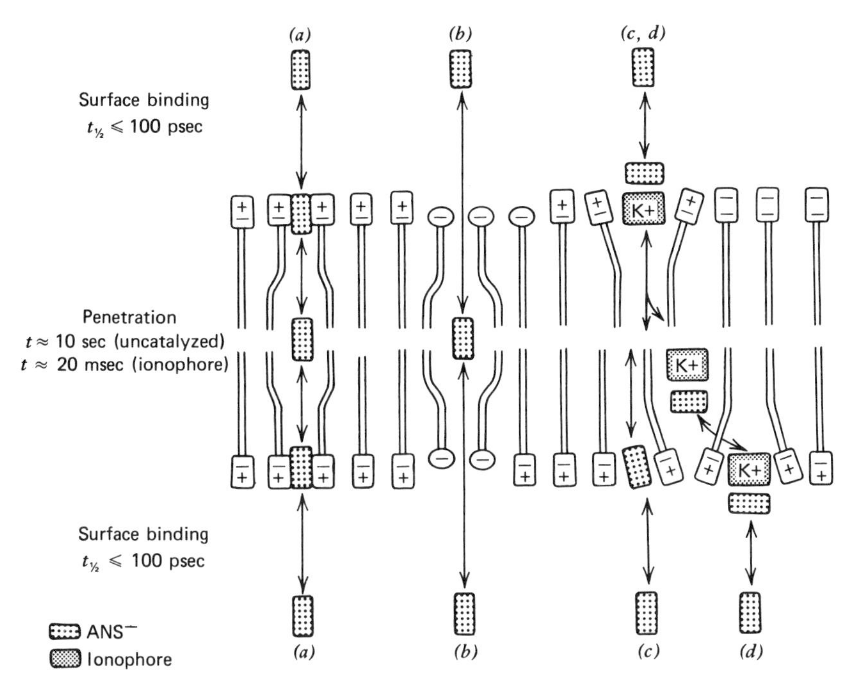

Figure 6.26. Permeability mechanisms for 1-anilino-8-naphthalenesulfonate (ANS$^-$). (*a*) ANS$^-$ binds to the polar headgroup region. Penetration of the membrane takes place concomitantly with a structural change in the hydrocarbon chain region. The ANS$^-$ is considered to have only a transient existence in this environment. The ANS$^-$ is transferred to another binding site (if present) on the other side of the membrane and is released to the inner aqueous phase. (*b*) Penetration takes place independently of the polar headgroup in cases where the latter does not support binding. For models (*a*) and (*b*) M$^+$ is considered to copermeate although its mechanism is not shown. (*c*) ANS$^-$ transport occurs by binding in a region perturbed by an ionophore followed by transport through this perturbed region. (*d*) ANS$^-$ transport occurs strictly by the mechanism of ion pairing with I—M$^+$, followed by cotransport of the ternary complex (taken from Haynes and Simkowitz, 1977).

Cationic carriers or ionophores create pathways additional to those related to substrate lipophilicities. These site mediated or *facilitated transport* pathways lower the energy barrier for moving molecules across the bilayer. Site facilitated transport can be mediated either by channel (pore) or by a carrier mechanism. The former is conceptualized as having a fixed opening through which the solutes pass. In the carrier mechanism the substrate binds to a site that alternates between the outer and inner interface of the bilayer by rotation, by conformation changes, by diffusion, or by a combination of these mechanisms. Selectivity of carrier mediated ion transport is accomplished by differential sequestering in the carrier. Cationic carriers or ionophores show a high degree of selectivities; in particular, they can distinguish between K^+ and Na^+ (Pressman, 1976). Cyclic peptides such as valinomycin, enniatin, or antamanide are typical ionophores. Crown ethers (see Chapter 7) have also been used as ionophores.

Different modes of ionophore mediated ion transports are shown schematically in Figure 6.25.

Conductance, fluorescence, and high resolution ^1H nmr spectroscopic techniques have provided insight into structures and conformation of ionophores in liposomes (Veatch and Stryer, 1977; Feigenson and Meers, 1980). Rate constants for the permeabilities of 1-anilino-8-naphthalanesulfonate across liposomes in the absence and in the presence of ionophores have been determined by stopped-flow spectrofluorometry (Haynes and Simkowitz, 1977). The two-phase kinetics have been discussed in terms of surface binding, penetration, and release (Figure 6.26; Haynes and Simkowitz, 1977). Binding occurs faster than 100 μsec while permeation is in the 5–100 sec range. Substrate incorporation often shows an induction period, as expected for cooperative processes (Sixl and Galla, 1980). Ionophores dramatically enhance the transit of the probe across the bilayer. Permeation of acetic acid across single-compartment liposomes has been determined to occur in the 0.001–10.0 sec timescale (Alger and Prestegard, 1979).

7. SYNTHETIC SURFACTANT VESICLES

Formation of closed bilayer structures from simple surfactants has been recognized for some time (Gebicki and Hicks, 1973). Exploitation of completely synthetic surfactant vesicles for mimicking membrane functions has been prompted by the ease of their formation and stabilities (Fendler, 1980a).

Vesicles obtained from different types of surfactants are collected in Table 6.8. Vesicles are seen to form from both cationic and anionic surfactants. Closed bilayer surfactant vesicles, contrary to geometric calculations (Israelachvili et al., 1976, 1980; Tanford, 1980), can apparently be obtained from surfactants having single-alkyl chains (Gebicki and Hicks, 1973, 1976; Hicks and Gebicki, 1977; Hargreaves and Deamer, 1978a, 1978b; Kunitake and Okahata, 1980; Kunitake et al., 1981). Those having rigid segments, for example, diphenylazomethine groups (Kunitake and Okahata, 1980), appear to be more stable than surfactant vesicles formed from long chain carboxylic acids. Relationships between amphiphile structures and vesicle morphologies have been delineated (Kunitake et al., 1981). Formation of single chain surfactant vesicles is significant since they may represent plausible prebiotic membrane precursors (Hargreave and Deamer, 1978b).

Formation of surfactant vesicles, rather than lamellar structures, depends on the nature of the surfactant and the length of the alkyl chains. At this point only generalizations can be made. Didodecyl dialkylammonium bromide, didodecyl and dioctadecyl sulfonates, decyl and didodecyl phosphates form vesicles, whereas their higher chain homologs give lamellar structures (Kunitake et al., 1977; Kunitake and Okahata, 1978). The larger the difference in chain lengths of the alkyl groups, the less tightly the surfactant vesicles pack (Nagamura et al., 1978).

Formation of surfactant vesicles from polymerizable surfactants is particularly noteworthy (Table 6.8; Day et al., 1979; Regen et al., 1980, 1981; Hub et al., 1980, 1981; Johnston et al., 1980; Bader et al., 1981; Akimoto et al., 1981; Tundo et al., 1982a, 1982b, 1982c, 1982d; O'Brien et al., 1981; Kunitake et al., 1981b; Lopez et al., 1982). Ultraviolet irradiation or addition of free radical initiators to vesicles, formed from vinyl or diacetylene surfactants, results in the formation of cross polymerization spherical bilayer surfactant vesicles having diameters between 200 and 700 Å. In contrast to other surfactant vesicles, or to liposomes, polymerized surfactant vesicles cannot be destroyed by the addition of up to 25% (v/v) alcohol (Regan et al., 1980; Hub et al., 1980; Tundo et al., 1982a, 1982b, 1982c, 1982d). Significantly, polymerized vesicles, like their unpolymerized counterparts, are able to organize molecules, retain their osmotic activities, and undergo thermotropic phase transitions (Tundo et al., 1982b). Polymerized surfactant vesicles open the door to creation of chemical dissymmetry (Figure 6.27; Tundo et al., 1982b) which, in turn, may lead to enhanced utility in photochemical energy transfer (see Chapters 12 and 13). Polymerized surfactant vesicles have been proposed to act as antitumor agents on a molecular and on a cellular level (Gros et al., 1981).

Surfactant vesicles, especially those containing two alkyl chains, behave quite analogously to liposomes. Once they are formed, they do not dissociate or disintegrate after weeks in aqueous solutions containing electrolytes in concentrations not exceeding 10^{-2} M. On standing they undergo, however, vesicle-vesicle fusion. Neutralization of charges on the exterior of vesicles, just as in the case of liposomes, facilitates fusion. Addition of alcohols or Triton X-100 destroys surfactant vesicles.

Unsonicated surfactant vesicles have onionlike multicompartment structures. Sonication of these results in the formation of fairly uniform single-walled vesicles. Increasing the sonication time, at a given power, results in an exponential decrease in the viscosity and turbidity of the solution, down to a point beyond which further sonication has no appreciable effect. This point can be considered to correspond to the appearance of single-compartment bilayer vesicles.

Weight-average molecular weights, $\overline{M}w$ values, of surfactant vesicles are appreciably greater than those for liposomes. Low-angle laser light scattering and photon correlation spectroscopy have established $\overline{M}w$ values for well-sonicated DODAC, and dihexadecylphosphate (DHP) surfactant vesicles to be $(13 \pm 5) \times 10^6$ and $(24 \pm 8) \times 10^6$ daltons, respectively (Herrmann and Fendler, 1979). In contrast, the $\overline{M}w$ value of single-compartment phosphatidylchlone liposomes is $(2 \pm 0.5) \times 10^6$ daltons (Huang and Mason, 1978). Hydrodynamic calculations favor prolate ellipsoidal or rodlike shapes for single-compartment DODAC and DHP surfactant vesicles (Herrmann and Fendler, 1979).

Surfactant vesicles, just like liposomes, undergo thermotropic phase transitions. Using a large variety of experimental techniques, pretransition and main transition temperatures for DODAC vesicles have been determined to be 30.0° and 36.2°C (Kano et al., 1979).

Table 6.8. Formation of Surfactant Vesicles

Surfactant	Method of Vesicle Formation	Characterization	Reference
Oleic acid	Thin oleic acid films, formed by evaporation of $CHCl_3$ solutions, are shaken with H_2O	Electron micrography	Gebicki and Hicks, 1973
Oleic and linoleic acid	Thin oleic acid films, formed by evaporation of $CHCl_3$ solutions, are shaken with H_2O and sonicated	Electron micrography, osmotic response, solute retention	Gebicki and Hicks, 1976
Dialkylammonium salts of the 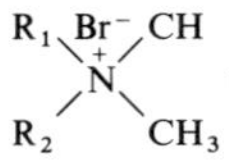 $R_1R_2N^+(CH)(CH_3)\ Br^-$ type $R_1 = R_2 = C_{14}H_{29}$ $R_1 = R_2 = C_{12}H_{25}$ $R_1 = C_{18}H_{37}, R_2 = C_{14}H_{29}$ $R_1 = C_{18}H_{37}, R_2 = C_{12}H_{25}$ $R_1 = C_{18}H_{37}, R_2 = C_{10}H_{21}$	Sonication	Electron micrography	Kunitake et al., 1977; Kunitake and Okahata, 1977; Deguchi and Mino, 1978
SDS dodecanol, $CH_3(CH_2)_{7-11}COOH$	Titrations, dilution from EtOH	Electron and phase contrast micrography, permeability determination	Hargreaves and Deamer, 1978a
Dihexadecylphosphate	Sonication	Electron micrography, gel filtration, substrate entrapment	Mortara et al., 1978
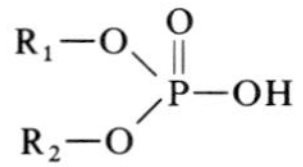 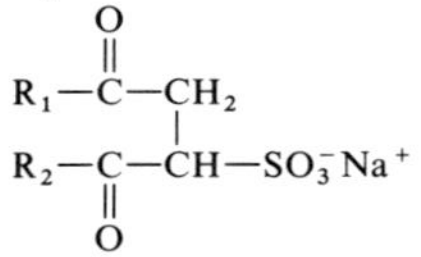 $(R_1{-}O)(R_2{-}O)P(=O){-}OH$ $R_1{-}C(=O){-}CH_2{-}CH(SO_3^-Na^+){-}C(=O){-}R_2$ $R_1 = R_2 = C_{10}H_{21}$ $R_1 = R_2 = C_{12}H_{25}$	Sonication	Electron micrography	Kunitake and Okahata, 1978

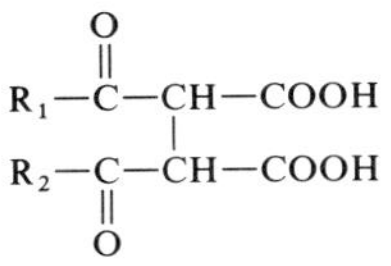 $R_1-C(=O)-CH(COOH)-CH(COOH)-C(=O)-R_2$; $R_1 = R_2 = C_{13}H_{27}$			
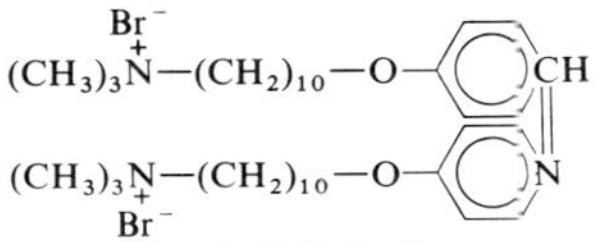 $(CH_3)_3\overset{+}{N}-(CH_2)_{10}-O-$ / $(CH_3)_3\overset{+}{N}-(CH_2)_{10}-O-$ (azobenzene-type aromatic core, $CH{=}N$), Br^- + cholesterol	Sonication	Electron micrography	Okahata and Kunitake, 1979
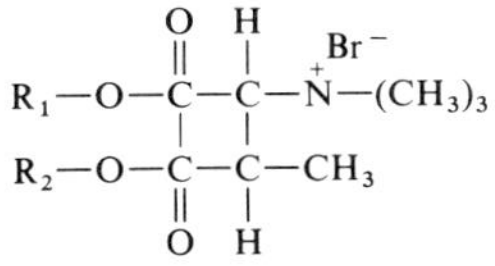 $R_1-O-C(=O)-CH-\overset{+}{N}(CH_3)_3\ Br^-$ / $R_2-O-C(=O)-CH-CH_3$; $R_1 = R_2 = C_{12}H_{25}$ (optically active)	Sonication	Electron micrography	Kunitake et al., 1979
Didodecyl phosphate	Sonication	Photochemistry	Czarniecki and Breslow, 1979
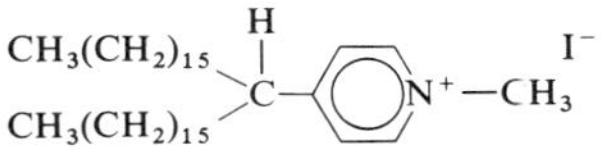 $(CH_3(CH_2)_{15})_2CH-C_5H_4N^+-CH_3\ I^-$	Sonication	Electron micrography	Sudhölter et al., 1980
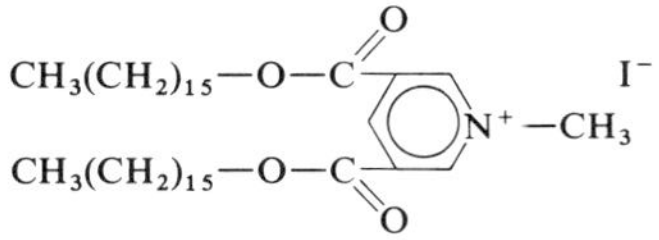 $(CH_3(CH_2)_{15}-O-C(=O))_2C_5H_3N^+-CH_3\ I^-$			
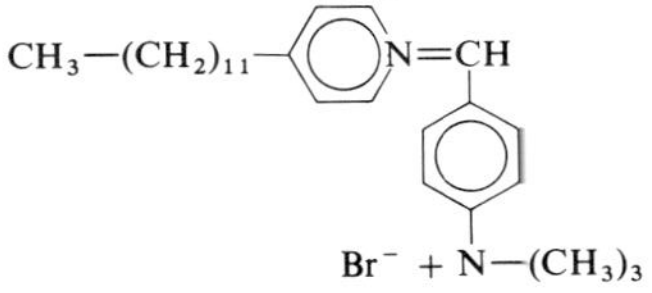 $CH_3-(CH_2)_{11}-C_5H_4N{=}CH-C_6H_4-\overset{+}{N}(CH_3)_3\ Br^-$	Sonication or alcohol injection	Electron micrography, light scattering, surface tension, differential scanning calorimetry	Kunitake and Okahata, 1980

Table 6.8 (*Continued*)

Surfactant	Method of Vesicle Formation	Characterization	Reference
$CH_3(CH_2)_n$–C₅H₄N=CH–C₆H₄–O–$(CH_2)_m$–$\overset{+}{N}$$(CH_3)_3$ Br^-; $n = 11, m = 3$; $n = 6, m = 9$			
$CH_3(CH_2)_{11}$–O–C₆H₄–C₆H₄–O–X			
$X = -(CH_2)_4-N^+(CH_3)_2-CH_3$ Br^- C_{12}—BPh—C_4—N^+			
$X = -(CH_2)_4-N^+(CH_3)_2-CH_2-(CH(OH))_4-CH_2OH$ C_{12}—BPh—C_4—N^+Glu			
$X = -(CH_2)_4-N^+(CH_3)_2-N^--COCH_3$ C_{12}—BPh—C_4—N^+N^-			
$X = -(CH_2CH_2-O)_n-H$ C_{12}—BPh—nG			
$X = -P(=O)(OH)-OH$ C_{12}—BPh—PO_4			
Br^- $HOOCCH_2CH_2\overset{+}{N}(CH_3)_2-(CH_2)_5CONHCH(CH_3)CON(C_{12}H_{25})_2$	Sonication	Electron micrography	Murakami et al., 1980

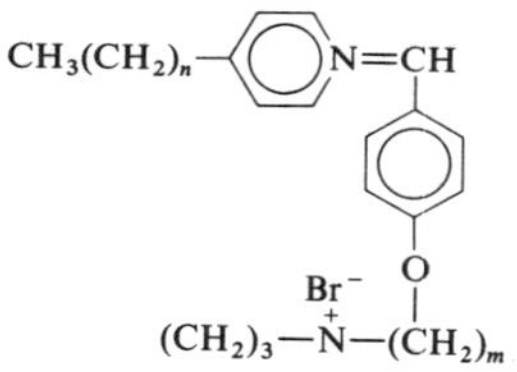

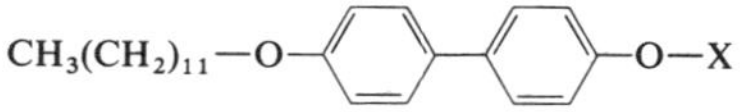

$H_3C-N(C_5H_4)(C_5H_4)N^{\oplus}-(CH_2)_2-NH-C(=O)-(CH_2)_{14}-C(=O)-NH-(CH_2)_2-{}^{\oplus}N(C_5H_4)(C_5H_4)N^{\oplus}-CH_3$	Sonication	Electron micrography	Baumgartner and Fuhrhop, 1980
$(CH_3)_2\overset{+}{N}[(CH_2)_5CH_3][(CH_2)_{11}-OC(=O)-C(CH_3)=CH_2]\ Br^-$	Sonication + free radical initiator to produce polymerized vesicles	Electron micrography	Regen et al., 1980
$[CH_3(CH_2)_{12}C\equiv C-C\equiv C(CH_2)_8COO(CH_2)_2]_2X$ $X = {}^-O(CH_2)_2O(CH_2)_2O^-$ $X = \overset{+}{N}(CH_3)_2,\ Br^-$ $X = \overset{+}{N}(CH_2)_2SO_3^-$ $X = NCH_3$ $CH_3(CH_2)_{12}C\equiv C-C\equiv C(CH_2)_8COOCH_2$ $CH_3(CH_2)_{12}C\equiv C-C\equiv(CH_2)_8COOCH$ $CH_2OP(=O)(O^-)O(CH_2)_2\overset{+}{N}(CH_3)_3$ $CH_3(CH_2)_{16}COOCH_2$ $CH_3(CH_2)_{12}C\equiv C-C\equiv C(CH_2)_8COOCH$ $CH_2OPO_3H_2$ $CH_3(CH_2)_{12}C\equiv C-C\equiv C-(CH_2)_8-R$ $R = COOH$ $R = CH_2OH$ $R = CH_2OPO_3H$	Sonication + ultraviolet irradiation to produce polymerized vesicles	Electron micrography	Hub et al., 1980

Table 6.8 (*Continued*)

Surfactant	Method of Vesicle Formation	Characterization	Reference
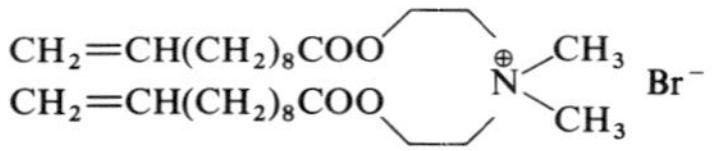 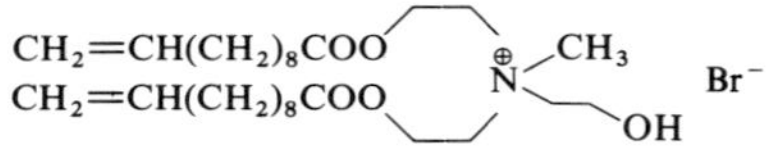 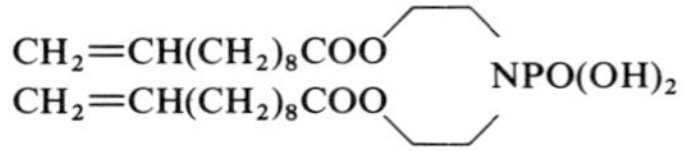 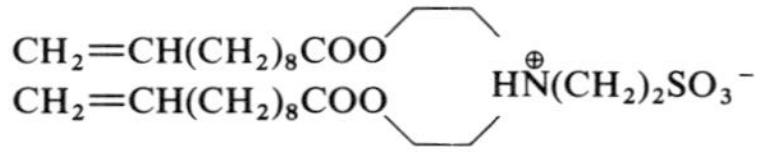 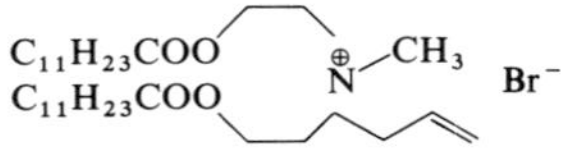 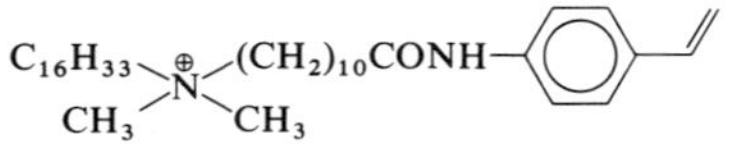 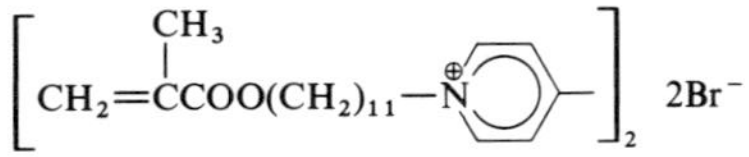 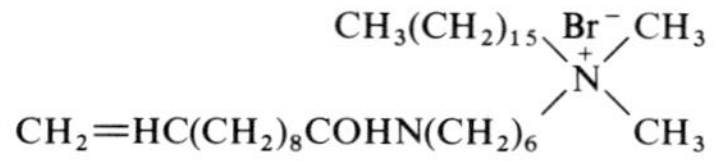 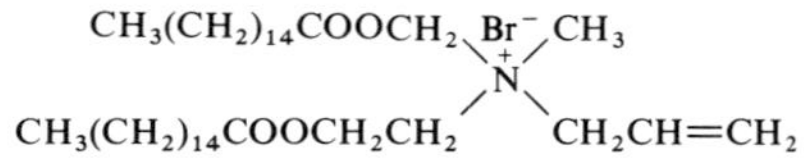 $[CH_2{=}CH(CH_2)_8COO(CH_2)_2]_2\overset{\oplus}{N}(CH_3)_2\ Br^-$	Sonication + uv irradiation on free radical initiator to produce polymerized vesicles	Electron micrography, gel filtration, turbidity measurements, substrate entrapment	Tundo et al., 1982a, 1982b, 1982c, 1982d
$[CH_2{=}CH(CH_2)_8COO(CH_2)_2]_2\overset{\oplus}{N}(CH_3)(CH_2CH_2OH)\ Br^-$			
$[CH_2{=}CH(CH_2)_8COO(CH_2)_2]_2NPO(OH)_2$			
$[CH_2{=}CH(CH_2)_8COO(CH_2)_2]_2H\overset{\oplus}{N}(CH_2)_2SO_3^-$			
$(C_{11}H_{23}COO(CH_2)_2)(C_{11}H_{23}COO(CH_2)_4CH{=}CH_2)\overset{\oplus}{N}CH_3\ Br^-$			
$C_{16}H_{33}\overset{\oplus}{N}(CH_3)_2(CH_2)_{10}CONH{-}C_6H_4{-}CH{=}CH_2$			
$[CH_2{=}C(CH_3)COO(CH_2)_{11}{-}\overset{\oplus}{N}C_5H_4{-}]_2\ 2Br^-$			
$CH_3(CH_2)_{15}(CH_2{=}HC(CH_2)_8COHN(CH_2)_6)\overset{+}{N}(CH_3)_2\ Br^-$			
$(CH_3(CH_2)_{14}COOCH_2)(CH_3(CH_2)_{14}COOCH_2CH_2)\overset{+}{N}(CH_3)(CH_2CH{=}CH_2)\ Br^-$			

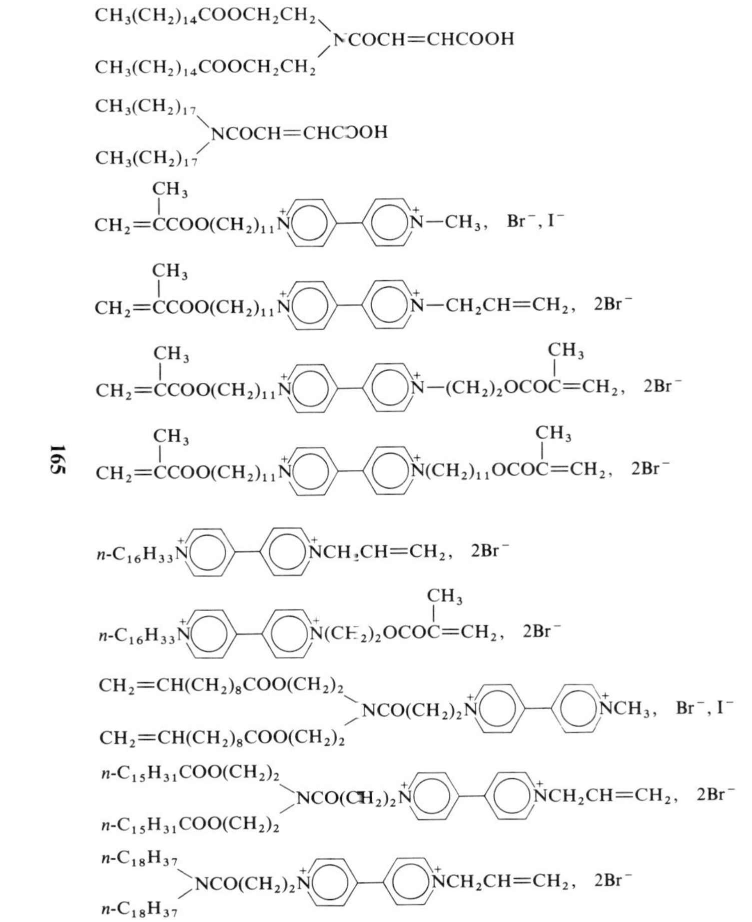
CH3(CH2)14COOCH2CH2
NCOCH=CHCOOH
CH3(CH2)14COOCH2CH2
CH3(CH2)17
NCOCH=CHCOOH
CH3(CH2)17
CH3
CH2=CCOO(CH2)11N
N—CH3, Br−, I−
CH3
CH2=CCOO(CH2)11N
N—CH2CH=CH2, 2Br−
CH3
CH2=CCOO(CH2)11N
CH3
N—(CH2)2OCOC=CH2, 2Br−
CH3
CH2=CCOO(CH2)11N
CH3
N(CH2)11OCOC=CH2, 2Br−
n-C16H33N
NCH2CH=CH2, 2Br−
CH3
n-C16H33N
N(CH2)2OCOC=CH2, 2Br−
CH2=CH(CH2)8COO(CH2)2
NCO(CH2)2N
NCH3, Br−, I−
CH2=CH(CH2)8COO(CH2)2
n-C15H31COO(CH2)2
NCO(CH2)2N
NCH2CH=CH2, 2Br−
n-C15H31COO(CH2)2
n-C18H37
NCO(CH2)2N
NCH2CH=CH2, 2Br−
n-C18H37

Table 6.8 *(Continued)*

Surfactant	Method of Vesicle Formation	Characterization	Reference
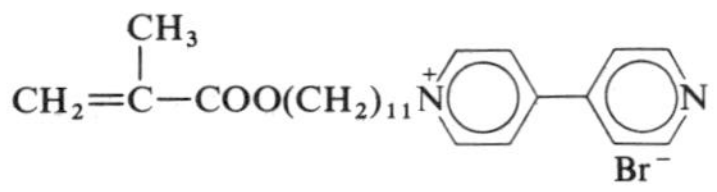 $CH_2{=}C(CH_3)-COO(CH_2)_{11}\overset{+}{N}C_5H_4-C_5H_4N$ Br^-			
 $[CH_3(CH_2)_{14}COO(CH_2)_2]_2NCOCH{=}CHCOO(CH_2)_2\overset{+}{N}C_5H_4-C_5H_4\overset{+}{N}CH_3$ Br^-, I^-			
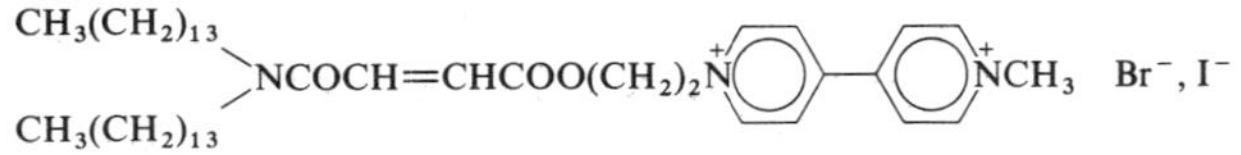 $[CH_3(CH_2)_{13}]_2NCOCH{=}CHCOO(CH_2)_2\overset{+}{N}C_5H_4-C_5H_4\overset{+}{N}CH_3$ Br^-, I^-			
$R = CH_2{=}C(CH_3)-CO$	Sonication + uv irradiation to produce polymerized vesicles	Electron micrography	Akimoto et al., 1981
$[R-NH-(CH_2)_{10}-CO-O-(CH_2)_2]_2\overset{\oplus}{N}(CH_3)_2$ $Br^{\ominus}$			
$HOOC-(CH_2)_8C{\equiv}C-C{\equiv}C-(CH_2)_8-COOH$			
$[H_3C-(CH_2)_{12}-CH{=}CH-CH{=}CH-CO-O]_2X$			
$X = -(CH_2)_2-N(CH_3)-(CH_2)_2-$			
$X = -(CH_2)_2-\overset{\oplus}{N}(CH_3)_2-(CH_2)_2-$ $Br^{\ominus}$			
$X = -CH_2-\overset{\vert}{C}H-CH_2OH$			
$X = -CH_2-\overset{\vert}{C}H-CH_2-O-P(=O)(O^{\ominus})-O-(CH_2)_2-\overset{\oplus}{N}(CH_3)_3$			

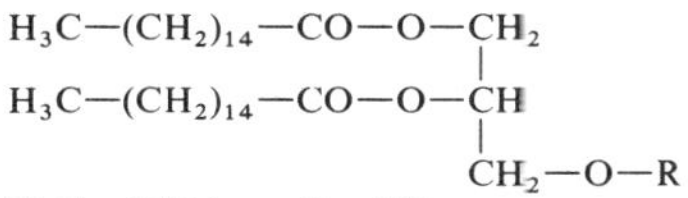 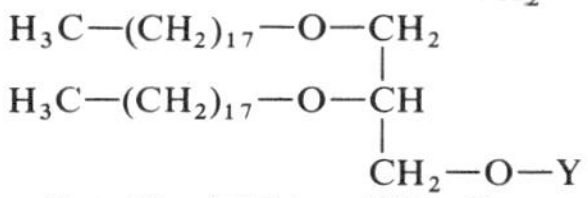 $H_3C-(CH_2)_{14}-CO-O-CH_2$ $H_3C-(CH_2)_{14}-CO-O-CH$ CH_2-O-R $H_3C-(CH_2)_{17}-O-CH_2$ $H_3C-(CH_2)_{17}-O-CH$ CH_2-O-Y $Y = Cl-(CH_2)_5-NH-R$ $Y = R$ 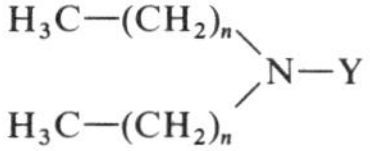 $H_3C-(CH_2)_n$ / $H_3C-(CH_2)_n$ $>N-Y$ $n = 11, Y = R$ $n = 17, Y = R$ $n = 11, Y = (CH_2)_3-NH-R$ $n = 17, Y = (CH_2)_3-NH-R$ 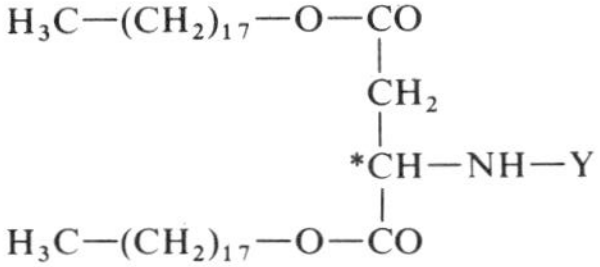 $H_3C-(CH_2)_{17}-O-CO$ CH_2 $^{*}CH-NH-Y$ $H_3C-(CH_2)_{17}-O-CO$ $Y = R$ $Y = CO-(CH_2)_5-NH-R$ Phosphatidylcholine diacetylene	Sonication + uv irradiation to produce polymerized vesicles	Electron micrography	O'Brien et al., 1981
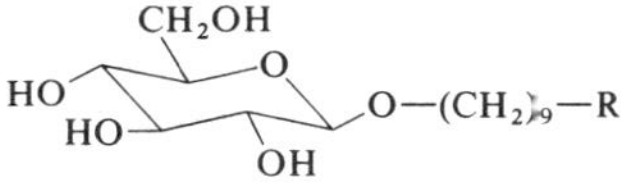 CH_2OH, HO, HO, OH (glucopyranoside ring), $O-(CH_2)_9-R$	Sonication + uv irradiation to produce polymerized vesicles		Bader et al., 1981
$R = C{\equiv}C-C{\equiv}C-(CH_2)_{12}-CH_3$ L-α-Lysophosphatidyl choline-1-palmitoyl L-α-glycerophosphoryl choline	Sonication + uv irradiation to produce polymerized vesicles	Electron micrography, substrate entrapment	Regen et al., 1981

Table 6.8 *(Continued)*

Surfactant	Method of Vesicle Formation	Characterization	Reference
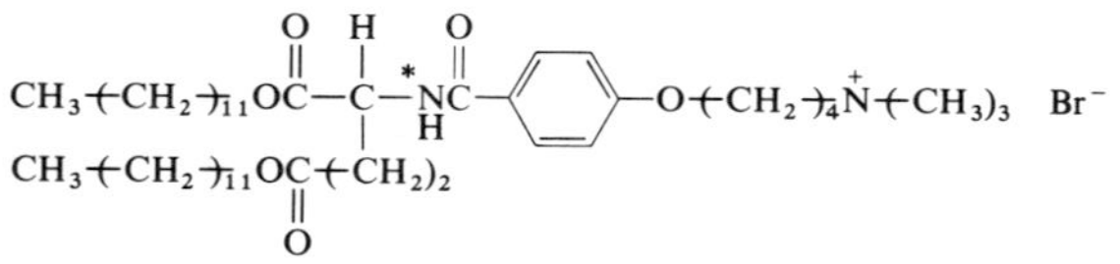 $CH_3(CH_2)_{11}OC(=O)-\overset{*}{C}H(-NHC(=O)-C_6H_4-O(CH_2)_4\overset{+}{N}(CH_3)_3\ Br^-)-(CH_2)_2C(=O)O(CH_2)_{11}CH_3$	Sonication	Circular dichroism, electron micrography	Kunitake et al., 1980a, 1980b
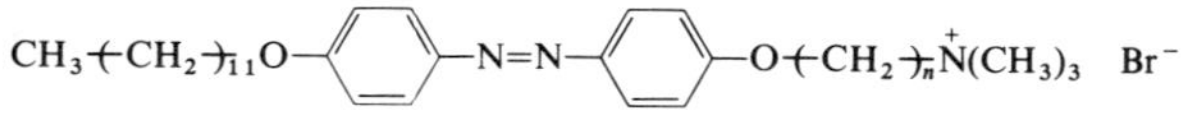 $CH_3(CH_2)_{11}O-C_6H_4-N{=}N-C_6H_4-O(CH_2)_n\overset{+}{N}(CH_3)_3\ Br^-$, $n = 2, 4, 10,$	Sonication	Electron micrography, photoisomerization	Kunitake et al., 1980a
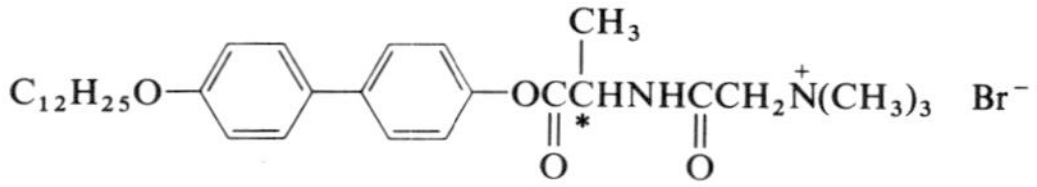 $C_{12}H_{25}O-C_6H_4-C_6H_4-OC(=O)\overset{*}{C}H(CH_3)NHC(=O)CH_2\overset{+}{N}(CH_3)_3\ Br^-$	Sonication	Electron micrography, circular dichroism	Kunitake et al., 1980b
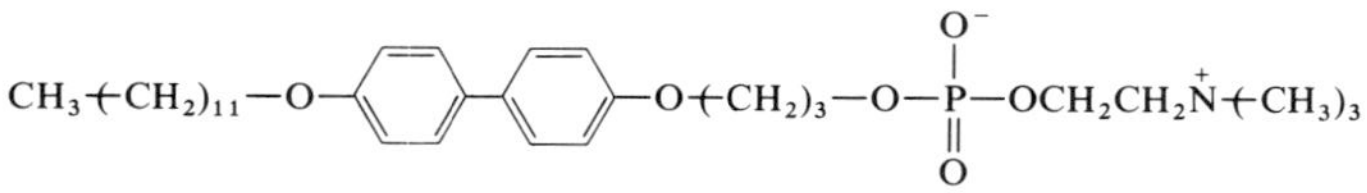 $CH_3(CH_2)_{11}-O-C_6H_4-C_6H_4-O(CH_2)_3-O-P(=O)(O^-)-OCH_2CH_2\overset{+}{N}(CH_3)_3$	Sonication	Electron micrography	Okahata et al., 1980
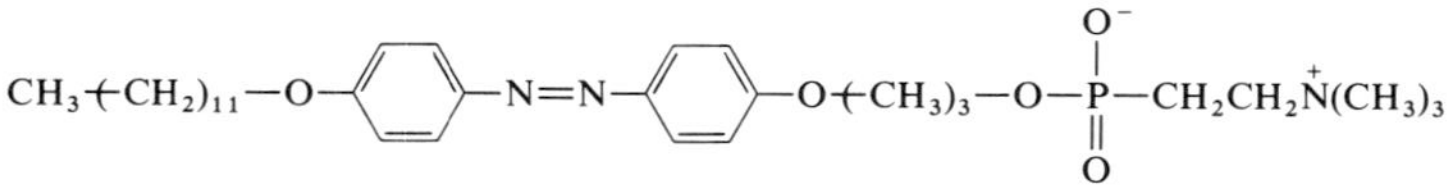 $CH_3(CH_2)_{11}-O-C_6H_4-N{=}N-C_6H_4-O(CH_3)_3-O-P(=O)(O^-)-CH_2CH_2\overset{+}{N}(CH_3)_3$	Sonication	Electron micrography, photoisomerization	Okahata et al., 1980
N,N,N-trimethyl-*N*-(3′-(6′-octyl-2′-naphthyloxy)propyl)ammonium bromide	Sonication	Electron microscopy, ^1H nmr and fluorescence spectroscopy	Nagamura et al., 1981

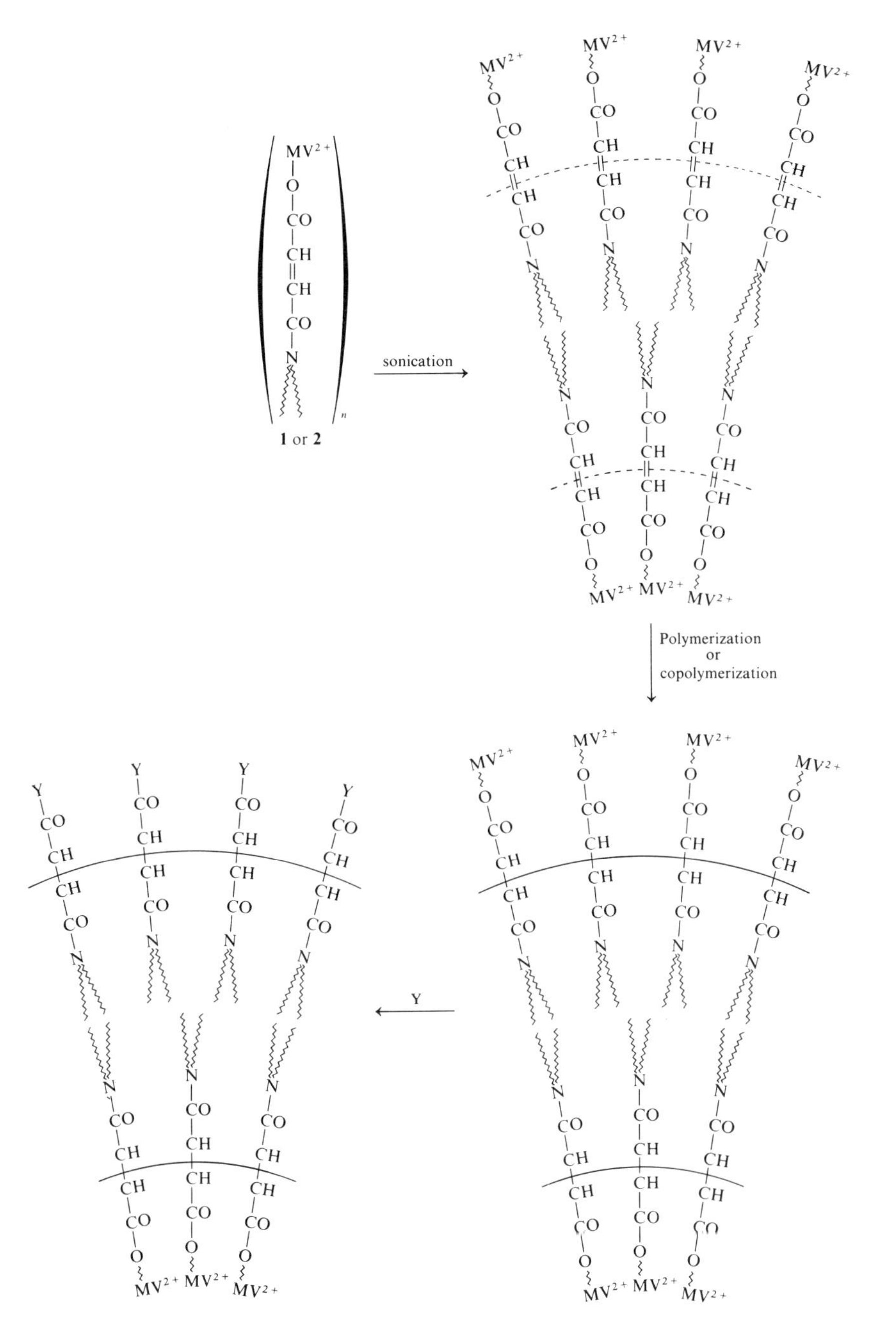

Figure 6.27. Scheme for preparing chemically dissymmetrical surfactant vesicles. **1** and **2** represent

$[CH_3(CH_2)_{14}COO(CH_2)_2]_2NCOCH{=}CHCO(CH_2)_2{-}\overset{+}{N}C_5H_4{-}C_5H_4\overset{+}{N}CH_3$, Br^-, I^- (**1**)

and $[CH_3(CH_2)_{17}]_2NCOCH{=}CHCOO(CH_2)_2\overset{+}{N}C_5H_4{-}C_5H_4\overset{+}{N}CH_3$, Br^-, I^- (**2**)

(taken from Tundo et al., 1981d).

Surfactant vesicles shrink in hyperosmolar and swell in hypoosmolar solutions (Kano et al., 1979). This osmotic activity is limited, however, to a narrower range of electrolytes than that used for liposomes. Electrolytes in excess of 0.10 *M* tend to precipitate surfactant vesicles. Initial shrinkage rates of DODAC vesicles increase with decreasing size of the alkali cation, implying some permeation of the alkali cations (Kano et al., 1979). This is a direct consequence of enhanced fluidities caused by electrostatic repulsions of the positively charged headgroups in DODAC vesicles. The observed increase in the initial shrinkage rates of phospholipid liposomes with increasing concentrations of added octadecylamine (Kano et al., 1979) is in accord with this postulate. Enhanced fluidities of surfactant vesicles are also manifested in decreased microviscosities in the bilayer compared with those in liposomes. The microviscosity for DODAC vesicles determined by means of fluorescence polarization of 2-methylanthracene was found to be 144 cP (Tran et al., 1978). Corresponding microviscosities of liposomes and aqueous micelles are 291 (Tran et al., 1978) and 30 cP (Shinitzky et al., 1971). Micropolarity in the electrical double layer of surfactant vesicles containing methyl pyridinium iodide headgroups was found to be comparable with that of dichloromethane (Sudhölter et al., 1980).

Surfactant vesicles interact with and entrap molecules. As in liposomes, polar molecules are entrapped in the aqueous interiors, charged species electrostatically bind to surfaces, and hydrophobic substrates are intercalated among the alkyl chains of surfactant vesicles. Protons and hydroxide ions move relatively freely across the bilayers of DODAC surfactant vesicles. In cholesterol containing DODAC vesicles, however, proton permeabilities become measurably slow and appreciable pH gradients can be maintained for some time (Fendler, 1980a). Substrate entrapment and release are comparable with those in lliposomes (Tran et al., 1978; Romero et al., 1978). These parameters depend, of course, on the nature of the substrate and of the surfactant vesicle. Interestingly, glucose is released at faster rates from didodecyldimethylammonium bromide than from DODAC (McNeil and Thomas, 1980). Although most of the properties of surfactant vesicles are analogous to those of phospholipid liposomes, unlike liposomes they are highly charged and have appreciable surface potentials. The possibility of building functional surfactant vesicles, of desired chemical composition and morphologies, renders them extremely interesting and potentially useful in a variety of applications.

REFERENCES

Akimoto, A., Dorn, K., Gross, L., Ringsdorf, H., and Schupp, H. (1981). *Angew. Chem. Int. Ed. Eng.* **20**, 90–91. Polymer Model Membranes.

Albon, N. and Sturtevant, J. M. (1978). *Proc. Natl. Acad. Sci. USA* **75**, 2258–2260. Nature of the Gel to Liquid Crystal Transition of Synthetic Phosphatidylcholines.

Alger, J. R. and Prestegard, J. H. (1979). *Biophys. J.* **28**, 1–14. Nuclear Magnetic Resonance Study of Acetic Acid Permeation of Large Unilamellar Vesicle Membranes.

Almgren, M. (1980). *Chem. Phys. Lett.* **71**, 539–543. Migration of Pyrene Between Lipid Vesicles in Aqueous Solution. A Stopped-Flow Study.

Almgren, M., Grieser, F., and Thomas, J. K. (1979). *J. Am. Chem. Soc.* **101**, 279–291. Dynamic and Static Aspects of Solubilization of Neutral Arenes in Ionic Micellar Solutions.

Aune, K. C., Gallagher, J. G., Gotto, A. M., and Morrisett, J. D. (1977). *Biochemistry* **16**, 2151–2156. Physical Properties of Dimyristoylphosphatidylcholine Vesicle and of Complexes Formed by its Interaction with Apolipoprotein C-III.

Bader, H., Ringsdorf, H., and Skura, J. (1981). *Angew. Chem. Int. Ed. Eng.* **20**, 91–92. Liposomes from Polymerizable Glycolipids.

Bakeeva, L. E., Grinius, L. L., Jasaitis, A. A., Kuliene, V. V., Levitsky, D. O., Liberman, E. A., Severina, I. I., and Skulachev, V. P. (1970). *Biochim. Biophys. Acta* **216**, 13–21. Conversion of Biomembrane-Produced Energy into Electrical Form. II. Intact Mitochondria.

Bangham, A. D. (1968). *Prog. Biophys. Mol. Biol.* **18**, 29–95. Membrane Models with Phospholipids.

Bangham, A. D., Standish, M. M., and Watkins, J. C. (1965). *J. Mol. Biol.* **13**, 238–252. Diffusion of Univalent Ions Across the Lamellae of Swollen Phospholipids.

Bangham, A. D., Standish, M. M., Watkins, J. C., and Weissmann, G. (1967a). *Protoplasma* **63**, 183–187. The Diffusion of Ions from a Phospholipid Model Membrane.

Bangham, A. D., de Gier, J., and Greville, G. D. (1967b). *Chem. Phys. Lipids* **1**, 225–246. Osmotic Properties and Water Permeability of Phospholipid Crystals.

Bangham, A. D., Hill, M. W., and Miller, N. G. A. (1974). In *Methods in Membrane Biology* (E. D. Korn, Ed.), Plenum Press, New York, pp. 1–68. Preparation and Use of Liposomes as Models of Biological Membranes.

Bansil, R., Day, J., Meadows, M., Rice, D., and Oldfield, E. (1980). *Biochemistry* **19**, 1938–1943. Laser Raman Spectroscopic Study of Specifically Deuterated Phospholipid Bilayers.

Barenholz, Y., Gibbes, D., Litman, B. J., Goll, J., Thompson, T. E., and Carlson, F. D. (1977). *Biochemistry* **16**, 2806–2810. A Simple Method for the Preparation of Homogeneous Phospholipid Vesicles.

Barenholz, Y., Amselem, S., and Lichtenberg, D. (1979). *FEBS Lett.* **99**, 210–214. A New Method for Preparation of Phospholipid Vesicles (Liposomes)—French Press.

Barrow, D. A. and Lentz, B. R. (1980). *Biochim. Biophys. Acta* **597**, 92–99. Large Vesicle Contamination in Small, Unilamellar Vesicles.

Batzri, S. and Korn, E. D. (1973). *Biochim. Biophys. Acta* **298**, 1015–1019. Single Bilayer Liposomes Prepared Without Sonication.

Baumgartner, E. and Fuhrhop, J. H. (1980). *Angew. Chem. Int. Ed. Eng.* **19**, 550–551. Vesicles with a Monolayer, Redox-Active Membrane.

Benz, R. and Zimmermann, U. (1981). *Biochim. Biophys. Acta* **640**, 169–178. The Resealing Process of Lipid Bilayers After Reversible Electrical Breakdown.

Berlinger, L. J. (1976). *Spin Labelling, Theory and Applications*, Academic Press, New York.

Bittman, R. and Blau, L. (1972). *Biochemistry* **11**, 4831–4839. The Phospholipid-Cholesterol Interaction. Kinetics of Water Permeability in Liposomes.

Blok, M. C., Van Der Neut-Kok, E. C. M., Van Deenen, L. L. M., and De Gier, J. (1975). *Biochim. Biophys. Acta* **406**, 187–196. The Effect of Chain Length and Lipid Phase Transitions on the Selective Permeability Properties of Liposomes.

Blok, M. C., Van Deenen, L. L. M., and De Gier, J. (1976). *Biochim. Biophys. Acta* **433**, 1–12. Effect of the Gel to Liquid Crystalline Phase Transition on the Osmotic Behaviour of Phosphatidylcholine Liposomes.

Blume, A. and Eibl, H. (1979). *Biochim. Biophys. Acta* **558**, 13–21. The Influence of Charge on Bilayer Membranes. Calorimetric Investigations of Phosphatidic Acid Bilayers.

Brendzel, A. M., and Miller, I. F. (1980). *Biochim. Biophys. Acta* **596**, 129–136. Liposome Filtration. Dependence on Transition Temperature.

Brown, M. F. and Seelig, J. (1978). *Biochemistry* **17**, 381–384. Influence of Cholesterol on the Polar Region of Phosphatidylcholine and Phosphatidylethanolamine Bilayers.

Brown, G. H. and Wolken, J. J. (1979). *Liquid Crystals and Biological Structures*, Academic Press, New York.

Browning, J. L. and Nelson, D. L. (1979). *J. Membr. Biol.* **49**, 75–103. Fluorescent Probes for Asymmetric Lipid Bilayers: Synthesis and Properties in Phosphatidyl Choline Liposomes and Erythrocyte Membranes.

Brunner, J., Skrabal, P., and Hauser, H. (1976). *Biochim. Biophys. Acta* **455**, 322–331. Single Bilayer Vesicles Prepared Without Sonication Physico-Chemical Properties.

Büldt, G. and Wohlgemuth, R. (1981). *J. Membr. Biol.* **58**, 81–100. The Headgroup Conformation of Phospholipids in Membranes.

Cafiso, D. S. and Hubbell, W. L. (1978). *Biochemistry* **17**, 187–195. Estimation of Transmembrane Potentials from Phase Equilibria of Paramagnetic Ions.

Cafiso, D. S. and Hubbell, W. L. (1980). *Biophys. J.* **30**, 243–263. Light Induced Interaction Potentials in Photoreceptor Membranes.

Chapman, D. (1980). In *Membrane Structure and Function* (E. E. Bittar, Ed.), John Wiley, New York, pp. 103–152. Studies Using Model Biomembrane Systems.

Chapman, D., Williams, R. M., and Ladbrooke, B. D. (1967). *Chem. Phys. Lipids* **1**, 445–475. Physical Studies of Phospholipids. VI. Thermotropic and Lyotropic Mesomorphism of Some 1,2-Diacylphosphatidylcholines (Lecithins).

Chen, R. F. and Edelhoch, H. (1975). *Biochemical Fluorescence: Concepts*, Marcel Dekker, New York.

Chen, L. A., Dale, R. E., Roth, S., and Brand, L. (1977). *J. Biol. Chem.* **252**, 2163–2169. Nanosecond Time-Dependent Fluorescence Depolarization of Diphenylhexatriene in Dimyristoyllecithin Vesicles and the Determination of "Microviscosity".

Chen, S. C., Sturtevant, J. M., and Gaffnen, B. J. (1980). *Proc. Natl. Acad. Sci. USA* **77**, 5060–5063. Scanning Calorimetric Evidence for a Third Phase Transition in Phosphatidylcholine Bilayers.

Cherry, R. J. (1979). *Biochim. Biophys. Acta* **559**, 289–327. Rotational and Lateral Diffusion of Membrane Proteins.

Cornell, B. A., Middlehurst, J., and Separovic, F. (1980). *Chem. Phys. Lett.* **73**, 569–571. Small Phospholipid Vesicles Cannot Undergo a Fluid-to-Crystalline Phase Transition.

Coster, H. G. L. and Zimmermann, V. (1975). *J. Membr. Biol.* **22**, 73–90. The Mechanism of Electrical Breakdown in the Membranes of Valonia utricularis.

Craig, I. F., Boyd, G. S., and Suckling, K. E. (1978). *Biochim. Biophys. Acta* **508**, 418–421. Optimum Interaction of Sterol Side Chains with Phosphatidylcholine.

Cullis, P. R. and De Kruijff, B. (1978). *Biochim. Biophys. Acta* **507**, 207–218. Polymorphic Phase Behaviour of Lipid Mixtures as Detected by ^{31}P NMR. Evidence that Cholesterol May Destabilize Bilayer Structure in Membrane Systems Containing Phosphatidylethanolamine.

Czarniecki, M. F. and Breslow, R. (1979). *J. Am. Chem. Soc.* **101**, 3675–3676. Photochemical Probes for Membrane Structures.

Darszon, A., Vandenberg, C. A., Schönfeld, M., Ellisman, M. H., Spitzer, N. C., and Montal, M. (1980). *Proc. Natl. Acad. Sci. USA* **77**, 239–243. Reassembly of Protein-Lipid Complexes into Large Bilayer Vesicles: Perspectives for Membrane Reconstitution.

Day, D., Hub, H. H., Ringsdorf, H. (1979). *Israel J. Chem.* **18**, 325–329. Polymerization of Mono and Bifunctional Diacetylene Derivatives in Monolayer at the Gas-Water Interface.

Deamer, D. W. (1978). In "Liposomes and Their Uses in Biology and Medicine" (D. Papahadjopoulos, Ed.), *Ann. New York Acad. Sci.* **308**, 250–257. Preparation and Properties of Ether Injection Liposomes.

Deamer, D. and Bangham, A. D. (1976). *Biochim. Biophys. Acta* **443**, 629–634. Large Volume Liposomes by an Ether Vaporization Method.

Deguchi, T. and Mino, J. (1978). *J. Colloid Interface Sci.* **65**, 155–161. Solution Properties of Long-Chain Dialkyldimethylammonium Salts. I. Formation of Vesicles by Dioctadecyldimethylammonium Chloride.

De Kruijff, B. (1978). *Biochim. Biophys. Acta* **506**, 173–182. ^{13}C Nmr Studies on [4-^{13}C]Cholesterol Incorporated in Sonicated Phosphatidylcholine Vesicles.

De Kruijff, B. and Baken, B. P. (1978). *Biochim. Biophys. Acta* **507**, 38–47. Rapid Transbilayer Movement of Phospholipids Induced by an Asymmetrical Perturbation of the Bilayer.

De Kruijff, B. and Wirtz, K. W. A. (1977). *Biochim. Biophys. Acta* **468**, 318–326. Induction of a Relatively Fast Transbilayer Movement of Phosphatidylcholine in Vesicles.

De Kruijff, B. and Zoelen, E. J. J. (1978). *Biochim. Biophys. Acta* **511**, 105–115. Effect of the Phase Transition on the Transbilayer Movement of Dimyristoyl Phosphatidylcholine in Unilamellar Vesicles.

De Kruyff, B., Van den Besselaar, A. M. H. P., and Van Deenen, L. L. M. (1977). *Biochim. Biophys. Acta* **465**, 443–453. Outside-Inside Distribution and Translocation of Lysophosphatidylcholine in Phosphatidylcholine Vesicles as Determined by ^{13}C nmr Using [N-$^{13}CH_3$] Enriched Lipids.

Demel, R. A. and De Kruijff, B. (1976). *Biochim. Biophys. Acta* **457**, 109–132. The Function of Sterols in Membranes.

Dody, M. C., Pownall, J. H., Kao, Y. J., and Smith, L. C. (1980). *Biochemistry* **19**, 108–116. Mechanism and Kinetics of Transfer of a Fluorescent Fatty Acid between Single-Walled Phosphatidylcholine Vesicles.

Edinin, M. (1974). *Annu. Rev. Biophys. Bioeng.* **3**, 179–201. Rotational and Transitional Diffusion in Membranes.

Eibl, H. and Blume, A. (1979). *Biochim. Biophys. Acta* **553**, 476–488. The Influence of Charge on Phosphatidic Acid Bilayer Membranes.

Feigenson, G. W. and Meers, P. R. (1980). *Nature* **283**, 313–314. ^{1}H nmr Study of Valinomycin Conformation in a Phospholipid Bilayer.

Fendler, J. H. (1980a). *Acc. Chem. Res.* **13**, 7–13. Surfactant Vesicles as Membrane Mimetic Agents: Characterization and Utilization.

Fendler, J. H. (1980b). In *Liposomes in Biological Systems* (G. Gregoriadis and A. C. Allison, Eds.), John Wiley, New York, pp. 87–100. Optimizing Drug Entrapment in Liposomes. Chemical and Biophysical Considerations.

Freire, E. and Snyder, B. (1980). *Biochemistry* **19**, 88–94. Estimation of the Lateral Distribution of Molecules in Two-Component Lipid Bilayers.

Gaffney, B. J. and Chen, S.-H. (1970). *Methods Membr. Biol.* **8**, 291–358. Spin-Label Studies in Membranes.

Galla, H.-J., Thielen, U., and Hartmann, W. (1979). *J. Membr. Biol.* **48**, 215–236. On Two-Dimensional Passive Random Walk in Lipid Bilayers in Biomembranes.

Gebhardt, C., Gruler, H., and Sackmann, E. (1977). *Z. Naturforsch.* **32c**, 581–596. On Domain Structure and Local Curvature in Lipid Bilayers and Biological Membranes.

Gebicki, J. M. and Hicks, M. (1973). *Nature* **243**, 232–234. Ufasomes Are Stable Particles Surrounded by Unsaturated Fatty Acid Membranes.

Gebicki, J. M. and Hicks, M. (1976). *Chem. Phys. Lipids* **16**, 142–160. Preparation and Properties of Vesicles Enclosed by Fatty Acid Membranes.

Georgescauld, D., Desmasez, J. P., Lapouyade, R., Babeau, A., Richard, H., and Winnik, M. (1980). *Photochem. Photobiol.* **31**, 539–545. Intramolecular Excimer Fluorescence: A New Probe of Phase Transitions in Synthetic Phospholipid Membranes.

Ginsberg, L. (1978). *Nature* **275**, 758–760. Does Ca^{2+} Cause Fusion or Lysis of Unilamellar Lipid Vesicles?

Gregoriadis, G. and Allison, A. C. (1980). *Liposomes in Biological Systems*, John Wiley, New York.

Griffin, R. G., Powers, L., and Pershan, P. S. (1978). *Biochemistry* **17**, 2718–2722. Head-Group Conformation in Phospholipids: A Phosphorus-31 Nuclear Magnetic Resonance Study of Oriented Monodomain Dipalmitoyl Phosphatidylcholine Bilayers.

Gros, L., Ringsdorf, H., Schupp, H. (1981). *Angew. Chem. Int. Ed. Eng.* **20**, 305–325. Polymeric Antitumor Agents on a Molecular and on a Cellular Level.

Gruenewald, B., Stankowski, S., and Blume, A. (1979). *FEBS Lett.* **102**, 227–229. Curvature Influence on the Cooperativity and the Phase Transition Enthalpy of Lecithin Vesicles.

Gruenewald, B., Blume, A., and Watanabe, F. (1980). *Biochim. Biophys. Acta* **597**, 41–52. Kinetic Investigations on the Phase Transition of Phospholipid Bilayers.

Hamilton, R. L., Jr., Goerke, J., Guo, L. S. S., William, M. C., and Havel, R. J. (1980). *J. Lipid Res.* **21**, 981–992. Unilamellar Liposomes Made with the French Pressure Cell: A Simple Preparative and Semiquantitative Technique.

Hare, F. and Lussan, C. (1978). *FEBS Lett.* **94**, 231–235. Mean Viscosities in Microscopic Systems and Membrane Bilayers.

Hare, F., Amiell, J., and Lussan, C. (1979). *Biochim. Biophys. Acta* **555**, 388–408. Is an Average Viscosity Tenable in Lipid Bilayers and Membranes? A Comparison of Semi-Empirical Equivalent Viscosities Given by Unbounded Probes: A Nitroxide and a Fluorophore.

Hargreaves, W. R. and Deamer, D. W. (1978a). *Biochemistry* **17**, 3759–3768. Liposomes from Ionic, Single-Chain Amphiphiles.

Hargreaves, W. R. and Deamer, D. W. (1978b). In *Light Transducing Membranes Structure, Function and Evolution* (D. W. Deamer, Ed.), Academic Press, New York, pp. 23–59. Origin and Early Evolution of Bilayer Membranes.

Hauser, H. (1975a). In *Water Relations of Foods* (R. B. Duckworth, Ed.), Academic Press, London, pp. 37–71. Water/Phospholipid Interactions.

Hauser, H. (1975b). In *Water, A Comprehensive Treatise* (F. Franks, Ed.), Vol. 4, Plenum Press, New York, pp. 209–303. Lipids.

Hauser, H., Guyer, W., Pascher, I., Skrabal, P., and Sundell, S. (1980a). *Biochemistry* **19**, 366–373. Polar Group Conformation of Phosphatidylcholine. Effect of Solvent and Aggregation.

Hauser, H., Pascher, I., and Sundell, S. (1980b). *J. Mol. Biol.* **137**, 249–264. Conformation of Phospholipids Crystal Structure of a Lysophosphatidylcholine Analogue.

Hauser, H., Guyer, W., and Spiess, M. (1980c). *J. Mol. Biol.* **137**, 265–282. The Polar Group Conformation of a Lysophosphatidylcholine Analogue in Solution. A High-Resolution Nuclear Magnetic Resonance Study.

Haynes, D. H. and Simkowitz, P. (1977). *J. Membr. Biol.* **33**, 63–108. 1-Aniliono-8-naphthalene-sulfonate: A Fluorescent Probe of Ion and Ionophore Transport Kinetics and Trans-membrane Asymmetry.

Haynes, D. H. and Westine, L. (1980). *J. Colloid Interface Sci.* **74**, 291–294. Why Divalent-Cation-Induced Aggregation of Phosphatidylserine and Phosphatidic Acid Vesicles Occurs at Less Than Diffusion-Controlled Rates.

Herrmann, U. and Fendler, J. H. (1979). *Chem. Phys. Lett.* **64**, 270–274. Low Angle Laser Light Scattering and Photon Correlation Spectroscopy in Surfactant Vesicles.

Hicks, M. and Gebicki, J. M. (1977). *Chem. Phys. Lipids* **20**, 243–252. Microscopic Studies of Fatty Acid Vesicles.

Hinz, H.-J. and Sturtevant, J. M. (1972a). *J. Biol. Chem.* **247**, 3697–3700. Calorimetric Investigation of the Influence of Cholesterol on the Transition Properties of Bilayers Formed from Synthetic L-α-Lecithin in Aqueous Suspensions.

Hinz, H.-J. and Sturtevant, J. M. (1972b). *J. Biol. Chem.* **247**, 6071–6075. Calorimetric Studies of Dilute Aqueous Suspensions of Bilayers Formed from Synthetic L-α-Lecithins.

Huang, C. (1969). *Biochemistry* **8**, 344–352. Studies of Phosphatidylcholine Vesicles. Formation and Physical Characteristics.

Huang, C. and Mason, J. T. (1978). *Proc. Natl. Acad. Sci. USA* **75**, 308–310. Geometric Packing Constraints in Egg Phosphatidylcholine Vesicles.

Hub, H. H., Hupfer, B., Koch, H., and Ringsdorf, H. (1980). *Angew. Chem. Int. Ed. Eng.* **19**, 938–940. Polymerizable Phospholipid Analogous New Stable Biomembrane and Cell Models.

Hub, H. H., Hupfer, B., Koch, H., and Ringsdorf, H. (1981). *J. Macromol. Sci. Chem.* **A15**, 701–705. Polymerization of Lipid and Lysolipid-Like Diacetylenes in Monolayers and Liposomes.

Hubbell, W. L. and McConnell, H. M. (1971). *J. Am. Chem. Soc.* **93**, 334–326. Molecular Motion in Spin Labelled Phospholipids and Membranes.

Hunt, G. R. and Tipping, L. R. H. (1978). *Biochim. Biophys. Acta* **507**, 242–261. A ^{1}H NMR Study of the Effects on Metal Ions, Cholesterol and *n*-Alkanes on Phase Transitions in the Inner and Outer Monolayers of Phospholipid Vesicular Membranes.

Ingolia, T. D. and Koshland, D. E., Jr. (1978). *J. Biol. Chem.* **253**, 3821–3829. The Role of Calcium in Fusion of Artificial Vesicles.

Israelachvili, J. N., Mitchell, D. J., and Ninham, B. W. (1976). *J. Chem. Soc. Faraday Trans. II* **72**, 1525–1538. Theory of Self-Assembly of Hydrocarbon Amphiphiles into Micelles and Bilayers.

Israelachvili, J. N., Marcelja, S., and Horn, R. G. (1980). *Quart. Rev. Biophys.* **13**, 121–200. Physical Principles of Membrane Organization.

Jain, M. K. (1972). *The Bimolecular Lipid Membrane: A System*, Van Nostrand Reinhold Co., New York.

Jain, M. K. and Wagner, R. C. (1980). *Introduction to Biological Membranes*, Wiley-Interscience, New York.

Jain, M. K. and White, H. B. (1977). *Adv. Lipid Res.* **15**, 1–60. Long Range Order in Biomembranes.

Jain, M. K., Tovissaint, D. G., and Cordes, E. H. (1973). *J. Membr. Biol.* **14**, 1–16. Kinetics of Water Penetration into Unsonicated Liposomes. Effects of *n*-Alkanols and Cholesterol.

Janiak, M. J., Small, D. M., and Shipley, G. G. (1976). *Biochemistry* **15**, 4575–4580. Nature of the Thermal Pretransition of Synthetic Phospholipids: Dimyristoyl- and Dipalmitoyllecithin.

Johnston, D. S., Sanghera, S., Pons, M., and Chapman, D. (1980). *Biochim. Biophys. Acta* **602**, 57–69. Phospholipid Polymers—Synthesis and Spectral Characteristics.

Kano, K. and Fendler, J. H. (1978). *Biochim. Biophys. Acta* **509**, 289–299. Pyranine as a Sensitive pH Probe for Liposome Interiors and Surfaces. pH Gradients Across Phospholipid Vesicles.

Kano, K. and Fendler, J. H. (1979). *Chem. Phys. Lipids* **23**, 189–200. Dynamic Fluorescence Investigations of the Effect of Osmotic Shocks on the Microenvironments of Charged and Uncharged Dipalmitoyl-D,L-α-Phosphatidylcholine Liposomes.

Kano, K., Romero, A., Djermouni, B., Ache, H., and Fendler, J. H. (1979). *J. Am. Chem. Soc.* **101**, 4030–4037. Characterization of Surfactant Vesicles as Membrane Mimetic Agents. II. Temperature-Dependent Changes of the Turbidity, Viscosity, Fluorescence Polarization of 2-Methylanthracene, and Positron Annihilation in Sonicated Dioctadecyldimethylammonium Chloride.

Kano, K., Yamaguchi, T., and Matsuo, T. (1980). *J. Phys. Chem.* **84**, 72–76. A Study of Intervesicle Exchange of Fluorescent Probes in Phospholipid Bilayer Membranes.

Kimelberg, H. K. and Mayhew, E. G. (1978). *CRC Crit. Rev. Toxicol.* **6**, 25–79. Properties and Biological Effects of Liposomes and Their Uses in Pharmacology and Toxicology.

Kinsky, S. C. (1974). In *Methods in Enzymology* (S. Fleischer and L. Parker, Eds.), Vol. 32, Academic Press, New York, pp. 501–513. Preparation of Liposomes and a Spectrophotometric Assay for Release of Trapped Glucose Marker.

Kirkland, J. J., Yau, W. W., Szoka, F. C. (1982). *Science* **215**, 296–298. Sedimentation Field Fractionation of Liposomes.

Korenbrot, J. I. (1977). *Annu. Rev. Physiol.* **39**, 19–49. Ion Transport in Membranes: Incorporation of Biological Ion-Translocating Proteins in Model Membrane Systems.

Kornberg, R. D., McNamee, M. G., and McConnell, H. M. (1972). *Proc. Natl. Acad. Sci. USA* **69**, 1508–1513. Measurements of Transmembrane Potentials in Phospholipid Vesicles.

Kotyk, A. and Janacek, K. (1977). *Membrane Transport, An Interdisciplinary Approach; Biomembranes*, Vol. 9, Plenum Press, New York.

Kremer, J. M. H. and Wiersema, P. H. (1977). *Biochim. Biophys. Acta* **471**, 348–360. Exchange and Aggregation in Dispersions of Dimyristoyl Phosphatidylcholine Vesicles Containing Myristic Acid.

Kremer, J. M. H., Esker, M. W. J. v. d., Pathmamanoharan, C., and Wiersema, P. H. (1977a). *Biochemistry* **16**, 3932–3935. Vesicles of Variable Diameter Prepared by a Modified Injection Method.

Kremer, J. M. H., Kops-Werkhoven, M. M., Pathmamanoharan, C., Gijzeman, O. L. J., and Wiersema, P. H. (1977b). *Biochim. Biophys. Acta* **471**, 177–188. Phase Diagrams and the Kinetics of Phospholipid Exchange for Vesicles of Different Composition and Radius.

Kroon, P. A., Kainosho, M., and Chan, S. I. (1976). *Biochim. Biophys. Acta* **433**, 282–293. Proton Magnetic Resonance Studies of Lipid Bilayer Membranes.

Kunitake, T. and Okahata, Y. (1977). *J. Am. Chem. Soc.* **99**, 3860. A Totally Synthetic Bilayer Membrane.

Kunitake, T. and Okahata, Y. (1978). *Bull. Chem. Soc. Jpn.* **51**, 1877–1879. Synthetic Bilayer Membranes with Anionic Head Groups.

Kunitake, T. and Okahata, Y. (1980). *J. Am. Chem. Soc.* **102**, 549–553. Formation of Stable Bilayer Assemblies in Dilute Aqueous Solution from Ammonium Amphiphiles with the Diphenylazomethine Segment.

Kunitake, T., Okahata, Y., Tamaki, K., Kumamura, F., and Takayanagi, M. (1977). *Chem. Lett.*, 387–390. Formation of the Bilayer Membrane from a Series of Quarternary Ammonium Salts.

Kunitake, T., Nakashima, N., Hayashida, S., and Yonemori, K. (1979). *Chem. Lett.*, 1413–1416. Chiral, Synthetic Bilayer Membranes.

Kunitake, T., Nakashima, N., Shimomura, M., and Okahata, Y. (1980a). *J. Am. Chem. Soc.* **102**, 6642–6442. Unique Properties of Chromophore Containing Bilayer Aggregates: Enhanced Chirality and Photochemically Induced Morphological Change.

Kunitake, T., Nakashima, N., and Morimitsu, K. (1980b). *Chem. Lett.*, 1347–1350. Enhanced Circular Dichroism and Fluidity of Disk-Like Aggregates of a Chiral, Single-Chain Amphiphile.

Kunitake, T., Nakashima, N., Takarabe, K., Nagai, M., Tsuge, A., and Yanagi, H. (1981b). *J. Am. Chem. Soc.* **103**, 5945–5947. Vesicles of Polymeric Bilayer and Monolayer Membranes.

Kunitake, T., Okahata, Y., Shimomura, M., Yasunami, S., and Takarabe, K. (1981a) *J. Am. Chem. Soc.* **103**, 5401–5413. Formation of Stable Bilayer Assemblies in Water from Single-Chain Amphiphiles. Relationship Between the Amphiphile Structure and the Aggregate Morphology.

Ladbroke, B. D., Williams, R. M., and Chapman, D. (1968). *Biochim. Biophys. Acta* **150**, 333–340. Studies on Lecithin-Cholesterol-Water Interactions by Differential Scanning Calorimetry and X-Ray Diffraction.

Lagaly, G. (1976). *Angew. Chem. Int. Ed. Eng.* **15**, 575–586. Kink-Block and Gauche-Block Structures of Bimolecular Films.

Lakowicz, J. R. and Knutson, J. R. (1980). *Biochemistry* **19**, 905–911. Hindered Depolarizing Rotations of Perylene in Lipid Bilayers. Detection by Lifetime-Resolved Fluorescence Anisotropy Measurements.

Lakshminarayanaiah, N. (1974). In *Electrochemistry*, Vol. 4, *Specialist Periodical Reports* (E. H. Thirsk, Ed.), The Chemical Society, London, pp. 167–276. Membrane Phenomena.

Lakshminarayanaiah, N. (1975). In *Electrochemistry*, Vol. 5, *Specialist Periodical Reports* (E. H. Thirsk, Ed.), The Chemical Society, London, pp. 132–219. Membrane Phenomena.

Lakshminarayanaiah, N. (1979). *Transport Phenomena in Membranes*, Academic Press, New York.

Lansman, J. and Haynes, D. H. (1975). *Biochim. Biophys. Acta* **394**, 335–347. Kinetics of a Ca^{2+}-Triggered Membrane Aggregation Reaction of Phospholipid Membranes.

Lawaczeck, R. (1978). *J. Colloid Interface Sci.* **66**, 247–256. Intervesicular Lipid Transfer and Direct Fusion of Phospholipid Vesicles: A Comparison on a Kinetic Basis.

Lawaczek, R. (1979). *J. Membr. Biol.* **51**, 229–261. On the Permeability of Water Molecules Across Vesicular Lipid Bilayers.

Lawaczeck, R., Kainosho, M., and Chan, S. I. (1976). *Biochim. Biophys. Acta* **443**, 313–330. The Formation and Annealing of Structural Defects in Lipid Bilayer Vesicles.

Lee, A. G. (1977a). *Biochim. Biophys. Acta* **472**, 237–281. Lipid Phase Transitions and Phase Diagrams. I. Lipid Phase Transitions.

Lee, A. G. (1977b). *Biochim. Biophys. Acta* **472**, 285–344. Lipid Phase Transitions and Phase Diagrams. II. Mixtures Involving Lipids.

Lelkes, P. I. (1979). *Biochem. Biophys. Res. Commun.* **90**, 656–662. Potential Dependent Rigidity Changes in Lipid Membrane Vesicles.

Lentz, B. R., Barenholz, Y., and Thompson, T. E. (1976). *Biochemistry* **15**, 4521–4528. Fluorescence Depolarization Studies of Phase Transitions and Fluidity in Phospholipid Bilayers. 1. Single Component Phosphatidylcholine Liposomes.

Lentz, B. R., Barrow, D. A., and Hoechli, M. (1980). *Biochemistry* **19**, 1944–1954. Cholesterol-Phosphatidylcholine Interactions in Multilamellar Vesicles.

Liao, M.-J. and Prestegard, J. H. (1979). *Biochem. Biophys. Res. Commun.* **90**, 1274–1279. Asymmetry Requirement for Ca^{2+} Induced Fusion of Phosphatidylcholine-Phosphatidic Acid Vesicles.

Likhtenschtein, G. I. (1976). *Spin Labelling Methods in Molecular Biology*, John Wiley, New York.

Lopez, E., O'Brien, D. F., Whitesides, T. H. (1982). *J. Am. Chem. Soc.* **104**, 305–307. Structural Effect on the Photopolymerization of Bilayer Membranes.

McIver, D. J. L. (1979). *Physiol. Chem. Phys.*, **11**, 289–302. Control of Membrane Fusion by Interfacial Water: A Model for the Actions of Divalent Cations.

McNeil, R. and Thomas, J. K. (1980). *J. Colloid Interface Sci.* **73**, 522–528. On the Nature of Surfactant Vesicle and Micelle Systems.

Mabrey, S. and Sturtevant, J. M. (1976). *Proc. Natl. Acad. Sci. USA* **73**, 3862–3866. Investigation of Phase Transitions of Lipids and Lipid Mixtures by High Sensitivity Differential Scanning Calorimetry.

Mabrey, S. and Sturtevant, J. M. (1978). In *Methods in Membrane Biology* (E. D. Korn, Ed.), Vol. 9, Plenum Press, New York, pp. 237–274. High Sensitivity Differential Calorimetry in the Study of Biomembranes and Related Model Systems.

Mabrey, S., Mateo, P. L., and Sturtevant, J. M. (1978). *Biochemistry* **17**, 2464–2468. High-Sensitivity Scanning Calorimetric Study of Mixtures of Cholesterol with Dimyristoyl- and Dipalmitoyl-phosphatidylcholines.

Mann, S., Skarnulis, A. J., and Williams, R. J. P. (1979). *J. Chem. Soc. Chem. Commun.*, 1067–1068. Location of Biological Compartments by High Resolution N.M.R. Spectroscopy and Electron Microscopy Using Magnetite-Containing Vesicles.

Mantsch, H. H., Saito, H., and Smith, I. C. P. (1977). *Prog. Nucl. Magn. Reson. Spectrosc.* **11**, 211–272. Deuterium Magnetic Resonance, Applications in Chemistry, Physics and Biology.

Marčelja, S. (1974a). *J. Chem. Phys.* **60**, 3599–3604. Chain Ordering in Liquid Crystals. I. Even-Odd Effect.

Marčelja, S. (1974b). *Biochim. Biophys. Acta* **367**, 165–176. Chain Ordering in Liquid Crystals. II. Structure of Bilayer Membranes.

Marsh, D. (1980). *Biochemistry* **19**, 1632–1637. Molecular Motion in Phospholipid Bilayers in the Gel Phase: Long Axis Rotation.

Martin, F. J. and MacDonald, R. C. (1976). *Biochemistry* **15**, 321–327. Phospholipid Exchange Between Bilayer Membrane Vesicles.

Mason, W. and Miller, N. G. A. (1979). *Biochem. Biophys. Res. Commun.* **91**, 878–885. Fusion of Charged and Uncharged Liposomes by *N*-Alkyl Bromides.

Merajver, S. D., Yorke, E. D., and DeRocco, A. G. (1981). *Phys. Rev.* **A23**, 897–907. Random-Walk Model of the Phase Transition of Hydrocarbon Chains on a Lattice.

Milsmann, M. H. W., Schwendener, R. A., and Weder, H. G. (1978). *Biochim. Biophys. Acta* **512**, 147–155. The Preparation of Large Single Bilayer Liposomes by a Fast and Controlled Dialysis.

Mitaku, S., Ikegami, A., and Sakanishi, A. (1978). *Biophys. Chem.* **8**, 295–304. Ultrasonic Studies of Lipid Bilayer. Phase Transition in Synthetic Phosphatidylcholine Liposomes.

Mortara, R. A., Quina, F. H., and Chaimovich, H. (1978). *Biochem. Biophys. Res. Commun.* **81**, 1080–1086. Formation of Closed Vesicles from a Simple Phosphate Diester. Preparation and Somc Properties of Vesicles of Dihexadecyl Phosphate.

Murakami, Y., Nakano, A., and Fukuya, K. (1980). *J. Am. Chem. Soc.* **102**, 4253–4254. Stable Single Compartment Vesicles with Zwitterionic Amphiphile Involving an Amino Acid Residue.

Nagamura, T., Mihara, S., Okahata, Y., Kunitake, T., and Matsuo, T. (1978). *Ber. Bunsenges. Phys. Chem.* **82**, 1093–1098. NMR and Fluorescence Studies on Self-Assembled Behavior of Dialkyldimethylammonium Salts in Aqueous Solutions.

Nagamura, T., Takeyama, N., and Matsuo, T. (1981). Unpublished results.

Nagle, J. F. and Scott, H. L. (1978). *Phys. Today* **31**, 38–47. Biomembrane Phase Transitions.

Nakagawa, Y., Inoue, K., and Nojima, S. (1979). *Biochim. Biophys. Acta* **553**, 307–319. Transfer of Cholesterol Between Liposomal Membranes.

Nichols, J. W. and Deamer, D. W. (1980). *Proc. Natl. Acad. Sci. USA* **77**, 2038–2042. Net Proton-Hydroxyl Permeability of Large Unilamellar Liposomes Measured by an Acid–Base Titration Technique.

Nichols, J. W., Hill, M. W., Bangham, A. D., and Deamer, D. W. (1980). *Biochim. Biophys. Acta* **596**, 393–403. Measurement of Net Proton-Hydroxyl Permeability of Large Unilamellar Liposomes with the Fluorescence pH Probe 9-Aminoacridine.

O'Brien, D. F., Whitesides, T. H., and Klingbiel, R. T. (1981). *J. Polym. Sci. Polym. Lett.* **19**, 95–101. The Photopolymerization of Lipid-Diacetylenes in Bimolecular-Layer Membranes.

Ohki, S. (1976). In *Progress in Surface and Membrane Science* (D. A. Cadenhead and J. F. Danielli, Eds.), Vol. 10, Academic Press, New York, pp. 117–252. Membrane Potential of Phospholipid Bilayer and Biological Membranes.

Okahata, Y. and Kunitake, T. (1979). *J. Am. Chem. Soc.* **101**, 5231–5234. Formation of Stable Monolayer Membranes and Related Structures in Dilute Aqueous Solution from Two-Headed Ammonium Amphiphiles.

Okahata, Y. and Kunitake, T. (1980). *Ber. Bunsenges. Phys. Chem.* **84**, 550–556. Self-Assembling Behavior of Single-Chain Amphiphiles with the Biphenyl Moiety in Dilute Aqueous Solution.

Okahata, Y., Ihara, H., Shimomura, M., Tawaki, S., and Kunitake, T. (1980). *Chem. Lett.*, 1169–1172. Formation of Disk-Like Aggregates from Single Chain Phosphocholine Amphiphiles in Water.

Oldfield, E. and Chapman, D. (1972). *Fed. Eur. Biochem. Soc. Lett.* **23**, 285–297. Dynamics of Lipids in Membranes: Heterogenity and the Role of Cholesterol.

Olson, F., Hunt, C. A., Szoka, F. C., Vail, W. J., and Papahadjopoulos, D. (1979). *Biochim. Biophys. Acta* **557**, 9–23. Preparation of Liposomes of Defined Size Distribution by Extrusion Through Polycarbonate Membranes.

Owen, C. S. (1980). *J. Membr. Biol.* **54**, 13–20. A Membrane-Bound Fluorescent Probe to Detect Phospholipid Vesicle-Cell Fusion.

Pagano, R. E. and Weinstein, J. N. (1978). *Annu. Rev. Biophys. Bioeng.* **7**, 435–468. Interactions of Liposomes with Mammalian Cells.

Papahadjopoulos, D. (1978). *Ann. New York Acad. Sci.* **308**. Liposomes and Their Uses in Biology.

Papahadjopoulos, D. and Kimelberg, H. K. (1974). In *Progress in Surface Science* (S. G. Davison, Ed.), Pergamon Press, Oxford, pp. 141–180. Phospholipid Vesicles (Liposomes) as Models for Biological Membranes, Their Properties and Interactions with Cholesterol and Proteins.

Papahadjopoulos, D. and Watkins, J. C. (1967). *Biochim. Biophys. Acta* **135**, 639–652. Phospholipid Model Membranes. II. Permeability Properties of Hydrated Liquid Crystals.

Papahadjopoulos, D., Vail, W. J., Jacobson, K., and Poste, G. (1975). *Biochim. Biophys. Acta* **394**, 483–491. Cochleate Lipid Cylinders: Formation by Fusion of Unilamellar Lipid Vesicles.

Papahadjopoulos, D., Vail, W. J., Newton, C., Nir, S., Jacobson, K., Poste, G., and Lazo, R. (1977). *Biochim. Biophys. Acta* **465**, 579–598. Studies on Membrane Fusion, III. The Role of Calcium Induced Phase Changes.

Parsegian, V. A. (1966). *Trans. Faraday Soc.* **72**, 848–860. Theory of Liquid-Crystal Phase Transition in Lipid and Water Systems.

Parsegian, V. A. (1968). In *Membrane Models and the Formation of Biological Membranes* (L. Bolis and B. L. Pethica, Eds.), North Holland, Amsterdam, pp. 307–317. An Energetic Model of Ionic Lipids in the Liquid Crystal State.

Parsegian, V. A., Fuller, N., and Rand, R. P. (1979). *Proc. Natl. Acad. Sci. USA* **76**, 2750–2754. Measured Work of Deformation and Repulsion of Lecithin Bilayers.

Pechold, W. and Blasenbrey, S. (1967). *Kolloid-Z. Z. Polym.* **216/17**, 235–244. Kooperative Rotationsisomerie in Polymeren. I. Schweltztheorie und Kinkenkonzehtrationen.

Pechhold, W. and Blasenbrey, S. (1970). *Kolloid-Z. Z. Polym.* **241**, 955–976. Molekülbewegung in Polymeren. III Teil: Mikrostructur and mechanische Eigenschaften.

Petersen, N. O. and Chan, S. I. (1977). *Biochemistry* **16**, 2657–2667. More on the Motional State of Lipid Bilayer Membranes: Interpretation of the Order Parameters Obtained from Nuclear Magnetic Resonance Experiments.

Pink, D. A., Green, T. J., and Chapman, D. (1980). *Biochemistry* **19**, 349–356. Raman Scattering in Bilayers of Saturated Phosphatidylcholines. Experiment and Theory.

Pope, J. M. and Cornell, B. A. (1978). In *Progress in Surface and Membrane Science* (D. A. Cadenhead, and J. F. Danielli, Eds.), Vol. 12, Academic Press, New York, pp. 183–243. Nmr Studies of Model Biological Membrane Systems: Unsonicated Surfactant-Water Dispersions.

Poste, G. and Nicolson, G. L. (1978). *Membrane Fusion*, North Holland, Amsterdam.

Pressman, B. D. (1976). *Annu. Rev. Biochem.* **45**, 501–530. Biological Applications of Ionophores.

Radda, G. (1975). *Methods Membr. Biol.* **4**, 97–188. Fluorescent Probes in Membrane Studies.

Radda, G. and Vanderkooi, J. (1972). *Biochim. Biophys. Acta* **265**, 509–549. Can Fluorescent Probes Tell Us Anything About Membranes?

Razin, S. (1972). *Biochim. Biophys. Acta* **265**, 241–296. Reconstitution of Biological Membranes.

Reeves, J. P. and Dowben, R. M. (1979). *J. Cell Physiol.* **73**, 49–60. Formation and Properties of Thin-Walled Phospholipid Vesicles.

Reeves, J. P. and Dowben, R. M. (1970). *J. Membr. Biol.* **3**, 123–141. Water Permeability of Phospholipid Vesicles.

Regen, S. L., Czech, B., and Singh, A. (1980). *J. Am. Chem. Soc.* **102**, 6638–6640. Polymerized Vesicles.

Regen, S. L., Singh, A., Oehme, G., and Singh, M. (1981). *J. Am. Chem. Soc.* **103**, to be published. Polymerized Phosphatidyl Choline Vesicles. Stabilized and Controllable Time-Release Carriers.

Rendi, R. (1967). *Biochim. Biophys. Acta* **135**, 333–346. Water Extrusion in Isolated Subcellar Fractions. VI. Osmotic Properties of Phospholipid Suspensions.

Rhoden, V. and Goldin, S. M. (1979). *Biochemistry* **18**, 4173–4176. Formation of Unilamellar Lipid Vesicles of Controllable Dimensions by Detergent Dialysis.

Roberts, M. F. and Dennis, E. A. (1977). *J. Am. Chem. Soc.* **99**, 6142–6143. Proton Nuclear Magnetic Resonance Demonstration of Conformationally Nonequivalent Phospholipid Fatty Acid Chains in Mixed Micelles.

Roberts, M. F., Bothner-By, A. A., and Dennis, E. A. (1978). *Biochemistry* **17**, 935–942. Magnetic Nonequivalence Within the Fatty Acyl Chains of Phospholipids in Membrane Models: ^{1}H Nuclear Magnetic Resonance Studies of the α-Methylene Groups.

Romero, A., Tran, C. D., Klahn, P. L., and Fendler, J. H. (1978). *Life Sci.* **22**, 1447–1450. Drug Entrapment in Surfactant Vesicles.

Sackman, E. (1978). *Ber. Bunsenges. Phys. Chem.* **82**, 891–909. Dynamic Molecular Organization in Vesicles and Membranes.

Schieren, H., Rudolph, S., Finkelstein, M., Coleman, P., and Weissmann, G. (1978). *Biochim. Biophys. Acta* **542**, 137–153. Comparison of Large Unilamellar Vesicles Prepared by a Petroleum Ether Vaporization Method with Multilamellar Vesicles. ESR, Diffusion and Entrapment Analyses.

Schindler, H. and Seelig, J. (1975). *Biochemistry* **14**, 2283–2287. Deuterium Order Parameters in Relation to Thermodynamic Properties of a Phospholipid Bilayer. A Statistical Mechanical Interpretation.

Seelig, J. (1977). *Quart. Rev. Biophys.* **10**, 353–418. Deuterium Magnetic Resonance: Theory and Application to Lipid Membranes.

Seelig, A. and Seelig, J. (1974). *Biochemistry* **13**, 4839–4845. The Dynamic Structure of Fatty Acyl Chains in a Phospholipid Bilayer Measured by Deuterium Magnetic Resonance.

Seelig, J. and Seelig, A. (1980). *Quart. Rev. Biophys.* **13**, 19–61. Lipid Conformation in Model Membranes and Biological Membranes.

Seelig, J., Gally, H. V., and Wohlgemuth, R. (1977). *Biochim. Biophys. Acta* **467**, 109–119. Orientation and Flexibility of the Choline Headgroup in Phosphatidylcholine Bilayers.

Shamoo, A. E. (1975). *Ann. New York Acad. Sci.* **264**, Carriers and Channels in Biological Systems.

Shinitzky, M., Dianoux, A. C., Gitler, C., and Weber, G. (1971). *Biochemistry* **10**, 2106–2113. Microviscosity and Order in the Hydrocarbon Region of Micelles and Membranes Determined with Fluorescent Probes. I. Synthetic Micelles.

Sixl, F. and Galla, H.-J. (1980). *Biochem. Biophys. Res. Commun.* **94**, 319–323. Evidence for a Nucleation Step in Lipid-Protein Interaction Kinetics of the Incorporation of Polymixin into Phosphatidic Acid Bilayer Vesicles.

Skarjune, R. and Oldfield, E. (1979). *Biochemistry* **18**, 5903–5909. Physical Studies of Cell Surface and Cell Membrane Structure. Determination of Phospholipid Headgroup Organization by Deuterium and Phosphorus Nuclear Magnetic Resonance Spectroscopy.

Slack, J. R., Anderton, B. H., and Day, W. A. (1973). *Biochim. Biophys. Acta* **323**, 547–559. A New Method for Making Phospholipid Vesicles and the Partial Reconstitution of the (Na^{+}, K^{+})-Activated ATPase.

Small, D. M. (1968). *J. Am. Oil. Chem. Soc.* **45**, 108–119. A Classification of Biological Lipids Based Upon Their Interaction in Aqueous Systems.

Smith, I. C. P. (1979). *Can. J. Biochem.* **57**, 1–14. Organization and Dynamics of Membrane Lipids as Determined by Magnetic Resonance Spectroscopy.

Smith, I. C. C., Tulloch, A. P., Stockton, G. W., Schreier, S., Joyce, A., Butler, K. W., Boulanger, Y., Blackwell, B., and Bennett, L. G. (1978). In "Liposomes and Their Uses in Biology and Medicine", (D. Papahadjopoulos, Ed.), *Ann. New York Acad. Sci.* **308**, 8–26. Determination of Membrane Properties at the Molecular Level by Carbon-13 and Deuterium Magnetic Resonance.

Snyder, B. and Freire, E. (1980). *Proc. Natl. Acad. Sci. USA* **77**, 4055–4059. Compositional Domain Structure in Phosphatidylcholine-Cholesterol and Sphingomyelin-Cholesterol Bilayers.

Stoeckenius, W. (1959). *J. Biophys. Biochem. Cytol.* **5**, 491–505. An Electron Microscope Study of Myelin Figures.

Strauss, G. and Ingenito, E. P. (1980). *Cryobiology* **17**, 508–515. Stabilization of Liposome Bilayers to Freezing and Thawing: Effects of Cryoprotective Agents and Membrane Proteins.

Sturtevant, J. M. (1974). In *Quantum Statistical Mechanics in the Natural Sciences* (B. Kursunoglu, S. Mintz, and S. Widmayer, Eds.), Plenum Press, New York, pp. 63–85. Phase Transitions of Phospholipids.

Sudhölter, E. J. R., Engberts, J. B. F. N., and Hoekstra, D. (1980). *J. Am. Chem. Soc.* **102**, 2467–2469. Vesicle Formation by Two Novel Synthetic Amphiphiles Carrying Micropolarity Reporter Head Groups.

Sunamoto, J., Kondo, H., and Yoshimatsu, A. (1978). *Biochim. Biophys. Acta* **510**, 52–62. Liposomal Membranes. I. Chemical Damage of Liposomal Membranes with Functional Detergent.

Sunamoto, J., Iwamoto, K., and Kondo, H. (1980). *Biochem. Biophys. Res. Commun.* **94**, 1367–1373. Liposomal Membranes VII. Fusion and Aggregation of Egg Liposomes as Promoted by Polysaccharides.

Sundaralingham, M. (1972). *Ann. New York Acad. Sci.* **195**, 324–355. Discussion Paper: Molecular Structures and Conformations of the Phospholipids and Sphingomyelins.

Szoka, F., Jr., and Papahadjopoulos, D. (1978). *Proc. Natl. Acad. Sci. USA* **75**, 4194–4198. Procedure for Preparation of Liposomes with Large Internal Aqueous Space and High Capture by Reverse-Phase Evaporation.

Szoka, F., Jr. and Papahadjopoulos, D. (1980). *Annu. Rev. Biophys. Bioeng.* **9**, 467–508. Comparative Properties and Methods of Preparation of Lipid Vesicles (Liposomes).

Szoka, F., Olson, F., Heath, T., Vail, W., Mayhew, E., and Papahadjopoulos, D. (1980). *Biochim. Biophys. Acta* **601**, 559–571. Preparation of Unilamellar Liposomes of Intermediate Size (0.1–0.2 μm) by a Combination of Reversed Phase Evaporation and Extrusion Through Polycarbonate Membranes.

Tanford, C. (1979). *Proc. Natl. Acad. Sci. USA* **76**, 3318–3319. Hydrostatic Pressure in Small Phospholipid Vesicles.

Tanford, C. (1980). *The Hydrophobic Effect. Formation of Micelles and Biological Membranes*, 2nd ed., Wiley-Interscience, New York.

Tien, H. T. (1974). *Bilayer Lipid Membranes (BLM) Theory and Practice*, Marcel Dekker, New York.

Tien, H. T. (1976). *Photochem. Photobiol.* **24**, 97–116. Electronic Processes and Photoelectric Aspects of Bilayer Lipid Membranes.

Tien, H. T. (1979). In *Photosynthesis in Relation to Model Systems* (J. Barber, Ed.), Elsevier, Amsterdam, pp. 115–173. Photoeffects in Pigmented Bilayer Lipid Membranes.

Tran, C. D., Klahn, P. L., Romero, A., and Fendler, J. H. (1978). *J. Am. Chem. Soc.* **100**, 1622–1624. Characterization of Surfactant Vesicles as Potential Membrane Models Effect of Electrolytes Substrates, and Fluorescence Probes.

Träuble, H. and Eibl, H. J. (1972). *Proc. Natl. Acad. Sci. USA* **71**, 214–219. Electrostatic Effects on Lipid Phase Transitions: Membrane Structure and Ionoc Environment.

Tsong, T. Y. (1974). *Proc. Natl. Acad. Sci. USA* **71**, 2684–2688. Kinetics of the Crystalline-Liquid Crystalline Phase Transition of Dimyristoyl-L-α-lecithin Bilayers.

Tundo, P., Kippenberger, D. J., Klahn, P. L., and Fendler, J. H. (1982a). *J. Am. Chem. Soc.* **103**, 456–461. Functionally Polymerized Surfactant Vesicles. Syntheses and Characterization.

Tundo, P., Kurihara, K., Kippenberger, D. J., Politi, M., Fendler, J. H. (1982b). *Angew. Chem. Int. Ed. Eng.* **21**, 81–82. Chemically Dissymmetrical Polymerized Surfactant Vesicles: Synthesis and Possible Utilization in Artificial Photosynthesis.

Tundo, P., Prieto, N. E., Kippenberger, D. J., Klahn, P. L., Kurihara, K., Politi, M., and Fendler, J. H. (1982c). *J. Am. Chem. Soc.* **103**, to be published. Redox Active Functionally Polymerized Surfactant Vesicles. Syntheses and Characterization.

Tundo, P., Kurihara, K., Prieto, N. E., Kippenberger, D. J., Politi, M., and Fendler, J. H. (1982d). *J. Am. Chem. Soc.* **103**, to be published. Chemically Dissymmetrical Polymerized Surfactant Vesicles. Syntheses and Utilization in Artificial Photosynthesis.

Tyrrell, D. A., Heath, T. D., Colley, C. M., and Ryman, B. E. (1976). *Biochim. Biophys. Acta* **457**, 259–302. New Aspects of Liposomes.

van Deenen, L. L. M. (1981). *FEBS Lett.* **123**, 3–15. Topology and Dynamics of Phospholipids in Membranes.

Vanderwerf, P. and Ullman, E. F. (1980). *Biochim. Biophys. Acta* **596**, 302–314. Monitoring of Phospholipid Vesicle Fusion by Fluorescence Energy Transfer between Membrane-Bound Dye Labels.

Van Dijck, P. W. M., Ververgaert, P., Verkleij, A. J., Van Deenen, L. L. M., and de Grier, J. (1975). *Biochim. Biophys. Acta* **406**, 465–478. Influence of Ca^{2+} and Mg^{2+} on the Thermotropic Behavior and Permeability Properties of Liposomes Prepared from Dimyristoyl Phosphatidylglycerol and Mixtures of Dimyristoyl Phosphatidylglycerol and Dimyristoyl Phosphatidylcholine.

Veatch, W. and Stryer, L. (1977). *J. Mol. Biol.* **113**, 89–102. The Dimeric Nature of the Gramicidin A Transmembrane Channel: Conductance and Fluorescence Energy Transfer Studies of Hybrid Channels.

Verkleij, A. J., De Kruyff, B., Ververgaert, P. H. J., Tocanne, J. F., and Van Deenen, L. L. M. (1974). *Biochim. Biophys. Acta* **339**, 432–437. The Influence of pH, Ca^{2+} and Protein on the Thermotropic Behavior of Negatively Charged Phospholipid, Phosphatidylglycerol.

Waggoner, A. S. (1976). *J. Membr. Biol.* **27**, 317–334. Optical Probes of Membrane Potential.

Wagner, R. (1980). In *Introduction to Biological Membranes* (M. K. Jain and R. C. Wagner, Eds.) John Wiley, New York, pp. 6–24. Membrane Biogenesis.

Wallach, D. F. H., Verma, S. P., and Fookson, J. (1979). *Biochim. Biophys. Acta* **559**, 153–208. Application of Laser Raman and Infrared Spectroscopy to the Analysis of Membrane Structure.

Watts, A., Marsh, D., and Knowles, P. F. (1978a). *Biochemistry* **17**, 1792–1800. Characterization of Dimyristoylphosphatidylcholine Vesicles and Their Dimensional Changes Through the Phase Transition. Molecular Control of Membrane Morphology.

Watts, A., Harlos, K., Maschke, W., and Marsh, D. (1978b). *Biochim. Biophys. Acta* **510**, 63–74. Control of the Structure and Fluidity of Phosphatidylglycerol Bilayers by pH Titration.

Wehry, E. L. (1976). *Modern Fluorescence Spectroscopy*, Plenum Press, New York.

Weinstein, J. N., Yoshikami, S., Henkart, P., Blumenthal, R., and Hagins, W. A. (1977). *Science* **195**, 489–491. Liposome-Cell Interaction: Transfer and Intracellular Release of a Trapped Fluorescence Marker.

White, S. H. (1980). *Proc. Natl. Acad. Sci. USA* **77**, 4048–4050. Small Phospholipid Vesicles: Internal Pressure, Surface Tension and Surface Free Energy.

Wilkins, M. H. F., Blaurock, A. E., and Engelman, D. M. (1971). *Nature New Biol.* **230**, 71–76. Bilayer Structure in Membranes.

Wilschut, J. and Papahadjopoulos, D. (1979). *Nature* **281**, 690–692. Ca^{2+}-Induced Fusion of Phospholipid Vesicles Monitored by Mixing of Aqueous Contents.

Zachariasse, K. A., Kühnle, W., and Weller, A. (1980). *Chem. Phys. Lett.* **73**, 6–11. Intramolecular Excimer Fluorescence as a Probe of Fluidity Changes and Phase Transitions in Phosphatidylcholine Bilayers.

Zaslavsky, B. Y., Borovskaya, A. A., Lavrinenko, A. K., Lisichkin, A. Y., Davidovich, Y. A., and Rogozhin, S. V. (1980). *Chem. Phys. Lett.* **26**, 49–55. Action of Surfactants on Egg Lecithin Liposomes.

Zimmermann, V., Pilwat, G., Beckers, F., and Riemann, F. (1976). *Bioelectrochem. Bioenerg.* **3**, 58–83. Effects of External Electric Fields on Cell Membranes.

Zingsheim, H. P. and Plattner, H. (1976). *Methods Membr. Biol.* **7**, 1–114. Electron Microscopic Methods in Membrane Biology.

Zumbuehl, O. and Weder, H. G. (1981). *Biochim. Biophys. Acta* **640**, 252–262. Liposomes of Controllable Size in the Range of 40 to 180 nm by Defined Dialysis/Detergent Mixed Micelles.

CHAPTER

7

HOST-GUEST SYSTEMS

Host–guest systems mimic the organizational ability of membranes and enzymes by bringing reactants together in highly structured specific microenvironments. This is accomplished by the noncovalent binding of substrates (the guests) into the cavities or sites provided by the hosts. Binding sites of the guests necessarily converge while those of the hosts diverge (Cram and Cram, 1978). The chemical challenge is to construct the simplest possible host that has suitable solubility, is easy to synthesize and stable, but at the same time binds appropriate guests selectively and reversibly. Linear (Vögtle and Weber, 1979) and macrocyclic polyethers (Cram and Cram, 1978; Lehn, 1978; Izatt and Christensen, 1978; Stoddart, 1979; Melson, 1979), and cyclodextrins (Griffiths and Bender, 1973; Bender and Komiyama, 1978) best fulfill these requirements. Each of these areas has been extensively investigated and adequately reviewed (*vide supra*). The purpose of the present chapter is to provide sufficient background for utilizing host-guest systems in membrane mimetic chemistry.

1 LINEAR AND MACROCYCLIC POLYETHERS

Crown ethers, coronands, cryptands, and podands are the terms used to describe polyether hosts containing repeating units of $(X—CH_2CH_2)_n$. According to the proposed nomenclature, coronands are taken to mean monocyclic compounds with any heteroatoms; cryptands are bi- and polycyclic compounds with any heteroatoms; podands are acyclic coronands and cryptands with any heteroatom, and the term crown ether is reserved for coronands having only oxygen as heteroatom (Vögtle and Weber, 1979). Topologically, coronands provide two-dimensional circular cavities while cryptands are capable of three-dimensional complexation. Structures **7.1**–**7.28** show typical linear and macrocyclic polyether hosts.

Complexation of cationic guests is the most important property of linear and macrocyclic polyethers. It provides means for investigating cation transport across hydrophobic regions of membranes and membrane models, allows anion solvation and activation, and finds applications in ion selective electrodes, phase transfer catalysis, organic synthesis, photochemical energy conversion, isomer and isotope separations. The magnitude of cation binding, the binding or

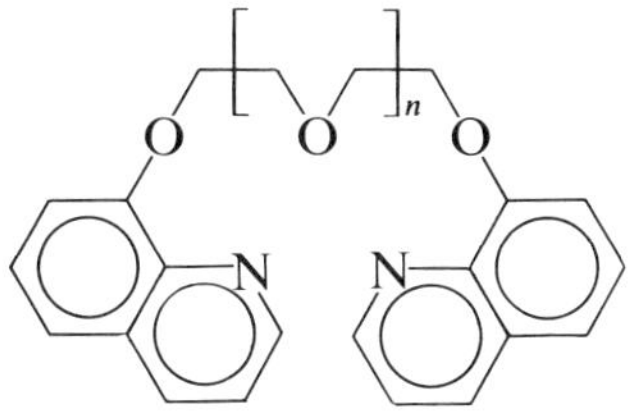

7.1

$n = 0\text{–}10$

Bis(8-quinolinoxy)podands

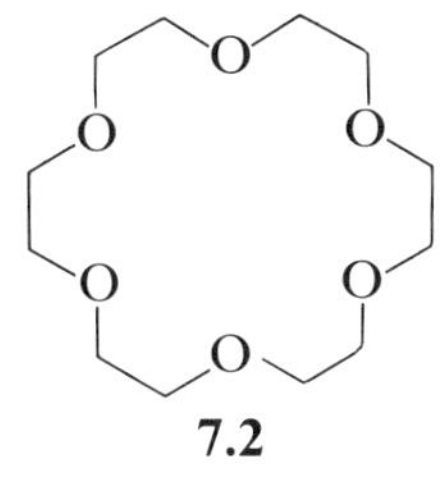

7.2

18-Crown-6
A crown ether

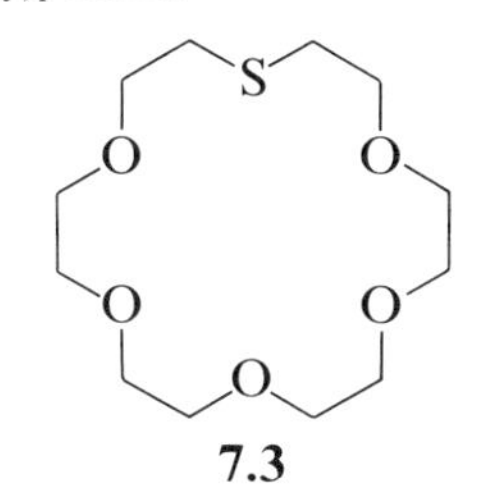

7.3

1-thio-4,7,10,13,16-pentaoxy
cyclooctadecane
A coronand

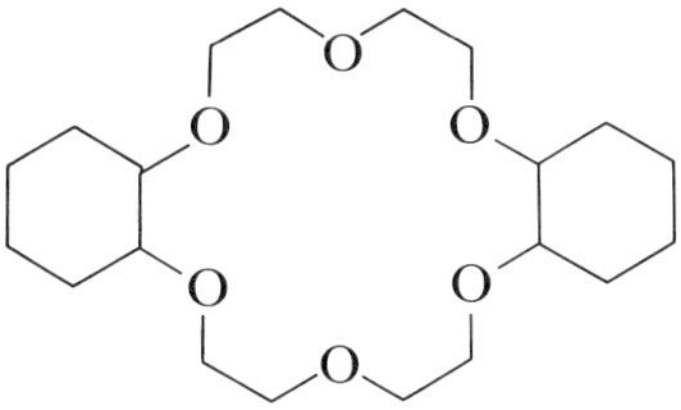

7.4

Perhydrodibenzo-18-crown-6
A crown ether

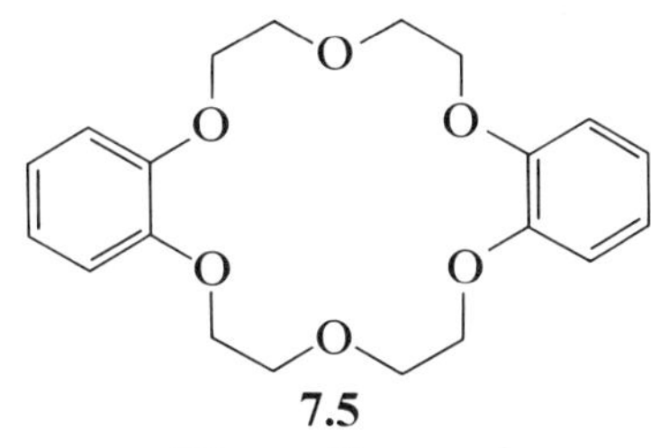

7.5

Dibenzo-18-crown-6
A crown ether

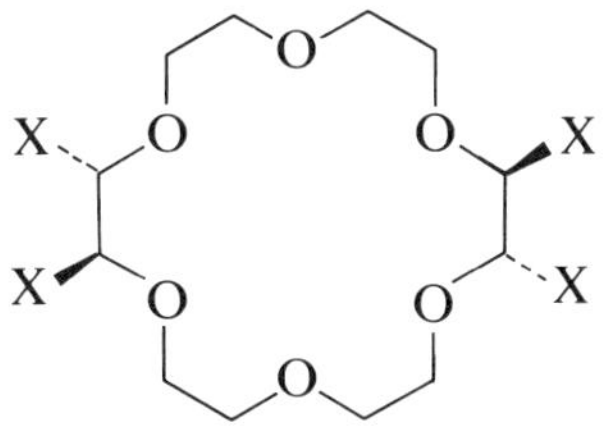

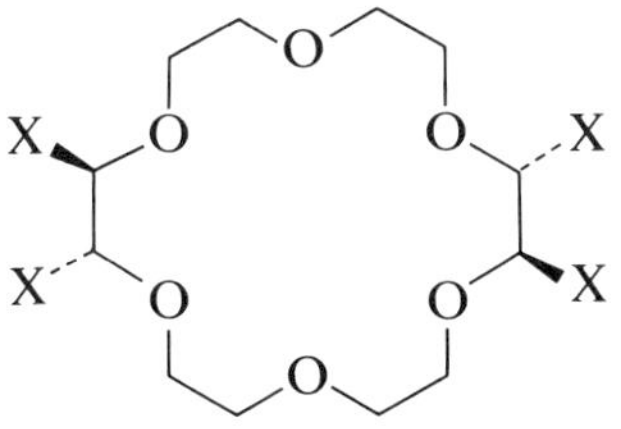

7.6$_{DD}$, X = D-CONHCH(CO_2Me)CH_2(3-indole)

7.6$_{LD}$, X = D-CONHCH(CO_2Me)CH_2(3-indole)

7.7$_{DL}$, X = L-CONHCH(CO_2Me)CH_2(3-indole)

7.7$_{LL}$, X = L-CONHCH(CO_2Me)CH_2(3-indole)

7.8$_{D}$, X = CONH-1-pyrene

7.8$_{L}$, X = CONH-1-pyrene

Optically active crown ethers (D and L) containing L- and D-tryptophan and -pyrene

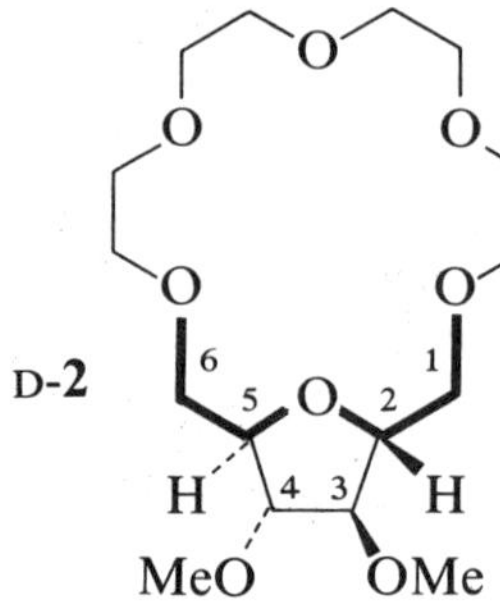

D-2

7.9

18-crown-6 derivative of 2,5-anhydro-D-mannitol
A chiral crown ether

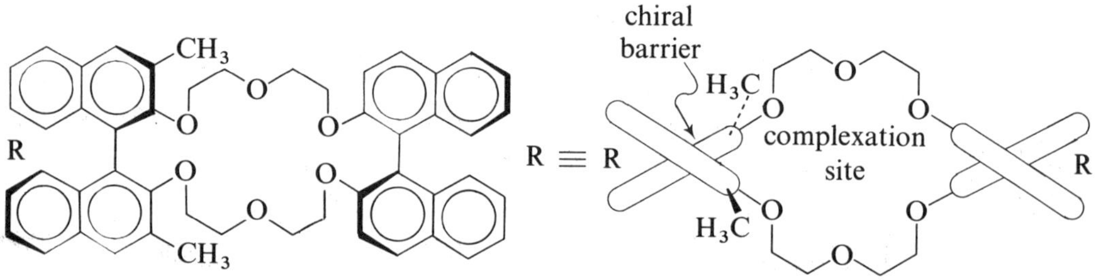

(RR)-7.10

Chiral crown ether containing binaphthyl units

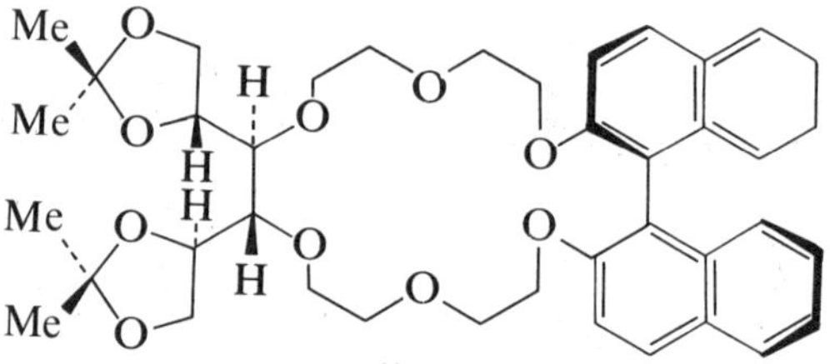

D-(S)-**7.11**

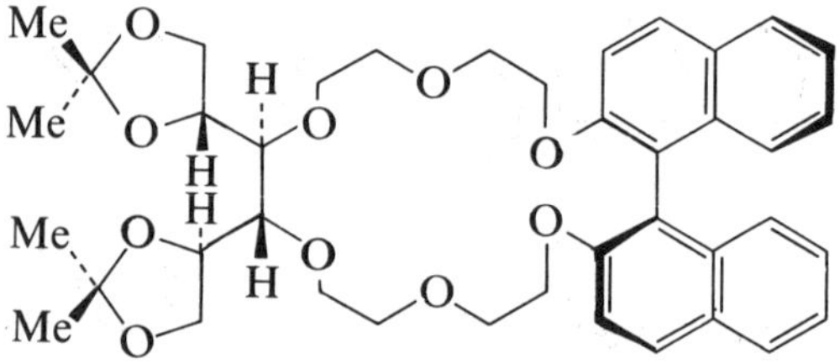

D-(R)-**7.11**

Chiral crown ethers containing di-*o*-isopropylydene-D-mannitol and binaphthyl units

7.12, [1.1.1], $m = n = 0$
7.13, [2.1.1], $m = 0, n = 1$
7.14, [2.2.1], $m = 1, n = 0$
7.15, [2.2.2], $m = n = 1$
7.16, [3.2.2], $m = 1, n = 2$
7.17, [3.3.2], $m = 2, n = 1$
7.18, [3.3.3], $m = n = 2$
7.19, $HC(CH_2OCH_2CH_2{-}OCH_2CH_2OCH_2)_3CH$

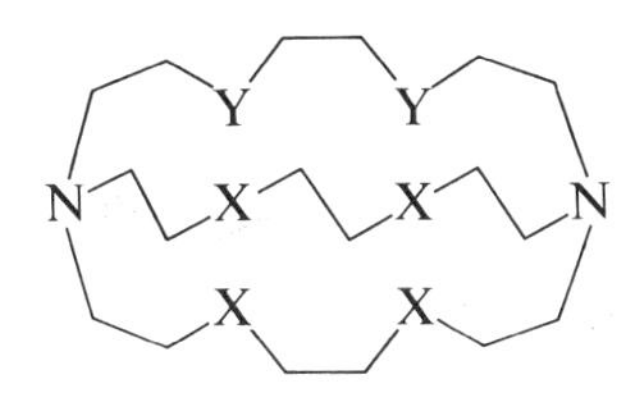

7.20, [2.2.C_8], X = 0, Y = CH_2
7.21, X = 0, Y = 0, NCH_3
7.22, X = 0, Y = NCH_3
7.23, X = NCH_3, Y = 0
7.24, X = 0, Y = S
7.25, X = S, Y = 0
7.26, X = Y = S
7.27, X = 0, Y = 0, NCH_2COOH
7.28, X = 0, Y = NCH_2CH_2COOH

Cryptands

stability constant, depends on the geometries and charges of the hosts and guests, on the number and kind of polar atoms in the host, and on the microenvironment of the guest in the host. Large numbers of stability constants have been determined for the binding of Li^+, Na^+, Rb^+, Cs^+, NH_4^+, Ba^{2+}, Sr^{2+}, Ca^{2+}, Mg^{2+}, RNH_3^+, and $Ar\overset{+}{N}{\equiv}N$ into structurally different hosts (Cram and Cram, 1978; Lehn, 1978; Izatt and Christensen, 1978; Melson, 1979; Vögtle and Weber, 1979; Stoddart, 1979).

Absorption spectrophotometry revealed the presence of two nonequivalent binding sites in **7.1** ($n = 3$) for alkali metal cations with binding constants of 10^3–10^4 M^{-1} and 10^2–10^3 M^{-1}, respectively (Tümmler et al., 1977, 1979). Stability constants were found to be independent of the ionic radii of the metal guest, indicating the flexibility of the tetraethylene glycol chain in the podand. Introduction of a bridging pyridine ring into the ether chain resulted in a modest degree of selectivity for Na^+ over K^+. More interestingly, the 8-quinolinoxy terminal groups in **7.1** render Mg^{2+} complexes considerably more stable than those of their alkali metal ion analogs (Tümmler et al., 1979). Structures of metal ion–podand complexes depend on the lengths of the ligand. X-ray crystallographic examination of RuI complexes of **7.1** indicated circular complexes for short chain ligands ($n \leq 4$), whereas helical structures predominate for longer chain ($n > 5$) complexes (Saenger et al., 1979). Several additional features emerged from the X-ray analysis of these podand complexes. The circular **7.1** ($n = 1$) complex of RbI possesses a plane of symmetry and is, therefore, achiral. The S-like KSCN complex of 1,5-bis{2-[5-(2-nitrophenoxy)-3-oxapentyloxy]-phenoxy}-3-oxapentane, **7.29**, on the other hand, has a C_2 symmetry (Figure 7.1) and is chiral.

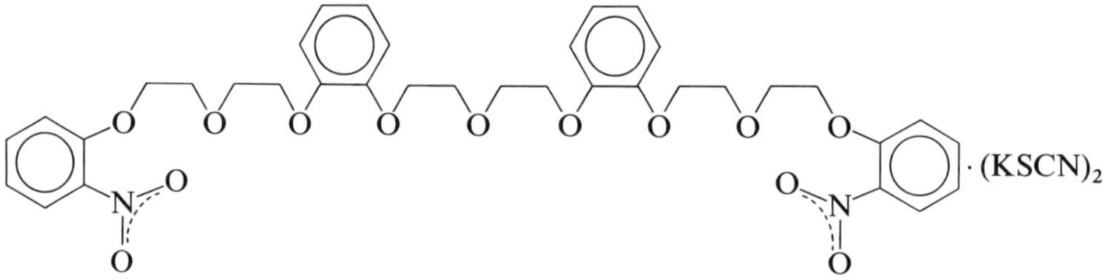

$\cdot (KSCN)_2$

7.29

In the crystal structure both enantiomers of **7.29** prevail. Metal ions in **7.1** and **7.29** coordinate to all heteroatoms of the podands and the coordination distance depends on the electronegativity of the heteroatom (Saenger et al., 1979). Further, the C—C—O—C and C—O—C—C torsion angles in the crystal were found to be trans and that of O—C—C—O to be gauche, with the trans forms being somewhat more flexible. The ligands wrapped preferentially in an "equatorial" zone around the cation while the "polar" regions are exposed to the terminal groups of the ligands (Saenger et al., 1979). Stronger binding and enhanced specificity are found in octopuslike ligands, or polypodands (Figure 7.2; Vögtle and Weber, 1974; Knöchel et al., 1975; MacNicol and Wilson, 1976; MacNicol et al., 1977; Weber et al., 1979).

As expected, crown ethers (**7.2**, **7.4**, **7.5**, **8.9**, **7.10**, **7.11**) bind cations more strongly than podands. A semiquantitative scale has been developed for assessing effectiveness of binding by distributing $t\text{-BuNH}_3^+X^-$ between D_2O and $CDCl_3$ and measuring its concentration, in the organic layer in the absence and in the presence of the host, spectrometrically (Timko et al., 1974, 1977). Association constants, Ka, and free energies of associations, ΔG^0, are defined by

$$\text{host} + t\text{-BuNH}_3^+X^- \underset{\text{CDCl}_3}{\overset{Ka}{\rightleftharpoons}} t\text{-BuNH}_3^+\text{host}X^- \tag{7.1}$$

$$\Delta G^0 = -\text{RT ln Ka} \tag{7.2}$$

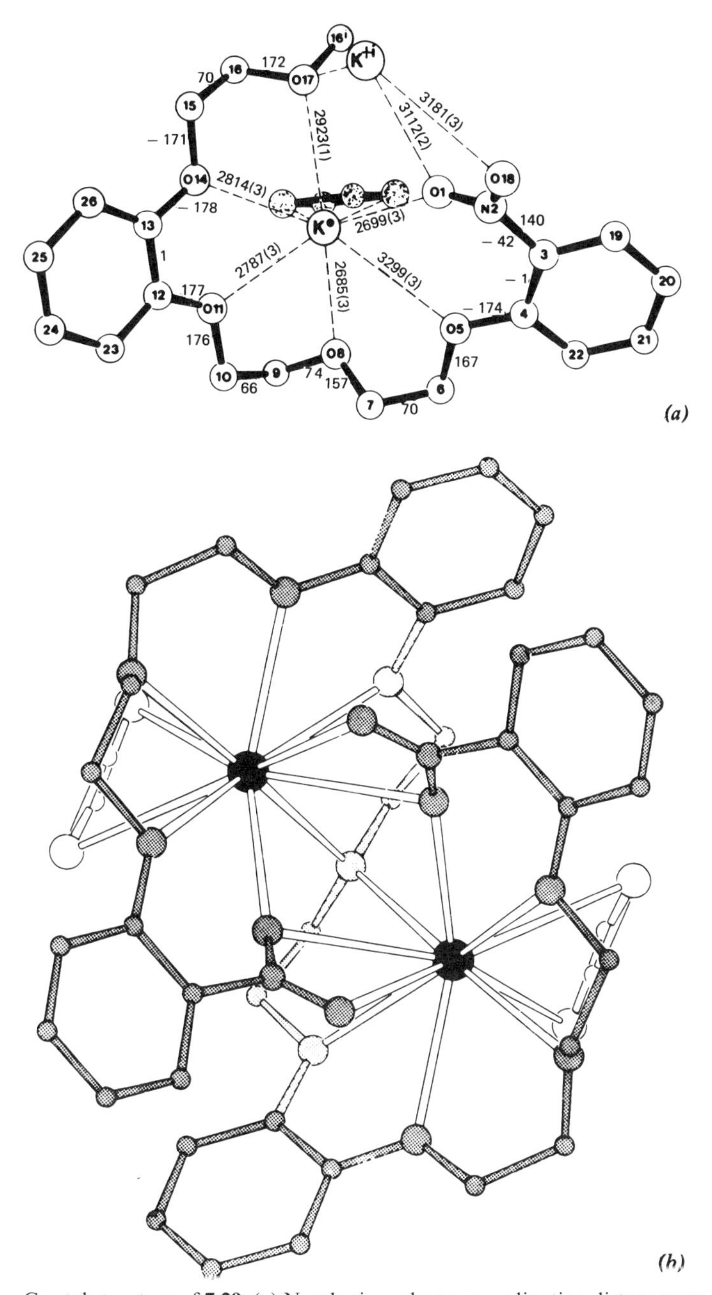

Figure 7.1. Crystal structure of **7.29**. (*a*) Numbering scheme, coordination distances, and torsion angles along the polyether chain. Atoms K^{+1} and C(16′) are related to K^+ and C(16), respectively, by twofold crystallographic symmetry. Both K^+ are threefold coordinated to nitro groups, twice to each O(1), 2.699 and 3.112 Å, and once to each O(18), 3.181 Å. In the figure only one of each bonding type is represented. The twofold disordered SCN^- anion is stippled. K^+ . . . S(N) distances, indicated by broken lines in the figure, are 3.387 and 3.256 Å, and represent K^+ . . . S rather than K^+ . . . N distances. (*b*) The coordination scheme of **7.29**. The twofold crystallographic axis passes through the central O(17). Parts of the molecule drawn more heavily are closer to the viewer than parts drawn lightly (taken from Saenger et al., 1979).

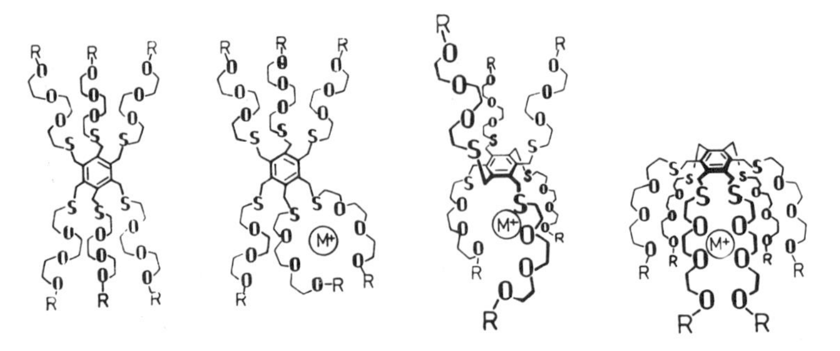

Figure 7.2. Metal–ion polypodand interactions (taken from Vögtle and Weber, 1979).

ΔG^0 values of -8.9 and -2.9 kcal mole^{-1} have been obtained for 18-crown-6, **7.2**, and its open chain podand analog, by means of t-$BuNH_3^+X^-$ distributions (Timko et al., 1977). The crown ether binds, therefore, some six orders of magnitude better than the open chain polyether. Free energies of associations are largely determined by the number of binding sites and they are, within a given series of crown ethers, additive (Cram and Cram, 1978). To a first approximation, stability constants are proportional to the fit of the cation in the cavity. Thus, Rb^+, with 2.96 Å diameter, is the most strongly bound cation in the 2.6–3.2 Å diameter cavity provided by dicyclohexyl-18-crown-6 (**7.4**; Pedersen and Frensdorff, 1972). Replacing some of the oxygen atoms in crown ethers by nitrogen leads to aza-oxa macrocyclic rings. The triaza-trioxa macrocycle[18]-N_3O_3 shows the highest stability and selectivity for binding primary ammonium cations discovered so far (Lehn and Vierling, 1980).

Guidelines for optimizing crown ether constructions as hosts have been delineated as follows (Cram and Cram, 1978):

1. Design hosts that combine maximum stability with minimum molecular weight.
2. Use repeating structural units wherever possible.
3. Design hosts that provide the maximum amount of symmetry compatible with desired properties.
4. Make maximum use of conformationally unambiguous molecular units.
5. Incorporate units that rigidly extend in three dimensions.
6. Maximally use molecular units that serve several purposes.
7. Use units with rigidly convergent and substitutable sites for binding and shaping.
8. Use units with rigidly divergent and substitutable sites for manipulation of gross physical properties (e.g., solubility), or for attachment to solid supports.
9. Avoid, in hosts, intramolecular complexation that competes with desired host–guest complexation.

10. Avoid hosts whose cavities can collapse by conformational reorganization.

11. Design series of hosts that can be synthesized from common intermediates.

12. Incorporate units that provide spectral probes for structures of the anticipated complexes.

13. Employ the maximum number of binding sites compatible with high rates of complexation-decomplexation.

14. Design hosts to avoid ambiguity as to which sites bind to the guest.

15. Design hosts that will avoid complexing unwelcome guests, and thus are able to differentiate between guests.

16. Remember that in host design strong entropic driving forces oppose selective complexation, and that the host will use all structural degrees of freedom available to avoid differentiating between guests.

17. In catalytic hosts design to stabilize the rate-limiting transition states.

18. Where covalent bonds are being made and broken between host and guest, treat the locus of the transition state as a binding site of low geometric tolerance.

These principles and detailed examination of CPK models have much aided the construction of crown ethers (e.g., **7.9** and **7.10**) capable of chiral recognition (Cram et al., 1975; Cram and Cram, 1978; Stoddart, 1979). Chiral crown ether, **7.10**, provides a good example for recognizing optically active guests (Kyba et al., 1973; Helgeson et al., 1974; Peacock and Cram, 1976; Peacock et al., 1980). Racemic amino acid or other amine (e.g., α-phenylethyl amine) guests are dissolved in an aqueous solution containing $LiClO_4$ or $LiPF_6$. The host, **7.10**, present in the $CHCl_3$ layer, extracts and complexes one of the enantiomers preferentially. The amount of guest extracted depends on the concentration of the lithium salt, the transfer agent. The difference in free energy between the diastereomeric complexes in the $CHCl_3$ layer, $\Delta(\Delta G^0)$, is given by

$$\Delta(\Delta G^0) = -RT \ln (EDC) \qquad (7.3)$$

where EDC is the enantiomeric distribution constant (Peacock et al., 1980). Under optimum conditions, using **7.10** as a chiral host, $\Delta(\Delta G^0)$ values of -0.48, -2.0, -1.95, and -2.15 kcal mole^{-1} have been obtained for α-phenylethylammonium hexafluorophosphate, methyl *p*-hydroxyphenylglycinate hexafluorophosphate, tryptophan perchlorate, and phenyl glycine perchlorate, respectively (Cram and Cram, 1978). These values represent quite remarkable chiral recognitions. Examination of molecular models indicates the importance of the methyl groups in **7.10**. The methyl group of the host, on the site remote from the face binding the guest, contacts the other naphthalene on the same side when the cavities on the binding side are enlarged by splaying motion. The consequence of this is that the guest in host (SS)-**7.10** is in closer chiral contact than that in (RR)-**7.10** (Peacock et al., 1980). A similar degree of chiral recognition has been reported with **7.11** (Curtis et al., 1976).

Photophysical techniques have been advantageously used for investigating chiral recognitions in crown ethers (Tundo and Fendler, 1980). Addition of $TbCl_3$ decreased the fluorescence yield of the tryptophan side arms of crown ethers **7.7** and **7.8**. Quantitative treatment of emission intensities led to binding constants of $(2.90 \pm 0.2) \times 10^4\ M^{-1}$ and $(2.76 \pm 0.1) \times 10^4\ M^{-1}$ for the Tb^{3+} complexes $\mathbf{7.7}_{DL}$ and $\mathbf{7.7}_{LL}$. Complexing an achiral guest into crown ethers is seen, therefore, to be unaffected by the chirality of the host, indicating similar cavity diameters for $\mathbf{7.7}_{DL}$ and $\mathbf{7.7}_{LL}$. Differences in the conformation of the tryptophan side arms in isomeric **7.7** are clearly indicated, however, by the different energy transfer efficiencies from $\mathbf{7.7}_{LL}$ and $\mathbf{7.7}_{DL}$ to Tb^{3+}, entrapped in the crown cavities (Figure 7.3; Tundo and Fendler, 1980). In the absence of crown ethers energy transfer is negligible. Conformational changes upon guest complexation have been inferred from changes in the monomer : excimer ratios in **7.8** (Tundo and Fendler, 1980).

Cryptands are the most efficient hosts. The stability constant for the K^+ complex of **7.15** is, for example, 10^5- and 10^6-fold greater than those for corresponding crown ethers and valinomycin (a naturally occurring macrocyclic antibiotic) K^+ complexes, respectively (Lehn, 1978). These large cryptate complex binding constants are paralleled by high degrees of selectivities.

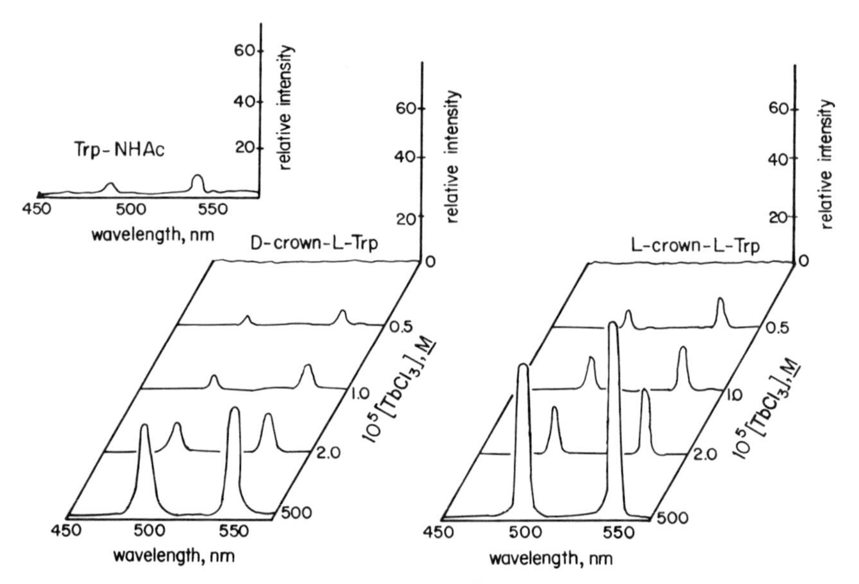

Figure 7.3. Relative emission intensities of $5.0 \times 10^{-3}\ M\ TbCl_3$ in the presence of $2.0 \times 10^{-4}\ M$ Trp-NHAc (upper left corner). Three-dimensional plots of emission spectra of varying amounts of $TbCl_3$ in the presence of $5.0 \times 10^{-5}\ M\ \mathbf{7.7}_{DL}$ (D-crown-L-Trp) and $\mathbf{7.7}_{LL}$ (L-crown-L-Trp) (excited at 290 nm) in MeOH (taken from Tundo and Fendler, 1980).

Selectivity coefficients for cation complexation are appreciably greater for cryptates than those for crown ethers. This is particularly important for distinguishing between di- and univalent cations. The Ba^{2+}/K^{+} binding ratio to **7.15** is, for example, approximately 10^4 (Dietrich et al., 1973). The efficient and specific cation bindings are largely due to spheroidal cryptand cavities, which are capable of optimal interactions with spherical cations (Figure 7.4).

Cryptands, containing two separate macrocyclic ligands, define three complexation sites: two lateral circular cavities inside the macrocycles and a center cavity. Binding of two metal ions into a cryptate is illustrated in Figure 7.5 (Lehn, 1978). Macrotricyclic cryptands are capable of binding guests in topo-selective manners (Kotzyba-Hibert et al., 1980). They can bring two substrates together within the cavities of a single cryptand and transport them simultaneously across barriers. Thus they can act as coreceptors, metallo-receptors, if one of the guests is a metal ion, and cocarriers. They can also function as cocatalysts by promoting reactions between the bound substrates with or without the participation of catalytic groups carried by side chains (Kotzyba-Hibert et al., 1980).

Metal ion binding to macrocyclic polyethers is a dynamic process. Relaxation studies for

$$\text{host} + M^{+} \underset{k_d}{\overset{k_f}{\rightleftharpoons}} \text{host complexed } M^{+} \tag{7.4}$$

indicated rate constants for k_f and k_d to be in the order of 10^8–10^9 M^{-1} sec^{-1} and 10^5–10^7 M^{-1} sec^{-1}, respectively (Liesegang and Eyring, 1978). Anion receptive polyammonium macrocyclic compounds have recently been developed (Dietrich et al., 1981).

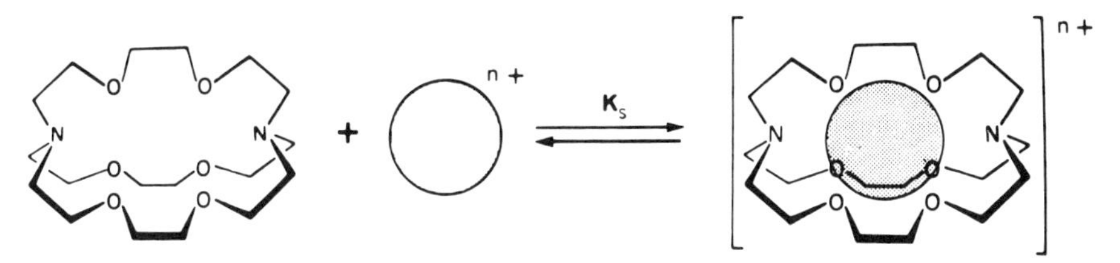

Figure 7.4. Formation equilibrium of a cryptate complex between **7.15** and a metal ion (taken from Lehn, 1978).

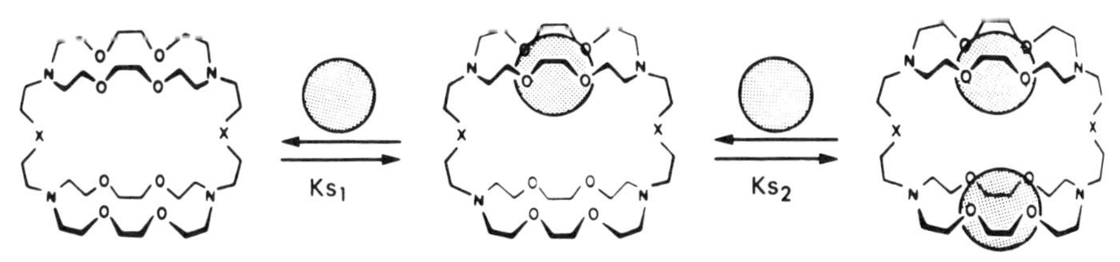

Figure 7.5. Equilibria for the successive formation of nonsymmetric mononuclear and symmetric binuclear cryptates (taken from Lehn, 1978).

2. CYCLODEXTRINS

Cyclodextrins are naturally occurring hosts (Cramer, 1954; French, 1957; Cramer and Hettler, 1967; Griffiths and Bender, 1973; Bergeron, 1977; Bender and Komiyama, 1977, 1978; Breslow, 1979; Saenger, 1980). Structurally, they are doughnut-shaped macrocyclic glucose polymers containing a minimum of 6-D(+)-glucopyranose units, attached by α-(1,4) linkages. Figure 7.6 shows space filling models of cyclohexaamylose (α-cyclodextrin), cycloheptaamylose (β-cyclodextrin), and cyclooctaamylose (γ-cyclodextrin), the formula for γ-cyclodextrin, and a schematic representation of the cavity of γ-cyclodextrin. The internal diameters of α, β-, and γ-cyclodextrins, estimated from X-ray analysis, are approximately 4.5, 7.0, and 8.5 Å, respectively (James et al., 1959). The interior of the doughnut, lined with CH groups, provides a relatively hydrophobic environment. One side of the torus contains primary hydroxyl groups, while the secondary hydroxyl groups on the C-2 and C-3 atoms of the glucose units are located on the other side (Figure 7.6). X-ray crystallography of α-cyclodextrin showed two water molecules in the cavity and another four on the outside (Figure 7.7; Saenger et al., 1976). These water molecules undergo circular homodromic hydrogen bond formation, which is stabilized by the cooperative effect (Saenger and Lindner, 1980). The six glucose

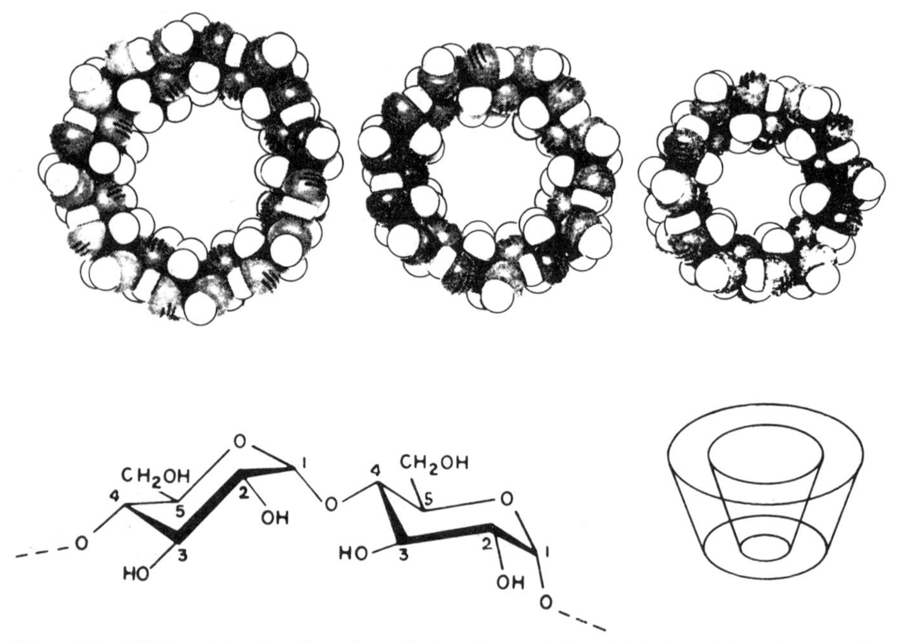

Figure 7.6. CPK models of α-, β-, and γ-cyclodextrins (top left to right) viewed from the secondary hydroxyl side of the torus. Schematic diagram of two glucopyranose units of a cyclodextrin molecule illustrating details of the α-(1,4)glycoside linkage and the numbering system (bottom left) and illustration of the conical cavity of γ-cyclodextrin.

units in α-cyclodextrin are nonequivalent; glucose 5 in Figure 7.7 is rotated more nearly normal to the axis of the molecule than are the other five glucose rings. This results in a distortion of the macrocyclic ring producing some steric strains. Interestingly, on complexation there is a conformational change resulting in the loss of macrocyclic distortion. The gross structural features of crystalline cyclodextrins have been confirmed in solution by nuclear magnetic resonance (nmr) (Takeo et al., 1973; Usui et al., 1973; Colson et al., 1974) and infrared (Casu et al., 1966) spectroscopy.

Treatment of starch with cyclodextrinase gives mixtures of α-, β-, γ-, and small amounts of higher cyclodextrins (Bender and Komiyama, 1978; Breslow, 1979). Guests entrapped in cyclodextrin cavities include organic compounds (substituted aromatic compounds, amino acids, nucleotides), small ions, and gases (Fendler and Fendler, 1975; Bender and Komiyama, 1978). To a first approximation the magnitude of binding correlates with the fit of the guest in the cyclodextrin cavity. Substrate-cyclodextrin binding constants range from 10^2 to 10^7 M^{-1}. Since cyclodextrins provide chiral cavities, complex formation can be stereoselective. Indeed circular dichroism is induced on complexing achiral molecules to cyclodextrins (Ikeda et al., 1975; Harata and Uedariva, 1975).

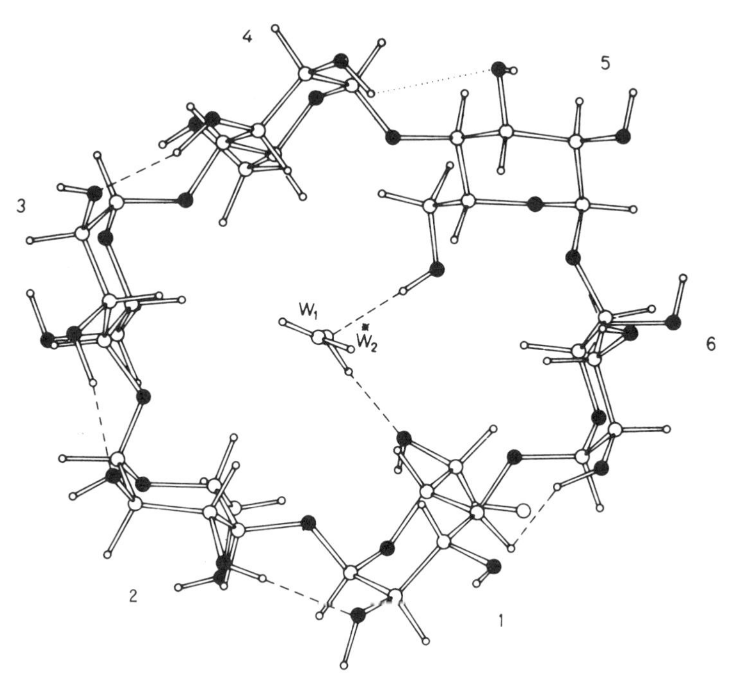

Figure 7.7. Structure of α-cyclodextrin hexahydrate; only two water molecules (W_1 and W_2) included in the cyclodextrin cavity are shown. Numbers refer to glucose units. The dotted line between the most rotated glucose 4 and glucose 5 represents a questionable hydrogen bond of 3.36 Å O---O distance. Water molecule W_2 is above W_1 and hydrogen bonded to W_1 and to two O(6) hydroxyl groups of α-cyclodextrin. The * mark shows the center of the cyclodextrin molecule (taken from Saenger et al., 1976).

Substrate entrapment in cyclodextrins is a dynamic process (Cramer et al., 1967; Rohrbach et al., 1977). There is an exchange between cyclodextrin-bound and -free substrates. The dynamics of quenching within the hydrophobic regions have been investigated (De Korte et al., 1980). Entrapped substrates are likely to undergo different types of motions within the cyclodextrin cavity (Behr and Lehn, 1976). Temperature-jump and ultrasonic studies indicate the presence of more than one process in cyclodextrin-bound substrate relaxations. The origins of these relaxations have not been unequivocally elucidated, however. In addition to substrate entry and exit, reorganization of water molecules and/or host conformational changes could be involved.

The driving force for substrate cyclodextrin complex formation has been discussed in terms of Van der Waals' interactions, hydrogen bondings, hydrophobic associations, and the release of high energy water molecules and of steric strain on complexation (Bender and Komiyama, 1978). Free energy changes accompanying the complexation of benzene, methyl orange, and *p*-iodoaniline in α-cyclodextrin have been calculated in terms of a four-step process (Figure 7.8; Tabushi et al., 1978). The release of the two water molecules from the cyclodextrin interior, the first step, is assompanied by losses of Van der Waals' interactions (H^{w}_{vdw}) and hydrogen bonding ($-2\Delta H_{\mathrm{H\text{-}bond}}$), gains of motional freedoms of two water molecules to translation ($2S^{w}_{\mathrm{trans}}$) and three-dimensional rotation ($2S^{w}_{\mathrm{rot(3\text{-}D)}}$), and a change in conformational energy of α-cyclodextrin. The second step, corresponding to the transformation of the expelled gaseous water molecules into the bulk liquid phase, is governed by an enthalpy ($-\Delta H^{w}_{\mathrm{vap}}$) and an entropy ($\Delta S^{w}_{\mathrm{gas}\rightarrow\mathrm{liq}}$) change. Concurrent with the water molecules vacating the interior of the cyclodextrin, desolvation of the guest needs to be considered. In this third step the hydrophobic guest is transferred from the aqueous solution to the ideal gaseous state, leaving a cavity behind. Free energy changes corresponding to this step are estimated by counting the number of water molecules surrounding the guest. The last step, the binding of the guest, is assessed in terms of the Van der Waals' stabilization (H^{g}_{vdw}) and a change in conformational energy

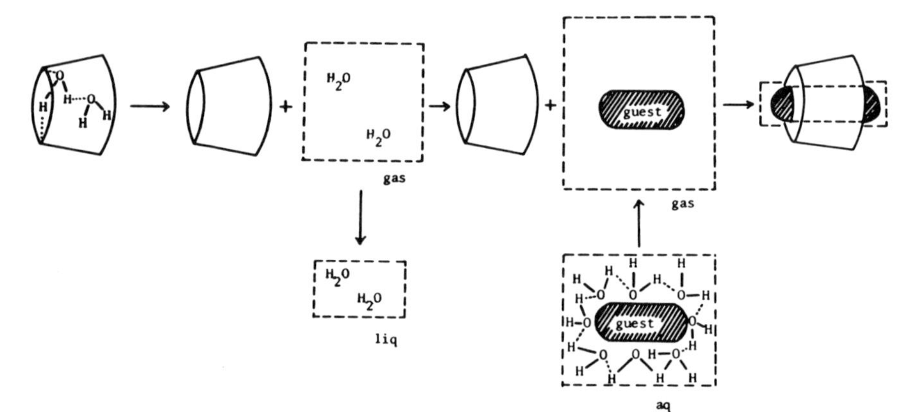

Figure 7.8. Schematic representation of the four-step thermodynamic process of inclusion of an organic guest molecule by α-cyclodextrin in water (taken from Tabushi et al., 1978).

of the cyclodextrin host. Enthalpy ($\Delta H_{\text{inclusion}}$) and entropy ($\Delta S_{\text{inclusion}}$) changes are calculated by means of (Tabushi et al., 1978)

$$\Delta H_{\text{inclusion}} = (H^{c}_{\text{vdw}} - H^{w}_{\text{vdw}}) + (H^{c}_{\text{conf}} - H^{w}_{\text{conf}}) - \Delta H^{g}_{\text{cluster}} - 2\Delta H^{w}_{\text{vap}} - 2H_{\text{H-bond}} \quad (7.5)$$

$$\Delta S_{\text{inclusion}} = (S^{g}_{\text{rot(1-D)}} - S^{g}_{\text{rot(3-D)}} - S^{g}_{\text{trans}}) - 2(-S^{w}_{\text{rot(3-D)}} - S^{w}_{\text{trans}}) + 2\Delta S^{w}_{\text{gas}\rightarrow\text{liq}} - \Delta S^{g}_{\text{cluster}} \quad (7.6)$$

where the superscripts c, w, and g refer to the cyclodextrin complex, the water molecule, and the guest, respectively. Table 7.1 lists the calculated thermodynamic parameters and observed free energy changes. The Van der Waals' stabilization is seen to play an important role in determining ΔH for inclusion. The large Van der Waals' stabilization for p-iodoaniline indicates this guest to have the best fit in α-cyclodextrin. Complex stabilization by Van der Waals' interaction is compensated for by breaking of water clusters and conformational changes (Tabushi et al., 1978).

In spite of substantial binding constants, rate enhancements for reactions of cyclodextrin-bound substrates are disappointingly meager (Bender and Komiyama, 1978). An investigation of substrate binding in cyclodextrin hosts

Table 7.1. Thermodynamic Parameters for α-Cyclodextrin Complexes[a]

Parameter	Guest		
	Benzene	p-Iodo Aniline	Methyl Orange
$H^{c}_{\text{vdw}} - H^{w}_{\text{vdw}}$, kcal mole^{-1}	−3.75	−10.09	−8.35
$H^{c}_{\text{conf}} - H^{w}_{\text{conf}}$, kcal mole^{-1}	+4.38	+4.38	+3.83
$-H^{g}_{\text{cluster}}$, kcal mole^{-1}	+4.10	+7.08	+6.71
$-2\Delta H^{w}_{\text{vap}}$, kcal mole^{-1}	−20.92	−20.92	−20.92
$-2H_{\text{H-bond}}$, kcal mole^{-1}	+12.20	+12.20	+12.20
$\Delta H_{\text{inclusion}}$, kcal mole^{-1}	−3.99	−7.35	−6.53
$S^{g}_{\text{rot(1-D)}} - S^{g}_{\text{rot(3-D)}} - S^{g}_{\text{trans}}$, cal deg^{-1} mole^{-1}	−52.00	−62.90	−67.60
$-\Delta S^{g}_{\text{cluster}}$, cal deg^{-1} mole^{-1}	+20.10	+34.80	−32.90
$-2(-S^{w}_{\text{rot(3-D)}} - S^{w}_{\text{trans}}) + 2S^{w}_{\text{gas}\rightarrow\text{liq}}$, cal deg^{-1} mole^{-1}	+33.60	+33.60	+33.60
$\Delta S_{\text{inclusion}}$, cal deg^{-1} mole^{-1}	+1.70	+5.50	−1.10
$\Delta H_{\text{inclusion}}$, kcal mole^{-1}	−3.99	−7.35	−6.53
$-T\Delta S_{\text{inclusion}}$, kcal mole^{-1}	−0.51	−1.64	+0.33
$\Delta G_{\text{inclusion}}$, kcal mole^{-1}	−4.50	−8.99	−6.20
$\Delta G_{\text{inclusion}}$, observed, kcal mole^{-1}		−5.90	−5.10

Source: Compiled from Tabushi et al., 1978.

[a] All values are calculated, unless stated otherwise.

showed appreciable decreases in the percentages of occupancy for most guests in going from the initial state, through the transition state, to tetrahedral intermediate and products (Breslow et al., 1980). Thus the observed inefficiencies are explicable in terms of greater ground state than transition state stabilization by cyclodextrin complexation. The primary requirement of a good catalyst is, of course, the opposite. It should stabilize the transition state more than the ground state. Attempts to bring about improved transition state stabilization included the functionalization of cyclodextrins, modification of guest structures, and solvent polarities (Breslow, 1979).

Cyclodextrin functionalization has been primarily directed toward decreasing ground state complexation. This can be accomplished by placing a cap on the floor of the host, which allows only partial guest penetration (Breslow, 1979). *N*-Formyl derivatives of β-cyclodextrin, **7.30** and **7.31**, provide hydrophobic caps, sufficient to observe enhanced hydrolysis of *m*-nitro and *m*-*tert*-butyl phenyl acetates (Emert and Breslow, 1975).

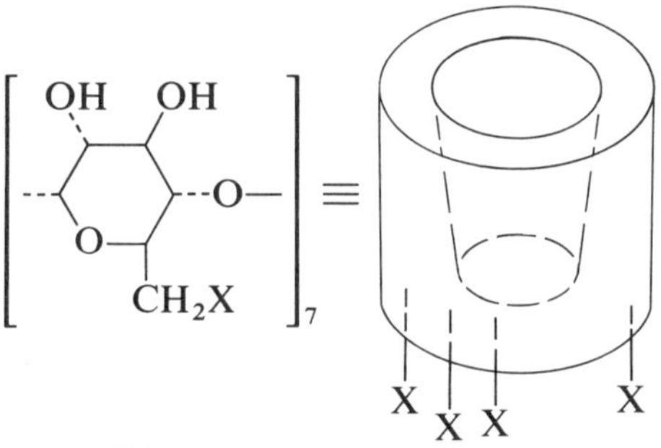

7.30, X = —N(CH$_3$)—C(H)=O

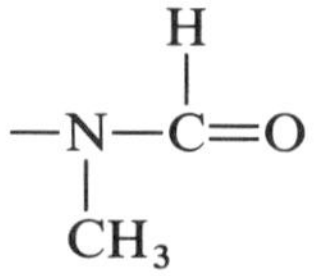

7.31, X = —N(CH$_2$CH$_3$)—C(H)=O

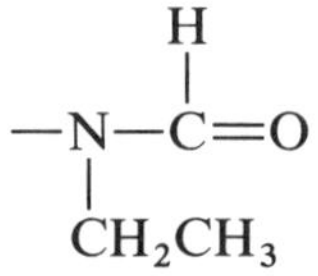

Rigid capping of β-cyclodextrin with diphenylmethansulfonate, **7.32**, or terephthalate, **7.33**, markedly increased the binding of hydrophobic guests (Tabushi et al., 1976).

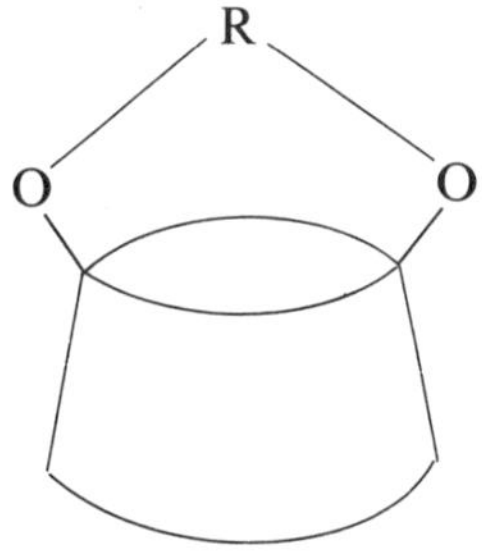

7.32, R = —SO$_2$—C$_6$H$_4$—CH$_2$—C$_6$H$_4$—SO$_2$—

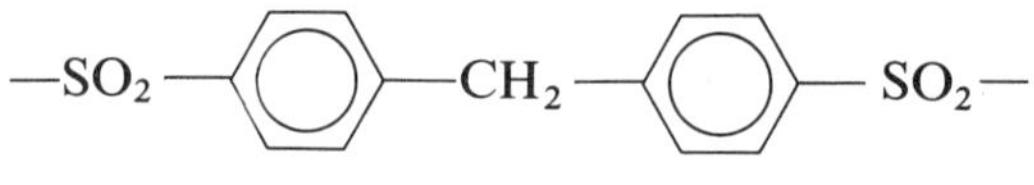

7.33, R = —CO—C$_6$H$_4$—CO—

Inclusion of suitable acceptors into the cavity of a β-cyclodextrin, rigidly capped by a benzophenone moiety, **7.34**, appreciably increased the triplet energy transfer between the photoexcited host and the entrapped guest (Tabushi et al., 1977a).

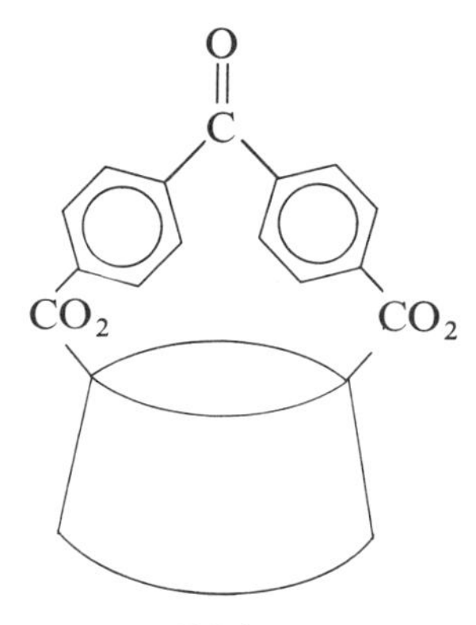

7.34

Azabenzene capped β-cyclodextrin undergoes photoinduced cis–trans isomerization with corresponding changes of the binding ability (Ueno et al., 1978). The trans configuration complexes one, and the cis configuration includes two guests:

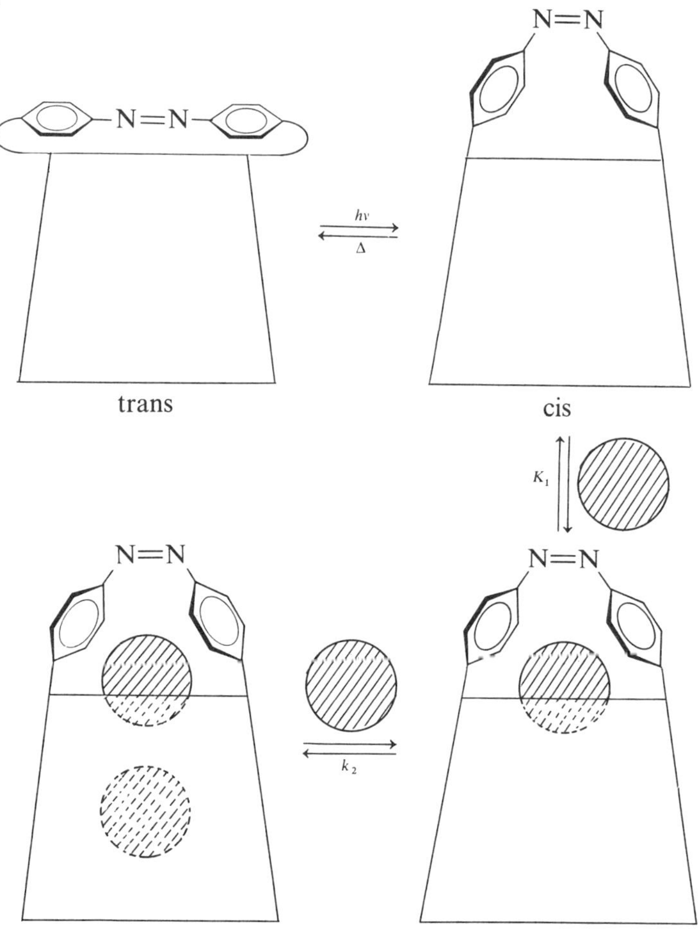

Polyamine functionalized cyclodextrins form Cu^{2+}, Zn^{2+}, or Mg^{2+} complexes as flexible caps and thus provide double recognitions, as illustrated for the adamantane carboxylate guest, **7.35** (Tabushi et al., 1977b).

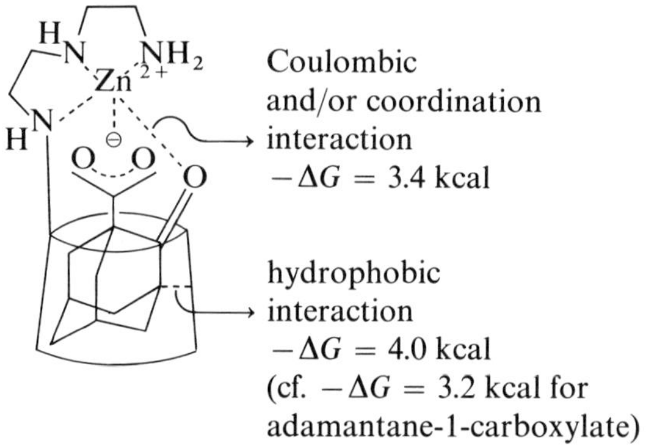

H N NH$_2$ Zn^{2+} H N O O O

Coulombic
and/or coordination
interaction
$-\Delta G = 3.4$ kcal

hydrophobic
interaction
$-\Delta G = 4.0$ kcal
(cf. $-\Delta G = 3.2$ kcal for
adamantane-1-carboxylate)

7.35

An interesting development is the formation of duplex cyclodextrins with multiple recognition sites (Tabushi et al., 1979).

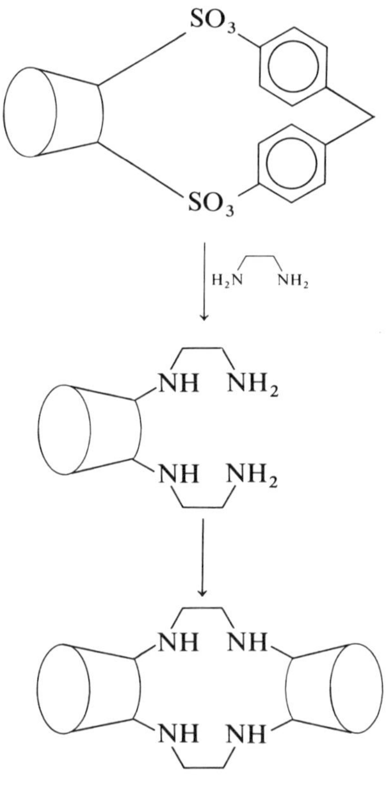

SO_3 SO_3

H_2N NH_2

NH NH_2 NH NH_2 (7.8)

NH NH NH NH

The most effective modification of the guest has been the introduction of a ferrocene nucleus (Siegel and Breslow, 1975). Acyl transfer,

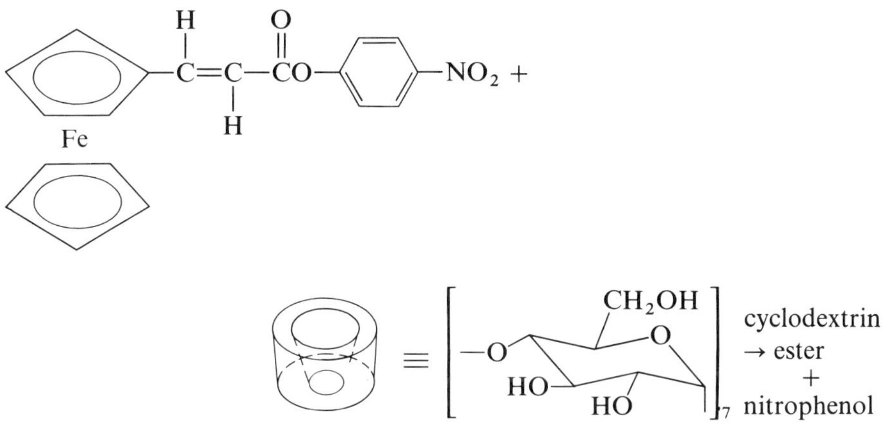

H O
—C=C—CO— —NO_2 +
H
Fe

CH_2OH
O
≡ —O
HO— HO
7
cyclodextrin
→ ester
+
nitrophenol

was found to be over 19 million-fold faster than the corresponding reaction in aqueous medium alone at the same pH (Breslow, 1979). Clearly, carefully designed host-guest systems mimic enzyme functions exceedingly well. Utilization of cyclodextrins, along with other enzyme models, is examined at length in Chapter 10.

3. RELATED MACROCYCLIC COMPOUNDS

Macrocyclic ligands, capable of complexing metal ions, have been known for some time (Busch, 1964, 1971; Curtis, 1968; Busch et al., 1971). They have been extensively used by coordination chemists to stabilize high oxidation states not normally available, to maintain constant stereochemistry about a metal ion while changing its oxidation state, and to increase metal ion stability constants by orders of magnitude over those seen in nonmacrocyclic complexes (Cotton and Wilkinson, 1980). These, in turn, provide opportunities for studying ligand field spectra and magnetic properties under a variety of different conditions (Urbach, 1979). Interest in macrocyclic compounds has also been stimulated by their potential as models for natural complexation in biological systems (Poon, 1973).

Typical examples of polyaza macrocycles are those of cyclic amines (**7.36**, **7.37**), cyclic imines (**7.38**, **7.39**), those containing pyridine (**7.40**), and benzene (**7.41**, **7.42**) fused rings. Unsubstituted and substituted paracyclophanes (**7.43**) have been extensively studied as enzyme models (Murakami et al., 1976, 1979; Odashima et al., 1980).

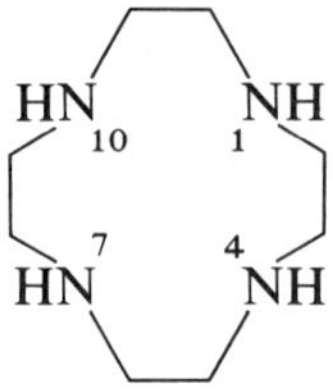

7.36

1,4,7,10-Tetraazacyclodecane
[12]aneN_4, cyclen

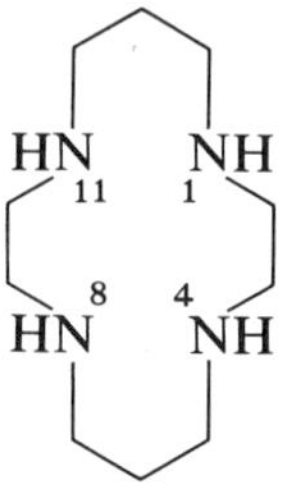

7.37

1,4,8,11-Tetraazacyclotetradecane
[14]aneN_4, cyclam

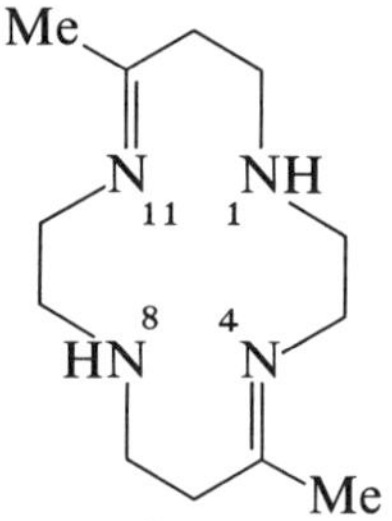

7.38

5,12-Dimethyl-1,4,8,11-tetraazacyclotetradec-4,11-diene
Me_2[14]dieneN_4

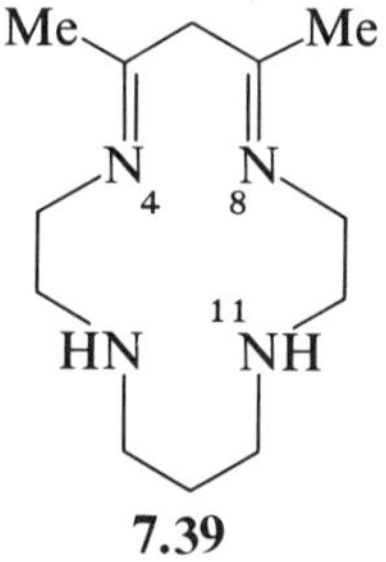

7.39

5,7-Dimethyl-1,4,8,11-tetraazacyclotetradec-4,7-diene
Me_2[14]dieneN_4

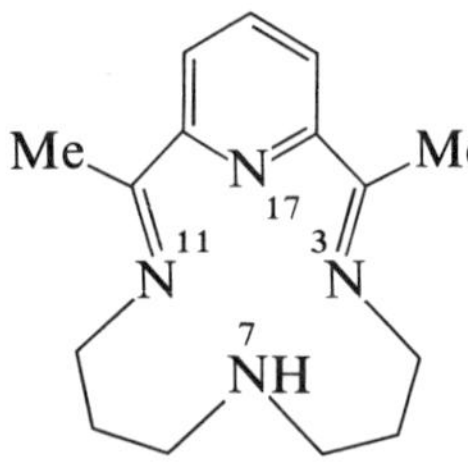

7.40

2,12-Dimethyl-3,7,11,17-tetraazabicyclo[11.3.1]septadec-1(17),2,11-13,15-pentaene
Me_2pyo[14]trieneN_4

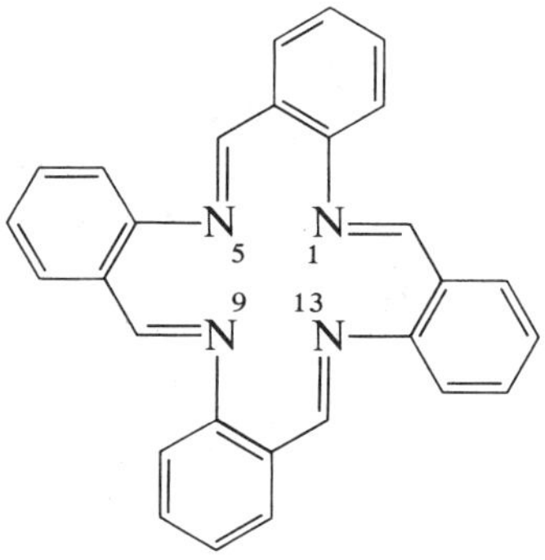

7.41

Tetrabenzo[*b,f,j,n*]-1,5,9,13-tetraazacyclohexadec-2,4,6,8,10,12,14,16-octane, tetrabenzo[*b,f,j,n*][1,5,9,13]tetraazacyclohexadecane
Bzo_4[16]octaeneN_4, TAAB

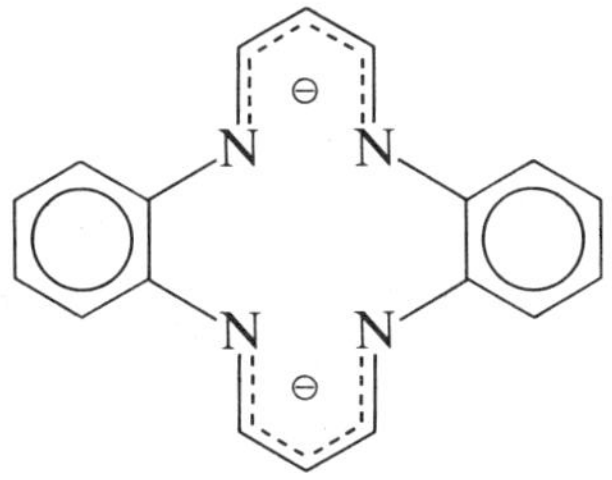

7.42

Dibenzo[*b,i*]-1,4,8,11-tetraazacyclotetradec-2,4,6,9,11-hexaenato(2-)

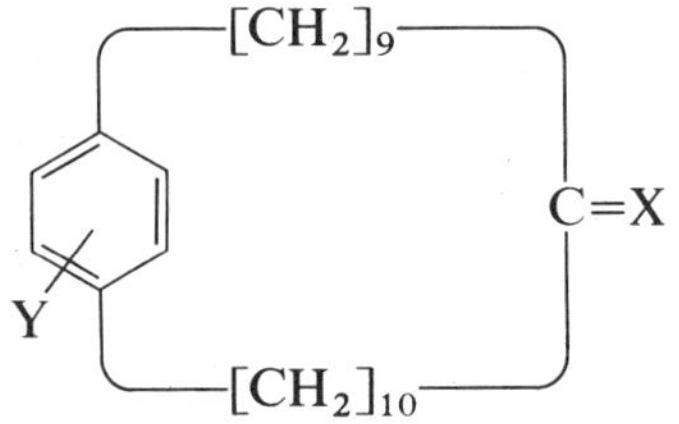

7.43a, X = NOH, Y = $CONH[CH_2]_2$–(imidazolyl, N/NH)

7.43b, X = O, Y = $CONH[CH_2]_2$–(imidazolyl, N/NH)

7.43c, X = NO^-, Y = H
7.43d, X = NOH, Y = $CH_2N^+[CH_3]_3Cl^-$

7.43e, X = NOH, Y = $COHN[CH_2]_2$–(imidazolium, HN/+/NH)

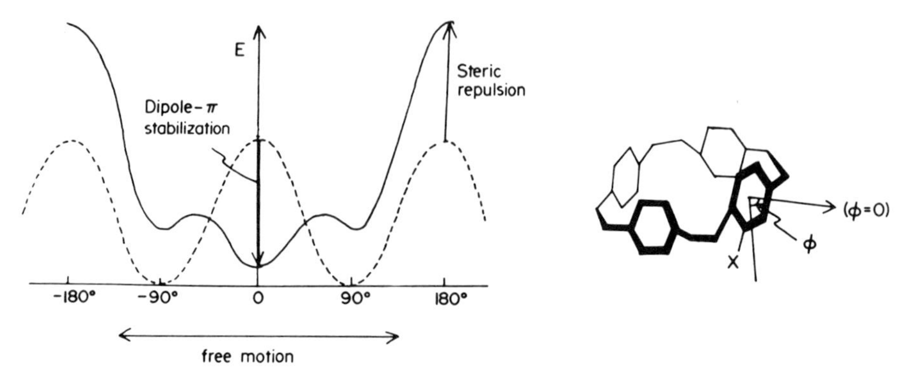

Figure 7.9. Assumed potential energy curve for **7.44** and proposed conformation for the face[3] · lateral structure (taken from Tabushi and Yamada, 1977).

Macrocyclic structures undergo, of course, a variety of conformational changes. For example, changes around the $C_2—C_3—C_6—C_9$ axis in 4-X-[2.2.2.2]paracyclophenes, **7.44**, are increasingly frozen at low temperatures if X is an electron withdrawing group (Tabushi and Yamada, 1977). A bulky X destabilizes the "face," relative to the "lateral" conformation of the benzene bearing the substituent, due to steric repulsions, leading to lateral, face, face, face

7.44

4-Substituted [2.2.2.2]paracyclophane

$N[(CH_2)_k][(CH_2)_l][(CH_2)_m]N$

7.45

$H^+—N$ — $N—\overset{+}{H}$	$H^+—N$ $H^+—N$	$N—\overset{+}{H}$ $H^+—N$
(out–out [o⁺o⁺])	(out–in [o⁺i⁺])	(in–in [i⁺i⁺])
7.46a	**7.46b**	**7.46c**

(face[3] · lateral) conformation. In the face[3] · lateral conformation of **7.44** the substituent turns toward the outside of the ring (Figure 7.9; Tabushi and Yamada, 1977). This results in a marked upfield shift of the *m*- and *p*-protons on the benzene ring. Electron withdrawing groups and decreased temperatures stabilize the face[3] · lateral conformation by dipole-π bond interactions.

An interesting property of macrobicyclic amines, **7.45**, is that in acid solution the protonated bridgehead nitrogens undergo isomerization (Park and Simmons, 1968a, 1968b; Simmons and Park, 1968; Simmons et al., 1970). In 50% aqueous trifluoroacetic acid the initial protonation sites are on the outside of the cavity (out–out, o^+o^+, **7.46a**). With time there is a slow change to form the macrobicyclic amine in which both NH^+ groups are in the inside of the cavity (in–in, i^+i^+, **7.46c**). For some compounds the out–in, o^+i^+, isomer (**7.46b**) has also been detected. **7.46a** $\rightleftharpoons$ **7.46c** isomerization has been discussed in terms of an eight-step mechanism (Simmons et al., 1970).

REFERENCES

Behr, J. P. and Lehn, J. M. (1976). *J. Am. Chem. Soc.* **98**, 1743–1747. Molecular Dynamics of α-Cyclodextrin Inclusion Complexes.

Bender, M. L. and Komiyama, M. (1977). In *Bioorganic Chemistry* (E. E. Van Tamelen, Ed.), Vol. 1, Academic Press, New York, pp. 14–176. Models for Hydrolytic Enzymes.

Bender, M. L. and Komiyama, M. (1978). *Reactivity and Structure Concepts in Organic Chemistry*, Vol. 6, Springer-Verlag, Berlin, Cyclodextrin Chemistry.

Bergeron, R. J. (1977). *J. Chem. Ed.* **54**, 204–207. Cycloamyloses.

Breslow, R. (1979). *Isr. J. Chem.* **18**, 187–191. Biomimetic Chemistry in Oriented Systems.

Breslow, R., Czarniecki, M. F., Emert, J., and Hamaguchi, H. (1980). *J. Am. Chem. Soc.* **102**, 762–770. Improved Acylation Rates Within Cyclodextrin Complexes from Flexible Capping of the Cyclodextrin and from Adjustment of the Substrate Geometry.

Busch, D. H. (1964). *Rec. Chem. Prog.* **25**, 107–126. The Significance of Complexes of Macrocyclic Ligands and Their Synthesis by Ligand Reactions.

Busch, D. H. (1971). *Science* **171**, 241–248. Metal Ion Control of Chemical Reactions.

Busch, D. H., Farmery, K., Goedken, V., Katovic, V., Melnyk, A. C., Sperati, C. R., and Tokel, N. (1971). In *Bioinorganic Chemistry* (R. F. Gould, Ed.), Advances in Chemistry Series, Vol. 100, American Chemical Society, Washington, D.C., pp. 44–78. Chemical Foundations for the Understanding of Natural Macrocyclic Complexes.

Casu, B., Reggiani, M., Gallo, G. G., and Vigevani, A. (1966). *Tetrahedron* **22**, 3061–3082. Hydrogen Bonding and Conformation of Glucose and Polyglucoses in Dimethylsulfoxide Solution.

Colson, P., Jennings, H. J., and Smith, I. C. P. (1974). *J. Am. Chem. Soc.* **96**, 8081–8087. Composition, Sequence, and Conformation of Polymers and Oligomers of Glucose as Revealed by Carbon 13 Nuclear Magnetic Resonance.

Cotton, F. A. and Wilkinson, G. (1980). *Advanced Inorganic Chemistry*, 4th ed, John Wiley, New York.

Cram, D. J. and Cram, J. M. (1978). *Acc. Chem. Res.* **11**, 8–14. Design of Complexes Between Synthetic Hosts and Organic Guests.

Cram, D. J., Helgelson, R. C., Koga, K., Kyba, E. P., Madan, K., Sousa, L. R., Siegel, M. G., Moreau, P., and Gokel, G. W. (1975). *J. Org. Chem.* **43**, 2758–2772. Host–Guest Complexation, 9. Macrocyclic Polyethers and Sulfides Shaped by One Rigid Dinaphthyl Unit and Attached Arms. Synthesis and Survey of Complexing Abilities.

Cramer, F. (1954). *Einschlussverbindungen*, Springer Verlag, Berlin.

Cramer, F. and Hettler, H. (1967). *Naturwissenschaften* **54**, 625–632. Inclusion Compounds of Cyclodextrins.

Cramer, F., Saenger, W., and Spatz, H.-C. (1967). *J. Am. Chem. Soc.* **89**, 14–20. Inclusion Compounds. XIX. The Formation of Inclusion Compounds of α-Cyclodextrin in Aqueous Solutions. Thermodynamics and Kinetics.

Curtis, N. F. (1968). *Coord. Chem. Rev.* **3**, 3–47. Macrocyclic Coordination Compounds Formed by Condensation of Metal-Amine Complexes with Aliphatic Carbonyl Compounds.

Curtis, W. D., King, R. M., Stoddart, J. F., and Jones, G. H. (1976). *J. Chem. Soc. Chem. Commun.*, 284–285. Enantiomeric Differentiation by Chiral Macrocyclic Polyethers Derived from D-Mannitol and Binaphthol.

De Korte, A., Langlois, R., and Cantor, C. R. (1980). *Biopolymers* **19**, 1281–1288. Fluorescence Quenching of Methyl-2-aminobenzoate in the Presence of Beta-cyclodextrin: A Model for the Dynamic Quenching of Chromophores in Hydrophobic Regions of Biopolymers.

Dietrich, B., Lehn, J. M., and Sauvage, J. P. (1973). *J. Chem. Soc. Chem. Commun.*, 15–16. Cryptates: Control over Bivalent/Monovalent Cation Selectivity.

Dietrich, B., Hosseini, M. W., Lehn, J. M., and Sessions, R. B. (1981). *J. Am. Chem. Soc.* **103**, 1282–1283. Anion Receptor Molecules. Synthesis and Anion-Binding Properties of Polyammonium Macrocycles.

Emert, J. and Breslow, R. (1975). *J. Am. Chem. Soc.* **97**, 670–672. Modification of the Cavity of β-Cyclodextrin by Flexible Capping.

Fendler, J. H. and Fendler, E. J. (1975). *Catalysis in Micellar and Macromolecular Systems*, Academic Press, New York.

French, D. (1957). *Adv. Carbohydrate Chem.* **12**, 189–260. The Schrardinger Dextrins.

Griffiths, D. W. and Bender, M. L. (1973). *Adv. Catal. Relat. Subj.* **23**, 209–261. Cycloamyloses as Catalysts.

Harata, K., and Uedaiva, H. (1975). *Bull. Chem. Soc. Jpn.* **48**, 375–378. Circular Dichroism Spectra of β-Cyclodextrin Complex with Naphthalene Derivatives.

Helgeson, R. C., Timko, P., Moreau, P., Peacock, S. C., Mayer, J. M., and Cram, D. J. (1974). *J. Am. Chem. Soc.* **96**, 6762–6763. Models for Chiral Recognition in Molecular Complexation.

Ikeda, K., Vekama, K., and Otagiri, M. (1975). *Chem. Pharm. Bull.* **23**, 201–208. Inclusion Complexes of β-Cyclodextrin with Antiinflamatory Drugs Fenamatesin Aqueous Solution.

Izatt, R. M. and Christensen, J. J. (1978). *Synthetic Multidentate Macrocyclic Compounds*, Academic Press, New York.

James, W. J., French, D., and Fundle, R. E. (1959). *Acta Cryst.* **12**, 385–389. Studies on the Schardinger Dextrins. IX. Structure of the Cycloxaamylose-Iodine Complex.

Knöchel, A., Dehler, J., and Rudolf, G. (1975). *Tetrahedron Lett.*, 3167–3170. Anionenaktivierung II. Vergleichende Übersicht über das anionenaktivierende Verhalten makrocyklischer Polyäther bei der Umzetzung von Benzylchlorid mit Kaliumacetat.

Kotzyba-Hibert, F., Lehn, J. M., and Vierling, P. (1980). *Tetrahedron Lett.*, 941–944. Multisite Molecular Receptors and Co-systems. Ammonium Cryptates of Macrocyclic Structures.

Kyba, E. P., Koga, K., Sousa, L. R., Siegel, M. G., and Cram, D. J. (1973). *J. Am. Chem. Soc.* **95**, 2692–2693. Chiral Recognition in Molecular Complexing.

Lehn, J. M. (1978). *Acc. Chem. Res.* **11**, 49–57. Cryptates: The Chemistry of Macropolycyclic Inclusion Complexes.

Lehn, J. M. and Vierling, P. (1980). *Tetrahedron Lett.*, 1323–1326. The [18]-N_3O_3 Aza-Oxa Macrocycle: A Selective Receptor for Primary Ammonium Cations.

Liesegang, G. W. and Eyring, E. M. (1978). In *Synthetic Multidentate Macrocyclic Compounds* (R. M. Izatt and J. J. Christensen, Eds.), Academic Press, New York, pp. 245–287. Kinetic Studies of Synthetic Multidentate Macrocyclic Compounds.

MacNicol, D. D. and Wilson, D. R. (1976). *J. Chem. Soc. Chem. Commun.*, 494–495. New Strategy for the Design of Inclusion Compounds: Discovery of the "Hexa-Hosts."

MacNicol, D. D., Hardy, A. D. V., and Wilson, D. R. (1977). *Nature* **266**, 611–612. Crystal and Molecular Structure of a "Hexa-Host" Inclusion Compound.

Melson, G. A. (1979). *Coordination Chemistry of Macrocyclic Compounds*, Plenum Press, New York.

Murakami, Y., Aoyama, Y., Ohno, K., Dobashi, K., Nakagawa, T., and Sunamoto, J. (1976). *J. Chem. Soc. Perkin Trans. I*, 1320–1326. Syntheses of Macrocyclic Enzyme Models. Part I. Preparation and Properties of [20]Paracyclophanes.

Murakami, Y., Aoyama, Y., Kida, M., Nakano, A., Dobashi, K., Tran, C. D., and Matsuda, Y. (1979). *J. Chem. Soc. Perkin Trans. I*, 1560–1567. Synthesis of Macrocyclic Enzyme Models. Part 2. Preparation and Substrate Binding Properties of [10.10]Paracyclophanes.

Odashima, K., Itai, A., Iitaka, Y., and Koga, K. (1980). *J. Am. Chem. Soc.* **102**, 2504–2505. Host–Guest Complex Formation Between a Water Soluble Polyparacyclophane and a Hydrophobic Guest Molecule.

Park, C. H. and Simmons, H. E. (1968a). *J. Am. Chem. Soc.* **90**, 2429–2431. Macrobicyclic Amines. II. out–out $\rightleftharpoons$ in–in Prototropy in 1, (k + 2)-Diazabicyclo[*k*,*l*,*m*]alkaneammonium Ions.

Park, C. H. and Simmons, H. E. (1968b). *J. Am. Chem. Soc.* **90**, 2431–2431. Macrobicyclic Amines. III. Encapsulation of Halide Ions by in, in-1(k + 2)-Diazabicyclo[*k*,*l*,*m*]alkaneammonium Ions.

Peacock, S. C. and Cram, D. J. (1976). *J. Chem. Soc. Chem. Commun.*, 282–284. High Chiral Recognition in α-Amino-Acid and -Ester Complexation.

Peacock, S. S., Walba, D. M., Gaeta, F. C. A., Helgelson, R. C., and Cram, D. J. (1980). *J. Am. Chem. Soc.* **102**, 2043–2052. Host–Guest Complexation. 22. Reciprocal Chiral Recognition Between Amino Acids and Dilocular Systems.

Pedersen, C. J. and Frensdorff, H. K. (1972). *Angew. Chem. Int. Ed. Eng.* **11**, 16–25. Macrocyclic Polyethers and Their Complexes.

Poon, C. K. (1973). *Coord. Chem. Rrv.* **10**, 1–35. Kinetics and Mechanisms of Substitution Reactions of Octahedral Macrocyclic Amine Complexes.

Rohrbach, R. P., Rodriquez, L. J., Eyring, E. M., and Wojcik, J. F. (1977). *J. Phys. Chem.* **81**, 944–948. An Equilibrium and Kinetic Investigation of Salt-Cycloamylose Complexes.

Saenger, W. (1980). *Angew. Chem. Int. Ed. Eng.* **19**, 344–362. Cyclodextrin Inclusion Compounds in Research and Industry.

Saenger, W. and Lindner, K. (1980). *Angew. Chem. Int. Ed. Eng.* **19**, 398–399. OH Clusters with Homodromic Circular Arrangement of Hydrogen Bonds.

Saenger, W., Noltemeyer, M., Manor, P. C., Hingerty, B., and Klar, B. (1976). *Bioorg. Chem.* **5**, 187–195. Topography of Cyclodextrin Inclusion Compounds. IX. "Induced-Fit"-Type Complex Formation of the Model Enzyme α-Cyclodextrin.

Saenger, W., Suh, I.-H., and Weber, G. (1979). *Isr. J. Chem.* **18**, 253–258. Wrapping of Metal Cations by Linear Polyethers.

Siegel, B. and Breslow, R. (1975). *J. Am. Chem. Soc.* **97**, 6869–6870. Lyophobic Binding of Substrates by Cyclodextrins in Nonaqueous Solvents.

Simmons, H. E. and Park, C. H. (1968). *J. Am. Chem. Soc.* **90**, 2428–2429. Macrobicyclic Amines, I. out–in Isomerism of 1, (k + 2)-Diazabicyclo 1, (k + 2)-Diazabicyclo[*k*.*l*.*m*]alkanes.

Simmons, H. E., Park, C. H., Uyeda, R. T., and Habibi, M. F. (1970). *Trans. New. Acad. Sci. Series II* **32**, 521–534. Macrobicyclic Molecules.

Stoddart, J. F. (1979). *Chem. Soc. Rev.* **7**, 85–142. From Carbohydrates to Enzyme Analogues.

Tabushi, I. and Yamada, H. (1977). *Tetrahedron* **33**, 1101–1104. Temperature Dependent Chemical Shift Change in 4-Substituted [2.2.2.2]Paracyclophanes.

Tabushi, I., Shimokawa, K., Shimizu, N., Shirakata, H., and Fujita, K. (1976). *J. Am. Chem. Soc.* **98**, 8755–7856. Capped Cyclodextrin.

Tabushi, I., Fujita, K., and Yuan, L. C. (1977a). *Tetrahedron Lett.*, 2503–2506. Specific Host-Guest Energy Transfer by Use of β-Cyclodextrin Rigidly Capped with a Benzophenone Moiety.

Tabushi, I., Shimizu, N., Sugimoto, T., Shiozuka, M., and Yamamura, K. (1977b). *J. Am. Chem. Soc.* **97**, 7100–7102. Cyclodextrin Flexibility Capped with Metal Ion.

Tabushi, I., Kiyosuke, Y., Sugimoto, T., and Yamamura, K. (1978). *J. Am. Chem. Soc.* **100**, 916–919. Approach to the Aspects of Driving Force of Inclusion by α-Cyclodextrin.

Tabushi, I., Kuroda, Y., and Shimokawa, K. (1979). *J. Am. Chem. Soc.* **101**, 1614–1615. Duplex Cyclodextrin.

Takeo, K., Hirose, K., and Kuge, T. (1973). *Chem. Lett.*, 1233–1236. Carbon-13 Nuclear Magnetic Resonance Spectra of Cycloamyloses and Their Peracetates.

Timko, J. M., Helgeson, R. C., Newcomb, M., Gokel, G. W., and Cram, D. J. (1974). *J. Am. Chem. Soc.* **96**, 7097–7099. Structural Parameters that Control Association Constants Between Polyether Hosts and Alkylammonium Guest Compounds.

Timko, J. M., Moore, S. S., Walba, D. M., Hiberty, P., and Cram, D. J. (1977). *J. Am. Chem. Soc.* **99**, 4207–4219. Host–Guest Complexation. 2. Structural Units that Control Association Constant Between Polyethers and *tert*-Butylammonium Salts.

Tümmler, B., Maass, G., Weber, E., Wehner, W., and Vögtle, F. (1977). *J. Am. Chem. Soc.* **99**, 4683–4690. Noncyclic Crown-Type Polyethers, Pyridinophane Cryptands, and Their Alkali Metal Complexes: Synthesis, Complex Stability and Kinetics.

Tümmler, B., Maass, G., Vögtle, F., Heimann, U., Sieger, H., and Weber, E. (1979). *J. Am. Chem. Soc.* **101**, 2588–2598. Open Chain Polyethers. Influence of Aromatic Donor End Groups on Thermodynamics and Kinetics of Alkali Metal Ion Complex Formation.

Tundo, P., and Fendler, J. H. (1980). *J. Am. Chem. Soc.* **102**, 1760–1762. Photophysical Investigations of Chiral Recognition in Crown Ethers.

Ueno, A., Yoshimura, H., Saka, R., and Osa, T. (1978). *J. Am. Chem. Soc.* **101**, 2779–2780. Photocontrol of Binding Ability of Capped Cyclodextrin.

Urbach, F. L. (1979). In *Coordination Chemistry of Macrocyclic Compounds* (G. A. Melson, Ed.), Plenum Press, New York, pp. 345–392. Ligand Field Spectra and Magnetic Properties of Synthetic Macrocyclic Complexes.

Usui, T., Yamaoka, N., Matsuda, K., Tuzimura, K., Sugiyama, H., and Seto, S. (1973). *J. Chem. Soc. Perkin Trans. I*, 2425–2432. Carbon-13 Nuclear Magnetic Resonance Spectra of Glucobioses, Glucotrioses and Glucans.

Vögtle, F. and Weber, E. (1974). *Angew. Chem. Int. Ed. Eng.* **13**, 814–816. Octopus Molecules.

Vögtle, F. and Weber, E. (1979). *Angew. Chem. Int. Ed. Eng.* **18**, 753–776. Multidentate Acyclic Ligands and Their Complexation.

Weber, E., Müller, W. M., and Vögtle, F. (1979). *Tetrahedron Lett.*, 2335–2338. Komplexe zwischen Neutralmoleculen: Neutralkomplexe von Krakenmolekülen mit Kronenethern und anderen Organischen Molekülen.

CHAPTER

8

POLYELECTROLYTES

Polymers having large numbers of ionized functional groups are referred to as polyelectrolytes or polyions. Polysoaps are polyelectrolytes in which the ionized groups are attached to the polymer backbone by long hydrocarbon chains. Hydrophobic interactions between the alkyl chains stabilize compact polysoap conformations in a manner analogous to micelle formation. Recognition of proteins and nucleic acids as polyelectrolytes has greatly stimulated research in this area. There are several books and reviews on polyelectrolytes (Rice and Nagasawa, 1961; Oosawa, 1971; Selegny et al., 1974; Dukhin and Shilov, 1974; Morawetz, 1975; Ise, 1975; Frisch et al., 1976) and on their utilization as media for reactivity control (Morawetz, 1970, 1978; Fendler and Fendler, 1975; Okubo and Ise, 1977; Overberger et al., 1978; Klotz, 1978). Physical and chemical properties of polyelectrolytes and theories governing their behavior are summarized in this chapter.

1. CHEMISTRY OF POLYELECTROLYTES

Typical synthetic polyelectrolytes are polymers containing ionizable carboxylic, sulfonic acid amine, imidazole, pyridine, and other functional groups (Table 8.1). The extent of ionization depends on the dissociation constant of the functional group and on the hydrogen (or hydroxide) ion concentration of the aqueous polymer solution. Weakly acidic (e.g., acrylic acid) or weakly basic (e.g., ethylene imine) functional groups can be partially neutralized by pH adjustments. This provides a control over the charge density of the polyelectrolyte. Alternative charge density control is accomplished in copolymers formed from appropriate ratios of ionizable and nonionizable monomers. Copolymerization of equivalent amounts of acidic and basic monomers results in amphoteric polyelectrolytes. More importantly, copolymerization provides means for introducing complex "tailor-made" functional groups at desired locations.

Conformations of macromolecules influence their charge densities. Most synthetic polymer chains are flexible and assume a variety of shapes. On dilution, expansion of polyelectrolytes causes some degree of charge separation. The maximum charge separation attainable is determined by the distances between the charge carrying groups on the backbone and on the elasticity of the

Table 8.1. Typical Polyelectrolytes

Polyelectrolyte	Structure	Polyelectrolyte	Structure
Poly(acrylate)	$(-CH_2-CH(CO_2^-)-)_n$	Poly-4(5)-(vinylimidazole)	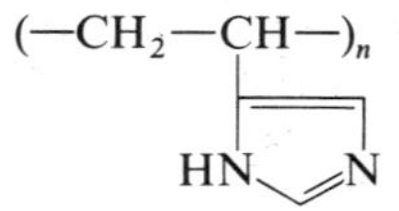 $(-CH_2-CH(C_3H_2N_2H)-)_n$ (imidazole ring: HN, N)
Poly(methylacrylate)	$(-CH_2-C(CH_3)(CO_2^-)-)_n$	Poly{1-[2-(adenin-9-yl)ethyl]-4-pyridinioethylene chloride}	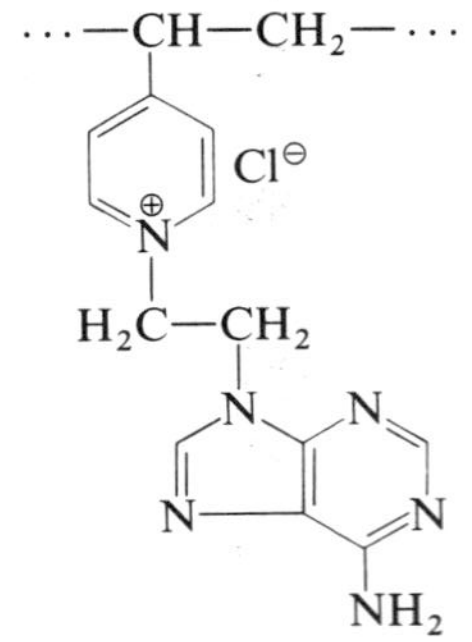 $\cdots-CH-CH_2-\cdots$ with pyridinium ring ($N^{\oplus}$, $Cl^{\ominus}$), $N-H_2C-CH_2-$adenine (N, N, N, N, NH_2)
Poly(vinylsulfonate)	$(-CH_2-CH(SO_3^-)-)_n$		

Poly(styrenesulfonate)	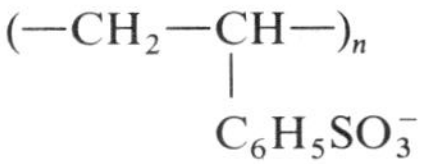 $(-CH_2-CH(C_6H_5SO_3^-)-)_n$	Poly{1-[2-(thymin-1-yl)ethyl]-4-pyridinioethylene chloride}	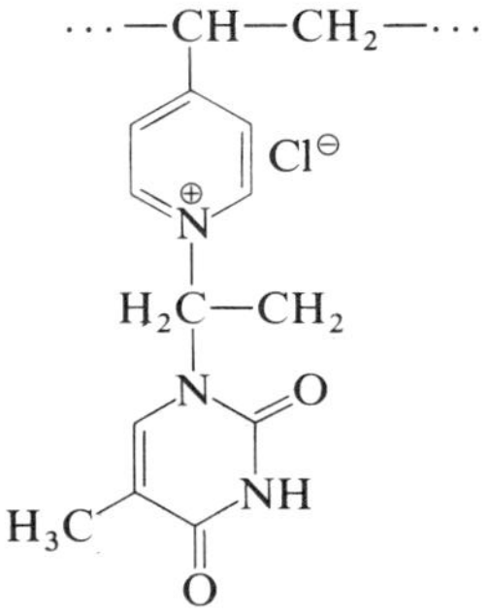 $\cdots-CH-CH_2-\cdots$ (4-pyridinio, $Cl^{\ominus}$; $N^{\oplus}-H_2C-CH_2-$thymin-1-yl, H_3C)
Poly(vinylaminehydrochloride)	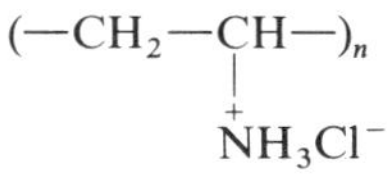 $(-CH_2-CH(\overset{+}{N}H_3Cl^-)-)_n$		
Poly(ethyleneimininehydrochloride)	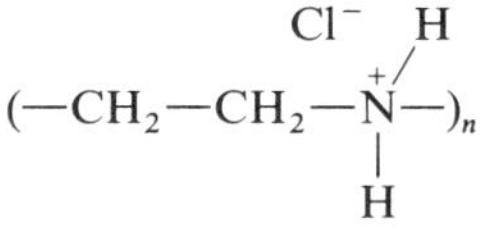 $(-CH_2-CH_2-\overset{+}{N}H_2-)_n$ Cl^-		
Poly(vinylpyridinehydrochloride)	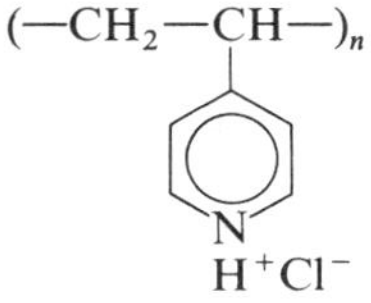 $(-CH_2-CH-)_n$ (4-pyridyl, N H^+Cl^-)	Poly-{1-[2-(theophyllin-7-yl)ethyl]-4-pyridinioethylene chloride}	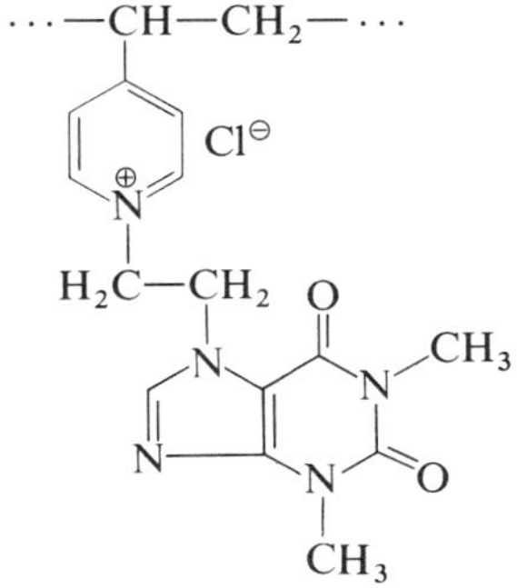 $\cdots-CH-CH_2-\cdots$ (4-pyridinio, $Cl^{\ominus}$; $N^{\oplus}-H_2C-CH_2-$theophyllin-7-yl, CH_3, CH_3)

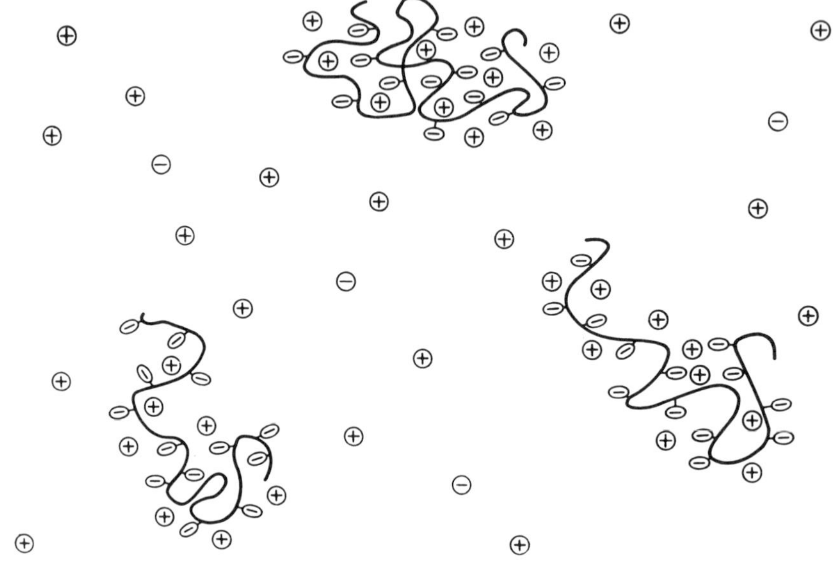

Figure 8.1. Schematic representation of the ion distribution in dilute solutions containing a salt of a polymeric acid composed of flexible chain molecules and a small concentration of added uni-univalent electrolyte (taken from Morawetz, 1975).

polymer. Even in extreme dilutions charge distribution is not uniform. Regions of the polyelectrolyte may contain high charge densities that selectively attract counterions (Figure 8.1). Proteins, nucleic acids, and their synthetic analogs have well-defined, usually compact, conformations in solutions. They are highly charged and have substantial surface potentials. Intrinsic viscosities reflect the conformational differences between synthetic and biological polyelectrolytes. The intrinsic viscosity of polyvinylpyrrolidine (22 ml g^{-1}) and serum albumin (4 ml g^{-1}) illustrates this point. Not surprisingly, binding constants of small molecules to serum albumin are appreciably greater than those for typical synthetic polyelectrolytes. Binding efficiency can be improved by creating more compact polyelectrolytes, by crosslinking, and/or by introducing highly branched ionizable groups.

2. THERMODYNAMICS OF POLYELECTROLYTE SOLUTIONS

Functional Group Dissociation at Polymer Backbones

Dissociation of large numbers of identical functional groups attached to macromolecular backbones cannot be readily expressed in terms of successive individual constants. Potentiometric titrations of polyelectrolytes give mean ionization constants, expressed as apparent dissociation constants, pK_{app}:

$$\text{p}K_{\text{app}} \equiv \text{pH} + \log \frac{1 - \alpha_i}{\alpha_i} \tag{8.1}$$

where α_i is the overall degree of ionization. Acidic polyelectrolytes electrostatically attract, whereas basic polyelectrolytes repel, hydrogen ions. Consequently, the pH at acidic polyelectrolyte surfaces is lower than that of bulk water. The opposite situation prevails, of course, for basic polyelectrolytes. Their surface pH will be higher than that of the surrounding bulk solution.

Apparent polyelectrolyte dissociation constants can also be expressed by

$$pK_{\text{app}} = pK^0 + \frac{\Delta G_{el}^i(\alpha_i)}{2 \cdot 3RT} \tag{8.2}$$

where pK^0 is the dissociation constant of the ionizing group in the absence of electrostatic interactions with other ionizing groups and $\Delta G_{el}^i(\alpha_i)$ is the electrostatic free energy for the removal of an equivalent of protons at a given degree of ionization. Combination of equations 8.1 and 8.2 leads to

$$\text{pH} = pK^0 + \log \frac{\alpha_i}{1 - \alpha_i} + \frac{\Delta G_{el}^i(\alpha_i)}{2 \cdot 3RT} \tag{8.3}$$

Attempts have been made to estimate ΔG_{el}^i and pK^0 values for spherical proteins. Estimated values differed considerably from pK^0 values determined by protein titrations (Morawtz, 1975). Effects of conformational changes on the ionization equilibria and uncertainties in the parameters used in the estimations are responsible for the discrepancies. Titrations of poly(α-L-glutamic acid) in the presence of added NaCl illustrates the complexities (Figure 8.2; Nagasawa and Holtzer, 1964). Four distinct regions are seen. The initial decrease of pK_{app} with increasing α_i (region I) corresponds to the association of the polypeptide at low charge densities. In region II the helical form predominates; there is a helix-coil transition in region III and in region IV the polypeptide assumes random coil conformations. Graphical treatment of the data allows the estimation of the fraction of polypeptide residues in the helical form at any given pH, f_h (Figure 8.2), and the standard free energy change, $\Delta G_{h \to c}^0$, corresponding to a helix-coil transition per amino acid residue, with both forms in the uncharged state from (Nagasawa and Holtzer, 1964)

$$\Delta G_{h \to c}^0 = -RT \ln \left(\frac{1 - f_h}{f_h} \right) \bar{\alpha}_i - 2.3RT \int_{\bar{\alpha}_i = 0}^{\bar{\alpha}_i} [(\alpha_i)_h - (\alpha_i)_c] \, d\text{pH} \tag{8.4}$$

where $f_h = [(\bar{\alpha}_i) - (\alpha_i)_c]/[(\alpha_i)_h - (\alpha_i)_c]$ and $(\alpha_i)_h$, $(\alpha_i)_c$, and $\bar{\alpha}_i$ are defined in Figure 8.2. The situation for synthetic polyelectrolytes undergoing chain expansions and conformational transitions is even more complex. Titrations of copolymers containing well-separated pairs of ionizable groups often show pronounced breaks that can be identified with successive intrinsic dissociation constants.

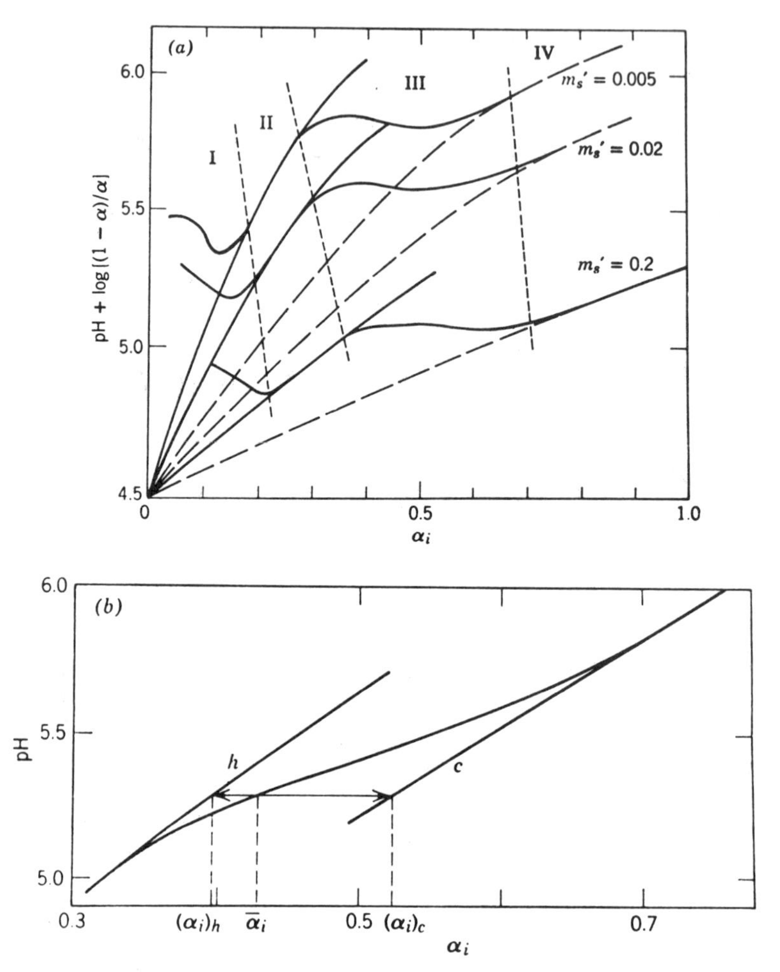

Figure 8.2. Interpretation of titration data on poly(α-L-glutamic acid). (*a*) Regions reflecting I, polymer association; II, titration of the helical form; III, the helix-coil transition; and IV, titration of the random coil form at various concentrations, m_s', of added NaCl. (*b*) Graphical method for estimating f_h. Titration curves *h* and *c* are estimated for systems where polymer could be kept fully helical or as a random coil over the entire range of α_i, respectively; $f_h = [(\bar{\alpha}_i) - (\alpha_i)_c]/[(\alpha_i)_h - (\alpha_i)_c]$ (taken from Nagasawa and Holtzer, 1964).

Ion Binding to Polyelectrolytes

The effective charge density of polyelectrolytes is much less than that based on the stoichiometric concentration of the ionized functional groups. This is a consequence of charge neutralization by counterion binding. Distribution of counterions is intimately involved with the polyelectrolyte effects and with the theories describing it (see Chapter 8.3). The fraction of counterions that remains "free" can be assessed from activity coefficient measurements (Ise, 1971;

Katchalsky, 1971). Chemical potentials for polyions (cations or anions) and their counterions (anions or cations), μ, are given by

$$\mu_c - \mu_c^0 = RT \ln a_c = RT \ln (\gamma_c m_c) \tag{8.5}$$

$$\mu_a - \mu_a^0 = RT \ln a_a = RT \ln (\gamma_a m_a) \tag{8.6}$$

where μ^0 is the respective chemical potential in the reference state, a is the single-ion activity, γ is the single-ion activity coefficient, m is the ionic concentration, and the subscripts c and a denote cations and anions, respectively. The mean activity, $a_{\pm}$, and the mean activity coefficient, $\gamma_{\pm}$, are expressed by

$$a = a_c^{\nu c} \cdot a_a^{\nu a} \equiv a_{\pm}^{\nu} \tag{8.7}$$

$$\gamma_{\pm}^{\nu} = \gamma_c^{\nu c} \cdot \gamma_a^{\nu a} \tag{8.8}$$

where ν_c and ν_a are the total numbers of anions and cations ($\nu = \nu_c + \nu_a$). Single-ion activity coefficients cannot be measured, of course. In principle, mean activity coefficients can be determined by (1) electromotive force (emf) measurements in concentration cells, (2) isopiestic vapor pressure, (3) osmotic pressure, (4) boiling point elevation or freezing point depression, (5) solubility, (6) sedimentation and diffusion measurements, and (7) by partitioning between two immiscible solvents. In practice, emf, isopiestic vapor pressure, and osmotic coefficient

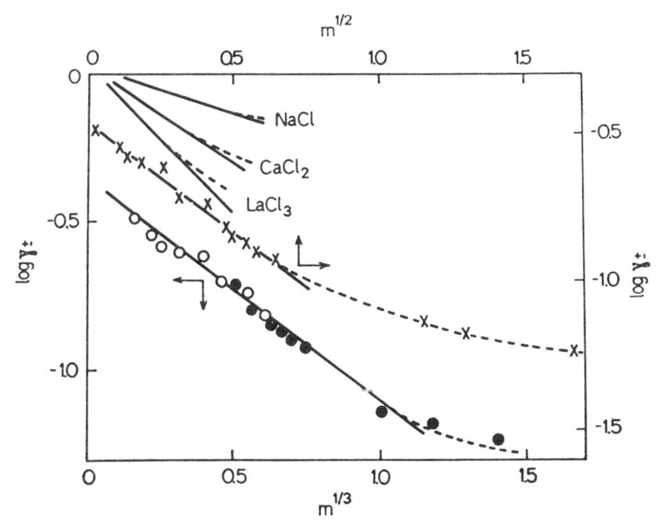

Figure 8.3. Mean ion activity coefficient depencency of sodium polyacrylate on the cube root and square root of their concentrations in aqueous solutions at 25°C, determined by the emf (○) and isopiestic (●) methods. Mean ion activity coefficient dependencies of NaCl, $CaCl_2$, and $LaCl_3$ on the square roots of their concentrations are given for comparison (taken from Ise, 1971).

measurements have been those most frequently used for determining polyelectrolyte mean activity coefficients.

Using sodium glass electrodes, Ise and Okubo (1965) measured the emf, E, of polyelectrolytes in a concentration cell of the type:

Na^+-selective glass electrode	Na-salt of polyelectrolyte $(a_\pm)_I$	Na-salt of polyelectrolyte $(a_\pm)_{II}$	Na^+-selective glass electrode

The passage of 1 faraday (F) of electricity results in the transfer of tp/vp moles of polyions and their tp counterions from a solution with mean ion activity $(a_\pm)_{II}$ to one with mean ion activity $(a_\pm)_I$ as governed by

$$E = \left(1 + \frac{1}{vp}\right)\frac{RT}{F}\int_{(a_\pm)_{II}}^{(a_\pm)_I} tp\, d\ln a \tag{8.9}$$

where t_p is the transference number of the polyion.

In isopiestic vapor pressure measurements the determined equilibrium osmotic coefficient, ϕ, is given by

$$\phi = \frac{v_{\text{ref}} \cdot m_{\text{ref}}}{(\alpha + 1)m}\phi_{\text{ref}} \tag{8.10}$$

where v_{ref} is the number of ions produced from the electrolyte molecule, m is the electrolyte concentration, and the subscript ref denotes the reference electrolyte, typically KCl, whose osmotic coefficient ϕ_{ref} is known. Substitution of ϕ values the Gibbs–Duheim relationship,

$$\ln \gamma_\pm = (\phi - 1) + \int_0^m (\phi - 1)\, d\ln m \tag{8.11}$$

gives the mean polyelectrolyte activity coefficients.

Intraion interactions prevail even at infinitely dilute polyelectrolyte solutions. The magnitude of these interactions depends, of course, on the average distances between the ionizable groups on the polymer backbone. In contrast to simple electrolytes, absolute mean activity coefficients do not exist in polyelectrolytes. The accessible parameter is the concentration dependence of $\gamma_\pm$. Figure 8.3 compares concentration dependencies of a typical polyelectrolyte with simple electrolytes. Cube root dependency has been rationalized in terms of the greater ease of ionic lattice formation in polyelectrolytes compared with that in simple salts (Ise, 1971).

An interesting property of polyelectrolytes is that, in salt free solutions, the product of the osmotic coefficient and the charge density remains constant as the charge density increases (Alexandrowicz and Katchalsky, 1963). Polyelectrolytes apparently act as osmotic buffers.

3. POLYELECTROLYTE THEORIES

Several theories have been postulated to account for the distribution of ions in polyelectrolyte solutions. The counterion condensation theory, developed by Manning (1969, 1972, 1974, 1978a, 1979, 1981; Record et al., 1978), is attractively descriptive. It rests upon two basic tenets. First, it assumes that the effective polyelectrolyte charge density, ξ_{eff}, has a definite maximum. Higher charge densities are reduced to the maximum by counterions "condensating" on the polyelectrolyte. Second, interactions between the polyelectrolyte and the counter- or coions are accurately treatable by the limiting form of the Debye–Hückel approximation assuming cylindrical shapes. Manning's theory appears to accommodate the observed dependence of fractional charges only on the valence of the counterion and on the axial charge density of the polyelectrolyte. Fractional charges of sodium polyacrylate, sodium polystyrene sulfonate, and sodium polyphosphate are equal (Manning, 1975; Kielman et al., 1976). The reduced polyelectrolyte charge density, ξ, is given by

$$\xi = \frac{b_B}{b} \tag{8.12}$$

where b_B and b are the Bjerrum length and the spacing between singly charged groups along the axes of the linear polyelectrolyte chain, respectively. The Bjerrum length is defined by

$$b_B = \frac{e^2}{DkT} \tag{8.13}$$

where e is the protonic charge, D is the bulk dielectric constant of water, k is Boltzmann's constant, and T is the absolute temperature. Operationally, counterion condensation is defined as the mode of binding of single counterion species of valence N to a polyelectrolyte, such that its charge fraction equals the constant value of $(N\xi)^{-1}$ over a broad concentration range (Manning, 1978a).

Mathematical derivation of the polyelectrolyte condensation theory (Manning, 1978a) is less straightforward. Indeed, Manning's assumptions and interpretations have been questioned (Stigter, 1978, Manning, 1978b; Gueron and Weisbuch, 1980). The volume of the region surrounding the polyelectrolyte within which the counterions are "bound," V_p, is expressed by (Manning, 1978a)

$$V_p = 4\pi(2.71)L_A N'(v + v')(\xi - N^{-1})b^3 \tag{8.14}$$

where L_A is Avogadro's number, v and v^1 are the numbers of counterions and coions, and N and N' are valencies of counterions and coions, respectively. The units of V_p and b are cubic centimeters per mole of polyion and angstroms, respectively. The number of associated counterions per fixed charge, Θ_1, in the presence of a uni-univalent electrolyte is

$$\Theta_1 = 1 - \xi^{-1} \tag{8.15}$$

A polyelectrolyte with $\xi > 1$ dissolved in a uni-univalent electrolyte of vanishingly small stoichiometric concentration, c_1, is surrounded by a local concentration, c_1^{loc}, of counterions:

$$c_1^{\text{loc}} = 10^3\Theta_1 V_p^{-1} = 24.3(\xi b^3)^{-1} \tag{8.16}$$

where c_1^{loc} has units of molarity. Table 8.2 gives values for the thickness of the cylindrical shell (Δx) of volume V_p, within which univalent counterions of local concentration c_1^{loc} are bound to polynucleotides (Record et al., 1976; Manning, 1978a). The data are explicable in terms of Manning's counterion condensation theory. It is seen, for example, that 76% of the phosphate charge in native DNA is neutralized by counterions.

Counterion distribution in polyelectrolytes is a somewhat elusive concept. Several modes of binding and interactions have been invoked. Manning's theory does not specify the precise significance of "bound" counterions. It simply treats all those falling within V_p. The cylindrical Gouy theory (Stigter, 1975, 1977, 1978), on the other hand, distinguishes between site and atmospherically or territorially bound (Spegt and Weill, 1976; Tivant et al., 1979) counterions. Site-bound counterions are believed to be coordinated to the oppositely charged groups on the polyion without interspacing water molecules of hydration. Territorially bound counterions are within the potential field of the polyelectrolyte, but retain the hydration shells and are free to move about. Using different terminology, site-bound counterions correspond to those retained in the Stern layer and territorially bound ones are analogous to those located within the Gouy–Chapman electrical double layer. Figure 8.4 shows potential-distance curves around a polyion (double-stranded DNA) according to the different theories (Schellman and Stigter, 1977).

A second distinguishing feature of the cylindrical Gouy theory is that it uses the full Poisson–Boltzmann equation. The attractive feature of the Debye–Hückel approximation, used in the condensation theory (Manning, 1978a), is that it leads to a linear differential equation for the potential. Conversely, the

Table 8.2. Parameters for Bound Univalent Cations in the Limit $c_1 \to 0$

Polynucleotide	b, Å	ξ	θ_1	V_P, $\text{cm}^3/\text{mole}^{-1}\ \text{P}^{-1}$	Δx, Å	c_1^{loc}, M
Poly A · 2 poly U	1.0	6.9	0.86	252	5	3.52
Poly A · poly U	1.6	4.5	0.78	589	7	1.32
Native DNA	1.7	4.2	0.76	656	7	1.18
Poly A	3.1	2.3	0.57	1592	12	0.35
Single-stranded DNA	4.0	1.8	0.44	2104	12	0.21
Poly U	4.5	1.6	0.38	2247	12	0.17

Source: Compiled from Manning, 1978a.

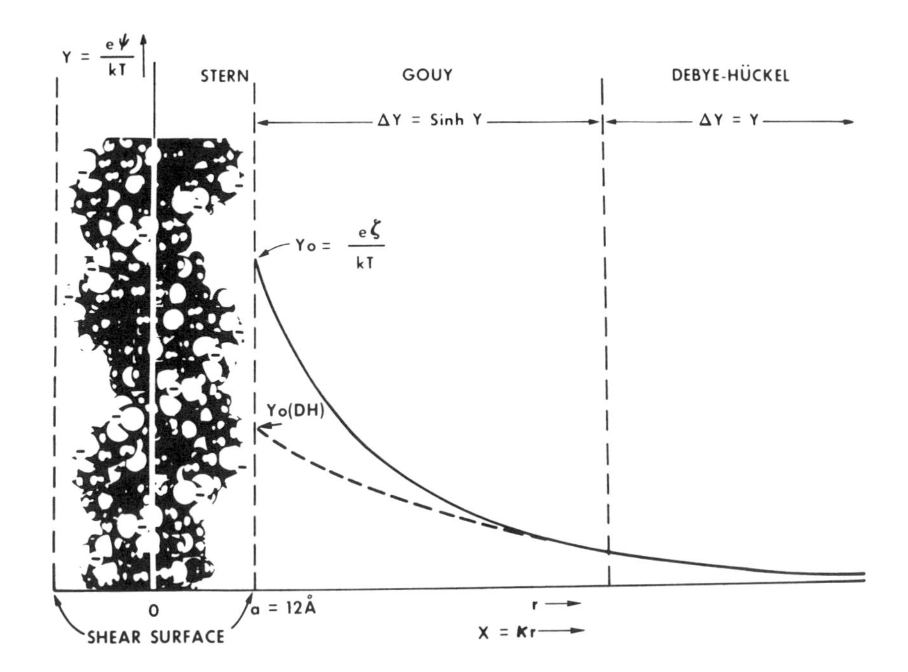

Figure 8.4. Schematic representation of the electrical double layer with Gouy-Chapman potential, $y = e\Psi/T$, versus distance, $x = \kappa r$, from the axis of cylinder with radius $x_0 = \kappa a$, for double-stranded DNA. The solid line, $y_0 = e\xi/kT$, is the Gouy potential from equation 8.17; the broken line is the Debye-Hückel potential from equation 8.17 taking $\beta = \gamma = 1$ and adjusting ξ for coincidence with the Gouy curve for large distance (taken from Schellman and Stigter, 1977).

nonlinear Poisson–Boltzmann equation must be solved numerically (Stigter, 1975, 1978; Fixman, 1979). The Gouy theory gives the interaction potential, y, for a uniformly charged polyelectrolyte cylinder with reduced surface potential, y_0, and linear charge density, ξ, in an ionic atmosphere, by (Schellman and Stigter, 1977)

$$y = \frac{2\xi}{\beta(x_0, y_0)} \frac{\gamma(x, y)}{\gamma(x_0 y_0)} \frac{K_0(X)}{x_0 K_1(x_0)} \tag{8.17}$$

where $x_0 = \kappa a$, $1/\kappa$ is the Debye distance, a is the radius of the cylinder, K_0 and K_1 are modified Bessel functions, and β and λ are numerical factors correcting the Debye–Hückel expression. Figure 8.4 illustrates the physical meaning of γ. The function $\gamma(x, y)$ converts the Debye–Hückel potential curve to the corresponding Gouy curve. The broken line in the figure indicates the Debye–Hückel approximation. At the surface of the cylinder

$$y_0 = \gamma(x_0, y_0)y_0(DH) \tag{8.18}$$

The function β corrects the Debye–Hückel approximation of the potential-charge relation. For $x = x_0$ equation 8.17 becomes

$$y_0 = \frac{2\xi}{\beta(x_0, y_0)} \frac{K_0(x_0)}{x_0 K_1(x_0)} \tag{8.19}$$

or for small x_0

$$y_0 = -\frac{2\xi \ln x_0}{\beta(x_0 y_0)} \tag{8.20}$$

If the Debye–Hückel approximation is exact for a cylindrical double layer, as required by the condensation theory (Manning, 1978b), then $\gamma = \beta = 1$. Depending on the systems and conditions selected, the condensation theory may give results similar to the Gouy model (Stigter, 1978; Wilson et al., 1980).

Poisson–Boltzmann distribution of both mono- and divalent counterions around polyelectrolytes of different shapes has been computed using tRNA as an example (Gueron and Weisbuch, 1980). At added 1–10 mM electrolyte concentrations polyelectrolyte properties are shape independent. Counterion distributions resemble those in the neighborhood of a plane with the same charge density. Three counterion regions have been described (Gueron and Weisbuch, 1980). The inner region, referred to as counterions in the immediate vicinity of the polyelectrolyte, CIV, extends up to a distance of $Rc/Z\xi$ from the surface (where Rc is the radius of the cylinder or equivalent sphere). The intermediate region extends to a distance where the electrostatic potential is kT/e. Beyond the intermediate region counterion distribution is unaffected by the polyelectrolyte surface and the Debye–Hückel approximation is valid (Figure 8.4). The CIV is the most important region of counterion distribution. In this region the counterion concentration can be as high as 10 M, and the charge density is remarkably insensitive to salts, Counterion binding in the CIV region can be calculated from the mass action law. Interestingly, these calculations result in a large and concentration dependent V_p, tends to infinity at dilute salt solutions (Gueron and Weisbuch, 1980).

REFERENCES

Alexandrowicz, Z. and Katchalsky, A. (1963). *J. Polym. Sci.* **A1**, 3231–3260. Colligative Properties of Polyelectrolyte Solutions in Excess of Salt.

Dukhin, S. S. and Shilov, V. N. (1974). *Dielectric Phenomena and the Double Layer in Disperse Systems and Polyelectrolytes*, John Wiley, New York.

Fendler, J. H. and Fendler, E. J. (1975). *Catalysis in Micellar and Macromolecular Systems*, Academic Press, New York.

Fixman, M. (1979). *J. Chem. Phys.* **70**, 4995–5005. The Poisson–Boltzmann Equation and Its Application to Polyelectrolytes.

Frisch, K. C., Klempner, D., and Patsis, A. V. (1976). *Polyelectrolytes*, Technomic, Westport, Connecticut.

Gueron, M. and Weisbuch, G. (1980). *Bipolymers* **19**, 353–382. Polyelectrolyte Theory, I. Counterion Accumulation, Site Binding, and Their Insensitivity to Polyelectrolyte Shape in Solutions Containing Finite Salt Concentrations.

Ise, N. (1971). *Adv. Polym. Sci.* **7**, 536–593. The Mean Activity Coefficient of Polyelectrolytes in Aqueous Solutions and Its Related Properties.

Ise, N. (1975). In *Polyelectrolytes and Their Applications* (A. Rembaum and E. Selegny, Eds.), D. Reidel Co., Dordrecht, Holland, pp. 71–96. Polyelectrolyte Catalysis of Ionic Reactions.

Ise, N. and Okubo, T. (1965). *J. Phys. Chem.* **69**, 4102–4109. Mean Activity Coefficients of Polyelectrolytes I. Measurements of Sodium Polyacrylates.

Katchalsky, A. (1971). *Pure Appl. Chem.* **26**, 327–373. Polyelectrolytes.

Kielman, H. S., Van der Hoeven, J. M. A. M., and Leyte, J. C. (1976). *Biophys. Chem.* **4**, 103–111. Nuclear Magnetic Relaxation of ^{23}Na and ^{7}Li Ions in Polyphosphate Solutions.

Klotz, I. M. (1978). In *Molecular Movements and Chemical Reactivity as Conditioned by Membranes, Enzymes and Other Macromolecules* (E. Lefever and A. Goldbeter, Eds.), Advan. Chem. Phys. **39**, Wiley-Interscience, New York, pp. 109–160. Synzymes: Synthetic Polymers with Enzyme-like Catalytic Activities.

Manning, G. S. (1969). *J. Chem. Phys.* **51**, 924–933. Limiting Laws and Counterion Condensation in Polyelectrolyte Solutions. I. Colligate Properties.

Manning, G. S. (1972). *Annu. Rev. Phys. Chem.* **23**, 117–140. Polyelectrolytes.

Manning, G. S. (1974). In *Polyelectrolytes* (E. Selegny, Ed.), D. Reidel Co., Dordrecht, Holland, pp. 39–55. Limiting Laws for Equilibrium and Transport Properties of Polyelectrolyte Solutions.

Manning, G. S. (1975). *J. Chem. Phys.* **62**, 748–749. Remarks on the Paper "Nucleic Magnetic Relaxation of ^{23}Na in Polyelectrolyte Solutions" by van der Klink, Zuiderweg, and Leyte.

Manning, G. S. (1978a). *Quart. Rev. Biophys.* **11**, 179–246. The Molecular Theory of Polyelectrolyte Solutions with Applications to the Electrostatic Properties of Polynucleotides.

Manning, G. S. (1978). *J. Phys. Chem.* **82**, 2349–2351. Comments on "A Comparison of Manning's Polyelectrolyte Theory with the Cylindrical Gouy Model" by D. Stigter.

Manning, G. S. (1979). *Acc. Chem. Res.* **12**, 443–449. Counterion Binding in Polyelectrolyte Theory.

Manning, G. S. (1981). *J. Phys. Chem.* **85**, 870–877. Limiting Laws and Counterion Condensation in Polyelectrolyte Solutions. 6. Theory of the Titration Curve.

Morawetz, H. (1970). *Acc. Chem. Res.* **3**, 354–360. Chemical Reaction Rates Reflecting Physical Properties of Polymer Solutions.

Morawetz, H. (1975). *Macromolecules in Solutions*, 2nd ed., Wiley-Interscience, New York.

Morawetz, H. (1978). *Isr. J. Chem.* **17**, 287–290. Some Recent Advances in the Study of the Reactivity of Systems Containing Polymers.

Nagasawa, M. and Holtzer, A. (1964). *J. Am. Chem. Soc.* **86**, 538–543. The Helix-Coil Transformation in Solution of Polyglutamic Acid.

Okubo, T. and Ise, N. (1977). *Adv. Polym. Sci.* **25**, 136–181. Synthetic Polyelectrolytes as Models of Nucleic Acids and Esterases.

Oosawa, F. (1971). *Polyelectrolytes*, Marcel Dekker, New York.

Overberger, C. G., Guterl, A. C., Jr., Kawakami, Y., Mathias, L. J., Meenakshi, A., and Tomono, T. (1978). *Pure Appl. Chem.* **50**, 309–313. Recent Developments in the Use of Polymers as Reactants in Organic Reactions.

Record, M. T., Jr., Woodbury, C. P., and Lohman, T. M. (1976). *Biopolymers* **15**, 893–915. Na^+ Effects on Transitions of DNA and Polynucleotides of Variable Linear Charge Density.

Record, M. T., Jr., Anderson, C. F., and Lohman, T. M. (1978). *Quart. Rev. Biophys.* **11**, 103–178. Thermodynamic Analysis of Ion Effects on Binding and Conformational Equilibria of Proteins and Nucleic Acids: The Roles of Ion Association and Release, Screening and Ion Effects on Water Activity.

Rice, S. A. and Nagasawa, M. (1961). *Polyelectrolyte Solutions*, Academic Press, London.

Schellman, J. A., and Stigter, D. (1977). *Biopolymers* **16**, 1415–1434. Electrical Double Layer, Zeta Potential, and Electrophoretic Charge of Double-Stranded DNA.

Selegny, E., Mandel, M., and Strauss, V. P. (1974). *Polyelectrolytes*, D. Reidel Co., Dordrecht, Holland.

Spegt, P. and Weill, G. (1976). *Biophys. Chem.* **4**, 143–149. Magnetic Resonance Distinction Between Site Bond and Atmospherically Bound Counterions in Polyelectrolyte Solutions.

Stigter, D. (1975). *J. Colloid Interface Sci.* **53**, 296–306. The Charged Colloidal Cylinder with a Gouy Double Layer.

Stigter, D. (1977). *Biopolymers* **16**, 1435–1448. Interactions of Highly Charged Colloidal Cylinders with Applications to Double-Stranded DNA.

Stigter, D. (1978). *J. Phys. Chem.* **82**, 1603–1606. A Comparison of Manning's Polyelectrolyte Theory with the Cylindrical Gouy Model.

Tivant, P., Turg, P., Chemla, M., Magdelenat, H., Spegt, P., and Weill, G. (1979). *Biopolymers* **18**, 1849–1857. Condensation of Co^{++} Ions on Chondroitin Sulfate from Transport Coefficients and Proton NMR Measurements in Aqueous Solutions.

Wilson, R. W., Rav, D. C., and Bloomfield, V. A. (1980). *Biophys. J.* **30**, 317–326. Comparison of Polyelectrolyte Theories of the Binding of Cations to DNA.

CHAPTER

9

COMPARISON OF THE DIFFERENT MEMBRANE MIMETIC AGENTS

Physical chemical properties of micelles, microemulsions, monolayers, organized multilayer assemblies, black lipid membranes (BLMs), vesicles, host–guest systems, and polyelectrolytes, collectively termed membrane mimetic agents, have been delineated in Chapters 2–8. The rational use of membrane mimetic agents requires a sufficient understanding of their physical chemical properties. In general, membrane mimetic agents are expected to (1) solubilize, concentrate, compartmentalize, organize, and localize reactants; (2) maintain proton and/or reactant gradients; (3) alter quantum efficiencies; (4) alter ionization potentials; (5) change oxidation and reduction properties; (6) change dissociation constants; (7) affect vectorial electron displacements; (8) alter photophysical pathways and rates; (9) alter chemical pathways and rates; (10) stabilize reactants, intermediates, transition states, and products; (11) separate products (charges); and (12) be chemically stable, optically transparent, and photochemically inactive. Not all models fulfill, of course, all these functions or provide unique media for all applications. It should be recognized that "organized assemblies" are notoriously disorganized. This is especially true for micelles and microemulsions. Even in organized multilayers there is a fair degree of disorder with respect to single monolayer contacts and schemes drawn for *X*, *Y*, and *Z* depositions (Figure 4.12) are purely speculative. This chapter provides the necessary understanding for choosing a given membrane model for a specific function. Properties of the different membrane mimetic agents, the various microenvironments they provide, and the way they solubilize and organize substrates (Fendler, 1980; Thomas, 1980; Israelachvili et al., 1980) are discussed comparatively.

1. PROPERTIES OF MEMBRANE MIMETIC AGENTS

Membrane mimetic agents can conveniently be divided into those assembled from surfactants, those acting as hosts, and those having ionized groups on polymer backbones. Aggregation behavior of surfactants depends on their chemical structures, on the nature of the media, and on the method of preparation. Opposing forces of repulsion between the polar headgroups and of association between the hydrocarbon chains of the surfactants are responsible for aggregation in water (Chapter 2). Dipole–dipole interactions provide the driving force for association in apolar solvents (Chapter 3). Formation of reversed micelles requires at least traces of water. These systems can be considered, therefore, to be surfactant entrapped water pools in hydrocarbon solvents. Increasing the concentration of entrapped water, that is, the size of water pools, at a given surfactant concentration, results in the formation of larger aggregates. If the water concentration is further increased, water-in-oil (w/o) microemulsions begin to appear (Chapter 3). Spreading an organic solution of a surfactant on water results in monolayer formation at the air-water interface (Chapter 4). Suitable surfactants undergo bilayer formation. There are two types of bilayers. The first type, the planar black (or bilayer) lipid membrane, the BLM, is formed on the orifice of a small pinhole (Chapter 5). The second type, the closed bilayer vesicle, is formed by the swelling of lipids. Bilayer vesicles are smectic mesophases of phospholipids (liposomes) or surfactants (surfactant vesicles) with

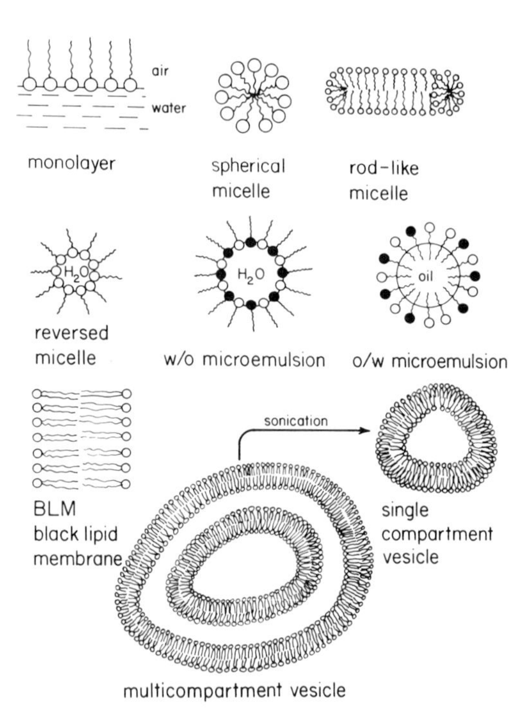

Figure 9.1. An oversimplified representation of organized structures of surfactants in different media (taken from Fendler, 1980).

water interspaced between them (Chapter 6). Figure 9.1 is a schematic representation of the structures formed from surfactants. These structures are, it must be realized, gross oversimplifications.

There are substantial morphological differences between the different classes of membrane mimetic agents. Aqueous micelles are spherical entities having 30–60 Å diameters. Reversed micelles have similar dimensions. Micellar dimensions can appreciably increase upon the solubilization of large molecules. Microemulsions and vesicles are considerably larger than micelles. Their diameters range from 50 to 1000 Å and from 300 to 5000 Å, respectively. Consequently, w/o microemulsions contain considerably larger water pools than reversed micelles (Kumar and Balasubramanian, 1980). Cavities provided by crown ethers, cryptands, cyclodextrins, and related hosts are small by comparison. They complex uni- and divalent metal ions, NH_4^+, $Ar\overset{+}{N}{\equiv}N$ and small aromatic compounds (Chapter 7). Dimensions of monolayers depend on the surface area of the subphase and on the surface pressure (Chapter 4); BLMs are confined within a relatively small pinhole (Chapter 5); and weight averaged molecular weights of polyions can be several million (Chapter 8).

The inherent stability of a given membrane mimetic agent is an important consideration. For all practical purposes polyions and synthetic and naturally occurring hosts are stable species. On the other hand, one can only talk about the kinetic stabilities of micelles, microemulsions, and vesicles. Micelles are in dynamic equilibrium with monomeric surfactants (see Section 2.4). The timescale for the release of a single surfactant molecule and its subsequent reincorporation, that is, for the dissociation of the micelle, is in the order of microseconds. The stepwise dissolution of micelles to monomers and the subsequent reassociation, that is, the dissolution of micelles, occurs on the millisecond timescale. Behavior of microemulsions is quite analogous. Conversely, surfactant residence times in vesicles are of the order of minutes to hours. Micelles, microemulsions, and vesicles can remain stable for weeks subsequent to their formation. Monolayers, under appropriate conditions, can be kept for an equally long time. BLMs, however, rarely last longer than a couple of hours.

Phase transition is an important property of monolayers, BLMs, and vesicles. Depending on the surface area–pressure isotherm, monolayers may be in a gaseous, fluid, or solid state (see Figure 4.5). Thermotropic phase transitions of BLMs and vesicles involve changes in the arrangements of lipids without altering the gross structural features of the bilayers (see Figure 6.14). Below the phase transition temperature the surfactant constituents of BLMs and vesicles are in highly ordered "solid" states, with their alkyl chains in all-trans conformations. Above the phase transition temperature lipids become "fluid" as the result of gauche rotations and kink formation. Micelles and host systems do not usually have temperature induced phase transitions. Polyions can undergo, however, conformational changes that may result in altered secondary and tertiary structures. There are additional motions of surfactants within the BLMs and vesicles. Surfactants may undergo segmental and rotational motions, lateral diffusion, and flip-flop (see Figure 6.20).

2. MICROENVIRONMENTS PROVIDED BY MEMBRANE MIMETIC AGENTS

Micelles, microemulsions, monolayers, organized multilayer assemblies, BLMs, vesicles, hosts, and polyelectrolytes provide regions of different polarities, acidities, and viscosities. Unambiguous determination of given regions, or indeed numerical assignment of values to given microenvironments, are nontrivial problems. Information is obtained through the use of spectroscopic (for example, absorption, emission, infrared, nuclear magnetic resonance (nmr), and electron paramagnetic resonance (epr)) probes. Such methods are plagued by difficulties. The first problem is the question of to what extent the probe perturbs the aggregates. Second, assumptions must be made on the location of probes. Third, decisions must be made on the solvents used as references. Microenvironments provided by aqueous micelles are the best understood.

Traditionally, aqueous micelles are considered to contain a hydrophobic "liquidlike" core and a polar surface. In light of more recent evidence (Fendler and Fendler, 1975; Mittal, 1977, 1979; Menger, 1979; Bunton, 1979), this picture should be somewhat modified. Micelles are dynamic species. The hydrocarbon tails are relatively mobile. Apparent microviscosities of approximately 8 cP (water = 1.0 cP, *n*-octanol = 8.9 cP, glycerol = 875 cP) have been determined from ^{13}C nmr spin-lattice relaxation times for the "interior" of aqueous micelles (Menger and Jerkunica, 1978). Polar headgroups of the surfactants are hydrated by, and are in contact with, a fair number of water molecules. The extent of water penetration into the micelles is still a matter of controversy (see Chapter 2).

There is a considerable amount of bound water at the inner surface of reversed micelles. By varying the size of the surfactant entrapped water pool, effective polarities of the environments can be substantially altered. Thus water in reversed micelles can attain polarities corresponding to that of benzene (Fendler, 1976). This, of course, represents a situation where all water molecules are tightly bound to the surfactant headgroups and corresponds to the region where extremely large rate enhancements (Fendler, 1976) and interesting photochemistry (Wong and Thomas, 1977) have been observed. There is no information, unfortunately, on the extent of water penetration into reversed micelles.

W/o microemulsions are expected to provide environments similar to those of reversed micelles. An important difference, however, is the considerably larger water pool in microemulsions than that in water. Thus, by definition, w/o microemulsions will always contain free and bound water molecules. The oil-in-water (o/w) microemulsion, in contrast to aqueous micelles, contains a sizeable hydrocarbon interior, This, in turn, provides highly apolar environments for entrapping substantial amounts of hydrophobic molecules in each aggregate.

Monolayers provide relatively polar environments close to their headgroups. Environments around the hydrocarbon tails depend on the surface pressure, that is, on the proximity of monolayer constituting surfactants to each other.

Vesicle bilayers are not as flexible as monolayers. They provide considerably more rigid interiors than micelles. Microviscosities in liposomes can be as high as 200 cP (Kano and Fendler, 1979). Rigidities of bilayers can be further increased by the addition of cholesterol. Vesicle entrapped water pool(s) provide additional unique environments. Since proton permeabilities across vesicle walls can be drastically reduced, it is possible to maintain substantial pH gradients between liposome entrapped and bulk water for several hours (Kano and Fendler, 1979; Cafiso and Hubbell, 1978). Vesicles are osmotically active. In hyperosmolar solutions they shrink and in hypoosmolar solutions they swell. An important consequence of osmotic shrinkage is that all of the liposome entrapped water becomes bound to the headgroups located at the inner surface of the vesicle.

Crown ethers and cyclodextrins have cavities of well-defined hydrophobicities. Structural alterations or changes in the dimensions of cavities cause only minor alterations in the microenvironments. Capping cyclodextrins provides, of course, additional binding sites and microenvironments.

Microenvironments around polyions can vary a great deal. The nature and the kind of functional groups and gross conformations are the influencing factors.

Ionizable membrane mimetic agents behave quite differently from simple electrolytes in dilute aqueous solutions. Micelles, monolayers, BLMs, vesicles, and polyions have appreciable surface potentials and charge densities. Charge distribution is treated by the various electrical double-layer theories. One important result is that negatively charged membrane mimetic agents attract positive counterions, while positively charged ones are surrounded by negative counterions. Consequently, the surface pH of anionic membrane mimetic agents is lower than that of the bulk solution. The reverse trend is observed for cationic membrane mimetic agents. The very nature of BLMs and vesicles allows the establishment of potential gradients across their bilayers. The dissymmetry thus created can be exploited in many practical applications.

3. SOLUBILIZATION AND ORGANIZATION IN MEMBRANE MIMETIC AGENTS

Substrate association with membrane mimetic agents is a dynamic process. Quantitative treatments for substrate solubilization dynamics in micelles have been provided in Chapter 2. Substrate binding constants and residence times in aqueous micelles are of the order of 10^3–10^6 M^{-1} and 10^3–10^5 sec^{-1}. Extensions of quenching studies to microemulsions resulted in similar values for substrate binding, entry, and exit rates.

The most likely substrate solubilization sites in micelles and microemulsions are the interface and Stern regions. There is no evidence for "deep" substrate penetration into the micellar core.

Substrates can penetrate into the hydrophobic regions of monolayers or associate with their headgroups. They can migrate laterally and indeed across monolayers in organized multilayer assemblies.

The considerable kinetic stabilities of BLMs and vesicles govern the dynamics of substrate interactions in these systems. Even the terminology is different from that used in micelles. Permeabilities and diffusion rates rather than residence times are determined. Substrate permeabilities are relatively slow. The timescale for the transport of water molecules across liposomes is of the order of seconds. Diffusion of larger molecules through liposomes occurs in the timescale of hours, days, and even weeks.

There are four extreme sites for substrate localization in vesicles. Hydrophobic molecules can be distributed among the hydrocarbon bilayers of the vesicles. Alternatively, they can be anchored by a long chain terminating in a polar headgroup. Polar molecules may move about relatively freely in vesicle entrapped water pools (particularly if they are electrostatically repelled from the

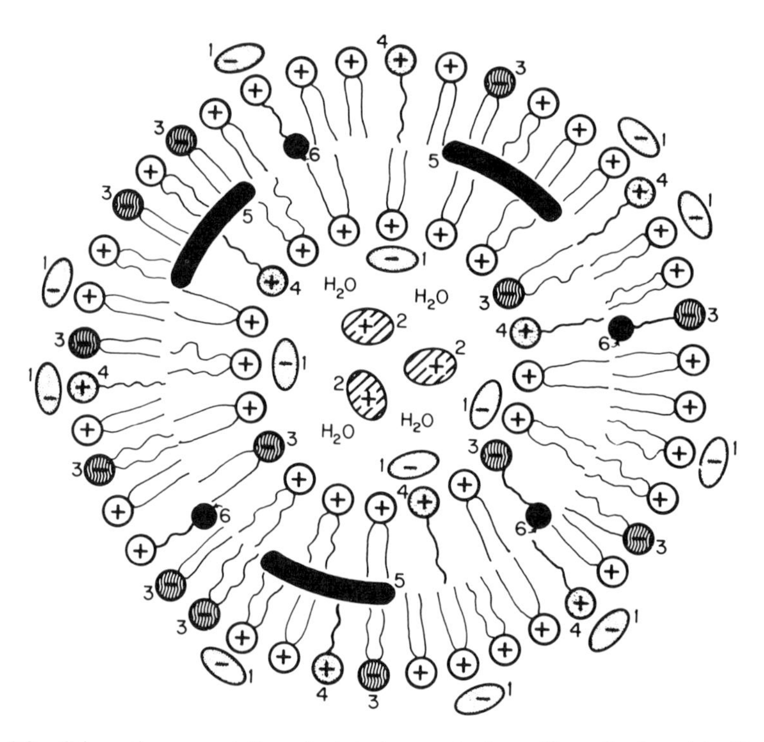

Figure 9.2. Schematic representation of substrate organization with a cationic vesicle. Negatively charged ionic compounds (1) may be electrostatically attracted to the inner and outer surface of the vesicle. Cations (2) are repelled electrostatically from vesicle headgroups and are located, therefore, in the middle of vesicle entrapped water pools. Charged molecules may, alternatively, be anchored by a long hydrocarbon tail (3 and 4). Hydrophobic molecules may be intercalated in the bilayer (5) or be anchored by ionic headgroups located on the end of a hydrocarbon chain (3–6) (taken from Fendler, 1980).

inner surface of the vesicles) or they may be associated with and bound to the inner and outer surface of vesicles. Polar molecules can also be anchored to the vesicle surface by a long hydrocarbon tail. Figure 9.2 illustrates different modes of substrate organization in vesicles.

Electrostatic interactions are of paramount importance. For example, cationic substrates associate strongly with negative charges located at the inner and outer surface of anionic vesicles. Gel filtration is used to separate free substrates from those associated with vesicles. Further charged substrates can be removed from the outer surface of the vesicles by passage through an appropriate ion-exchange resin.

Substrate organization in hosts or in polyions is much more restricted. The usual mode of host-guest interaction is the intercalation of the guest in the cavity of the host. Covalently linked sidearms on crown ethers or cyclodextrins may provide additional sites for interactions. Rate constants for entry and exit of guests into macrocyclic host cavities are in the order of 10^8–10^9 sec^{-1} and 10^5–10^7 M^{-1} sec^{-1}, respectively. Substrates are usually bound to polyion backbones. Quite often substrates themselves are the constituents of membrane mimetic agents.

4. SELECTION OF MEMBRANE MIMETIC AGENTS

Different features of a given membrane mimetic agent should be considered prior to its exploitation. Table 9.1 compares the features of different membrane mimetic agents. It should be realized that micelles and liposomes have been investigated in considerably greater detail than have microemulsions or monolayers. Similarly, host-guest interactions are better understood than those in polyions. Catalytic activities of micelles and polyelectrolytes have been compared recently (Klotz et al., 1981). The available sites for interaction are often an important consideration. For example, in energy transfer experiments there is often need to compartmentalize no more than one donor per aggregate and, at the same time, to concentrate a large number of acceptors. The relatively small ionic micelles meet these requirements best. A point of illustration is the efficient energy transfer from micellar sodium dodecyl sulfate (SDS) solubilized naphthalene to terbium chloride (Escabi-Perez et al., 1977). In the absence of micelles there is no energy transfer. Since there is less than one naphthalene molecule in each micelle, triplet-triplet annihilation, preventing energy transfer, is precluded (Escabi-Perez et al., 1977). Micelles are also the best media for altering the path of photochemical reactions (Turro and Cherry, 1978; Turro and Kraeutler, 1978). Microemulsions and vesicles, on the other hand, serve well the need for organizing high concentrations of polar and apolar molecules in each aggregate. The large compartments are also needed for carrying maximum amounts of drugs. Vesicles appear to serve this purpose best. Hosts and polyions are the preferred media for organic sythesis, since they can be readily separated from the reaction products.

Table 9.1. Comparison of Membrane Mimetic Agents

Characteristic	Aqueous Micelles	Reversed Micelles	Microemulsions	Monolayers
Constituents	Various surfactants	Various surfactants	Surfactants, cosurfactants, polar and apolar solvents	Various surfactants
Method of preparation	Dissolving appropriate concentration (> critical micelle concentration (CMC)) of the surfactant in water	Dissolving appropriate concentration of surfactant in the apolar solvents	Dissolving appropriate concentration of the surfactant and the cosurfactant in the appropriate solvent	Spreading surfactants, dissolved in an organic solvent, on water surface
Weight averaged molecular weights	2000–6000	2000–6000	10^5–10^6	Depends on area covered and density of coverage
Diameter, Å	30–60	40–80	50–1000	Depends on area covered and density of coverage
Stability	Weeks, months	Weeks, months	Weeks, months	Hours, days
Dilution by H_2O	Destroyed	Water pool enlarged → w/o microemulsion formed	o/w + water → aqueous micelles; w/o + water → phase separation	Destroyed
Effects of temperature changes	$T\uparrow$ → Kraft point	$T\uparrow$ → Kraft point; $T\downarrow$ → possibility of looking at supercooled water	$T\uparrow$ → Kraft point; $T\downarrow$ → possibility of looking at supercooled water	Phase transition
Number of solubilizates taken up	Few	Few	Large	Large
Available solubilization sites	Surface, Stern layer, vicinity of headgroups	Aqueous pool, inner surface, surfactant tail	Inner pool, inner surface, surfactant tail	On surface, around the hydrocarbons

BLMs	Vesicles	Host–Guest Systems	Polyions
Mostly lipids of biological origins	Lipids and surfactants	Macrocyclic polyethers, cyclodextrins	Various ionizable groups attached to polymer backbones
Painting surfactant, dissolved in an organic solvent, on a pinhole	Shaking thin lipid films in water, ultrasonication, injecting ether or alcohol solutions of lipids into water, gel filtration lipid-detergent micelles	Dissolving hosts in appropriate solvent	Dissolving polymers in water
Depends on area covered and density of coverage	$> 10^7$	500–5000	Depends on polymer
Depends on area covered and density of coverage	300–10,000	4–10 (cavity size)	Depends on polymer
Hours	Weeks	Weeks, months	Weeks, months
Destroyed	Unaltered	Unaltered in water	Unaltered
Phase transition	Phase transition	Possible conformational changes	Possible conformational changes
Large	Large	Typically 1 : 1	Large
Either or both sides of the surface, among the bilayer	Aqueous pool, either or both sides of the surface, among the bilayer	In cavities	Around charged surfaces or backbones

REFERENCES

Bunton, C. A. (1979). *Catal. Rev. Sci. Eng.* **20**, 1–56. Reaction Kinetics in Aqueous Surfactant Solutions.

Cafisco, D. S. and Hubbell, W. L. (1978). *Biochemistry* **17**, 187–195. Estimation of Transmembrane Potentials from Phase Equilibria of Paramagnetic Ions.

Escabi-Perez, J. R., None, F., and Fendler, J. H. (1977). *J. Am. Chem. Soc.* **99**, 7749–7754. Energy Transfer in Micellar Systems. Steady State and Time Resolved Luminescence of Aqueous Micelle Solubilized Naphthalene and Terbium Chloride.

Fendler, J. H. (1976). *Acc. Chem. Res.* **9**, 153–161. Interactions and Reactions in Reversed Micellar Systems.

Fendler, J. H. (1980). *J. Phys. Chem.* **84**, 1485–1491. Microemulsions, Micelles, and Vesicles as Media for Membrane Mimetic Photochemistry.

Fendler, J. H., and Fendler, E. J. (1975). *Catalysis in Micellar and Macromolecular Systems*, Academic Press, New York.

Israelachvili, J. M., Marcelja, S., and Horn, R. G. (1980). *Quart. Rev. Biophys.* **13**, 121–200. Physical Principles of Membrane Organization.

Kano, K. and Fendler, J. H. (1979). *Chem. Phys. Lipids* **23**, 189–200. Dynamic Fluorescence Investigations of the Effect of Osmotic Shocks on the Microenvironment of Charged and Uncharged Dipalmitoyl-D,L-α-phosphatidylcholine Liposomes.

Klotz, I. M., Drake, E. N., and Sisido, M. (1981). *Bioorg. Chem.* **10**, 63–74. Comparison of Biomimetic Catalytic Properties of Modified Polyethylenimines with Those of Micelles.

Kumar, C. and Balasubramanian, D. (1980). *J. Phys. Chem.* **84**, 1895–1899. Structural Features of Water-in-Oil Microemulsions.

Menger, F. M. (1979). *Acc. Chem. Res.* **12**, 111–117. On the Structure of Micelles.

Menger, F. M. and Jerkunica, J. M. (1978). *J. Am. Chem. Soc.* **100**, 688–691. Anisotropic Motion Inside a Micelle.

Mittal, K. L. (1977). *Micellization, Solubilization, and Microemulsions*, Plenum Press, New York.

Mittal, K. L. (1979). *Solution Chemistry of Surfactants*, Plenum Press, New York.

Thomas, J. K. (1980). *Chem. Rev.* **80**, 283–299. Radiation Induced Reactions in Organized Assemblies.

Turro, N. J. and Cherry, W. R. (1978). *J. Am. Chem. Soc.* **100**, 7431–7432. Photoreactions in Detergent Solutions. Enhancement of Regioselectivity Resulting from the Reduced Dimensionality of Substrates Sequestered in a Micelle.

Turro, N. J. and Kraeutler, B. (1978). *J. Am. Chem. Soc.* **100**, 7432–7434. Magnetic Isotope and Magnetic Field Effects on Chemical Reactions. Sunlight and Soap for the Efficient Separation of ^{13}C and ^{12}C Isotopes.

Wong, M. and Thomas, J. K. (1977). In *Micellization, Solubilization and Microemulsions* (K. L. Mittal, Ed.), Plenum Press, New York, pp. 647–664. Some Kinetic Studies in Reversed Micellar Systems—Aerosol-OT (Diisoctylsulfosuccinate)/H_2O/Heptane Solution.

PART 2

APPLICATION OF MEMBRANE MIMETIC AGENTS

CHAPTER 10

ENZYMES AND MEMBRANE MIMETIC SYSTEMS

Investigations of interactions and reactions in micelles, polyions, and host–guest systems have been mainly prompted by their potential as enzyme models. Analogies have been drawn between the structure of micelles and globular proteins and between reaction kinetics in enzymes and micellar or other macromolecular systems (Morawetz, 1970, 1978; Bender, 1971; Cordes and Gitler, 1973; Bunton, 1973, 1979; Jencks, 1975; Fendler and Fendler, 1975; Fendler, 1976; Piszkiewicz, 1977a, 1977b; Bender and Komiyama, 1977, 1978; Cram and Cram 1978; Lehn, 1978; Klotz, 1978; Breslow, 1979; O'Connor et al., 1981). Validities of these analogies are reexamined in this chapter. Two distinct levels of enzyme modeling have been approached experimentally. Simpler models have been designed to examine the catalytic consequences of substrate bindings. Most of the pre-1975 work in micellar and macromolecular catalysis falls into this category. More recently, attention has been paid for mimicking the active sites of given enzyme-substrate interactions. Details of these more advanced modelings are discussed in this chapter. Kinetic treatments of reactivities in micellar and related systems are given in Chapter 11. The data accumulated since the 1975 compilation (Fendler and Fendler) are provided in Chapter 12.

Interactions of enzymes themselves with membrane mimetic agents have been extensively investigated. Enzymes have been entrapped in aqueous pools provided by reversed micelles or vesicles, incorporated among the hydrophobic moieties, or attached to the surface of monolayers, black lipid membranes (BLMs), and vesicles. Results of these investigations are also examined in this chapter. An attempt is made to exhaustively tabulate pertinent data relating to enzymes in membrane mimetic systems. These investigations not only are inherently interesting and provide valuable insight into enzyme structures and catalysis, but they also have potential applications in enzyme mediated synthesis.

1. MEMBRANE MIMETIC AGENTS AS ENZYME MODELS

Theories of Enzyme Catalysis

Enzymes are highly specific and extremely efficient catalysts. They enhance reaction rates 10^5–10^{10}-fold in relatively dilute aqueous solutions around neutral pH at ambient temperatures (Koshland, 1962; Bruice, 1962; Jencks, 1969; Fife, 1975; Fendler, 1976). Not surprisingly, there is a growing interest in developing a comprehensive theory of enzyme catalysis (Koshland, 1956, 1962; Bruice, 1962, 1976; Page and Jencks, 1971; Wolfenden, 1972; Page, 1973; Lienhard, 1973; Jencks, 1975; Fersht, 1977; Luisi, 1979). Given the multitude of enzyme catalyzed reactions and their specificity, it may never be possible to postulate a detailed theory encompassing all systems. Information available is sufficient, however, for discussing gross features.

Enzymes are proteins whose molecular weights are in the tens of thousands (Boyer, 1970). Structures of several enzymes, their active sites, and substrate interactions therein have been determined by X-ray crystallography. Active sites of many enzymes are situated near the *N*-terminal end of an α-helix. The dipole field, generated by the α-helix, was suggested to play an important part in binding substrates and in enhancing reactivities (Hol et al., 1978; Van Duijnen et al., 1979, 1980). The microenvironment for the substrate at the active site is quite different from that in bulk water. Well-recognized manifestations of this unique environment are the "superreactivity" (Cohen, 1970; Shaw, 1970) and the substantially altered p*Ka* values (Tanford, 1962) of functional groups in proteins. Electrostatic, hydrophobic, steric, dielectric interactions and neighboring group participation (Page, 1973) contribute to altered substrate reactivities at enzyme active sites. *A priori* this requires relatively rigid organizations. Enzymes, however, undergo pH, temperature, or substrate induced conformational changes, referred to as allosteric interactions or isomerizations (Citri, 1973). Conformational changes can be quite subtle; alteration of a few bond angles often suffices. This apparent dichotomy between rigidity and structural changes is an essential feature of enzyme functions. Thus the catalytic site has to be able to shift from one relatively rigid conformation into another rigid structure on receiving a specific chemical signal. Allosteric changes are mediated by cooperative protein subunit interactions often extending to large distances. Small molecules cannot provide all the requirements of enzyme catalysis. Large flexible macromolecules, such as proteins, are needed for creating active sites with the required microenvironments, rigidity, stereochemistry, and interactions (Luisi, 1979).

Several attempts have been made to account for the superior catalytic powers of enzymes in terms of known physical organic-chemical principles (Jencks, 1975; Fife, 1975; Fersht, 1977). In its simplest form an enzyme catalyzed reaction,

$$E + S \underset{k_{-1}}{\overset{k_1}{\rightleftharpoons}} ES \xrightarrow{k_2} P \tag{10.1}$$

$$\frac{k_2 + k_{-1}}{k_1} = K_m \tag{10.2}$$

is contrasted with the uncatalyzed reaction,

$$S \overset{k_u}{\rightleftharpoons} P \tag{10.3}$$

or with that catalyzed by a nonenzymatic (acid or base, for example) catalyst, C,

$$C + S \rightleftharpoons CS \longrightarrow P \tag{10.4}$$

where E, C, S, and P refer to the enzyme, the nonenzyme catalyst, the substrate, and the product, respectively; ES and CS are the enzyme-substrate and the nonenzyme catalyst-substrate complexes; K_m is the Michaelis-Menten constant. Quantitative discussion of enzyme catalysis requires the selection of a set of standard reaction conditions in terms of thermodynamic standard states of free energies. The choice of "uncatalyzed" (equation 10.3) or nonenzymatically catalyzed (equation 10.4) standard reaction is quite arbitrary, of course.

Free energies of reactions can be lowered by destabilizing the ground state, stabilizing the transition state, or affecting both of these states by the enzyme. The enzyme-substrate binding energy is also utilized for the catalysis. Figure 10.1 illustrates free energy diagrams for uncatalyzed and enzyme catalyzed reactions (Schowen, 1978). Two extreme cases are considered. In the first case (Figure 10.1*a*) the concentration of the substrate, and hence the steady state concentration of the enzyme-substrate complex, is much lower than that of the enzyme. Under these conditions the catalytic free energy, ΔG_{cat}, arises from the free energy released upon lowering the transition state of the uncatalyzed reaction as a result of forming the enzyme-substrate complex, ES. Catalysis arises, therefore, solely from the free energy of transition state binding, ΔG_b^*,

$$\Delta G_{\text{cat}} = \Delta G_u^* - \Delta G_c^* = -\Delta G_b^* \tag{10.5}$$

ΔG_{cat} is defined as the difference in free energies of activation for the uncatalyzed reaction (subscript u) and the catalyzed reaction (subscript c). In the second case concentrations of S and ES are sufficiently high compared with that of E. Further, ES is destabilized with respect to the ground states of E and S (Figure 10.1*b*). Under these conditions the free energy expended by the enzyme in stabilizing the reactant state, ΔG_{ES}, is to be subtracted from ΔG_b^* to give the overall catalytic free energy,

$$\Delta G_{\text{cat}} = -\Delta G_b^* + \Delta G_{\text{ES}} \tag{10.6}$$

Free energies due to enzyme catalysis can, of course, be dissected into as many contributing terms as one wishes. Attempts have been made at various times to estimate such factors as (1) the juxtaposition, in correct stereochemical fashion,

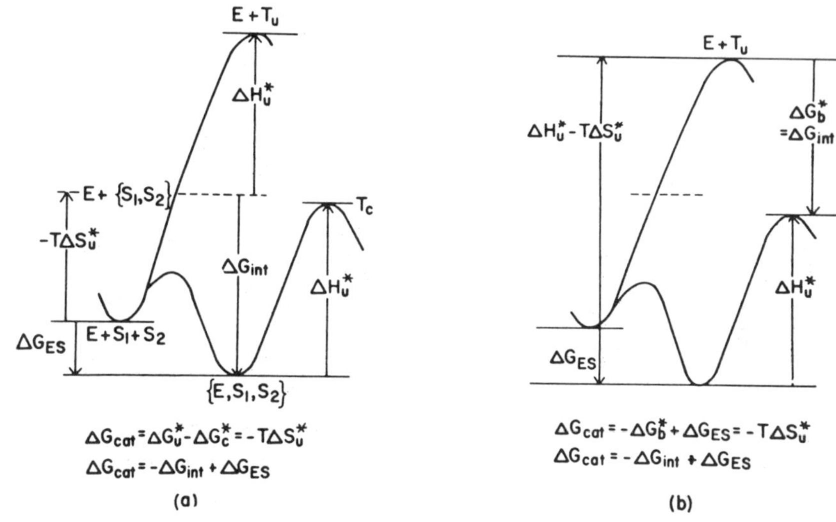

Figure 10.1 Free energy diagrams for uncatalyzed and enzyme catalyzed reactions of the substrate S at different standard state concentrations of S. (*a*) Concentration of S low compared with K_m, so that no substantial concentration of enzyme-substrate complex is present. The entire reduction in free energy of activation effected by the enzyme originates in stabilization of the uncatalyzed transition state T_U with formation of the catalyzed transition state T_C. The whole binding or stabilization energy appears as catalytic free energy. (*b*) Concentration of S high compared with K_m so that the enzyme is mainly complexed. The inhibitory complexation of the substrate by the enzyme countervails against the transition state stabilization and the substrate stabilization energy must be subtracted from the transition state stabilization energy to obtain the catalytic free energy (Taken from Schowen, 1978).

of all nucleophilic, general acid, and base groups (proximity or propinquity effect); (2) the ground state desolvation–transition state solvation (milieu effect); and (3) the conformation optimization (Bruice, 1970). Kinetic investigations of model reactions in dilute aqueous solutions lead to numerical assignments of catalytic contribution to the proximity effect. A comparison of lactonization rates illustrates the "number game" being played (Storm and Koshland, 1972a, 1972b). Relative rates of lactonization are seen to increase with increasing proximity of the reacting group (Table 10.1). The large difference in reactivity between lactonization and thiolactonization (compare **10.1** with **10.2** in Table 10.1), attributed to the small change in orientation between the oxygen and sulfur atoms, has been rationalized in terms of orbital steering (Storm and Koshland, 1970). Orbital steering requires the correct alignment of the reactive orbitals for optimal catalysis and has been considered to be responsible for rate enhancements as large as 10^4 (Storm and Koshland, 1970; Koshland, 1972). Orbital steering has been called an unnecessary concept (Capon, 1971). The methods used for proximity corrections, as well as the choice of standard reactions, have been severely criticized (Bruice, 1970; Bruice et al., 1971; Page and Jencks, 1971; Jencks and Page, 1974).

Table 10.1. Effect of Structure on Lactonization Rates

Structure	X = O		X = S	
	Relative Rates	Corrected Relative Rates[a]	Relative Rates	Corrected Relative Rates[a]
$CH_3COOH + CH_3CH_2-XH$	1	1	1	1
 COOH, XH	80	413	384	2,020
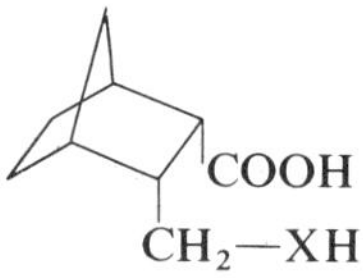 COOH, CH_2-XH	6,620	1,660	90	5
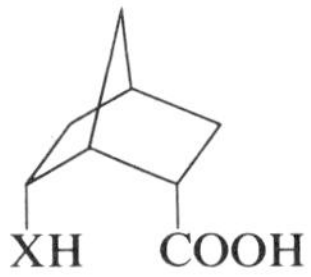 XH, COOH **10.1**	1,030,000	18,700	821,000	15,000
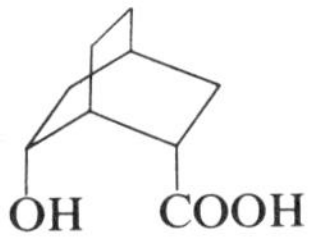 OH, COOH **10.2** $R_2 = H$	873			
$R_2 = CH_3$	98,700			

Source: Compiled from Gandour, 1978.
[a] Storm and Koshland, 1972a.

Free energies due to enzyme catalysis can also be expressed in terms of enthalpy and entropy contributions:

$$\Delta G_{cat} = \Delta H_{cat} - T\Delta S_{cat} \tag{10.7}$$

The extent of enthalpy versus entropy contribution to enzyme catalysis is a much debated and as yet unsettled problem (Jencks, 1975). Reduction of the number of translational degrees of freedom, by restriction of rotation or by conformational change, has been termed the "induced fit" theory (Koshland, 1960; Koshland et al., 1966). Jencks (1975) favors entropy loss upon substrate binding to the enzyme, the "Circe-effect," as a major cause for rate enhancements.

It is evident from this brief discussion that no universally accepted theory is available for rationalizing enzyme reactivities even in the simplest form (equation 10.1). Hopefully, investigation of models will calrify some issues that, at present, are incompletely understood and controversial.

Modeling of Substrate Binding

Micelles, cyclodextrins, and polyelectrolytes have been extensively used for modeling substrate-enzyme binding and concomitant free energy reductions (Morawetz, 1970, 1978; Jencks, 1975; Fendler and Fendler, 1975; Bender and Komiyama, 1977, 1978; Klotz, 1978; Bunton, 1979; Breslow, 1979).

Rate enhancements of unimolecular reactions in the presence of mimetic agents are the simplest to rationalize. Catalysis is the primary consequence of substrate destabilization–transition state stabilization in the polar environments of the aggregates. Investigations of unimolecular ractions provide, therefore, a means to establish the extent to which membrane mimetic agents enhance reactivities by virtue of altering the microenvironments at the site of reactions. Decarboxylations have been investigated most extensively since the stoichiometry and the nature of substrate binding are known or can be determined, the reaction is unimolecular, devoid of complications, and the rates are highly solvent dependent and easy to monitor. Rate constants for the decarboxylation of 6-nitrobenzisoxazole-3-carboxylate ions (**10.3**),

$$\textbf{10.3} \longrightarrow [\text{transition state}]^{\ominus} \longrightarrow CO_2 + \text{2-cyano-5-nitrophenoxide } (O_2N\text{-}C_6H_3(C{\equiv}N)O^-) \quad (10.8)$$

have been determined in water, in apolar aprotic solvents (Kemp and Paul, 1970, 1975; Kemp et al., 1955), in aqueous micelles (Bunton and Minch, 1970; Bunton et al., 1973, 1975), in polyelectrolytes (Suh et al., 1976), and in β-cyclodextrin (Suh et al., 1976). Apolar environments are seen to increase substantially the

Table 10.2. Relative Rates of 6-Nitrobenzisoxazole-3-carboxylate Ion Decarboxylation

Medium	k'_M, sec^{-1} / k'_w, sec$^{-1\,a}$(T, °C)	Reference
Methanol	33 (30)	Kemp and Paul, 1975
Chloroform	10^2 (30)	Kemp and Paul, 1975
Carbon tetrachloride	2×10^2(30)	Kemp and Paul, 1975
Benzene	6×10^2(30)	Kemp and Paul, 1975
Dioxane	5×10^3(30)	Kemp and Paul, 1975
Diethylether	1×10^4(30)	Kemp and Paul, 1975
Acetonitrile	4×10^5(30)	Kemp and Paul, 1975
N-Methylpyrrolidine	3×10^7(30)	Kemp and Paul, 1975
Aqueous micellar hexadecyltrimethyl-ammonium bromide (CTAB)	2×10^2(25)	Bunton et al., 1973, 1975
Aqueous micellar cysolecithin	3×10^2(25)	Bunton et al., 1975
Cycloheptaamylose (β-cyclodextrin)	3.5 (60.4)	Straub and Bender, 1972
Poly(vinylbenzo-18-crown-6)	6×10^2(25)	Smid et al., 1975
$[(C_2H_4\overset{+}{N})_m(C_{12}H_{25})_{0.25m}(CH_3)_{1.75m}]Cl^-$, $m = 1{,}400$	4×10^2(25)	Suh et al., 1976
$[(C_2H_4\overset{+}{N})_m(C_{12}H_{25})_{0.25m}(C_2H_5)_{1.75m}]Cl^{-1}$ $m = 1{,}400$	1×10^3(25)	Suh et al., 1976

[a] $k'_w = 3 \times 10^{-6}$ sec^{-1} at 25°C (Bunton et al., 1975).

decarboxylation rate of **10.3** (Table 10.2). The rate profiles are similar in all media. Increasing concentrations of surfactants, polyions, or β-cyclodextrin increase the rate constants sigmoidally to a plateau, beyond which there is no more rate enhancement. This behavior is characteristic for unimolecular reactions and is described by

$$\begin{array}{ccc} S + D_n & \underset{}{\overset{K_s}{\rightleftharpoons}} & SD_n \\ \downarrow k'_w & & \downarrow k'_M \\ P & & P \end{array} \tag{10.9}$$

where S and P stand for the substrate and the product; D_n represents micelles (containing n monomers), polyions, or cyclodextrins (containing n binding sites), SD_n is the substrate-model complex, K_s is the binding constant, k'_w and k'_M are the rate constants in water and in the presence of micelles or cyclodextrins or polyions. Provided that the concentration of S is sufficiently small (compared with

D_n) not to effect changes in structural and catalytic properties, the observed rate constant, k_ψ, can be expressed by

$$k_\psi = \frac{k'_w + k'_M K_s[D_n]}{1 + K_s[D_n]} \tag{10.10}$$

Rate enhancements, manifested in $k'_M > k'_w$, can be attributed to ground state destabilization–transition state stabilization, due to solvation change of the bound reactant. Rate enhancements caused by reduced polarities provided by enzyme active sites are quite analogous. Both in enzymes and in models the driving force for rate enhancement is the sole consequence of a change in the reaction milieu. Enzyme solvent effects in pyruvate decarboxylase catalyzed decarboxylations have been considered to cause 10^5–10^6-fold rate enhancements (Straub and Bender, 1972). None of the complexing agents caused such large rate enhancements for the decarboxylation of **10.3** (Table 10.2). Interestingly, dipolar aprotic solvents alone caused rate enhancements similar to that manifested by the enzyme (Table 10.2). Changes of reaction milieu alone (both in enzyme and in models) are insufficient to bring about catalytic specificity.

Treatments of bimolecular reactions are more complex (see Section 11.1). The primary function of membrane mimetic agents is to take up both reactants in their environments. The substrate-model binding energy is then utilized to overcome the entropy requirements involved in bringing the reacting functional groups together. Electrostatic, hydrophobic, and steric forces come into play.

Ionic reactants lacking hydrophobic groups are concentrated on the surface of oppositely charged micelles, microemulsions, vesicles, or polyions. The greatest catalytic effect is expected between reactants of the same charge, localized on oppositely charged surfaces. Rate constants for the Hg^{2+} induced aquation of $Co(NH_3)_5Cl_2^{2+}$,

$$Co(NH_3)_5Cl^{2+} + Hg^{2+} + H_2O \longrightarrow Co(NH_3)_5H_2O^{3+} + HgCl^{+} \tag{10.11}$$

are enhanced by factors of 140,000 in aqueous anionic micellar sodium dodecyl sulfate (SDS) (Cho and Morawetz, 1972) and 176,000-fold in anionic polyvinylsulfonate polyions (Morawetz and Vogel, 1969), for example Rate enhancements of comparable magnitude have been observed for a number of electron transfer reactions on charged surfaces (see Chapter 12). Altered rates of electron transfer are significant in photochemical solar energy conversion (see Chapter 13). Rate increases between nonhydrophobic ions of oppositely charged surfaces are the consequence of reactant localization by electrostatic attractions and transition state stabilization. Rate constants increase very quickly with increasing reagent concentrations up to a point beyond which they decline. This behavior is explicable in terms of the reagent saturating the available catalytic surface. Enzyme-substrate kinetic rate profiles are quite analogous, of course.

Reaction rates between a neutral and a charged molecule depend, once again, on the uptake of both reactants by the aggregate. Associations of molecules are governed by hydrophobic and electrostatic interactions. The longer the hydrocarbon tail on an ester, for example, the greater the extent of its incorporation in a cationic micelle and hence the faster the rate of its base catalyzed hydrolysis (see Chapter 12 for examples and data).

Bimolecular reactions show little substrate selectivity in nonfunctional membrane mimetic agents. Reactions in reversed micelles (Fendler, 1976) provide a possible exception. Aquation of tris(oxalato)chromate(III) anion is up to 5.4 million-fold faster in octadecyltrimethylammonium tetradecanoate solubilized water pools in benzene than it is in water (O'Connor et al., 1973, 1974). Furthermore, such small changes in the substrate as replacing the chromium metal by cobalt or replacing the oxalato by azido ligand results in a drastic decrease of the rate enhancements (10^2–10^3-fold) in the same reversed micellar system (O'Connor et al., 1973). An important function of reversed micelles is to provide sites of variable polarities and microenvironments for ionic species (see Chapter 3). Metal ion desolvation in reversed micelles has been suggested to resemble the active sites of metalloenzymes (Sunamoto and Hamada, 1978). Similarly, investigations of the hydrolysis of adenosine 5′-triphosphate (ATP) (Seno et al., 1975), and the iodination of ketones (Seno et al., 1976) in dodecylammonium propionate (DAP) solubilized water pools in hexane was considered to model the ATPase action and the active sites of aldolase and enolose enzymes. Similarly to reversed micelles, polarities provided by cyclodextrin hosts can be controlled by carrying out reactions in aqueous DMSO mixtures of different composition (Siegel and Breslow, 1975).

Jencks (1975) has considered that micellar (and presumably related) catalysis is the consequence of a reduction of the activity coefficients and the standard free energy of the reactant in the micellar phase and that substrate confinement in this phase results in favorable entropy gains. The observed noncovalent substrate binding, saturation type kinetics, and competitive inhibition, in nonfunctional model systems, are similar to the behavior of enzymes (Fendler and Fendler, 1975). Disappointingly, however, rate enhancements in these systems rarely exceed a thousandfold and only show a modest degree of substrate specificity. Presumably, the noncovalent binding and the dynamic nature of the substrate–nonfunctional model systems do not approximate well the rigid configurational requirements of enzyme-substrate interactions.

A promising recent approach is to reproduce the surface properties of proteins in small oligopeptides capable of forming amphiphatic helices (Fukushima et al 1979, 1980; Yokoyama et al., 1980; DeGrado et al., 1981). The properties of synthetic rationally designed docosapeptide, **10.4**, are remarkably similar to those of plasma apolipoprotein A-1 (Apo-I).

Both **10.4** and Apo-I bind to single-compartment egg lecithin liposomes; with dissociation constant $K_d = 1.92 \times 10^{-6}\ M$ and binding capacity $N = 1.51 \times 10^{-2}$ peptide/lecithin for **10.4** and $K_D = 9.0 \times 10^{-7}\ M$ and $N = 1.74 \times 10^{-3}$ proteins/lecithin for Apo-I (Yokoyama et al., 1980). Both **10.4** and Apo-I

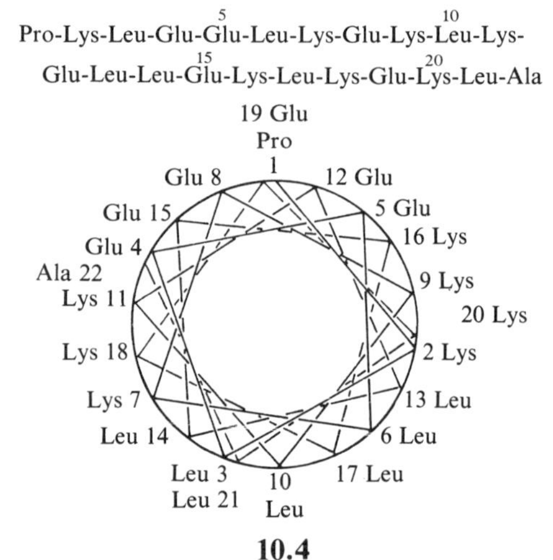

10.4

activate the reactions of lecithin : cholesterol acyltransferase. The rates of enzyme reaction are linear with respect to liposome-bound **10.4** or liposome-bound Apo-I. Both **10.5** and Apo-I form monolayers with similar surface properties (Fukushima et al., 1979). Plasma apoprotein–lipid association has been modeled by investigating the intrinsic fluorescence of a 20-residue synthetic lipid-associating protein **10.5** prior and subsequent to binding to dimyristoyl phosphatidylcholine (Pownall et al., 1980).

5 10
Val-Ser-Ser-Leu-Leu-Ser-Ser-Leu-Lys-Glu-

15 20
Tyr-Trp-Ser-Ser-Leu-Lys-Glu-Ser-Phe-Ser

10.5

These results substantiate the proposal that functions of lipoproteins are related to their potential for forming "amphiphilic" α-helical structures (Fukushima et al., 1980) and open the door to systematic studies of lipid-protein interactions at the molecular level (Fukushima et al., 1979).

Functional Active Site Modeling

Enzyme catalyses are mediated by amino acid moieties localized far apart in the linear sequence of proteins. They are brought together in tertiary structures to form active sites with required configurations, microenvironments, and functionalities. Not surprisingly, modeling enzyme actions in dilute aqueous solutions by small molecules resembling the active sites resulted in meager catalyses. Functionalized micelles, polyions, and cyclodextrins are expected to mimic more faithfully tertially enzyme structures. Investigations have been directed toward modeling given enzyme environments and actions.

Micellar pyridoxal-5′-P Schiff bases represent, for example, good models for the vitamin B_6 site of glycogen phosphorylase (Kupfer et al., 1977). Pyridoxal-5′-P in phosphorylase is embedded in a hydrophobic environment while the substrates, glycogen, glucose-1-phosphate are water soluble. Absorption and fluorescence spectra of the micellar pyrodoxal-5′-P Schiff base are entirely analogous to those of the enzyme system (Kupfer et al., 1977).

Functionalized cyclodextrin, **10.6**, formed from all-trans retinal and ω-aminoethylamino-β-cyclodextrin,

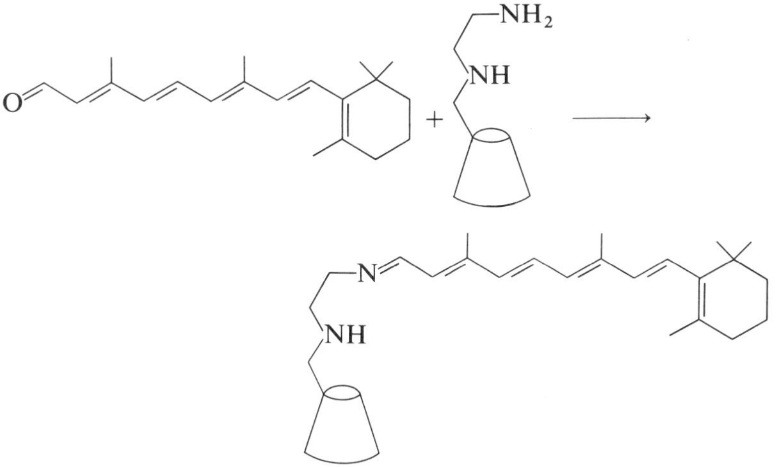

NH_2 NH O + ⟶ N NH (10.12)

10.6

reproduced remarkably well the spectral behavior of bovin rhodopsin (Tabushi et al., 1979). The absorption maximum of **10.6** at pH = 1.16, 497 nm, is quite similar to that of bovin rhodopsin, 498 nm, or bovin luminorhodopsin, 497 nm (Figure 10.2). Conversely, nonfunctionalized Schiff bases **10.7** and **10.8** behaved quite differently.

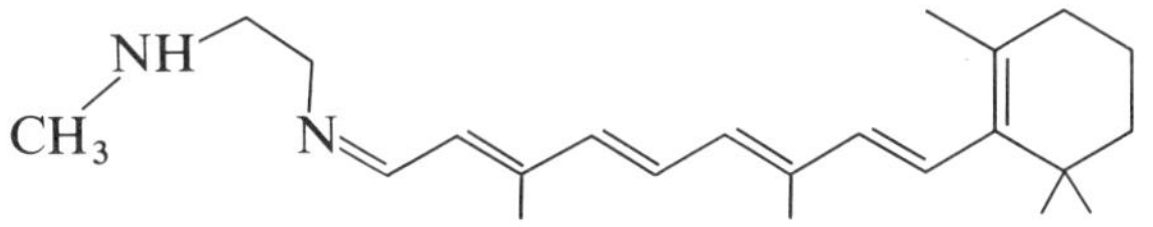

NH CH_3 N

10.7

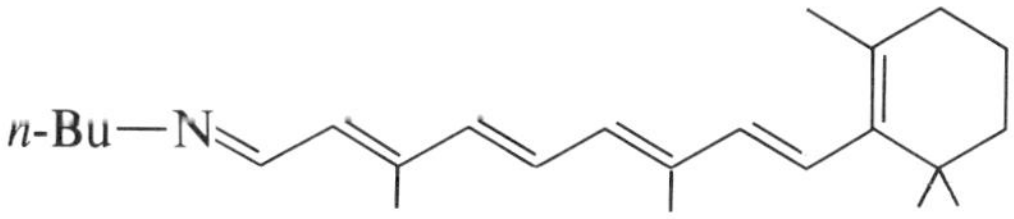

n-Bu—N

10.8

Absorption maxima of **10.7** and **10.8** in cavities of nonfunctionalized β-cyclodextrin showed appreciable blue shifts to 444 nm and 437 nm, respectively. It has been suggested that the shift for **10.6** was caused by covalent combination of the CO recognition site and hydrophobic binding. The proposed structure,

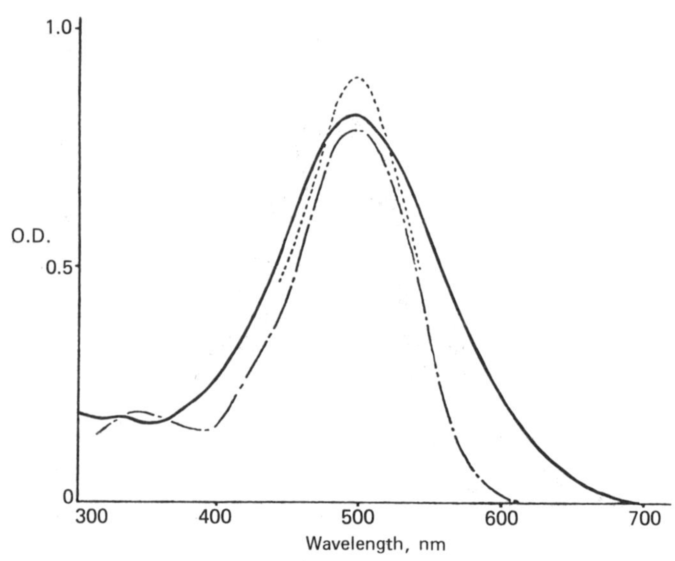

Figure 10.2 Electronic spectra of **10.6** · $(H^+)_2$ (—) at pH 1.16 in the aqueous HCl solution and bovin rhodopsin (–·–) and lumirhodopsin (----) (G. Wald, J. Durell, and R. C. C. St. George, *Science* **111**, 179(1950)). Concentrations of all compounds are 2×10^{-5} *M* (taken from Tabushi et al., 1979).

10.9

10.9, was considered to be a good approximation of rhodopsin (Tabushi et al., 1979).

Metal ion capped cyclodextrins (**7.35**, for example) also provide double recognition sites and are considered to be models for metalloenzymes (Tabushi et al., 1977).

Functional cyclodextrins have also been utilized for modeling enzyme mediated transaminations (Breslow et al., 1980a). In a typical process pyridoxamine phosphate, **10.10**, reacts with an α-keto acid such as pyruvic acid, **10.11**, to form a Schiff base that tautomerizes and cleaves to pyrodoxal phosphate, **10.12**, and an amino acid, in this case alanine, **10.13**. The cycle is completed by using a different amino acid such as phenylalanine, **10.14**, to convert the pyridoxal coenzyme back to **10.10**, while the amino acid **10.13** is converted to the keto acid **10.15** (Walsh, 1979). Although this reaction sequence can be carried out with the coenzyme alone, the rates are much slower than those in the presence of enzyme and the process is quite unselective. Substantial rate enhancements and appreciable degrees of selectivity were obtained on using pyridoxal-pyrodoxamine coenzymes ocvalently linked to β-cyclodextrin, **10.16** (Breslow et al., 1980a). Indolepyruvic acid, **10.17**, was converted to tryptophan, **10.18**, approximately 200 times faster by **10.16** than by pyridoxamine, **10.19**. Conversely, alanine formation, **10.13**, from pyruvic acid, **10.11**, occurred at the same rate from **10.19** or **10.16**. Competitive reaction of **10.16** with **10.11** and **10.17** leads to 97% tryptophan at the beginning of the reaction, but the selectivity decreased with increasing completion of the reaction. This finding was rationalized in terms of binding the

CH_2NH_2, HO, CH_2X, CH_3, N + $R-C(=O)-COOH$

10.10 $X = OPO_3H_2$ **10.11** $R = CH_3$
10.19 $X = OH$ **10.15** $R = CH_2C_6H_5$
10.16 $X = SCH_2$ **10.17** $R = CH_2$– NH

(10.13)

CHO, HO, CH_2X, CH_3, N + $R-CH(NH_2)-COOH$

10.12 $X = OPO_3H_2$ **10.13** $R = CH_3$
10.14 $R = CH_2C_6H_5$
10.18 $R = CH_2$– NH

aromatic rings of **10.15** or **10.17** into the cyclodextrin cavity during transamination (**10.20**) with the resultant increase of both the forward and the reversed rate

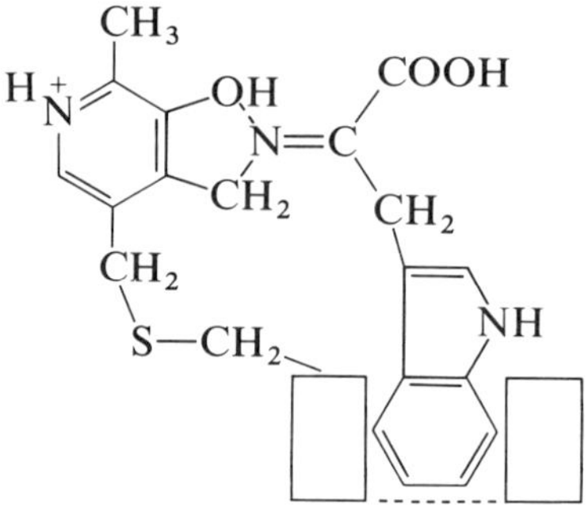

10.20

constants, and hence with a loss of selectivity with time due to rapid equilibration of the keto acid with the amino acid (Breslow et al., 1980a). Significantly, induction of optical activity has also been realized on using **10.16** (Breslow et al., 1980a).

A high degree of functional cyclodextrin mediated selectivity was obtained for the cleavage of cyclic phosphate **10.21** (Breslow et al., 1978, 1980b):

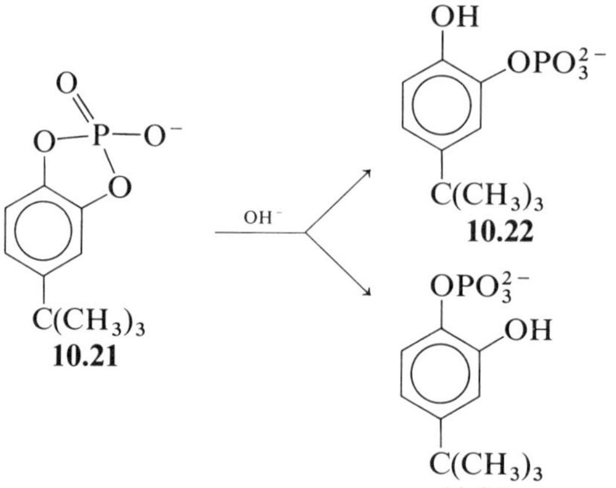

10.21 $\xrightarrow{OH^-}$ **10.22** + **10.23** (10.14)

In water distribution of products **10.22** and **10.23** is 50 : 50. In the presence of β-cyclodextrilyl-6,6′-bisimidazole, **10.24**, only the P—O bond was cleaved, giving exclusively **10.22**. Conversely, in the presence of **10.25** adn **10.26** the product of reaction 10.14 is exclusively **10.23**. These results were rationalized in terms of the different structural arrangements of substrate cyclodextrin complexes (Breslow et al., 1980b). The substrate, **10.21**, occupies a position in **10.24** such that attacking and leaving oxygens are 180° apart (**10.27**). In this arrangement water, brought in by the imidazole group attached directly to cyclodextrin, must approach in line with the P—O(1) bond, and cannot get far enough to align

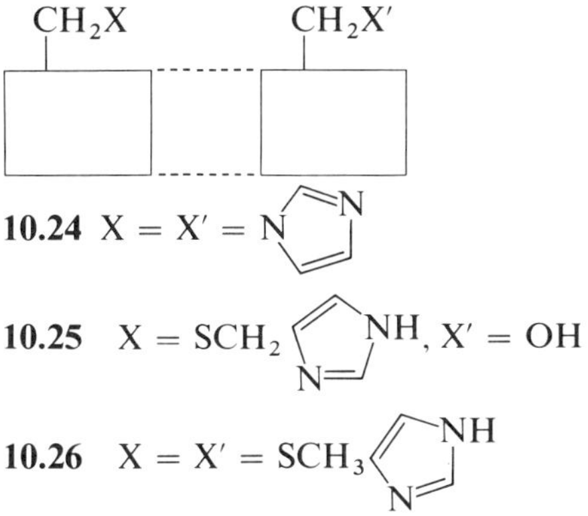

CH_2X CH_2X'

10.24 X = X′ = N-imidazolyl

10.25 X = SCH_2-imidazolyl, X′ = OH

10.26 X = X′ = SCH_3-imidazolyl

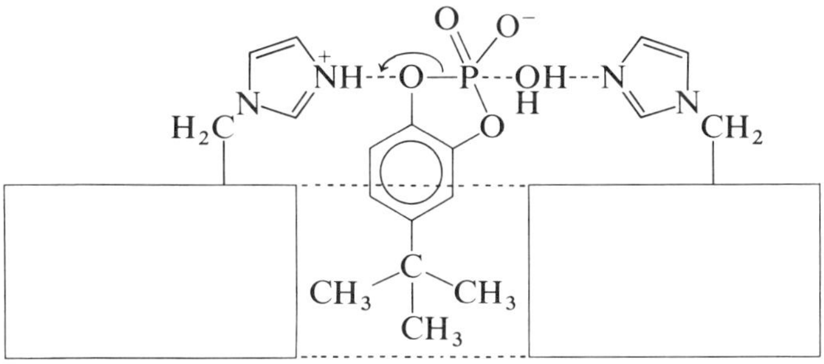

10.27

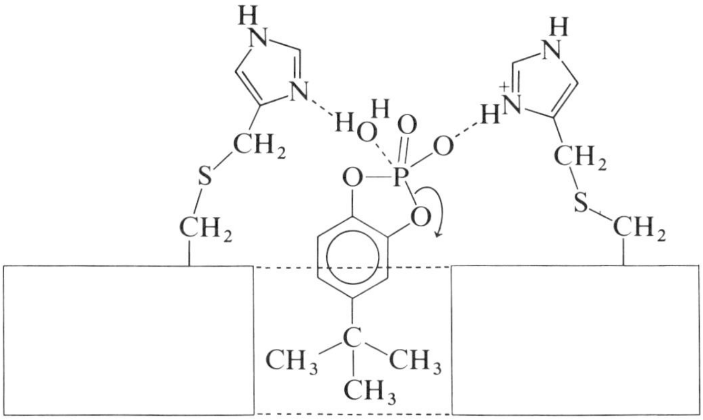

10.28

with the P—O(2) bond. In **10.26** the imidazole moieties are sufficiently removed from the cyclodextrin to deliver water to the P—O(2) position (**10.28**). These results provide strong evidence that subtle geometric changes in enzyme active sites are capable of altering product distributions dramatically, and that imidazole capped cyclodextrins model ribonuclease functions quite well (Breslow et al., 1980b).

Hydrolytic enzymes, particularly chymotrypsin, have been extensively modeled. The hydroxyl group of serine 195 residue and the imidazole group of hystidine 57 are the functional groups believed to be responsible for the catalysis. The mechanism involves the acylation of the enzyme bound substrate by the serine hydroxyl-group, and subsequent deacylation:

$$\underset{\boxed{\text{enzyme}}}{\text{OH}} + \text{R}\overset{\text{O}}{\overset{\|}{\text{C}}}\text{X} \underset{k_{-1}}{\overset{k_1}{\rightleftharpoons}} \underset{\text{enzyme-substrate complex}}{\underset{\boxed{\text{enzyme}}}{\text{OH}} \cdot \text{R}\overset{\text{O}}{\overset{\|}{\text{C}}}\text{X}} \xrightarrow[\text{acylation}]{k_2} \underset{\boxed{\text{enzyme}}}{\text{OCOR}} + \text{HX}$$

$$\underset{\boxed{\text{enzyme}}}{\text{OCOR}} \xrightarrow[\text{H}_2\text{O deacylation}]{k_3} \text{R}\overset{\text{O}}{\overset{\|}{\text{C}}}\text{OH} + \underset{\boxed{\text{enzyme}}}{\text{OH}} \qquad (10.15)$$

The imidazole group on histidine acts as a general base for both the acylation and the deacylation steps. Aspartic acid 101, hydrogen bonded to histidine 57, assists the base catalysis by "charge relay."

Reaction 10.15 has been modeled by functionalized micelles (Tonellato, 1979; Moss et al., 1980), polyions (Morawetz, 1978), and cyclodextrins (Breslow, 1979). Typically, rate constants for the hydrolysis of *p*-nitrophenyl esters have been determined in the absence and in the presence of enzyme models containing hydroxyl, oxime, imidazole, histidine, amine, and thiol moieties. Table 10.3 collects the results of selected investigations. The different conditions and standards used do not warrant quantitative comparisons. A number of qualitative conclusions can be drawn, however, Catalytic efficiency depends on the extent that the enzyme model, the host, is capable of destabilizing the ground state and stabilizing the transition state. The importance of these factors is clearly illustrated for ester hydrolyses in cyclodextrins (Breslow et al., 1980c). Rate enhancements in nonfunctional cyclodextrins are negligible since the ground state is stabilized more effectively than the transition state. Flexible capping of the cyclodextrin resulted in an improved geometry. Guided by molecular models, substrates were selected that remained bound as much as possible in the cyclodextrin cavity during the progress of the reaction. As expected, the greatest rate enhancement was observed for the hydrolysis of ferrocenylacrylate ester (see equation 7.9), since the geometry of binding of this substrate was almost the same as that of its hydrolytic tetrahedral intermediate. Conversely, other substrates penetrated cyclodextrin much more effectively than their tetrahedral intermediates and, hence, their ground states gained greater stabilization than their transition states (Breslow et al., 1980c).

Substrate binding in micelles is much less rigid than in cyclodextrins. Nevertheless, catalytic efficiencies similar to that of chymotrypsin have been demonstrated in well-designed bifunctional and polyfunctional micelles and comicelles (see Table 10.3). To date, no truly bifunctional catalysis has been observed for

Table 10.3. Enzyme Modeling in Functional Systems[a]

System	Reaction Substrate	Result	Reference
Aqueous $n\text{-}C_{11}H_{23}CON(CH_3)OH$ and CTAB comicelles	Hydrolysis of PNPA	Acylation rate is comparable to that of chymotrypsin	Tabushi et al., 1974
Aqueous $CH_3(CH_2)_{15}\overset{+}{N}(CH_3)_3Br^-$ and $CH_3(CH_2)_{10}CON(CH_3)OH$ comicelles	Hydrolysis of PNPA	Acylation rate is comparable to that of chymotrypsin; there is an efficient turnover	Tabushi and Kuroda, 1974
Aqueous $CH_3(CH_2)_{11}N(OH)C(O)Im$ + CTAB and $CH_3(CH_2)_{11}N(OH)C(O)C_6H_5$ + CTAB comicelles	Hydrolysis of PNPA	Bifunctional comicelles are much better catalysts than nonfunctional micelles	Kunitake et al., 1976
Aqueous micellar $n\text{-}C_{16}H_{33}\overset{+}{N}(CH_3)(R_1)R_2Cl^-$ $R_1 = CH_3$, $R_2 = CH_3$ $R_1 = CH_3$, $R_2 = CH_2CH_2OH$ $R_1 = CH_2C_6H_5$, $R_2 = CH_2CH_2OH$ $R_1 = CH_3$, $R_2 = CH_2Im$ $R_1 = CH_2CH_2OH$, $R_2 = CH_2Im$	Hydrolysis of PNPA and PNPH	Functional micelles are efficient catalysts	Moss et al., 1975
Aqueous $C_{13}H_{27}CONH(CH)(COOH)CH_2Im$ + CTAB and $(C_{16}H_{33})\overset{+}{N}(CH_3)(C_2H_5)_2CH_2Im$, Cl^- + CTAB comicelles	Hydrolysis of PNPA	Structural differences determine catalytic efficiencies	Tonellato, 1976
Aqueous micellar CTAB or SDS or POOA	RS^- + PNPA → $RSCOCH_3$ + *p*-nitrophenol RSH = coenzyme A or glutathione	Reactive ion pairs formed between cationic CTAB and coenzyme A or glutathione are responsible for the large rate enhancements	Shinkai and Kunitake, 1976

Table 10.3 (*Continued*)

System	Reaction Substrate	Result	Reference
Aqueous micellar $C_{16}H_{33}\overset{+}{N}(CH_3)_2(CH_2CH_2OH)CH_2Im, Cl^-$, $C_{16}H_{33}\overset{+}{N}(CH_3)_2CH_2Im, Cl^-$, $C_{16}H_{33}\overset{+}{N}(CH_3)(C_2H_5)CH_2CH_2OH, Br^-$	Hydrolysis of PNPA, PNPH, CH_3COIm, $C_5H_{11}Im$, $CF_3CON(Bu)$-$C_6H_4(NO_2)_2$	Effective nucleophilic site changes from the imidazole to the hydroxyl function ongoing from the hydrolysis of esters to that of amides	Anoardi and Tonellato, 1977
Aqueous micellar n-$C_{16}H_{33}\overset{+}{N}(CH_3)(CH_2Im)CH_2CH_2OH$	Hydrolysis of PNPA and PNPH	Sequential rather than bifunctional catalysis	Moss et al., 1977
Aqueous $C_{18}H_{37}\overset{+}{N}(CH_3)_2(CH_2CH_2OH), Br^-$ + $C_{13}H_{27}CONHCH(COOH)CH_2Im$ comicelles	Hydrolysis of PNPA	Hydrolysis occurs by a fast acylation on the imidazole followed by a fast quantitative transfer of the acyl group to the hydroxyl group	Tagaki et al., 1977
Aqueous micellar n-$C_{12}H_{25}\overset{+}{N}(CH_3)_2CH_2C(O)C{=}(NOH)Ph, Cl^-$, n-$C_{16}H_{25}\overset{+}{N}(CH_3)_2CH_2C(O)C{=}(NOH)Ph, Cl^-$	Hydrolysis of PNPA and PNPH	Acylation rate is comparable to that of chymotrypsin	Anoardi et al., 1978a
Aqueous micellar $R\overset{+}{N}(CH_3)_2CH_2SH, Br^-$ $R = n$-$C_{12}H_{25}$ or $R = n$-$C_{16}H_{33}$	Hydrolysis of PNPA and PNPH	Reaction involves the nucleophilic attack of the thiolate ion at the carbonyl carbon; functional micelles approach activities comparable to the related enzyme, ficin	Anoardi et al., 1978b
Aqueous micellar n-$C_{16}H_{33}\overset{+}{N}(CH_3)_2CH_2CH_2SH, Cl^-$, n-$C_{16}H_{33}\overset{+}{N}(CH_3)_2CH_2Im, Cl$	Hydrolysis of PNPA and PNPH	Acylation with respect to that by thiocholine is enhanced 251(PNPA)- and 2850(PNPH)-fold. Deacylation of the product	Moss et al., 1980

		n-$C_{16}H_{33}\overset{+}{N}(CH_3)_2CH_2CH_2SCOCH_3$, Cl^-, is 271-fold faster in micellar $C_{16}H_{33}\overset{+}{N}(CH_3)_2CH_2Im$, Cl^- than in micellar CTACl	
Aqueous $CH_3(CH_2)_{12}-\underset{\underset{\text{histMe}}{\vert}}{\underset{C=O}{\vert}}{C}-H(CH_2)_4\overset{+}{N}(CH_3)_3$, Cl^- **10.29**	Hydrolysis of PNPA	Mechanism involves acyl transfer to imidazole with no concurrent general-base catalysis; the resulting acylimidazole breaks down to acetate ion; no charge relay observed	Brown et al., 1979
Aqueous micellar $CH_3(CH_2)_{13}\underset{O=CN(CH_3)OH}{\underset{\vert}{C}}H(CH_2)_2\overset{+}{N}(CH_3)_3$, Cl^- $CH_3(CH_2)_9\underset{O=CN(CH_3)OH}{\underset{\vert}{C}}H(CH_2)_2\overset{+}{N}(CH_3)_3$, Cl^- and their comicelles with **10.29**	Hydrolysis of PNPH, phenylalanine esters	Effective catalysis by noncooperative mechanism	Brown and Lynn, 1980
β-Cyclodextrin with and without flexible capping in water and in DMSO:H_2O = 60:40 (v/v)	Acylation of nitrophenyl acetates and esters attached to cinnamic acid, adamantate, and ferrocene frameworks	Host–guest ground state–transition state optimization leads to an acceleration of acylations relative to hydrolysis of 10^6–10^7-fold, exceeding that for chymotrypsin with PNPA	Breslow et al., 1980c

[a] Abbreviations: PNPA = p-nitrophenylacetate; PNPH = p-nitrophenylhexanoate; Im = imidazole; CTAB = hexadecyltrimethylammonium bromide, $C_{16}H_{33}\overset{+}{N}(CH_3)_3$, Br^-; CTACl = hexadecyltrimethylammonium chloride, $C_{16}H_{33}\overset{+}{N}(CH_3)_3$, Cl^-; SDS = sodium dodecyl sulfate, $CH_3(CH_2)_{11}OSO_3^-Na^+$; POOA = $CH_3(CH_2)_7CH=CH(CH_2)_7CH_2(CH_2CH_2O)_{10}OH$; histMe = histidine methyl ester.

ester hydrolysis in micellar systems (Tonellato, 1979). Experimental results for the hydrolysis of *p*-nitrophenyl acetate in bifunctional micelles do not support a cooperative one-step reaction but are compatible with a two-step mechanism; initial slow *N*-acylation followed by a rapid hydroxyl group mediated deacylation (Moss et al., 1977). The catalytic cycle involves acylation of the imidazole ring, acyl transfer to the hydroxyl function, and regeneration:

$$\text{micelle}\begin{matrix}\text{Im}\\ \text{OH}\end{matrix} \underset{}{\overset{\text{RCOOAr}}{\rightleftharpoons}} \text{micelle}\begin{matrix}\text{Im}\\ \text{OH}\end{matrix}\ \text{RCOOAr} \xrightarrow{-\text{HOAr}} \text{micelle}\begin{matrix}\text{Im}-\overset{\overset{\text{O}}{\|}}{\text{C}}-\text{R}\\ \text{OH}\end{matrix} \xrightarrow{\text{fast}} \text{micelle}\begin{matrix}\text{Im}\\ \text{O}-\underset{\underset{\text{O}}{\|}}{\text{C}}-\text{R}\end{matrix} \xrightarrow{-\text{RCO}_2\text{H}} \text{micelle}\begin{matrix}\text{Im}\\ \text{OH}\end{matrix} \qquad (10.16)$$

Interestingly, comicelles show the greatest catalytic efficiency (Kunitake et al., 1976). No functional vesicles have been utilized yet for modeling enzyme catalysis. Rapid advances in this fascinating area of chemistry are fully expected.

2. ENZYMES IN MEMBRANE MIMETIC SYSTEMS

Surfactant-protein interactions have been extensively investigated. In the simplest form there is competition between surfactant self-association (micelle formation) and surfactant protein binding. The latter may lead to denaturation and/or conformational changes. Investigations of the effects of surfactants on protein stability and conformational changes have provided insight into the structure of proteins (Nemethy, 1967; Jencks, 1969). Examinations of the effects of micellar surfactants on enzyme catalyzed reactions have been aimed at obtaining better understanding of the mechanisms and the active sites involved in enzyme catalyses. These types of studies have been discussed previously (Fendler and Fendler, 1975; Tanfor, 1980). Attention is focused here on modeling membrane-protein interactions in monolayers, bilayers, and vesicles and on exploiting reversed micelles as media for investigating properties and reactivities of enzymes.

Proteins in Monolayers

Monolayers provide conceptually, if not experimentally (see Chapter 4), the simplest framework for the investigation of protein-lipid interactions (Shah,

1972; Demel, 1974; Rothfield and Fried, 1975). They represent the first membrane mimetic agents investigated. Introduction of enzymes into monolayers dates to 1935 (Schulman and Hughes, 1935). Table 10.4 summarizes the data obtained. Enzymatic reactions manifest time dependent surface pressure changes. Substrates are typically located within the monolayers, while the enzyme is distributed in the subphase. Depending on the reaction, the product(s) may be retained in the monolayer or enter into the subphase.

A two-phase model has been proposed to account for pancreatic phospholipase catalyzed hydrolysis of phospholipids in monolayers (Figure 10.3; Verger et al., 1973). It involves two successive equilibria. In the first equilibrium the water-soluble enzyme enters into the monolayer, and thereby it is stabilized (indicated by E* in Figure 10.3). The second equilibrium is the formation of the enzyme-substrate complex, E*S, within the subphase of the monolayer. In the catalytic step the product (P) is formed and the enzyme is regenerated. If the product is water-soluble, it is expelled into the aqueous subphase. Rate constants, treated kinetically in terms of two-dimensional Michaelis–Menten kinetics, agreed well with this model (Verger et al., 1973). The advantage of investigating enzymatic reactions in monolayers is the relative ease with which the concentration and organization of substrates and enzymes can be controlled. A distinct disadvantage is the experimental difficulties in handling monolayers (see Chapter 4).

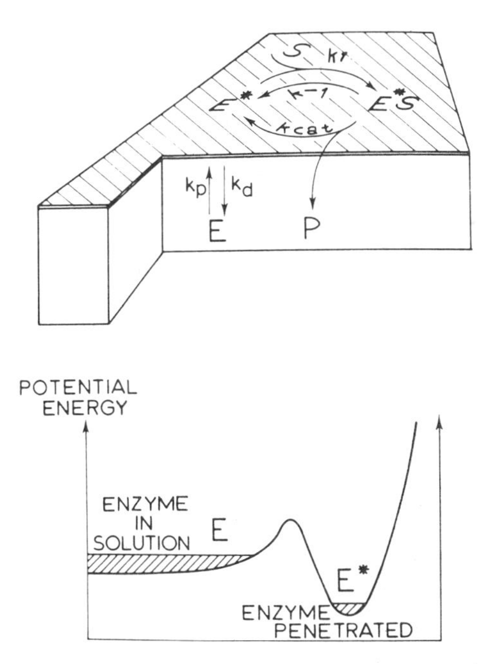

Figure 10.3 Proposed model for the action of a soluble enzyme at an interface (taken from Verger et al., 1973).

Table 10.4. Proteins in Monolayers

Monolayer	Protein	Method of Investigation	Results	Reference
3-*sn*-Phosphatidylcholines, 1,2-didocenaoyl-3-*sn*-phosphatidylethanolamine, 1,2-didocanoyl-3-*sn*-phosphatidylglycerol, 1,2-didodecanoyl-3-*sn*-phosphatidyl-3′-*sn*-lysyl glycerol	Pancreatic phospholipase	Time dependent surface pressure changes are determined	Hydrolysis involves an induction time, related to the reversible penetration of the enzyme into the monolayer; mechanism rationalized in terms of a two-dimensional Michaelis-Menten model	Verger et al., 1973
Trioctanoi, 1,2-dioctanoin monolayer on water	Pancreatic lipase	Surface pressure is determined as a function of reaction time in a Langmuir trough	Hydrated enzyme molecules act upon substrates lying within the monolayer; enzymatic hydrolysis is first order with respect to the surface concentration of glyceride substrate, independent of pressure	Lagocki et al., 1970, 1973
1,2-Dioctanoyl-*sn*-glycero-3-phosphonylcholine monolayer on water	α-Phospholipase A_2 (snake venom *C. adamanteus*)	Time dependent changes of surface pressure are determined in a Langmuir trough	Dimeric forms of the enzyme and presence of bound Ca^{2+} are required; reaction is first order with respect to the surface concentration of the lipid substrate and of the enzyme dimer	Shen et al., 1975
1,3-Didecanoylglycerol monolayer on water	Pancreatic lipase (containing colipase)	Time dependent changes of surface pressure are determined in a Langmuir trough	Effects of enzyme, added Ca^{2+}, concentrations, and pH are investigated	Dervichian and Barque, 1979; Barque and Dervichian, 1979

Trihexanoylglycerol, 1,3-dihexanoyl-2-butylether glycerol, *p*-chlorobenzoyl decanoate monolayers on water	Glycerol-ester hydrolase	Time dependent surface pressure changes are determined	Enzyme activity depends on the nature of the monolayer and on surface pressure	Esposito et al., 1973
Dipalmitoylphosphatidyl choline + methylstearate + eicosylsodiumsulfate + eicosyltrimethylammonium bromide + arachiolic acid monolayers on water	Malate dehydrogenase	Quantity of adsorbed protein and enzyme activity are determined	Enzyme activity in monolayers is substantially lower than that in bulk water	Peters and Fromherz, 1975
Dipalmitoylphosphatidyl choline and chloroform monolayers on water	Acetylcholine receptor protein (*Torpedo marmorata*)	Surface pressure is determined in the absence and presence of protein	Interaction is proportional to surface area	Popot et al., 1977
Silochrome, lecithin, and cholesterol monolayers deposited on silochrome	α-Chymotrypsin, albumin, lysozyme, methemoglobin, cytochrome C	Epr spectroscopy of spin probes covalently bound to the enzyme or adsorbed on monolayer	Mobility of probes depends on the structure of probe and the nature of the support; proteins adsorbed on silochrome and cholesterol monolayers are not very mobile	Genkin et al., 1977
Dipalmitoylphosphatidyl-choline, phosphatidic acid + dansylphosphatidyl-ethanolamine (donor)	Cytochrome C	Fluorescence spectroscopy, energy transfer from donor to heme in protein (acceptor)	Orientation of protein in membrane assessed	Teissie, 1981

Proteins in BLMs

A number of proteins have been incorporated into BLMs (Table 10.5). By their very nature BLMs represent a more advanced membrane model than micelles. The advantage of BLM systems is the ease with which electrical properties can be monitored. Difficulties in handling BLMs, their unknown composition, and inherent instabilities (see Chapter 5) are distinct disadvantages.

Proteins in Liposomes

The relative ease of liposome preparation (see Chapter 6) has prompted the numerous investigations of protein-liposome interactions (Table 10.6). A large variety of experimental techniques has been used to establish the extent and nature of protein binding. Proteins can be adsorbed on the outer surface of vesicles (Kimelberg and Papahadjopoulos, 1971; Hebdon et al., 1979; Dombrose et al., 1979), where they may be immobilized in domains or move laterally. Alternatively, proteins may penetrate liposomes, span the bilayers, or be completely entrapped in the aqueous interiors. Entrapment or penetration depends on the size of the protein and the morphology of the liposome as well as on electrostatic and hydrophobic interactions. Trapping efficiencies of ^{125}I labeled globular proteins (lysozyme, α-chymotrypsogen, peroxidase, conalbumin, α-amylase) in small and large single -and multicompartment liposomes have been determined by gel electrophoresis and autoradiography (Adrian and Huang, 1979). The trapped proteins were separated from untrapped ones by gel filtration and ultrafiltration. True entrapments, rather than surface adsorption, were demonstrated by trypsin digestion. Liposome entrapped proteins were unaffected by externally added trypsin, but upon their release by the action of detergents they were all digested. Proteins with molecular weight of up to 96,920 (α-amylase) could be entrapped in large unilamellar and multilamellar liposomes with high efficiency. Conversely, only the smallest two proteins (lysozyme and α-chymotrypsinogen) could be entrapped with high efficiency in single-compartment liposomes. The structure of water, bound at the inner surface liposomes, and steric factors were considered to be important in determining trapping efficiencies. The probability, P, of a given protein of radius R_P being trapped inside the liposome is given by (Adrian and Huang, 1979)

$$P = 1 - \frac{D^3 - (D - A^3) + [(R_P - R_w)/(D - A - R_w)](D - A)^3}{D^3} \qquad (10.17)$$

or simplified to

$$P = \left(\frac{D - A}{D}\right)^3 \left(\frac{D - A - R_P}{D - A - R_w}\right) \qquad (10.18)$$

where D is the radius of the total internal volume of a spherical single-compartment liposome, A is the thickness of the bound water layer, and R_w is the radius water (~ 1.65 A). The probability of solute exclusion by seteric hindrance is given by $(R_P - R_w)/(D - A - R_w)$ and the radius of the protein is taken from Stokes' radius:

$$R_p = \frac{Mr(1 - \bar{v}\rho)}{N6\pi\eta S_{20,w}} \tag{10.19}$$

where Mr is the molecular weight, $S_{20,w}$ is the sedimentation coefficient of the protein, $\bar{v}$ is the partial specific volume of the protein, ρ is the density, and η is the viscosity of water. Calculations gave satisfactory trapping efficiencies (Table 10.7).

Proteins in general do not alter gross liposome morphologies, although they mediate vesicle-vesicle fusions (Dawidowicz and Rothman, 1976) and alter phase transitions (see Chapter 6). Interactions of proteins with liposomes have been treated theoretically by assuming that the protein is a rigid body in the bilayer and that the perturbation, described by an order parameter, decays smoothly in space and is related to the free energy of the system (Owicki et al., 1978; Owicki and McConnell, 1979). Permeabilities of protons, ions, and small molecules are substantially influenced by proteins (Brunner et al., 1978).

Utilization of large unilamellar liposomes in protein entrapment is particularly significant. This system has the stability associated with liposomes and the potential for making direct electrical measurements. Indeed penetration of a microelectrode into the interior of a large bovine rhodopsin containing single-compartment liposomes has recently been described (Darszon et al., 1980).

The entrapment of proteins in liposomes provides opportunities for modeling such membrane mediated functions as cell interactions and recognition, as well as metabolically or photosynthetically driven proton and electron translocations. It also opens the door, via chemical transformations (Heath et al., 1980), to immunochemistry and target directed drug delivery (see Chapter 14).

Proteins in Reversed Micelles

The importance of investigating reversed micelle solubilized enzymes are increasingly recognized (Luisi and Wolf, 1982). Sodium bis(2-ethylhexyl) sulfosuccinate, sodium di-(2-ethylhexyl)sulfosuccinate (aerosol-OT), and phospholipid solubilized water pools in hydrocarbon solvents are the most frequently used systems. aerosol-OT is the favored surfactant since it can solubilize substantial amounts of water (see Chapter 3), which in turn is able to accommodate large proteins ($\overline{\text{MW}} > 500{,}000$). Proteins are introduced as concentrated aqueous solutions into the reversed micelle–hydrocarbon system. Alternatively, they are extracted from aqueous solutions or solids. The solubility of proteins in reversed micelles is governed by the surfactant concentration, the water to surfactant ratio, the temperature, and the type and concentration of buffers, co-, and counterions present. Given the right condition any protein tested up to now could be solubilized in reversed micelles.

Table 10.5. Proteins in BLMs

BLM	Protein	Method of Investigation	Results	Reference
Egg lecithin, bovine phosphatidylserine, cardiolipin	Cytochrome-C (horse heart), bovine acetylcholinesterase, cytochrome oxidase	Electrical properties determined	Addition of enzymes did not alter resistance, but it altered the capacitance	Gitler and Montal, 1972b
Asymmetric BLM, containing reconstituted rhodopsin + lipoprotein, glycerol dioleate, cholesterol or retina phospholipids + glycerol dioleate, cholesterol	Rhodopsin	Electrical properties determined	Rhodopsin containing BLM is very short-lived ($\sim$1 min)	Montal and Korenbrot, 1973
Oxidized cholesterol: lecithin = 10:1 (v/v)	Bovine serum, human serum, rabbit anti-human serum	Resistance, voltage, and capacity of BLM measured during the immune reaction	BLMs are developed as models to study immunology	Mountz and Tien, 1978
Reconstituted photosynthetic reaction center	Photosynthetic center from *Rhodopseudomonas sphaeroides* R-26, cytochrome-C	Steady state and transient photovoltaic effects investigated in the presence of donors and acceptors	Photoelectric signal arises from transfer of electrons from a secondary donor (cytochrome-C) to a secondary acceptor (ubiquinone-0) located on opposite sides of the bilayer	Schönfeld et al., 1979b

Phosphatidylcholine, monogaloctosyl diglyceride, + chlorophyll-a and β-carotene	G-phycocyanin, plastocyanin	Photovoltaic effects investigated as a function of additives	Phycocyanine and plastocyanine enhanced the photosensitivity	Chen and Berns, 1979
Reconstituted acetylcholine receptor	Acetylcholine receptor (from *Torpedo californica*)	Conductance measurements on spreading reconstituted vesicles at an air–water interface	Fluctuations of membrane conductance, corresponding to opening and closing of receptor channels, are observed; single channel conductance of 16 ± 3 pS (in 0.1 *M* NaCl) with a mean channel opening time of 35 ± 5 msec estimated	Nelson et al., 1980
Oxidized cholesterol	Hepatic binding protein	Conductance measurements	Voltage dependent conductance increases reflect protein penetration and translocation	Blumenthal et al., 1980
Lecithin-2,6-di(*t*-butyl)*p*-cresol	Melittin	Surface potential measurements	Melittin binds strongly to the interface, where it remains localized	Schoch and Sargent, 1980
Bovine phosphatidylcholine, phosphatidylethanolamine, phosphatidylserine, cholesterol (BLM on a porous disk)	Rhodopsin	Illumination induces permeabilities	Illumination increases the permeability of K^+, Na^+, and Ca^+ with a high selectivity for Ca^{2+}	Kossi and Leblanc, 1981

Table 10.6. Proteins in Liposomes

Liposome	Protein	Method of Investigation	Results	Reference
Sonicated and mechanically dispersed phosphatidylcholine, phosphatidylserine, and their mixtures	Lysozyme, cytochrome-C, ribonuclease, poly-L-lysine	Electrophoretic mobility	Surface interaction (no entrapment); protein bound as ion-pair in liposome surface	Kimelberg and Papahadjopoulos, 1971
Sonicated cardiolipin lecithin	Cytochrome-C (horse heart)	Epr using doxyl stearic acid spin label	Data are consistent with electrostatic binding of cytochrome-C to the charged groups of the phospholipid; presence of extrinsic proteins does not interfere with measurement of boundary lipids in intact membranes	Van and Griffith, 1975
Sonicated soy bean phospholipid + potassium cholate + cytochrome-C oxidase (reconstituted liposome)	Ferrocytochrome-C	Oxygen consumption and absorption, spectrophotometry, pH measurements	Ferrocytochrome-C induces proton ejection, $[H^+]$ release depends on enzyme–lipid ratio; cytochrome-C oxidase acts as a proton pump; $\sim 0.9\ H^+$ released per ferrocytochrome-C molecule oxidized	Casey et al., 1979
Sonicated 1-palmitoyl-2-oleoyl-*sn*-glycero-3-phosphorylcholine, 1,2-bis(9,10-dibrostearoyl)-*sn*-glycerol-3-phosphorylcholine	Cytochrome-b_5	Intervesicle exchange is studied by fluorescence spectroscopy	Kinetics of cytochrome-b_5 exchange between liposomes are consistent with the transfer of the enzyme through the aqueous phase rather than with vesicle-vesicle collisions	Leto et al., 1979; Leto and Holloway, 1979

Cholate filtrated microsomal phospholipids	Cytochrome-b_5 (rabbit liver), epoxide hydrase, cytochrome-P-450 (liver microsomal)	High pressure liquid chromatographic analysis of the different benzo(a)pyrene metabolites	Substrate oxygenation is stereospecific; cytochrome-P-450_{LM2} efficiently converts benzo(a)pyrene in the presence of epoxide hydrase to 4,5-dihydroxy-4,5-dihydrobenzo(a)pyrene; cytochrome-P-450_{LM4} primarily forms 9,10-dihydroxy-9,10-dihydrobenzo(a)pyrene; different cytochromes affect oxygenations at different rates.	Brunström and Ingelman-Sundberg, 1980; Ingelman-Sundberg and Johansson, 1980
Mechanically dispersed L-α-dipalmitoyllecithin, L-α-dimyristoyllecithin, and L-α-dioleoyllecithin	Cytochrome-C	Fluorescence polarization intensities and lifetimes of lipid bound *N*-(1-alinonaphthyl-4)-maleimide and lipophilic 1,6-diphenyl-1,3,5-hexatriene	Temperature dependent conformational changes in oxidized cytochrome oxidase only, at 20°C are due to intrinsic enzyme conformational change, at 38°C are due to lipid phase transition	Kawato et al., 1980, 1981
Dialized sodium cholate phospholipid (egg yolk lecithin, dimyristoyl-phosphatidylcholine, dioleoylphosphatidyl-choline) mixtures	Cytochrome-C and cytochrome-C_2 + bacterial photosynthetic reaction centers	Differential scanning calorimetry, steady state and laser spectroscopy	Oxidation of cytochromes by bacterial photosynthetic centers are investigated	Overfield and Wraight, 1980
Mechanically dispersed dipalmitoylphosphatidyl-choline, dimyristoylphos-phatidylcholine, and diolcoylphosphatidyl-choline	Cytochrome-C oxidase (from bovine heart)	Oxygen consumption measurements following addition of sodium as corbate	Breaks in Arrhenius plots correspond to structural transition of cytochrome oxidase	Yoshida et al., 1979

Table 10.6 (*Continued*)

Liposome	Protein	Method of Investigation	Results	Reference
Sonicated soy bean phospholipids	Cytochrome-C oxidase (beef heart, spin labeled)	Light induced pH changes	Enzyme rotates rapidly with a correlation time of 34 μsec at 4°C	Ariano and Azzi, 1980
Dialized or sonicated sodium cholate phosphatidylcholine mixtures	Cytochrome-C oxidase (bovine heart, spin labeled)	Epr	Rotational mobility depends on the method of preparation	Swanson et al., 1980
Sonicated dimyristoyl-L-lecithin	Rhodium(III)-protophorphyrin IX derivative of cytochrome-b_5	Laser flash-induced transient dichroism	Triplet decay is perturbed in liposomes; the ratio of the limiting anisotropy to the initial anisotropy is 0.6, implying a cone of restricted motion of 34° for the protein in the bilayer	Vaz et al., 1979b
Sonicated retinal outer segment phospholipids	RDS membranes with ~25% bleached rhodopsin (bovine)	^1H nmr relaxations at 100 and 360 MHz	Interaction of rhodopsin with the more fluid membrane phospholipids affect high frequency segmental motions; trans–gauche isomerizations are significantly restricted	Brown et al., 1977
Sonicated phosphatidyl-ethanolamine, phosphatidylcholine	Rhodopsin	^1H and ^{31}P nmr spectroscopy to follow Mn^{2+}, Co^{2+}, and Eu^{3+} equilibration across bilayer in presence and absence of light	In dark vesicles are sealed to metal ions, in light they are translocated ∴ rhodopsin is a transmembrane protein	O'Brien et al., 1977

Sonicated soy bean phospholipids	Rhodopsin	Column chromatography using $^{22}Na^+$, $^{134}Cs^+$, $^{36}Cl^-$, ^{3}N-glycerol, ^{3}H-glucose, ^{3}H- or ^{14}C-inulin	Light exposure increases Na^+, Cs^+, Ca^+, glycerol, and glucose permeabilities, does not affect those for Cl^-, sucrose, and inulin.	Darszon et al., 1977
Egg yolk phosphatidylcholine + cholic acid	Rhodopsin	Proteolysis of reconstituted membrane with papain and thermolysin, iodination catalyzed by lactoperoxidase	Rhodopsin polypeptide spans the thickness of the membrane	Fung and Hubbell, 1978
Rhodopsin-egg lecithin (detergent removal)	Rhodopsin (+ covalently attached spin labeled fatty acid)	Epr	Rhodopsin boundary layer under physiological conditions is associated with low microviscosity; however, low temperatures, low lipid to protein ratios, or both, can induce dramatic modifications of the physical state of the boundary lipids	Davoust et al., 1979
Soy bean phospholipids, egg phosphatidylcholine	Bacteriorhodopsin (*Halobacterium halobrium*)	Fluorescence, flow dialysis	Equations predicting ion flows are verified experimentally	Hellingwerf et al., 1979
Dimyristoylphosphatidyl-choline, egg phosphatidylcholine, asolecithin	Bacteriorhodopsin	Light induced pH changes	Efficient proton translocation from inside to outside on illumination	Dencher and Heyn, 1979

Table 10.6 (*Continued*)

Liposome	Protein	Method of Investigation	Results	Reference
Sonicated soy bean phospholipid, dioleoylphosphatidylcholine, dimyristoylphosphatidylcholine + sodium cholate	Bacteriorhodopsin (99% free from endogenous lipids)	Light induced pH changes	Reconstituted bacteriorhodopsin efficiently translocates protons from outside to inside on illumination	Huang et al., 1980
Sonicated dimyristoylphosphatidylcholine + cholesterol	Bacteriorhodopsin	Laser flash-induced transient dichroism	Above phase transition temperature in the absence of cholesterol bacteriorhodopsin exhibits rotational mobility in liposomes; cholesterol incorporation results in loss of of rotational mobility	Cherry et al., 1980
Reconstituted (removal of apolar solvent) protein lipid complexes	Bovine and squid rhodopsin, reaction centers from *Rhodopseudomas sphaeroides*, beef heart cytochrome-C oxidase, acetylcholine receptors from *Torpedo californica*	Electron microscopy, absorption spectroscopy, light induced proton translocation, electrical measurements	Reconstituted membrane proteins show biological activity	Darszon et al., 1980
Sonicated egg yolk phosphatidylcholine, dimyristoyl phosphatidylcholine, dipalmitoylphosphatidylcholine	Phospholipase A_2 (cobra venom)	Enzyme activity measurements	Gel states hydrolized 2–3 times faster than the liquid crystalline state	Kensil and Dennis, 1979

Sonicated phosphatidylcholine, phosphatidylethanol-amine, phosphatidylinositol, rendered asymmetric by amidation of the outer surface	Phospholipase A_2 (snake venom)	Rate of phospholipid degradation is determined	Selective cleavage of the lipid at the outer and inner surface of liposomes	Sundler et al., 1978
Dispersed or sonicated L-α-dipalmitoylphospha-tidylcholine	Phospholipase A_2 (bee venom)	Hydrolysis of phospholipids	Phospholipase A_2 preferentially catalyzes hydrolysis of the substrate available at or near defect structures	Upreti and Jain, 1980
Sonicated dimyristoyl lecithin	Apoprotein (from porcine high density lipoprotein)	Epr, nmr, CD	Nature of lipid–protein interaction is elucidated	Andres et al., 1976
Hydrated dipalmitoyl-phosphatidylcholine and egg yolk phosphatidylcholine (1:1 molar ratio) on slides	Apolipoprotein C ApoC-III (fluorescein-labeled)	Fluorescence recovery after photobleaching	Apoprotein is bound to the bilayer near the polar groups as suggested by its lateral diffusion coefficient; binding of the protein does not alter gel-to-liquid; crystalline phase transition but mediates a reversible aggregation of vesicles about 33°C	Vaz et al., 1979a
Sonicated dimyristoyl-phosphatidylcholine	A-1 Apoproteins (human and bovine)	^{31}P nmr, fluorescence spectra of dansylated apoproteins, and 1,6-diphenylhexatriene probes	Apoprotein binding results in apoprotein–liposome and apoprotein–micelle complexes; formation of the former is very much faster than the latter.	Jonas and Drengler, 1980a, 1980b

Table 10.6 (*Continued*)

Liposome	Protein	Method of Investigation	Results	Reference
Unilamellar phosphatidylcholine liposome (French press)	Apolipoproteins (A–I, apoE), HDL, and VLDL lipoproteins	Electron microscopy, fluorescence spectroscopy of marker release	Proteins induce marker release and disc formation	Guo et al., 1980
Mechanically shaken egg lecithin + cholesterol in the presence and absence of hydrocortisone	Heat aggregated and native immunoglobulins (G-type)	Epr, using 3-doxylcholestane or 4-[3β-hydroxylcholest-5-enamido]-tempo spin labels and solute diffusions	Heat aggregated immunoglobulin mobilized hydrophobic spin probe within bilayers of anionic liposomes; native immunoglobulin ineffective; hydrocortisone prevented the perturbation of the nonpolar membrane regions and reduced leakage provoked by aggregated immunoglobulin	Schieren et al., 1978
Sonicated egg lecithin, dimyristoylphosphatidyl-choline	High density (HDL_3) lipoprotein	^1H nmr spectroscopy (270 MHz)	Weak, nonspecific; surface interactions apoprotein peptide chains are intercalated between lipid headgroups	Hauser, 1975
Sonicated egg phosphatidylcholine + cholesterol	Bovine high density serum lipoprotein	Electron microscopy, fluorescence spectroscopy (steady state, time resolved, and polarization)	Direct addition of excess lipids from membranes or other lipoproteins is a possible mechanism for lipid transfer	Jonas, 1979

Sonicated (small single-compartment), ether injected (large single-compartment), and mechanically agitated (multicompartment) dicetyl phosphate, egg phosphatidylcholine, brain gangliosides, L-α-dipalmitoyl phosphatidylcholine, sphingomyelin, and gangliosides	Peroxidase (horseradish), inulin (carbohydrate)	Gel filtration using isotopically labeled inulin and horseradish peroxidase	Liposome disruption by human serum is concentration and time dependent; loss is reduced by treating the serum to remove the complement by incorporating sphingomyelin into multilamellar liposomes; leakage of inulin is also reduced; cholesterol has the same effect	Finkelstein and Weissmann, 1979
Vortexed cardiolipin, phosphatidylserine, phosphatidylglycerol, phosphatidylcholine, or phosphatidylethanolamine	Cytochrome-C	^{31}P nmr, electron microscopy	Cytochrome-C induces hexagonal H_{11} phase and inverted micelle structure	DeKruijff and Cullis, 1980
Sonicated phosphatidylethanolamine, phosphatidylcholine, phosphatidylserine	Horseradish peroxidase	Electron microscopy, isoelectric focusing, binding studies	Horseradish peroxidase is covalently attached to the liposome surface	Heath et al., 1980
Sonicated phosphatidylcholine + sodium cholate	(Na^+, K^+)ATPase (lamb kidney medulla)	Gel filtration (Sephadex G-50) to separate free and liposome entrapped $^{22}Na^+$ and $^{86}Rb^+$	Transport is catalyzed with a ratio of $3Na^+$ to $2K^+$, oubain inhibited transport.	Anner et al., 1977
Dialized deuterated phospholipids	ATPase, human brain lipophilin, beef brain myelin, proteolipid apoprotein	^2H FT nmr spectroscopy	Protein incorporation, in general, has little or no effect on the segmental motion of lipid bilayers	Rice et al., 1979

Table 10.6 (*Continued*)

Liposome	Protein	Method of Investigation	Results	Reference
Sonicated or detergent diluted asolectin (soy phospholipid)	CFo-CF, ATP-synthetase (extracted from spinach)	Pulse radiolysis	Pulse radiolytically driven phosphorylation; ADP + ^{32}Pi + $nH^+_{in} \rightleftharpoons {}^{32}ATP + H_2O + nH^+_{out}$ is observed	Gould et al., 1979
Sonicated egg yolk phosphatidylcholine	Bovine heart transhydrogenase	Fluorescence (using 9-aminoacridine) and pH measurements of proton translocation	Proton translocation demonstrated in reconstituted vesicles; one or less proton is translocated for each hydride ion equivalent transferred between the substrates	Earle and Fisher, 1980
Sonicated egg yolk phosphatidylcholine	γ-Glutamyl transpeptidase	L-^{14}[C]-glutamate uptake determined	Selective glutamate uptake observed in glutathione carrying vesicles, but not in empty ones; glutamate uptake is sensitive to the temperature of incubation, to inhibitors, and to enzyme activity	Sikka and Kalra, 1980
Vortexed L-α-dioleylphosphatidylcholine + cholesterol (French press)	Lecithin: cholesterol acyltransferase, apolipoprotein A-1	Determining rate of ^{3}[H] labeled cholesteryl ester formation from liposomes containing ^{3}[H] cholesterol	Relationship between lecithin: cholesterol acyltransferase an apoprotein A-1 has been investigated	Chajek et al., 1980

Sonicated phosphatidylcholine and mixtures of phosphatidylcholine, lysophosphatidylcholine	C-Reactive protein (CRP)	Gel filtration of ^{125}I labeled protein	Binding of protein requires the incorporation of lysophosphatidylcholine into phosphatidylcholine liposomes; proposed as model for the binding CRP to damaged but not to intact cells	Volanakis and Wirtz, 1979
Sonicated diacylphosphatidylserine, phosphatidylcholine, lysophosphatidylcholine, deoxycholate	Myelin (from bovine brain)	Circular dichroic spectra	Conformational changes upon binding is interpreted in terms of an increased α-helix folding in the protein caused hydrophobic interactions	Keniry and Smith, 1979
Dioleoyl-*sn*-glycerophosphorylcholine + dioleoyl-*sn*-glycerophosphorylglycerol	Bovine protrombin fragment 1	Quantitative gel filtration using ^{14}[C] labeled liposomes	Protein binds reversibly to liposome surface, binding is a function of [Ca^{2+}] between 0.5 and 2.7 m*M*, but independent above; proposed model for binding suggested to be similar to vitamin K dependent coagulation factors	Dombrose et al., 1979
Vortexed and centrifuged (300 g) dimyristoylphosphatidylcholine, dipalmitoylphosphatidylcholine + cholesterol	Anti-nitroxide IgG (from rabbit antisera)	Laser-fluorescence-photobleaching recovery	Rates of lateral diffusion of fluorescent labeled specific anti-nitroxide IgG bound with both combining sites to nitroxide containing liposomes are measured; diffusion coefficients range from 10^{-11} to 10^{-8} $cm^2 sec^{-1}$; antibody dependent immune responses are observed	Smith et al., 1979
Sonicated phospholipids	Microsomes	Enzyme activity determinations	Surface potential changes correlate with enzyme activity.	Nalecz et al., 1980

Table 10.6 (*Continued*)

Liposome	Protein	Method of Investigation	Results	Reference
Sonicated and mechanically agitated phosphatidyl glycerol and phosphatidylcholine	Myelin (proteolipid) apoprotein	Electron microscopy and differential scanning calorimetry	Fusion of vesicles observed by using myelin or dimethylsulfoxide but not by using myristic acid or lysolecithin	Papahadjopoulos et al., 1976
Sonicated L-α-dimyristoyl phosphatidylcholine, L-α-dilauroyl phosphatidylcholine, L-α-dioleoyl phosphatidylcholine, L-α-dilinoleyl phosphatidylcholine, L-α-diarachidonoyl phosphatidylcholine	Amphotericin B (polyene antibiotic)	Stop-flow kinetic measurements	Shortening the fatty acyl chain length of saturated phosphatidylcholines and increasing the number of double bonds enhance the rate of initial uptake of antibiotic; cholesterol reduces it; initial rate of uptake depends on competition of lipid–lipid and lipid–amphotericin B interactions	Chen and Bittman, 1977
Sonicated L-3-dipalmitoyl phosphatidylcholine, L-3-dioleoyl phosphatidylcholine, L-3-dilinoleoyl	Polylysine (MW = 70,000)	Absorption and CD spectroscopy, dilatometry, and electron microscopy	Above phase transition temperature, hydrophobic (temperature reversible) below polar surface (temperature irreversible) interactions	Campbell and Pawagi, 1979
Sonicated egg yolk phosphatidylcholine, phosphatidylethanolamine, phosphatidylserine	Adenylate cyclase (rat brain)	Enzyme activity determinations	Enzyme is "almost quantitatively" adsorbed at outer surface, being partially exposed to water	Hebdon et al., 1979
Mechanically agitated or sonicated phosphatidylethanolamine, phosphatidylglycerol cardiolipin	Cyclopropane fatty acid synthetase (*E. coli*)	Lipid analysis	Methylenation of C=C in the lipid is catalyzed by the enzyme both at the inner and outer layers of liposomes	Taylor and Cronan, 1979

Sonicated egg yolk phosphatidylcholine	Sucrase · isomaltase enzyme complex from brush border membrane	^{1}H and ^{31}P nmr, epr	90% protein incorporated; incorporation is asymmetric, being oriented toward external aqueous phase with minimal perturbation of packing; permeabilities of Na^{+}, D-fructose, and D-glucose are increased by the protein nonspecifically	Brunner et al., 1978
Sonicated or injected (ether solution into warm water) phosphatidylcholine, [^{3}H]-dipalmitoylphoshatidylcholine	Lysozyme, α-chymotrypsinogen, peroxidase, conalbumin, α-amylase	^{125}I labeling	Trapping efficiencies (lysozyme, 100%; α-chymotrypsinogen, 97.5%; perioxidase, 77.5%, conalbumin, 59.5%; α-amylase, 49.5%) are related to the size of proteins	Adrian and Huang, 1979
Mechanically agitated lipids isolated from membranes	Colchicine	Fluorescence polarization of 1,6-diphenyl-1,3,5-hexatriene	Alterations in microviscosity are due to changes in the lipid composition when vesicles are phagocited	Berlin and Ferra, 1977
Sonicated dipalmitoyl-phosphatidylcholine	Polymyxin-B (peptide antibiotic)	Fluorescence polarization of 1,6-diphenyl-1,3,5-hexatriene; electron paramagnetic resonance spectroscopy using fatty acid spin labels; electron microscopy	Polymyxin binds cooperatively producing a new phase transition and domain structures	Hartmann et al., 1978
Mechanically dispersed dipalmitoylphosphatidylcholine, dimyristoylphosphatidylcholine; dielaidoylphosphatidylcholine, dipalmitoylphosphatidylethanolamine	Glycophorin	Electron microscopy (freeze-fracture); epr spectroscopy (spin label); phase diagrams	Lipids are immobilized near the protein clusters	Grant and McConnell, 1974

Table 10.6 (*Continued*)

Liposome	Protein	Method of Investigation	Results	Reference
Sonicated dioleoylphosphatidylcholine (^3H labeled), (bis-9,10-dibromostearoyl)-phosphatidylcholine	Beef liver, phospholipid exchange protein	Density gradient centrifugation	Protein mediated phospholipid exchange and fusion are investigated; no spontaneous intervesicle exchange, but fusion over several hours.	Dawidowicz and Rothman, 1976
Single-compartment egg yolk phosphatidylcholine (detergent removal from octyl glucoside comicelles)	Glycoporin A	Electron microscopy, ion-flux measurements	Reconstitution process is surveyed	Mimms et al., 1981

Table 10.7. Trapping Efficiencies of Proteins in Single-Compartment Liposomes[a]

Protein	M_r	$S_{20,w}$	$\bar{v}$	*Av* Radius, Å, Calculated[b]	Probability of Trapping, Calculated[c]	Relative Trapping Efficiency, %[d]	
						Calculated	Experimental[e]
Lysozyme	14,100	1.91	0.703	17.5	0.466	100	100
α-Chymotrypsinogen	23,650	2.58	0.721	21.0	0.435	93.5	97.5
Peroxidase	39,780	3.48	0.699	26.2	0.390	83.7	77.5
Conalbumin	86,180	5.05	0.732	39.1	0.278	59.6	59.5
α-Amylase	96,920	6.47	0.717	44.3	0.233	59.9	49.5

Source: Taken from Adrian and Huang, 1979.

[a] See text for definitions.

[b] Calculated by equation 10.19.

[c] Calculated by equation 10.18.

[d] Trapping efficiency relative to lysozyme.

[e] Trapping efficiency $= (A \times b)/(B \times a) \times 100\%$, where A and a are the relative amounts of lysozyme in the original solution and in the vesicle entrapped form, respectively, and B and b are the relative amounts of a given protein marker in the original solution and in the vesicle entrapped form, respectively.

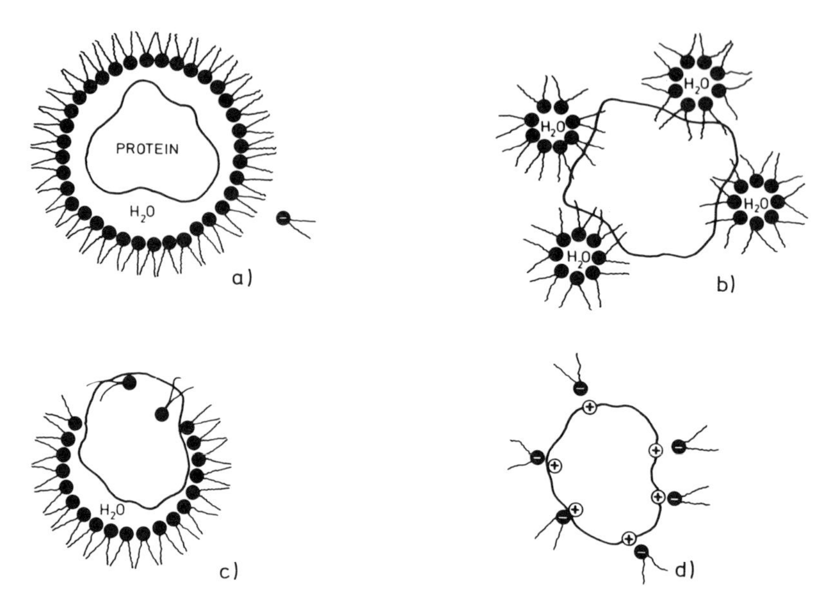

Figure 10.4 Possible situations for a biopolymer solubilized in a micellar hydrocarbon solution (cross section). (*a*) Water-shell model. (*b*) Concerted-micelles mechanisms. (*c*) Protein partially exposed to the hydrocarbon solvent. (*d*) Ion pair interactions between the biopolymer and the ionized surfactant heads (taken from Luisi and Wolf, 1982).

There is relatively meager structural information on reversed micelle solubilized proteins. Depending on their sizes, polarities, and charges several modes of protein–reversed micelle interactions have been suggested (Figure 10.4; Luisi and Wolf, 1982). Reversed micelles are known to increase their volumes upon entrapping large polar substrates (Fendler, 1976). Several hundred surfactant molecules have been considered to wrap around a hydrated vitamin B_{12} molecule, thereby protecting it from the bulk hydrocarbon solvent (Fendler et al., 1974). Relatively small hydrophilic proteins are likely to be solubilized analogously (Figure 10.4*a*). Alternatively, highly charged proteins might be brought into solution by ion-pair formation (Figure 10.4*d*). Proteins whose surfaces are predominantly hydrophobic might be solubilized by several small micelles (Figure 10.4*b*) or partially exposed to the organic solvent (Figure 10.4*c*). Most proteins investigated to date are completely surrounded by surfactants in reversed micelles. Table 10.8 collects the calculated structural parameters for reversed micelle entrapped proteins.

Activities of enzymes entrapped in the aqueous pools of reversed micelles appeared to be similar to those in bulk water. Indeed this has been found for phospholipase, α-chymotrypsin, cytochrome C, horseradish peroxidase, and catalase etc. (Table 10.9). Effects of pH on the α-chymotryspin catalyzed hydrolysis of *N*-acetyl-L-tryptophan methyl ester have been examined in detail in aerosol-OT solubilized water in heptane (Menger and Yamada, 1979). The

Table 10.8. Calculated Structure Parameters for Protein-Containing Reversed Micelles

System	w_0	n^a	j^b	r_0, Å^c
RNase isooctane/aerosol-OT (0.05 *M*)	5.7	91	521	19.5
RNase isooctane/AOT (0.05 *M*)	11.1	193	2128	26.6
RNase isooctane/AOT (0.05 *M*)	22.3	395	8823	40.6
Lysozyme *n*-octane/AOT (0.05 *M*)	22.5	364	8195	39.9
Lysozyme *n*-octane/AOT (0.05 *M*)	27.4	636	17452	50.6
LADH isooctane/AOT (0.05 *M*)	46.4	2022	93844	89.6

Source: Compiled from Bonner et al., 1980.
[a] Surfactant aggregation number.
[b] Number of water molecules in the water pool (which contains only one protein molecule).
[c] Inner core radius of the protein-containing micelle. To obtain the total radius, add approximately 12 Å (the dimension of an AOT molecule).

pH rate profile of the catalysis in reversed micelles resembled that in water except for a 1.5 pH unit shift to more basic values. The pH of aqueous buffers did not change upon solubilization in heptane by aerosol-OT provided that the water to surfactant ratio, *R*, remained 25 or greater. When *R* decreased to 4 or less the apparent pH in the reversed micelle–buffer system decreased by 3 units from its value in bulk water. Since the 1.5 pH unit change occurred even in the larger surfactant solubilized water pools ($R > 25$), the changed hydrolysis rates were the consequence of altered pKa values at the active site. Apparently, the sulfonate groups of aerosol-OT stabilize the imidazole conjugate acid, believed to be responsible for chymotrypsin catalysis, and thereby render it more acidic (Menger and Yamada, 1979). It is tempting to conclude that hydrogen ion concentrations in large surfactant entrapped water pools in organic solvents ($R > 25$) remain the same as those in bulk water. Addition of proteins profoundly affects, however, the availability of free water molecules in given pools and alters the structures of reversed micelles. Expressing proton or substrate concentrations in reversed micelles is a nontrivial problem (Smith and Luisi, 1980; El Seoud et al., 1982). The total, overall, or stoichiometric concentration refers to the number of molecules present in the total volume (bulk apolar solvent + water). Alternatively, concentrations of completely hydrocarbon-insoluble substrates can be expressed in terms of those dissolved in the water pools.

Kinetics of second order reactions depend, of course, on the local or "real" concentration of both reactants (see Chapter 11). Distributions of enzymes and substrates within the available surfactant solubilized water pools need to be considered in the kinetic treatments of the data in reversed micelles. If the number of water pools is larger than the number of enzyme and/or substrate molecules, then the rate determining step for the reaction may become the

Table 10.9. Proteins in Reversed Micelles

Surfactant/Solvent	Protein	Method of Investigation	Results	Reference
Phosphatidyl serine 0.2% (w/v), phosphatidylcholine 0.2% (w/v) in decane	Cytochrome-C	Absorption spectrophotometry	Extraction of protein into decane strongly depends on the concentration of dated H^+, Na^+, Li^+, K^+, Mg^{2+}, Ca^{2+}; Ca^{2+} does not denature the protein	Gitler and Montal, 1972a
Methyltrioctylammonium chloride (12 m*M*) in cyclohexane	α-Chymotrypsin, trypsin, pepsin, and glucagon	Absorption, fluorescence, and circular dichroic spectra	Enzymes are readily transferred from water into cyclohexane phases and into a second water phase; gross features of enzymes remain unaltered	Luisi et al., 1979
Methyltrioctylammonium chloride (0.48% w/v) in cyclohexane	α-Chymotrypsin, trysin, pepsin, glucagone	Absorption, fluorescence ^1H nmr spectroscopy, optical rotary dispersion	Structural integrity of proteins is retained; there is one protein per reversed micelles; one molecule of α-chymotrypsin is solubilized by circa 18 molecules of surfactants and 40 molecules of H_2O	Luisi et al., 1977
Soybean phospholipids in hexane	Rhodopsin	Absorption spectrophotometry extraction	Enzyme remains configurationally stable	Darszon et al., 1978
Sodium bis(2-ethylhexyl)sulfosuccinate in octane or benzene	α-Chymotrypsin, peroxidase	Active center titration substrate hydrolyses	Enzymes remain stable; kinetics of enzyme catalyzed reaction are different in reversed micelles from that in H_2O	Martinek et al., 1978
Sorbitantristearate in heptane or corn or sunflower oil	Bacterial cytochrome P-450	Absorption spectrophotometric measurements of heme group oxidation rates at temperature down to $-30°C$	Enzyme–substrate complexes and intermediates can be trapped in supercooled water	Douzou et al., 1978

Sodium bis(2-ethylhexyl)sulfosuccinate (0.055–0.20 *M*) in heptane	α-Chymotrypsin	Kinetics of hydrolysis of *N*-acetyl-L-tryptophan methyl ester determined at different pH values	At high pH enzyme is twice as active in reversed micelles as in water due to altered p*Kb* values	Menger and Yamada, 1979
Sodium bis(2-ethylhexyl)sulfosuccinate 1.5% (w/v) in *n*-heptane or silicone oil	Cytochrome-C, trypsin	Enzyme activity determined spectrophotometrically using benzoyl-L-arginine as substrate, down to −38°C	Enzymes remain active; intermediates can be studied in supercooled water	Douzou et al., 1979
Sodium bis(2-ethylhexyl)sulfosuccinate (0.55–0.94 *M*) in *n*-octane	Ribonuclease	Absorption and circular dicroic spectra; enzyme activity determined using cytidine 2′3′-phosphate as substrate	Natural folding of enzyme and activity are retained or enhanced	Wolf and Luisi, 1979
Phosphatidylserine + M^{2+} (Cd^{2+} or Ba^{2+} or Sr^{2+} or Sr^{2+} or Ca^{2+}) in hexane	Reaction centers isolated from the R26 mutant of *Rhodopseudomonas sphaeroides*	Flash photolysis in presence of electron donors and acceptors absorption spectroscopy	Reaction center remains active for electron transfer in reversed micelles	Kendall-Tobias and Crofts, 1979
Soybean phospholipids in hexane	Reaction centers isolated from the R26 mutant of *Rhodopseudomonas sphaeroides*	Absorption and epr spectra at low temperature in the presence of electron donors and acceptors	Reaction center remains active for electron transfer in reversed micelles	Schönfeld et al., 1979a, 1980
Sodium bis(2-ethylhexyl)sulfosuccinate in *n*-heptane or silicon oils	Cytochrome-C, peroxidase (horseradish), catalase, cytochrome P450 (bacterial)	Absorption spectra and enzyme activity measurements down to −20°C	Enzymes remain active; intermediates can be studied in supercooled water	Balny and Douzou, 1979

Table 10.9 (*Continued*)

Surfactant/Solvent	Protein	Method of Investigation	Results	Reference
Sodium bis(2-ethylhexylsulfosuccinate (0.6–2.5%, v/v) in isooctane	α-Chymotrypsin	Enzymatic activity monitored by using *N*-glutaryl-L-phenylalanine *p*-nitroanilide, fluorescence, and circular dichroic spectra	Enhanced activity (up to five-fold) in reversed micellar solubilized water pools; enzyme has a more ordered conformation in reversed micelles	Barbaric and Luisi, 1980
Phosphatidylcholine in diethylether:methanol (95:5, v/v)	Phospholipase A2	Kinetic investigations of lipid hydrolysis	Enzyme activity is a complex function of substrate, water, and added Ca^{2+} concentration	Misiorowski and Wells, 1974; Poon and Wells, 1974; Wells, 1974
Sodium bis(2-ethylhexyl)sulfosuccinate (50 m*M*) in isooctane	Lysozyme	Absorption, fluorescence, and circular dichroic spectra	Enzyme activity depends on the water content and pH; helical contents and thermal stability of lysozyme increases in reversed micelles	Grandi et al., 1980
Sodium bis(2-ethylhexyl)sulfosuccinate (0.1–0.5 *M*), CTAB (0.1–0.5 *M*), Brij-56 (0.1–0.5 *M*) in benzene, $CHCl_3$, octane, or cyclohexane	α-Chymotrypsin, trypsin, pyrophosphatase, peroxidase, lactate dehydrogenase, pyruvate kinase	Enzyme activity measurements by following rates of reactions spectrophotometrically	Enzymatic activity retained, kinetics for rate enhancements are rationalized, mechanisms discussed	Martinek et al., 1981a, 1981b; Martinek and Semenov, 1981
Sodium bis(2-ethylhexyl)sulfosuccinate (50 m*M*) in isooctane	Horse liver alcohol dehydrogenase	Kinetics active site tiration, fluorescence and circular dichroic spectra	The enzyme has a turnover number and spectroscopic properties similar to those in water solution	Meier and Luisi, 1980

encounter between the enzyme and the substrate containing micelles (Luisi and Wolf, 1982). Otherwise, the Michaelis-Menten kinetics, expressing concentrations in a defined manner, describe well enzyme catalysis in surfactant solubilized water pools in hydrocarbon solvents. A theory, based on concepts developed for aqueous micellar catalysis, has been developed to account for the kinetics of enzyme catalyzed reaction in reversed micelles (Martinek et al., 1981a, 1981b). For α-chymotrypsin (Menger and Yamada, 1979) and ribonuclease (Wolf and Luisi, 1979), the maximal enzyme catalyzed rates are enhanced in reversed micelles with respect to those in water. These "superactivities" have been interpreted in terms of conformational changes (Luisi and Wolf, 1981).

Conformations of α-chymotrypsin (Barbaric and Luisi, 1981), ribonuclease (Wolf and Luisi, 1979), lysozyme (Smith et al., 1981), and horse liver alcohol dehydrogenase (Meier and Luisi, 1980) have been investigated in reversed micelles spectroscopically. The gross conformations of these enzymes do not alter appreciably. The circular dichroic spectra of the enzymes in the 260–320 nm region show very similar patterns in bulk and surfactant entrapped water. Conversely, there is an increased ellipticity in the 190–230 nm region for the enzymes in reversed micelles, indicating increased helicities. Elucidation of structural details, the nature and the type of hydrogen bonding between the surfactant entrapped protein and water molecules must await the outcome of further investigations.

Cytochrome-C and cytochrome-C_3 have retained their ability to mediate electron transfer in aerosol-OT reversed micelles; they reacted rapidly with electrons, generated by pulse radiolysis or, alternatively formed in the laser photoionization of pyrene (Visser and Fendler, 1982).

Reversed micelles provide convenient media for the investigation of enzyme catalyzed reactions at subzero temperatures. Decreased reaction rates at low temperatures render feasible the detection and characterization of many transients. Douzou (1977), Fink (1977), and their co-workers have pioneered cryoenzymology in aqueous : nonaqueous ($MeOH : H_2O$, $DMSO : H_2O$, $DMF : H_2O$, ethyleneglycol : H_2O) solvent mixtures. The high concentration of organic cosolvents needed alters, however, far too drastically the environments of enzymes. The use of sufficiently large surfactant solubilized water pools alleviates this problem. Surfactants stabilize the supercooled water pools surrounding the enzyme against freezing due to heterogeneous nucleation. Experimentally accessible temperatures of supercooled water are limited only by the onset of homogeneous nucleation, approximately $-40°C$ (Douzou et al., 1978). Cytochrome and trypsin were shown to retain their activities in surfactant entrapped supercooled water in hydrocarbon solvents (Douzou et al., 1978, 1979). Activation energies for the trypsin catalyzed hydrolysis of benzoylarginine ethyl ester were found to be essentially identical in bulk water, determined between 4° and 30°C, and in surfactant entrapped supercooled water, determined at temperatures down to $-38°C$ (Douzou et al., 1979).

Reversed micelles are potentially applicable to enzyme mediated synthesis using hydrocarbon soluble substrates. Decanal was, for example, reduced to decanol by horse liver dehydrogenenase and NADH in the aerosol-OT–isooctance reversed micelle system (Meier and Luisi, 1980). Surfactant entrapped water pools in organic solvents offer ideal media for the enzymatic synthesis of peptides. In water, particularly at low pH values, competing hydrolysis prevents oligomerization. Water activity can be kept low in reversed micelles. Formation of oligopeptides containing up to 40 monomer units, in yields as high as 95%, have been observed in the spontaneous polymerization of alanyl adenylate in aerosol-OT–hexadecyltrimethylammonium bicarbonate reversed micelles in benzene (Armstrong et al., 1978). Reversed micelles can also be exploited in the selective transport of enzymes (Luisi et al., 1979), small molecules, electrons, and protons (see Chapter 13) as well as in the determination of the molecular weight and sites of proteins by ultracentrifugation (Levashov et al., 1981).

REFERENCES

Adrian, G. and Huang, L. (1979). *Biochemistry* **18**, 5610–5614. Entrapment of Proteins in Phosphatidylcholine Vesicles.

Andrews, A. L., Atkinson, D., Barratt, M. D., Finer, E. G., Hauser, H., Henry, R., Leslie, R. B., Owens, N. L., Phillips, M. C., and Robertson, R. N. (1976). *Eur. J. Biochem.* **64**, 549–563. Interaction of Apoprotein from Porcine High Density Lipoprotein with Dimyristoyllecithin.

Anner, B. M., Lane, L. K., Schwartz, A., and Pitts, B. (1977). *Biochim. Biophys. Acta* **467**, 340–345. A Reconstituted $Na^+ + K^+$ Pump in Liposomes Containing Purified $(Na^+ + K^+)$ATPase from Kidney Medula.

Anoardi, L. and Tonellato, U. (1977). *J. Chem. Soc. Chem. Commun.*, 401–402. Catalysis of Amide Hydrolysis due to Micelles Containing Imidazole and Hydroxy Functional Groups.

Anoardi, L., Buzzacarini, F., Fornasier, R., and Tonellato, U. (1978a). *Tetrahedron Lett.*, 3945–3948. A Powerful Nucleophilic Micellar Reagent. Synthesis and Properties of α-Oximino-Ketone-Functionalized Surfactants.

Anoardi, L., Fornasier, R., Sostero, D., and Tonellato, U. (1978b). *Gazz. Chim. Ital.* **108**, 707–708. Functional Micellar Catalysis. Synthesis and Properties of Thiocholine-Type Surfactants.

Ariano, B. H. and Azzi, A. (1980). *Biochem. Biophys. Res. Commun.* **93**, 478–485. Rotational Motion of Cytochrome C Oxidase in Phospholipid Vesicles.

Armstrong, D. W., Seguin, R., McNeal, C. J., Macfarlane, R. D., and Fendler, J. H. (1978). *J. Am. Chem. Soc.* **100**, 4605–4606. Spontaneous Polypeptide Formation from Amino Acyl Adenylates in Surfactant Aggregates.

Balny, C. and Douzou, P. (1979). *Biochimie* **61**, 445–452. New Trends in Cryoenzymology: II—Aqueous Solutions of Enzymes in Apolar Solvents.

Barbaric, S. and Luisi, P. L. (1981). *J. Am. Chem. Soc.* **103**, 4239–4245, Micellar Solubilization of Biopolymers in Organic Solvents. 5. Activity and Conformation of α-Chymotripsin in Isooctane in AOT.

Barque, J. P. and Dervichian, D. G. (1979). *J. Lipid Res.* **20**, 447–455. Enzyme-Substrate Interaction in Lipid Monolayers. II. Binding and Activity of Lipase in Relation to Enzyme and Substrate Concentration and to Other Factors.

Bender, M. L. (1971). *Mechanisms of Homogeneous Catalysis from Protons to Proteins*, Wiley-Interscience, New York.

Bender, M. L. and Komiyama, M. (1977). In *Bioorganic Chemistry* (E. E. VanTamelen, Ed.), Vol. 1, Academic Press, New York, pp. 14–176. Models for Hydrolytic Enzymes.

Bender, M. L. and Komiyama, M. (1978). In *Reactivity and Structure Concepts in Organic Chemistry*, Vol. 6, Springer-Verlag, Berlin. Cyclodextrin Chemistry.

Berlin, R. D. and Fera, J. P. (1977). *Proc. Natl. Acad. Sci. USA* **74**, 1072–1076. Changes in Membrane Microviscosity Associated with Phagocytosis: Effects of Colchicine.

Blumenthal, R., Klausner, R. D., and Weinstein, J. N. (1980). *Nature* **288**, 333–337. Voltage-Dependent Translocation of the Asialoglycoprotein Receptor Across Lipid Membranes.

Bonner, F. J., Wolf, R., and Luisi, P. L. (1980). *J. Solid Phase Biochem.* **5**, 255–268. Micellar Solubilization of Biopolymers in Hydrocarbon Solvents. I. A Structural Model for Protein Containing Reversed Micelles.

Boyer, P. D. (1970). *The Enzymes*, Vol. 2, Academic Press, New York.

Breslow, R. (1979). *Isr. J. Chem.* **18**, 187–191. Biomimetic Chemistry in Oriented Systems.

Breslow, R., Doherty, J. B., Guillot, G., and Lipsey, C. (1978). *J. Am. Chem. Soc.* **100**, 3227–3229. β-Cyclodextrinylbisimidazole, A Model for Ribonuclease.

Breslow, R., Hammond, M., and Laver, M. (1980a). *J. Am. Chem. Soc.* **102**, 421–422. Selective Transamination and Optical Induction by a β-Cyclodextrin-Pyridoxamine Artificial Enzyme.

Breslow, R., Bovy, P., and Hersh, C. L. (1980b). *J. Am. Chem. Soc.* **102**, 2115–2117. Reversing the Selectivity of Cyclodextrin Bisimidazole Ribonuclease Mimics by Changing the Catalyst Geometry.

Breslow, R., Czarniecki, M. F., Emert, J., and Hamaguchi, H. (1980c). *J. Am. Chem. Soc.* **102**, 762–770. Improved Acylation Rates within Cyclodextrin Complexes from Flexible Capping of the Cyclodextrin and from Adjustment of the Substrate Geometry.

Brown, J. M. and Lynn, J. L., Jr., (1980). *Ber. Bunsenges. Phys. Chem.* **84**, 95–100. Structural and Catalytic Aspects of Functional Micelles. Ester Hydrolysis by Hydroxamic Acids Bound to Cationic Surfactants.

Brown, M. F., Miljanich, G. P., and Dratz, E. A. (1977). *Proc. Natl. Acad. Sci. USA* **74**, 1978–1982. Proton Spin-Lattice Relaxation of Retinal Rod Outer Segment Membranes and Liposomes of Extracted Phospholipids.

Brown, J. M., Chaloner, P. A., and Colens, A. (1979). *J. Chem. Soc. Perkin Trans. II*, 71–76. Acyl Transfer Reactions in Functional Micelles Studied by Proton Magnetic Resonance at 270 MHz.

Bruice, T. C. (1962). *Brookhaven Symp. Biol.* **15**, 52–84. Intramolecular Catalysis and the Mechanism of Chymotrypsin Actions.

Bruice, T. C. (1970). In *The Enzymes* (P. D. Boyer, Ed.), Vol. 2, Academic Press, New York, pp. 217–279. Proximity Effects and Enzyme Catalysis.

Bruice, T. C. (1976). *Ann. Rev. Biochem.* **45**, 331–373. Some Pertinent Aspects of Mechanisms as Determined with Small Molecules.

Bruice, T. C., Brown, A., and Harris, D. O. (1971). *Proc. Natl. Acad. Sci. USA* **68**, 658–661. On the Concept of Orbital Steering in Catalytic Reactions.

Brunner, J., Hauser, H., and Semenza, G. (1978). *J. Biol. Chem.* **253**, 7538–7546. Single Bilayer Lipid-Protein Vesicles Formed from Phosphatidylcholine and Small Intestinal Sucrase · Isomaltase.

Brunström, A. and Ingelman-Sunberg, M. (1980). *Biochem. Biophys. Res. Commun.* **95**, 431–439. Benzo(A)pyrene Metabolism by Purified Forms of Rabbit Liver Microsomal Cytochrome P-450, Cytochrome b_5 and Epoxide Hydrase in Reconstituted Phospholipid Vesicles.

Bunton, C. A. (1973). *Prog. Solid State Chem.* **8**, 239–281. Micellar Catalysis and Inhibition.

Bunton, C. A. (1979). *Catal. Rev. Sci. Eng.* **20**, 1–56. Reaction Kinetics in Aqueous Surfactant Solutions.

Bunton, C. A. and Minch, M. J. (1970). *Tetrahedron Lett.*, 3881–3884. Micellar Catalyzed Decarboxylation of 6-Nitrobenzisoxazole-3-carboxylate Ion.

Bunton, C. A., Minch, M. J., Hidalgo, J., and Sepulveda, L. (1973). *J. Am. Chem. Soc.* **95**, 3262–3273. Electrolyte Effects on the Cationic Micelle Catalyzed Decarboxylation of 6-Nitrobenzisoxazole-3-carboxylate Anion.

Bunton, C. A., Kamego, A., Minch, M. J., and Wright, J. L. (1975). *J. Org. Chem.* **40**, 1321–1337. Effect of Changes in Surfactant Structure on Micellarly Catalyzed Spontaneous Decarboxylations and Phosphate Ester Hydrolysis.

Campbell, I. M. and Pawagi, A. B. (1979). *Can. J. Biochem.* **57**, 1099–1109. Temperature-Dependent Interactions between Poly-L-lysine and Phosphatidylcholine Vesicles.

Capon, B. (1971). *J. Chem. Soc. B*, 1207–1209. Orbital Steering: An Unnecessary Concept.

Casey, R. P., Chappell, J. B., and Azzi, A. (1979). *Biochem. J.* **182**, 149–156. Limited-Turnover Studies on Proton Translocation in Reconstituted Cytochrome C Oxidase-Containing Vesicles.

Chajek, T., Aron, L., and Fielding, C. J. (1980). *Biochemistry* **19**, 3673–3677. Interaction of Lecithin: Cholesterol Acyltransferase and Cholesteryl Ester Transfer Protein in the Transport of Cholesteryl Ester into Sphingomyelin Liposomes.

Chen, S. S. and Berns, D. S. (1979). *J. Membr. Biol.* **47**, 113–127. Effect of Plastocyanin and Phycocyanin on the Photosensitivity of Chlorophyll-Containing Bilayer Membranes.

Chen, W. C. and Bittman, R. (1977). *Biochemistry* **16**, 4145–4149. Kinetics of Association of Amphotericin B with Vesicles.

Cherry, R. J., Müller, V., Holenstein, C., and Heyn, M. P. (1980). *Biochim. Biophys. Acta* **596**, 145–151. Lateral Segregation of Proteins Induced by Cholesterol in Bacteriorhodopsin-Phospholipid Vesicles.

Cho, J.-R. and Morawetz, H. (1972). *J. Am. Chem. Soc.* **94**, 375–377. Catalysis of Ionic Reactions by Micelles Reaction of $Co(NH_3)_5Cl^{2+}$ with Hg^{2+} in Sodium Alkyl Sulfate Solutions.

Citri, N. (1973). *Adv. Enzymol.* **37**, 397–648. Conformational Adaptibility in Enzymes.

Cohen, A. (1970). In *The Enzymes* (P. D. Boyer, Ed.), Vol. 1, Academic Press, New York, pp. 147–211. Chemical Modification as a Probe of Structure and Function.

Cordes, E. H. and Gitler, C. (1973). In *Progress in Bioorganic Chemistry* (E. T. Kaiser and F. J. Kezdy, Eds.), Vol. 2, Wiley, New York, pp. 1–53. Reaction Kinetics in the Presence of Micelle Forming Surfactants.

Cram, D. J. and Cram, J. M. (1978). *Acc. Chem. Res.* **11**, 8–14. Design of Complexes Between Synthetic Hosts and Organic Guests.

Darszon, A., Montal, M., and Zarco, J. (1977). *Biochem. Biophys. Res. Commun.* **76**, 820–827. Light Increases the Ion and Non-electrolyte Permeability of Rhodopsin–Phospholipid Vesicles.

Darszon, A., Philipp, M., Zarco, J., and Montal, M. (1978). *J. Membr. Biol.* **43**, 71–90. Rhodopsin-Phospholipid Complexes in Apolar Solvents: Formation and Properties.

Darszon, A., Vanderberg, C. A., Schönfeld, M., Ellisman, M. H., Spitzer, N. C., and Montal, M. (1980). *Proc. Natl. Acad. Sci. USA* **77**, 239–243. Reassembly of Protein-Lipid Complexes into Large Bilayer Vesicles: Perspectives for Membrane Reconstitution.

Davoust, J., Schoot, B. M., and Devaux, P. (1979). *Proc. Natl. Acad. Sci. USA* **76**, 2755–2759. Physical Modifications of Rhodopsin Boundary Lipids in Lecithin–Rhodopsin Complexes: A Spin Label Study.

Dawidowicz, E. A. and Rothman, J. E. (1976). *Biochim. Biophys. Acta* **455**, 621–630. Fusion and Protein-Mediated Phospholipid Exchange Studied with Single Bilayer Phosphatidylcholine Vesicles of Different Density.

DeGrado, W. F., Kezdy, F. J., and Kaiser, E. T. (1981). *J. Am. Chem. Soc.* **103**, 679–681. Design, Synthesis and Characterization of a Cytotoxic Peptide with Melittin-Like Activity.

DeKruijff, B. and Cullis, P. R. (1980). *Biochim. Biophys. Acta* **602**, 477–490. Cytochrome C Specifically Induces Non-Bilayer Structures in Cardiolipin-Containing Model Membranes.

Demel, R. A. (1974). *Methods Enzymol.* **32B**, 539–544. Monolayers. Use and Interaction.

Dencher, N. A. and Heyn, M. P. (1979). *FEBS Lett.* **108**, 307–310. Bacteriorhodopsin Monomers Pump Protons.

Dervichian, D. G. and Barque, J. P. (1979). *J. Lipid Res.* **20**, 437–446. Enzyme-Substrate Interaction in Lipid Monolayers. I. Experimental Conditions and Fundamental Kinetics.

Dombrose, F. A., Gitel, S. N., Zawalich, K., and Jackson, G. M. (1979). *J. Biol. Chem.* **254**, 5027–5040. The Association of Bovine Prothrombin Fragment I with Phospholipid.

Douzou, P. (1977). *Cryobiochemistry*, Academic Press, New York.

Douzou, P., Debey, P., and Franks, F. (1978). *Biochim. Biophys. Acta* **523**, 1–8. Supercooled Water as Medium for Enzyme Reactions at Subzero Temperatures.

Douzou, P., Keh, E., and Balny, C. (1979). *Proc. Natl. Acad. Sci. USA* **76**, 681–684. Cryoenzymology in Aqueous Media: Micellar Solubilized Water Clusters.

Earle, S. R. and Fisher, R. R. (1980). *J. Biol. Chem.* **255**, 827–830. A Direct Demonstration of Proton Translocation Coupled to Transhydrogenation in Reconstituted Vesicles.

El Seoud, O. A., Chinelatto, A. N., Shimizv, M. R. (1982). *J. Colloid Interface Sci.* July. Acid–Base Indicator Equilibria in the Presence of Aerosol-OT Aggregates in Heptane. Ion Exchange in Reversed Micelles.

Esposito, S., Semeriva, M., and Desnuelle, P. (1973). *Biochim. Biophys. Acta* **302**, 293–304. Effect of Surface Pressure on the Hydrolysis of Ester Monolayers by Pancreatic Lipase.

Fendler, J. H. (1976). *Acc. Chem. Res.* **9**, 153–161. Interactions and Reactions in Reversed Micellar Systems.

Fendler, J. H. and Fendler, E. J. (1975). *Catalysis in Micellar and Macromolecular Systems*, Academic Press, New York.

Fendler, J. H., Nome, F., and Van Woert, H. C. (1974). *J. Am. Chem. Soc.* **96**, 6745–6753. Effects of Surfactants on Ligand Exchange Reactions in Vitamin B_{12a} in Water and in Benzene. Influence of Water and of Solvent Restriction.

Fersht, A. (1977). *Enzymes. Structure and Mechanism*, Freeman, San Francisco.

Fife, T. H. (1975). *Adv. Phys. Org. Chem.* **11**, 1–122. Physical Organic Model Systems and the Problem of Enzymatic Catalysis.

Fink, A. L. (1977). *Acc. Chem. Res.* **10**, 233–239. Cryoenzymology: The Study of Enzyme Mechanisms at Subzero Temperatures.

Finkelstein, M. C. and Weissmann, G. (1979). *Biochim. Biophys. Acta* **587**, 202–216. Enzyme Replacement via Liposomes. Variations in Lipid Composition Determine Liposomal Integrity in Biological Fluids.

Fukushima, D., Kupferberg, J. P., Yokoyama, S., Kroon, D. J., Kaiser, E. T., and Kezdy, F. J. (1979). *J. Am. Chem. Soc.* **101**, 3703–3704. A Synthetic Amphiphilic Docasapeptide with the Surface Properties of Plasma Apolipoprotein A-1.

Fukushima, D., Kaiser, E. T., Kezdy, F. J., Kroon, D. J., Kupferberg, J. P., and Yokoyama, S. (1980). *Ann. New York Acad. Sci.* **348**, 365–377. Rational Design of Synthetic Models for Lipoproteins.

Fung, B. K.-K. and Hubbell, W. L. (1978). *Biochemistry* **21**, 4403–4410. Organization of Rhodopsin in Photoreceptor Membranes. 2. Transmembrane Organization of Bovine Rhodopsin: Evidence from Proteolysis and Lactoperoxidase-Catalyzed Iodination of Native and Reconstituted Membranes.

Gandour, R. D. (1978). In *Transition States of Biochemical Processes* (R. D. Gandour and R. L. Schowen, Eds.), Plenum Press, New York, pp. 529–552. Intramolecular Reactions and the Relevance of Models.

Genkin, M. V., Darydov, R. M., Shapiro, A. B., and Krylov, O. V. (1977). *Zh. Fiz. Khim.* **53**, 1282 **CA 87**, 73915h. Study of the Interaction of Adsorbed Proteins with the Surface of a Solid Support Area with Lipid Monolayers by a Spin Label Method.

Gitler, C. and Montal, M. (1972a). *FEBS Lett.* **28**, 329–332. Formation of Decane-Soluble Proteolipids: Influence of Monovalent and Divalent Cations.

Gitler, C. and Montal, M. (1972b). *Biochem. Biophys. Res. Commun.* **47**, 1486–1491. Thin-Proteolip Films: A New Approach to the Reconstitution of Biological Membranes.

Grandi, C., Smith, R. E., and Luisi, L. P. (1981). *J. Biol. Chem.* **256**, 837–843. Activity and Conformation of Lysozyme in Isooctane Reverse Micelles.

Grant, C. W. M. and McConnell, H. J. (1974). *Proc. Natl. Acad. Sci. USA* **71**, 4653–4657. Glycophorin in Lipid Bilayers.

Gould, J. M., Patterson, L. K., Ling, E., and Winget, G. D. (1979). *Nature* **280**, 607–609. Phosphorylation in a Simple System of Lipids and Chloroplast ATP Synthetase Driven by Pulsed Ionising Radiation.

Guo, L. S., Hamilton, R. L., Hamilton, G., Goerke, J., Weinstein, J. N., and Havel, R. J. (1980). *J. Lipid Res.* **21**, 993–1003. Interaction of Unilamellar Liposomes with Serum Lipoproteins and Apolipoproteins.

Hartman, W., Galla, H.-J., and Sackmann, E. (1978). *Biochim. Biophys. Acta* **510**, 124–139. Polymyxin Binding to Charged Lipid Membranes. An Example of Lipid-Protein Interactions.

Hauser, H. (1975). *FEBS Lett.* **60**, 71–75. Lipid-Protein Interaction in Porcine High-Density (HDL_3) Lipoprotein.

Heath, T. D., Robertson, D., Birbeck, M. S. C., and Davies, A. J. S. (1980). *Biochim. Biophys. Acta* **599**, 42–62. Covalent Attachment of Horseradish Peroxidase to the Outer Surface of Liposomes.

Hebdon, G. M., LeVine, H., III, Minand, R. B., Sahyoun, N. E., Schmitges, C. J., and Cuatrecasas, P. (1979). *J. Biol. Chem.* **254**, 10459–10465. Incorporation of Rat Brain Adenylate Cyclase into Artificial Phospholipid Vesicles.

Hellingwerf, K. J., Arents, J. C., Scholte, B. J., and Westerhoff, H. V. (1979). *Biochim. Biophys. Acta* **547**, 561–582. Bacteriorhodopsin in Liposomes. II. Experimental Evidence in Support of a Theoretical Model.

Hol, W. G. J., Van Duijnen, P. T., and Berendsen, H. J. C. (1978). *Nature* **273**, 443–446. The α-Helix Dipole and the Properties of Proteins.

Huang, K.-S., Bayley, H., and Khorana, H. G. (1980). *Proc. Natl. Acad. Sci. USA* **77**, 323–327. Delipidation of Bacteriorhodopsin and Reconstitution with Exogenous Phospholipids.

Ingelman-Sundberg, M. and Johansson, I. (1980). *Biochemistry* **19**, 4004–4011. Catalytic Properties of Purified Forms of Rabbit Liver Microsomal Cytochrome P-450 in Reconstituted Phospholipid Vesicles.

Jencks, W. P. (1969). *Catalysis in Chemistry and Enzymology*, McGraw-Hill, New York.

Jencks, W. P. (1975). *Adv. Enzymol. Relat. Areas Mol. Biol.* **43**, 219–410. Binding Energy, Specificity and Enzymic Catalyses: The Circe Effect.

Jencks, W. P. and Page, M. I. (1974). *Biochem. Biophys. Res. Commun.* **57**, 887–892. "Orbital Steering," Entropy, and Rate Accelerations.

Jonas, A. (1979). *J. Lipid Res.* **20**, 817–824. Interaction of Bovine Serum High Density Lipoprotein with Mixed Vesicles of Phosphatidylcholine and Cholesterol.

Jonas, A. and Drengler, S. M. (1980a). *J. Biol. Chem.* **255**, 2183–2189. Two Types of Complexes Formed by the Interaction of Apolipoprotein A-I with Vesicles of L-α-Dimyristoylphosphatidylcholine.

Jonas, A. and Drengler, S. M. (1980b). *J. Biol. Chem.* **255**, 2190–2194. Kinetics and Mechanism of Apoliprotein A–I Interaction with L-α-Dimyristoylphosphatidylcholine Vesicles.

Kawato, S., Ikegami, A., Yoshida, S., and Orii, Y. (1980). *Biochemistry* **19**, 1598–1603. Fluorescent Probe Study of Temperature-Induced Conformational Changes in Cytochrome Oxidase in Lecithin Vesicle and Solubilized Systems.

Kawato, S., Yoshida, S., Orii, Y., Ikegami, A., and Kinosita, K. (1981). *Biochim. Biophys. Acta* **634**, 85–92. Nanosecond Time Resolved Fluorescence Investigations of Temperature-Induced Conformational Changes in Cytochrome Oxidase in Phosphatidylcholine Vesicles and Solubilized Systems.

Kemp, D. S. and Paul, K. (1970). *J. Am. Chem. Soc.* **92**, 2553–2554. Decarboxylation of Benzisoxazole-3-carboxylic Acids. Catalysis by Extraction of Possible Relevance to the Problem of Enzymatic Mechanism.

Kemp, D. S. and Paul, K. (1975). *J. Am. Chem. Soc.* **97**, 7305–7312. The Physical Organic Chemistry of Benzisoxazoles. III. The Mechanism and the Effects of Solvents on Rates of Decarboxylation of Benzisoxazole-3-carboxylic Acids.

Kemp, D. S., Cox, D. D., and Paul, K. G. (1975). *J. Am. Chem. Soc.* **97**, 7312–7318. The Physical Organic Chemistry of Benzisoxazoles. IV. The Origins and Catalytic Nature of the Solvent Rate Acceleration for the Decarboxylation of 3-Carboxybenzisoxazoles.

Kendall-Tobias, M. and Crofts, A. R. (1979). *Biophys. J.* **25**, 54a. Reaction Centres in Hexane.

Keniry, M. A. and Smith, R. (1979). *Biochim. Biophys. Acta* **578**, 381–391. Circular Dichroic Analysis of the Secondary Structure of Myelin Basic Protein and Derived Peptides Bound to Detergents and to Lipid Vesicles.

Kensil, C. R. and Dennis, E. A. (1979). *J. Biol. Chem.* **254**, 5843–5848. Action of Cobra Venom Phospholipase A_2 on the Gel and Liquid Crystalline States of Dimyristoyl and Dipalmitoylphosphatidylcholine Vesicles.

Kimelberg, H. K. and Papahadjopoulos, D. (1971). *J. Biol. Chem.* **46**, 1142–1148. Interactions of Basic Proteins with Phospholipid Membranes.

Klotz, I. M. (1978). In *Molecular Movements and Chemical Reactivity as Conditioned by Membranes, Enzymes and Other Macromolecules* (E. Lefever and A. Goldbeter, Eds.), Advances in Chemical Physics, Vol. **39**, Wiley-Interscience, New York, pp. 109–160. Synzymes. Synthetic Polymers with Enzymelike Catalytic Activities.

Koshland, D. E., Jr. (1956). *J. Cell. Comp. Physiol.* **47**, Supp. 1, 217–234. Molecular Geometry in Enzyme Action.

Koshland, D. E., Jr. (1960). *Adv. Enzymol.* **22**, 45–97. The Active Site and Enzyme Action.

Koshland, D. E., Jr. (1962). *J. Theor. Biol.* **2**, 75–86. The Comparison of Non-Enzymic and Enzymic Reaction Velocities.

Koshland, D. E., Jr. (1972). *Proc. Robert A. Welch Found. Conf. Chem. Res.* **15**, 53–91. The Catalytic Power of Enzymes.

Koshland, D. E., Jr., Nemethy, G., and Filmer, D. (1966). *Biochemistry* **5**, 365–385. Comparison of Experimental Binding Data and Theoretical Models in Proteins Containing Subunits.

Kossi, C. N. and Leblanc, R. M. (1981). *J. Colloid Interface Sci.* **80**, 426–436. Rhodopsin in a New Bilayer Membrane.

Kunitake, T., Okahata, Y., and Sakamoto, T. (1976). *Chem. Lett.*, 459–462. The Hydrolysis of *p*-Nitrophenyl Acetate by a Micellar Bifunctional Catalyst.

Kupfer, A., Gani, V., and Shaltiel, S. (1977). *Biochem. Biophys. Res. Commun.* **79**, 1004–1010. Micelles of Pyridoxal-5′-Phosphate Schiff Bases—An Improved Model for the B_6 Site of Glycogen Phosphorylase.

Lagocki, J. W., Boyd, N. D., Law, J. H., and Kezdy, F. J. (1970). *J. Am. Chem. Soc.* **92**, 2923–2925. Kinetic Analysis of the Action of Pancreatic Lipase on Lipid Monolayers.

Lagocki, J. W., Law, J. H., and Kezdy, F. J. (1973). *J. Biol. Chem.* **248**, 580–587. The Kinetic Study of Enzyme Action on Substrate Monolayers.

Lehn, J.-M. (1978). *Acc. Chem. Res.* **11**, 49–57. Cryptates: The Chemistry of Macropolycyclic Inclusion Complexes.

Leto, T. L. and Holloway, P. W. (1979). *J. Biol. Chem.* **254**, 5015–5019. Mechanism of Cytochrome b_5 Binding to Phosphatidylcholine Vesicles.

Leto, T. L., Roseman, M. A., and Holloway, P. W. (1979). *Biochemistry* **19**, 1911–1916. Mechanism of Exchange of Cytochrome b_5 between Phosphatidylcholine Vesicles.

Levashov, A. V., Khmelnitsky, Y. L., Klyachko, N. L., Chernyak, V. Y., Martinek, K. (1981). *Anal. Biochem.* Ultracentrifugation of Reversed Micelles in Organic Solvents: New Approach to Determination of Molecular Weight and Effective Size of Proteins.

Lienhard, G. E. (1973). *Science* **180**, 149–154. Enzymatic Catalysis and Transition State Theory.

Luisi, P. L. (1979). *Naturwissenshaften* **66**, 498–504. Why Are Enzymes Macromolecules?

Luisi, P. L. and Wolf, R. (1982). In *Solution Behavior of Surfactants. Theoretical and Applied Aspects* (E. J. Fendler and K. L. Mittal, Eds.), Plenum Press, New York, in press. Micellar Solubilization of Enzymes in Hydrocarbon Solvents.

Luisi, P. L., Henninger, F., Joppich, M., Dossena, A., and Casnati, G. (1977). *Biochem. Biophys. Res. Commun.* **74**, 1384–1389. Solubilization and Spectroscopic Properties of α-Chymotrypsin in Cyclohexane.

Luisi, P. L., Bonner, F. J., Pellegrini, A., Wiget, P., and Wolf, K. (1979). *Helv. Chim. Acta* **62**, 740–753. Micellar Solubilization of Proteins in Aprotic Solvents and Their Spectroscopic Characterization.

Martinek, K. and Semenov, A. N. (1981). *Biochim. Biophys. Acta* **658**, 90–101. Enzymatic Synthesis in Biphasic Aqueous-Organic Systems. II. Shift of Ionic Equilibria.

Martinek, K., Levashov, A. V., Klyachko, N. J., and Berezin, I. V. (1978). *Dok. Akad. Nauk SSSR, Eng.* Ed. **236**, 951–953. Catalysis by Water-Soluble Enzymes in Organic Solvents. Stabilization of Enzymes Against Denaturation (Inactivation) Through Their Inclusion in Reversed Micelles of Surfactants.

Martinek, K., Levashov, A. V., Klyachko, N. L., Pantin, V. I., and Berezin, I. V. (1981a). *Biochim. Biophys. Acta* **657**, 277–294. The Principles of Enzyme Stabilization. VI. Catalysis by Water-Soluble Enzymes Entrapped into Reversed Micelles of Surfactants in Organic Solvents.

Martinek, K., Semenov, A. N., and Berezin, I. V. (1981b). *Biochim. Biophys. Acta* **658**, 76–89. Enzymatic Synthesis in Biphasic Aqueous-Organic Systems. I. Chemical Equilibrium Shift.

Meier, P. and Luisi, P. L. (1980). *J. Solid Phase Biochem.* **5**, 269–282. Micellar Solubilization of Biopolymers in Hydrocarbon Solvents.

Menger, F. M. and Yamada, K. (1979). *J. Am. Chem. Soc.* **101**, 6731–6734. Enzyme Catalysis in Water Pools.

Mimms, L. T., Zampighi, G., Nozaki, Y., Tanford, C., and Reynolds, J. A. (1981). *Biochemistry* **20**, 833–840. Phospholipid Vesicle Formation and Transmembrane Protein Incorporation Using Octyl Glucoside.

Misiorowski, R. L. and Wells, M. A. (1974). *Biochemistry* **13**, 4921–4927. The Activity of Phospholipase A_2 in Reverses Micelles of Phosphatidylcholine in Diethyl Ether: Effect of Water and Cations.

Montal, M. and Korenbrot, J. I. (1973). *Nature* **246**, 219–221. Incorporation of Rhodopsin Proteolipid into Bilayer Membranes.

Morawetz, H. (1970). *Acc. Chem. Res.* **3**, 354–360. Chemical Reaction Rates Reflecting Physical Properties of Polymer Solutions.

Morawetz, H. (1978). *Isr. J. Chem.* **17**, 287–290. Some Recent Advances in the Study of the Reactivity of Systems Containing Polymers.

Morawetz, H. and Vogel, B. (1969). *J. Am. Chem. Soc.* **91**, 563–568. Catalysis of Ionic Reactions by Polyelectrolytes. Reaction of $Co(NH_3)_5Cl^{2+}$ and $Co(NH_3)_5Br^{2+}$ with Hg^{2+} in Poly(sulfonic acid) Solution.

Moss, R. A., Nahas, R. C., Ramaswami, S., and Sanders, W. J. (1975). *Tetrahedron Lett.*, 3379–3382. A Comparison of Hydroxyl and Imidazole Functionalized Micellar Catalysis in Ester Hydrolyses.

Moss, R. A., Nahas, R. C., and Ramaswami, S. (1977). *J. Am. Chem. Soc.* **99**, 627–629. Sequential Bifunctional Micellar Catalysis.

Moss, R. A., Bizzogotti, G. O., and Huang, C.-W. (1980). *J. Am. Chem. Soc.* **102**, 754–762. Nucleophilic Esterolytic and Displacement Reactions of a Micellar Thiocholine Surfactant.

Mountz, J. D. and Tien, H. T. (1978). *J. Bioenerg. Biomembr.* **10**, 139–151. Bilayer Lipid Membranes (BLM): Study of Antigen-Antibody Interactions.

Nalecz, M., Zborowski, J., Famulski, K. S., and Wojtczak, L. (1980). *Eur. J. Biochem.* **112**, 75–80. Effect of Phospholipid Composition on the Surface Potential of Liposomes and the Activity of Enzymes Incorporated into Liposomes.

Nelson, N., Anholt, R., Lindstron, J., and Montal, M. (1980). *Proc. Natl. Acad. Sci. USA* **77**, 3057–3061. Reconstitution of Purified Acetylcholine Receptors with Functional Ion Channels in Planar Lipid Bilayers.

Nemethy, G. (1967). *Angew. Chem. Int. Ed. Eng.* **6**, 195–206. Hydrophobic Interactions.

O'Brien, D. F., Zumbulyadis, N., Michaels, F. M., and Ott, R. A. (1977). *Proc. Natl. Acad. Sci. USA* **74**, 5222–5226. Light-Regulated Permeability of Rhodopsin: Egg Phosphatidylcholine Recombinant Membranes.

O'Connor, C. J., Fendler, E. J., and Fendler, J. H. (1973). *J. Am. Chem. Soc.* **95**, 600–602. Dramatic Rate Enhancement of the Aquation of the Tris(oxalato)chromate(III) Anion by Surfactant Solubilized Water in Benzene.

O'Connor, C. J., Fendler, E. J., and Fendler, J. H. (1974). *J. Am. Chem. Soc.* **96**, 370–375. Catalysis by Reversed Micelles in Nonpolar Solvents. Trans–Cis Isomerization of Bis(oxalato)diaquochromate(III).

O'Connor, C. J., Ramage, R. E., and Porter, A. J. (1981). *Adv. Colloid Surf. Sci.*, to be published. Surfactant Systems as Enzyme Models.

Overfield, R. E. and Wraight, C. A. (1980). *Biochemistry* **19**, 3322–3327. Oxidation of Cytochromes C and C_2 by Bacterial Photosynthetic Reaction Centers in Phospholipid Vesicles. I. Studies with Neutral Membranes.

Owicki, J. C. and McConnell, H. M. (1979). *Proc. Natl. Acad. Sci. USA* **76**, 4750–4754. Theory of Protein-Lipid and Protein-Protein Interactions in Bilayer Membranes.

Owicki, J. C., Springgate, M. W., and McConnell, H. M. (1978). *Proc. Natl. Acad. Sci. USA* **76**, 1616–1619. Theoretical Study of Protein-Lipid Interactions in Bilayer Membranes.

Page, M. I. (1973). *Chem. Soc. Rev.* **2**, 295–323. Energetics of Neighboring Group Participation.

Page, M. I. and Jencks, W. P. (1971). *Proc. Natl. Acad. Sci. USA* **68**, 1678–1683. Entropic Contributions to Rate Accelerations in Enzymic and Intramolecular Reactions and the Chelate Effect.

Papahadjopoulos, D., Hui, S., Vail, W. J., and Poste, G. (1976). *Biochim. Biophys. Acta* **448**, 245–264. Studies on Membrane Fusion. I. Interactions of Pure Phospholipid Membranes and the Effect of Myristic Acid, Lysolecithin, Proteins and Dimethylsulfoxide.

Peters, J. and Fromherz, P. (1975). *Biochim. Biophys. Acta* **394**, 111–119. Interaction of Electrically Charged Lipid Monolayers with Malate Dehydrogenase.

Piszkiewicz, D. (1977a). *J. Am. Chem. Soc.*, **99**, 1550–1557. Positive Cooperativity in Micelle-Catalyzed Reactions.

Piszkiewicz, D. (1977b). *J. Am. Chem. Soc.* **99**, 7695–7696. Cooperativity in Bimolecular Micelle-Catalyzed Reactions. Inhibition of Catalysts by High Concentrations of Detergent.

Poon, P. H. and Wells, M. A. (1974). *Biochemistry* **13**, 4928–4936. Physical Properties of Egg Phosphatidylcholine in Diethyl Ether–Water Solutions.

Popot, J. L., Demel, R. A., Sobel, A., Van Deenen, L. L. M., and Changeux, J. P. (1977). *C. R. Hebd. Seances Acad. Sci. Ser. D* **285**, 1005–1008. Preferential Affinity of the Acetylcholine Receptor Protein for Certain Lipids as Studied by the Monolayer Method.

Pownall, J. H., Hu, A., Gotto, A. M., Jr., Albers, J. J., and Sparrow, J. T. (1980). *Proc. Natl. Acad. Sci. USA* **77**, 3154–3158. Activation of Lecithin: Cholesterol Acyltransferase by a Synthetic Model Lipid-Associating Peptide.

Rice, D. M., Meadows, M. D., Scheinman, A. O., Goñi, F. M., Gómez-Fernández, J. C., Moscarello, M. A., Chapman, D., and Oldfield, E. (1979). *Biochemistry* **18**, 5893–5903. Protein-Lipid Interactions. A Nuclear Magnetic Resonance Study of Sarcoplasmic Reticulum Ca^{2+}, Mg^{2+}-ATPase, Lipophilin, and Proteolipid Apoprotein-Lecithin Systems and a Comparison with the Effects of Cholesterol.

Rothfield, L. I. and Fried, V. A. (1975). *Methods Membr. Biol.* **4**, 277–292. Use of Monolayer Techniques in Reconstitution of Biological Activity.

Schieren, H., Weissmann, G., Seligman, M., and Coleman, P. (1978). *Biochem. Biophys. Res. Commun.* **82**, 1160–1167. Interactions of Immunoglobulins with Liposomes. An ESR and Diffusion Study Demonstrating Protection by Hydrocortisone.

Schoch, P. and Sargent, D. F. (1980). *Biochim. Biophys. Acta* **602**, 234–247. Quantitative Analysis of the Binding of Melittin to Planar Lipid Bilayers Allowing for the Discrete-Charge Effect.

Schönfeld, M., Montal, M., and Feher, G. (1979a). *Biophys. J.* **25**, 203a. Formation and Characterization of a Reaction Center-Lipid Complex in an Organic Solvent.

Schönfeld, M., Montal, M., and Feher, G. (1979b). *Proc. Natl. Acad. Sci. USA* **76**, 6351–6355. Functional Reconstitution of Photosynthetic Reaction Centers in Planar Lipid Bilayers.

Schönfeld, M., Montal, M., and Feher, G. (1980). *Biochemistry* **19**, 1535–1542. Reaction Center-Phospholipid Complex in Organic Solvents: Formation and Properties.

Schowen, R. L. (1978). In *Transition States of Biochemical Processes* (R. D. Gandour and R. L. Schowen, Eds.), Plenum Press, New York, pp. 77–114. Catalytic Power and Transition State Stabilization.

Schulman, J. H. and Hughes, A. M. (1935). *Biochem. J.* **29**, 1243–1252. Monolayers of Proteolytic Enzymes and Proteins. IV. Mixed Unimolecular Films.

Seno, M., Shiraishi, S., Araki, K., and Kise, H. (1975). *Bull. Chem. Soc. Jpn.* **48**, 3678–3681. Nonenzymatic Hydrolysis of Adenosine 5′-Triphosphate in Micellar and Reversed Micellar Systems.

Seno, M., Araki, K., and Shiraishi, S. (1976). *Bull. Chem. Soc. Jpn.* **49**, 1901–1905. Iodination Reactions of Ketones in the Reversed Micellar Systems of Dodecylammonium Propionate in Hexane.

Shah, D. O. (1972). *Prog. Surf. Sci.* **3**, 221–278. Monolayers of Lipids in Relation to Membranes.

Shaw, E. (1970). In *The Enzymes* (P. D. Boyer, Ed.), Vol. 1, Academic Press, New York, pp. 91–146. Chemical Modification by Active-Site-Directed Reagents.

Shen, B. W., Tsao, F. H. C., Law, J. H., and Kezdy, F. J. (1975). *J. Am. Chem. Soc.* **97**, 1205–1208. Kinetic Study of the Hydrolysis of Lecithin Monolayers by *Crotalus Adamanteus* α-Phospholipase A_2. Monomer-Dimer Equilibrium.

Shinkai, S. and Kunitake, T. (1976). *Bull. Chem. Soc. Jpn.* **49**, 3219–3223. Micellar Activation of Nucleophilic Reactivity of Coenzyme A and Glutathione toward *p*-Nitrophenyl Acetate.

Siegl, B. and Breslow, R. (1975). *J. Am. Chem. Soc.* **97**, 6869–6870. Lyophobic Binding of Substrates by Cyclodextrins in Nonaqueous Solvents.

Sikka, S. C. and Kalra, V. K. (1980). *J. Biol. Chem.* **255**, 4399–4402. γ-Glutamyl Transpeptidase Mediated Transport of Amino Acid in Lecithin Vesicles.

Smid, J., Shah, S., Wong, L., and Hurley, J. (1975). *J. Am. Chem. Soc.* **97**, 5932–5933. Solute Binding and Catalytic Effects in Aqueous Solutions of Poly(vinyl crown ethers).

Smith, R. E. and Luisi, P. L. (1980). *Helv. Chim. Acta* **63**, 2302–2311. Micellar Solubilization of Biopolymers in Hydrocarbon Solvents. III. Empirical Definition of an Acidity Scale in Reversed Micelles.

Smith, L. M., Parce, W. P., Smith, B. A., and McConnell, H. M. (1979). *Proc. Natl. Acad. Sci. USA* **76**, 4177–4179. Antibodies Bound to Lipid Haptens in Model Membranes Diffuse as Rapidly as the Lipids Themselves.

Storm, D. R. and Koshland, D. E., Jr. (1970). *Proc. Natl. Acad. Sci. USA* **66**, 445–452. A Source for Special Catalytic Power of Enzymes: Orbital Stirring.

Storm, D. R. and Koshland, D. E., Jr. (1972a). *J. Am. Chem. Soc.* **94**, 5805–5814. An Indication of the Magnitude of Orientation Factors in Esterification.

Storm, D. R. and Koshland, D. E., Jr. (1972b). *J. Am. Chem. Soc.* **94**, 5815–5825. Effect of Small Changes in Orientation on Reaction Rates.

Straub, T. S. and Bender, M. L. (1972). *J. Am. Chem. Soc.* **94**, 8875–8881. Cycloamylases as Enzyme Models. The Decarboxylation of Phylacetate Anions.

Suh, J., Scarpa, I. S., and Klotz, I. M. (1976). *J. Am. Chem. Soc.* **98**, 7060–7064. Catalysis of Decarboxylation of Nitrobenzisoxazolecarboxylic Acid and of Cyanophenylacetic Acid by Modified Polyethylenimines.

Sunamoto, J. and Hamada, T. (1978). *Bull. Chem. Soc. Jpn.* **51**, 3130–3135. Solvochromism and Thermochromism of Cobalt(II) Complexes Solubilized in Reversed Micelles.

Sundler, R., Alberts, A. W., and Vagelos, P. R. (1978). *J. Biol. Chem.* **253**, 5299–5304. Phospholipases as Probes for Membrane Sidedness.

Swanson, M. S., Quintanilha, A. T., and Thomas, D. D. (1980). *J. Biol. Chem.* **255**, 7494–7502. Protein Rotational Mobility and Lipid Fluidity of Purified and Reconstituted Cytochrome C Oxidase.

Tabushi, I. and Kuroda, Y. (1974). *Tetrahedron Lett.*, 3613–3616. A Mixed Micelle Catalyst of Extremely Great Acitivity.

Tabushi, I., Kuroda, Y., and Kita, S. (1974). *Tetrahedron Lett.*, 643–646. Organic Catalyst of Enzyme Activity, *N*-Methyl-*N*-Laurylhydroxamic Acid in CTAB Micelle.

Tabushi, I., Shimizu, N., Sugimoto, T., Shiozuka, M., and Yamamura, K. (1977). *J. Am. Chem. Soc.* **97**, 7100–7102. Cyclodextrin Flexibly Capped with Metal Ion.

Tabushi, I., Kuroda, Y., and Shimokawa, K. (1979). *J. Am. Chem. Soc.* **101**, 4759–4760. Cyclodextrin Having an Amino Group as a Rhodopsin Model.

Tagaki, W., Kobayashi, S., and Fukushima, D. (1977). *J. Chem. Soc. Chem. Commun.*, 29–30. Facile Acyl Transfer from Imidazole to the Hydroxy-group in a Cationic Micelle in the Hydrolysis of *p*-Nitrophenyl Acetate.

Tanford, C. (1962). *Adv. Protein Chem.* **17**, 69–168. The Interpretation of Hydrogen Ion Titration Curves of Proteins.

Tanford, C. (1980). *The Hydrophobic Effect. Formation of Micelles and Biological Membranes*, 2nd ed., Wiley-Interscience, New York.

Tayłor, F. R., and Cronan, J. E., Jr. (1979). *Biochemistry* **18**, 3292–3300. Cyclopropane Fatty Acid Synthase of *Escherichia coli.* Solubilization, Purification, and Interaction with Phospholipid Vesicles.

Teissie, J. (1981). *Biochemistry* **20**, 1554–1560. Interaction of Cyctochrome C with Phospholipid Monolayers. Orientation and Penetration as Functions of the Packing Density of Film, Nature of the Phospholipids and Ionic Content of the Aqueous Phase.

Tonellato, U. (1976). *J. Chem. Soc. Perkin Trans. II*, 771–776. Catalysis of Ester Hydrolysis by Cationic Micelles of Surfactants Containing the Imidazole Ring.

Tonellato, U. (1979). In *Solution Chemistry of Surfactants* (K. L. Mittal, Ed.), Plenum Press, New York, pp. 541–558. Functional Micellar Catalysis.

Upreti, G. C. and Jain, M. K. (1980). *J. Membr. Biol.* **55**, 113–121. Action of Phospholipase A_2 on Unmodified Phosphatidylcholine Bilayers: Organizational Defects are Preferrred Sites of Action.

Van, S. P. and Griffith, O. H. (1975). *J. Membr. Biol.* **20**, 155–170. Bilayer Structure in Phospholipid-Cyctochrome C Model Membranes.

Van Duijnen, P. T., Thole, B. T., and Hol, W. G. J. (1979). *Biophys. Chem.* **9**, 273–280. On the Role of the Active Site Helix in Papain, An Ab Initio Molecular Orbital Study.

Van Duijnen, P. T., Thole, B. T., Broer, R., and Nievwpoort, W. C. (1980). *Int. J. Quantum Chem.* **17**, 651–671. Active-Site α-Helix in Papain and the Stability of the Ion Pairs RS^- ... ImH^+. Ab initio Molecular Orbital Study.

Vaz, W. L. C., Jacobson, K., Wu, E., and Derzko, Z. (1979a). *Proc. Natl. Acad. Sci. USA* **76**, 5645–5649. Lateral Mobility of an Amphipathic Apolipoprotein, APOC-III, Bound to Phosphatidylcholine Bilayers with and Without Cholesterol.

Vaz, W. L. C., Austin, R. H., and Vogel, H. (1979b). *Biophys. J.* **26**, 415–426. The Rotational Diffusion of Cytochrome b_5 in Lipid Bilayer Membranes. Influence of the Lipid Physical State.

Verger, R., Mieras, M. C. E., and De Haas, G. H. (1973). *J. Biol. Chem.* **248**, 4023–4034. Action of Phospholipase A at Interfaces.

Visser, A. J. G. and Fendler, J. H. (1982). *J. Phys. Chem.* **86**, 947–950. Reduction of Reversed Micelle Entrapped Cytochrome-C and Cytochrome-C_3 by Electrons Generated by Pulse Radiolysis or Pyrene Photoionization.

Volanakis, J. E. and Wirtz, K. W. (1979). *Nature* **281**, 155–157. Interaction of C-reactive Protein with Artificial Phosphatidylcholine Bilayers.

Walsh, C. (1979). *Enzymatic Reaction Mechanisms*, Freeman, San Francisco.

Wells, M. A. (1974). *Biochemistry* **13**, 4937–4942. The Nature of Water Inside Phosphatidylcholine Micelles in Diethyl Ether.

Wolf, R. and Luisi, P. L. (1979). *Biochem. Biophys. Res. Commun.* **89**, 209–217. Micellar Solubilization of Enzymes in Hydrocarbon Solvents, Enzymatic Activity and Spectroscopic Properties of Ribonuclease in *N*-Octane.

Wolfenden, R. (1972). *Acc. Chem. Res.* **5**, 10–18. Analog Approaches to the Structure of the Transition State in Enzyme Reactions.

Yokoyama, S., Fukushima, D., Kupferberg, J. P., Kezdy, F. J., and Kaiser, E. T. (1980). *J. Biol. Chem.* **255**, 7333–7339. The Mechanism of Activation of Lecithin:Cholesterol Acyltransferase by Apolipoprotein A-I and an Amphiphilic Peptide.

Yoshida, S., Orii, Y., Kawato, S., and Ikegami, A. (1979). *J. Biochem.* **86**, 1443–1450. Effects of Temperature on Cytochrome Oxidase Activity in Solubilized Form and in Lipid Vesicle Systems.

CHAPTER

11

REACTIVITY CONTROL AND SYNTHETIC APPLICATIONS

Alteration of reaction rates in the environments of micelles, vesicles, polyelectrolytes, and macrocyclic hosts has been recognized for some time (Fendler and Fendler, 1975). Utilization of membrane mimetic agents for modifying reaction pathways, stereochemistry and isotopic composition, as well as for syntheses, is of more recent development. Basic principles of these applications are delineated in the present chapter. Exhaustive data tabulation on altered reactivities is provided in Chapter 12.

1. KINETIC TREATMENTS OF REACTIVITIES IN MICELLAR AND RELATED SYSTEMS

Kinetic treatments have been developed most extensively for reactions in the presence of aqueous micelles. Processes occurring at rates comparable with micelle dissociation or dissolution have already been treated in terms of substrate entry and exit in Chapter 2. Additionally, fast intramicellar reactions have been successfully treated in terms of a stochastic approach (Hatlee and Kozak, 1980, 1981a, 1981b). The principal qualitative features found upon constructing and solving the stochastic master equation for irreversible (Hatlee and Kozak, 1980) and reversible (Hatlee and Kozak, 1981a) intramicellar kinetic processes were the apparent changes of kinetic order upon compartmentalization and the need for introducing an apparent equilibrium constant. The ensuing discussion assumes statistical substrate distributions and reactivities at timescales very much longer than those involved in the micellar equilibria.

In most treatments (Berezin et al., 1973; Romsted, 1975, 1977; Martinek et al., 1977; Bunton et al., 1978, 1979, 1980a, 1980b, 1980c, 1980d, 1981; Bunton, 1979a, 1979b; Moroi, 1980; Chaimovich et al., 1982) reactivities are considered to be sums of reactions occurring in the bulk aqueous (R_w) and the micellar pseudophases (R_M):

$$R_{total} = R_w + R_M \tag{11.1}$$

For unimolecular reactions partitioning of only one substrate needs to be considered. Assuming that the substrate does not perturb the micellar equilibria and that the monomer concentration remains constant above the critical micelle concentration (CMC), unimolecular reactions in aqueous micelles are described by (Menger and Portnoy, 1967)

$$\begin{array}{ccc} S + Dn & \xrightleftharpoons{K_s} & SM \\ \downarrow k'_w & & \downarrow k'_M \\ \text{products} & & \text{products} \end{array} \qquad (10.9)\ (11.2)$$

where S is the substrate, Dn is the detergent, M is the micelle, SM is the substrate-micelle complex, k'_w and k'_M are the first order rate constants in the bulk and micellar pseudophases, respectively, and K_s is the binding constant given by

$$K_s = \frac{[SM]}{[S][Dn]} \qquad (11.3)$$

The observed first order rate constant for the reaction, k_Ψ, is described by

$$k_\Psi = \frac{k'_w + k'_M K_s[M]}{1 + K_s[M]} \qquad (10.10)\ (11.4)$$

where the micelle concentration, [M], is expressed as

$$[M] = \frac{[Dn] - CMC}{n} \qquad (11.5)$$

where n is the aggregation number. Recognizing the analogy between equation 11.4 and the Michaelis–Menten equation for enzyme catalyzed reaction allowed the treatment of data by

$$\frac{1}{k'_w + k_\Psi} = \frac{1}{k'_w + k'_M} + \frac{1}{(k'_w - k'_M)K_s[M]} \qquad (11.6)$$

which is similar to the Lineweaver–Burke equation of enzyme kinetics. Equation 11.6 well describes the kinetic of micelle inhibited and catalyzed unimolecular processes. Its validity is substantiated by the good agreement between kinetically and independently determined substrate-micelle binding constants (Bunton, 1979a, 1979b).

Equation 11.4 does not adequately describe the kinetics of bimolecular reactions in micellar solutions. Partitioning of both reactants (A and B) between the bulk and aqueous phases has to be considered (Berezin et al., 1973; Martinek et al., 1977):

$$\begin{aligned} (A + B)_w &\xrightarrow{k'_w} \text{products} \\ (A + B)_M &\xrightarrow{k'_M} \text{products} \end{aligned} \tag{11.7}$$

where the subscripts w and M refer to the bulk water and micellar pseudophases. At reagent concentrations low enough not to affect the CMC, the overall rate, described by equation 11.1, is modified to

$$R_{\text{total}} = k'_M[A]_M[B]_M[Dn]\bar{V} + k'_w[A]_w[B]_w(1 - [Dn]\bar{V}) \tag{11.8}$$

where [Dn] is the stoichiometric detergent concentration, $\bar{V}$ is the molar volume of the surfactant, and the concentrations of reagents A and B are given by material balances

$$[A]_{\text{total}} = [A]_M[Dn]\bar{V} + [A]_w(1 - [Dn]\bar{V}) \tag{11.9}$$

$$[B]_{\text{total}} = [B]_M[Dn]\bar{V} + [B]_w(1 - [Dn]\bar{V}) \tag{11.10}$$

If the chemical reaction 11.7 does not affect the partition equilibria,

$$\begin{aligned} [A]_w &\rightleftharpoons [A]_M \\ [B]_w &\rightleftharpoons [B]_M \end{aligned} \tag{11.11}$$

then the observed second order rate constant for reactions in aqueous micelles is given by

$$k_{2\Psi} = \frac{k_M P_A P_B[Dn]\bar{V} + k_w(1 - [Dn]\bar{V})}{\{1 + (P_A - 1)[Dn]\bar{V}\}\{1 + (P_B - 1)[Dn]\bar{V}\}} \tag{11.12}$$

For dilute surfactant solutions the volume fraction of the micellar phase is small ($[Dn]\bar{V} \ll 1$) and, if both A and B bind strongly to the micelles ($P_A \gg 1$ and $P_B \gg 1$), then equation 11.12 simplifies to

$$k_{2\Psi} = \frac{(k_M/\bar{V})K_A K_B[Dn] + k_w}{(1 + K_A[Dn]\bar{V})(1 + K_B[Dn])} \tag{11.13}$$

where the binding constants are expressed by

$$\begin{aligned} K_A &= (P_A - 1)\bar{V} \\ K_B &= (P_B - 1)\bar{V} \end{aligned} \tag{11.14}$$

Equations 11.12 and 11.13 have been found to describe well several bimolecular reactions in aqueous (Bunton 1979a, 1979b; Bhalekar and Engberts, 1978) and reversed (Lim and Fendler, 1978) micelles as well as in surfactant vesicles (Lim and Fendler, 1979; Fendler and Hinze, 1981). The experimental data on the dependence of $k_{2\Psi}$ on [Dn] can be used to calculate k_M, K_A, and K_B. For this purpose equation 11.13 is transformed to

$$\frac{[\mathrm{Dn}]}{k_{2\Psi} - k_w} = \alpha + \beta[\mathrm{Dn}]\frac{k_{2\Psi}}{k_{2\Psi} - k_w} + \gamma[\mathrm{Dn}]^2\frac{k_{2\Psi}}{k_{2\Psi} - k_w} \tag{11.15}$$

where

$$\alpha = \frac{\bar{V}}{k_M K_A K_B} \tag{11.16}$$

$$\beta = \alpha(K_A + K_B) \tag{11.17}$$

$$\gamma = \alpha K_A K_B \tag{11.18}$$

A plot of the data according to equation 11.15 gives a value for the intercept, α. This value allows further analysis in terms of rearranged equation 11.15:

$$\left(\frac{1}{k_{2\Psi}} - \frac{\alpha}{\mathrm{Dn}}\right)\left(1 - \frac{k_w}{k_{2\Psi}}\right) = \beta + \gamma[\mathrm{Dn}] \tag{11.19}$$

which, in turn, provides numerical values for β and γ and hence for k_M, K_A, and K_B. Figure 11.1 illustrates the treatment of the kinetic data for the reaction of sodium ascorbate with a stable free radical on the surface of dioctadecyldimethylammonium chloride surfactant vesicles (Lim and Fendler, 1979).

Useful as it may be, equation 11.13 is inadequate for treating ionic reactions in the presence of charged micelles and for accounting for electrolyte effects on the rates of micelle catalyzed reactions (Bunton 1979a, 1979b). The pseudophase model is apparently insufficient for these systems. There are two recent modifications of equation 11.12. The first, developed by Romsted (1975, 1977), incorporates counterion distributions (see Chapter 8) in the micellar Stern layer in unbuffered solutions. The second, the ion exchange model (Quina and Chaimovich, 1979; Chaimovich et al., 1979, 1982; Bonilha et al., 1979; Quina et al., 1980), allows for estimations of micelle-bound and free ions in the presence of buffers and electrolytes.

The model developed by Romsted (1977) considers the micellar Stern layer to be saturated by hydrophilic counterions. The degree of ionization (α), reflecting the counterion distribution between the aqueous and micellar pseudophases, is assumed to be independent of the surfactant concentration and the ionic strength at constant temperature. The hydrophilic ionic reagent,

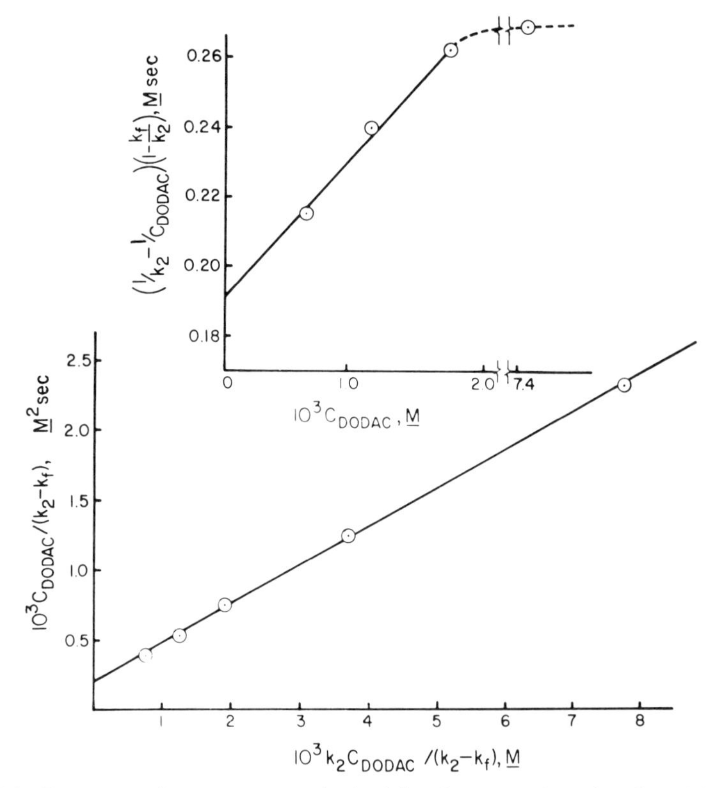

Figure 11.1 Treatment of rate constants obtained for the scavenging of sodium 2,2,5,5-tetramethyl-1-pyrrolidinyloxy-3-carboxylate spin probe by sodium ascorbate in the presence of dioctadecyldimethylammonium chloride (DoDAC) surfactant vesicles according to equation 11.15 and 11.19 (insert) (taken from Lim and Fendler, 1979).

X, and the nonreactive micellar counterion, Y, exchange rapidly between the two phases:

$$X_M + Y_w \rightleftharpoons X_w + Y_M \tag{11.20}$$

where the subscripts M and w refer, as before, to micelles and water. The selectivity coefficient for the ionic reagent at the Stern layer,

$$K_{X/Y} = \frac{X_w Y_M}{X_M Y_w} \tag{11.21}$$

determines the concentration of the hydrophilic reagent in the micellar phase. It is quite feasible, therefore, to obtain high X_M concentrations even when the stoichiometric concentration of Y far exceeds that of X. The observed second order rate constant between a neutral reagent A and a hydrophilic reagent in the presence of an ionic micelle is described by the modified equation 11.13 (Romsted, 1977):

$$k_2 = \frac{k_M \beta \Pi K_A[\mathrm{Dn}]}{(1 + K_A[\mathrm{Dn}])(X_t + Y_t K_{X/Y})} + \frac{k_w}{K_A[\mathrm{Dn}] + 1} \qquad (11.22)$$

where β is the degree of counterion binding to the Stern layer ($\beta = 1 - \alpha$), Π is the molar density of the micellar phase expressed in moles of surfactant per liter of micellar phase, X_t is the total concentration of the hydrophilic reagent, $Y_t = [\mathrm{Dn}] + [\mathrm{BY}]$, and [BY] is the concentration of added salts. Equation 11.22 has successfully predicted the kinetic behavior of a large number of second order reactions in ionic micelles as well as salt effects therein (Romsted, 1977; Bunton, 1979a; Bunton et al., 1982). One critical test using reactive counterion surfactants succeeded initially, then produced some interesting failures of the model (Bunton et al., 1979, 1980a, 1980c, 1982). This model cannot, however, be applied to buffered systems.

The ion exchange model has provided quantitative rationalizations for (1) the binding of a reactive ion to the micelle in the absence or presence of buffers; (2) the first order reaction of an ionic substrate in the micelle; (3) the second order reaction of a neutral substrate with an ionic reagent; (4) the effect of micelles on dissociation of weak acids and bases; and (5) the binding of OH^- to cationic micelles (Quina and Chaimovich, 1979; Chaimovich et al., 1979, 1981; Bonilha et al., 1979; Quina et al., 1980). This model assumes that (1) the distribution of aggregate sizes can be presented in terms of most probable aggregation number $\bar{N}$; (2) ion-ion and ion-headgroup-headgroup interactions are noncooperative; (3) degrees of ionization (α's) of the individual micellar species are the same; (4) ion-ion exchange rates are rapid compared to the lifetime of the micelle; and (5) activities of micellar and ionic species can be treated in terms of their concentrations. With these assumptions the selectivity coefficient for a reactive counterion X^- in B^+X^-, in micelle forming detergent Dn^+Y^-, in the absence or the presence of an added common salt, B^+Y^-, is given by

$$K_{X/Y} = \frac{X_M^-}{(X_T^- - X_M^-)} \frac{\alpha[\mathrm{Dn}] + \mathrm{CMC} + X_M^- + [B^+Y^-]_T}{(1 - \alpha)[\mathrm{Dn}] - X_M^-)} \qquad (11.23)$$

where the subscripts T and M refer to total concentrations of the appropriate species and to those present in the micellar pseudophase. At high detergent concentration

$$\lim_{[\mathrm{Dn}] \to \infty} K_{X/Y} = \frac{\alpha}{1 - \alpha} \frac{X_M^-}{X_w^-} \qquad (11.24)$$

equation 11.24 predicts that X_M^-/X_w^- tends to a limiting value and it allows, therefore, the assessment of K_{X^-/Y^-}. Addition of a buffer maintains X_w^- rather than X_M^- (Quina and Chaimovich, 1979). Rate constants for the reaction of a univalent ionic substrate, S^-, in an oppositely charged micelle, Dn^+Y^-, are given by

$$k_{\Psi} = \frac{k'_M K_{S^-/Y^-}(Y^-_M/Y^-_w) + k'_w}{1 + K_{S^-/Y^-}(Y^-_M/Y^-_w)} \tag{11.25}$$

Equation 11.25 should be compared to equation 11.4. The selectivity of micelle-bound ions rather than substrate partitioning is expressed in equation 11.25. Similarly, incorporation of the concepts of ion exchange theory in treating bimolecular reactions in micelles (see equation 11.13) leads to equations 11.26 and 11.27 for reactions between S and an oppositely charged X in the absence and in the presence of buffers, respectively (Quina and Chaimovich, 1979):

$$k_{2\Psi} = \frac{X_T[(k_M/\overline{V})(K_S K_{X/Y})(Y^-_M/Y^-_w) + k_w]}{(1 + K_S[Dn])[1 + K_{X/Y}(Y^-_M/Y^-_w)]} \tag{11.26}$$

$$k_{2\Psi} = \frac{X_w[(k_M/\overline{V})(K_S K_{X/Y})(Y^-_M/Y^-_w) + k_w]}{(1 + K_S[Dn])} \tag{11.27}$$

Elucidation of the effects of micelles on the dissociation constants of substrates is an important contribution of the ion exchange theory (Chaimovich et al., 1981). Dissociation of weakly acidic substrate, HA, in a cationic micelle, Dn^+Y^-, is described by the following equations (Quina and Chaimovich, 1979):

$$HA_w \overset{K_{aw}}{\rightleftharpoons} A^-_w + H^+_w \tag{11.28}$$

$$K_{aw} = \frac{[A^-]_w[H^+_w]}{[HA_f]} \tag{11.29}$$

$$HA_w \overset{K_{HA}}{\rightleftharpoons} HA_M \tag{11.30}$$

$$K_{HA} = \frac{[HA_M]}{([HA_w][Dn])} \tag{11.31}$$

$$A^-_w + Y^-_M \overset{K_{^-A/Y^-}}{\rightleftharpoons} A^-_M + Y^-_w \tag{11.32}$$

$$K_{^-A/Y^-} = \frac{[A^-_M][Y^-_w]}{([A^-_w][Y^-_M])} \tag{11.33}$$

$$^-OH_w + Y^-_M \overset{K_{^-OH/Y^-}}{\rightleftharpoons} {}^-OH_M + Y^-_w \tag{11.34}$$

$$K_{^-OH/Y^-} = \frac{[^-OH_M][Y^-_w]}{([^-OH_w][Y^-_M])} \tag{11.35}$$

$$HA_M + {}^-OH_M \overset{K_{aM}}{\rightleftharpoons} A^-_M + H_2O \tag{11.36}$$

$$K_{aM} = \frac{[^-A_M][D]}{([HA_M][^-OH_M])} \tag{11.37}$$

The apparent pK_a, pK_{app}, in the presence of Dn^+Y^-, is defined as the value of the intermicellar pH ($-\log[H_w^+]$) at which

$$[A_w^-] + [A_M^-] \equiv [HA_w] + [HA_M] \tag{11.38}$$

and K_{app} is given by

$$K_{app} = [H_w^+] = K_a \frac{1 + K_{-A/Y^-}(Y_M^-/Y_w^-)}{1 + K_{HA}[Dn]} \tag{11.39}$$

Equation 11.39 predicts the shape of the pK_{app} versus [Dn] curve to be a function of the relative magnitudes of K_{-A/Y^-} and k_{HA}. Conversely, the analogous equation, derived by Berezin and co-workers (1973),

$$K_{app} = [H^+] = K_a \frac{1 + K_A[D]}{1 + K_{HA}[D]} \tag{11.40}$$

cannot directly account for electrolyte effects on K_{app}.

The ion exchange theory accounts for the observed rate constant for the reaction between a neutral substrate, S, and the nucleophile, A^-, derived from the weak acid, HA, in the presence of cationic surfactant, D^+Y^-, by equation 11.41 (Quina and Chaimovich, 1979):

$$k_{2\Psi} = [HA_T]K_a \frac{(k_M/\bar{V})(K_S K_{-A/Y^-})(Y_M^-/Y_w^-) + k_w}{(1 + K_{HA}[Dn])([H_w^+] + K_{app})(1 + K_S[Dn])} \tag{11.41}$$

The behavior of $k_{2\Psi}$ as a function of a surfactant concentration has been computed for equations 11.26, 11.27, and 11.41 under a variety of experimental conditions (Quina and Chaimovich, 1979; Chaimovich et al., 1982). Experimental results substantiate well the model calculations (Bonilha et al., 1979; Quina et al., 1980).

It should be noted that the assumptions in the Romsted–Bunton (Bunton et al., 1979) and the ion exchange (Chaimovich et al., 1982) treatments are the same. The approaches differ only in the mathematical treatments and experimental tests. The main difference between the two approaches is that equations in the former are derived in terms of stoichiometric quantities of materials including hyperophilic ions, while the ion exchange equations are expressed in terms of the concentration of the hydrophilic ions in the aqueous phase, which has to be determined independently.

Distribution of reactive counterions, discussed in terms of micellar surface potentials (Almgren and Rydholm, 1979; Funasaki, 1979; Dung et al., 1980), has led to equations similar to those based on the ion exchange model.

Regardless of the model used, rate enhancements for bimolecular reactions appear to be the mere consequence of concentrating the reagents on the micellar surface. Treatments of reactions involving functional micelles have led to similar conclusions (Bunton, 1979a; Fornasier and Tonellato, 1980).

Treatments derived for analyzing reactivities in the presence of charged micelles are likely to be applicable to those in reversed micelles, microemulsions, polyions, monolayers, and vesicles. Indeed kinetics of ionic reactions in polyelectrolytes have been analyzed in terms of the counterion condensation theory (see Chapter 8 and Manning, 1979) as early as 1975 (Mita et al., 1975, 1976). Comparison of reactivities in different membrane mimetic systems (Fendler, 1980; Fendler and Hinze, 1981) will allow the selection of the best system for a given application.

In closing this section, it is appropriate to point out the need for care in treating reactivities in membrane mimetic systems. It is advisable to follow the methodologies developed for investigating reaction kinetics in homogeneous solutions (Bunnett, 1961); Frost and Pearson, 1962). Purity of substrates and surfactants is mandatory. It is important to realize that some surfactants can affect reactions as general acids or general bases. This effect has been recognized for hydrolyses in reversed micelles (O'Connor et al., 1973, 1981).

2. ALTERATION OF REACTION PATHS

Reagent organization in, and the different microenvironments provided by, membrane mimetic agents open the possibility of controlling pathways leading to product formation:

$$\text{reagent} \begin{cases} \longrightarrow \text{product I} \\ \longrightarrow \text{product II} \end{cases} \tag{11.42}$$

Aqueous micelles and cyclodextrins have been used most extensively for altering reaction pathways (Table 11.1). The principle is illustrated by the efficient energy transfer from micellar sodium dodecyl sulfate (SDS) solubilized naphthalene to terbium chloride (Escabi-Perez et al., 1977). Micelles compartmentalize no more than one donor into each micellar interior and at the same time concentrate large numbers of acceptor molecules at the surface. Subsequent to excitation, naphthalene singlets intersystem cross into the triplet domain. These species, in turn, transfer their energies to terbium chloride. In the absence of micelles, competing triplet-triplet annihilation obviates electron transfer.

Molecular organization in micelles has been advantageously exploited in photochemical energy conversion (see Chapter 13) as well as in photochemistry (Turro et al., 1980a; Fendler, 1980). Photodecarbonylation of dissymmetrical

Table 11.1. Effects of Membrane Mimetic Agents on Reaction Paths[a]

System	Reaction	Effects on Product Distribution	Reference
Aqueous micellar CTAB, SDS, Igepal, DDAS	OSO_3^-, NO_2, NO_2 (substituted benzene) + RNH_2 → hydrolysis → O^-, NO_2, NO_2 (substituted benzene); → nucleophilic substitution → NHR, NO_2, NO_2 (substituted benzene) RNH_2 = nicotinamide, pyridine, 2-picoline, 4-picoline, imidazole, hydrazine, morpholine, glycine piperidine	Micellar CTAB, DDAPS and Igepal + DDAPS completely suppress the nucleophilic substitution by hydrazine and morpholine	Fendler et al., 1972
Aqueous micelle forming *n*-decyl phosphate	$ROPO_3H_2 \xrightarrow{Cl^-} RCl + Pi$ $ROPO_3H_2 \underset{}{\overset{H^+}{\rightleftharpoons}} ROPO_3H_3^+$ $ROPO_3H_3^+ \xrightarrow{H_2O} ROH + Pi$ (crossed arrows: $ROPO_3H_2$ → ROH + Pi; $ROPO_3H_3^+$ → RCl + Pi)	Micellization eliminates the Cl^- ion promoted reaction	Bunton et al., 1976
Aqueous micellar Brij-35, Triton X-100	$K_3[CO(CN)_5H]$ catalyzed hydrogenation of dienes	Micelles substantially alter product distribution	Reger and Habib, 1978
Aqueous micellar CTAB	$ArCHO \underset{k_{-1}}{\overset{k_1}{\rightleftharpoons}} Ar{-}C(H)(CN){-}O^- \underset{k_{-2}}{\overset{k_2}{\rightleftharpoons}} Ar{-}\overset{\ominus}{C}(CN){-}OH$ $Ar{-}\overset{\ominus}{C}(CN){-}OH \xrightarrow{ArCHO}$ benzoin condensation $Ar{-}\overset{\ominus}{C}(CN){-}OH \xrightarrow{\text{Flavin(Fl)} \rightarrow FlH^-} ArCOCN$ $ArCOCN \rightarrow ArCO_2^-$ (CN^- returned to k_1 step) or analogous reaction with $ArCOCO_2^-$	Flavin trapping of carbanion, diverting condensation to oxidation, is greatly facilitated by micellar CTAB	Shinkai et al., 1980

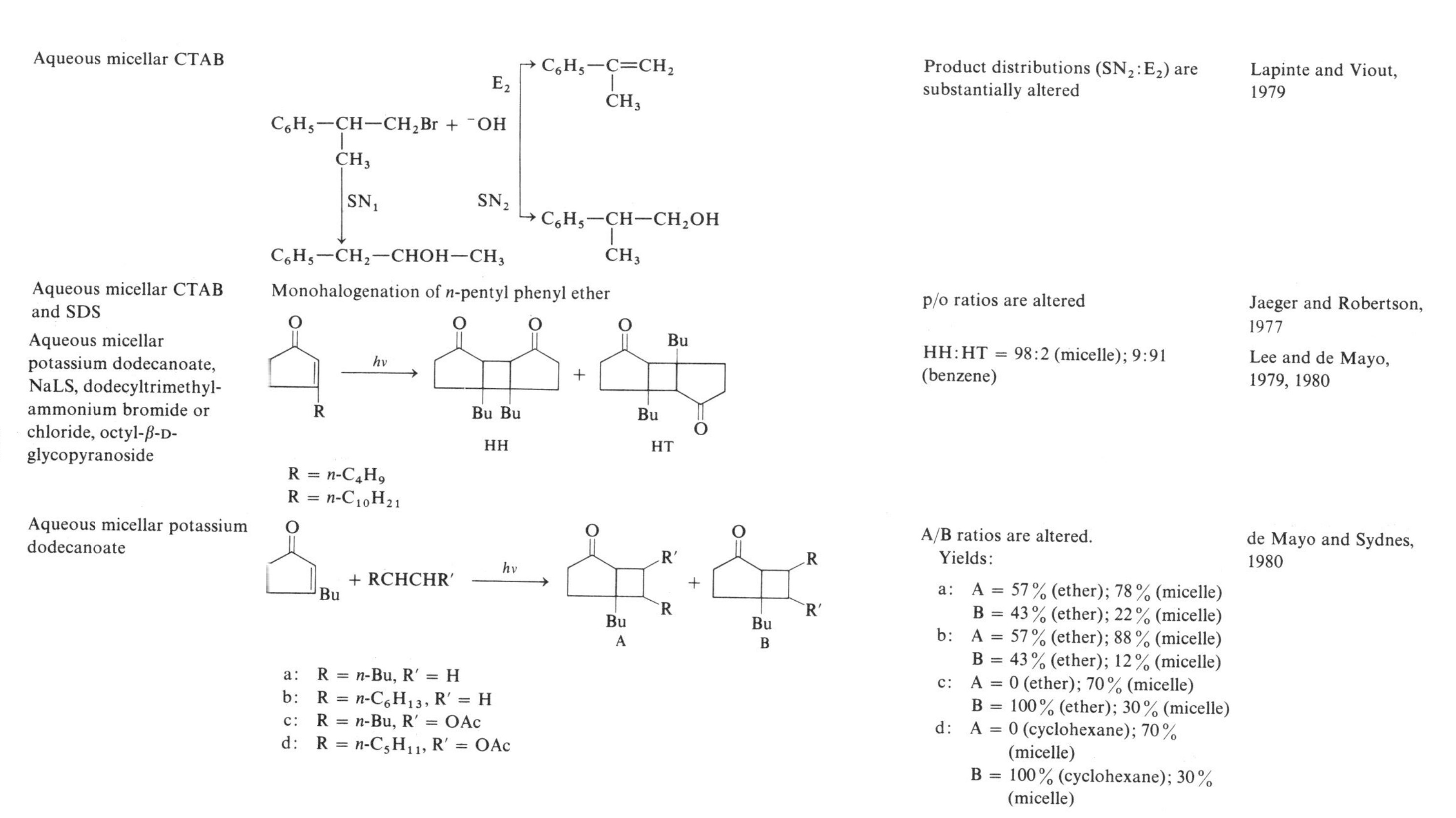

Medium	Reaction	Effect	Reference
Aqueous micellar CTAB	$C_6H_5-CH(CH_3)-CH_2Br + {}^{-}OH$ (E_2, SN_2, SN_1)	Product distributions ($SN_2:E_2$) are substantially altered	Lapinte and Viout, 1979
Aqueous micellar CTAB and SDS	Monohalogenation of *n*-pentyl phenyl ether	p/o ratios are altered	Jaeger and Robertson, 1977
Aqueous micellar potassium dodecanoate, NaLS, dodecyltrimethyl-ammonium bromide or chloride, octyl-β-D-glycopyranoside	Photodimerization, $R = n\text{-}C_4H_9$; $R = n\text{-}C_{10}H_{21}$	HH:HT = 98:2 (micelle); 9:91 (benzene)	Lee and de Mayo, 1979, 1980
Aqueous micellar potassium dodecanoate	+ RCHCHR′ ($h\nu$) → A + B a: R = *n*-Bu, R′ = H b: $R = n\text{-}C_6H_{13}$, R′ = H c: R = *n*-Bu, R′ = OAc d: $R = n\text{-}C_5H_{11}$, R′ = OAc	A/B ratios are altered. Yields: a: A = 57% (ether); 78% (micelle) B = 43% (ether); 22% (micelle) b: A = 57% (ether); 88% (micelle) B = 43% (ether); 12% (micelle) c: A = 0 (ether); 70% (micelle) B = 100% (ether); 30% (micelle) d: A = 0 (cyclohexane); 70% (micelle) B = 100% (cyclohexane); 30% (micelle)	de Mayo and Sydnes, 1980

Table 11.1 (*Continued*)

System	Reaction	Effects on Product Distribution	Reference
Aqueous micellar CTAB	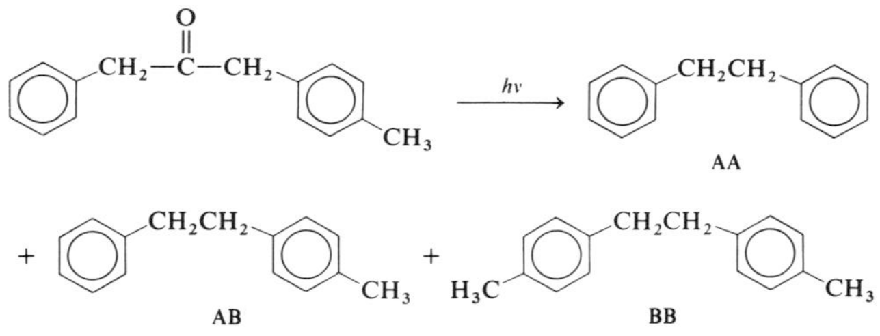 $C_6H_5CH_2COCH_2C_6H_4CH_3 \xrightarrow{h\nu}$ $C_6H_5CH_2CH_2C_6H_5$ (AA) + $C_6H_5CH_2CH_2C_6H_4CH_3$ (AB) + $H_3CC_6H_4CH_2CH_2C_6H_4CH_3$ (BB)	If there is less than one ketone per micelle, the photoproduct is exclusively AB	Turro and Cherry, 1978
Aqueous micellar 1-hydroxyethyl-2-dimethyl-alkylammonium bromide (alkyl = n-$C_{12}H_{25}$, n-$C_{16}H_{33}$)	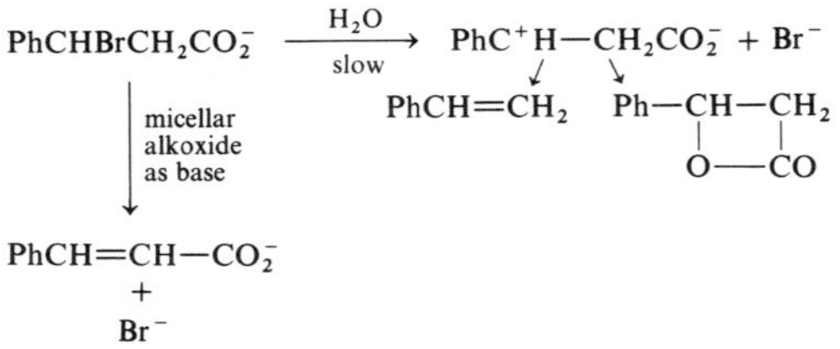 $PhCHBrCH_2CO_2^- \xrightarrow[slow]{H_2O} PhC^+H{-}CH_2CO_2^- + Br^-$ → $PhCH{=}CH_2$ and $Ph{-}CH{-}CH_2$ (with O—CO ring); $PhCHBrCH_2CO_2^-$ →(micellar alkoxide as base) $PhCH{=}CH{-}CO_2^-$ + Br^-	In micellar solution SN_1 process is completely replaced by E_2 elimination to give transcinnamate ion	Bunton et al., 1974
Aqueous micellar sodium tetradecanoate	Photolysis of 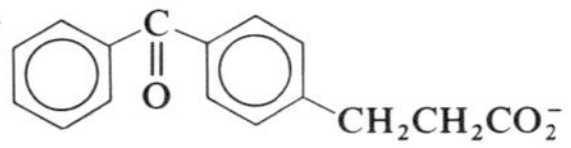 $C_6H_5C(=O)C_6H_4CH_2CH_2CO_2^-$	Hydrogen is abstracted only from position 5–8 of the surfactant	Mitani et al., 1979
Aqueous micellar SDS	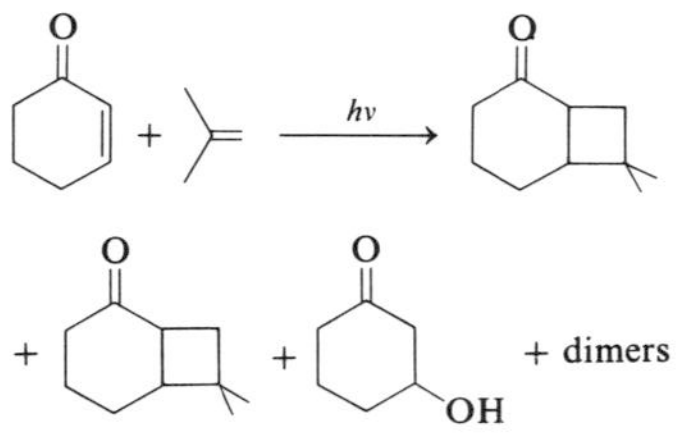 cyclohexenone + isobutylene $\xrightarrow{h\nu}$ bicyclic cyclobutane ketone + isomeric bicyclic ketone + 3-hydroxycyclohexanone + dimers	In micelles cycloaddition prevails, dimerization is suppressed	Braun, 1981

Aqueous micellar SDS	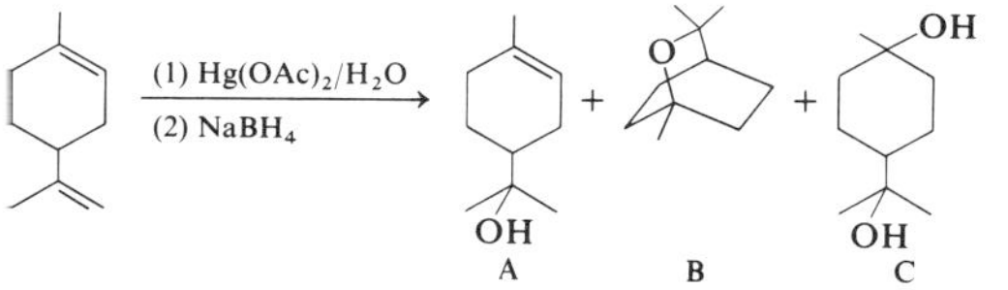 (1) $Hg(OAc)_2/H_2O$; (2) $NaBH_4$ → A + B + C (structures; OH labels)	In micelles yield of A is 97%; in THF/H_2O there is a mixture of A (70%), B + C (14%), and starting material (19%)	Link et al., 1980
Aqueous micellar SDS	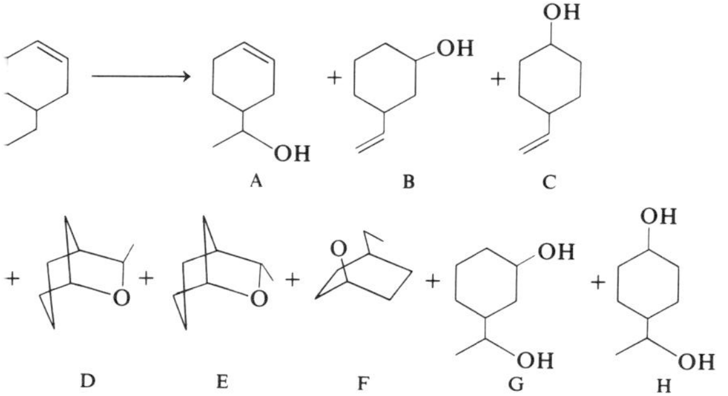 → A + B + C + D + E + F + G + H (structures; OH labels)	In micelles yield of A is 90%; in THF/H_2O there is a mixture of A, B, C, D, E, F, G, and H, with only 20% of A	Link et al., 1980
Aqueous micellar dodecyltrimethyl ammonium chloride	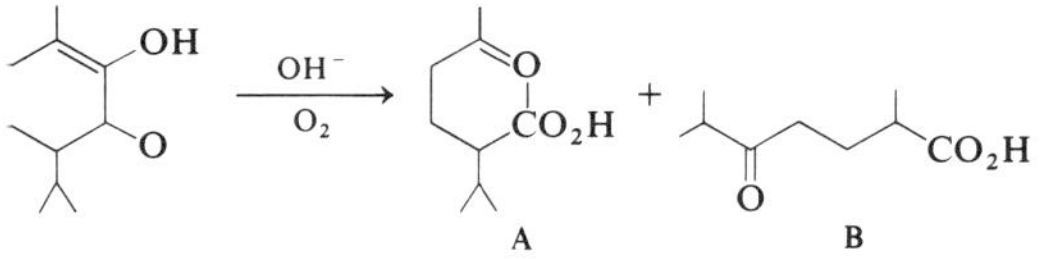 $\xrightarrow[O_2]{OH^-}$ A (CO_2H) + B (CO_2H) (structures)	In micelles yield of A is 76%, B is 11%, only traces of byproducts; in absence of micelles yield of A is 47%, B is 9%, and several byproducts are in 35%.	Utaka et al., 1980
Aqueous micellar CTACl	$C_6H_5COAd \xrightarrow{h\nu} C_6H_5CHO$ (A) $+ C_6H_5COC_6H_5$ (B) $+ AdH$ (C) $+ 1\text{-}C_6H_5Ad$ (D) $+ AdAd$ (E) $+ AdOH$ (F) $+ C_6H_5COCOC_6H_5$ (G) $+ AdCl$ (H) $+ C_6H_5COOH$ (I) Ad = 1-adamantyl	Yields are: A = 75% (C_6H_6), 7% (CTACl); B = 6% (C_6H_6); C = 33% (C_6H_6), 60% (CTACl), 5% (CTACl + $CuCl_2$); D = 35% (C_6H_6); E = 3% (C_6H_6); F = 3% (CTACl), 38% (CTACl + $CuCl_2$); G = 3% (CTACl); H = 33% (CTACl + $CuCl_2$); I = 45% (CTACl + $CuCl_2$)	Turro and Tung, 1980

Table 11.1 *(Continued)*

System	Reaction	Effects on Product Distribution	Reference
Aqueous micellar polyoxyethylenepolyoxypropylene cetyl ether	Acenaphthylene + $CH_2{=}CH{-}CN$ $\xrightarrow{h\nu}$ cyclobutane CN adduct + photodimer	The photodimer is readily formed in micelles; dimer:CN adduct = 0%–30% (benzene); 30%:70% (micelle)	Nakamura et al., 1977, 1978
Aqueous micellar CTAB, reversed micellar dodecylammonium acetate in cyclohexane	Photodimerization of *N*-ω-carboxylalkyl-2-pyridones and their 4-alkyl derivatives	Ratio of cis:trans dimer increases with decreasing length of alkyl chain in micelles; in 4-alkyl derivatives only cis dimer is obtained	Nakamura et al., 1981
Reversed micellar CTAB	Benzidinine rearrangement of hydrazobenzene-3,3′-disulfonic acid and concomitant azobenzene-3,3′-disulfonic acid formation	Exclusive azobenzene-3,3′-disulfonic acid formation Horiguchi, 1980	Sunamoto and
Reversed comicellar hexadecyltrimethylammonium bicarbonate di-2-ethylhexyl sodium sulfosuccinate in benzene	Alanyl adenylate → polypeptides; Alanyl adenylate → hydrolysis products	High yield (up to 95%) polypeptide formation containing up to 41 amino acid units	Armstrong et al., 1978
Cyclodextrin	$Ph{-}CH_2\overset{+}{S}(Me){-}CH_2Ph\ BF_4^-$ $\xrightarrow{OH^-}$ $Ph\overset{-}{C}H\overset{+}{S}(Me)CH_2Ph$; Stevens → $Ph{-}CH(CH_2Ph){-}S{-}Me$; Sommelet → o-$CH_3C_6H_4{-}CH(Ph){-}S{-}Me$	Stevens:Sommelet = 0.38:1 (H_2O); 2:3 (cyclodextrin)	Mitani et al., 1974

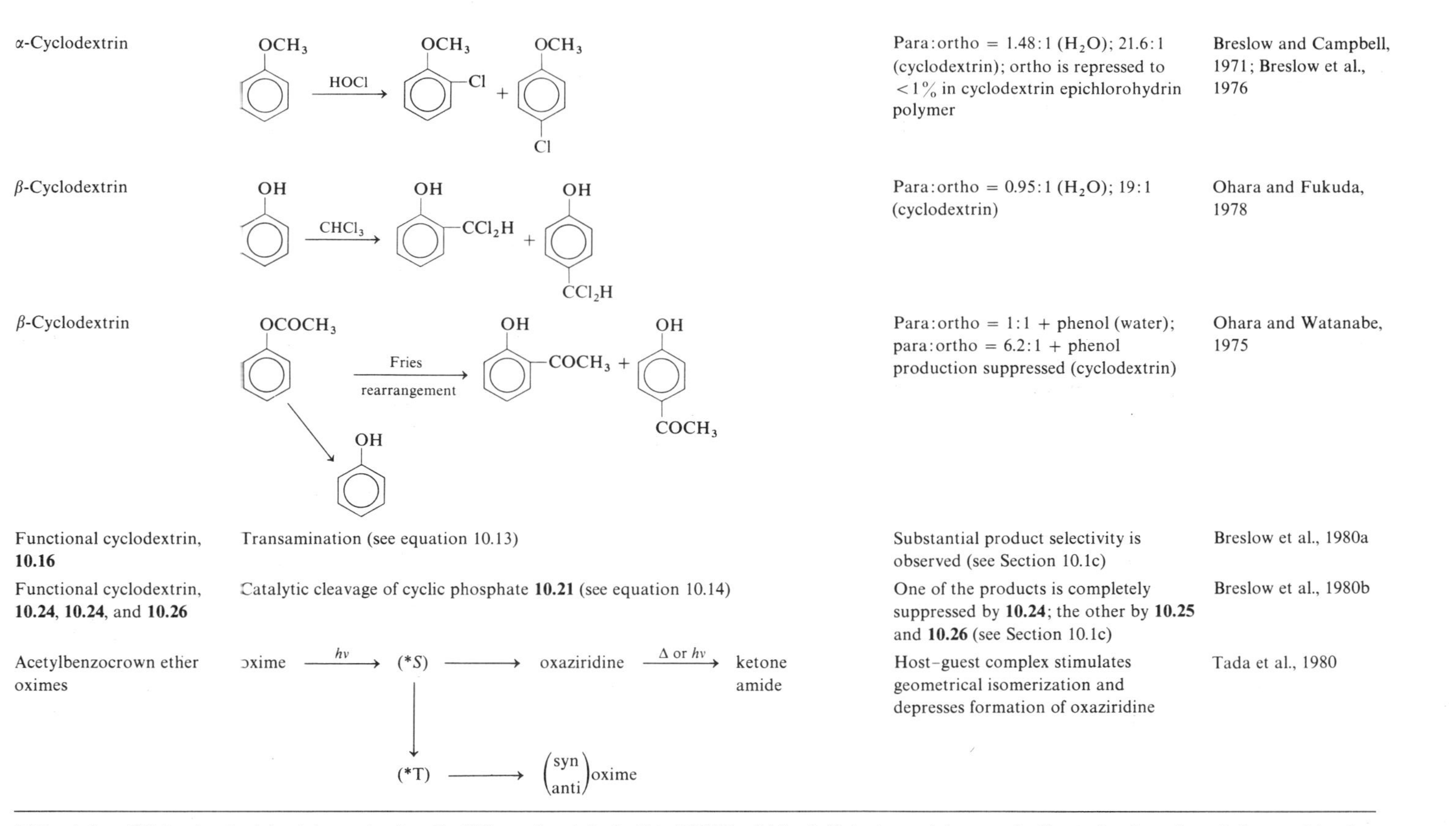

α-Cyclodextrin	Anisole (OCH_3) $\xrightarrow{HOCl}$ o-chloroanisole + p-chloroanisole	Para:ortho = 1.48:1 (H_2O); 21.6:1 (cyclodextrin); ortho is repressed to <1% in cyclodextrin epichlorohydrin polymer	Breslow and Campbell, 1971; Breslow et al., 1976
β-Cyclodextrin	Phenol (OH) $\xrightarrow{CHCl_3}$ o-CCl_2H phenol + p-CCl_2H phenol	Para:ortho = 0.95:1 (H_2O); 19:1 (cyclodextrin)	Ohara and Fukuda, 1978
β-Cyclodextrin	Phenyl acetate ($OCOCH_3$) $\xrightarrow[\text{rearrangement}]{\text{Fries}}$ o-$COCH_3$ phenol + p-$COCH_3$ phenol; also → phenol (OH)	Para:ortho = 1:1 + phenol (water); para:ortho = 6.2:1 + phenol production suppressed (cyclodextrin)	Ohara and Watanabe, 1975
Functional cyclodextrin, **10.16**	Transamination (see equation 10.13)	Substantial product selectivity is observed (see Section 10.1c)	Breslow et al., 1980a
Functional cyclodextrin, **10.24**, **10.24**, and **10.26**	Catalytic cleavage of cyclic phosphate **10.21** (see equation 10.14)	One of the products is completely suppressed by **10.24**; the other by **10.25** and **10.26** (see Section 10.1c)	Breslow et al., 1980b
Acetylbenzocrown ether oximes	oxime $\xrightarrow{h\nu}$ (*S) ⟶ oxaziridine $\xrightarrow{\Delta \text{ or } h\nu}$ ketone, amide; (*S) ↓ (*T) ⟶ (syn/anti) oxime	Host–guest complex stimulates geometrical isomerization and depresses formation of oxaziridine	Tada et al., 1980

[a] Abbreviations: CTAB = hexadecyltrimethylammonium bromide; SDS = sodium dodecyl sulfate; DDAPS = 3-(dimethyldodecylammonio)-propane-1-sulfonate; Igepal = polyoxyethylene nonylphenol.

dibenzylketones in homogeneous solution results in statistical product formation (Turro and Cherry, 1978):

$$\underset{\substack{\text{O}\\ \|\\ \text{ACB}}}{\text{PhCH}_2\overset{\text{O}}{\overset{\|}{\text{C}}}\text{CH}_2\text{Ar}} \xrightarrow{h\nu} \underset{\substack{\text{AA}\\ 25\%}}{\text{PhCH}_2\text{CH}_2\text{Ph}} + \underset{\substack{\text{AB}\\ 50\%}}{\text{PhCH}_2\text{CH}_2\text{Ar}} + \underset{\substack{\text{BB}\\ 25\%}}{\text{ArCH}_2\text{CH}_2\text{Ar}} \quad (11.43)$$

Product distribution is dramatically altered in the presence of hexadecyltrimethylammonium chloride (CTACl). Increasing the surfactant to ketone ratios results in a sigmoidal increase of the cross products, AB, at the expense of coupled products, AA and BB (Figure 11.2). When there is less than one reactant per micelle AB becomes the exclusive photoproduct. Under this condition the photolytically generated A· and ·B readily react with each other prior to their escape from the micellar cage (Turro and Cherry, 1978). Conversely, in homogeneous solution there is nothing to prevent the radicals from reacting with each other in a statistical manner (equation 11.43).

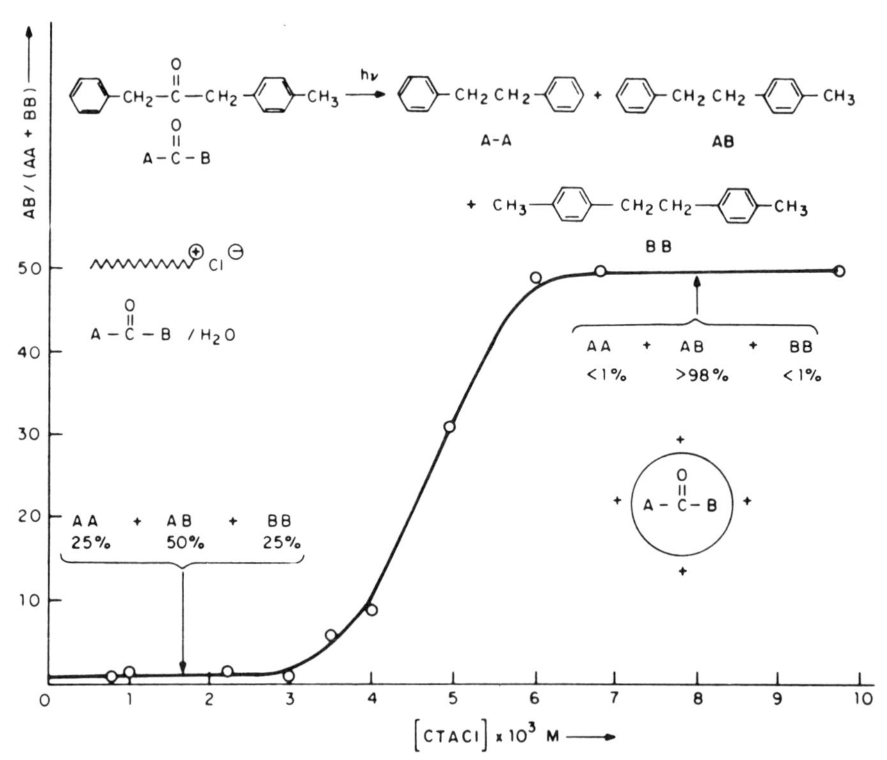

Figure 11.2 Product distribution in the photolysis of dissymmetrical dibenzylketones as a function of CTACl concentration (taken from Turro and Cherry, 1978).

Aqueous micellar SDS has been shown to control the selective monofunctionalization of nonconjugated dienes (Link et al., 1980). Limonene and 4-vinylcyclohexene gave high yields of monools in the presence of micelles, while hosts of other products were formed in the usual THF:H_2O mercuration media (see Table 11.1). These results were rationalized in terms of mercuration at the micellar surface. Strong stabilization of the cationic mercury species by the anionic surfactant headgroups forces out the water that would otherwise have solvated the cationic complex. This effectively lowers the water concentration in the immediate vicinity of the reaction site and favors, therefore, nucleophilic reactivity (Link et al., 1980).

Micelles can also change reaction mechanisms. An example is provided by the decomposition of 3-bromo-3-phenylpropionate ion (Bunton et al., 1974). The rate determining step for this reaction in aqueous solution is the formation of a zwitterionic carbonium ion. This carbonium ion rapidly decarboxylates with elimination to give styrene as the major product and a β-lactone as the minor product. In the presence of micellar 1-hydroxyethyl-2-dimethylalkylammonium bromide the SN_1 reaction is replaced by an E_2 mechanism and *trans*-cinnamate ion becomes the major product (Bunton et al., 1974). Effects of micelles on altering ratios of SN_1 to SN_2-E_2 mechanisms have been discussed (Lapinte and Viout, 1979).

Substrate inclusion into cyclodextrin cavities has also resulted in altered product distributions (see Table 11.1). Monolayers, bilayers, and vesicles have not yet been exploited for altering reaction pathways.

An important utilization of micelles, cyclodextrins, and crown ethers is to suppress or eliminate undesirable side reactions and hence stabilize drugs, foodstuffs, herbicides, pesticides, and light or oxygen sensitive materials (Elworthy et al., 1968; Saenger, 1980).

3. ALTERATION OF STEREOCHEMISTRY

Stereochemistry plays a central role in natural product synthesis, molecular recognition, and biochemical and pharmacological processes. Most enzyme catalyzed reactions are highly stereospecific. Interest in exploiting membrane mimetic systems for promoting stereoselectivities, chiral recognitions, and asymmetric inductions is hardly surprising, therefore. Maximum stereospecificities are to be expected for systems that provide relatively rigid and specific binding sites for a given substrate and its reactive transition state. Free energies of binding and/or reactivities of the enantiomers should be sufficiently different to bring about chiral discriminations. Cavities of functionalized optically active cyclodextrins and crown ethers are expected to provide more rigid and hence energetically more favorable binding sites for chiral discrimination than those available in micelles. This expectation is borne out by the stereoselectivities observed in host–guest systems, which are substantially greater than those observed in micelles (see Table 11.2).

Table 11.2. Altered Stereochemistry in Membrane Mimetic Systems

System	Reaction	Stereoselectivity	Reference
Aqueous optically active $PhCH(OH)CHMe\overset{+}{N}Me_2R\ Br^-$, $R = C_{10}H_{21}, C_{12}H_{25}$	Hydrolysis of optically active 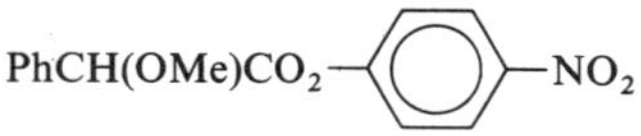 $PhCH(OMe)CO_2-C_6H_4-NO_2$	Modest degree of stereoselective catalysis.	Bunton et al., 1971
Aqueous micellar 2-octylammonium and 2-octyltrimethylammonium ions	Nitrous acid deamination of optically active 2-aminooctane	24% net inversion (H_2O) → 6% retention (micelles); effects are highly dependent on micellar couterion	Moss et al., 1973
Aqueous micellar CTAB, *N*-benzyl-*N*-cetyldimethylammonium bromide, *d*-, *l*-, and *dl*-*N*-α-methylbenzyl-*N*,*N*-dimethylcetylammonium bromide, *d*-, *l*-, and *dl*-*N*-α-methyl-*p*-methoxybenzyl-*N*,*N*-dimethylcetylammonium bromide	Hydrolysis of *d*-, *l*-, and *dl*-*p*-nitrophenyl-α-methoxyphenyl acetates	No substantial stereoselectivity observed	Moss and Sunshine, 1974
Aqueous micellar $Me(CH_2)_{11}-(CH(C(=O)R))-(CH_2)_4\overset{+}{N}Me_3$; $R = OH$; $R = MeO_2C\sim(CH)NH-$ with CH_2-imidazolium (HN, NH, +) Cl^- 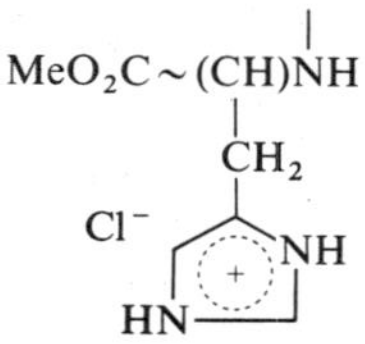	Hydrolysis of R—(−)— or S—(+)— 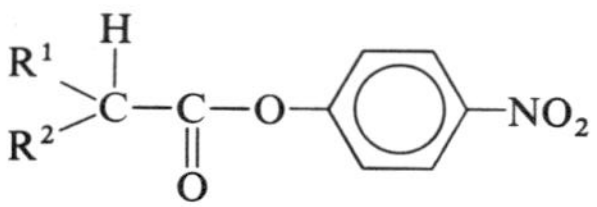 $R^1R^2C(H)-C(=O)-O-C_6H_4-NO_2$; $R^1 = Ph, R^2 = Me$; $R^1 = PhCH_2, R^2 = MeCONH$	Enantioselective ($k(R)/k(S)$ = 3:1) deacylation	Brown and Bunton, 1974

Aqueous micellar CTAB, SDS	Hydrolysis of optically active 1-methylheptyltrifluoromethanesulfonate	70% inversion (H_2O), 47% retention (CTAB); concentration dependent inversion → retention (SDS)	Okamoto et al., 1975
Aqueous micellar (+)—(R)-*N*-dodecyl-*N*,*N*-dimethyl-α-phenylethylammonium bromide	$PhC(=O)-R \xrightarrow[H_2O]{NaBH_4} PhCH(OH)R$ R = Me, Et, *N*-Pr, *i*-Pr, *t*-Bu	Modest degree of asymmetry (up to 1.5%)	Goldberg et al., 1978
Aqueous sodium cholate	$PhC(=O)-R \xrightarrow[H_2O]{NaBH_4} PhCH(OH)R$	Modest degree of asymmetry (up to 1.7%)	Sugimoto et al., 1978
Optically active and racemic $RCONHCH(CH_2CH_3)COONa$ and $RCONHCH(CHMe_2)COONa$ R = capryl, lauryl, myristoyl, palmitoyl, lauroyl	Micelle formation	CMC values of pure enantiomers are smaller than those of their racemates	Miyagishi and Nishida, 1978
Aqueous mixed micelles of CTAB and optically active $R—CONHCH(CO_2H)CH_2$—(imidazolyl) R = Me, Ph, $Me(CH_2)_8$, $Me(CH_2)_{16}$	Deacylation of optically active $R—OCONHCH(CH_2Ph)—CO_2C_6H_4NO_2$-*p*	Modest degree of stereoselective hydrolysis ($k_L/k_D \sim 2$)	Ihara, 1978, 1980; Ihara et al., 1980b

Table 11.2 *(Continued)*

System	Reaction	Stereoselectivity	Reference
Aqueous mixed micelles of CTAB and optically active  $CH_3(CH_2)_{10}CONHCH(CO_2H)$–(imidazolyl)	Deacylation of optically active $R_2OCONHCH(R_1)COO—C_6H_4NO_2$-*p*	Modest degree of stereoselective hydrolysis ($k_L/k_D \sim 2$)	Yamada et al., 1979a, 1979b
Aqueous micellar n-$C_{16}H_{33}N^+(CH_3)_2(R)$, Cl^-; R = CH_3(CTAB), CH_2CH_2SH, 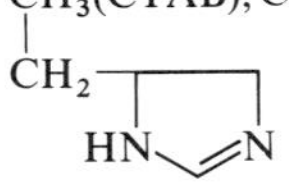 CH_2–(imidazolyl)	Cleavage of LL- and DL-*N*-carbobenzyloxyalanylproline-*p*-nitrophenylester	Stereoselectivity ($k_{LL}/k_{DD} \doteq$ 0.6–4.3) increases with increasing catalytic efficiency of the surfactant	Moss et al., 1979
Aqueous micellar n-$C_{16}H_{33}N^+(CH_3)_2(R)$, Cl^-; R = CH_3(CTAB), CH_2CH_2SH	Cleavage of LL- and DL- 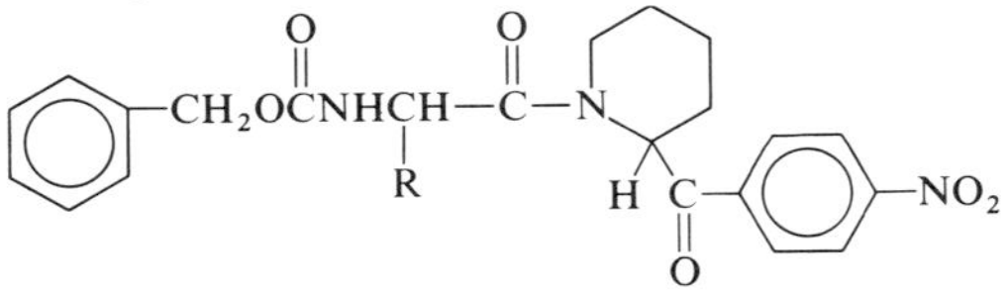 $C_6H_5CH_2OC(O)NHCH(R)C(O)$–N(piperidyl)–CH–C(O)–$C_6H_4$–$NO_2$; R = CH_3, $CH_2C_6H_5$, CH_2—(3-indoyl), $CH_2CH(CH_3)_2$, $CH(CH_3)_2$	Stereoselectivities (k_{LL}/k_{DL} = 0.6–5.0) depend on specific substrate–surfactant interactions within the substrate–micelle complex	Moss et al., 1980a
Aqueous micellar n-$C_{16}H_{33}\overset{+}{N}R(CH_3)_2$, Cl^-; R = CH_3(CTAB), CH_2CH_2SH, $CH_2CH_2SCH_3$, $CH_2CH_2NHCO(CH)NH_2$ with CH_2SH on CH	Cleavage of diastereometric tripeptide esters, *Z*-(D or L)-Phe-(D or L)-Phe-(L)-Pro-*p*-nitrophenyl	Stereoreactivities depend on the surfactant–substrate interactions	Moss et al., 1980b

Aqueous micellar optically active *N*-acylglumates (carbon number in acyl group = 12, 14, 16)	Formation of chiral aggregates (CD band at 220 nm)	Chiral hydrogen bonds at concentrations > CMC	Sakamoto and Hatano, 1980
Aqueous comicelles of *N*-(*N*-dodecanoyl-L-histidyl)-L-leucine and R-(+)-*N*-α-methylbenzyl-*N*,*N*-dimethyloctadecylammonium bromide	Deacylation of $H(CH_2)_{n-1}COCHCH(CH_2Ph)$ with p-$NO_2C_6H_4OOC$ on the CH; n = 10–16	Stereoselective (k_L/k_D = 1.0–5.5) catalysis	Ohkubo et al., 1980
Aqueous micellar CTAB	Hydrolysis of L or D $C_6H_5CH_2OCONHCHCOOC_6H_4NO_2$-$p$ (with $CH_2C_6H_5$ on the CH) by optically active $C_6H_5CH_2OCONHCHCONOH$ (R on the N; $CH_2C_6H_5$ on the CH), R = H, CH_3; $C_6H_5CH_2OCONHCHCONOH$ (R on the N; $(CH_2)_4NHCO_2CH_2C_6H_4$ on the CH), R = H, CH_3; $H_2NCHCONH(CH_2)_{11}CH_3$ (R on the CH), R = H, CH_3; $C_6H_5CH_2OCONHCHCOOH$ (with $(CH_2)_4NH_2$ on the CH), R = H, CH_3	Modest stereoselectivity (k_D/k_L = 0.9–2.6)	Ihara et al., 1980a

Table 11.2 (*Continued*)

System	Reaction	Stereoselectivity	Reference
Aqueous micellar CTAB + D- or L- $CH_3(CH_2)_{10}CONHCH(CH_2\text{-imidazolyl})COOH$	Oxidation of 3,4-dihydroxyphenyl-CH_2—CH(NH_2)—COOH (L-DOPA)	Modest degree of stereoselectivity ($k_L/k_D \pm 1.4$)	Yamada et al., 1980
Reversed micellar optically active (+) and (−) α-phenylethylammonium dodecanoates in hexane	Hydrolysis of optically active L- and D-*p*-nitrophenyl-α-methoxyphenyl acetate	Modest degree of stereoselectivity ($k_L/k_D \sim 1.4$)	Kon-no et al., 1981
Enantiomeric and racemic *N*-α-methylbenzylstearamide	Monolayer formation	Force-area curves of the enantiomers are different from that of their racemate	Arnett et al., 1978; Arnett and Thompson, 1981
sn-2- and *sn*-3-dipalmitoyl phosphatidylcholine	Monolayer formation	At a given surface pressure the *sn*-2-phosphatidylcholine is more expanded than its *sn*-3 analogue	Seelig et al., 1980
Lecithin emulsions (reversed micelles?)	$RC(=O)—CH_3 \xrightarrow{NaBH_4} RCH(OH)CH_3$ $R = C_2H_5, n\text{-}C_5H_{11}, n\text{-}6_6H_{13}, n\text{-}C_9H_{19}$	Enhanced rate of reduction and extremely modest asymmetric induction	Doiuchi and Minoura, 1976/77
sn-3- and *sn*-1-dipalmitoyl phosphatidylcholine (sonicated) liposomes	Effect of cholesterol on CH_2 proton linewidth	^1H nmr linewidths are affected differently; stereospecific phospholipid–cholesterol interactions in the hydrogen belt region	Chatterjie and Brockerhoff, 1978
sn-3- and *sn*-2-dipalmitoyl phosphatidylcholine (unsonicated) liposomes	^2H and ^{31}P nmr spectroscopy	Hydrocarbon interior of *sn*-2 phosphatidylcholines is more disordered than that of the *sn*-3 analogue	Seelig et al., 1980

N,N-didodecyl-N^{α}-[6-(trimethylammonio)hexanoyl]-histinamide bromide surfactant vesicle	Hydrolysis of L- and D-N-benzyloxycarbonylphenylalanine p-nitrophenyl esters	Modest degree of stereoselectivity ($k_L/k_D \sim 2.7–4.4$)	Murakami et al., 1981
β-Cyclodextrin	Acylation of β-cyclodextrin by $^{H}C—CO_2pNP$ (*1′*), $pNPO_2C—C^{H}$ (*1″*), CO_2pNP (*2*) [ferrocene structures; *1′* and *1″* are the enantiomers of *1*] 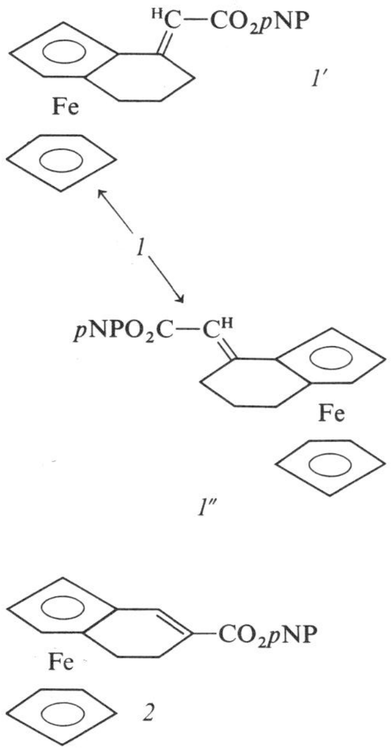	20- and 7-fold enantioselectivity in the acylation of β-cyclodextrin by *1′* versus *1″*, and by analogous enantiomers of *2*.	Trainor and Breslow, 1981
α- and β-cyclodextrin	Diastereomeric complexes with racemates	Partial resolution of mandelic, phenylchloroacetic, and atrolactic acid, isopropyl methylphosphinates, and their ethyl analogues, sulfoxides, and sulfonates	Cramer and Dietsche, 1959; Benschop and Van den Berg, 1970; Mikolajcczyk and Drabowicz, 1978

Table 11.2 (*Continued*)

System	Reaction	Stereoselectivity	Reference
α-Cyclodextrin	Hydrolysis of (S)-(+)- and (R)-(−)-savin	Partial resolution of racemate	Van Hooidonk and Breebart-Hansen, 1970
α-Cyclodextrin	Binding of chiral benzene derivatives	Small chiral recognition	Cooper and MacNicol, 1978
β-Cyclodextrin	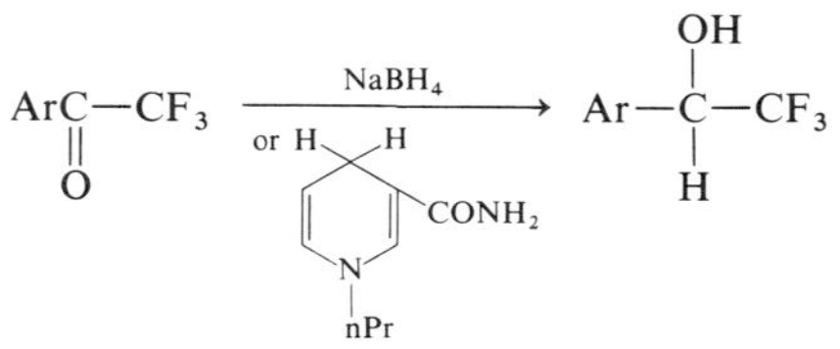 $ArC(=O)-CF_3 \xrightarrow[\text{or 1-nPr-1,4-dihydronicotinamide}]{NaBH_4} Ar-CH(OH)-CF_3$ Ar = phenyl, 1-naphthyl, 2-naphthyl	Modest degree of asymmetric induction	Baba et al., 1978
β-Cyclodextrin	Complexation of (+) and (−) fenchone, monitored by epr using spin labels	(+) fenchone is more complexed than (−) fenchone	Michon and Rassat, 1979
α- and β-cyclodextrin	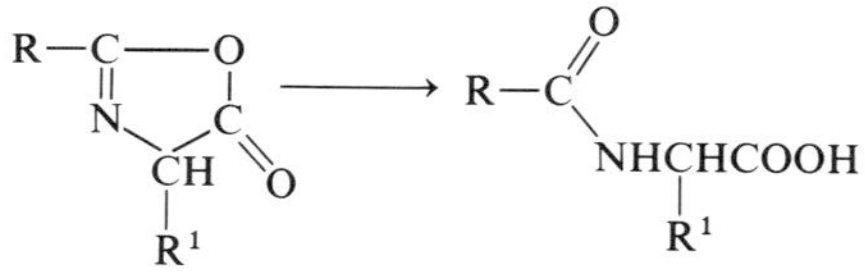 oxazolone (R—C=N—CH(R^1)—C(=O)—O) ⟶ $R-C(=O)-NHCH(R^1)COOH$ $R = C_6H_5, R^1 = C_6H_5CH_2$ $R = CH_3, R^1 = C_6H_5CH_2$ $R = C_6H_5, R^1 = CH_3$ $R = CH_3, R^1 = CH_3$	Substantial enantiomeric selectivity	Daffe and Fastrez, 1980

Optically active multiheteromacrocyclic crown ethers containing 1,1′-dinaphthyl units	Complexation primary amine salts	Optical resolution of salts, amines, amino acids, amino acid esters by chromatography and selective extraction (amino acid resolving machine)	Cram et al., 1975; Kyba et al., 1978b; Sousa et al., 1978; Newcomb et al., 1979; Peacock et al., 1980; Lingenfelter et al., 1981
11.9	Thiolysis of *p*-nitrophenyl ester salts of chiral dipeptides	High degree ($k_L/k_D = 50$–90) of chiral recognition	Lehn and Sirlin, 1978
$\mathbf{7.6}_{DD}$, $\mathbf{7.6}_{LD}$, $\mathbf{7.6}_{DL}$, $\mathbf{7.6}_{LL}$, $\mathbf{7.8}_{D}$, $\mathbf{7.8}_{L}$	Energy transfer from tryptophan side arms to crown entrapped Tb^{3+}, energy transfer from crown entrapped Gly-L-Trp salts by pyrene side arms (see Section 7.1)	Chiral recognition of up to 0.5 kcal $mole^{-1}$	Tundo and Fendler, 1980
11.10	$\mathbf{11.10} + Mg^{2+} \rightleftharpoons \mathbf{11.10}\cdot Mg^{2+}$; $\xrightarrow{H^+,\ RC(=O)R^1}$ $RCH(OH)R^1 + Mg^{2+}$ + pyridinium salt of **11.10** ($\xrightarrow{Na_2S_2O_4}$ back to **11.10**)	Up to 86% asymmetric induction in high yields of products	de Vries and Kellogg, 1979
Chiral ammonium salts as phase transfer agents in benzene-water two-phase systems	$PhCOR \longrightarrow PhCH(OH)R$ R = Me, Et, *i*-Pr, *t*-Bu	Up to 32% asymmetric induction	Colonna and Fornasier, 1978
Chiral thiazolium salts in two-phase systems	$2ArCHO \longrightarrow ArC(H)(OH)—C(=O)—Ar$	Highest optical purity is observed at the lowest yield of benzoin	Tagaki et al., 1980

Selective complexation of cyclodextrins has been advantageously utilized for racemate resolutions (Cramer and Dietsche, 1959; Benschop and Van den Berg, 1970; Mikolajczyk and Drabowitcz, 1978).

Crown ethers carrying chirals side arms have been shown to be even more useful in recognizing optically active primary amine salts (Cram et al., 1975). Strategies for designing optimal host–guest interactions have been delineated in Section 7.1. Further details on the synthesis and characterization of crown ether host–guest compounds can be found in a series of elegant papers by Cram and his co-workers (Kyba et al., 1977, 1978a, 1978b; Timko et al., 1977; Newcomb et al., 1977a, 1977b, 1979; Moore et al., 1977; Helgelson et al., 1977; Cram et al., 1978). Hosts **11.1–11.7** were surveyed for their power of recognition of $LMSC^{*}NH_3^{+}X^{-}$,

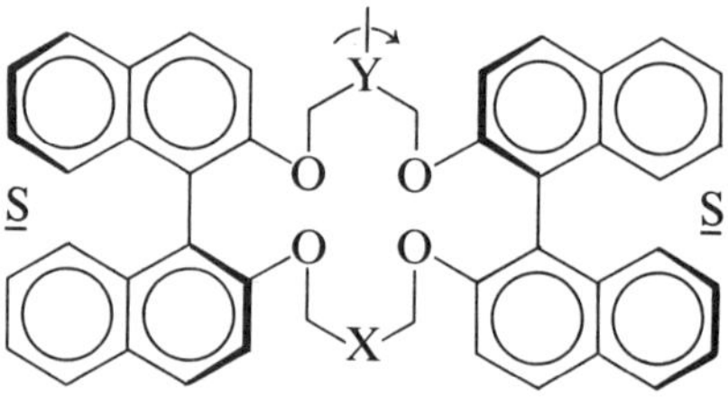

	X	Y
(SS)-**11.1**	CH_2OCH_2	CH_2OCH_2
(SS)-**11.2**	CH_2OCH_2	$CH_2CH_2CH_2$
(SS)-**11.3**	CH_2OCH_2	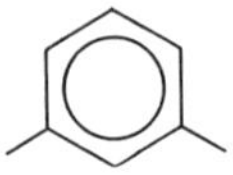
(SS)-**11.4**	CH_2OCH_2	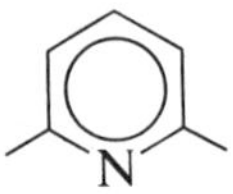 N
(SS)-**11.5**	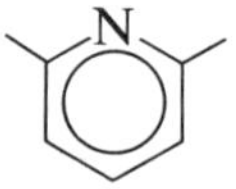 N	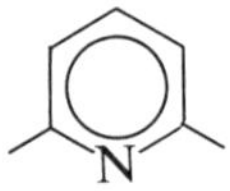 N
(SS)-**11.6**	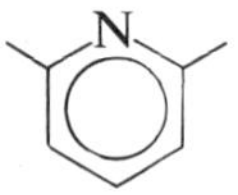 N	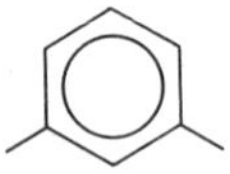
(SS)-**11.7**	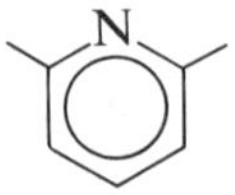 N	$CH_2CH_2CH_2$

where L, M, and S are large, medium, and small groups, respectively (Kyba et al., 1978b):

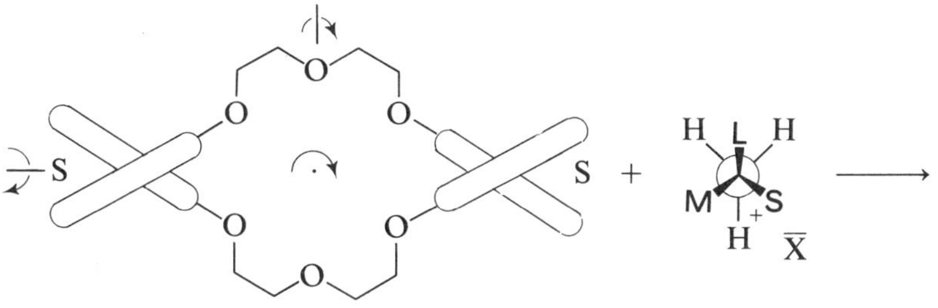

(SS)-**11.1**

(11.44)

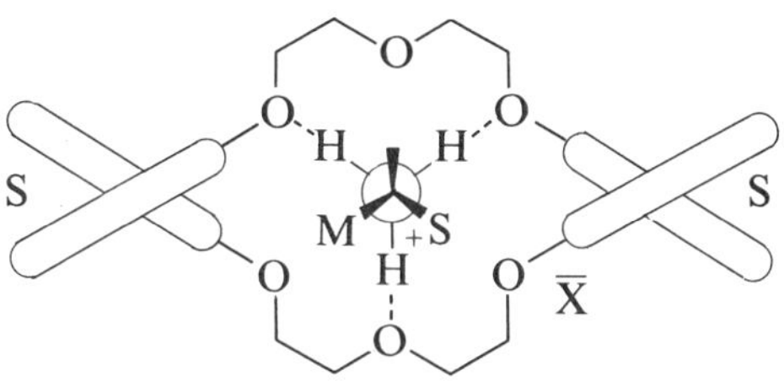

(SS)-**11.8**

The six oxygen atoms in (SS)-**11.8** define a plane. Perpendicular to this plane are the four naphthalene rings of the host and the —NC* bond axis of the guest. Two of the naphthalene rings are below the plane of the crown, while the other two, along with LMSC* portion of the guest, are above it. The symmetry of (SS)-**11.1** ensures the formation of the same complex regardless of whether the guest approaches from above or from below the plane of the host. Portions of the guests are entrapped in different chiral cavities. In the suggested arrangement, L was considered to be located in one of two identical chiral cavities, and M and S in the other. Discrimination was proposed to originate in the arrangements of M and S. In one diastereomer S was considered to align with one naphthalene wall, providing space for M with an orientation parallel to the opposite wall. The other diastereomer was presumed to have inverted geometry (Kyba et al., 1978b).

Several methods of optical resolutions are based on stereoselective guest complexation to **11.1** or to related crown ethers. Both liquid-liquid and liquid-solid chromatography have been used (Sousa et al., 1978). In the liquid-liquid chromatography racemic amine salts, absorbed on minimum amounts of celite, are eluted by optically pure **11.1** in $CHCl_3$ or CH_2Cl_2 on a celite column saturated with $NaPF_6$ or $LiPF_6$. Solid-liquid chromatography is carried out by immobilizing **11.1** on silica gel through Ar—$Si(CH_3)_2OSi$ bonds at the 6 position on the naphthalene ring. Amine salts, dissolved in $CHCl_3$ or CH_2Cl_2, and 18-crown-6 carrier, dissolved in isopropanol, were allowed to pass through a column containing immobilized **11.1**. Racemates of α-phenylethylamine, methyl phenylglycinate, methyl *p*-hydroxyphenylglycinate, methyl valinate, methyl phenylalanine, isopropyl phenylglycinate, methyl *p*-hydroxyglycinate, methyl valinate, methyl tyrosinate, and methyl tryptophanete have been resolved by these chromatographic techniques (Sousa et al., 1978).

Racemic amino acids have been resolved by the selective transport of amino acid salts in a "catalytic amino acid resolving machine" (Figure 11.3; Newcomb et al., 1979) using (RR)- and (SS)-dimethyl substituted **11.1**, **10.7**. The left-hand arm of the W-tube contained (SS)-**10.7** in $CHCl_3$ while (RR)-**10.7** was placed in the right-hand arm of the W-tube. Aqueous solutions of racemic amino acid salts were placed in central arms of the W-tube above the $CHCl_3$ phase. Equal volumes of 10^{-3} *M* HCl were introduced above the $CHCl_3$ phase in the left and right arms of the W-tube. Stirring facilitated the transport of the enantiomeric guest salts from the central aqueous compartment to those on either side of the W-tube. Resolution of racemic

$$C_6H_5CH(CH_3)NH_3Br, \; C_6H_5CH(CO_2CH_3)NH_3Cl,$$

and

$$p\text{-}HOC_6H_4(CO_2CH_3)NH_3Cl$$

resulted in enantiomers with 70–90% optical purities (Newcomb et al., 1979). Crown ethers containing carbohydrate residues (Stoddart, 1979, 1980) and cryptates (Lehn, 1978) also have the potential of being constituents of "amino acid resolving machines." Polymers containing chiral cavities have been utilized for resolving racemates (Wulff et al., 1977a, 1977b).

Rigid confinements of optically active dipeptides in chiral crown ether, **11.9**, led to dramatic enantiomeric differences in their reactivities (Lehn and Sirlin, 1978). The rate constant for the release of *p*-nitrophenol from Gly-L-

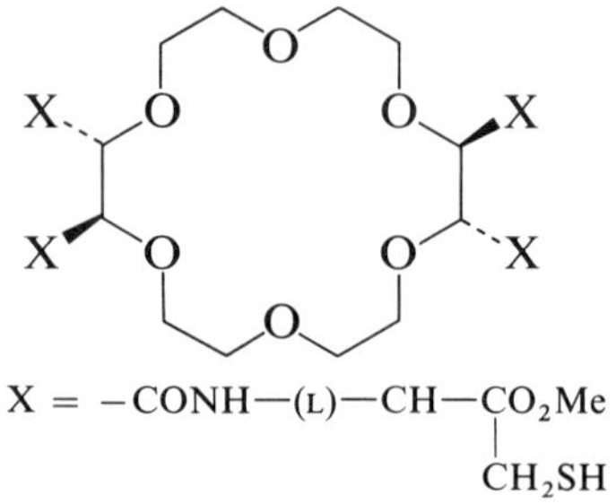

$X = -CONH-(L)-CH(CH_2SH)-CO_2Me$

11.9

Phe-$OC_6H_4NO_2p \cdot HBr$ was found to be fiftyfold greater than that from its enantiomer, Gly-D-Phe-$OC_6H_4NO_2p \cdot HBr$ in

$$CH_2Cl_2{:}MeOH{:}H_2O = 97.9{:}2.0{:}0.1.$$

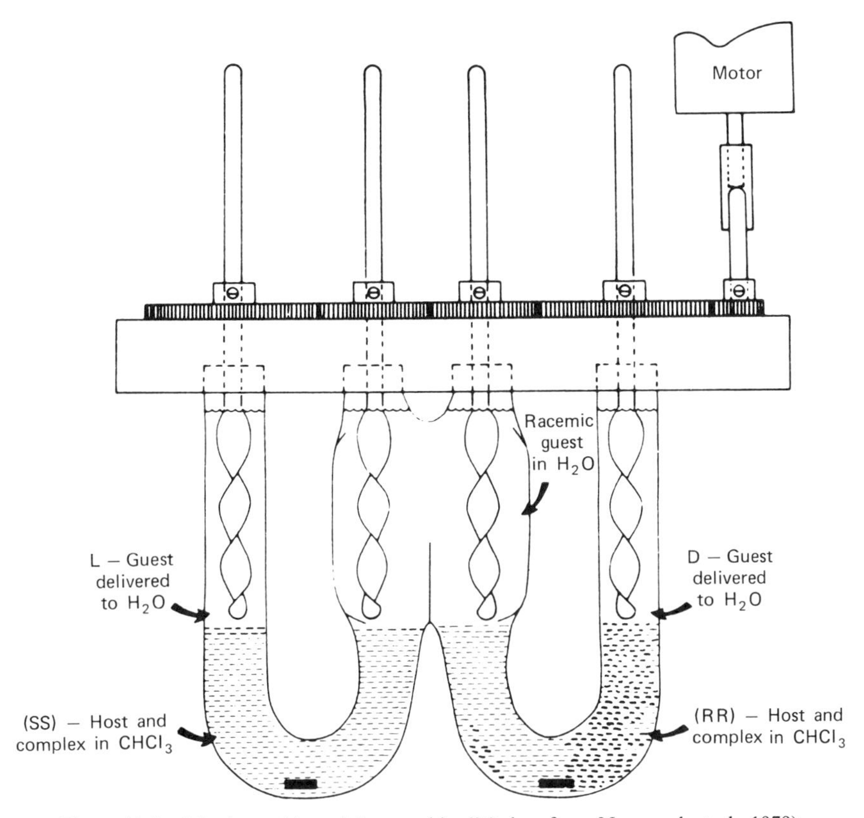

Figure 11.3 "Amino acid resolving machine" (taken from Newcomb et al., 1979).

An even greater discrimination (ninetyfold) was found in CH_2Cl_2:EtOH = 95:5 (Lehn and Sirlin, 1978). Substrate entrapment in **11.9** resulted in a substantially enhanced hydrolysis rate compared with that in the absence of the crown ether. Significantly, the SH side groups on the crown ether in **11.9** are essential for the catalyst. No catalysis was observed in the S-benzyl analog of **11.9**. These facts, as well as the observed saturation-type kinetics, render **11.9** a good enzyme model (Lehn and Sirlin, 1978).

The synthetic utility of functionalized crown ethers is beautifully demonstrated by the asymmetric reduction of ketones (de Vries and Kellogg, 1979; Table 11.2). Carbonyl group reduction was facilitated by binding to the magnesium complex of an optically active 1,4-dihydropyridine containing crown ether, **11.10**. $C_6H_5COCF_3$, $C_6H_5COCO_2C_2H_5$, $C_6H_5COCONH_2$, and $C_6H_5COCONHC_2H_5$ were reduced to the corresponding S-alcohols in 82, 80, 67, and 37% chemical and 68, 86, 64, and 78% optical yields. The formation of (S)-enantiomers was rationalized by assuming that the oxygen of the carbonyl carbon oriented toward the 4 position of the 1,4-dihydropyridine in **11.10**, that the phenyl group is the largest group, and that complexed α-dicarbonyl compounds assume cis conformations (de Vries and Kellogg, 1979). Since the

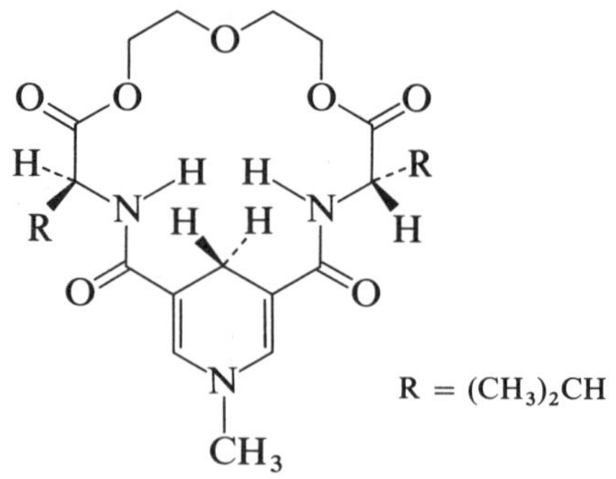

R = $(CH_3)_2CH$

11.10

pyridinium salt of **11.10** is reconverted by $Na_2S_2O_4$, **11.10** acts as a cyclic catalyst for asymmetric ketone reductions (Table 11.2).

Chiral functional groups attached to polymer backbones (Inove, 1976; Luisi, 1979) or acting as phase transfer agents (Colonna and Fornasier, 1978; Starks and Liotta, 1979) have also been utilized in asymmetric synthesis and in optical resolution.

Until recently, only meager stereoselectivities had been observed in aqueous micelles (Table 11.2). It appears that strong and specific substrate micelle interactions are necessary requirements for chiral discriminations. Placing chiral centers in hydrophobic regions leads to more pronounced discriminations than those elicited at or around the micellar headgroups (Sugimoto and Baba, 1979). Pronounced differences have been observed in the reactivities between diastereomeric dipeptides (Moss et al., 1979, 1980a, 1980b; see Table 11.2). These differences are the results of interactions between diastereomers that discriminate between like (DD with LL or DL with LD) and unlike (DD with LD or LL with DL) pairs. Diastereomeric and enantiomeric (interactions between two chiral molecules that discriminate between like, + with + or − with −, and unlike, − with + or + with −, pairs) recognitions have been treated theoretically in terms of short and long range nonbinding interactions (Craig and Mellor, 1976; Wynberg, 1976). Enantiomeric and diastereomeric discriminations are expected to be most prevalent in solids and in concentrated solutions. The function of micelles is to bring together pairs of interacting molecules. Accordingly, both binding and functionalization are expected to contribute to the observed discriminations. Chirality of substrates, rather than that of the surfactants, determine the magnitude of selectivity (Moss et al., 1979). Enantiomeric discrimination is also manifested in different CMC values (Miyagishi and Nishida, 1978) and monolayer packing (Arnett et al., 1978) of optically pure and racemic surfactants.

Highly structured micelles (Tachibana and Kurihara, 1976), crown ethers (Kaneko et al., 1979), and liquid crystals (Pollman et al., 1976; Stegemeyer et al., 1979; Seuron and Solladie, 1980) have been shown to induce chirality of achiral compounds.

4. ISOTOPE ENRICHMENT

A significant recent development is the utilization of micelles and other organized assemblies for the enrichment of magnetic isotopes (Turro and Kraeutler, 1980). The underlying principle is based on the radical pair model of chemically induced dynamic nuclear polarization (CIDNP) (Buchachenko and Zhidomirov, 1977; Ward, 1972; Lawler, 1972; Kaptein, 1975; Muus, 1977). According to this theory, the chemical reactivity of radical pairs is expected to depend on the hyperfine interactions of the orbitally uncoupled electrons of the radical pair with nuclear spins (magnetic isotope effect) or laboratory magnets (magnetic field effect). It should be possible, therefore, to separate magnetic isotopes from nonmagnetic ones (Atkins and Lambert, 1975; Buchachenko et al., 1976; Buchachenko, 1977; Sagdeev et al., 1977a, 1977b) in photochemical reactions involving suitable radical pairs. Criteria for establishing the existence of nuclear spin isotope effects have been delineated (Lawler, 1980). The function of micelles is to compartmentalize appropriate radicals and to alter their microenvironments and reactivities (Fendler, 1980).

Isotope enrichment in the photolysis of dibenzyl ketone in aqueous micelles has been investigated most extensively (Turro and Kraeutler, 1978; Turro et al., 1979, 1980b, 1980c, 1980d; Kraeutler and Turro, 1980a, 1980b). Photolysis of dibenzyl ketone (**11.11**) in aqueous CTAB micelles results in the formation of 1,2-diphenylethane and carbon monoxide in high quantum yields ($\Phi \sim 0.8$).

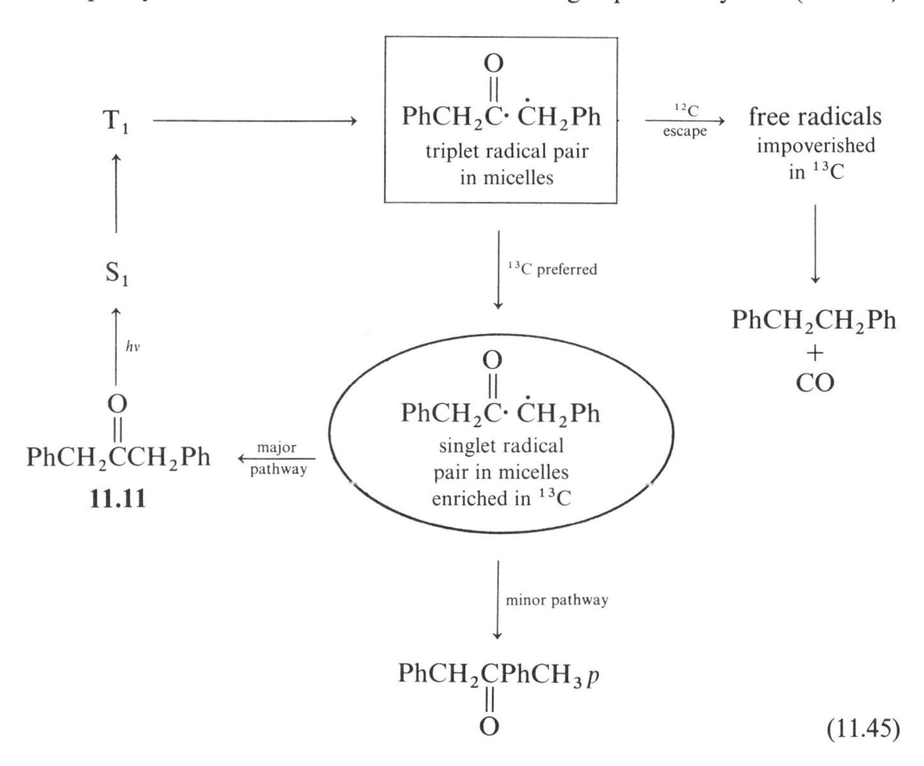

(11.45)

The mechanism involves the formation of a triplet radical pair within the micellar cage (equation 11.45). The triplet radical pair, containing ^{13}C nuclei, undergoes nuclear hyperfine coupling induced intersystem crossing to a singlet radical pair that, in turn, regenerates **11.11** or forms phenyl-*p*-acetophenone. Since the ^{12}C nuclei is nonmagnetic, it cannot undergo nuclear hyperfine coupling induced triplet radical pair to singlet radical pair conversion. Consequently, cage products are formed faster for ^{13}C than for ^{12}C nuclei with the resultant ^{13}C enrichment of **11.11**. The isotope enrichment factor, α, is expressed by

$$\alpha = \frac{\text{rate of disappearance of } ^{12}\text{C ketone}}{\text{rate of disappearance of } ^{13}\text{C ketone}} \equiv \frac{^{12}\Phi}{^{13}\Phi} \tag{11.46}$$

where $^{12}\Phi/^{13}\Phi$ is the ratio of quantum yields of disappearance of ^{12}C and ^{13}C in **11.11**. It is related to the separation factor S_f by

$$\log S_f = \left(\frac{\alpha - 1}{\alpha}\right) \log \left(\frac{1 - F_0^*}{1 - F_f^*}\right) \tag{11.47}$$

where F_0^* and F_f^* are the molar fractions of ^{13}C in the starting ketone before the reaction and after the fraction f of ketone has been converted, respectively, and

$$S_f = \frac{^{13}\text{C}_{\text{final}}/^{12}\text{C}_{\text{final}}}{^{13}\text{C}_{\text{initial}}/^{12}\text{C}_{\text{initial}}} \tag{11.48}$$

where $^{13}C_{final}/^{12}C_{final}$ is the ratio of concentrations of ^{13}C and ^{12}C in **11.11** after a fractional conversion of f, and $^{13}C_{initial}/^{12}C_{initial}$ is the corresponding ratio in the starting ketone. For small ^{13}C contents and small values of α equation 11.47 can be simplified to

$$\log S_f \approx -\left(\frac{\alpha - 1}{\alpha}\right) \log (1 - f) \tag{11.49}$$

$$\log S_f \approx \left(\frac{^{12}\Phi - {}^{13}\Phi}{^{12}\Phi}\right) \log (1 - f) \tag{11.50}$$

Figure 11.4 shows plots of the separation factor as a function of the fraction of reacted **11.11** (f) in benzene and in aqueous micellar CTACl solutions at two different magnetic field strengths (Turro and Kraeutler, 1978). The larger the value of α, the larger the ^{13}C enrichment remaining in **11.11**. The beneficial effect of micellar compartmentalization is clearly shown in Figure 11.5 (Kraeutler and Turro, 1980b). Increasing the surfactant concentration results in a sigmoidal increase of ^{13}C enrichment. Interestingly, variation of the average occupancy of the micelle from 1 to 10 molecules of **11.11** per aggregate did not appreciably affect the α values. The important function of the micelles is, therefore, to provide

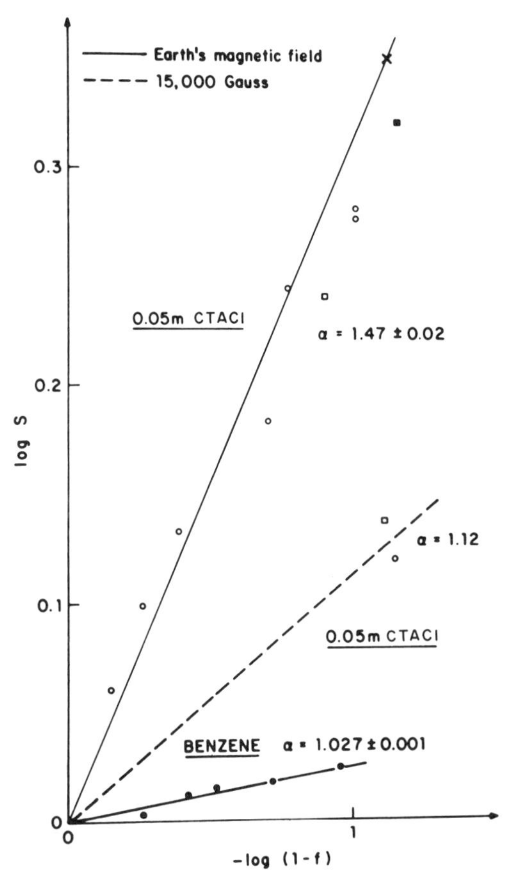

Figure 11.4 Plot of data for isotope enrichment in the photolysis of $PhCH_2COCH_2Ph$ according to equation 11.50 (taken from Turro and Kraeutler, 1978).

dynamic "super cages" that preserve the spin correlation of radical pairs in high local concentration for a time sufficient to boost the magnetic isotope effect.

The probability of radical recombination can be calculated from (Kraeutler and Turro, 1980b):

$$P \approx \varepsilon \int_0^\infty F(t)|C(t)|^2 \exp(-kt)\, dt \tag{11.51}$$

where ε is the probability that the cage singlet radical pair will recombine to a specific product; $F(t)|C(t)|^2 \exp(-kt)$ is the probability that a radical pair generated at time $t = 0$ re-encounters to give the cage singlet radical pair at time t; $|C(t)|^2$ describes the spin dynamics of the pair; $F(t)$ gives its diffusional motion; and $\exp(-kt)$ relates the pair disappearance due to a first order chemical reaction.

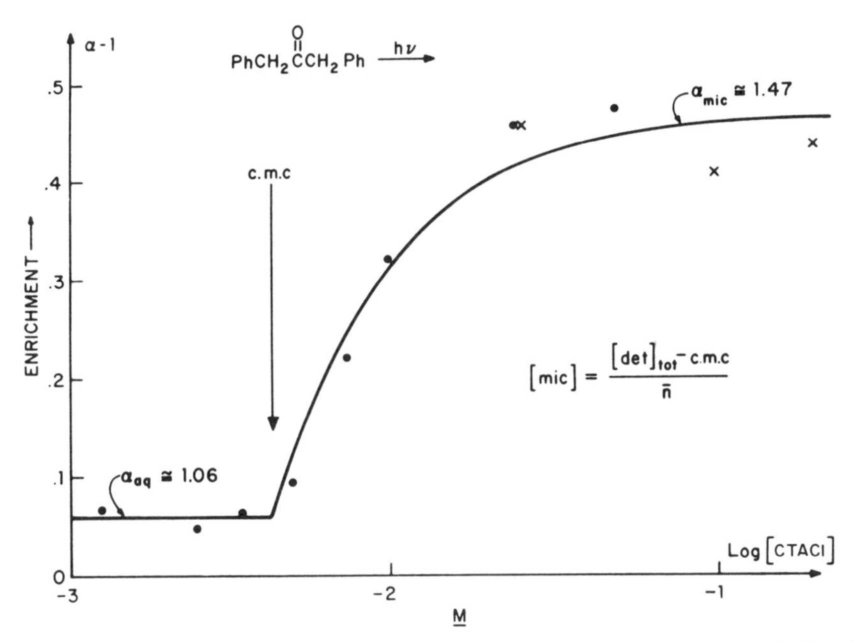

Figure 11.5 Effect of increasing surfactant concentration on the isotope enrichment obtained in the photolysis of $PhCH_2COCH_2Ph$ (taken from Kraeutler and Turro, 1980b).

The mechanism for isotope enrichment (equation 11.45) has been discussed in terms of the progress of **11.11** along different potential energy surfaces (Figure 11.6; Turro and Kraeutler, 1978). Breaking the OC—CH_2 bond corresponds to the energetically downhill movement of **11.11** on the electronically repulsive triplet surface. Hyperfine interaction of the magnetic moment of the ^{13}C with the odd electron of the radical generated is responsible for providing a "hole" on the triple energy surface, which leads to the regeneration of **11.11** along the singlet surface. In the absence of electron nuclear hyperfine coupling, nonmagnetic ^{12}C cannot "jump" to the singlet potential surface. ^{13}C impoverished **11.11** molecules will continue to move along their predetermined path to complete the bond breaking, yielding reactive $PhCH_2\dot{C}O$ and $\dot{C}H_2Ph$ radicals. The micellar supercage acts as a boundary that "reflects" "overshoot" **11.11** molecules to the "crucial geometries" required for intersystem crossing from the triplet to the singlet potential surface (Turro and Kraeutler, 1980). The observation of different CIDNP effects for 1-(*p*-tolyl)-3-phenylacetone in homogeneous and in micellar solutions (Hutton et al., 1979) substantiates this interpretation.

Under optimal conditions several thousand per unit enrichment in ^{13}C isotope of **11.11** is possible (Turro and Kraeutler, 1980). Similar isotopic enrichments have been reported in the photolysis of 1,2-diphenyl-2,2-dimethylpropanone (Turro and Mattay, 1980) in micellar solutions and in emulsion polymerization using **11.11** as an initiator (Turro et al., 1980e).

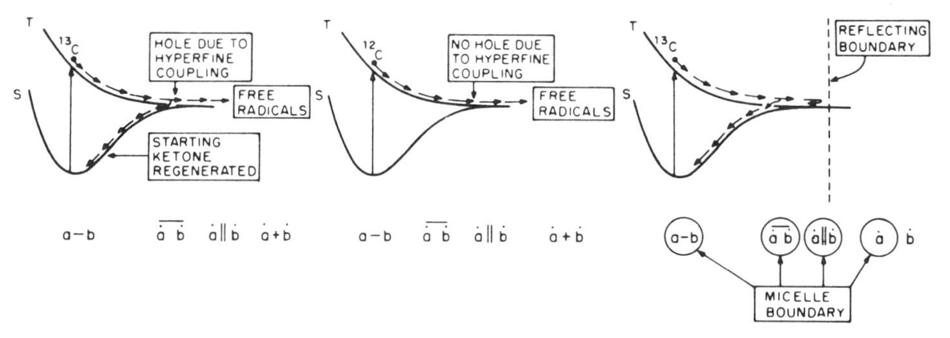

Figure 11.6 Proposed mechanism for isotope enrichment in the photolysis of ketones in micellar solutions (taken from Turro and Kraeutler, 1978).

5. SYNTHETIC APPLICATIONS

Reactions between water-insoluble substrates and water-soluble reactants present a challenging synthetic problem. Dipolar aprotic solvents provide good media for many reactions. They are quite expensive, however. Carrying out reactions in stirred aqueous nonaqueous apolar two-phase solvent systems in the presence of phase transfer agents (Weber and Gokel, 1977; Starks and Liotta, 1979; Dehmlow and Dehmlow, 1980) offers many advantages. The need for expensive organic solvents is obviated in phase transfer catalysis. Furthermore, it becomes possible to use less expensive nucleophiles, lower temperatures, and shorter reaction times. Phase transfer catalysis has been utilized in a large number of laboratory and industrial syntheses (Keller, 1979). Relatively little attention was placed, however, on elucidating mechanistic intricacies for phase transfer catalysis. According to the accepted dogma (Starks, 1971), an organic soluble quarternary ammonium or phosphonium cation, Q^+, transports the nucleophile, Y^-, into the organic phase, the site of reaction:

$$\begin{array}{llll} RX + QY & \longrightarrow & RY + QX & \text{(organic phase)} \\ QY\uparrow & \rightsquigarrow & \downarrow QX & \text{(interphase)} \\ NaX + QY & \rightleftharpoons & NaQ + QX & \text{(aqueous phase)} \end{array} \quad (11.52)$$

Subsequent to reaction, Q^+, along with the leaving group, is transported back to the aqueous phase. Most of the mechanistic studies have been carried out on nucleophilic substitutions at an alkyl carbon atom. Different mechanism(s) may well prevail for other reactions (Makosza, 1979). Furthermore, the importance of many factors, such as anion desolvation, catalyst aggregation both in water and in the organic phase, and interfacial interactions and reactions, have not been quantitatively elucidated.

The situation for micelles is quite different. Although the mechanism for micellar catalysis is well understood (see Section 11.1), there have been limited applications of micelles in preparative chemistry. The advantage of large rate enhancements is offset by the limited amounts of substrate solubilization, the restricted temperature range, and the difficulty of separating products from the surfactants.

Addition of catalytic amounts of cationic CTAB increased the yield of $KMnO_4$ mediated oxidation of piperonal to piperonylic acid from 37 to 65% (Menger et al., 1975). Yields of benzoic acid in the hydrolysis of α,α,α-trichlorotoluene increased even more appreciably in the presence of CTAB or Brij-35 surfactants. The catalysis may involve either emulsions or micelles (Menger et al., 1975).

Unique environments of micelles have been utilized advantageously in peptide syntheses. Thus while the base catalyzed condensation of micelle forming thioalanine S-dodecyl ester gave a good yield of 3,6-dimethyl-2,5-piperazinedione, only traces of cyclic dipeptides were formed from nonmicelle thioalanine S-ethyl ester (Kawabata and Kinoshita, 1975). Polycondensation of amino acyl adenylates in surfactant entrapped water pools in hydrocarbon solvents also has potential synthetic applicability (Armstrong et al., 1978). High molecular weight peptides, containing up to 41 amino acid units, were formed in 85% yield in reversed comicellar hexadecyltrimethylammonium bicarbonate di-2-ethylhexyl sodium sulfosuccinate in benzene at room temperature (see Table 11.1; Armstrong et al., 1978).

β-Cyclodextrin was reported to increase the yield of vitamin K_1 or K_2 analogs in a one-step preparation from 20% to approximately 60% (Tabushi et al., 1977).

Immobilization of surfactants (Brown and Jenkins, 1976) and phase transfer agents (Regen, 1975, 1976, 1977, 1979; Molinary et al., 1977; Tundo, 1977, 1978; Chiellini and Solaro, 1977; Komeili–Zadeh et al., 1978; Colonna et al., 1978; Regen and Besse, 1979; Tundo and Venturello, 1979) provides highly versatile and efficient catalysts. The principles of triphase catalysis (Regen, 1979) are illustrated by reaction of cyanide ion with 1-halooctane in benzene-water two-phase solvent systems using a crosslinked polystyrene resin bearing quaternary ammonium groups as the catalyst (Regen, 1975, 1976):

$$n\text{-}C_8H_{17}X(\text{org}) + NaCN(\text{aq}) \longrightarrow n\text{-}C_8H_{17}CN(\text{org}) + NaBr(\text{aq})$$

$$\text{(polystyrene)}\text{-}C_6H_4\text{-}CH_2\text{-}\overset{\oplus}{N}(CH_3)_2\text{-}n\text{-}C_4H_9\,X^- \qquad (11.53)$$

No reaction took place in the absence of polymer immobilized catalyst. Large catalytic surfaces and the possibility of continuous operation, as well as catalyst recovery, are the advantages of this method. Attractive synthetic procedures have been developed for nucleophilic substitution, dihalocarbene generation, oxidation and reduction (Regen, 1979). The mechanism of polymer immobilized phase transfer catalysis has been discussed in terms of the diffusion of reactants across the hydration shell of the catalyst, followed by reaction at the particle surface (Regen and Besse, 1979). More extensive synthetic exploitation of immobilized membrane mimetic agents is fully expected in the immediate future.

REFERENCES

Almgren, M. and Rydholm, R. (1979). *J. Phys. Chem.* **83**, 360–364. Influence of Counterion Binding on Micellar Reaction Rates. Reaction between *p*-Nitrophenyl Acetate and Hydroxide Ion in Aqueous Cetyltrimethylammonium Bromide.

Armstrong, D. W., Seguin, R., McNeal, C. J., Macfarlane, R. D., and Fendler, J. H. (1978). *J. Am. Chem. Soc.* **100**, 4605–4606. Spontaneous Polypeptide Formation from Amino Acyl Adenylates in Surfactant Aggregates.

Arnett, E. M. and Thompson, O. (1981). *J. Am. Chem. Soc.* **103**, 968–970. Chiral Aggregation Phenomena. 2. Evidence for Partial "Two-Dimensional Resolution" in a Chiral Monolayer.

Arnett, E. M., Chao, J., Kinzig, B., Stewart, M., and Thompson, O. (1978). *J. Am. Chem. Soc.* **100**, 5575–5576. Chiral Aggregation Phenomena. I. Acid Dependent Chiral Recognition in a Monolayer.

Atkins, P. A. and Lambert, T. P. (1975). *Annu. Rep. Chem. Soc. Sec. A* **72**, 67–88. The Effect of a Magnetic Field on Chemical Reactions.

Baba, N., Matsumura, Y., and Sugimoto, T. (1978). *Tetrahedron Lett.* **44**, 4281–4284. Asymmetric Reduction of Aryl Trifluoromethyl Ketones with an Achiral NADH Model Compound in a Chiral Hydrophobic Binding Site of Sodium Cholate Micelle, β-Cyclodextrin and Bovine Serum Albumin.

Benschop, H. P. and Van den Berg, G. R. (1970). *Chem. Commun.*, 1431–1432. Stereospecific Inclusion in Cycloamyloses: Partial Resolution of Isopropyl Methylphosphinate and Related Compounds.

Berezin, I. V., Martinek, K., and Yatsimirskii, A. K. (1973). *Russ. Chem. Rev. Engl. Transl.* **42**, 787–802. Physicochemical Foundations of Micellar Catalysis.

Bhalekar, A. A. and Engberts, J. B. F. N. (1978). *J. Am. Chem. Soc.* **100**, 5914–5920. Electron Transfer Reactions of Transition Metal Amino Carboxylates in the Presence of Micelle-forming Surfactants. Catalysis by Cetyl Trimethyl Bromide. Reduction of $Mn(cydta)^{-}$ by $Co(edta)^{2-}$ and $Co(cydta)^{2-}$.

Bonilha, J. B. S., Chaimovich, H., Toscano, V. G., and Quina, F. (1979). *J. Phys. Chem.* **83**, 2463–2470. Photophenomena in Surfactant Media. 2. Analysis of the Alkaline Photohydrolysis of 3,5-Dinitroanisole in Aqueous Micellar Solutions of *N*-tetradecyl-*N,N,N*-trimethylammonium Chloride.

Braun, A. M. (1981). Unpublished results, cited in N. J. Turro, M. Grätzel, and A. M. Braun, *Angew. Chem. Int. Ed. Eng.* **19**, 675–696 (1980).

Breslow, R. and Campbell, P. (1971). *Bioorg. Chem.* **1**, 140–150. Selective Aromatic Substitution by Hydrophobic Binding of a Substrate to a Simple Cyclodextrin Catalyst.

Breslow, R., Kohn, H., and Siegel, B. (1976). *Tetrahedron Lett.*, 1645–1646. Methylated Cyclodextrin and a Cyclodextrin Polymer as Catalysts in Selective Anisole Chlorination.

Breslow, R., Hammond, M., and Lauer, M. (1980a). *J. Am. Chem. Soc.* **102**, 421–422. Selective Transamination and Optical Induction by a β-Cyclodextrin-pyridoxamine Artificial Enzyme.

Breslow, R., Bovy, P., and Hersh, C. L. (1980b). *J. Am. Chem. Soc.* **102**, 2115–2117. Reversing the Selectivity of Cyclodextrin Bisimidazole Tibonuclease Mimics by Changing the Catalyst Geometry.

Brown, J. M. and Bunton, C. (1974). *J. Chem. Soc. Chem. Commun.*, 969–971. Stereoselective Micelle-Promoted Ester Hydrolysis.

Brown, J. M. and Jenkins, J. A. (1976). *J. Chem. Soc. Chem. Commun.*, 458–459. Micelle Related Heterogeneous Catalysis Anion-Activation by Polymer-Linked Cationic Surfactants.

Buchachenko, A. L. (1977). *Russ. J. Phys. Chem. Eng. Transl.* **51**, 1445–1451. Enrichment of Magnetic Isotopes—New Method of Investigation of Chemical Reaction Mechanisms.

Buchachenko, A. L. and Zhidomirov, F. M. (1971). *Russ. Chem. Rev. Eng. Transl.* **40**, 801–818. Chemically Induced Polarization of Electrons and Nuclei.

Buchachenko, A. L., Galimov, E. M., Ershov, V. V., Nikoforov, G. A., and Pershin, A. D. (1976). *Dok. Phys. Chem. Eng. Transl.* **228**, 451–453. Isotope Enrichment Induced by Magnetic Interactions in Chemical Reactions.

Bunnett, J. F. (1961). In *Technique of Organic Chemistry* (S. L. Friess, E. S. Lewis, and A. Weissberger, Eds.), Vol. 8, 2nd ed., Interscience, New York, pp. 177–283. The Interpretation of Rate Data.

Bunton, C. A. (1979a). In *Solution Chemistry of Surfactants* (K. L. Mittal, Ed.), Vol. 2, Plenum Press, New York, pp. 519–540. Micellar Catalysis and Inhibition Sources of Rate Enhancements in Functional and Nonfunctional Micelles.

Bunton, C. A. (1979b). *Catal. Rev. Sci. Eng.* **20**, 1–56. Reaction Kinetics in Aqueous Surfactant Solutions.

Bunton, C. A., Robinson, L., and Stam, M. F. (1971). *Tetrahedron Lett.*, 121–124. Stereospecific Micellar Catalyzed Ester Hydrolysis.

Bunton, C. A., Kamego, A. A., and Ng, P. (1974). *J. Org. Chem.* **39**, 3469–3471. Micellar Effects upon the Decomposition of 3-Bromo-3-phenylpropionic Acid. Effect of Changes in Surfactant Structure.

Bunton, C. A., Diaz, S., Romsted, S., and Valenzuela, O. (1976). *J. Org. Chem.* **41(18)**, 307–40. The Effect of Substrate Micellization on the Hydrolysis of *n*-Decyl Phosphate.

Bunton, C. A., Romsted, L. S., and Smith, J. H. (1978). *J. Org. Chem.* **43**, 4299–4303. Quantitative Treatement of Micellar Catalysts of Reactions Involving Hydrogen Ions.

Bunton, C. A., Romsted, L. S., and Savelli, G. (1979). *J. Am. Chem. Soc.* **101**, 1253–1259. Test of the Pseudophase Model of Micellar Catalysis: Its Partial Failure.

Bunton, C. A., Frankson, J., and Romsted. L. S. (1980a). *J. Phys. Chem.* **84**, 2607–2611. Reaction of *p*-Nitrophenyl Diphenyl Phosphate in Cetyltrimethylammonium Fluoride. Apparent Failure of the Pseudophase Model for Kinetics.

Bunton, C. A., Romsted, L. S., and Sepulveda, L. (1980b). *J. Phys. Chem.* **84**, 2611–2618. A Quantitative Treatment of Micellar Effects upon Deprotonation Equilibria.

Bunton, C. A., Romsted, L. S., and Thamavit, C. (1980c). *J. Am. Chem. Soc.* **102**, 3900–3903. The Pseudophase Model of Micellar Catalysis. Addition of Cyanide Ion to *N*-Alkylpyridinium Ions.

Bunton, C. A., Hamed, F., and Romsted, L. S. (1980d). *Tetrahedron Lett.* **21**, 1217–1218. Predicted Rate Enhancements of Dephosphorylation in Functional Oximate Comicelles.

Bunton, C. A., Hong, Y.-S., Romsted, L. S. (1982). In *Solution Behavior of Surfactants, Theoretical and Applied Aspects* (E. J. Fendler and K. L. Mittal, Eds.), Plenum Press, New York, in press.

A Quantitative Treatment of the Deprotonation Equilibria of Benzimidazole in Basic Solutions of Cetyltrimethylammonium Ion (CTAX) Surfactants.

Chaimovich, H., Bonilha, J. B. S., Politi, M. J., and Quina, F. H. (1979). *J. Phys. Chem.* **83**, 1851–1854. Ion Exchange in Micellar Solutions. 2. Binding of Hydroxide to Positive Micelles.

Chaimovich, H., Aleixo, R. M., Cuccovia, I. M., Zanette, D., and Quina, F. (1982). In *Solution Behavior of Surfactants, Theoretical and Applied Aspects* (E. J. Fendler and K. L. Mittal, Eds.), Plenum Press, New York, in press. The Quantitative Analysis of Micellar Effects on Chemical Reactivity and Equilibria: An Evolutionary Overview.

Chatterjie, N. and Brockerhoff, H. (1978). *Biochim. Biophys. Acta* **511**, 116–119. Evidence for Stereospecific Phospholipid–Cholesterol Interaction in Lipid Bilayers.

Chiellini, E. and Solaro, R. (1977). *J. Chem. Soc. Chem. Commun.*, 231–232. Stereo-Ordered Macromolecular Matrixes Bearing Ammonium Groups as Catalysts in Alkylation and Carbenation Reactions.

Colonna, S. and Fornasier, R. (1978). *J. Chem. Soc. Perkin Trans. I*, 371–373. Asymmetric Induction in the Borohydride Reduction of Carbonyl Compounds by Means of Chiral Phase Transfer Catalysts 2.

Colonna, S., Fornasier, R., and Pfeiffer, V. (1978). *J. Chem. Soc. Perkin Trans. I*, 8–12. Asymmetric Induction in the Darzens Reaction by Means of Chiral Phase Transfer in a Two Phase System. The Effect of Binding the Catalyst to a Solid Polymeric Support.

Cooper, A. and MacNicol, D. D. (1978). *J. Chem. Soc. Perkin Trans. II*, 760–763. Chiral Host–Guest Complexes: Interaction of α-Cyclodextrin with Optically Active Benzene Derivatives.

Craig, D. P. and Mellor, D. P. (1976). *Fortschr. Chem. Forsch.* **63**, 1–48. Discriminating Interactions between Chiral Molecules.

Cram, D. J., Helgelson, R. C., Sousa, L. R., Timko, J. M., Newcomb, M., Moreau, P., de Jong, F., Gokel, G. W., Hoffman, D. H., Domeier, L. A., Peacock, S. C., Madan, K., and Kaplan, L. (1975). *Pure Appl. Chem.* **43**, 327–349. Chiral Recognition in Complexation of Guest by Designed Host Molecules.

Cram, D. J., Helgelson, R. C., Koga, K., Kyba, E. P., Madan, K., Sousa, L. R., Siegel, M. G., Moreau, P., and Gokel, G. W. (1978). *J. Org. Chem.* **43**, 3758–3772. Host–Guest Complexation, 9. Macrocyclic Polyethers and Sulfides Shaped by One Rigid Dinaphthyl Unit and Attached Arms. Synthesis and Survey of Complexity Abilities.

Cramer, F., and Dietsche, W. (1959). *Chem. Ber.* **92**, 378–384. Über Einschlussverbindungen XV. Spaltung von Racematen mit Cyclodextrinen.

Daffe, V. and Fastrez, J. (1980). *J. Am. Chem. Soc.* **102**, 3601–3605. Enantiomeric Enrichment in the Hydrolysis of Oxazolones Catalyzed by Cyclodextrins on Proteolytic Enzymes.

Dehmlow, E. V. and Dehmlow, S. S. (1980). *Phase Transfer Catalysis*, Verlag Chemie, New York.

de Mayo, P. and Sydnes, L. K. (1980). *J. Chem. Soc. Chem. Commun.*, 994–995. Biphasic Photochemistry: Micelles Control of Regioselectivity in Enone in Photoannelations.

de Vries, J. G. and Kellog, R. M. (1979). *J. Am. Chem. Soc.* **101**, 2759–2761. Asymmetric Reductions with a Chiral 1,4-Dihydropyridine Crown Ether.

Doiuchi, T. and Minoura, Y. (1976/77). *Isr. J. Chem.* **15**, 84–88. Asymmetric Reduction of Ketones with $NaBH_4$.

Dung, M. H., Knox, D. G., and Kozak, J. J. (1980). *Ber. Bunsenges. Phys. Chem.* **84**, 789–795. Electrostatic Focusing in Micellar and Microelectrode Kinetic Processes.

Elworthy, P. H., Florence, A. T., and Macfarlane, C. B. (1968). *Solubilization of Surface Active Agents*, Chapman & Hall, London.

Escabi-Perez, J. R., Nome, F., and Fendler, J. H. (1977). *J. Am. Chem. Soc.* **99**, 7749–7754. Energy Transfer in Micellar Systems. Steady State and Time Resolved Luminescence of Aqueous Micelle Solubilized Naphthalene and Terbium Chloride.

Fendler, J. H. (1980). *J. Phys. Chem.* **84**, 1485–1491. Microemulsions, Micelles, and Vesicles as Media for Membrane Mimetic Photochemistry.

Fendler, J. H. and Fendler, E. J. (1975). *Catalysis in Micellar and Macromolecular Systems*, Academic Press, New York.

Fendler, J. H. and Hinze, W. L. (1981). *J. Am. Chem. Soc.* **103**, 5439–5447. Reactivity Control in Micelles and Surfactant Vesicles. Kinetics and Mechanism of Base Catalyzed Hydrolysis of 5,5′-Dithiobis(2-Nitrobenzoic Acid) in Water, in Hexadecyltrimethylammonium Bromide Micelles, and in Dioctadecyldimethylammonium Chloride Surfactant Vesicles.

Fendler, J. H., Fendler, E. J., and Smith, L. W. (1972). *J. Chem. Soc. Perkin Trans. II*, 2097–2104. Micellar Effects on the Hydrolysis and Aminolysis of 2,4-Dinitrophenyl Sulphate.

Fornasier, R. and Tonellato, V. (1980). *J. Chem. Soc. Faraday Trans. I* **76**, 1301–1310. Functional Micellar Catalysis. Part 3—Quantitative Analysis of the Catalytic Effects due to Functional Micelles and Comicelles.

Frost, A. A. and Pearson, R. G. (1962). *Kinetics and Mechanisms*, John Wiley, New York.

Funasaki, N. (1979). *J. Phys. Chem.* **83**, 1998–2003. Micellar Effects on the Kinetics and Equilibrium of Chemical Reactions in Salt Solutions.

Goldberg, S. I., Baba, N., Green, R. L., Pandian, R., Stowers, J., and Dunlap, R. B. (1978). *J. Am. Chem. Soc.* **100**, 6768–6779. Micelle-Enzyme Analogy: Stereochemical and Substrate Selectivity.

Hatlee, M. D. and Kozak, J. J. (1980). *J. Chem. Phys.* **72**, 4358–4367. A Stochastic Approach to the Theory of Intramicellar Kinetics. I. Master Equation for Irreversible Reactions.

Hatlee, M. D. and Kozak, J. J. (1981a). *J. Chem. Phys.* **74**, 1098–1109. A Stochastic Approach to the Theory of Intramicellar Kinetics. II. Master Equation for Reversible Reactions.

Hatlee, M. D. and Kozak, J. J. (1981b). *J. Chem. Phys.* **74**, 5627–5635. A Stochastic Approach to the Theory of Intramicellar Kinetics. III. The Homogeneous System Limit.

Helgelson, R. C., Tarnowski, T. L., Timko, J. M., and Cram, D. J. (1977). *J. Am. Chem. Soc.* **99**, 6411–6418. Host-Guest Complexation 6. The [2.2]Paracyclophanyl Structural Unit in Host Compounds.

Hutton, R. S., Roth, H. D., Kraeutler, B., Cherry, W. R., and Turro, N. J. (1979). *J. Am. Chem. Soc.* **101**, 2227–2228. Chemically Induced Dynamic Nuclear Polarization from the Selective Recombination of Radical Pairs in Micelles.

Ihara, Y. (1978). *J. Chem. Soc. Chem. Commun.*, 984–985. Stereoselective Reaction of *p*-Nitrophenyl *N*-Acyl-phenylalanines with *N*-Acyl-L-histidine in Mixed Micelles.

Ihara, Y. (1980). *J. Chem. Soc. Perkin Trans. II*, 1483–1487. Stereoselective Micellar Catalysis. Reactions of Amino-Acid Ester Derivatives with *N*-Acyl-L-histidine in Micelles.

Ihara, H., Ono, S., Shosenji, H., and Yamada, K. (1980). *J. Org. Chem.* **45**, 1623–1625. Enantioselective Ester Hydrolysis by Hydroxamic Acids of *N*-Benzylcarboxyl-L-amino Acids on Optically Active Amines in Cetyltrimethylammonium Bromide Micelles.

Ihara, Y., Nango, M., and Kuroki, N. (1980b). *J. Org. Chem.* **45**, 5009–5011. Stereoselective Micellar Bifunctional Catalysis.

Inove, S. (1976). *Adv. Polym. Sci.* **21**, 77–106. Asymmetric Reactions of Synthetic Polypeptides.

Jaeger, D. A. and Robertson, R. E. (1977). *J. Org. Chem.* **42**, 3298–3301. Micellar Effects on the Monohalogenation of *n*-Pentyl Phenyl Ether.

Kaneko, O., Matsuura, N., Kimura, K., and Shono, T. (1979). *Chem. Lett.*, 369–372. Induced Circular Dichroism of Substituted Benzophenones by Complexation with Chiral Crown Ethers.

Kaptein, R. (1975). *Adv. Free-Radical Chem.* **5**, 319–380. Chemically Induced Dynamic Nuclear Polarization: Theory and Applications in Mechanistic Chemistry.

Kawabata, Y. and Kinoshita, M. (1975). *Makromol. Chem.* **176**, 49–56. Functional Micelles. 2. Micellar Effect on the Preparation of Cyclic Dipeptide from Thioanaine S-Dodecyl Ester.

Keller, W. E. (1979). "Compendium of Phase-Transfer Reactions and Related Synthetic Methods, Comprehensive Register of Syntheses by Phase Transfer Catalysis, Ion-Pair Extraction and Reactions of Quaternary Ammonium Salts in Homogeneous Phase of Nonaqueous Solvents," Fluka AG, CH-9470, Switzerland.

Komeili-Zadeh, H., Dou, H. J. M., and Metzger, D. J. (1978). *J. Org. Chem.* **43**, 156–159. Triphase Catalysis. C-Alkylation of Nitriles.

Kon-no, K., Toska, M., and Kitahara, A. (1981). *J. Colloid Interface Sci.* **79**, 581–583. Enantioselectivity of Ester Aminolysis by Optically Active Surfactants in Hexane.

Kraeutler, B. and Turro, N. J. (1980a). *Chem. Phys. Lett.* **70**, 266–269. Photolysis of Dibenzyl Ketone in Micellar Solution. Correlation of Isotopic Enrichment Factors with Photochemical Efficiency Parameters.

Kraeutler, B. and Turro, N. J. (1980b). *Chem. Phys. Lett.* **70**, 270–275. Probes for Micellar Cage Effect. The Magnetic ^{13}C Isotope Effect and a New Cage Product in the Photolysis of Dibenzyl Ketone.

Kyba, E. P., Helgelson, R. C., Madan, K., Gokel, G. W., Tarnowski, S., Moore, S., and Cram, D. J. (1977). *J. Am. Chem. Soc.* **99**, 2564–2571. Host–Guest Complexation 1. Concept and Illustration.

Kyba, E. P., Gokel, G. W., de Jong, F., Koga, K., Sousa, L. R., Siegel, M. G., Kaplan, L., Sogah, G. D. Y., and Cram, D. J. (1978a). *J. Org. Chem.* **42**, 4173–4184. Host–Guest Complexation. 7. The Binaphthyl Structural Unit in Hosts Compounds.

Kyba, E. P., Timko, J. M., Kaplan, L. J., de Jong, F., Gokel, G. W., and Cram. D. J. (1978b). *J. Am. Chem. Soc.* **100**, 4555–4568. Host–Guest Complexation. 11. Survey of Chiral Recognition of Amino Ester Salts by Dilocular Bisdinaphthyl Hosts.

Lapinte, C. and Viout, P. (1979). *Tetrahedron* **35**, 1931–1944. Catalyse Micellaire—I. Etude de la Réactivite en Milieu Micellaire A L'Aide de Réactions Compétitives.

Lawler, R. G. (1972). *Acc. Chem. Res.* **5**, 25–33. Chemically Induced Dynamic Nucelar Polarization (CIDNP) II. The Radical Pair Model.

Lawler, R. (1980). *J. Am. Chem. Soc.* **102**, 430–431. Criteria for Establishing the Existence of Nuclear Spin Isotope Effects.

Lee, K.-H. and de Mayo, P. (1979). *J. Chem. Soc. Chem. Commun.*, 493–495. Biphasic Photochemistry: Micellar Regioselectivity in Enone Dimerisation.

Lee, K.-H. and de Mayo, P. (1980). *Photochem. Photobiol.* **31**, 311–314. Biphasic Photochemistry: Photochemical Regiospecificity and Critical Micelle Concentration Determination.

Lehn, J.-M. (1978). *Acc. Chem. Res.* **11**, 49–57. Cryptates: The Chemistry of Macropolycyclic Inclusion Complexes.

Lehn, J.-M. and Sirlin, C. (1978). *J. Chem. Soc. Chem. Commun.*, 949–951. Molecular Catalysis: Enhanced Rates of Thiolysis with High Structural and Chiral Recognition in Complexes of a Reactive Macrocyclic Receptor Molecule.

Lim, Y. Y. and Fendler, J. H. (1978). *J. Am. Chem. Soc.* **100**, 7490–7494. Spin Probes in Reversed Micelles. Electron Paramagnetic Resonance Spectra of 2,2,5,5,-Tetramethylpyrrolidine-1-oxyl Derivatives in Benzene in the Presence of Dodecylammonium Propionate Aggregates.

Lim, Y. Y. and Fendler, J. H. (1979). *J. Am. Chem. Soc.* **101**, 4023–4029. Characterization of Surfactant Vesicles as Membrane Mimetic Agents. 1. Interactions and Reactions of Sodium 2,2,5,5-Tetramethyl-1-pyrrolidinyloxy-3-carboxylate Spin Probe in Sonicated Dioctadecyldimethylammonium Chloride.

Lingenfelter, D. S., Helgelson, R. C., and Cram, D. J. (1981). *J. Org. Chem.* **46**, 393–406. Host–Guest Complexation. 23. High Chiral Recognition of Amino Acid and Ester Guests by Hosts Containing One Chiral Element.

Link, C. M., Jansen, D. K., and Sukenik, C. N. (1980). *J. Am. Chem. Soc.* **102**, 7798–7799. Hydroxymercuration of Nonconjugated Dienes in Aqueous Micelles.

Luisi, P. L. (1979). In *Optically Active Polymers* (E. Sélégny, Ed.), D. Reidel Co., Amsterdam, pp. 357–401. Synthetic Optically Active Polymers as Catalysts for Asymmetric Synthesis.

Makosza, M. (1979). *Surv. Prog. Chem.* **9**, 1–50. Two Phase Reactions in Organic Chemistry.

Manning, G. S. (1979). *Acc. Chem. Res.* **12**, 443–449. Counterion Binding in Polyelectrolyte Theory.

Martinek, K., Yatsimirski, A. K., Levashov, A. V., and Berezin, I. V. (1977). In *Micellization, Solubilization, and Microemulsions* (K. L. Mittal, Ed.), Plenum Press, New York, pp. 489–508. The Kinetic Theory and the Mechanisms of Micellar Effects on Chemical Reactions.

Menger, F. M. and Portnoy, C. E. (1967). *J. Am. Chem. Soc.* **89**, 4698–4703. On the Chemistry of Reactions Proceeding Inside Molecular Aggregates.

Menger, F. M., Rhee, J. V., and Rhee, H. K. (1975). *J. Org. Chem.* **40**, 3803–3805. Application of Surfactants to Synthetic Organic Chemistry.

Michon, J. and Rassat, A. (1979). *J. Am. Chem. Soc.* **101**, 995–996. Nitroxides, 73. Electron Spin Resonance Study of Chiral Recognition by Cyclodextrin.

Mikolajczyk, M. and Drabowicz, J. (1978). *J. Am. Chem. Soc.* **100**, 2510–2515. Optical Resolution of Chiral Sulfinyl Compounds with β-Cyclodextrin Inclusion Complexes.

Mita, K., Kunugi, S., Okudo, T., and Ise, N. (1975). *J. Chem. Soc. Faraday Trans. I* **71**, 936–945. Theoretical Studies of Polyelectrolyte "Catalysis" of Ionic Reactions, Part 1. Interionic Reactions Between Oppositely Charged Species.

Mita, K., Okubo, T., and Ise, N. (1976). *J. Chem. Soc. Faraday Trans. I* **72**, 1033–1042. Theoretical Studies of Polyelectrolyte "Catalysis" of Ionic Reactions, Part 2. Interionic Reactions Between Similarly Charged Species.

Mitani, M., Tsuchida, T., and Koyama, K. (1974). *J. Chem. Soc. Chem. Commun.*, 869. Perturbation of Rearrangement Causes by Cyclohepta-amylose.

Mitani, M., Suzuki, T., Takeuchi, H., and Koyama, K. (1979). *Tetrahedron Lett.*, 803–804. Regioselective Functionalization of Unactivated Methylene Groups Induced by Photochemistry in Micellar Solutions.

Miyagashi, S. and Nishida, M. (1978). *J. Colloid Interface Sci.* **65**, 380–386. Influence of Chirality on Micelle Formation of Sodium *N*-Acylalanates and Sodium *N*-Lauroylvalinates.

Molinari, H., Montanary, F., and Tundo, P. (1977). *J. Chem. Soc. Chem. Commun.*, 639–641. Heterogeneous Phase-Transfer Catalysis: High Efficiency of Catalysts Bonded by a Long Chain to a Polymer Matrix.

Moore, S. S., Tarnowski, T. L., Newcomb, M., and Cram, D. J. (1977). *J. Am. Chem. Soc.* **99**, 6398–6405. Host–Guest Complexation 4. Remote Substituent Effects on Macrocyclic Polyether Binding to Metal and Ammonium Ions.

Moroi, Y. (1980). *J. Phys. Chem.* **84**, 2186–90. Distribution of Solubilizates among Micelles and Kinetics of Micelle-Catalyzed Reactions.

Moss, R. A. and Sunshine, W. L. (1974). *J. Org. Chem.* **39**, 1083–1089. Micellar Catalysts of Ester Hydrolysis. The Influence of Chirality and Head Group Structure in "Simple" Surfactants.

Moss, R. A., Talkowski, C. J., Reger, D. W., and Powell, C. E. (1973). *J. Am. Chem. Soc.* **95**, 5215–5224. Micellar Control of Stereochemistry and Kinetics in the Nitrous Acid Deamination Reaction.

Moss, R. A., Lee, Y., and Lukas, T. J. (1979). *J. Am. Chem. Soc.* **101**, 2499–2301. Micellar Stereoselectivity. Cleavage of Diastereomeric Substrates by Functional Surfactant Micelles.

Moss, R. A., Lee, Y.-S., and Alvis, K. W. (1980a). *J. Am. Chem. Soc.* **102**, 6646–6648. Excpetional Micellar Stereoselectivity in Esterolysis Reactions: The Micelle–Enzyme Analogy.

Moss, R. A., Lee, Y.-S., and Alwis, K. W. (1980b). *Tetrahedron Lett.* **22**, 283–286. Micellar Diastereoselectivity—Tripeptide Substrates.

Murakami, Y., Nakano, A., Yoshimatsu, A., and Fukuya, K. (1981). *J. Am. Chem. Soc.* **103**, 728–730. Functionalized Vesicular Assembly. Enantioselective Catalysis of Ester Hydrolysis.

Muus, L. T. (1977). *Chemically Induced Magnetic Polarization*, D. Reidel Co., Dordrecht, The Netherlands.

Nakamura, Y., Imakura, Y., Kato, T., and Morita, Y. (1977). *J. Chem. Soc. Chem. Commun.*, 887–888. Reactions Using Micellar System: Photochemical Dimerization of Acenaphthylene.

Nakamura, Y., Imakura, Y., and Morita, Y. (1978). *Chem. Lett.*, 965–968. Reactions Using Micellar System: Photocycloadditions of Acenaphthylene with Acrylonitrile and Methyl Acrylate.

Nakamura, Y., Kato, T., and Morita, Y. (1981). *Tetrahedron Lett.* **22**, 1025–1028. A Micellar Alignment Effect in the Photodimerization of *N*-ω-carboxyalkyl-1-pyridones and their 4-Alkyl Derivatives in Micelle or Reversed Micelle.

Newcomb, M., Timko, J. M., Walba, D. M., and Cram, D. J. (1977a). *J. Am. Chem. Soc.* **99**, 6392–6398. Host–Guest Complexation 3. Organization of Pyridyl Binding Sites.

Newcomb, M., Moore, S. S., and Cram, D. J. (1977b). *J. Am. Chem. Soc.* **99**, 6405–6410. Host–Guest Complexation 5. Convergent Functional Groups in Macrocyclic Polyethers.

Newcomb, M., Toner, J. L., Helgelson, R. C., and Cram, D. J. (1979). *J. Am. Chem. Soc.* **101**, 4941–4947. Host–Guest Complexation 20. Chiral Recognition in Transport as a Molecular Basis for a Catalytic Resolving Machine.

O'Connor, C. J., Fendler, E. J., and Fendler, J. H. (1973). *J. Org. Chem.* **38**, 3371–3375. Hydrolysis of 2,4-Dinitrophenyl Sulfate in Benzene in the Presence of Alkylammonium Carboxylate Surfactants.

O'Connor, C. J., Lomax, T. D., and Ramage, R. E. (1981). In *Solution Behavior of Surfactants, Theoretical and Applied Aspects* (E. J. Fendler and K. L. Mittal, Eds.), Plenum Press, New York, in press. Kinetic Concepts in Reversed Micellar Systems.

Ohara, M. and Fukuda, J. (1978). *Pharmazie* **33**, 467. Selective Reimer–Tiemann Formylation of Phenol via Cyclodextrin Inclusion Complex.

Ohara, M. and Watanabe, K. (1975). *Angew. Chem. Int. Ed. Eng.* **14**, 820. Selective Photochemical Tries Rearrangement of Phenyl Acetate in the Presence of β-Cyclodextrin.

Ohkubo, K., Sugahara, K., Yoshinaga, K., and Veoka, R. (1980). *J. Chem. Soc. Chem. Commun.*, 637–639. Enantioselective Deacylation of Long Chain *p*-Nitrophenyl *N*-Acylphenylalanates by *N*-(*N*-Dodecanoyl-L-histidyl)-L-leucine and a Cationic Surfactant.

Okamoto, K., Kinoshita, T., and Yoneda, H. (1975). *J. Chem. Soc. Chem. Commun.*, 922–923. Micellar Control of the Rate and Stereochemical Course of the Hydrolysis of 1-Methylheptyl Trifluoromethanesulphonate.

Peacock, S. S., Walba, D. M., Gaeta, F. C. A., Helgeson, R. C., and Cram, D. J. (1980). *J. Am. Chem. Soc.* **102**, 2043–2052. Host–Guest Complexation. 22. Reciprocal Chiral Recognition Between Amino Acids and Dilocular Systems.

Pollman, P., Mainusch, K. J., and Stegemeyer, H. (1976). *Z. Physik. Chem. Neue Folge* **103**, 295–309. Circularpolarisation der Fluoreszent achiraler Moleküle in cholesterischen Flüssigkristallphasen.

Quina, F. H. and Chaimovich, H. (1979). *J. Phys. Chem.* **83**, 1844–1850. Ion Exchange in Micellar Solutions. I. Conceptual Framework for Ion Exchange in Micellar Solutions.

Quina, F. H., Politi, M. J., Cuccovia, I. M., Baumgarten, E., Martins-Franchetti, S. M., and Chaimovich, H. (1980). *J. Phys. Chem.* **84**, 361–365. Ion Exchange in Micellar Solutions. 4. "Buffered" Systems.

Regen, S. L. (1975). *J. Am. Chem. Soc.* **97**, 5956–5957. Triphase Catalysis.

Regen, S. L. (1976). *J. Am. Chem. Soc.* **98**, 6270–6274. Triphase Catalysis. Kinetics of Cyanide Displacement on 1-Bromooctane.

Regen, S. L. (1977). *J. Am. Chem. Soc.* **99**, 3838–3840. Microenvironment Within a Solid Phase Cosolvent.

Regen, S. L. (1979). *Angew. Chem. Int. Ed. Eng.* **18**, 421–429. Triphase Catalysis.

Regen, S. L. and Besse, J. J. (1979). *J. Am. Chem. Soc.* **101**, 4059–4063. Liquid-Solid-Liquid Triphase Catalysis. Consideration of the Rate-Limiting Step, Role of Stirring, and Catalyst Efficiency for Simple Nucleophilic Displacement.

Reger, D. L. and Habib, M. M. (1978). *J. Mol. Catal.* **4**, 315–324. Effects of Micelles on the $K_3[Co(CN)_5H]$ Catalyzed Hydrogenation of 2-Methylbutadiene, 2,3-Dimethylbutadiene and Trans-1,3-Pentadiene.

Romsted, L. S. (1975). Ph.D. Thesis, Indiana University, Bloomington, Indiana. Kinetic Treatment of Micellar Catalysis.

Romsted, L. S. (1977). In *Micellization, Solubilization, and Microemulsions* (K. L. Mittal, Ed.), Plenum Press, New York, pp. 509–530. A General Kinetic Theory of Rate Enhancements for Reactions Between Organic Substrates and Hydrophilic Ions in Micellar Systems.

Saenger, W. (1980). *Angew. Chem. Int. Ed. Eng.* **19**, 344–362. Cyclodextrin Inclusion Compounds in Research and Industry.

Sagdeev, R. Z., Leshina, T. V., Kamkha, M. A., Molin, Y. N., and Rezvukhin, A. I. (1977a). *Chem. Phys. Lett.* **48**, 89–90. A Magnetic Isotopic Effect in the Triplet Sensitized Photolysis of Dibenzoyl Peroxide.

Sagdeev, R. Z., Salikhov, K. M., and Molin, Y. M. (1977b). *Russ. Chem. Rev. Eng. Transl.* **46**, 297–315. The Influence of the Magnetic Field on Processes Involving Radicals and Triplet Molecules in Solutions.

Sakamoto, K. and Hatano, M. (1980). *Bull. Chem. Soc. Jpn.* **53**, 339–343. Formation of Chiral Aggregates of Acylamino Acids in Solution.

Seelig, J., Dijkman, R., and de Haas, G. H. (1980). *Biochemistry* **19**, 2215–2219. Thermodynamic and Conformational Studies on *sn*-2-Phosphatidylcholines in Monolayers and Bilayers.

Seuron, P. and Solladie, G. (1980). *J. Org. Chem.* **45**, 715–719. Asymmetric Induction in Liquid Crystals: Optically Active *trans*-Cyclooctane from Hofmann Elimination in New Cholesteric Mesophases.

Shinkai, S., Yamashita, T., Kusano, Y., Ide, T., and Manabe, O. (1980). *J. Am. Chem. Soc.* **102**, 2335–2340. Flavin Trapping of Carbanion Intermediates as Catalyzed by Cyanide Ion and Cationic Micelle.

Sousa, L. R., Sogah, G. D. Y., Hoffman, D. H., and Cram, D. J. (1978). *J. Am. Chem. Soc.* **100**, 4569–4576. Host–Guest Complexation. 12. Total Optical Resolution of Amine and Amino Ester Salts by Chromatography.

Starks, C. M. (1971). *J. Am. Chem. Soc.* **93**, 195–199. Phase-Transfer Catalysis, I. Heterogeneous Reactions Involving Anion Transfer by Quaternary Ammonium and Phosphonium Salts.

Starks, C. M. and Liotta, C. (1979). *Phase Transfer Catalysis: Principles and Techniques*, Academic Press, New York.

Stegemeyer, H., Stille, W., and Pollman, P. (1979). *Isr. J. Chem.* **18**, 312–317. Circular Fluorescence Polarization of Achiral Molecules in Cholesteric Liquid Crystals.

Stoddart, J. F. (1979). *Chem. Soc. Rev.* **7**, 85–142. From Carbohydrates to Enzyme Analogues.

Stoddart, J. F. (1980). In *Bioenergetics and Thermodynamics: Model Systems* (A. Braibanti, Ed.), D. Reidel Co., Amsterdam, pp. 43–62. Holes, Handedness, Handles, and Hopes: Meeting the Requirements of Primary Binding, Chirality, Secondary Interactions and Functionality in Enzyme Analogues.

Sugimoto, T. and Baba, N. (1979). *Isr. J. Chem.* **18**, 214–219. Asymmetric Reactions in Chiral Hydrophobic Binding Sites.

Sugimoto, T., Matsumura, Y., Imanishi, T., Tanimoto, S., and Okano, M. (1978). *Tetrahedron Lett.* **37**, 3431–3434. Asymmetric Reduction of Aromatic Ketones Bound in the Chiral Hydrophobic Interior of Sodium Cholate Micelle.

Sunamoto, J. and Horiguchi, D. (1980). *Nippon Kagaku Kaishi*, 475–480. Control of Reaction Path in Reversed Micelles—Acid Catalyzed and Non-Catalyzed Reactions of Hydraazobenzene-3,3′-disulfonic Acid.

Tabushi, I., Fujita, K., and Kawakubo, H. (1977). *J. Am. Chem. Soc.* **99**, 6456–6457. One Step Preparation of Vitamin K_1 or K_2 Analogues by Cyclodextrin Inclusion Catalysis.

Tachibana, T. and Kurihara, K. (1976). *Naturwissenschaften* **11**, 532. Induced Circular Dichroism of Achiral Dye Solubilizates in Aqueous Micellar Solutions of a Chiral Surfactant.

Tada, M. Hirano, H., and Suzuki, A. (1980). *Bull. Chem. Soc. Jpn.* **53**, 2304–2308. Photochemistry of Host–Guest Complex. III. Effect of Guest Cation on the Photoreactivity of Acetophenone Oxime Derivatives Having Crown Ether Moiety.

Tagaki, W., Tamura, Y., and Yano, Y. (1980). *Bull. Chem. Soc. Jpn.* **53**, 478–480. Asymmetric Benzoic Condensation Catalyzed by Optically Active Thiazolium Salts in Micellar Two-Phase Media.

Timko, J. M., Moore, S. S., Walba, D. M., Hiberty, P., and Cram, D. J. (1977). *J. Am. Chem. Soc.* **99**. 4207–4219. Host-Guest Complexation. 2. Structural Units that Control Association Constants Between Polyethers and *tert*-Butylammonium Salts.

Trainor, G. L. and Breslow, R. (1981). *J. Am. Chem. Soc.* **103**, 154–158. High Acylation Rates and Enantioselectivity with Cyclodextrin Complexes of Rigid Substrates.

Tundo, P. (1977). *J. Chem. Soc. Chem. Commun.*, 641–642. Silica Gel as a Polymeric Support for Phase-Transfer Catalysts.

Tundo, P. (1978). *Synthesis*, 315–316. Easy and Economical Synthesis of Widely Porous Resins: Very Efficient Support for Immobilized Phase Transfer Catalysis.

Tundo, P. and Fendler, J. H. (1980). *J. Am. Chem. Soc.* **102**, 1760–1762. Photophysical Investigations of Chiral Recognition in Crown Ethers.

Tundo, P. and Venturello, P. (1979). *J. Am. Chem. Soc.* **101**, 6606–6613. Synthesis, Catalytic Activity, and Behavior of Phase-Transfer Catalysts Supported on Silica Gel. Strong Influence of Substrate Adsorption on the Polar Polymeric Matrix on the Efficiency of the Immobilized Phosphonium Salts.

Turro, N. J. and Cherry, W. R. (1978). *J. Am. Chem. Soc.* **100**, 7431–7432. Photoreactions in Detergent Solutions. Enhanced Regioselectivity Resulting from the Reduced Dimensionality of Substrates Sequestered in a Micelle.

Turro, N. J. and Kraeutler, B. (1978). *J. Am. Chem. Soc.* **100**, 7432–7434. Magnetic Isotope and Magnetic Field Effects on Chemical Reactions. Sunlight and Soap for the Efficient Separation of ^{13}C and ^{12}C Isotopes.

Turro, N. J. and Kraeutler, B. (1980). *Acc. Chem. Res.* **13**, 369–377. Magnetic Field and Magnetic Isotope Effects in Organic Photochemical Reactions. A Novel Probe of Reaction Mechanisms and a Method for Enrichment of Magnetic Isotopes.

Turro, N. J. and Mattay, J. (1980). *Tetrahedron Lett.* **21**, 1799–1802. Photochemistry of 1,2-Diphenyl-2,2-dimethyl-propanone-1 in Micellar Solutions. Cage Effects, Isotope Effects and Magnetic Field Effects.

Turro, N. J. and Tung, C.-H. (1980). *Tetrahedron Lett.* **21**, 4321–4322. Photochemistry of Phenyladamantyl Ketone in Homogeneous Organic and in Micellar Solution.

Turro, N. J., Kraeutler, B., and Anderson, D. R. (1979). *J. Am. Chem. Soc.* **101**, 7435–7437. Magnetic and Micellar Effects on Photoreactions. Micellar Cage and Magnetic Isotope Effects on Quantum Yields. Correlation of ^{13}C Enrichment Parameters with Quantum Yield Measurements.

Turro, N. J., Grätzel, M., and Braun, A. M. (1980a). *Angew. Chem. Int. Ed. Eng.* **19**, 675–696. Photophysical and Photochemical Processes in Micellar Systems.

Turro, N. J., Anderson, D. R., and Kraeutler, B. (1980b). *Tetrahedron Lett.* **21**, 3–6. Photochemistry of Ketones in Micellar Solution: Structural and Viscosity Effects on Carbon-13 Isotopic Enrichment.

Turro, N. J., Chow, M. F., and Kraeutler, B. (1980c). *Chem. Phys. Lett.* **73**, 545–549. The Dynamics of the Photodecarbonylation of Dibenzylketone in a Micellar Detergent Solution: Effect of Temperature on the Absolute Quantum Yields and on ^{13}C Enrichment.

Turro, N. J., Chow, M. F., Chung, C. J., Weed, G. C., and Kraeutler, B. (1980d). *J. Am. Chem. Soc.* **102**, 4843–4845. Magnetic Field and Magnetic Isotope Effects on Cage Reactions in Micellar Solutions.

Turro, N. J., Chow, M.-F., Chung, C.-J., and Tung, C. H. (1980e). *J. Am. Chem. Soc.* **102**, 7391–7393. An Efficient, High Conversion Photoinduced Emulsion Polymerization. Magnetic Field Effects on Polymerization Efficiency and Polymer Molecular Weight.

Utaka, M., Matsushita, S., Yamasaki, H., and Takeda, A. (1980). *Tetrahedron Lett.* **21**, 1063–1064. Micellar Catalysis in the Base-Catalyzed Autoxidation of Diosphenol.

Van Hooidonk, C. and Breebart-Hansen, C. A. E. (1970). *Recl. Trav. Chim. Pays-Bas. Belg.* **89**, 289–299. Stereospecific Reaction of Isopropyl Methylphosphonofluoridate (Sarin) with α-Cyclodextrin. A Model for Enzyme Inhibition.

Ward, H. R. (1972). *Acc. Chem. Res.* **5**, 18–24. Chemically Induced Dynamic Nuclear Polarization (CIDNP). I. The Phenomenon, Examples and Applications.

Weber, W. P. and Gokel, G. W. (1977). *Phase Transfer Catalysis in Organic Synthesis*, Springer, Heidelberg.

Wulff, G., Vesper, W., Grobe-Einsler, R., and Sarhan, A. (1977a). *Makromol. Chem.* **178**, 2799–2816. Enzyme-Analogue Built Polymers, 4. On the Synthesis of Polymers Containing Chiral Cavities and Their Use for the Resolution of Racemates.

Wulff, G., Grobe-Einsler, R., Vesper, W., and Sarhan, A. (1977b). *Makromol. Chem.* **178**, 2817–2825. Enzyme-Analogue Built Polymers, 5. On the Specificity of Chiral Cavities Prepared in Synthetic Polymers.

Wynberg, H. (1976). *Chimia* **30**, 445–451. Asymmetric Catalysis in Oxidation Reactions. The Consequences of Interactions Between Enantiomers.

Yamada, K., Shosenji, H., and Ihara, H. (1979a). *Chem. Lett.*, 491–494. The Hydrolysis of *p*-Nitrophenyl Esters of α-Amino Acids by *N*-Lauryl L- or D-Histidine in Cationic Micelles.

Yamada, K., Shosenji, H., Ihara, H., and Otsubo, Y. (1979b). *Tetrahedron Lett.* **27**, 2529–2532. Enantioselectivity Catalyzed Hydrolysis of *p*-Nitrophenyl Esters of *N*-Protected L-Amino Acids by *N*-Lauroyl L- or D-Histidine in CTABr Micelles.

Yamada, K., Shosenji, H., Otsubo, Y., and Ono, S. (1980). *Tetrahedron Lett.* **21**, 2649–2652. Enantioselectively Catalyzed Oxidation of 3,4-Dihydroxy-L-Phenylalanine by *N*-Lauroyl L- or D-Histidine-Cu(II) Complex in CTABr Micelles.

CHAPTER

12

DATA TABULATION FOR REACTIVITIES IN MEMBRANE MIMETIC SYSTEMS

An attempt is made in this chapter to compile the literature data in membrane mimetic chemistry exhaustively. Sufficient information has been provided to allow the reader to compare critically the published original research. Further experimental details and interpretations should be obtained from the cited references. No data prior to 1975 have been included. Tables 12.1–12.10 update the previous compilation (Fendler, and Fendler, 1975). The remaining tables emphasize the very much increased scope of membrane mimetic chemistry. Due to their importance in solar energy conversion, emphasis has been placed on energy and electron transfer and charge separation in micelles, microemulsions, black lipid membranes (BLMs), vesicles, and polyions (Tables 12.7, 12.9, 12.12, 12.14, 12.15, and 12.19).

The following list of abbreviations has been used in the tables:

PNPA	*p*-Nitrophenylacetate
PNPH	*p*-Nitrophenylhexanoate
PNPP	*p*-Nitrophenylphosphate
CEAB	$CH_3(CH_2)_{15}\overset{+}{N}(Me_2)(Et)$, Br^-
CBzAC	$CH_3(CH_2)_{15}\overset{+}{N}(Me_2)(CH_2C_6H_5)$, Cl^-
Im	Imidazole
DMOG	*N*,*N*-Dimethyl-*N*-octadecylglycine
CTAB	Hexadecyltrimethylammonium bromide
CTACl	Hexadecyltrimethylammonium chloride
SDS	Sodium dodecyl sulfate
SDOS	Sodium 1-dodecane sulfonate
POOA	Polyoxyethylene(n = 10) oleyl alcohol
k_0	Rate in absence of surfactant but in presence of all components stated in the reaction column
k_Ψ	Maximum rate in the rate constant versus surfactant concentration profile

k_Ψ/k_0	+ = rate enhancements
k_Ψ/k_0	− = rate retardation
DTAC	$CH_3(CH_2)_{11}\overset{+}{N}(CH_3)_3, Cl^-$
OTAC	$CH_3(CH_2)_{17}\overset{+}{N}(CH_3)_3, Cl^-$
PLE	Polyoxyethylene lauryl ether
NDB	*N*-Dodecyl betaine
DTAB	Decyltrimethylammonium bromide
TTAB	Tetradecyltrimethylammonium bromide
CPyB	Cetylpryidinium bromide
DOTAC	Dodecyltrimethylammonium chloride
HPC	Hexadecylpyridinium chloride

Table 12.1. Hydrolysis and Solvolysis of Carboxylic Esters in Aqueous Micelles

Reaction	Effect of Surfactant	Reference
O^-–N(~~~)–C(=O)Im + CTAB $\xrightarrow[k_a]{pNO_2C_6H_4OAc}$ AcO–N(~~~)–C(=O)Im + $^-OC_6H_4NO_2p$; $\xrightarrow{k_d}$ CH_3CO_2H + ^-O–N(~~~)–C(=O)Im; $\xrightarrow[pNO_2C_6H_4OAc]{k_m}$ $^-OC_6H_4NO_2p$ 30°C, pH = 8.1, 3 v/v % EtOH, 0.15 *M* tris buffer	Multifunctional surfactant containing hydroxamate and imidazole moieties N-Lauryl benzohydroxamic acid + CTAB $k_{cat} = 0.06\ M^{-1}$ sec, N-methyl benzohydroxamatic acid $k_{cat} = 1.3$; N-lauryl-4-imidazoylcarbonydroxamic acid + CTAB, $k_{cat} = 32$ $k_{cat} = \dfrac{k_a k_d}{k_a[pNO_2C_6H_4OAC] + k_d}$	Kunitake *et al.*, 1975
Hydrolysis of *p*-nitrophenyl benzoyl choline + OH^-, 25.0°C, 0.01 *M* NaOH	CTAB, $k_\Psi/k_0 \sim 1$	Bunton and McAneny, 1976
Hydrolysis of hexadecyl(2-hydroxyethyl)diethyl ammonium bromide *p*-nitrobenzoyl ester, 25.0°C, 0.01 *M* NaOH	CTAB, $k_\Psi/k_0 \sim 41$; $n\text{-}C_{16}H_{33}\overset{+}{N}Me_2CH_2CH_2OHBr^-$, $k_\Psi/k_0 \sim 33$	

Table 12.1 (*Continued*)

Reaction	Effect of Surfactant	Reference
Hydrolysis of *p*-nitrophenylheptanoate, 30.0°C, pH = 8.5, 0.02 *M* KCl	Octadecyldiethylhydroxyethylammonium bromide, $k_\Psi/k_0 \sim 10^5$	Martinek et al., 1975a, 1975b
$\text{benzimidazole (N-}R_1\text{)} + R_2COOC_6H_4NO_2p \longrightarrow \text{1-}R_1\text{-3-(}O{=}C{-}R_2\text{)benzimidazolium}^{+} + {}^{-}OC_6H_4NO_2p$		Martinek et al., 1975c
$R_1 = H, R_2 = CH_3(CH_2)_5$	SDS, $k_\Psi/k_0 \sim 0.03$;	
$R_1 = CH_3, R_2 = CH_3(CH_2)_5$	CTAB, $k_\Psi/k_0 \sim 0.009$	
$R_1 = CH_3, R_2 = CH_3(CH_2)_5$	SDS, $k_\Psi/k_0 \sim 0.007$	
Benzimidazole anion, $R_2 = CH_3$	CTAB, $k_\Psi/k_0 \sim 10$; $k_\Psi/k_0 \sim 11$; $k_\Psi/k_0 \sim 14$	
Benzimidazole anion, $R_2 = CH_3(CH_2)_2$	Kinetic treatment for micelles catalysis is derived (see Chapter 11)	
Benzimidazole anion, $R_2 = CH_3(CH_2)_5$		
30.0°C, pH varied, 0.02 *M* borate buffer		
Hydrolysis of *p*-nitrophenyl acetate	$n\text{-}C_{16}H_{33}\overset{+}{N}CH_3R_1R_2Cl^-$	Moss et al., 1975
25.0°C, pH = 8.0, 0.01 *M* phosphate buffer	$R_1 = CH_3, R_2 = CH_3$, $k_\Psi/k_0 \sim 9$	
	$R_1 = CH_3, R_2 = CH_2CH_2OH$, $k_\Psi/k_0 \sim 110$	
	$R_1 = CH_2C_6H_5, R_2 = CH_2CH_2OH$, $k_\Psi/k_0 \sim 250$	
	$R_1 = CH_3, R_2 = CH_2Im$, $k_\Psi/k_0 \sim 11{,}000$	
	$R_1 = CH_2CH_2OH, R_2 = CH_2Im$, $k_\Psi/k_0 \sim 7200$	

25.0°C, pH = 8.0, 0.4 *M* phosphate buffer	$R_1 = CH_3$, $R_2 = CH_2CH_2OH$, $k_\Psi/k_0 \sim 25$ $R_1 = CH_3$, $R_2 = CH_2Im$, $k_\Psi/k_0 \sim 1400$ $R_1 = EtOH$, $R_2 = CH_2Im$, $k_\Psi/k_0 \sim 930$ Mechanism is sequential; initial slow acylation is followed by a rapid hydroxyl group mediated deacylation (see equation (10.16))	
Hydrolysis of *p*-nitrophenyl hexanoate	$n\text{-}C_{16}H_{33}\overset{+}{N}CH_3R_1R_2Cl^-$	
25.0°C, pH = 8.0, 0.01 *M* phosphate buffer	$R_1 = CH_3$, $R_2 = CH_3$, $k_\Psi/k_0 \sim 13$ $R_1 = CH_3$, $R_2 = CH_2Im$, $k_\Psi/k_0 \sim 20{,}000$ $R_1 = CH_2CH_2OH$, $R_2 = CH_2Im$, $k_\Psi/k_0 \sim 10{,}000$	
25.0°C, pH = 8.0, 0.4 *M* phosphate buffer	$R_1 = CH_3$, $R_2 = CH_2CH_2OH$, $k_\Psi/k_0 \sim 48$ $R_1 = CH_3$, $R_2 = CH_2Im$, $k_\Psi/k_0 \sim 6300$ $R_1 = CH_2CH_2OH$, $R_2 = CH_2Im$, $k_\Psi/k_0 \sim 5600$ Mechanism is sequential; initial slow acylation is followed by a rapid hydroxyl group mediated deacylation (see equation 10.16).	
Hydrolysis of *p*-nitrophenyl acetate, 30°C, pH = 8.0, 3 v/v % EtOH, 0.01 *M* borate buffer, $\mu = 0.01$ (KCl) (see scheme above, Kunitake et al., 1975)	Multifunctional surfactants containing hydroxamate and imidazole moieties Turnover rate for hydroxamate group = $k_{\text{turnover}} = k_a k_d[pNO_2C_6H_4OAc]_\circ[(Co)]/k_a[pNO_2C_6H_4OAc]_\circ + k_d = k_{\text{turnover}}[Co]$ *N*-Lauryl(4-imidazolecarbo)hydroxamic acid + CTAB, $10^4 k_{\text{turnover}} = 380\ \text{sec}^{-1}$ *N*-Lauryl[β-(4-imidazole) propio]hydroxamic acid + CTAB, $10^4 k_{\text{turnover}} = 9.4\ \text{sec}^{-1}$; *N*-laurylbenzohydroxamic acid + CTAB, $10^4 k_{\text{turnover}} = 0.13\ \text{sec}^{-1}$ (Imidazole, $10^4 k_{\text{turnover}} = 0.07\ \text{sec}^{-1}$; α-Chymotrypsin, $10^4 k_{\text{turnover}} = 150\ \text{sec}^{-1}$)	Kunitake et al., 1976

Table 12.1 (*Continued*)

Reaction	Effect of Surfactant	Reference
p-$NO_2C_6H_4OAc + RS^- \xrightarrow{d} RSCOCH_3 + p$-$NO_2C_6H_4OH$, 30°C, pH = varied, 2 v/v% EtOH, 0.02 M buffer; RS^- glutathionate coenzyme A(CoASH)	Nonionic and anionic micelles are ineffective CTAB, $k_\Psi/k_0 \sim 100$; $k_\Psi/k_0 \sim 290$ Rate enhancements explicable in terms of hydrophobic ion-pair formation	Shinkai and Kunitake, 1976a
Nucleophilic attack of tetraethylammonium N-methylmyristohydroxamate on p-nitrophenyl acetate, 22°C, pH = 9.99	CTAB, $k_\Psi/k_0 \sim +$ Nucleophilicity of hydroxamate anion is greatly enhanced on desolvation	Shinkai and Kunitake 1976b
Nucleophilic attack of N-methylmyristohydroxamic acid on p-nitrophenyl acetate, 30°C, pH = 8.5, $\mu = 0.01$ (KCl)	SDS, $k_\Psi/k_0 \sim 0.3$; CTAB, $k_\Psi/k_0 \sim 205$; Tween 80, $k_\Psi/k_0 \sim 5$; Triton X-100, $k_\Psi/k_0 \sim 9$; POOA ($n = 10$), $k_\Psi/k_0 \sim 9$; Brij-35, $k_\Psi/k_0 \sim 5$	Kunitake et al., 1976
Nucleophilic attack of N-methylmyristohydroxamic acid on p-nitrophenyl acetate in the presence of trioctylmethylammonium chloride, 30°C, pH = 8.5, $\mu = 0.01$ (KCl)	Tween 80, $k_\Psi/k_0 \sim 21$; Triton X-100, $k_\Psi/k_0 \sim 70$; POOA ($n = 10$), $k_\Psi/k_0 \sim 53$; Brij-35, $k_\Psi/k_0 \sim 29$	
Nucleophilic attack of dodecyl(2-hydroxyimidophenethyl) dimethyl ammonium bromide on p-nitrophenyl acetate, 30°C, pH = 8.1, $\mu = 0.01$ (KCl)	SDS, $k_\Psi/k_0 \sim 0.1$; CTAB, $k_\Psi/k_0 \sim 194$; Tween 80, $k_\Psi/k_0 \sim 14$; Triton X-100, $k_\Psi/k_0 \sim 11$; POOA, $k_\Psi/k_0 \sim 58$; Brij-35, $k_\Psi/k_0 \sim 17$ Nucleophilicity of hydroxamate anions is greatly enhanced on desolvation; pH independent rate constants are given	
Hydrolysis of p-nitrophenyl acetate, 25.0°C	Cetyl(imidazol-4-ylmethyl)dimethyl ammonium chloride (*1*) pH — $k_\Psi(M^{-1}\ sec^{-1})/k_{Im}(M^{-1}\ sec^{-1})$ 7.2 — ~1 7.7 — ~3.7 8.7 — ~9 9.2 — ~20	Tonellato, 1976

Reaction	Micellar effect	Reference
Hydrolysis of *p*-nitrophenyl hexanoate, 25.0°C	*1*,	
	pH — $\frac{k_{\Psi}(M^{-1}\ sec^{-1})}{k_{Im}(M^{-1}\ sec^{-1})}$: 7.2 — ~36; 7.7 — ~260; 8.7 — ~1000; 9.2 — ~1300	
Hydrolysis of $XCH_2CO_2C_6H_4NO_2$-*p*, 25.0°C, pH = 8.99	CTAB, k_{Ψ}/k_0: X = PhO 1.2; X = PhS 19; X = pMeOC$_6$H$_4$ 1.7; X = pNO$_2$C$_6$H$_4$ 298. Increased micellar catalysis with change of acidity of the CH_2 groups interpreted in terms of change to E_1cB mechanism	Tagaki et al., 1976
Hydrolysis of *p*-nitrophenyl acetate, 25.0°C, pH = 8.0, μ = 0.05 (KCl)	CTACl, $k_{\Psi}/k_0 \sim 1$; $R\overset{+}{N}(CH_3)_2CH_2CH_2NH_2, Cl^-$, $k_{\Psi}/k_0 \sim 91$ (this buffer); $R\overset{+}{N}(CH_3)_2CH_2CH_2NHC(=O)-CH(CH_3)-NH_2, Cl^-$ (*1*), $k_{\Psi}/k_0 \sim 45, 23$ (tris phosphate buffer); $R\overset{+}{N}(CH_3)_2CH_2CH_2NHC(=O)-CH(CH_2Im)-NH-C-Boc, Cl^-$ (*2*); $R = n\text{-}C_{16}H_{33}$	Moss et al., 1977a

Table 12.1 (*Continued*)

Reaction	Effect of Surfactant	Reference
	$k_\Psi/k_0 \sim 130$ (phosphate buffer); $CH_3(CH_2)_{11}CH(CONHCH(CH_2Im)COOCH_3)-(CH_2)_4\overset{+}{N}(CH_3)_3, Cl^-$ (*3*), $k_\Psi/k_0 \sim 360$	
Hydrolysis of D- or L-*N*-acetylphenylalanine *p*-nitrophenyl ester, 25.0°C, pH = 8.0, $\mu = 0.05$ (KCl)	CTACl, $k_\Psi/k_0 \sim 1$; *2*, $k_\Psi/k_0 \sim 0.77$, $k_D \sim k_L$; *3*, $k_\Psi/k_0 \sim 2.7$; *4*, $k_\Psi/k_0 \sim 42$	
Hydrolysis of *p*-nitrophenyl acetate, 60.0°C, pH = 9.56	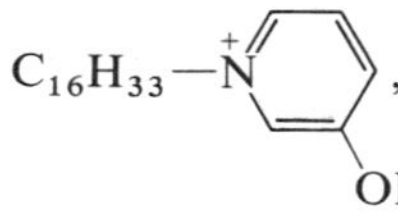 $C_{16}H_{33}-\overset{+}{N}$(3-hydroxypyridinium), $k_\Psi/k_0 \sim +$	Gani, 1977
Hydrolysis of *p*-nitrophenyl acetate;	CTAB, $C_{16}H_{33}\overset{+}{N}(CH_3)_2CH_2CH_2OH, Br^-$, $k_\Psi/k_0 \sim 1$	Meyer and Viout, 1977
Hydrolysis of $CH_3CO_2(CH_3)_2\overset{+}{N}(CH_3)_2C_{16}H_{33}, Br^-$, pH = 8.5, 9.5, T = ?	S_N2 versus E_2 mechanism	
Hydrolysis of *p*-nitrophenyl acetate, 30°C, pH = 7.10, 0.05 *M* phosphate buffer	$C_{18}H_{37}\overset{+}{N}Me_3, Br^-$, $k_\Psi/k_0 \sim 10$; $C_{18}H_{37}\overset{+}{N}(CH_2CH_2OH)Me_2$, Br^-, $k_\Psi/k_0 \sim 28$; $C_{18}H_{37}\overset{+}{N}(Me_3), Br^-$ 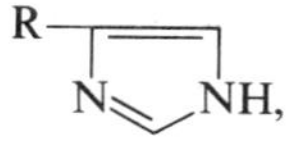 + R–(imidazolyl, N, NH), $R = C_{13}H_{27}CONHCH(CO_2H)CH_2$	Tagaki et al., 1977

Reaction	Surfactant / Results	Reference
	$k_\Psi/k_0 \sim 382$	
	$C_{18}H_{37}\overset{+}{N}(CH_2CH_2OH)Me_2, Br^- + R$–(imidazol-4-yl)	
	$R = C_7H_{15}CONHCH_2CH_2$	
	$k_\Psi/k_0 \sim 474$	
	In mixed micelles, fast acylation of the imidazole is followed by quantitative and fast transfer of the acyl group to the hydroxy group	
Hydrolysis of *p*-nitrophenyl acetate, 25.0°C	Cetyl(imidazol-4-ylmethyl)dimethylammonium chloride (*1*)	Tonellato, 1977

pH	$\frac{k_\Psi(M^{-1}\,sec^{-1})}{k_{Im}(M^{-1}\,sec^{-1})}$
7.2	~1
7.9	~8
8.65	~30
8.86	~119

Cetylethyl(2-hydroxyethyl)methylammonium bromide (*2*)

pH	$\frac{k_\Psi(M^{-1}\,sec^{-1})}{k_{Im}(M^{-1}\,sec^{-1})}$
7.9	~0.25
8.65	~1.4
8.86	~3.7

Diethyl-(2-hydroxyethyl)methylammonium bromide (*3*)

pH	$\frac{k_\Psi(M^{-1}\,sec^{-1})}{k_{Im}(M^{-1}\,sec^{-1})}$
8.86	~0.006

Table 12.1 *(Continued)*

Reaction	Effect of Surfactant	Reference
	Cetyl(2-hydroxyethyl)(imidazol-4-ylmethyl)methylammonium chloride hydroxychloride (*4*)	
	pH — $\frac{k_{\Psi}(M^{-1}\ sec^{-1})}{k_{Im}(M^{-1}\ sec^{-1})}$	
	7.2 — ~2.3	
	7.9 — ~13	
	8.65 — ~45	
	8.86 — ~209	
	Ethyl-(2-hydroxyethyl)(imidazol-4-ylmethyl)methylammonium chloride hydrochloride (*5*)	
	pH — $\frac{k_{\Psi}(M^{-1}\ sec^{-1})}{k_{Im}(M^{-1}\ sec^{-1})}$	
	7.9 — ~0.1	
	8.65 — ~0.2	
Hydrolysis of *p*-nitrophenyl hexanoate, 25.0°C	5,	
	pH — $\frac{k_{\Psi}(M^{-1}\ sec^{-1})}{k_{Im}(M^{-1}\ sec^{-1})}$	
	7.2 — ~70	
	7.9 — ~460	
	8.65 — ~1846	
	8.86 — ~3928	

6,

pH	$\frac{k_{\Psi}(M^{-1}\ \text{sec}^{-1})}{k_{\text{Im}}(M^{-1}\ \text{sec}^{-1})}$
7.9	$\sim$6.5
8.65	$\sim$52
8.86	$\sim$96

7,

pH	$\frac{k_{\Psi}(M^{-1}\ \text{sec}^{-1})}{k_{\text{Im}}(M^{-1}\ \text{sec}^{-1})}$
8.65	$\sim$0.006
8.86	$\sim$0.008

8,

pH	$\frac{k_{\Psi}(M^{-1}\ \text{sec}^{-1})}{k_{\text{Im}}(M^{-1}\ \text{sec}^{-1})}$
7.2	$\sim$211
7.9	$\sim$675
8.65	$\sim$3076
8.86	$\sim$9285

9,

pH	$\frac{k_{\Psi}(M^{-1}\ \text{sec}^{-1})}{k_{\text{Im}}(M^{-1}\ \text{sec}^{-1})}$
7.2	$\sim$0.17
7.9	$\sim$0.22
8.65	$\sim$0.38

Mechanism for bifunctional micellar catalysis involves the acylation of the imidazole ring followed by a relatively rapid acyl transfer to the hydroxy function in the micellar phase

Table 12.1 *(Continued)*

Reaction	Effect of Surfactant	Reference
Hydrolysis of *p*-nitrophenyl acetate, 30°C, pH = varied, $\mu = 0.8$(KBr)	*n*-Decyldimethyl(2-hydroxyethyl)ammonium bromide (*1*) $k_\Psi/k_0 \sim 23$; *n*-decyldimethyl(3-hydroxypropyl)ammonium bromide (*2*), $k_\Psi/k_0 \sim 3.7$; *n*-decyldiethyl(2-hydroxyethyl)ammonium bromide (*3*), $k_\Psi/k_0 \sim 33$; *n*-decyldimethyl(1,1-dimethyl-2-hydroxyethyl) ammonium bromide (*4*), $k_\Psi/k_0 \sim 11$; *n*-decyldimethyl(2,3-dihydroxypropyl)ammonium bromide (*5*), $k_\Psi/k_0 \sim 12$	Shiffman et al., 1977
Hydrolysis of *p*-nitrophenyl hexanoate, 30°C, pH = varied, $\mu = 0.8$(KBr)	*1*, $k_\Psi/k_h \sim 14$; *2*, $k_\Psi/k_0 \sim 1.8$; *3*, $k_\Psi/k_0 \sim 18$; *4*, $k_\Psi/k_0 \sim 6$; *5*, $k_\Psi/k_0 \sim 6.6$	
Hydrolysis of 2,4-dinitrophenyl acetate, 30°C, pH = varied, $\mu = 0.8$(KBr)	*1*, $k_\Psi/k_0 \sim 33$; *2*, $k_\Psi/k_0 \sim 4$; *3*, $k_\Psi/k_0 \sim 53$; *4*, $k_\Psi/k_0 \sim 19.3$; *5*, $k_\Psi/k_0 \sim 10$ Similarities to esterases are discussed	
Hydrolysis of *p*-nitrophenyl acetate, 25.0°C, pH = 8.4, 0.02 *M* borate buffer	$C_{16}H_{33}\overset{+}{N}(Me)(CH_2CH_2OH)Im$, Cl^- (*1*), $k_\Psi = 22\ M^{-1}\ sec^{-1}$; $C_{16}H_{33}\overset{+}{N}(Me_2)CH_2Im$, Cl^- (*2*), $k_\Psi = 11\ M^{-1}\ sec^{-1}$; $C_{16}H_{33}\overset{+}{N}(Me)(Et)CH_2CH_2OH$, Br^- (*3*), $k_\Psi = 0.32\ M^{-1}\ sec^{-1}$	Anoardi and Tonellato, 1977
Hydrolysis of *p*-nitrophenyl hexanoate, 25.0°C, pH = 8.4, 0.02 *M* borate buffer	*1*, $k_\Psi = 640\ M^{-1}\ sec^{-1}$; *2*, $k_\Psi = 390\ M^{-1}\ sec^{-1}$; *3*, $k_\Psi = 7\ M^{-1}\ sec^{-1}$	
Hydrolysis of MeC(=O)—N-imidazole (1-acetylimidazole), 25.0°C, pH = 8.4, 0.02 *M* borate buffer	*1*, $k_\Psi = 7.8\ M^{-1}\ sec^{-1}$; *2*, $k_\Psi = 13.5\ M^{-1}\ sec^{-}1$	

Hydrolysis of $C_5H_{11}C(=O)-N$ (imidazole) 25.0°C, pH = 8.4, 0.02 M borate buffer	*1*, $k_\Psi = 31\ M^{-1}\ sec^{-1}$; *2*, $k_\Psi = 95\ M^{-1}\ sec^{-1}$	
Hydrolysis of $CF_3-C(=O)-N(Bu^n)-C_6H_3(NO_2)_2$ (2,4-dinitrophenyl) 25.0°C, pH = 8.4, 0.02 M borate buffer	*1*, $k_\Psi = 510\ M^{-1}\ sec^{-1}$; *2*, $k_\Psi = 40\ M^{-1}\ sec^{-1}$; *3*, $k_\Psi = 910\ M^{-1}\ sec^{-1}$ Effective nucleophilic site changes from the imidazole to the hydroxy function on going from the hydrolysis of ester to that of amides	
Hydrolysis of $HO_2C-C_6H_3(NO_2)-OCO(CH_2)_{n-1}H$ 31.0°C, pH = 9.06, $\mu = 0.083\ M$ tris(hydroxymethyl)aminomethane buffer $n = 2$	$H-(CH_2)_{m-1}CONHOH$; PhCH(OH)CONHOH (MHA) $m = 4$, $k_\Psi = 26.8\ M^{-1}\ sec^{-1}$; $m = 4$ + CTAB, $31.7\ M^{-1}\ sec^{-1}$; $m = 6$, $k_\Psi = 25.3\ M^{-1}\ sec^{-1}$; $m = 6$ + CTAB, $k_\Psi = 96.3\ M^{-1}\ sec^{-1}$; $m = 8$, $k_\Psi = 27.2\ M^{-1}\ sec^{-1}$; $m = 8$ + CTAB, $k_\Psi = 325\ M^{-1}\ sec^{-1}$; $m = 10$, $k_\Psi = 31.3\ M^{-1}\ sec^{-1}$; $m = 10$ + CTAB, $k_\Psi = 413\ M^{-1}\ sec^{-1}$; MHA, k_Ψ, $40.8\ M^{-1}\ sec^{-1}$; MHA + CTAB, $k_\Psi = 116\ M^{-1}\ sec^{-1}$	Ueoka et al., 1977

Table 12.1 *(Continued)*

Reaction	Effect of Surfactant	Reference
$n = 4$	$m = 4$, $k_\Psi = 8.3\ M^{-1}\ sec^{-1}$; $m = 4$ + CTAB, $k_\Psi = 8.7\ M^{-1}\ sec^{-1}$; $m = 6$, $k_\Psi = 8\ 6\ M^{-1}\ sec^{-1}$; $m = 6$ + CTAB, $k_\Psi = 45.9\ M^{-1}\ sec^{-1}$; $m = 8$, $k_\Psi = 8.8\ M^{-1}\ sec^{-1}$; $m = 8$ + CTAB, $k_\Psi = 185\ M^{-1}\ sec^{-1}$; $m = 10$, $k_\Psi = 12.2\ M^{-1}\ sec^{-1}$; $m = 10$ + CTAB, $k_\Psi = 480\ M^{-1}\ sec^{-1}$; MHA, $k_\Psi = 13.2\ M^{-1}\ sec^{-1}$; MHA + CTAB, $k_\Psi = 49.2\ M^{-1}\ sec^{-1}$	
$n = 6$	$m = 4$, $k_\Psi = 8.3\ M^{-1}\ sec^{-1}$; $m = 4$ + CTAB, $k_\Psi = 7.5\ M^{-1}\ sec^{-1}$; $m = 6$, $k_\Psi = 8.6\ M^{-1}\ sec^{-1}$; $m = 6$ + CTAB, $k_\Psi = 33.8\ M^{-1}\ sec^{-1}$; $m = 8$, $k_\Psi = 9.9\ M^{-1}\ sec^{-1}$; $m = 8$ + CTAB, $k_\Psi = 190.3\ M^{-1}\ sec^{-1}$; $m = 10$, $k_\Psi = 13.8\ M^{-1}\ sec^{-1}$; $m = 10$ + CTAB, $k_\Psi = 497.4\ M^{-1}\ sec^{-1}$; MHA, $k_\Psi = 12.9\ M^{-1}\ sec^{-1}$; MHA + CTAB, $k_\Psi = 50.7\ M^{-1}\ sec^{-1}$	
$n = 10$	$m = 4$, $k_\Psi = 7.3\ M^{-1}\ sec^{-1}$; $m = 4$ + CTAB, $k_\Psi = 6.2\ M^{-1}\ sec^{-1}$; $m = 6$, $k_\Psi = 8.7\ M^{-1}\ sec^{-1}$; $m = 6$ + CTAB, $k_\Psi = 30.9\ M^{-1}\ sec^{-1}$; $m = 8$, $k_\Psi = 8.7\ M^{-1}\ sec^{-1}$; $m = 8$ + CTAB $k_\Psi = 144\ M^{-1}\ sec^{-1}$; $m = 10$, $k_\Psi = 20.3\ M^{-1}\ sec^{-1}$; $m = 10$ + CTAB, $k_\Psi = 298\ M^{-1}\ sec^{-1}$; MHA, $k_\Psi = 12.8\ M^{-1}\ sec^{-1}$; MHA + CTAB, $k_\Psi = 43.5\ M^{-1}\ sec^{-1}$	
$n = 12$	$m = 4$, $k_\Psi = 7.0\ M^{-1}\ sec^{-1}$; $m = 4$ + CTAB, $k_\Psi = 5.3\ M^{-1}\ sec^{-1}$; $m = 6$, $k_\Psi = 7.7\ M^{-1}\ sec^{-1}$; $m = 6$ + CTAB, $k_\Psi = 28.5\ M^{-1}\ sec^{-1}$; $m = 8$, $k_\Psi = 8.5\ M^{-1}\ sec^{-1}$; $m = 8$ + CTAB, $k_\Psi = 145.7\ M^{-1}\ sec^{-1}$; $m = 10$, $k_\Psi = 14.9\ M^{-1}\ sec^{-1}$; $m = 10$, $k_\Psi = 14.9\ M^{-1}\ sec^{-1}$; $m = 10$ + CTAB, $k_\Psi = 346.1\ M^{-1}\ sec^{-1}$; MHA, $k_\Psi = 11.2\ M^{-1}\ sec^{-1}$; MHA + CTAB, $k_\Psi = 39\ M^{-1}\ sec^{-1}$	

$n = 16$	$m = 4$, $k_\Psi = 0.2\ M^{-1}\ sec^{-1}$; $m = 4$ + CTAB, $k_\Psi = 5\ M^{-1}\ sec^{-1}$; $m = 6$, $k_\Psi = 0.4\ M^{-1}\ sec^{-1}$; $m = 6$ + CTAB, $k_\Psi = 25.7\ M^{-1}\ sec^{-1}$; $m = 8$, $k_\Psi = 1.6\ M^{-1}\ sec^{-1}$; $m = 8$ + CTAB, $k_\Psi = 157\ M^{-1}\ sec^{-1}$; $m = 10$, $k_\Psi = 9.7\ M^{-1}\ sec^{-1}$; $m = 10$ + CTAB, $k_\Psi = 317\ M^{-1}\ sec^{-1}$; MHA, $k_\Psi = 0.05\ sec^{-1}$; MHA + CTAB, $k_\Psi = 36.7\ M^{-1}\ sec^{-1}$	
Hydrolysis of p-$NO_2C_6H_4OCO(CH_2)_{n-1}H$, 31.°C, pH = 9.06, 0.083 M tris(hydroxymethyl)aminoethane buffer	$H(CH_2)_{m-1}CONHOH$, BHA=benzohydroxamic acid	Ueoka and Ohkubo, 1978
$n = 2$	BHA, $k_\Psi = 25\ M^{-1}\ sec^{-1}$; $m = 10$, $k_\Psi = 40\ M^{-1}\ sec^{-1}$; $m = 12$, $k_\Psi = 29\ M^{-1}\ sec^{-1}$; BHA + DTAC, $k_\Psi = 58.8\ M^{-1}\ sec^{-1}$; BHA + CTAB, $k_\Psi = 56.8\ M^{-1}\ sec^{-1}$; BHA + OTAC, $k_\Psi = 68.1\ M^{-1}\ sec^{-1}$; $m = 10$ + DTAC, $k_\Psi = 113\ M^{-1}\ sec^{-1}$; $m = 10$ + CTAB, $k_\Psi = 1032\ M^{-1}\ sec^{-1}$; $m = 10$ + OTAC, $k_\Psi = 1001\ M^{-1}\ sec^{-1}$; $m = 12$ + DTAC, $k_\Psi = 132\ M^{-1}\ sec^{-1}$; $m = 12$ + CTAB, $k_\Psi = 664\ M^{-1}\ sec^{-1}$; $m = 12$ + OTAC, $k_\Psi = 1396\ M^{-1}\ sec^{-1}$	
$n = 4$	BHA, $k_\Psi = 9.7\ M^{-1}\ sec^{-1}$; $m = 10$, $k_\Psi = 40.1\ M^{-1}\ sec^{-1}$; $m = 12$, $k_\Psi = 17.5\ M^{-1}\ sec^{-1}$; BHA + DTAC, $k_\Psi = 34\ M^{-1}\ sec^{-1}$; BHA + CTAB, $k_\Psi = 44.9\ M^{-1}\ sec^{-1}$; BHA + OTAC, $k_\Psi = 51.1\ M^{-1}\ sec^{-1}$; $m = 10$ + DTAC, $k_\Psi = 74.6\ M^{-1}\ sec^{-1}$; $m = 10$ + CTAB, $k_\Psi = 1296\ M^{-1}\ sec^{-1}$; $m = 10$ + OTAC, $k_\Psi = 1053\ M^{-1}\ sec^{-1}$; $m = 12$ + DTAC, $k_\Psi = 91\ M^{-1}\ sec^{-1}$; $m = 12$ + CTAB, $k_\Psi = 694\ M^{-1}\ sec^{-1}$; $m = 12$ + OTAC, $k_\Psi = 1511\ M^{-1}\ sec^{-1}$	

Table 12.1 *(Continued)*

Reaction	Effect of Surfactant	Reference
$n = 6$	BHA = 9.1 M^{-1} sec^{-1}; $m = 10$, $k_{\Psi} = 20$ M^{-1} sec^{-1}; $m = 12$, $k_{\Psi} = 38.7$ M^{-1} sec^{-1}; BHA + DTAC, $k_{\Psi} = 40.1$ M^{-1} sec^{-1}; BHA + CTAB, $k_{\Psi} = 68.9$ M^{-1} sec^{-1}; BHA + OTAC, $k_{\Psi} = 90$ M^{-1} sec^{-1}; $m = 10$ + DTAC, $k_{\Psi} = 96.1$ M^{-1} sec^{-1}; $m = 10$ + CTAB, $k_{\Psi} = 1554$ M^{-1} sec^{-1}; $m = 10$ + OTAC, $k_{\Psi} = 1805$ M^{-1} sec^{-1}; $m = 12$ + DTAC, $k_{\Psi} = 123$ M^{-1} sec^{-1}; $m = 12$ + CTAB, $k_{\Psi} = 1074$ M^{-1} sec^{-1}; $m = 12$ + OTAC, $k_{\Psi} = 2110$ M^{-1} sec^{-1}	
$n = 10$	BHA, $k_{\Psi} = 0.90$ M^{-1} sec^{-1}; $m = 10$, $k_{\Psi} = 5.8$ M^{-1} sec^{-1}; $m = 12$, $k_{\Psi} = 54.2$ M^{-1} sec^{-1}; BHA + DTAC, $k_{\Psi} = 38.8$ M^{-1} sec^{-1}; BHA + CTAB, $k_{\Psi} = 70.8$ M^{-1} sec^{-1}; BHA + OTAC, $k_{\Psi} = 91.2$ M^{-1} sec^{-1}; $m = 10$ + DTAC, $k_{\Psi} = 104$ M^{-1} sec^{-1}; $m = 10$ + CTAB, $k_{\Psi} = 2292$ M^{-1} sec^{-1}; $m = 10$ + OTAC, $k_{\Psi} = 1960$ M^{-1} sec^{-1}; $m = 12$ + DTAC, $k_{\Psi} = 140$ M^{-1} sec^{-1}; $m = 12$ + CTAB, $k_{\Psi} = 1680$ M^{-1} sec^{-1}; $m = 12$ + OTAC, $k_{\Psi} = 2299$ M^{-1} sec^{-1}	
$n = 12$	BHA, $k_{\Psi} = 0.72$ M^{-1} sec^{-1}; $m = 10$, $k_{\Psi} = 4.7$ sec^{-1}; $m = 12$, $k_{\Psi} = 89.3$ M^{-1} sec^{-1}; BHA + DTAC, $k_{\Psi} = 41$ M^{-1} sec^{-1}; BHA + OTAC, $k_{\Psi} = 86.7$ M^{-1} sec^{-1}; $m = 10$ + DTAC, $k_{\Psi} = 106$ M^{-1} sec^{-1}; $m = 10$ + CTAB, $k_{\Psi} = 1382$ M^{-1} sec^{-1}; $m = 10$ + OTAC, $k_{\Psi} = 1867$ M^{-1} sec^{-1}; $m = 12$ + DTAC, $k_{\Psi} = 146$ M^{-1} sec^{-1}; $m = 12$ + CTAB, $k_{\Psi} = 1138$ M^{-1} sec^{-1}; $m = 12$ + OTAC, $k_{\Psi} = 2017$ M^{-1} sec^{-1}	

Reaction and conditions	Results	Reference
$n = 16$	$m = 10$, $k_\Psi = 4.4\ M^{-1}\ sec^{-1}$; $m = 12$, $k_\Psi = 104\ M^{-1}\ sec^{-1}$; BHA + DTAC, $k_\Psi = 25.8\ M^{-1}\ sec^{-1}$; BHA + CTAB, $k_\Psi = 18.3\ M^{-1}\ sec^{-1}$; BHA + OTAC, $k_\Psi = 53.5\ M^{-1}\ sec^{-1}$; $m = 10$ + DTAC, $k_\Psi = 103\ M^{-1}\ sec^{-1}$; $m = 10$ + CTAB, $k_\Psi = 451\ M^{-1}\ sec^{-1}$; $m = 10$ + OTAC, $k_\Psi = 1364\ M^{-1}\ sec^{-1}$; $m = 12$ + DTAC, $k_\Psi = 101\ M^{-1}\ sec^{-1}$; $m = 12$ + CTAB, $k_\Psi = 531\ M^{-1}\ sec^{-1}$; $m = 12$ + OTAC, $k_\Psi = 1065\ M^{-1}\ sec^{-1}$	
As—Cys + PNPA $\xrightarrow{k_1}$ AsCys—SAc $\xrightarrow{k_2}$ As—Cys—NAc; As—Cys—N_1S—Ac_2 $\underset{k_\Psi(+OH^-)}{\overset{k_3(+PNPA)}{\rightleftharpoons}}$ As—Cys—NAc; PNPA, 25.0°C, pH = 8, $\mu = 0.05$ (KCl)	As—Cys = $n\text{-}C_{16}H_{33}\overset{+}{N}(Me_2)CH_2CH_2NHC(=O)—C(H)(CH_2SH)—\overset{+}{N}H_3$, $2Cl^-$; $k_1 = 1.04\ sec^{-1}$, $26\ M^{-1}\ sec^{-1}$; $k_2 = 0.44\ sec^{-1}$; $k_3 = 1.45\ sec^{-1}$, $36.3\ M^{-1}\ sec^{-1}$; $k_4 = (1\text{–}5)10^{-4}\ sec^{-1}$; $\dfrac{k_1(\text{As—Cys})^1 M^{-1}\ sec^{-1}}{k_1(\text{CTAC})\ M^{-1}\ sec^{-1}} = 1860$	
Hydrolysis of $CH_3(CH_2)_9$—$OCOC_6H_4NO_2p$, 30°C, pH = 9.5–10.8, $\mu = 0.8$	$CH_3(CH_2)_9\overset{+}{N}(Me_2)(CH_2)_2OH$ (*1*), $k_\Psi = 0.16\ M^{-1}\ sec^{-1}$; $CH_3(CH_2)_9\overset{+}{N}(Me_2)(CH_2)_3OH$ (*2*), $k_\Psi = 0.10\ M^{-1}\ sec^{-1}$; $CH_3(CH_2)_9\overset{+}{N}(Et_2)(CH_2)_2OH$ (*3*), $k_\Psi = 0.19\ M^{-1}\ sec^{-1}$; $CH_3(CH_2)_9\overset{+}{N}(Me_2)C(Me_2)CH_2OH$ (*4*), $k_\Psi = 0.13\ M^{-1}\ sec^{-1}$; $CH_3(CH_2)_9\overset{+}{N}(Me_2)(CH_2)_2OMe$ (*5*), $k_\Psi = 0.022\ M^{-1}\ sec^{-1}$; $CH_3(CH_2)_9\overset{+}{N}(Me_2)(CH_2)_2OEt$ (*6*), $k_\Psi = 0.022\ M^{-1}\ sec^{-1}$; $CH_3(CH_2)_9\overset{+}{N}(Me_2)CH_2CH_3$ (*7*), $k_\Psi = 0.025\ M^{-1}\ sec^{-1}$	Wexler et al., 1978

Table 12.1 *(Continued)*

Reaction	Effect of Surfactant	Reference
Hydrolysis of $CH_3(CH_2)_9\overset{+}{N}(Me_2)CH_2OCOC_6H_4NO_2$-*p* 30°C, pH = 9.5–10.8, $\mu = 0.8$	*1*, $k_\Psi/k_0 \sim 1$, $k_\Psi = 13\ M^{-1}\ sec^{-1}$; *2*, $k_\Psi/k_0 \sim 6$; *3*, $k_\Psi/k_0 \sim 1$; *4*, $k_\Psi/k_0 \sim 1$; *5*, $k_\Psi/k_0 \sim 1$; *6*, $k_\Psi/k_0 \sim 1$; *7*, $k_\Psi/k_0 \sim 1$	
Hydrolysis of $CH_3(CH_2)_9\overset{+}{N}(Me_2)(CH_2)_3OCOC_6H_4NO_2$-*p*, 30°C, pH = 9.5–10.8, $\mu = 0.8$	*1*, $k_\Psi/k_0 \sim 6.5$, $k_\Psi = 1.7\ M^{-1}\ sec^{-1}$; *2*, $k_\Psi/k_0 \sim 1$; *3*, $k_\Psi/k_0 \sim 5.9$; *4*, $k_\Psi/k_0 \sim 4$; *5*, $k_\Psi/k_0 \sim 1$; *6*, $k_\Psi/k_0 \sim 1$; *7*, $k_\Psi/k_0 \sim 1$;	
Hydrolysis of $CH_3(CH_2)_9COO-C_6H_3(NO_2)_2$ (2,4-dinitrophenyl ester; structure) 30°C, pH = 9.5–10.8, $\mu = 0.8$	*1*, $k_\Psi = 505\ M^{-1}\ sec^{-1}$; *2*, $k_\Psi = 63\ M^{-1}\ sec^{-1}$; *3*, $k_\Psi = 790\ M^{-1}\ sec^{-1}$; *4*, $k_\Psi = 405\ M^{-1}\ sec^{-1}$; *5*, $k_\Psi = 30\ M^{-1}\ sec^{-1}$; *6*, $k_\Psi = 11.2\ M^{-1}\ sec^{-1}$; *7*, $k_\Psi = 11.7\ M^{-1}\ sec^{-1}$	
Hydrolysis of $CH_3(CH_2)_9\overset{+}{N}(Me_2)(CH_2)_3COO-C_6H_3(NO_2)_2$ (2,4-dinitrophenyl ester; structure) 30.0°C, pH = 9.5–10.8, $\mu = 0.8$	*1*, $k_\Psi/k_0 \sim 3.6$, $k_0 = 1689\ M^{-1}\ sec^{-1}$; *2*, $k_\Psi/k_0 \sim 1$; *3*, $k_\Psi/k_0 \sim 7$; *4*, $k_\Psi/k_0 \sim 1.9$; *5*, $k_\Psi/k_0 \sim 1$; *6*, $k_\Psi/k_0 \sim 0.8$; *7*, $k_\Psi/k_0 \sim 1.3$	
Hydrolysis of $CH_3(CH_2)_9COO—C_6H_4NO_2$-*p*, 30°C, pH = 9.5–10.8, $\mu = 0.8$	*1*, $k_\Psi = 78\ M^{-1}\ sec^{-1}$; *2*, $k_\Psi = 8.3\ M^{-1}\ sec^{-1}$; *3*, $k_\Psi = 86\ M^{-1}\ sec^{-1}$; *4*, $k_\Psi = 39\ M^{-1}\ sec^{-1}$; *5*, $k_\Psi = 2.9\ M^{-1}\ sec^{-1}$; *6*, $k_\Psi = 1.4\ M^{-1}\ sec^{-1}$; *7*, $k_\Psi = 1.6\ M^{-1}\ sec^{-1}$ Proximity effects control the power of micelles Br^- and F^-; salt effects are examined	

Hydrolysis of *p*-nitrophenyl acetate, 25.0°C, pH = 9.87, 0.01 *M* bicarbonate buffer	CTAB, $k_\Psi/k_0 \sim 16.8$	Funasaki, 1978
Hydrolysis of *p*-nitrophenyl propionate, 25.0°C, pH = 9.87, 0.01 *M* bicarbonate buffer	CTAB, $k_\Psi/k_0 \sim 14.2$	
Hydrolysis of *p*-nitrophenyl butyrate, 25.0°C, pH = 9.87, 0.01 *M* bicarbonate buffer	CTAB, $k_\Psi/k_0 \sim 14.1$ Kinetic data are used to derive equations for micelle catalyzed reactions	
Hydrolysis of *p*-nitrophenyl acetate, 23.0°C, pH = 8.0, $\mu = 0.02$ *M*	$n\text{-}C_{16}H_{33}N(CH_3)_2CH_2CH_2SH, Cl^-$, $k_\Psi(M^{-1}\ sec^{-1})/k_\Psi(CTACl) \simeq 34{,}600$	Moss et al., 1978b
Hydrolysis of *p*-$NO_2C_6H_4OCOCH_3$, 31.0°C, pH = 9.06, tris buffer, $\mu = 0.083$ (KCl), 10 vol % CH_3CN	CTAB + $H(CH_2)_3CONHOH$ (*1*), $k_\Psi(M^{-1}\ sec^{-1})/k_{CTAB}(M^{-1}\ sec^{-1}) = 1.7$; CTAB + $H(CH_2)_5CONHOH$ (*2*), $k_\Psi(M^{-1}\ sec^{-1})/k_{CTAB} = 4.4$; CTAB + $H(CH_2)_7CONH$ (*3*), $k_\Psi(M^{-1}\ sec^{-1})/k_{CTAB}(M^{-1}\ sec^{-1}) = 18.6$; CTAB + $H(CH_2)_9CONH$ (*4*), $k_\Psi(M^{-1}\ sec^{-1})/k_{CTAB}(M^{-1}\ sec^{-1}) = 25.8$	Ueoka et al., 1978
Hydrolysis of *p*-$NO_2C_6H_4OCO(CH_2)_3H$, 31°C, pH = 9.06, tris buffer $\mu = 0.083$ (KCl), 10 vol. % CH_3CN	*1*, $k_\Psi(M^{-1}\ sec^{-1})/k_{CTAB}(M^{-1}\ sec^{-1}) = 2.1$; *2*, $k_\Psi(M^{-1}\ sec^{-1})/k_{CTAB}(M^{-1}\ sec^{-1}) = 9.2$; *3*, $k_\Psi(M^{-1}\ sec^{-1})/k_{CTAB}(M^{-1}\ sec^{-1}) = 37.6$; *4*, $k_\Psi(M^{-1}\ sec^{-1})/k_{CTAB}(M^{-1}\ sec^{-1}) = 64.2$	
Hydrolysis of *p*-$NO_2C_6H_4OCO(CH_2)_5H$, 31°C, pH = 9.06, tris buffer, $\mu = 0.083$ (KCl), 10 vol. % CH_3CN	*1*, $k_\Psi(M^{-1}\ sec^{-1})/k_{CTAB}(M^{-1}\ sec^{-1}) = 3.1$; *2*, $k_\Psi(M^{-1}\ sec^{-1})/k_{CTAB}(M^{-1}\ sec^{-1}) = 12.2$; *3*, $k_\Psi(M^{-1}\ sec^{-1})/k_{CTAB}(M^{-1}\ sec^{-1}) = 47.6$; *4*, $k_\Psi(M^{-1}\ sec^{-1})/k_{CTAB}(M^{-1}\ sec^{-1}) = 77.7$	
Hydrolysis of *p*-$NO_2C_6H_4OCO(CH_2)_9H$, 31°C, pH = 9.06, tris buffer, $\mu = 0.083$ (KCl), 10 vol. % CH_3CN	*1*, $k_\Psi(M^{-1}\ sec^{-1})/k_{CTAB}(M^{-1}\ sec^{-1}) = 59$; *2*, $k_\Psi(M^{-1}\ sec^{-1})/k_{CTAB}(M^{-1}\ sec^{-1}) = 243$; *3*, $k_\Psi(M^{-1}\ sec^{-1})/k_{CTAB}(M^{-1}\ sec^{-1}) = 542$; *4*, $k_\Psi(M^{-1}\ sec^{-1})/k_{CTAB}(M^{-1}\ sec^{-1}) = 395$	
Hydrolysis of *p*-$NO_2C_6H_4OCO(CH_2)_{11}H$, 31°C, pH = 9.06, tris buffer, $\mu = 0.083$ (KCl), 10 vol. % CH_3CN	*1*, $k_\Psi(M^{-1}\ sec^{-1})/k_{CTAB}(M^{-1}\ sec^{-1}) = 204$; *2*, $k_\Psi(M^{-1}\ sec^{-1})/k_{CTAB}(M^{-1}\ sec^{-1}) = 404$; *3*, $k_\Psi(M^{-1}\ sec^{-1})/k_{CTAB}(M^{-1}\ sec^{-1}) = 610$; *4*, $k_\Psi(M^{-1}\ sec^{-1})/k_{CTAB}(M^{-1}\ sec^{-1}) = 294$	

Table 12.1 *(Continued)*

Reaction	Effect of Surfactant	Reference
Hydrolysis of $p\text{-}NO_2C_6H_4OCO(CH_2)_{15}H$, 31°C, pH = 9.06, tris buffer, $\mu = 0.083$ (KCl), 10 vol. % CH_3CN	*1*, $k_\Psi(M^{-1}\ sec^{-1})k_{CTAB}(M^{-1}\ sec^{-1}) = 66$; *2*, $k_\Psi(M^{-1}\ sec^{-1})/k_{CTAB}(M^{-1}\ sec^{-1}) = 141$; *3*, $k_\Psi(M^{-1}\ sec^{-1})/k_{CTAB}(M^{-1}\ sec^{-1}) = 312$; *4*, $k_\Psi(M^{-1}\ sec^{-1})/k_{CTAB}(M^{-1}\ sec^{-1}) = 103$	
Hydrolysis of $HO_2C\text{-}C_6H_3(NO_2)\text{-}OCOCH_3$ 31°C, pH = 9.06, tris buffer, $\mu = 0.083$ (KCl), 10 vol. % CH_3CN	*1*, $k_\Psi(M^{-1}\ sec^{-1})/k_{CTAB}(M^{-1}\ sec^{-1}) = 1$; *2*. $k_\Psi(M^{-1}\ sec^{-1})/k_{CTAB}(M^{-1}\ sec^{-1}) = 4$; *3*, $k_\Psi(M^{-1}\ sec^{-1})/k_{CTAB}(M^{-1}\ sec^{-1}) = 12$; *4*, $k_\Psi(M^{-1}\ sec^{-1})/k_{CTAB}(M^{-1}\ sec^{-1}) = 13$	
Hydrolysis of $HO_2C\text{-}C_6H_3(NO_2)\text{-}OCO(CH_2)_3H$ 31°C, pH = 9.06, tris buffer, $\mu = 0.083$ (KCl), 10 vol. % CH_3CN	*1*, $k_\Psi(M^{-1}\ sec^{-1})/k_{CTAB}(M^{-1}\ sec^{-1}) = 1$; *2*, $k_\Psi(M^{-1}\ sec^{-1})/k_{CTAB}(M^{-1}\ sec^{-1}) = 5.3$; *3*, $k_\Psi(M^{-1}\ sec^{-1})/k_{CTAB}(M^{-1}\ sec^{-1}) = 21$; *4*, $k_\Psi(M^{-1}\ sec^{-1})/k_{CTAB}(M^{-1}\ sec^{-1}) = 39$	
Hydrolysis of $HO_2C\text{-}C_6H_3(NO_2)\text{-}OCO(CH_2)_5H$ 31°C, pH = 9.06, tris buffer, $\mu = 0.083$ (KCl), 10 vol. % CH_3CN	*1*, $k_\Psi(M^{-1}\ sec^{-1})/k_{CTAB}(M^{-1}\ sec^{-1}) = 0.9$; *2*, $k_\Psi(M^{-1}\ sec^{-1})/k_{CTAB}(M^{-1}\ sec^{-1}) = 4$; *3*, $k_\Psi(M^{-1}\ sec^{-1})/k_{CTAB}(M^{-1}\ sec^{-1}) = 19$; *4*, $k_\Psi(M^{-1}\ sec^{-1})/k_{CTAB}(M^{-1}\ sec^{-1}) = 36$	

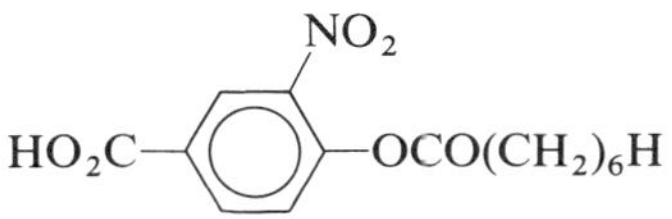 Hydrolysis of $HO_2C-C_6H_3(NO_2)-OCO(CH_2)_6H$ 31°C, pH = 9.06, tris buffer, μ = 0.083 (KCl), 10 vol. % CH_3CN	*1*, $k_\Psi(M^{-1}\,sec^{-1})/k_{CTAB}(M^{-1}\,sec^{-1}) = 0.9$; *2*, $k_\Psi(M^{-1}\,sec^{-1})/k_{CTAB}(M^{-1}\,sec^{-1}) = 3.6$; *3*, $k_\Psi(M^{-1}\,sec^{-1})/k_{CTAB}(M^{-1}\,sec^{-1}) = 16.5$; *4*, $k_\Psi(M^{-1}\,sec^{-1})/k_{CTAB}(M^{-1}\,sec^{-1}) = 14.7$	
Hydrolysis of 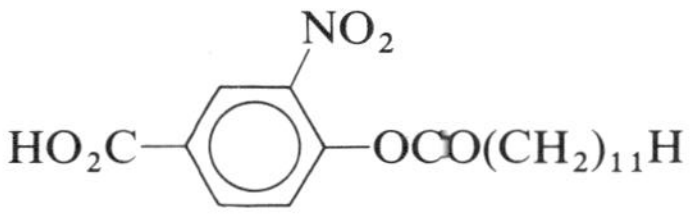 $HO_2C-C_6H_3(NO_2)-OCO(CH_2)_{11}H$ 31°C, pH = 9.06, tris buffer, μ = 0.083 (KCl), 10 vol. % CH_3CN	*1*, $k_\Psi(M^{-1}\,sec^{-1})/k_{CTAB}(M^{-1}\,sec^{-1}) = 0.8$; *2*, $k_\Psi(M^{-1}\,sec^{-1})/k_{CTAB}(M^{-1}\,sec^{-1}) = 3.7$; *3*, $k_\Psi(M^{-1}\,sec^{-1})/k_{CTAB}(M^{-1}\,sec^{-1}) = 17.1$; *4*, $k_\Psi(M^{-1}\,sec^{-1})/k_{CTAB}(M^{-1}\,sec^{-1}) = 23.2$	
Hydrolysis of 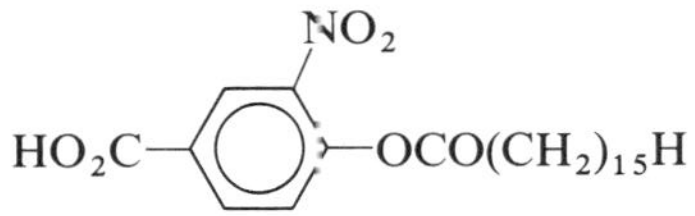 $HO_2C-C_6H_3(NO_2)-OCO(CH_2)_{15}H$ 31°C, pH = 9.06, tris buffer, μ = 0.083 (KCl), 10 vol. % CH_3CN	*1*, $k_\Psi(M^{-1}\,sec^{-1})/k_{CTAB}(M^{-1}\,sec^{-1}) = 25$; *2*, $k_\Psi(M^{-1}\,sec^{-1})/k_{CTAB}(M^{-1}\,sec^{-1}) = 64$; *3*, $k_\Psi(M^{-1}\,sec^{-1})/k_{CTAB}(M^{-1}\,sec^{-1}) = 98$; *4*, $k_\Psi(M^{-1}\,sec^{-1})/k_{CTAB}(M^{-1}\,sec^{-1}) = 32$	
Hydrolysis of *p*-nitrophenyl acetate, 25.0°C, tris buffer	CTAB + $n\text{-}C_{12}H_{25}\overset{+}{N}(CH_3)_2CH_2COC(NOH)PH, Cl^-$ (*1*), $k_\Psi(pH = 7.15) = 920\ M^{-1}\,sec^{-1}$, $k_\Psi(pH = 7.95) - 2570\ M^{-1}\,sec^{-1}$; CTAB + $n\text{-}C_{16}H_{33}\overset{+}{N}(CH_3)_2CH_2CH_2SH, Br^-$, $k_\Psi(pH = 7.15) = 10\ M^{-1}\,sec^{-1}$, $k_\Psi(pH = 7.95) = 615\ M^{-1}\,sec^{-1}$; CTAB + $n\text{-}C_{16}H_{33}\overset{+}{N}(CH_3)_2CH_2C(NOH)Ph, Br^-$ (*2*), $k_\Psi(pH = 7.15) = 55\ M^{-1}\,sec^{-1}$, $k_\Psi(pH = 7.95) = 325\ M^{-1}\,sec^{-1}$ Chymothrypsin (pH = 7.95) = 560 $M^{-1}\,sec^{-1}$	Anoardi et al., 1978a

Table 12.1 *(Continued)*

Reaction	Effect of Surfactant	Reference
Hydrolysis of *p*-nitrophenyl hexanoate, 25.0°C, tris buffer	*1*, $k_\Psi(\mathrm{pH} = 7.15) = 8.4 \times 10^3\ M^{-1}\ \mathrm{sec}^{-1}$, $k_\Psi(\mathrm{pH} = 7.95) = 2.3 \times 10^4\ M^{-1}\ \mathrm{sec}^{-1}$ (30°C); *2*, $k_\Psi(\mathrm{pH} = 7.15) = 4.8 \times 10^2\ M^{-1}\ \mathrm{sec}^{-1}$, $k_\Psi(\mathrm{pH} = 7.95) = 2.5 \times 10^3\ M^{-1}\ \mathrm{sec}^{-1}$	
Hydrolysis of *p*-nitrophenyl acetate, 25.0°C, tris buffer, $\mu = 0.1$ (KCl), 1% CH_3CN	CTAB + $n\text{-}C_{12}H_{25}\overset{+}{N}(Me_2)CH_2CH_2SH$, Br^-, $k_\Psi(\mathrm{pH} = 7.15) = 120\ M^{-1}\ \mathrm{sec}^{-1}$; CTAB + $n\text{-}C_{16}H_{33}\overset{+}{N}(Me_2)CH_2CH_2SH$, Br^-, $k_\Psi(\mathrm{pH} = 7.15) = 210\ M^{-1}\ \mathrm{sec}^{-1}$, $k_\Psi(\mathrm{pH} = 7.95) = 580\ M^{-1}\ \mathrm{sec}^{-1}$	Anoardi et al., 1978b
Thiolysis of *p*-nitrophenyl acetate, 30.0°C, pH = 8.5,		Cuccovia et al., 1978
p-Chlorothiophenoxide	CTAB, $k_\Psi/k_0 \sim 42$	
Thiophenoxide	CTAB, $k_\Psi/k_0 \sim 69$	
p-Methylthiophenoxide	CTAB, $k_\Psi/k_0 \sim 55$	
p-Methoxythiophenoxide	CTAB, $k_\Psi/k_0 \sim 57$	
Hydrolysis of *p*-nitrophenyl acetate, 25.0°C, pH = 8.0, $\mu = 0.05$ (KCl)	$n\text{-}C_{16}H_{33}\overset{+}{N}(Me_2)CH_2CH_2NHCOC(H)(CH_2Im)\text{—NH—C—tBOC}$, Cl^- (*1*), $k_\Psi(M^{-1}\ \mathrm{sec}^{-1})/k_{\mathrm{CTACl}}(M^{-1}\ \mathrm{sec}^{-1}) = 130$; $n\text{-}C_{16}H_{33}\overset{+}{N}(Me_2)CH_2Im$, Cl^-, $k_\Psi(M^{-1}\ \mathrm{sec}^{-1})/k_{\mathrm{CTACl}}(M^{-1}\ \mathrm{sec}^{-1}) = 360$; $n\text{-}C_{16}H_{33}\overset{+}{N}(Me_2)CH_2CH_2NHCO\text{—}C(H)(CH_2SH)\text{—}\overset{+}{N}H_3$, $2Cl^-$ (*2*), $k_\Psi(M^{-1}\ \mathrm{sec}^{-1})/k_{\mathrm{CTACl}}(M^{-1}\ \mathrm{sec}^{-1}) = 1860$	Moss et al., 1978c

Hydrolysis of L-*N*-Acetylphenylalanine *p*-nitrophenyl ester, 25.0°C, pH = 8.0, $\mu = 0.05$ (KCl)	*1*, $k_D \sim k_L$, $k_\Psi(M^{-1}\ sec^{-1})/k_{CTACl}(M^{-1}\ sec^{-1}) = 2.7$; $n\text{-}C_{16}H_{33}\overset{+}{N}(Me_2)CH_2Im, Cl^-$, $k_\Psi(M^{-1}\ sec^{-1})/k_{CTACl}(M^{-1}\ sec^{-1}) = 42$; *2*, $k_\Psi(M^{-1}\ sec^{-1})/k_{CTACl}(M^{-1}\ sec^{-1}) = 180$	
Hydrolysis of *p*-nitrophenyl dodecanoate, 30.0°C, pH = 10.8, dioxan:0.025 *M* phosphate buffer = 40:60 (v/v)	Bis-2-[2-(3α,7α,12α-trihydroxy-5β-cholanamidoethoxy)ethyl ether] $k_\Psi/k_0 \sim +$	Cairns-Smith and Rasool, 1978
Hydrolysis of 2,4-dinitrophenyl acetate, 60.0°C, pH varied, $\mu = 0.8\ M$ (KCl)	$CH_3(CH_2)_9\overset{+}{N}(Me_2)CH_2COOH, Br^-$ (*1*), $10^3\ k_\Psi = 42\ min^{-1}$ (pH ~ 3), $k_{OH} = 20{,}343\ M^{-1}\ min^{-1}$	Rav-Acha et al., 1978
Hydrolysis of 2,4-dinitrophenyl hexanoate, 60.0°C, pH varied, $\mu = 0.8\ M$ (KCl)	*1*, $10^3\ k_\Psi = 27\ min^{-1}$ (pH ~ 3), $k_{OH} = 6850\ M^{-1}\ min^{-1}$	
Hydrolysis of 2,4-dinitrophenyl decanoate, pH varied, $\mu = 0.8\ M$ (KCl)	*1*, $10^3\ k_\Psi$ (pH ~ 3) = $2.5\ min^{-1}$ (30.0°C), $27\ min^{-1}$ (60.0°C), $k_{OH} = 590\ M^{-1}\ min^{-1}$ (30.0°C), $8600\ M^{-1}\ min^{-1}$ (60.0°C); $CH_3(CH_2)_{15}\overset{+}{N}(Me_2)CH_2COOH, Br^-$, $10^3\ k_\Psi$ (pH ~ 3) = $82\ min^{-1}$ (60.0°C); $CH_3(CH_2)_9\overset{+}{N}(Me_2)CH_2CH_2COOH, Br^-$, $10^3\ k_\Psi$ (pH ~ 3) = $13\ min^{-1}$ (30.0°C), $k_{OH} = 500\ M^{-1}\ min^{-1}$ (30.0°C); $CH_3(CH_2)_9\overset{+}{N}(Me_2)(CH_2)_3COOH, Br^-$, $10^3\ k_\Psi$ (pH ~ 3) = $60\ min^{-1}$ (30.0°C), $650\ min^{-1}$ (60.0°C), $k_{OH} = 480\ M^{-1}\ min^{-1}$ (30.0°C), $30{,}000\ M^{-1}\ min^{-1}$ (60.0°C); $CH_3(CH_2)_{15}\overset{+}{N}(Me_2)(CH_2)_3COOH, Br^-$, $10^3\ k_\Psi$ (pH ~ 3) = $1000\ min^{-1}$ (60.0°C)	
Hydrolysis of *p*-nitrophenyl decanoate, 30.0°C, $[OH^-]$ varied	$CH_3(CH_2)_9\overset{+}{N}(Me_2)(CH_2)_2OH$ (*1*), $k_\Psi = 375\ M^{-1}\ sec^{-1}$; $CH_3(CH_2)_{11}\overset{+}{N}(Me_2)(CH_2)_2OH$ (*2*), $k_\Psi = 505\ M^{-1}\ sec^{-1}$; $CH_3(CH_2)_{13}\overset{+}{N}(Me_2)(CH_2)_2OH$ (*3*), $k_\Psi = 812\ M^{-1}\ sec^{-1}$; $CH_3(CH_2)_{15}\overset{+}{N}(Me_2)(CH_2)_2OH$ (*4*), $k_\Psi = 977\ M^{-1}\ sec^{-1}$; $CH_3(CH_2)_9\overset{+}{N}(Me_2)(CH_2)_3OH$ (*5*), $k_\Psi = 57\ M^{-1}\ sec^{-1}$;	Pillersdorf and Katzhendler, 1979a

Table 12.1 *(Continued)*

Reaction	Effect of Surfactant	Reference
	$CH_3(CH_2)_{13}\overset{+}{N}(Me_2)(CH_2)_3OH$ (*6*), $k_\Psi = 79\ M^{-1}\ sec^{-1}$; $CH_3(CH_2)_{15}\overset{+}{N}(Me_2)(CH_2)_3OH$ (*7*), $k_\Psi = 91\ M^{-1}\ sec^{-1}$; $CH_3(CH_2)_9\overset{+}{N}(Et_2)(CH_2)_2OH$ (*8*), $k_\Psi = 446\ M^{-1}\ sec^{-1}$; $CH_3(CH_2)_{15}\overset{+}{N}(Et_2)(CH_2)_2OH$ (*9*), $k_\Psi = 1070\ M^{-1}\ sec^{-1}$; $CH_3(CH_2)_9\overset{+}{N}(Me_2)C(CH_3)_2CH_2OH$ (*10*), $k_\Psi = 208\ M^{-1}\ sec^{-1}$; $CH_3(CH_2)_9\overset{+}{N}(Me_2)CH_2CH(CH_3)OH$ (*11*), $k_\Psi = 59\ M^{-1}\ sec^{-1}$; $CH_3(CH_2)_{13}\overset{+}{N}(Me_2)CH_2CH(CH_3)OH$ (*12*), $k_\Psi = 72\ M^{-1}\ sec^{-1}$; $CH_3(CH_2)_{15}\overset{+}{N}(Me_2)CH_2CH(CH_3)OH$ (*13*), $k_\Psi = 147\ M^{-1}\ sec^{-1}$; $CH_3(CH_2)_9\overset{+}{N}(Me_2)CH_2CH_3$ (*14*), $k_\Psi = 13.5\ M^{-1}\ sec^{-1}$; $CH_3(CH_2)_{15}\overset{+}{N}(Me_2)CH_2CH_3$ (*15*), $k_\Psi = 18.5\ M^{-1}\ sec^{-1}$	
Hydrolysis of 2,4-dinitrophenyl decanoate, 30.0°C, $[^-OH]$ varied	*1*, $k_\Psi = 3466\ M^{-1}\ sec^{-1}$; *2*, $k_\Psi = 5248\ M^{-1}\ sec^{-1}$; *3*, $k_\Psi = 7547\ M^{-1}\ sec^{-1}$; *4*, $k_\Psi = 9036\ M^{-1}\ sec^{-1}$; *5*, $k_\Psi = 57\ M^{-1}\ sec^{-1}$; *6*, $k_\Psi = 761\ M^{-1}\ sec^{-1}$; *7*, $k_\Psi = 846\ M^{-1}\ sec^{-1}$; *8*, $k_\Psi = 5250\ M^{-1}\ sec^{-1}$; *9*, $k_\Psi = 7333\ M^{-1}\ sec^{-1}$; *10*, $k_\Psi = 2666\ M^{-1}\ sec^{-1}$; *11*, $k_\Psi = 545\ M^{-1}\ sec^{-1}$; *12*, $k_\Psi = 615\ M^{-1}\ sec^{-1}$; *13*, $k_\Psi = 1388\ M^{-1}\ sec^{-1}$; *14*, $k_\Psi = 142\ M^{-1}\ sec^{-1}$; *15*, $k_\Psi = 189\ M^{-1}\ sec^{-1}$	

Hydrolysis of $NO_2-C_6H_4-C(=O)-O(CH_2)_2\overset{+}{N}(Me_2)(CH_2)_9CH_3$ 30.0°C, [^-OH] varied	*1*, $k_\Psi = 82\ M^{-1}\ sec^{-1}$; *2*, $k_\Psi = 80\ M^{-1}\ sec^{-1}$; *3*, $k_\Psi = 87\ M^{-1}\ sec^{-1}$; *4*, $k_\Psi = 94\ M^{-1}\ sec^{-1}$; *5*, $k_\Psi = 350\ M^{-1}\ sec^{-1}$; *6*, $k_\Psi = 430\ M^{-1}\ sec^{-1}$; *7*, $k_\Psi = 530\ M^{-1}\ sec^{-1}$; *8*, $k_\Psi = 101\ M^{-1}\ sec^{-1}$; *9*, $k_\Psi = 115\ M^{-1}\ sec^{-1}$; *10*, $k_\Psi = 85\ M^{-1}\ sec^{-1}$; *11*, $k_\Psi = 80\ M^{-1}\ sec^{-1}$; *12*, $k_\Psi = 80\ M^{-1}\ sec^{-1}$; *13*, $k_\Psi = 75\ M^{-1}\ sec^{-1}$; *14*, $k_\Psi = 52\ M^{-1}\ sec^{-1}$; *15*, $k_\Psi = 78\ M^{-1}\ sec^{-1}$	
Hydrolysis of $NO_2-C_6H_4-C(=O)-O-(CH_2)_2-\overset{+}{N}(Me_2)(CH_2)_9CH_3$ 30.0°C, [^-OH] varied	*1*, $k_\Psi = 42\ M^{-1}\ sec^{-1}$; *2*, $k_\Psi = 51\ M^{-1}\ sec^{-1}$; *3*, $k_\Psi = 54\ M^{-1}\ sec^{-1}$; *4*, $k_\Psi = 62\ M^{-1}\ sec^{-1}$; *5*, $k_\Psi = 6.3\ M^{-1}\ sec^{-1}$; *6*, $k_\Psi = 7.1\ M^{-1}\ sec^{-1}$; *7*, $k_\Psi = 7.7\ M^{-1}\ sec^{-1}$; *8*, $k_\Psi = 51\ M^{-1}\ sec^{-1}$; *9*, $k_\Psi = 60\ M^{-1}\ sec^{-1}$; *10*, $k_\Psi = 27\ M^{-1}\ sec^{-1}$; *11*, $k_\Psi = 7\ M^{-1}\ sec^{-1}$; *12*, $k_\Psi = 9.1\ M^{-1}\ sec^{-1}$; *13*, $k_\Psi = 19\ M^{-1}\ sec^{-1}$; *14*, $k_\Psi = 6.1\ M^{-1}\ sec^{-1}$; *15*, $k_\Psi = 8\ M^{-1}\ sec^{-1}$ Reactivity of cationic hydroxylic micelles toward ester hydrolysis is attributed to electrostatic parameters affecting the basicity of the head group	
Hydrolysis of phenyl decanoate, 30.0°C, $\mu = 0.15\ M$ (KBr)	$CH_3(CH_2)_{15}\overset{+}{N}(Me_2)(CH_2)_3CONHOH, Br^-$, $10^3\ k_0 = 2.5\ sec^{-1}$, $10^3\ k_c = 16\ sec^{-1}$; $CH_3(CH_2)_{15}\overset{+}{N}(Me_2)C(CH_2)_3CONHOH$, $Br^- + CH_3(CH_2)_{10}\overset{+}{N}(Me_2)(CH_2)_2H, Br^-$ (1:4), $10^3\ k_0 = 0.08\ sec^{-1}$, $10^3\ k_c = 5.4\ sec^{-1}$; $CH_3(CH_2)_{15}\overset{+}{N}(Me_2)C(CH_2)_3CONHOH$, $+ CH_3\overset{+}{N}(Me_2)CH_2, COOH, Br^-$ (1:4), $10^3\ k_0 = 0.25\ sec^{-1}$, $10^3\ k_c = 3\ sec^{-1}$;	Pillersdorf and Katzhendler, 1979b

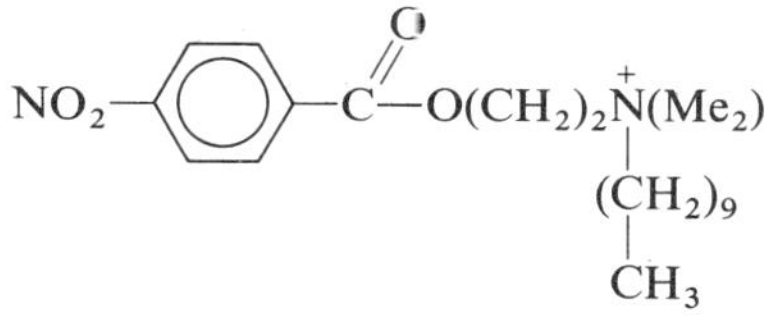

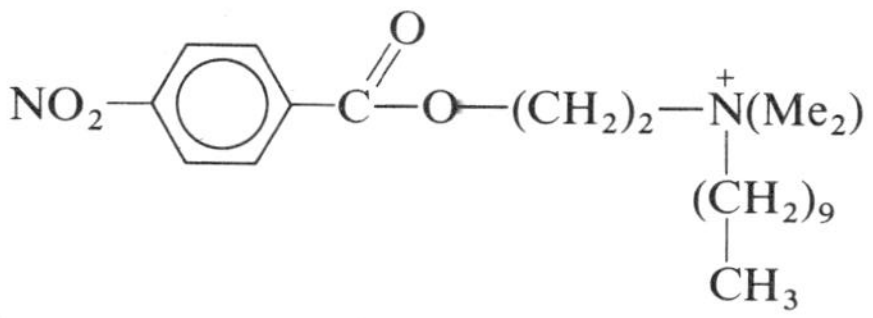

Table 12.1 *(Continued)*

Reaction	Effect of Surfactant	Reference
	$CH_3(CH_2)_{15}\overset{+}{N}(Me_2)C(CH_2)_3CONHOH$, Br^- + $CH_3(CH_2)_9\overset{+}{N}(Me_2)(CH_2)_2COOH$, Br^- (1:4), $10^3\ k_0 = 0.16\ sec^{-1}$, $10^3\ k_c = 8.1\ sec^{-1}$; $CH_3(CH_2)_{15}\overset{+}{N}(Me_2)C(CH_2)_3CONHOH$, Br^- + $CH_3(CH_2)_{15}\overset{+}{N}(Me_2)(CH_2)_3COOH$, Br^- (1:4), $10^3\ k_0 = 0.4\ sec^{-1}$, $10^3\ k_c = 5.5\ sec^{-1}$	
Hydrolysis of $CH_3(CH_2)_9\overset{+}{N}(Me_2)(CH_2)_2OCOOC_6H_4CH_3$-*p*, 30.0°C, $\mu = 0.8M$	None, $k_{OH} = 0.43\ M^{-1}\ sec^{-1}$ $CH_3(CH_2)_9\overset{+}{N}(Me_2)(CH_2)_2OH$, Br^- (*1*), $k'_{OH} = 0.25\ M^{-1}\ sec^{-1}$ ($0.83\ M^{-1}\ sec^{-1}$, $\mu = 0.15$); $CH_3(CH_2)_9\overset{+}{N}(Me_2)(CH_2)_3OH$, Br^- (*2*), $(k'_1 + k'_3) = 1.14\ M^{-1}\ sec^{-1}$ ($4.3\ M^{-1}\ sec^{-1}$, $\mu = 0.15$), $k_c = 6.5\ sec^{-1}$; $CH_3(CH_2)_9\overset{+}{N}(Et_2)CH_2CH_3$, Br^- (*3*), $k'_{OH} = 0.32\ M^{-1}\ sec^{-1}$; $CH_3(CH_2)_9\overset{+}{N}(Me_2)CH_2CH_3$, Br^- (*4*), $k'_{OH} = 0.13\ M^{-1}\ sec^{-1}$; $CH_3(CH_2)_9\overset{+}{N}(Me_2)CH_2COOH$, Br^- (*5*), $k'_{OH} = 0.12\ M^{-1}\ sec^{-1}$; $CH_3(CH_2)_9\overset{+}{N}(Me_2)(CH_2)_3COOH$, Br^- (*6*), $k'_{OH} = 0.12\ M^{-1}\ sec^{-1}$	Pillersdorf and Katzhendler, 1979c
Hydrolysis of $CH_3(CH_2)_9\overset{+}{N}(Me_2)(CH_2)_2OCOOC_6H_5$, Br^-, 30.0°C, $\mu = 0.8\ M$	None, $k'_{OH} = 0.87\ M^{-1}\ sec^{-1}$; *1*, $k'_{OH} = 0.53\ M^{-1}\ sec^{-1}$ ($2.0\ M^{-1}\ sec^{-1}$, $\mu = 0.15$); *2*, $(k'_1 + k'_3) = 2.3\ M^{-1}\ sec^{-1}$ ($8.3\ M^{-1}\ sec^{-1}$, $\mu = 0.15$), $k_c = 12.6\ sec^{-1}$ ($46.4\ sec^{-1}$, $\mu = 0.15$); *3*, $k'_{OH} = 0.69\ M^{-1}\ sec^{-1}$; *4*, $k'_{OH} = 0.32\ M^{-1}\ sec^{-1}$; *5*, $k'_{OH} = 0.26\ M^{-1}\ sec^{-1}$; *6*, $k'_{OH} = 0.25\ M^{-1}\ sec^{-1}$	

Hydrolysis of $CH_3(CH_2)_9\overset{+}{N}(Me_2)(CH_2)_2\ OCOOC_6H_4Br$-*p*, 30.0°C, $\mu = 0.80\ M$	None, $k'_{OH} = 1.8\ M^{-1}\ sec^{-1}$; *1*, $k'_{OH} = 1.32\ M^{-1}\ sec^{-1}$ ($5.4\ M^{-1}\ sec^{-1}$, $\mu = 0.15$); *2*, $(k'_1 + k'_3) = 5.5\ M^{-1}\ sec^{-1}$ ($24.0\ M^{-1}\ sec^{-1}$, $\mu = 0.15$), $k_c = 31\ sec^{-1}$; *3*, $k'_{OH} = 1.66\ M^{-1}\ sec^{-1}$; *4*, $k'_{OH} = 0.85\ M^{-1}\ sec^{-1}$; *5*, $k'_{OH} = 0.68\ M^{-1}\ sec^{-1}$; *6*, $k'_{OH} = 0.69\ M^{-1}\ sec^{-1}$
Hydrolysis of $CH_3(CH_2)_9\overset{+}{N}(Me_2)(CH_2)_2OCOOC_6H_4NO_2$-*p* 30.0°C, $\mu = 0.80\ M$	None, $k'_{OH} = 13\ M^{-1}\ sec^{-1}$; *1*, $k'_{OH} = 18\ M^{-1}\ sec^{-1}$ ($81\ M^{-1}\ sec^{-1}$, $\mu = 0.15$); *2*, $(k'_1 + k'_3) = 82\ M^{-1}\ sec^{-1}$ ($350\ M^{-1}\ sec^{-1}$, $\mu = 0.15$), $k_c = 471\ sec^{-1}$; *3*, $k'_{OH} = 17.6\ M^{-1}\ sec^{-1}$; *4*, $k'_{OH} = 12\ M^{-1}\ sec^{-1}$; *5*, $k'_{OH} = 9.1\ M^{-1}\ sec^{-1}$; *6*, $k'_{OH} = 9.3\ M^{-1}\ sec^{-1}$
Hydrolysis of $CH_3(CH_2)_9\overset{+}{N}(Me_2)(CH_2)_3OCOOC_5H_5CH_3$-*p*, 30.0°C, $\mu = 0.8\ M$	None, $k'_{OH} = 0.11\ M^{-1}\ sec^{-1}$; *1*, $(k'_1 + k'_3) = 0.22\ M^{-1}\ sec^{-1}$ ($0.51\ M^{-1}\ sec^{-1}$, $\mu = 0.15$), $k_c = 0.072\ sec^{-1}$; *2*, $k'_{OH} = 0.023\ M^{-1}\ sec^{-1}$ ($0.063\ M^{-1}\ sec^{-1}$, $\mu = 0.15$); *4*, $k'_{OH} = 0.016\ M^{-1}\ sec^{-1}$
Hydrolysis of $CH_3(CH_2)_9\overset{+}{N}(Me_2)(CH_2)_3OCOOC_6H_5$, $\mu = 0.8\ M$	None, $k'_{OH} = 0.17\ M^{-1}\ sec^{-1}$; *1*, $(k'_1 + k'_3) = 0.37\ M^{-1}\ sec^{-1}$ ($1{,}25\ M^{-1}\ sec^{-1}$, $\mu = 0.15$), $k_c = 0.12\ sec^{-1}$; *2*, $k'_{OH} = 0.05\ M^{-1}\ sec^{-1}$ ($0.16\ M^{-1}\ sec^{-1}$, $\mu = 0.15$); *4*, $k'_{OH} = 0.04\ M^{-1}\ sec^{-1}$
Hydrolysis of $CH_3(CH_2)_9\overset{+}{N}(Me_2)(CH_2)_3OCOOC_6H_4Br$-*p*, $\mu = 0.8\ M$	None, $k'_{OH} = 0.36\ M^{-1}\ sec^{-1}$; *1*, $(k'_1 + k'_3) = 1.21\ M^{-1}\ sec^{-1}$ ($3.7\ M^{-1}\ sec^{-1}$, $\mu = 0.15$), $k_c = 0.4\ sec^{-1}$; *2*, $k'_{OH} = 0.13\ M^{-1}\ sec^{-1}$ ($0.44\ M^{-1}\ sec^{-1}$, $\mu = 0.15$); *4*, $k'_{OH} = 0.11\ M^{-1}\ sec^{-1}$
Hydrolysis of $CH_3(CH_2)_9\overset{+}{N}(Me_2)(CH_2)_3OCOOC_6H_4CN$-*p*, 30°C, $\mu = 0.8\ M$	None, $k'_{OH} = 1.19\ M^{-1}\ sec^{-1}$; *1*, $(k'_1 + k'_3) = 6.31\ M^{-1}\ sec^{-1}$ ($27.6\ M^{-1}\ sec^{-1}$, $\mu = 0.15$), $k_c = 2.0\ sec^{-1}$; *2*, $k'_{OH} = 0.83\ M^{-1}\ sec^{-1}$ ($3.46\ M^{-1}\ sec^{-1}$, $\mu = 0.15$); *4*, $k'_{OH} = 0.80\ M^{-1}\ sec^{-1}$
Hydrolysis of $CH_3(CH_2)_9\overset{+}{N}(Me_2)(CH_2)_3OCOOC_6H_4Br$-*p*, 30.0°C, $\mu = 0.8\ M$	None, $k'_{OH} = 1.7\ M^{-1}\ sec^{-1}$; *1*, $(k'_1 + k'_3) = 11.0\ M^{-1}\ sec^{-1}$ ($42\ M^{-1}\ sec^{-1}$, $\mu = 0.15$), $k_c = 3.6\ sec^{-1}$; *2*, $k'_{OH} = 1.3\ M^{-1}\ sec^{-1}$ ($6.3\ M^{-1}\ sec^{-1}$, $\mu = 0.15$); *4*, $k'_{OH} = 1.7\ M^{-1}\ sec^{-1}$

Table 12.1 *(Continued)*

Reaction	Effect of Surfactant	Reference
Hydrolysis of $CH_3(CH_2)_9OCOOC_6H_4CH_3$-*p*, 30.0°C, $\mu = 0.8\ M$	*1*, $10^3(k'_1 + k'_3) = 0.52\ M^{-1}\ sec^{-1}$, $10^3\ k_c = 0.13\ sec^{-1}$; *2*, $10^3(k'_1 + k'_3) = 0.40\ M^{-1}\ sec^{-1}$, $10^3\ k_c = 1.77\ sec^{-1}$; *4*, $10^3\ k'_{OH} = 0.16\ M^{-1}\ sec^{-1}$	
Hydrolysis of $CH_3(CH_2)_9OCOOC_6H_5$, 30.0°C, $\mu = 0.8\ M$	*1*, $10(k'_1 + k'_3) = 1.45\ M^{-1}\ sec^{-1}$, $10^3\ k_c = 0.38\ sec^{-1}$; *2*, $10^3(k'_1 + k'_3) = 1.0\ M^{-1}\ sec^{-1}$, $10^3 k_c = 4.41\ sec^{-1}$; *4*, $10^3 k'_{OH} = 0.4\ M^{-1}\ sec^{-1}$	
Hydrolysis of $CH_3(CH_2)_9OCOOC_6H_4Br$-*p*, 30.0°C, $\mu = 0.8\ M$	*1*, $10^3(k'_1 + k'_3) = 5.8\ M^{-1}\ sec^{-1}$, $10^3 k_c = 1.66\ sec^{-1}$; *2*, $10^3(k'_1 + k'_3) = 3.6\ M^{-1}\ sec^{-1}$, $10^3 k_c = 16.9\ sec^{-1}$; *4*, $10^3 k'_{OH} = 1.3\ M^{-1}\ sec^{-1}$	
Hydrolysis of $CH_3(CH_2)_9OCOOC_6H_4CN$-*p*, 30.0°C, $\mu = 0.8\ M$	*1*, $10^3(k'_1 + k'_3) = 60\ M^{-1}\ sec^{-1}$, $10^3 k_c = 17.3\ sec^{-1}$; *2*, $10^3(k'_1 + k'_3) = 45\ M^{-1}\ sec^{-1}$, $10^3 k_c = 235\ sec^{-1}$; *4*, $10^3 k'_{OH} = 13\ M^{-1}\ sec^{-1}$	
Hydrolysis of $CH_3(CH_2)_9OCOOC_6H_4NO_2$-*p*, 30.0°C, $\mu = 0.8\ M$	*1*, $10^3(k'_1 + k'_3) = 160\ M^{-1}\ sec^{-1}$, $10^3 k_c = 50\ sec^{-1}$; *2*, $10^3(k'_1 + k'_3) = 100\ M^{-1}\ sec^{-1}$, $10^3 k_c = 552\ sec^{-1}$; *4*, $10^3 k'_{OH} = 25\ M^{-1}\ sec^{-1}$ Hydrocyclic micelles are ineffective catalysts; data fits the Hammett's equation	
$PhCH_2C(O)OC_6H_4NO_2p$ $+ OH^-$ $+ O_2H^-$ $+ PhC(Me_2)O_2^-$ 30.0°C, *p*H = 9.45, 0.05 *M* borate buffer	$CH_3(CH_2)_{15}\overset{+}{N}Me$, Cl^- $k_\Psi/k_0 \sim 31$ $k_\Psi/k_0 \sim 91$ $k_\Psi/k_0 \sim 9 \times 10^3$ Hydrophobic effects in micellar catalysis are demonstrated	Brown and Darwent, 1979a

$Me(CH_2)_9CH(CH_2)_6\overset{+}{N}Me_3, Cl^-$ with $O{=}C{-}OC_6H_4NO_2\text{-}p$ on CH	$CH_3(CH_2)_{15}\overset{+}{N}Me_3, Cl^-$, $k_\Psi/k_0 \sim +$	Brown and Darwent, 1979b
$Me(CH_2)_{13}CH{-}(CH_2)_2\overset{+}{N}Me_3, Cl^-$ with $O{=}COC_6H_4NO_2\text{-}p$ on CH	Rate enhancements occur below concentrations associated with micelle formation; hydrophobic ion pairs enhance effective anion concentration	
$+ OH^-, O_2H^-, PhCMe_2OO^-$, 30.0°C, pH = 9.45, 0.05 *M* borate buffer		
Hydrolysis of *p*-nitrophenyl acetate, 313°K, p*D* varied, (followed by 270 MHz ^1H nmr)	$CH_3(CH_2)_{12}CH(CH_2)_4\overset{+}{N}Me_3, Cl^-$ with $O{=}C{-}N(H)CH(CO_2CH_3)$ on CH, and $CH_2{-}Im$ on CH(CO$_2$CH$_3$)	Brown et al., 1979
	Rapid reaction with *p*-nitrophenyl acetate in phosphate buffer, p*D* 7–8; entirely by acyl transfer to the imidazole with no concurrent general-base catalysis; the resulting acyl-imidazole breaks down by a pH independent route between p*D* 7.0–7.9, forming acetate ion and monoacetyl phosphate	
Hydrolysis of *p*-nitrophenyl *N*-acetylphenylalanine, 313 K, p*D* varied	No acylimidazole intermediate is observed	
Hydrolysis of *p*-nitrophenyl acetate, 313 K, p*D* varied	$CH_3(CH_2)_{14}C(H)(CH_2)_2\overset{+}{N}Me_3, Cl^-$ with $O{=}CN(OH)CH_3$ on C(H)	
	Ready acylation to give acylhydroxamate that is unstable to hydrolysis	
p-Nitrophenylacetate + OH^-, 25.0°C, no buffers	CTAB, $k_\Psi/k_0 \sim +$; k_Ψ determined in series I: $[OH]_{tot} = 10$ m*M*, $[CTAB] = [Br^-]_{tot}$ varied; series II: $[CTAB] = 10$ m*M*, $[^-OH]_{tot} + [Br^-]_{tot} = 26$ m*M*, ratio $[OH^-]/[Br^-]$ varied; series III: $[CTAB] = [Br^-]_{tot} = 10$ m*M*, $[^-OH]_{tot}$ varied. Kinetic data are quantitatively treated in terms of counterion binding and micellar surface potentials (see Section 11.1)	Almgren and Rydholm, 1979

Table 12.1 (*Continued*)

Reaction	Effect of Surfactant	Reference
Hydrolysis of	Dodecylhydrogen sulfate	Taniguchi et al., 1979
Methyl acetate	$\Delta\Delta V^{\ddagger} = 4.0$ cm^3 mole^{-1}, $\Delta V^{\ddagger} = -5.3$ cm^3 mole^{-1}	
Ethylacetate	$\Delta\Delta V^{\ddagger} = 9.8$ cm^3 mole^{-1}, $\Delta V^{\ddagger}_{\Psi} = 0.40$ cm^3 mole^{-1}	
n-Propylacetate	$\Delta\Delta V^{\ddagger} = 18.7$ cm^3 mole^{-1}, $\Delta V^{\ddagger}_{\Psi} = 9.7$ cm^3 mole^{-1}	
n-Butylacetate	$\Delta\Delta V^{\ddagger} = 27.9$ cm^3 mole^{-1}, $\Delta V^{\ddagger}_{\Psi} = 20$ cm^3 mole^{-1}	
40.0°C, pressure varied up to 2 kbar	$\Delta\Delta V^{\ddagger} = \Delta V^{\ddagger}_{\Psi} - \Delta V^{\ddagger}_{HCl}$	
Hydrolysis of *p*-$NO_2C_6H_4OCOR$, 30.0°C, $\mu = 0.10$ (KCl), 10.8% (v/v) EtOH–1% (v/v) dioxane	*N*-Hexadecyl-N^{α}-glutanyl-L-cysteinamide (*1*)	Murakami et al., 1979a
R = CH_3	$k_{\Psi}/k_0 \sim 1$ (pH = 9.30)	
R = $CH_3(CH_2)_4$	$k_{\Psi}/k_0 \sim 9.8$ (pH = 9.30), 3.0 (pH = 12.0)	
R = $CH_3(CH_2)_8$	$k_{\Psi}/k_0 \sim 575$ (pH = 9.30), 43 (pH = 12.0)	
R = $CH_3(CH_2)_{10}$	$k_{\Psi}/k_0 \sim 1509$ (pH = 9.30), 267 (pH – 12.0)	
R = $CH_3(CH_2)_{14}$	$k_{\Psi}/k_0 \sim 2397$ (pH = 9.30), 1432 (pH = 12.0)	
R = C_6H_{11}— (cyclohexyl, H)	$k_{\Psi}/k_0 \sim 11.3$ (pH = 9.30)	
R = C_6H_{11}—CH_2	$k_{\Psi}/k_0 \sim 13.6$ (pH = 9.30)	
R = C_6H_{11}—$CH(CH_3)$—	$k_{\Psi}/k_0 \sim 1.75$ (pH = 9.30)	
R = C_6H_5—	$k_{\Psi}/k_0 \sim 140$ (pH = 9.30)	
R = C_6H_4—$CH_2OCONHCH_2$	$k_{\Psi}/k_0 \sim 530$ (pH = 9.30)	

	N-Hexadecanoyl-L-cysteine (*2*)	
R = CH_3	$k_\Psi/k_0 \sim 1$ (pH = 9.30)	
R = $CH_3(CH_2)_4$	$k_\Psi/k_0 \sim 4.4$ (pH = 9.30), 1.9 (pH = 12.0)	
R = $CH_3(CH_2)_8$	$k_\Psi/k_0 \sim 199$ (pH = 9.30), 19.9 (pH = 12.0)	
R = $CH_3(CH_2)_{10}$	$k_\Psi/k_0 \sim 441$ (pH = 9.30), 138 (pH = 12.0)	
R = $CH_3(CH_2)_{14}$	$k_\Psi/k_0 \sim 801$ (pH = 9.30), 292 (pH = 12.0)	
R = C_6H_{11}—	$k_\Psi/k_0 \sim 5.2$ (pH = 9.30)	
R = C_6H_{11}—CH_2—	$k_\Psi/k_0 \sim 3.6$ (pH = 9.30)	
R = C_6H_5—$CH(CH_3)$—	$k_\Psi/k_0 \sim 5.7$ (pH = 9.30)	
R = C_6H_5—	$k_\Psi/k_0 \sim 3.5$ (pH = 9.30)	
R = C_6H_5—$CH_2OCONHCH_2$	$k_\Psi/k_0 \sim 6.6$ (pH = 9.30)	
	1 + CTAB (1:20)	
R = CH_3	$k_\Psi/k_0 \sim 3320$ (pH = 7.60)	
R = $CH_3(CH_2)_4$	$k_\Psi/k_0 \sim 3080$ (pH = 7.60)	
	2 + CTAB (1:16)	
R = CH_3	$k_\Psi/k_0 \sim 611$ (pH = 7.60)	
R = $CH_3(CH_2)_4$	$k_\Psi/k_0 \sim 1360$ (pH = 7.60)	
	2 + CTAB (1:20)	
R = CH_3	$k_\Psi/k_0 \sim 368$ (pH = 7.60)	
R = $CH_3(CH_2)_4$	$k_\Psi/k_0 \sim 1160$ (pH = 7.60)	
Hydrolysis of *p*-nitrophenyl acetate, 30.0°C, pH = 8.65, $\mu = 0.10\ M$ (KCl), 9.8% v/v EtOH, 1.0% v/v dioxane, 1.0% MeOH	*N*-Hexadecyl-N^α-(3-trimethylammoniopropionyl)-L-cysteinamide bromide, $k_\Psi/k_0 \sim 376$; *N*-Dodecyl-N^α-(6-trimethylammoniohexanoyl)-L-cysteinamide bromide, $k_\Psi/k_0 \sim 326$	Murakami et al., 1979b

Table 12.1 *(Continued)*

Reaction	Effect of Surfactant	Reference
Hydrolysis of *p*-nitrophenyl acetate, 30.0°C, pH = 8.65, $\mu = 0.10\ M$ (KCl), 9.8% v/v EtOH, 1.0% v/v dioxane, 1.0% v/v MeOH	*N*-Hexadecyl-N^{α}-(3-trimethylammoniopropionyl)-L-cysteinamide bromide, $k_{\Psi}/k_0 \sim 2560$; *N*-Dodecyl-N^{α}-(6-trimethylammoniohexanoyl-L-cysteinamide bromide, $k_{\Psi}/k_0 \sim 884$	
Hydrolysis of *p*-nitrophenyl acetate, 30.0°C, pH = 8.9, 3 v/v % EtOH, $\mu = 0.01\ M$ (KCl) (see scheme above, Kunitake et al., 1975)	Multifunctional surfactants containing hydroxamate and imidazole moieties $CH_3(CH_2)_{11}$—N^+(imidazolium)N—CH_2—C(=O)—N(O^-)—CH_2—C_6H_5 R = H: $C_{12}Im^+$—HA R = CH_3: $C_{12}MIm^+$—HA $CH_3(CH_2)_{N-1}$—(pyridinium N^+)—CH_2—C(=O)—N(O^-)—CH_2—C_6H_5 n = 8: C_8Py^+—HA n = 13: $C_{13}Py^+$—HA $CH_3(CH_2)_{11}$N(O^-)—C(=O)—C_6H_5 C_{12}BHA	Kunitake et al., 1979

$CH_3(CH_2)_{11}-C(=O)-N(O^-)-CH_3$ $C_{13}MHA$

$C_6H_5-CH_2N(O^-)-C(=O)-C_6H_5$ BBHA

$C_{12}IM^+HA$, $k_a = 12\ M^{-1}\ sec^{-1}$; $C_{12}Im^+HA$ + SDS, $k_a = 1.2\ M^{-1}\ sec^{-1}$; $C_{12}IM^+HA$ + CTAB, $k_a = 1000\ M^{-1}\ sec^{-1}$, $10^2 k_d = 3.5\ sec^{-1}$; $C_{12}Im^+HA$ + POOA, $k_a = 270\ M^{-1}\ sec^{-1}$, $10^3 k_d = 0.21\ sec^{-1}$; $C_{12}Im^+HA$ + DMOG, $k_a = 1200\ M^{-1}\ sec^{-1}$, $10^2 k_d = 0.37\ sec^{-1}$; $C_{12}MIm^+HA$, $k_a = 13\ M^{-1}\ sec^{-1}$; $C_{12}MIM^+HA$ + SDS, $k_a = 0.8\ M^{-1}\ sec^{-1}$; $C_{12}MIm^+HA$ + CTAB, $k_a = 1300\ M^{-1}\ sec^{-1}$; $C_{12}MIm^+HA$ + POOA, $k_a = 250\ M^{-1}\ sec^{-1}$; $C_{12}MIm^+HA$ + DMOG, $k_a = 1100\ M^{-1}\ sec^{-1}$; $C_{13}Py^+HA$, $k_a = 14\ M^{-1}\ sec^{-1}$; $C_{13}Py^+HA$ + SDS, $k_a = 1.0\ M^{-1}\ sec^{-1}$; $C_{13}Py^+HA$ + CTAB, $k_a = 1100\ M^{-1}\ sec^{-1}$; $C_{13}Py^+HA$ + POOA, $k_a = 310\ M^{-1}\ sec^{-1}$; $C_{13}Py^+HA$ + DMOG, $k_a = 1300\ M^{-1}\ sec^{-1}$; C_8Py^+HA, $k_a = 12\ M^{-1}\ sec^{-1}$; C_8Py^+HA + CTAB, $k_a = 320\ M^{-1}\ sec^{-1}$; C_8Py^+HA + POOA, $k_a = 45\ M^{-1}\ sec^{-1}$; C_8Py^+HA + DMOG, $k_a = 420\ M^{-1}\ sec^{-1}$; $C_{12}BHA$ + SDS, $k_a = 0.9\ M^{-1}\ sec^{-1}$; $C_{12}BHA$ + CTAB, $k_a = 1500\ M^{-1}\ sec^{-1}$; $C_{12}BHA$ + POOA, $k_a = 25\ M^{-1}\ sec^{-1}$; $C_{12}BHA$ + DMOG, $k_a = 75\ M^{-1}\ sec^{-1}$; $C_{13}MHA$ + SDS, $k_a = 1.3\ M^{-1}\ sec^{-1}$;

Table 12.1 *(Continued)*

Reaction	Effect of Surfactant	Reference
	$C_{13}MHA$ + CTAB, $k_a = 1300\ M^{-1}\ sec^{-1}$; $C_{13}MHA$ + POOA, $k_a = 32\ M^{-1}\ sec^{-1}$; $C_{13}MHA$ + DMOG, $k_a = 45\ M^{-1}\ sec^{-1}$; BBHA, $k_a = 12\ M^{-1}\ sec^{-1}$; BBHA + CTAB, $k_a = 1500\ M^{-1}\ sec^{-1}$; BBHA + POOA, $k_a = 13\ M^{-1}\ sec^{-1}$; BBHA + DMOG, $k_a = 23\ M^{-1}\ sec^{-1}$ Rate enhancements are rationalized in terms of hydrophobic ion pairs	
Hydrolysis of $CH_3COO-C_6H_4-X$ 25.0°C, pH = 9.2, 0.05 *M* carbonate buffer	$C_{18}H_{37}\overset{+}{N}(Me_2)CH_2CH_2Im, Cl^-$ X = pNO$_2$, $10^3k_\Psi = 450\ sec^{-1}$ X = mNO$_2$, $10^3k_\Psi = 69\ sec^{-1}$ X = pCOCH$_3$, $10^3k_\Psi = 100\ sec^{-1}$ X = pCl, $10^3k_\Psi = 12.2\ sec^{-1}$ X = H, $10^3k_\Psi = 4.95\ sec^{-1}$ X = p-CH$_3$, $10^3k_\Psi = 4.45\ sec^{-1}$ X = m-CH$_3$, $10^3k_\Psi = 4.61\ sec^{-1}$ X = pt-Bu, $10^3k_\Psi = 12.8\ sec^{-1}$ X = pOCH$_3$, $10^3k_\Psi = 3.14\ sec^{-1}$	Tagaki et al., 1979
Hydrolysis of $C_5H_{11}COO-C_6H_4-X$ 25.0°C, pH = 9.2, 0.05 *M* carbonate buffer	$C_{18}H_{37}\overset{+}{N}(Me_2)CH_2CH_2Im, Cl^-$ X = p-NO$_2$, $10^3k_\Psi = 1250\ sec^{-1}$ X = pCOCH$_3$, $10^3k_\Psi = 313\ sec^{-1}$ X = pCl, $10^3k_\Psi = 33.3\ sec^{-1}$ X = H, $10^3k_\Psi = 23.3\ sec^{-1}$ X = pCH$_3$, $10^3k_\Psi = 9.8\ sec^{-1}$ X = ptBu, $10^3k_\Psi = 9.52\ sec^{-1}$ X = pOCH$_3$, $10^3k_\Psi = 9.42\ sec^{-1}$	

Hydrolysis of D- or L- $CH_3CH_2OCONHCH(CH_2Ph)OCCOC_6H_4NO_2$-*p* 25.0°C, pH = 7.17, 0.05 *M* tris buffer, μ = 0.2 *M* (KCl), 10.0–6.67 v/v % MeOH—CH_3CN	CTAB + L-$CH_3(CH_2)_{10}CONHCH(CH_2Im)$—COOH (*1*), $k_L = 56.3\ M^{-1}\ sec^{-1}$, $k_D = 36.2\ M^{-1}\ sec^{-1}$; CTAB + D-*1*, $k_L = 36.2\ M^{-1}\ sec^{-1}$, $k_D = 55.8\ M^{-1}\ sec^{-1}$; CTAB + $CH_3(CH_2)_{10}CONHOH$ (*1*), $k_L = 110.7\ M^{-1}\ sec^{-1}$, $k_D = 111.0\ M^{-1}\ sec^{-1}$ L-$CH_3(CH_2)_{11}NHCOCH(CH_3)NMe_3, I^-$ (*3*) L-*3* + L-*1*, $k_L = 20.5\ M^{-1}\ sec^{-1}$, $k_D = 13.5\ M^{-1}\ sec^{-1}$; L-*3* + D-*1*, $k_L = 16.5\ M^{-1}\ sec^{-1}$, $k_D = 23.5\ M^{-1}\ sec^{-1}$; L-*3* + *2*, $k_L = 65.7\ M^{-1}\ sec^{-1}$, $k_D = 70\ M^{-1}\ sec^{-1}$	Yamada et al., 1979a
Hydrolysis of $R_2OCONHCH(R_1)COOC_6H_4$—NO_2p 25.0°C, pH = 7.4, 0.05 *M* tris buffer, μ = 0.2 *M* (KCl)		Yamada et al., 1979b
$R_1 = H, R_2 = C_6H_5CH_2$	CTAB + L-$CH_3(CH_2)_{10}CONHCH(CH_2Im)COOH$ (*60*), $k_\Psi = 37.4\ M^{-1}\ sec^{-1}$ CTAB + D-*1*, $k_\Psi = 37.4\ M^{-1}\ sec^{-1}$	
$R_1 = CH_3, R_2 = C_6H_5CH_2$ (L)	CTAB + L-*1*, $k_\Psi = 50.8\ M^{-1}\ sec^{-1}$; CTAB + D-*1*, $k_\Psi = 33.8\ M^{-1}\ sec^{-1}$	
$R_1 = CH(CH_3)_2, R_2 = C_6H_5CH_2$ (L)	CTAB + L-*1*, $k_\Psi = 12.2\ M^{-1}\ sec^{-1}$; CTAB + D-*1*, $k_\Psi = 6.51\ M^{-1}\ sec^{-1}$	
$R_1 = CH_2CH(CH_3)_2, R_2 = C_6H_5CH_2$ (L)	CTAB + L-*1*, $k_\Psi = 92.2\ M^{-1}\ sec^{-1}$; CTAB + D-*1*, $k_\Psi = 39.9\ M^{-1}\ sec^{-1}$	

Table 12.1 *(Continued)*

Reaction	Effect of Surfactant	Reference
$R_1 = CH_2C_6H_5, R_2 = C_6H_5CH_2$ (L)	CTAB + L-*1*, $k_\psi = 149.6\ M^{-1}\ sec^{-1}$; CTAB + D-*1*, $k_\psi = 68\ M^{-1}\ sec^{-1}$	
$R_1 = CH_2C_6H_5, R_2 = C_6H_5CH_2$ (D)	CTAB + L-*1*, $k_\psi = 65.4\ M^{-1}\ sec^{-1}$;	
$R_1 = CH_2$–(3-indolyl), $R_2 = C_6H_5CH_2$(L)	CTAB + L-*1*, $k_\psi = 21.4\ M^{-1}\ sec^{-1}$; CTAB + D-*1*, $k_\psi = 17.4\ M^{-1}\ sec^{-1}$	
Hydrolysis of $C_6H_5CH_2OCON$—$CHOOC_6H_4NO_2p$ (pyrrolidine ring: H_2C, CH_2, $C H_2$) 25.0°C, pH = 7.4, 0.05 M phosphate buffer, $\mu = 0.2\ M$ (KCl)	CTAB + L-*1*, $k = 3.57\ M^{-1}\ sec^{-1}$; CTAB + D-*1*, $k = 2.40\ M^{-1}\ sec^{-1}$	
Hydrolysis of p-$NO_6C_6H_4OCO(CH_2)_{n-1}H$, 31.0°C, pH = 9.06, $\mu = 0.083\ M$ (KCl) tris buffer, 10 % v/v CH_3CN		Ueoka and Matsumoto, 1980
$n = 2$	$CH_3(CH_2)_8CONHOH$ (*1*) + CTAB, $k_\psi = 1.033\ M^{-1}\ sec^{-1}$; *1* + CEAB, $k_\psi = 1,364\ M^{-1}\ sec^{-1}$; *1* + CBzAc, $k_\psi = 987\ M^{-1}\ sec^{-1}$; $CH_3(CH_2)_8CONHC(CONHOH)HCH_2C_6H_5$ (*2*) + CTAB, $k_\psi = 355\ M^{-1}\ sec^{-1}$; *2* + CEAB, $k_\psi = 587\ M^{-1}\ sec^{-1}$; *2* + CBzAc, $k_\psi = 706\ M^{-1}\ sec^{-2}$; $CH_3(CH_2)_7CH[CH_3(CH_2)_5]CONHOH$ (*3*) + CTAB, $k_\psi = 1960\ M^{-1}\ sec^{-1}$; *3* + CEAB, $k_\psi = 1720\ M^{-1}\ sec^{-1}$; *3* + CBzAc, $k_\psi = 1647\ M^{-1}\ sec^{-1}$	

$n = 4$	*1* + CTAB, $k_\Psi = 1296\ M^{-1}\ \text{sec}^{-1}$; *1* + CEAB, $k_\Psi = 1368\ M^{-1}\ \text{sec}^{-1}$; *1* + CBzAc, $k_\Psi = 1057\ M^{-1}\ \text{sec}^{-1}$; *2* + CTAB, $k_\Psi = 256\ M^{-1}\ \text{sec}^{-1}$; *2* + CEAB, $k_\Psi = 408\ M^{-1}\ \text{sec}^{-1}$; *2* + CBzAc, $k_\Psi = 486\ M^{-1}\ \text{sec}^{-1}$; *3* + CTAB, $k_\Psi = 2610\ M^{-1}\ \text{sec}^{-1}$; *3* + CEAB, $k_\Psi = 2106\ M^{-1}\ \text{sec}^{-1}$; *3* + CBzAc, $k_\Psi = 2039\ M^{-1}\ \text{sec}^{-1}$
$n = 6$	*1* + CTAB, $k = 1556\ M^{-1}\ \text{sec}^{-1}$; *1* + CEAB, $k_\Psi = 2129\ M^{-1}\ \text{sec}^{-1}$; *1* + CBzAc, $k_\Psi = 1710\ M^{-1}\ \text{sec}^{-1}$; *2* + CTAB, $k_\Psi = 331\ M^{-1}\ \text{sec}^{-1}$; *1* + CEAB, $k_\Psi = 481\ M^{-1}\ \text{sec}^{-1}$; *2* + CBzAc, $k_\Psi = 625\ M^{-1}\ \text{sec}^{-1}$; *3* + CTAB, $k_\Psi = 3930\ M^{-1}\ \text{sec}^{-1}$; *3* + CEAB, $k_\Psi = 3487\ M^{-1}\ \text{sec}^{-1}$; *3* + CBzAc, $k_\Psi = 2954\ M^{-1}\ \text{sec}^{-1}$
$n = 10$	*1* + CTAB, $k_\Psi = 2292\ M^{-1}\ \text{sec}^{-1}$; *1* + CEAB, $k_\Psi = 2322\ M^{-1}\ \text{sec}^{-1}$; *1* + CBzAc, $k_\Psi = 2213\ M^{-1}\ \text{sec}^{-1}$; *2* + CTAB, $k_\Psi = 352\ M^{-1}\ \text{sec}^{-1}$; *2* + CEAB, $k_\Psi = 541\ M^{-1}\ \text{sec}^{-1}$; *2* + CBzAc, $k_\Psi = 635\ M^{-1}\ \text{sec}^{-1}$; *3* + CTAB, $k_\Psi = 3815\ M^{-1}\ \text{sec}^{-1}$; *3* + CEAB, $k_\Psi = 3207\ M^{-1}\ \text{sec}^{-1}$; *3* + CBzAc, $k_\Psi = 3206\ M^{-1}\ \text{sec}^{-1}$
$n = 12$	*1* + CTAB, $k_\Psi = 1383\ M^{-1}\ \text{sec}^{-1}$; *1* + CEAB, $k_\Psi = 2419\ M^{-1}\ \text{sec}^{-1}$; *1* + CBzAc, $k_\Psi = 1827\ M^{-1}\ \text{sec}^{-1}$; *2* + CTAB, $k_\Psi = 318\ M^{-1}\ \text{sec}^{-1}$; *2* + CEAB, $k_\Psi = 495\ M^{-1}\ \text{sec}^{-1}$; *2* + CBzAc, $k_\Psi = 654\ M^{-1}\ \text{sec}^{-1}$; *3* + CTAB, $k_\Psi = 3659\ M^{-1}\ \text{sec}^{-1}$; *3* + CEAB, $k_\Psi = 3053\ M^{-1}\ \text{sec}^{-1}$; *3* + CBzAc, $k_\Psi = 3068\ M^{-1}\ \text{sec}^{-1}$
$n = 16$	*1* + CTAB, $k_\Psi = 451\ M^{-1}\ \text{sec}^{-1}$; *1* + CEAB, $k_\Psi = 558\ M^{-1}\ \text{sec}^{-1}$; *1* + CBzAc, $k_\Psi = 1699\ M^{-1}\ \text{sec}^{-1}$; *2* + CTAB, $k_\Psi = 254\ M^{-1}\ \text{sec}^{-1}$; *2* + CEAB, $k_\Psi = 380\ M^{-1}\ \text{sec}^{-1}$; *2* + CBzAc, $k_\Psi = 654\ M^{-1}\ \text{sec}^{-1}$; *3* + CTAB, $k_\Psi = 1524\ M^{-1}\ \text{sec}^{-1}$; *3* + CEAB, $k_\Psi = 1885\ M^{-1}\ \text{sec}^{-1}$; *3* + CBzAc, $k_\Psi = 3379\ M^{-1}\ \text{sec}^{-1}$

Table 12.1 *(Continued)*

Reaction	Effect of Surfactant	Reference
Hydrolysis of HOOC—$C_6H_3(NO_2)$—$OCO(CH_2)_{n-1}H$	H—$(CH_2)_{m-1}$CONHOH, Cm (m = 6, 8, 10, 12) $RCONHCH(CH_2H)CH_2$—(imidazolyl)	Ueoka et al., 1980
31.0°C, pH = 7.45, 0.083 M tris buffer 10% (v/v) CH_3CN	AcetHis (R = CH_3), LauHis (R = $C_{11}H_{23}$)	
$n = 2$	C_6, $k_\Psi = 0.9\ M^{-1}\ sec^{-1}$; C_6 + DTAC, $k_\Psi = 1.83\ M^{-1}\ sec^{-1}$; C_6 + CTAC, $k_\Psi = 2.30\ M^{-1}\ sec^{-1}$; C_6 + OTAC, $k_\Psi = 2.15\ M^{-1}\ sec^{-1}$; C_8, $k_\Psi = 0.99\ M^{-1}\ sec^{-1}$; C_8 + DTAC, $k_\Psi = 4.30\ M^{-1}\ sec^{-1}$; C_8 + DTAB, $k_\Psi = 13.6\ M^{-1}\ sec^{-1}$; C_8 + OTAC, $k_\Psi = 13.6\ M^{-1}\ sec^{-1}$; C_{10}, $k_\Psi = 0.94\ M^{-1}\ sec^{-1}$; C_{10} + DTAC, $k_\Psi = 6.22\ M^{-1}\ sec^{-1}$; C_{10} + CTAC, $k_\Psi = 41.5\ M^{-1}\ sec^{-1}$; C_{10} + OTAC, $k_\Psi = 11.7\ M^{-1}\ sec^{-1}$; C_{12}, $k_\Psi = 4.91\ M^{-1}\ sec^{-1}$; C_{12} + DTAC, $k_\Psi = 5.82\ M^{-1}\ sec^{-1}$; C_{12} + CTAB, $k_\Psi = 33.6\ M^{-1}\ sec^{-1}$; C_{12} + OATAC, $k_\Psi = 52.3\ M^{-1}\ sec^{-1}$; AcetHis, $k_\Psi = 0.45\ M^{-1}\ sec^{-1}$; AcetHis + DTAC, $k_\Psi = 0.18\ M^{-1}\ sec^{-1}$; AcetHis + CTAB, $k_\Psi = 0.27\ M^{-1}\ sec^{-1}$; AcetHis + OTAC, $k_\Psi = 0.24\ M^{-1}\ sec^{-1}$; LauHis, $k_\Psi = 0.73\ M^{-1}\ sec^{-1}$; LauHis + DTAC, $k_\Psi = 3.31\ M^{-1}\ sec^{-1}$; LauHis + CTAB, $k_\Psi = 6.40\ M^{-1}\ sec^{-1}$; LauHis + OTAC, $k_\Psi = 5.20\ M^{-1}\ sec^{-1}$	

$n = 4$	C_6, $k_\Psi = 0.27\ M^{-1}\ sec^{-1}$; C_6 + DTAC, $k_\Psi = 0.46\ M^{-1}\ sec^{-1}$; C_6 + CTAB, $k_\Psi = 0.82\ M^{-1}\ sec^{-1}$; C_6 + OTAC, $k_\Psi = 0.67\ M^{-1}\ sec^{-1}$; C_8, $k_\Psi = 0.99\ M^{-1}\ sec^{-1}$; C_8 + DTAC, $k_\Psi = 1.45\ M^{-1}\ sec^{-1}$; C_8 + CTAB, $k_\Psi = 5.6\ M^{-1}\ sec^{-1}$; C_8 + OTAC, $k_\Psi = 6.13\ M^{-1}\ sec^{-1}$; C_{10}, $k_\Psi = 0.37\ M^{-1}\ sec^{-1}$; C_{10} + DTAC, $k_\Psi = 2.57\ M^{-1}\ sec^{-1}$; C_{10} + CTAB, $k_\Psi = 21.4\ M^{-1}\ sec^{-1}$; C_{10} + OTAC, $k_\Psi = 6.8\ M^{-1}\ sec^{-1}$; AcetHis, $k_\Psi = 0.41\ M^{-1}\ sec^{-1}$; AcetHis + DTAC, $k_\Psi = 0.05\ M^{-1}\ sec^{-1}$; AcetHis + CTAB, $k_\Psi = 0.47\ M^{-1}\ sec^{-1}$; LauHis + DTAC, $k_\Psi = 2.37\ M^{-1}\ sec^{-1}$; LauHis, $k_\Psi = 0.47\ M^{-1}\ sec^{-1}$; LauHis + DTAC, $k_\Psi = 2.37\ M^{-1}\ sec^{-1}$; LauHis + CTAB, $k_\Psi = 6.0\ M^{-1}\ sec^{-1}$; LauHis + OTAC, $k_\Psi = 4.83\ M^{-1}\ sec^{-1}$
$n = 6$	C_6, $k_\Psi = 0.27\ M^{-1}\ sec^{-1}$; C_6 + DTAC, $k_\Psi = 0.44\ M^{-1}\ sec^{-1}$; C_6 + CTAB, $k_\Psi = 0.8\ M^{-1}\ sec^{-1}$; C_6 + OTAC, $k_\Psi = 0.93\ M^{-1}\ sec^{-1}$; C_8, $k_\Psi = 0.3\ M^{-1}\ sec^{-1}$; C_8 + DTAC, $k_\Psi = 1.32\ M^{-1}\ sec^{-1}$; C_8 + CTAB, $k_\Psi = 5.9\ M^{-1}\ sec^{-1}$; C_8 + OTAC, $k_\Psi = 6.3\ M^{-1}\ sec^{-1}$; C_{10}, $k_\Psi = 0.4\ M^{-1}\ sec^{-1}$; C_{10} + DTAC, $k_\Psi = 2.53\ M^{-1}\ sec^{-1}$; C_{10} + CTAB, $k_\Psi = 31.4\ M^{-1}\ sec^{-1}$; C_{10} + OTAC, $k_\Psi = 8.77\ M^{-1}\ sec^{-1}$; C_{12}, $k_\Psi = 0.82\ M^{-1}\ sec^{-1}$; C_{12} + DTAC, $k_\Psi = 2.56\ M^{-1}\ sec^{-1}$; C_{12} + CTAB, $k_\Psi = 20.0\ M^{-1}\ sec^{-1}$; C_{12} + OTAC, $k_\Psi = 26.5\ M^{-1}\ sec^{-1}$; AcetHis, $k_\Psi = 0.34\ M^{-1}\ sec^{-1}$; AcetHis + DTAC, $k_\Psi = 0.03\ M^{-1}\ sec^{-1}$; AcetHis + CTAB, $k_\Psi = 26.5\ M^{-1}\ sec^{-1}$; AcetHis + OTAC, $k_\Psi = 0.01\ M^{-1}\ sec^{-1}$; LauHis, $k_\Psi = 0.59\ M^{-1}\ sec^{-1}$; LauHis + DTAC, $k_\Psi = 2.52\ M^{-1}\ sec^{-1}$; LauHis + CTAB, $k_\Psi = 7.14\ M^{-1}\ sec^{-1}$; LauHis + OTAC, $k_\Psi = 5.07\ M^{-1}\ sec^{-1}$

Table 12.1 *(Continued)*

Reaction	Effect of Surfactant	Reference
$n = 10$	C_6, $k_\Psi = 0.26\ M^{-1}\ sec^{-1}$; C_6 + DTAC, $k_\Psi = 0.39\ M^{-1}\ sec^{-1}$; C_6 + CTAB, $k_\Psi = 0.7\ M^{-1}\ sec^{-1}$; C_6 + OTAC, $k_\Psi = 0.66\ M^{-1}\ sec^{-1}$; C_8, $k_\Psi = 0.41$; C_8 + DTAC, $k_\Psi = 1.07\ M^{-1}\ sec^{-1}$; C_8 + CTAB, $k_\Psi = 4.76\ M^{-1}\ sec^{-1}$; C_8 + OTAC, $k_\Psi = 4.96\ M^{-1}\ sec^{-1}$; C_{10}, $k_\Psi = 0.47\ M^{-1}\ sec^{-1}$; C_{10} + DTAC, $k_\Psi = 2.44\ M^{-1}\ sec^{-1}$; C_{10} + CTAB, $k_\Psi = 20.6\ M^{-1}\ sec^{-1}$; C_{10} + OTAC, $k_\Psi = 5.91\ M^{-1}\ sec^{-1}$; C_{12}, $k_\Psi = 2.27\ M^{-1}\ sec^{-1}$; C_{12} + DTAC, $k_\Psi = 1.77\ M^{-1}\ sec^{-1}$; C_{12} + CTAB, $k_\Psi = 16.8\ M^{-1}\ sec^{-1}$; C_{12} + OTAC, $k_\Psi = 23.8\ M^{-1}\ sec^{-1}$; AcetHis, $k_\Psi = 0.23\ M^{-1}\ sec^{-1}$; AcetHis + DTAC, $k_\Psi = 0.03\ M^{-1}\ sec^{-1}$; AcetHis + CTAB, $k_\Psi = 0.01\ M^{-1}\ sec^{-1}$; AcetHis + OTAC, $k_\Psi = 0.01\ M^{-1}\ sec^{-1}$; LauHis, $k_\Psi = 1.49\ M^{-1}\ sec^{-1}$; LauHis + DTAC, $k_\Psi = 2.27\ M^{-1}\ sec^{-1}$; LauHis + CTAB, $k_\Psi = 6.0\ M^{-1}\ sec^{-1}$; LauHis + OTAC, $k_\Psi = 4.5\ M^{-1}\ sec^{-1}$	
$n = 12$	C_6, $k_\Psi = 0.21\ M^{-1}\ sec^{-1}$; C_6 + DTAC, $k_\Psi = 0.37\ M^{-1}\ sec^{-1}$; C_6 + CTAB, $k_\Psi = 0.67\ M^{-1}\ sec^{-1}$; C_6 + OTAC, $k_\Psi = 0.7\ M^{-1}\ sec^{-1}$; C_8, $k_\Psi = 0.32\ M^{-1}\ sec^{-1}$; C_8 + DTAC, $k_\Psi = 1.01\ M^{-1}\ sec^{-1}$; C_8 + CTAB, $k_\Psi = 4.03\ M^{-1}\ sec^{-1}$; C_8 + OTAC, $k_\Psi = 4.3\ M^{-1}\ sec^{-1}$; C_{10}, $k_\Psi = 0.42\ M^{-1}\ sec^{-1}$; C_{10} + DTAC, $k_\Psi = 2.54\ M^{-1}\ sec^{-1}$; C_{10} + CTAB, $k_\Psi = 14.5\ M^{-1}\ sec^{-1}$; C_{10} + OTAC, $k_\Psi = 6.94\ M^{-1}\ sec^{-1}$; C_{12} $k_\Psi = 4.15\ M^{-1}\ sec^{-1}$; C_{12} + DTAC, $k_\Psi = 1.77\ M^{-1}\ sec^{-1}$; C_{12} + CTAB, $k_\Psi = 13.8\ M^{-1}\ sec^{-1}$; C_{12} + OTAC, $k_\Psi = 21.3\ M^{-1}\ sec^{-1}$; AcetHis, $k_\Psi = 0.35\ M^{-1}\ sec^{-1}$;	

	AcetHis + DTAC, $k_\Psi = 0.02\ M^{-1}\ sec^{-1}$; AcetHis + CTAB, $k_\Psi = 0.02\ M^{-1}\ sec^{-1}$; AcetHis + OTAC, $k_\Psi = 0.02\ M^{-1}\ sec^{-1}$; LauHis, $k_\Psi = 1.99\ M^{-1}\ sec^{-1}$; LauHis + DTAC, $k_\Psi = 2.48\ M^{-1}\ sec^{-1}$; LauHis + CTAB, $k_\Psi = 5.59\ M^{-1}\ sec^{-1}$; LauHis + OTAC, $k_\Psi = 3.98\ M^{-1}\ sec^{-1}$	
$n = 16$	C_6, $k_\Psi \sim 0$; C_6 + DTAC, $k_\Psi = 0.34\ M^{-1}\ sec^{-1}$; C_6 + CTAB, $k_\Psi = 0.54\ M^{-1}\ sec^{-1}$; C_6 + OTAC, $k_\Psi = 0.52\ M^{-1}\ sec^{-1}$; C_8, $k_\Psi \sim 0$; C_8 + DTAC, $k_\Psi = 1.0\ M^{-1}\ sec^{-1}$; C_8 + CTAB, $k_\Psi = 3.28\ M^{-1}\ sec^{-1}$; C_8 + OTAC, $k_\Psi = 3.85\ M^{-1}\ sec^{-1}$; C_{10}, $k_\Psi = 0.56\ M^{-1}\ sec^{-1}$; C_{10} + DTAC, $k_\Psi = 2.21\ M^{-1}\ sec^{-1}$; C_{10} + CTAB, $k_\Psi = 17.6\ M^{-1}\ sec^{-1}$; C_{10} + OTAC, $k_\Psi = 4.48\ M^{-1}\ sec^{-1}$; C_{12}, $k_\Psi = 7.73\ M^{-1}\ sec^{-1}$; C_{12} + DTAC, $k_\Psi = 2.26\ M^{-1}\ sec^{-1}$; C_{12} + CTAB, $k_\Psi = 5.1\ M^{-1}\ sec^{-1}$; C_{12} + OTAC, $k_\Psi = 15.8\ M^{-1}\ sec^{-1}$; AcetHis, $k_\Psi = 0.06\ M^{-1}\ sec^{-1}$; AcetHis + DTAC, $k_\Psi = 0.07\ M^{-1}\ sec^{-1}$; AcetHis + CTAB, $k_\Psi = 0.07\ M^{-1}\ sec^{-1}$; AcetHis + OTAC, $k_\Psi = 0.06\ M^{-1}\ sec^{-1}$; LauHis, $k_\Psi = 3.47\ M^{-1}\ sec^{-1}$; LauHis + DTAC, $k_\Psi = 2.62\ M^{-1}\ sec^{-1}$; LauHis + CTAB, $k_\Psi = 5.45\ M^{-1}\ sec^{-1}$; LauHis + OTAC, $k_\Psi = 3.73\ M^{-1}\ sec^{-1}$	
Hydrolysis of *p*-nitrophenyl acetate, 25.0°C, pH = 7.95	$C_{16}H_{33}\overset{+}{N}Me_2CH_2Im, Cl^-$ (I), $k_m/k_0 = 1.9$; $C_{16}H_{33}\overset{+}{N}(Me)(Et)CH_2CH_2OH, Br^-$ (II), $k_m/k_0 = 0.4$; $C_{16}H_{33}\overset{+}{N}(Me)(CH_2CH_2OH)CH_2Im, Cl^-$ (III), $k_m/k_0 = 1.9$; $C_{16}H_{33}\overset{+}{N}(Me_2)CH_2CH_2SH, Br^-$ (IV) $k_m/k_0 = 1.8$ k_m = "true" micellar rate constant, data are used to calculate kinetic rate equations for functional micellar catalysis	Fornasier and Tonellato, 1980

Table 12.1 (*Continued*)

Reaction	Effect of Surfactant	Reference
Hydrolysis of *p*-nitrophenyl hexanoate, 25.0°C, pH = 7.95	I, $k_m/k_0 = 0.8$; II, $k_m/k_0 = 0.15$; III, $k_m/k_0 = 1.4$; IV, $k_m/k_0 = 0.4$ k_m = "true" micellar rate constant; data are used to calculate kinetic rate equations for functional micellar catalysis.	
Hydrolysis of *p*-nitrophenyl acetate, 25.0°C, pH = 8.0, $\mu = 0.05$ (KCl)	$n\text{-}C_{16}H_{33}\overset{+}{N}(Me_2)CH_2CH_2OOH$, $Cf_3SO_3^-$, $k_\Psi/k_0 \sim 20{,}000$	Moss and Alwis, 1980
Hydrolysis of *p*-nitrophenyl acetate, 25.0°C, $\mu = 0.05$ (KCl)	$n\text{-}C_{16}H_{33}\overset{+}{N}Me_2CH_2CH_2SH$, Cl^-, $k_\Psi/k_{thiocholine} = 251$ (pH = 7), 485 (pH = 8.0); k_Ψ/k_0(deacylation) $\sim +$	Moss et al., 1980
Hydrolysis of *p*-nitrophenyl hexanoate, 25.0°C, pH = 7, $\mu = 0.05$ (KCl)	$n\text{-}C_{16}H_{33}\overset{+}{N}Me_2CH_2CH_2SH$, Cl^-, $k_\Psi/k_{thiocholine} = 2850$; k_Ψ/k_0(deacylation) $\sim +$	
Hydrolysis of *p*-nitrophenyl hexanoate, 30.0°C, pH = 7.7, 0.05 *M* phosphate buffer	CTAB + $CH_3(CH_2)_{13}C(CONMeOH)CH_2CH_2\overset{+}{N}Me_3$, Cl^- (*1a*), $k_\Psi = 2.3\ M^{-1}\ sec^{-1}$; CTAB + $CH_3(CH_2)_9C(CONMeOH)(CH_2)_6\overset{+}{N}Me_3$, Cl^- (*2b*), $k_\Psi = 5.1\ M^{-1}\ sec^{-1}$ Mixed micelles of *1a* and *2b* and the surfactant histidine **10.29** (see Table 10.3) react with *p*-nitrophenyl hexanoate by separate noncooperative mechanisms	Brown and Lynn, 1980
Hydrolysis of optically active $R_1OCONHCH(R_2)COOC_6H_4NO_2\text{-}p$ 25.0°C, pH = 7.30, 0.02 *M* phosphate buffer, 2.0×10^{-3} *M* CTAB	RCONHCH(COOH)CH$_2$–(imidazol-4-yl; HN, N ring)	Ihara, 1978, 1980

$R_1 = R_2 = PhCH_2$	R = Me, $k_\Psi(L) = 2.75\ M^{-1}\ sec^{-1}$; $k_\Psi(D) = 1.96\ M^{-1}\ sec^{-1}$ R = Ph, $k_\Psi(L) = 174\ M^{-1}\ sec^{-1}$; $k_\Psi(D) = 101\ M^{-1}\ sec^{-1}$ R = $CH_3(CH_2)_8$, $k_\Psi(L) = 1730\ M^{-1}\ sec^{-1}$; $k_\Psi(D) = 690\ M^{-1}\ sec^{-1}$ R = $CH_3(CH_2)_{16}$, $k_\Psi(L) = 2740\ M^{-1}\ sec^{-1}$; $k_\Psi(D) = 985\ M^{-1}\ sec^{-1}$	
$R_1 = PhCH_2$, $R^2 = CH_3$	R = Me, $k_\Psi(L) = 3.07\ M^{-1}\ sec^{-1}$; $k_\Psi(D) = 2.25\ M^{-1}\ sec^{-1}$ R = Ph, $k_\Psi(L) = 83.5\ M^{-1}\ sec^{-1}$; $k_\Psi(D) = 61.5\ M^{-1}\ sec^{-1}$ R = $CH_3(CH_2)_8$, $k_\Psi(L) = 783\ M^{-1}\ sec^{-1}$; $k_\Psi(D) = 358\ M^{-1}\ sec^{-1}$ R = $CH_3(CH_2)_{16}$, $k_\Psi(L) = 1150\ M^{-1}\ sec^{-1}$; $k_\Psi(D) = 544\ M^{-1}\ sec^{-1}$	
$R_1 = PhCH_2$, $R_2 = H$	R = Me, $k_\Psi = 1.9\ M^{-1}\ sec^{-1}$ R = Ph, $k_\Psi = 71.3\ M^{-1}\ sec^{-1}$ R = $CH_3(CH_2)_8$, $k_\Psi = 469\ M^{-1}\ sec^{-1}$ R = $CH_3(CH_2)_{16}$, $k_\Psi = 649\ M^{-1}\ sec^{-1}$	
$R_1 = Me$, $R_2 = PhCH_2$	R = Me, $k_\Psi(L) = 3.64\ M^{-1}\ sec^{-1}$; $k_\Psi(D) = 2.64\ M^{-1}\ sec^{-1}$ R = PH, $k_\Psi(L) = 119\ M^{-1}\ sec^{-1}$; $k_\Psi(D) = 83.5\ M^{-1}\ sec^{-1}$ R = $CH_3(CH_2)_8$, $k_\Psi(L) = 877\ M^{-1}\ sec^{-1}$; $k_\Psi(D) = 403\ M^{-1}\ sec^{-1}$ R = $CH_3(CH_2)_{16}$, $k_\Psi(L) = 1370\ M^{-1}\ sec^{-1}$; $k_\Psi(D) = 622\ M^{-1}\ sec^{-1}$	
Hydrolysis of optically active $R_1OCONHCH(R_2)COOC_6H_4NO_2$-*p*	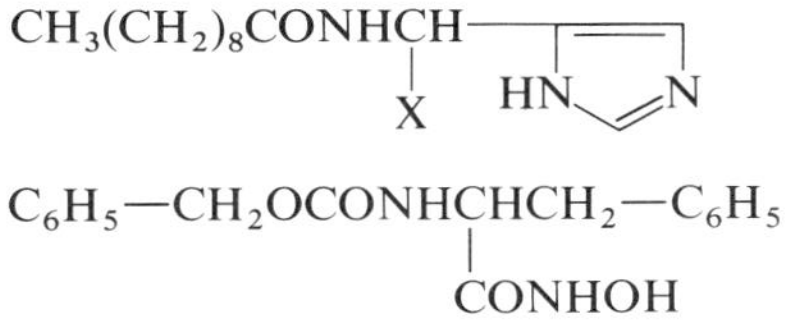 $CH_3(CH_2)_8CONHCH(X)$–(imidazolyl, HN / N) $C_6H_5{-}CH_2OCONHCH(CONHOH)CH_2{-}C_6H_5$	Ihara et al., 1980

Table 12.1 (*Continued*)

Reaction	Effect of Surfactant	Reference
25.0°C, pH = 7.30, 0.02 M phosphate buffer + 6.0×10^{-3} M CTAB	*1a*, X = COOH *1b*, X = CONHOH	*71*
$R_1 = R_2 = PhCH_2$	*1a*, $k_\Psi(L) = 572\ M^{-1}\ sec^{-1}$, $k_\Psi(D) = 231\ M^{-1}\ sec^{-1}$; *1b*, $k_\Psi(L) = 2880\ M^{-1}\ sec^{-1}$, $k_\Psi(D) = 2060\ M^{-1}\ sec^{-1}$; *2*, $k_\Psi(L) = 1530\ M^{-1}\ sec^{-1}$, $k_\Psi(D) = 1770\ M^{-1}\ sec^{-1}$	
$R_1 = PhCH_2$, $R_2 = CH_3$	*1a*, $k_\Psi(L) = 281\ M^{-1}\ sec^{-1}$, $k_\Psi(D) = 140\ M^{-1}\ sec^{-1}$; *1b*, $k_\Psi(L) = 2350\ M^{-1}\ sec^{-1}$, $k_\Psi(D) = 1920\ M^{-1}\ sec^{-1}$; *2*, $k_\Psi(L) = 1630\ M^{-1}\ sec^{-1}$, $k_\Psi(D) = 1670\ M^{-1}\ sec^{-1}$	
$R_1 = CH_3$, $R_2 = PhCH_2$	*1a*, $k_\Psi(L) = 314\ M^{-1}\ sec^{-1}$, $k_\Psi(D) = 145\ M^{-1}\ sec^{-1}$; *1b*, $k_\Psi(L) = 2230\ M^{-1}\ sec^{-1}$, $k_\Psi(D) = 1610\ M^{-1}\ sec^{-1}$; *2*, $k_\Psi(L) = 1450\ M^{-1}\ sec^{-1}$, $k_\Psi(D) = 1810\ M^{-1}\ sec^{-1}$	

Table 12.2. Hydrolysis and Solvolyses of Phosphate and Sulfate Esters in Aqueous Micelles

Reaction	Effect of Surfactant	Remarks	Reference
Hydrolysis of diethyl *p*-nitrophenyl phosphate, 25.0°C, pH = 12	CTAB, $k_\Psi/k_0 = 9$; $C_{16}\overset{+}{N}Me_2CH_2CH_2OH$ (*1*), $k_\Psi/k_0 = 3$	D_2O solvent isotope effect CTAB $= H_2O \neq 1$	Bunton and Diaz, 1976
Hydrolysis of di-*n*-hexyl *p*-nitrophenol phosphate, 25.0°C, pH = 12	CTAB, $k_\Psi/k_0 = 18$; *1*, $k_\Psi/k_0 = 256$	$1 \leftrightharpoons C_{16}\overset{+}{N}Me_2CH_2CH_2O^- + H^+$	
Hydrolysis of *n*-decyl phosphate, 100°C, pH = 4.5 and [HCl or $HClO_4$] = 1–3 *M*	Monoanion hydrolysis unaffected by self-micellization; undissociated *n*-decylphosphoric acid hydrolysis is affected by self-micellization (altered mechanism, see Table 11.1)		Bunton et al., 1976a
Hydrolysis of diethyl *p*-nitrophenyl phosphate, 25.0°C, pH = 8.5–13.5	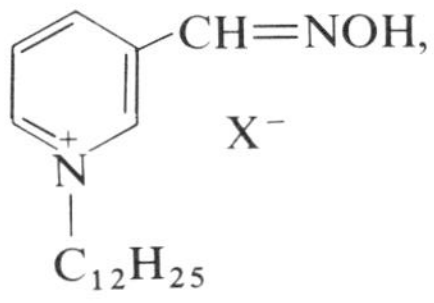 [3-(CH=NOH)-1-$C_{12}H_{25}$-pyridinium X^- structure] CH=NOH, $k_\Psi/k_0 \sim 10^4$; X^-; $\overset{+}{N}$–$C_{12}H_{25}$; X = I, Cl, Br, CH_3SO_3, $CH_3C_6H_4SO_3$, 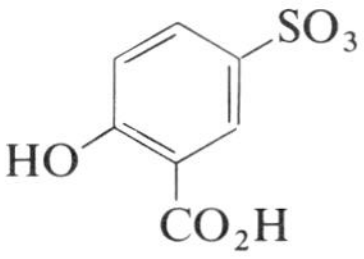 [structure: HO, CO_2H, SO_3 substituted benzene]	Micellar catalysis is effective only for the reaction with *neutral* substrates; hydrolysis of protonated substrates unaffected by micelles	Epstein et al., 1978
Hydrolysis of O-ethyl S-2-diisopropylaminoethyl methylphosphonothioate, 25.0°C, pH = 8.5–13.5	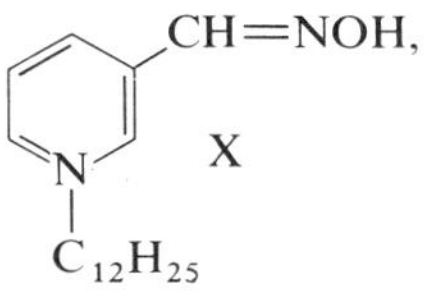 [pyridine structure] CH=NOH, $k_\Psi/k_0 \sim 10^4$; X; N–$C_{12}H_{25}$; X = I, Cl, Br, CH_3SO_3, $CH_3C_6H_4SO_3$,	Micellar catalysis is effective only for the reaction with *neutral* substrates; hydrolysis of protonated substrates unaffected by micelles 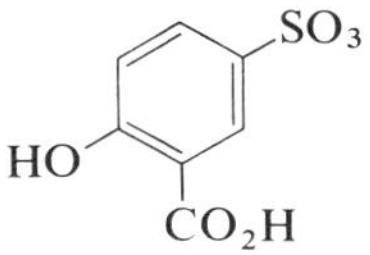 [structure: HO, CO_2H, SO_3 substituted benzene]	

Table 12.2 *(Continued)*

Reaction	Effect of Surfactant	Remarks	Reference
Hydrolysis of *p*-nitrophenyl diphenyl phosphate, 25.0°C, pH = 8.0, 0.02 *M* borate buffer	CTAB, $k_{\Psi}/k_0 \simeq 46$; {5-[(α-*N*-L-histidinyl methyl ester) carbonyl] *n*-heptadecyl} trimethylammonium chloride (*1*), $k_{\Psi}/k_0 = 95$ Dimethylhexadecyl [(4-imidazolyl)-methyl]ammonium chloride (*2*), $k_{\Psi}/k_0 = 400$	Imidazole moieties of the surfactant act as general bases; micelle catalysis is almost wholly due to increased concentration of reactive groups at the micellar surface	Brown et al., 1980
	2 + Brij(10^{-3} *M*), [substrate] > functional micelle	No buildup of phosphorylated imidazole	
	1 immobilized on polystyrene ion exchange resin, $k_{\Psi}/k_0 = +$	Reusable catalyst	
Hydrolysis of *d*-ethyl *p*-nitrophenyl phosphate, 25.0°C, pH = 8.0, 0.015 *M* borate buffer	*1*, $k_{\Psi}/k_0 \sim 5$		
Hydrolysis of di-*n*-hexyl *p*-nitrophenyl phosphate, 25.0°C, pH = 8.0, 0.015 *M* borate buffer	*1*, $k_{\Psi}/k_0 \sim 110$		
Hydrolysis of 2,4-dinitrophenyl phosphate, 25.0°C, pH = 8.0, 0.015 *M* borate buffer	*1*, $k_{\Psi}/k_0 \sim 8$		
Hydrolysis of *p*-nitrobenzoyl phosphate 25.0°C, pH = 2, monoanion dianion	CTAB, $k_{\Psi}/k_0 \sim 2$; CTAB, $k_{\Psi}/k_0 \sim 5$; CTAB + octyloxyamine, $k_{\Psi}/k_0 \sim 14$; CTAB + decylguanidinium bromide, $k_{\Psi}/k_{CTAB} \sim 2$		Bunton and McAneny, 1977

Hydrolysis of $CH_3COOPO_3^{2-}$, 39.0°C, pH = 7, $\mu = 0.6\ M$, Mn^{2+}, Co^{2+}, Zn^{2+}, Ni^{2+}, Cu^{2+}	5-Alkyltriethylene amine, $k_\Psi/k_0 \sim ++$; dodecylammonium chloride, $k_\Psi/k_0 \sim 160$	Melhaldo and Gutsche, 1978
Hydrolysis of 2,4-dinitrophenyl sulfate, 30.0°C, pH = 8–10	$k_0 = 5 \times 10^{-7}\ M^{-1}\ sec^{-1}$(25.0°C); *N*-benzylbenzohydroxamic acid, 30 vol. % EtOH, $k_\Psi < 0.001\ M^{-1}\ sec^{-1}$; $C_{12}H_{25}ImCH_2CO(NO)CH_2H_6H_5$ (*1*) + DMOG, $k_\Psi = 6.1\ M^{-1}\ sec^{-1}$; *1* + POOA, $k_\Psi = 6.8\ M^{-1}\ sec^{-1}$; *1* + CTAB, $k_\Psi = 11\ M^{-1}\ sec^{-1}$	Kunitake and Sakamoto, 1979
Hydrolysis of phenyl-, *p*-butyl-, and *p*-decyl phenyl phosphatosulfates, 39.5°C, pH varied	SDS, $k_\Psi/k_0 \sim +$	Eiki et al., 1980

Table 12.3. Miscellaneous Hydrolyses and Solvolyses in Aqueous Micelles

Reaction	Effect of Surfactant	Reference
Hydrolysis of $n\text{-}C_{16}H_{33}\overset{+}{N}Me_2CH_2CH_2OCOC_6H_4NO_2\text{-}p$, 25°C, 0.005 *M* NaOH	CTAB, $k_\Psi/k_0 = 45$; $n\text{-}C_{16}H_{33}\overset{+}{N}Me_2CH_2CH_2OH$, Br^-, $k_\Psi/k_0 \sim 43$	Bunton and McAneny, 1976
Hydrolysis of 4-methylphenylurea, 101.0°C, 0.09–1.0 *M* NaOH	CTAB, $k_\Psi/k_0 \sim 1$; SDS, $k_\Psi/k_0 \sim 1$; A-730, $k_\Psi/k_0 \sim 1$	Mollett and O'Connor, 1976
Hydrolysis of 4-nitrophenylurea, 101.0°C, 0.09–1.0 *N* NaOH	CTAB, $k_\Psi/k_0 \sim 1$; SDS, $k_\Psi/k_0 \sim 1$; A-730, $k_\Psi/k_0 \sim 1$	
Hydrolysis of *n*-decyloxalate, 40°C, 45°C, 50°C, pH = 50 $[\omega^{2+}] = 2 \times 10^{-3}$ *M*, $\mu = 0.1$ *M*	SDS, $k_\Psi/k_0 \sim 50$ $\Delta H^\ddagger_\Psi/\Delta H^\ddagger_0 \sim 1$, $\Delta S^\ddagger_\Psi/S^\ddagger_0 \sim 2$	Ong and Kostenbauder, 1976
Hydrolysis of isopropyl methylphosphonofluoride and diisopropylphosphonofluoridate, 30°C, pH = 9.3 (borate buffer, 10^{-3} *M* 1-oxamino-2-ketononane	CTAB, $k_\Psi/k_0 \sim 2$	Epstein et al., 1977
Hydrolysis of *p*-tosylmethylperchlorate, 25.0°C, ph = 5.5–5.7	NLS, $k_\Psi/k_0 \sim 0.7$; CTAB, $k_\Psi/k_0 \sim 40$; Ingepal, $k_\Psi/k_0 \sim 1.9$	Jagt and Engberts, 1977
Hydrolysis of *p*-nitrophenylsulfonylmethyl nitrate, 25.0°C, pH ~ 4.4	CTAB, $k_\Psi/k_0 \sim 20$; rates in CTAB increase upon stirring	
$pCH_3C_6H_4SO_2CH_2OClO_3 + B \xrightarrow{\text{slow}} pCH_3C_6H_4SO_2CH^-OClO_3 + BH^+ \xrightarrow[\text{fast}]{H_2O} pCH_3C_6H_4SO_2H + HCOOH + ClO_3^-$		Van de Langkruis and Engberts, 1979

25.0°3, pH = 3.5		
B = p-$CH_3C_6H_4SO_2^-$	CTAB, $k_\Psi/k_0 \sim 14 \times 10^3$	
B = $C_6H_5CO_2^-$	CTAB, $k_\Psi/k_0 \sim 10 \times 10^3$	
B = p-$BrC_6H_4SO_2^-$	CTAB, $k_\Psi/k_0 \sim 10 \times 10^3$	
B = p-$NO_2C_6H_4SO_2^-$	CTAB, $k_\Psi/k_0 \sim 0.7 \times 10^3$	
B = HCO_2^-	CTAB, $k_\Psi/k_0 \sim 3 \times 10^3$	
B = $CH_3(CH_2)_2CO^-$	CTAB, $k_\Psi/k_0 \sim 9 \times 10^3$	
Hydrolysis of $CH_3(CH_2)_6CONHOH$, 50.7°C, 0.203 *N* HCl	SDS, $k_\Psi/k_0 \sim 8.7$	Bernt and Sendelbach, 1977
Hydrolysis of $PhCH_2CONHOH$, 55.0°C, 0.314 *N* HCl	SDS, $k_\Psi/k_0 \sim 3.9$	
Hydrolysis of procaine, 40°C, pH = 7–12	PLE, $k_\Psi/k_0 \sim -0.17$; SDS, $k_\Psi/k_0 \sim 0.11$; CTAB, $k_\Psi/k_0 \sim 0.15$; NDB, $k_\Psi/k_0 \sim 0.21$	Tomida et al., 1978
Hydrolysis of *N*-alkyl-4-cyanopyridinium ions, 30°C, pH varied	CTAB, $k_\Psi/k_0 \sim +$ Micelles alter selectivity of reactions involving small hydrophilic ions	Politi et al., 1978
Hydrolysis of Ar(Me)N—C(=O)—R		Broxton et al., 1978
65.5°C, 0.00537 *M* NaOH		
Ar = $pNO_2C_6H_4$, R = $(CH_2)_6Me$	CTAB, $k_\Psi/k_0 \sim 77$	
Ar = $pNO_2C_6H_4$, R = Ph	CTAB, $k_\Psi/k_0 \sim 36$	
Ar = $pNO_2C_6H_4$, R = Me	CTAB, $k_\Psi/k_0 \sim 6$	
Ar = pyridine-4-yl, R = Ph	CTAB, $k_\Psi/k_0 \sim 8$	
Hydrolysis of 1,3,5-triaza-1,3,5-trinitrocyclohexane	Ethylhexadenyldimethylammonium bromide (*1*), $k_\Psi/k_0 \sim 100$	Croce and Okamoto, 1979
Hydrolysis of 1,3,5,7-tetraaza-1,3,5,7-tetranitrocyclooctane, 25°C, NaOH varied	*1*, $k_\Psi/k_0 \sim 27$	

Table 12.3 *(Continued)*

Reaction	Effect of Surfactant	Reference
Hydrolysis of $pNO_2C_6H_4CH(OR)_2$, 25°C		Bunton et al., 1979a
R = Me	$pC_8H_{17}OC_6H_4SO_3H$ (*1*), $k_\Psi/k_0 \sim 0.055$; $pC_{12}H_{25}OC_6H_4SO_3H$, (*2*), $k_\Psi/k_0 \sim 0.076$	
R = Et	$C_{14}H_{29}SO_3H$, $k_\Psi/k_0 \sim 0.020$; *1*, $k_\Psi/k_0 \sim 0.049$; *2*, $k_\Psi/k_0 \sim 0.056$; $C_{12}H_{25}SO_4Na$, $k_\Psi/k_0 \sim 0.046$	
Hydrolysis of $XC_6H_4CH_2CONHOH$, 50.1°C, 0.101 *N* HCl		Berndt et al., 1979
X = H	SDOS, $k_\Psi/k_0 \sim 6.72$	
X = mCH$_3$	SDOS, $k_\Psi/k_0 \sim 9.15$	
X = pCH$_3$	SDOS, $k_\Psi/k_0 \sim 10.4$	
X = pBr	SDOS, $k_\Psi/k_0 \sim 15$	
X = pCH_3CH_2	SDOS, $k_\Psi/k_0 \sim 12$	
X = pCH_3CH_2	SDS, $k_\Psi/k_0 \sim 11.8$	
Hydrolysis of 4-nitroacetanilide, 370 K°, pH = 9	CTAB, $k_\Psi/k_0 \sim 5.3$; SDS, $k_\Psi/k_0 \sim 0.60$ Micellar catalysis is inhibited by added salts	O'Connor and Tan, 1980
Hydrolysis of 5,5′-dithiobis-(2-nitrobenzoic acid), 25°C, pH = 9.20	TTAB, $k_\Psi/k_0 \sim +$	Hiramatsu, 1977

Table 12.4. Nucleophilic Substitutions in Aqueous Micelles

Reaction	Effect of Surfactant	Reference
2-Chloroquinoxaline + OH^-, 25.0°C	CTAB, $k_\Psi/k_0 \sim 4$; SDS, $k_\Psi/k_0 \sim 0.20$	Flamini et al., 1975
2,4-Dinitrofluorobenzene + PhO^-, 25.0°C	CTAB, $k_\Psi/k_0 \sim 230$	Chaimovich et al., 1975
2,4-Dinitrofluorobenzene + PhS^-, 25.0°C	CTAB, $k_\Psi/k_0 \simeq 1100$; Brij-58, $k_\Psi/k_0 \sim 36$	
2,6-Dinitro-4-trifluoromethylbenzene sulfonate + H_2NR, 25.0°C		Bunton and Wright, 1975
H_2NR = glycinate (*1*)	CTAB, $k_\Psi/k_0 \sim 6$	
H_2NR = glycylglycinate (*2*)	CTAB, $k_\Psi/k_0 \sim 6$	
H_2NR = leucinate (*3*)	CTAB, $k_\Psi/k_0 \sim 93$	
H_2NR = phenylalaninate (*4*)	CTAB, $k_\Psi/k_0 \sim 740$	
H_2NR = α-phenylglycinate (*5*)	CTAB, $k_\Psi/k_0 \sim 247$	
H_2NR = glycinamide (*6*)	CTAB, $k_\Psi/k_0 \sim 1$	
H_2NR = OH (*7*)	CTAB, $k_\Psi/k_0 \sim 3$	
2,4-Dinitrofluorobenzene +		
+*1*	CTAB, $k_\Psi/k_0 \sim 28$	
+*2*	CTAB, $k_\Psi/k_0 \sim 26$	
+*3*	CTAB, $k_\Psi/k_0 \sim 34$	
+*4*	CTAB, $k_\Psi/k_0 \sim 107$	
+*5*	CTAB, $k_\Psi/k_0 \sim 65$	
+*6*	CTAB, $k_\Psi/k_0 \sim 5.5$	
+*7*	CTAB, $k_\Psi/k_0 \sim 60$	
2,4-Dinitrochlorobenzene + OH^-, 43.5°C	Addition of alcohols decrease the CTAB catalyzed rate	Blandamer and Reid, 1975

Table 12.4 (*Continued*)

Reaction	Effect of Surfactant	Reference
$(Me_2N\text{-}C_6H_5)_2C^+Ph + OH^-$ (malachite green), 25.0°C	$n\text{-}C_{16}H_{33}\overset{+}{N}Me_2CH_2CH_2OH$, $k_\Psi/k_0 \sim 600$	Bunton and Paik, 1976
$(MeOC_6H_5)_3C^+ + OH^-$ (trianisylmethylcation), 25.0°C	$n\text{-}C_{16}H_{33}\overset{+}{N}Me_2CH_2CH_2OH$, $k_\Psi/k_0 \sim 10^3$	
2,4,6-Trinitrobenzene sulfonate ion + x, 25.0°C		Bunton et al., 1976b
x = glycinate	CTAB, $k_\Psi/k_0 \sim 6$	
x = phenylglycinate	CTAB, $k_\Psi/k_0 \sim 174$	
x = leucinate	CTAB, $k_\Psi/k_0 \sim 38$	
x = glycineamide	CTAB, $k_\Psi/k_0 < 1$	
x = aniline	CTAB, $k_\Psi/k_0 \sim 30$	
x = OH^-	CTAB, $k_\Psi/k_0 \sim 22$	
x = PhO^-	CTAB, $k_\Psi/k_0 \sim 1900$	
2,6-Dinitro-4-trifluoromethylbenzene sulfonate + x		
x = glycinate	CTAB, $k_\Psi/k_0 \sim 6$	
x = phenylglycinate	CTAB, $k_\Psi/k_0 \sim 247$	
x = leucinate	CTAB, $k_\Psi/k_0 \sim 93$	
x = glycineamide	CTAB, $k_\Psi/k_0 < 1$	
x = aniline	CTAB, $k_\Psi/k_0 \sim 96$	
x = OH^-	CTAB, $k_\Psi/k_0 \sim 15$	
p-Nitrophenyl diphenyl phosphate + RCH=NOH		Bunton and Ihara, 1977
p-Nitrobenzaldoximate	CTAB, $k_\Psi/k_0 \sim 2340$	
2-Pyridinecarbaldoximate	CTAB, $k_\Psi/k_0 \sim 254$	
3-Pyridinecarbaldoximate	CTAB, $k_\Psi/k_0 \sim 142$	
4-Pyridinecarbaldoximate	CTAB, $k_\Psi/k_0 \sim 275$	
2-Quinolinecarbaldoximate	CTAB, $k_\Psi/k_0 \sim 3700$	
25.0°C, pH = 10	Metal cations do not assist reaction in the presence of CTAB	

p-Nitrophenyl-3-phenylpropionate + RCH=NOH		
p-Nitrobenzaldoximate	CTAB, $k_\Psi/k_0 \sim 2210$	
2-pyridinecarbaldoximate	CTAB, $k_\Psi/k_0 \sim 307$	
3-pyridinecarbaldoximate	CTAB, $k_\Psi/k_0 \sim 152$	
4-pyridinecarbaldoximate	CTAB, $k_\Psi/k_0 \sim 283$	
2-quinolinecarbaldoximate	CTAB, $k_\Psi/k_0 \sim 1910$	
25.0°C, pH = 10	Metal cations do not assist reaction in the presence of CTAB	
2,4,6-Trinitrotoluene + 3,3′-diamino-*N*-methyldipropylalamine, 25.0°C	$k_{\Psi(\text{micelle})}/k_{\Psi(\text{monomer})} \sim 55$ Alkylphenoxypolyethanol, $k_\Psi/k_0 \sim 8$ CTAB, $k_\Psi/k_0 \sim 130$	Okamoto and Wang,
Malachite green + OH^-, 25.0°C, $\mu = 0.1\ M$ (NaCl)	$k_\Psi = k_0 + k_1[OH]$; Brij, $k_0 = 9.2 \times 10^{-4}\ \text{sec}^{-1}$, $k_1 = 1.1 \times 10^3\ M^{-1}\ \text{sec}^{-1}$; SDS, $k_0 = 0.97 \times 10^{-4}\ \text{sec}^{-1}$, $k_1 = 2.0 \times 10^{-2}\ M^{-1}\ \text{sec}^{-1}$ *N*-Dodecyl-*β*-alanine, concerted catalysis; p*K* of due are affected (see Table 12.5)	Funasaki, 1977
p-Nitrophenyldiphenylphosphate + *n*-dodecylphosphate, 25.0°C, pH = 9.1, 0.01 *M* borate buffer	$k_{\Psi(\text{micelle})}/k_{\Psi(\text{monomer})} \sim 2$; CTAB, $k_\Psi/k_0 \sim 100$	Bunton et al., 1978a
2,4-Dinitrophenyl-3-phenylpropionate		
+*n*-Butylphosphate	CTAB, $k_\Psi/k_0 \sim 1.5$	
+*n*-Dodecylphosphate	CTAB, $k_\Psi/k_0 \sim -$	
25.0°C, pH = 9.1, 0.01 *M* borate buffer		
2,4-Dinitrofluorobenzene		Bunton et al., 1978b
+Glycylglycine	Glucose, $k_\Psi/k_{CTAB} \sim 0.47$	
$+OH^-$ 2.50°C, 0.01 *M* NaOH, 0.025 *M* CTAB 25.0°C, 0.01 *M* NaOH	Glucose, $k_\Psi^{OH}/k_{CTAB} \sim 0.19$, $K_\Psi^{glu}/k_{CTAB} \sim 3.5$; −methylglucoside, $k_\Psi^{OH}/k \sim 1.69$, $k_\Psi^{glu}/k_0 \sim 1$; sorbose, $k_\Psi/k_0 \sim 2.3$; sorbitol, $k_\Psi/k_0 \sim 2.3$; *p*-Octyloxybenzyltrimethylammonium bromide (*1*), $k_\Psi/k_0 \sim 28$; *1* + glucose, $k_\Psi/k_0 \sim 106$	

Table 12.4 *(Continued)*

Reaction	Effect of Surfactant	Reference
Malachite green + OH^-, 25.0°C, 0.01 *M* NaOH	SDS, $k_\Psi/k_0 \sim 9.3$; CTAB, $k_\Psi/k_0 \sim 13$; Igepal, $k_\Psi/k_0 \simeq 1$	Bunton et al., 1978c
Malachite green + 1-benzyldihydronicotinamide, 25.0°C	CTAB, $k_\Psi/k_0 \sim 272$, NaLS, $k_\Psi/k_0 \sim 15$	
Malachite green + BH_4^-	CTAB, $k_\Psi/k_0 \sim 274$	
2,4-Dinitrofluorobenzene + PhO^-	CTAB, $k_\Psi/k_0 \sim 750$	Bunton et al., 1979b
p-Nitrophenyl diphenyl phosphate	Rates are quantitatively treated in terms of distribution of both reactants between aqueous and micellar phases	
$+PhO^-$	CTAB, $k_\Psi/k_0 \sim 3000$	
$+pMeC_6H_4O^-$	CTAB, $k_\Psi/k_0 \sim 4000$	
25.0°C, pH = 10	Rates are quantitatively treated in terms of distribution of both reactants between aqueous and micellar phases	
p-Nitrophenyldiphenyl phosphate		Bunton and Sepulveda, 1979
$+C_6H_5O^-$	CTAB, $10^3k_\Psi = 10\ sec^{-1}$, $10^2k_0 = 3.2\ M^{-1}\ sec^{-1}$	
$+pMeC_6H_4O^-$	CTAB, $10^3k_\Psi = 13\ sec^{-1}$, $10^2k_0 = 3.4\ M^{-1}\ sec^{-1}$	
$+pEtC_6H_4O^-$	CTAB, $10^3k_\Psi = 20\ sec^{-1}$	
$+pPrC_6H_4O^-$	CTAB, $10^3k_\Psi = 22\ sec^{-1}$	
$+ptBUC_6H_4O^-$	CTAB, $10^3k_\Psi = 22\ sec^{-1}$	
$+ptAmC_6H_4O^-$	CTAB, $10^3k_\Psi = 24\ sec^{-1}$	
$+2C_{10}H_7O^-$	CTAB, $10^3k_\Psi = 27\ sec^{-1}$	
25.0°C, pH = 10		
Smiles rearrangement of DL-2-aminododecanoic acid *N*-methyl-*p*-nitroanilide to DL-2-(*p*-nitrophenyl)aminododecanoic acid, 25.0°C, 1% CH_3CN, pH = 10	CTAB, $k_\Psi/k_0 \sim 4$	Sunamoto et al., 1980a

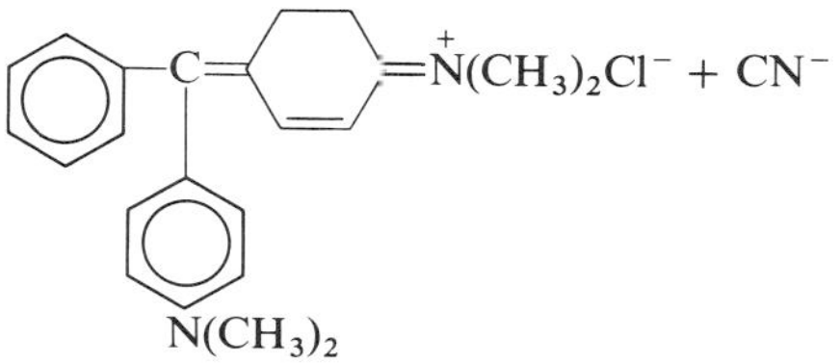 C_6H_5–C(=C_6H_4=$\overset{+}{N}(CH_3)_2Cl^-$)($C_6H_4N(CH_3)_2$) + CN^- 25.0°C, pH 9.5	CTAB, $k_\Psi/k_0 \sim 357$, SDS, $k_\Psi/k_0 \sim 0.03$	Srivastava and Katiyar 1980
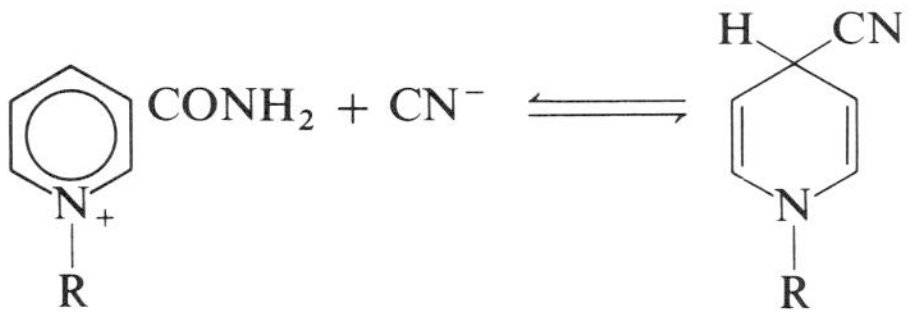 3-$CONH_2$-pyridinium(N^+–R) + CN^- $\rightleftharpoons$ 4-H,4-CN-1,4-dihydropyridine(N–R) $R = C_{12}H_{25}$ $R = C_{14}H_{29}$ $R = C_{16}H_{33}$	CTACN, $k_\Psi/k_0 \sim +$; CTABr, $k_\Psi/k_0 \sim +$ Relative efficiencies of CN^- and Br^- for a cationic micelle are evaluated; catalysis is rationalized in terms of the pseudophase model	Bunton et al., 1980

Table 12.5. Dissociation Constants in Aqueous Micelles

Reaction	Effect of Surfactant	Reference
$pNO_2C_6H_4C(H)(CN)(R) + H_2O \overset{K_a}{\rightleftharpoons} pNO_2C_6H_4C^{\ominus}(CN)(R) + H_3O^+$		Minch et al., 1975
R = H	$pK_a(CTAB) = 11.3$, $pK_a(H_2O) = 13.3$	
$R = C_6H_5$	$pK_a(CTAB) = 8.3$, $pK_a(H_2O) = 10.9$	
$R = p\text{-}ClC_6H_4$	$pK_a(CTAB) = 8.0$, $pK_a(H_2O) = 10.7$	
$R = p\text{-}NO_2C_6H_4$	$pK_a(CTAB) = 5.75$, $pK_a(H_2O) = 10.4$	
R = C≡N	$pK_a(CTAB) = -0.18$, $pK_a(H_2O) = 1.89, 2.35$	
1-nitroindene $\overset{K_a}{\rightleftharpoons}$ 1-nitroindenyl anion ($\ominus$) $+ H^+$	$pK_a(CTAB) = 1.5$, $pK_a(H_2O) = 2.0$	
$pNO_2C_6H_4OH \overset{K_a}{\rightleftharpoons} pNO_2C_6H_4O^- + H^+$	$pK_a(CTAB) = 6.35$, $pK_a(H_2O) = 7.15$	
2,6-di-*t*Bu-4-NO^2-phenol (OH) $\overset{K_a}{\rightleftharpoons}$ 2,6-di-*t*Bu-4-NO_2-phenoxide (O)	$pK_a(CTAB) = 5.13$, $pK_a(H_2O) = 6.92$	

Methylcobalamin	$pK_a(H_2O) = 2.63$, $pK_a(SDS) = 5.73$, $pK_a(CTAB) = 2.65$, $pK_a(\text{Triton X}) = 2.65$	Beckmann and Brown, 1976
Cyanocobalamin	$pK_a(H_2O) \simeq 0$, $pK_a(SDS) = 3.2$, $pK_a(CTAB) \simeq 0$, $pK_a(\text{Triton X}) \simeq 0$	
Aquocobalamin	$pK_a(H_2O) < 0$, $pK_a(SDS) = 2.8$, $pK_a(CTAB) < 0$, $pK_a(\text{Triton X}) \simeq 0$	
Malachite green	$pK_a(H_2O) = 6.90$, $pK_a(\text{N-dodecyl-}\beta\text{-alanine})\ (1) = 5.3$, $pK_a(\text{Brij}) = 4.90$	
Brilliant green	$pK_a(H_2O) = 7.90$, $pK_a(1) = 5.55$, $pK_a(\text{Brij}) = 5.8$	Funasaki, 1977
Crystal violet	$pK_a(H_2O) = 9.36$, $pK_a(1) = 7.2$	
Benzoic acid, 2,4-dimethylbenzoic acid, 2,6-dimethylbenzoic acid, 4(1,1′-dimethylethyl)benzoic acid, 4-dodecyl oxybenzoic acid	$pK_a(\text{HPC, CTAB}) < 1$; $\Delta pK_a(SDS) = 0.5–3$	Pelizetti and Pramauro, 1980
Dimethyl yellow	$pK_a(H_2O) = 3.35$, $pK_a(SDS) = 4.85$	James and Robinson, 1978
Pyridine-2-azo-*p*-dimethylaniline	$pK_a(H_2O) \sim 4.5$, $pK_a(SDS) = 5.98$	

Table 12.6. Miscellaneous Ionic Reactions in Aqueous Micelles

Reaction	Effect of Surfactant	Reference
Acid catalyzed hydration of 1-benzyl-1,4-dihydronicotinamide, 30°C, pH = 6.3, 0.01 *M* phosphate buffer	NaLS, $k_\Psi/k_0 \sim 1.7$; CTAB, $k_\Psi/k_0 \sim 0.12$	Shinkai et al., 1975
Acid catalyzed hydration of 1-lauryl-1,4-dihydronicotinamide, 30°C, pH = 6.3, 0.01 *M* phosphate buffer	NaLS, $k_\Psi/k_0 \sim 33$; CTAB, $k_\Psi/k_0 \sim 0.08$; 1-lauryl-3-carbamoylpyridinium ion, $k_\Psi/k_0 \sim 0.02$	
HSO_3^- catalyzed polymerization of acrylamide and methylacrylate, 30°C, pH = 5.30	CTAB, $k_\Psi/k_0 \sim 3$	Kim and Griffith, 1976; Kim, 1977
Acid catalyzed benzidine rearrangements of:		Bunton and Rubin, 1976
1,2-Diphenylhydrazine (2 proton)	CTAB, $k_\Psi/k_0 \sim 0.06$; Brij-58, $k_\Psi/k_0 \sim 0.1$	
1,2-Di-*o*-tolylhydrazine	NaLS, $k_\Psi/k_0 \sim 4300$ (2 proton)	
1,2-Di-*o*-anisyl hydrazine	NaLS, $k_\Psi/k_0 \sim 5000$ (2 proton); NaLS, $k_\Psi/k_0 \sim 50$ (1 proton)	
25.0°C		
Acid (HA) catalyzed hydration of NADH (reduced form of nicotinamide adenine dinucleotide)		Shinkai et al., 1976a
HA = $ClCH_2COOH$	SDS, $k_H = 521\ M^{-1}\ sec^{-1}$, $k_{HA} = 12.3\ M^{-1}\ sec^{-1}$; CTAB, $k_0 = 5 \times 10^{-5}\ sec^{-1}$, $k_{HA} = 0.08\ M^{-1}\ sec^{-1}$	
HA = $C_6H_5CONHCH_2COOH$	SDS, $k_H = 494\ M^{-1}\ sec^{-1}$, $k_{HA} = 7.67\ M^{-1}\ sec^{-1}$	
HA = CH_3COOH	SDS, $k_H = 341\ M^{-1}\ sec^{-1}$, $k_{HA} = 0.58\ M^{-1}\ sec^{-1}$; CTAB, $k_0 = 5 \times 10^{-5}\ sec^{-1}$, $k_H \sim 0$, $k_{HA} = 0.038\ M^{-1}\ sec^{-1}$	
HA = $CH_3(CH_2)_8COOH$	SDS, $k_H = 304\ M^{-1}\ sec^{-1}$, $k_{HA} = 0.905\ M^{-1}\ sec^{-1}$	
HA = iso-C_3H_7COOH	SDS, $k_H = 311\ M^{-1}\ sec^{-1}$, $k_{HA} = 0.64\ M^{-1}\ sec^{-1}$, CTAB, $k_0 = 5 \times 10^{-5}\ sec^{-1}$, $k_H \sim 0$, $k_{HA} = 0.0023\ M^{-1}\ sec^{-1}$	
HA = $^-OOCCH_2COOH$	SDS, $k_H = 225\ M^{-1}\ sec^{-1}$, $k_{HA} = 0.0368\ M^{-1}\ sec^{-1}$, CTAB, $k_0 = 3 \times 10^{-5}\ sec^{-1}$, $k_H \sim 0$, $k_{HA} = 0.0008\ M^{-1}\ sec^{-1}$	
30°C		

Peroxidisulfate oxidation of I^-, 25.0°C	CTAB, $k_\Psi/k_0 \sim 25$	Blandamer et al., 1976
Peroxidisulfate oxidation of $Fe(CN)_6^{4-}$, 25.0°C, 6×10^{-4} *M* KOH	CTAB, $k_\Psi/k_0 \sim 32$	
Cobaloxime($Co(CH)_2^-$) + EtBr, n-$C_5H_{11}Cl$, $ClCH_2CO_2^-$, $ClCH_2CH_2CO_2^-$	CTAB, $k_\Psi/k_0 \sim 1200, 420, 373$	Allen and Bunton, 1976
Propane derivative of cobaloxime(Co(COH)CO*pn*°)		
+ EtBr	CTAB, $k_\Psi/k_0 \sim 8.5$	
+ $nC_5H_{11}Cl$	CTAB, $k_\Psi/k_0 \sim 200$; SDS, $k_\Psi/k_0 \sim 1$	
+ $ClCH_2CO_2^-$	CTAB, $k_\Psi/k_0 \sim 173$; SDS, $k_\Psi/k_0 \sim 0.19$	
25.0°C, 0.05 *M* NaOH		
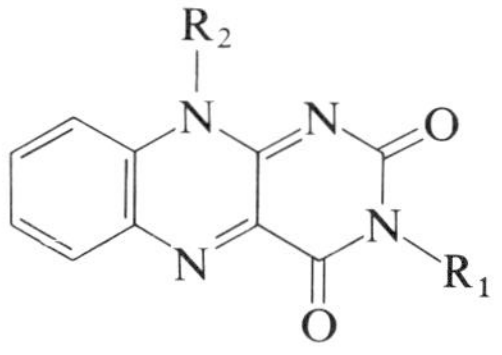 [isoalloxazine, N-R_2, N-R_1] + nitroethane → $CH_3COH + NO_2^-$		Shinkai et al., 1976b
$R_1 = Me(CH_2)_{15}$, $R_2 = n$Bu 30°C, pH = 8.0–9.2, μ = 0.15 *M* (KCl) 3 vol.% EtOH	CTAB, $k_\Psi/k_0 \sim 10^3$ For R_1 = Me, R_2 = Et or R_1 = Me, $R_2 = n$Bu, no oxidation in absence of micelles, only hydrolysis in the presence of CTAB	
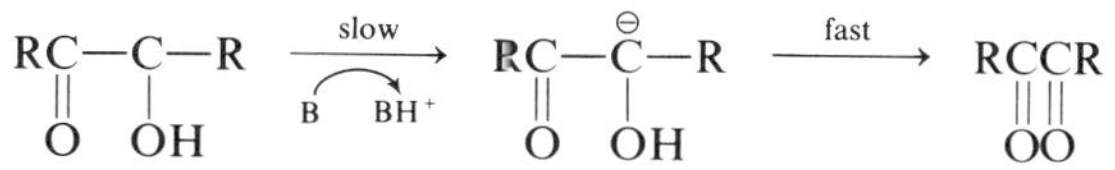 RC(=O)—C(OH)—R $\xrightarrow[\text{B} \to \text{BH}^+]{\text{slow}}$ RC(=O)—$\overset{\ominus}{\text{C}}$(OH)—R $\xrightarrow{\text{fast}}$ RC(=O)C(=O)R		Shinkai and Kunitake, 1976c
RCOCOHR = furoin, B = OH^-	CTAB, $k_\Psi/k_0 \sim 1.4$; POOA, $k_\Psi/k_0 \sim 0.6$	
RCOCOHR = furoin, B = *N*-benzylbenzohydroxamic acid (*1*)	POOA, $k_\Psi/k_0 \sim 3.4$	
RCOCOHR = Furoin, B = *N*-methylmyristohydroxamic acid (*2*)	POOA, $k_\Psi/k_0 \sim 10$; CTAB, $k_\Psi/k_0 \sim 413$	

Table 12.6 (*Continued*)

Reaction	Effect of Surfactant	Reference
RCOCOHR = benzoin, B = OH	CTAB, $k_\Psi/k_0 \sim 7.0$; POOA, $k_\Psi/k_0 \sim 0.2$	
RCOCOHR = benzoin, B = *1*	CTAB, $k_\Psi/k_0 \sim 376$	
RCOCOHR = benzoin, B = *2*	POOA, $k_\Psi/k_0 \sim 30$; CTAB, $k_\Psi/k_0 \sim 3500$	
30°C, pH = 11.0, 1% v/v CH_3CN		
Dye formation between		Tong and Glesmann, 1977
	$k_I = M^{-1}$ sec^{-1} in H_2O, k_{II} = sec^{-1} in micelle	
I, R = C_2H_5; II, R = $C_{18}H_{37}$ and:		
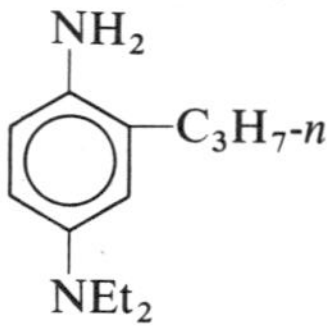	$\log k_I = 3.68$, $\log k_{II} = 1.95$	
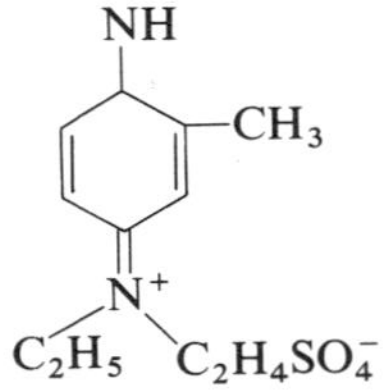	$\log k_I = 4.87$, $\log k_{II} > 3.0$	

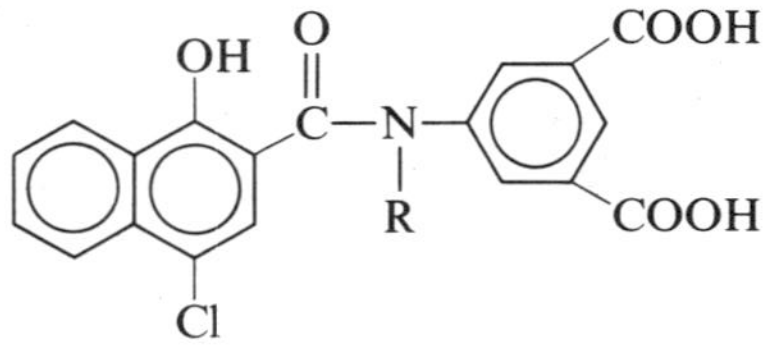

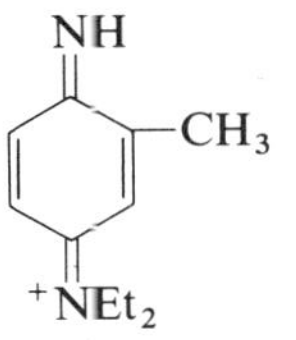

NH, CH_3, $^{+}NEt_2$ — $\log k_{I} = 3.70$, $\log k_{II} = 1.9$

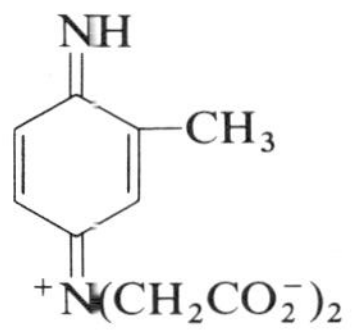

NH, CH_3, $^{+}N(CH_2CO_2^-)_2$ — $\log k_{I} = 4.4$, $\log k_{II} > 1.1$

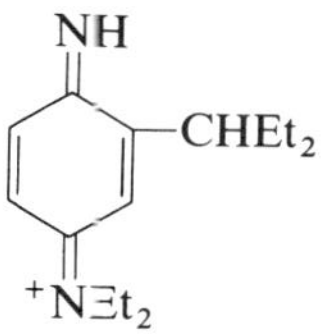

NH, $CHEt_2$, $^{+}N\Xi t_2$ — $\log k_{I} = 3.2$, $\log k_{II} = 1.6$

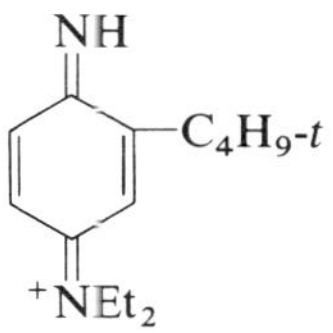

NH, C_4H_9-*t*, $^{+}NEt_2$ — $\log k_{I} = 0.40$, $\log k_{II} = -1.1$

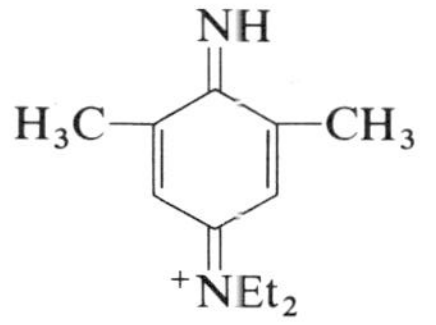

NH, H_3C, CH_3, $^{+}NEt_2$ — $\log k_{I} = 1.56$, $\log k_{II} = -0.2$

$T = ?$, pH = 10–11

Table 12.6 (*Continued*)

Reaction	Effect of Surfactant	Reference
Nitrosation of dihexylamine by sodium nitrite in the presence of amines, 25.0°C, pH = 3.5, citrate-phosphate buffer	DTAB, $k_\Psi/k_0 \sim 3.3$ (diethylamine), 7.3 (di-*sec*-butylamine), 11 (methylbutylamine), 53 (dibutylamine), 350 (dipentyl), 800 (dihexyl), 1 (morpholine), 3.8 (pyrrolidine), 4.8 (piperidine)	Okun and Archer, 1977
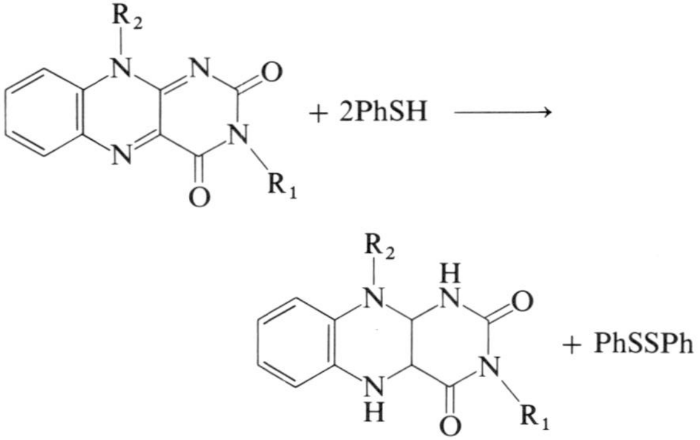 R_2 N N O N O R_1 + 2PhSH ⟶ R_2 N H N O N H O R_1 + PhSSPh $R_1 = CH_3(CH_2)_{15}$, $R_2 = CH_3(CH_2)_3$(I) $R_1 = CH_3$, $R_2 = CH_3CH_2$(II) 30°C, $\mu = 0.02$ phosphate buffer	$k_\Psi = k_3'[PhSH]_T^2[\text{isoalloxazine}]$; $k_3' = k_3[PhSH][PhS^-] = [k_3 K_{app} a_H/(K_{app} + a_H)^2][PhSH]_T^2$ I + PhSH, $pk_{app} = 6.6$, $k_3 = 5.78 \times 10^4\ M^{-2}\ sec^{-1}$ II + PhSH, $pk_{app} = 6.6$, $k_3 = 3.2 \times 10^4\ M^{-2}\ sec^{-1}$	Shinkai et al., 1977
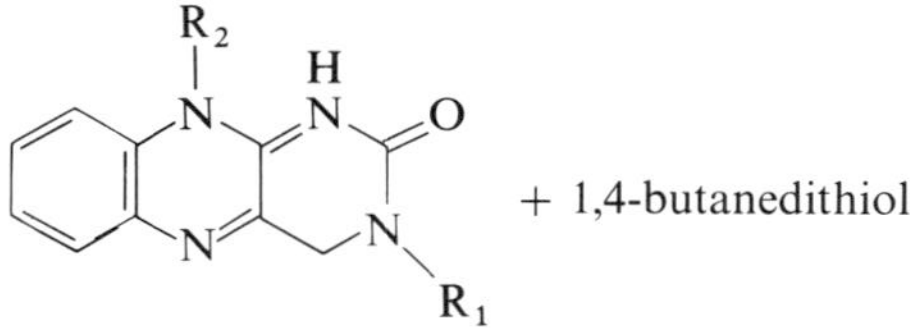 R_2 N H N O N N R_1 + 1,4-butanedithiol		Shinkai and Kunitake, 1977

Reaction	Surfactant, k_Ψ/k_0	Reference
$R_1 = CH_3$, $R_2 = CH_3CH_2$(*1*)	CTAB, $k_\Psi/k_0 \sim 22$; SDS, $k_\Psi/k_0 = -$; Brij-35, $k_\Psi/k_0 = -$	
$R_1 = CH_3(CH_2)_{15}$, $R_2 = CH_3(CH_2)_3$(*2*) 30°C, pH = 10.05, $\mu = 0.06$ *M* KCl	CTAB, $k_\Psi/k_0 \sim 400$	
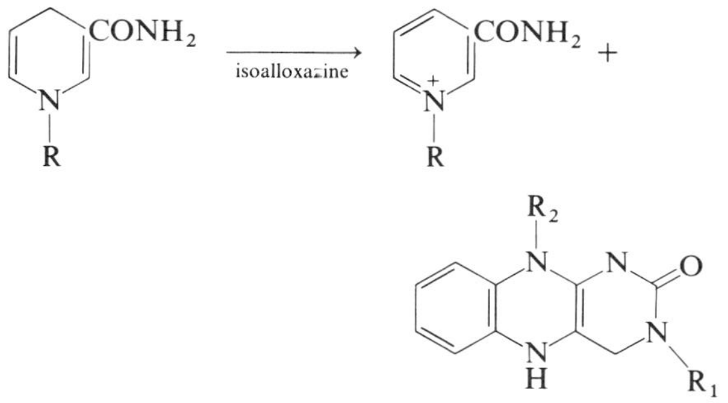 1-R-1,4-dihydronicotinamide ($CONH_2$) $\xrightarrow{\text{isoalloxazine}}$ 1-R-pyridinium-3-carboxamide ($CONH_2$) + reduced isoalloxazine (N-R_2, N-R_1, C=O, N-H)		Shinkai et al., 1978a
$R = CH_2C_6H_5$, $R_1 = CH_3$, $R_2 = C_2H_5$	CTAB, $k_\Psi/k_0 \sim 2$; CPyB, $k_\Psi/k_0 \sim 1.4$; SDS, $k_\Psi/k_0 \sim 2.3$; POOA, $k_\Psi/k_0 \sim 1$	
$R = CH_2C_6H_5$, $R_1 = C_{16}H_{33}$, $R_2 = C_4H_9$	CTAB, $k_\Psi/k_0 \sim 1$; CPyB, $k_\Psi/k_0 \sim 1.4$; SDS, $k_\Psi/k_0 \sim 0.11$; POOA, $k_\Psi/k_0 \sim 0.74$	
$R = C_{12}H_{25}$, $R_1 = CH_3$, $R_2 = C_2H_5$	CTAB, $k_\Psi/k_0 \sim 0.08$; CPyB, $k_\Psi/k_0 \sim 0.05$; SDS, $k_\Psi/k_0 \sim 2$	
Pyridine-2-azo-*p*-dimethylamine + Ni^{2+}, 25.0°C, pH varied	SDS, $k_\Psi/k_0 \sim 10^3$	James and Robinson, 1978; Robinson and White, 1978; Reinsborough and Robinson, 1979; Holzwarth et al., 1978

Table 12.6 (*Continued*)

Reaction	Effect of Surfactant	Reference
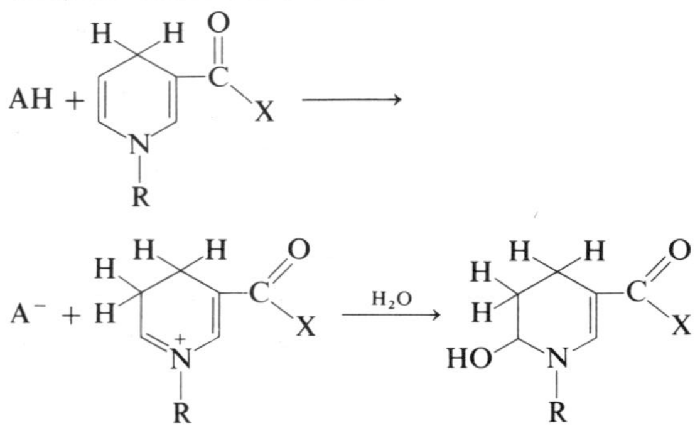 AH + [1-R-1,4-dihydropyridine-3-C(O)X] ⟶ A⁻ + [1-R-1,4,5,6-tetrahydro-5-H-pyridinium-3-C(O)X] $\xrightarrow{H_2O}$ [6-HO-1-R-1,4,5,6-tetrahydropyridine-3-C(O)X] 25.0°C, 0.001–0.005 M HCl	CTAB, $k_\Psi/k_0 \sim 1$; NaLS, $k_\Psi/k_0 \sim 6.6$ ($R = CH_2Ph$, $X = NH_2$); NaLS, $k_\Psi/k_0 \sim 14$ ($R = CH_2Ph$, $X = CH_3$)	Bunton et al., 1978d
Base catalyzed isomerization of prostaglandins 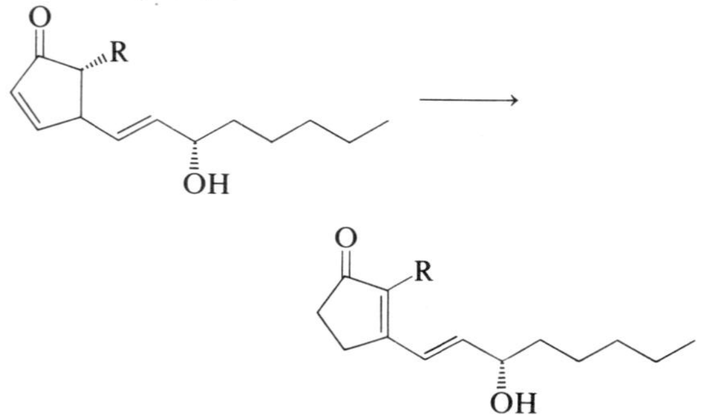 [prostaglandin A structure] ⟶ [prostaglandin B structure] $R = (CH_2)_6COOH$ or $CH_2CH{=}CH(CH_2)_3COOH$, 60°C, pH = 11.2	DOTAC, $k_\Psi/k_0 \sim 3$; SDS, $k_\Psi/k_0 \sim 1$; Brij-35, $k_\Psi/k_0 \sim 1.5$	Uekama et al., 1978

$RCHO \underset{k_{-1}}{\overset{k_1}{\rightleftharpoons}} RC(O^-)(CN) \underset{k_{-2}}{\overset{k_2}{\rightleftharpoons}} RC^-(OH)(CN) \xrightarrow{k_3} RCOCN \xrightarrow{H_2O} RCOOH$ (with CN^- regenerated; I ⇌ IH_2, H_2O_2, O_2) $K = k_1/k_{-1}$ C_4H_9 … R_3 $R_3 = CH_3$ (*5*) $R_3 = C_{16}H_{33}$ (*6*) 30°C	RCHO = $pClC_6H_4CHO$, $10^6 k_2 K = 0.49$ (H_2O), 107 (I + CTAB), 126 (II + CTAB)M^{-1} sec^{-1}; RCHO = C_6H_5CHO, $10^6 k_2 K = 0.10$ (H_2O), 25 (I + CTAB), 17 (II + CTAB)M^{-1} sec^{-1}; RCHO = C_3H_7CHO, $10^6 k_2 K = 0.01$ (H_2O), 0.23 (I + CTAB), 0.57 (II + CTAB)M^{-1} sec^{-1}; RCHO = HCHO, $10^6 k_2 K = 0.001$ (H_2O), 0.001 (I + CTAB), 0.04 (II + CTAB)M^{-1} sec^{-1}	Shinkai et al., 1978b
Acid catalyzed epimerization of 15(5)-15-methyl prostaglandin $F_{2\alpha}$, 25.0°C, pH = 2.5	Myristyl-γ-picolinium chloride, $k_\Psi/k_0 \sim 0.003$	Cho and Allen, 1978
$p\text{-}NO_2C_6H_4CH_2CH_2\overset{+}{N}Me_3, I^- \xrightarrow{OH^-}$ $p\text{-}NO_2C_6H_4CH{=}CH_2 - NMe_3$, 39.0°C, [OH] varies	SDS, $k_\Psi/k_0 \sim 0.2$; CTAB, $k_\Psi/k_0 \sim 3$; *N*,*N*-dimethyl-*N*-hexadecyl-*N*-(2-hydroxy ethyl)ammonium bromide, $k_\Psi/k_0 \sim 3$	Minch et al., 1978
Decarboxylation of 5-amino-1,3,4-thiadiazole-2-carboxylic acid, pH = 0.46–4.6, 40.0°C, 50.6°C, 59.3°C	CTAB, $k_\Psi/k_0 \sim 5$; Triton X-100, $k_\Psi/k_0 \sim 5.7$; NaLS, $k_\Psi/k_0 \sim 1.3$	Spinelli et al., 1978
Methylation of thiophenoxide ions by surfactants, 25.0°C, pH = 8.0, phosphate buffer	$n\text{-}C_{16}H_{33}\overset{+}{N}(Me_2)CH_2CH_2\overset{+}{S}(Me_2)$, $2CF_3SO_3^-$ (*7*), $k_\Psi = 0.12$ sec^{-1}; *7* + CTACl, $k_\Psi = 0.16$ sec^{-1}; $(Me_3)\overset{+}{N}CH_2\overset{+}{S}(Me_2)$, $2CF_3SO_3^-$, $k_\Psi = 0.04$ sec^{-1}; $n\text{-}C_{16}H_{33}\overset{+}{N}(Me_2)CH_2CH_2OSO_2CH_3$, Cl^-, $k_\Psi = 0.00162$ sec^{-1}	Moss and Sanders, 1979

Table 12.6 (*Continued*)

Reaction	Effect of Surfactant	Reference
Oximation of methyl ketones ($CH_3(CH_2)_nC{=}OCH_3$, $0 \leq n \leq 5$ and $n = 8$), cycloalkanones ($n = 4$–8), substituted cyclohexanones, 30.0°C, pH = 8.5, $\mu = 0.5$ (NaCl)	SDS, $k_\Psi/k_0 \sim 1$–12 k_Ψ/k_0 increases linearly from $n = 0$ to $n = 4$ methyl ketones, thereafter independent of n; k_Ψ/k_0 increases with ring size for cycloalkanones; no correlation for substituted cyclohexanones	Finiels and Geneste, 1979
Hydration of 3-formyl-*N*-tetradecyl pyridinium bromide as a function of surfactant concentration, ambient temperature	Between 0–0.5 *M* in the presence of *N*-tetradecyltrimethylammonium bromide and SDS spectra is the same as in H_2O; activity of H_2O at the surface of ionic micelles is, therefore, not very different from that in bulk aqueous environment	De Albrizzio and Cordes, 1979
Cyclization of salicylaldoximate to benzisoxazole, 20.0°C, pH = 10.15	CTAB, $k_\Psi/k_0 \sim 2.4$	Meyer, 1979
Complexation of pyridine-2-azopodidimethylaniline with Ni^{2+}	SDES, $k/k_0 \sim 200$ Local ion concentrations at charged micellar surfaces are calculated	Diekmann and Frahm, 1979
Reduction of methylene blue with L-ascorbic acid and L-cysteine	CTAB, $k/k_0 = +$ (alters dissociation constant)	Seno et al., 1979
Base catalyzed elimination from $pXC_6H_4CR_2CH_2Br$, 25.0°C, [NaOH] = 0.5 *M*		Yano et al., 1979
X = NO_2, R = H	$n\text{-}C_{18}H_{37}\overset{+}{N}Me_2CH_2CH_2OH$ (*1*), $k_\Psi/k_0 \sim 182$, CTAB, $k_\Psi/k_0 \sim 8$	
X = NO_2, R = D	*1*, $k_\Psi/k_0 \sim 153$	
X = CN, R = H	*1*, $k_\Psi/k_0 \sim 144$; CTAB, $k_\Psi = 7.36 \times 10^{-3}$ sec^{-1}	
X = Cl, R = H	*1*, $k_\Psi/k_0 \sim 129$; CTAB, $k_\Psi = 3.24 \times 10^{-4}$ sec^{-1}	
X = H, R = H	*1*, $k_\Psi/k_0 \sim 51$; CTAB, $k_\Psi = 1.16 \times 10^{-4}$ sec^{-1}	
X = OMe, R = H	*1*, $k_\Psi/k_0 \sim 35$; CTAB, $k_\Psi = 3.43 \times 10^{-5}$ sec^{-1} Hammett correlations suggest the transition states to be more carbanion like in micelles than in water	

Reversible complex formation between murexide and Ni^{2+}, 25.0°C, pH = 4.5–5.6	$K_\Psi/K_0 \sim -$	Fischer et al., 1979
Acetone + Ce(IV) $\underset{}{\overset{k}{\rightleftharpoons}}$ complex $\xrightarrow{k}$ products 30.0°C, 1.0 M $HClO_4$	SDS, $k_\Psi/k_0 \sim +$	Maruthamuthu and Usha, 1979
Carboxylate ion catalyzed oxidation of diethyl sulfide, 15°C, 25°C, 42°C	SDS, $k_\Psi/k_0 \sim 100$ Rate constants exhibit two linear first order regions with respect to carboxylate ion concentration in SDS micelles, but in water they are described by first and second order terms	Young and Hou, 1979
Flavin mediated oxidation of aldehydes (30.0°C) and oxidative decarboxylation of α-keto acids (50.0°C) (see Shinkai et al., 1978b above for scheme) 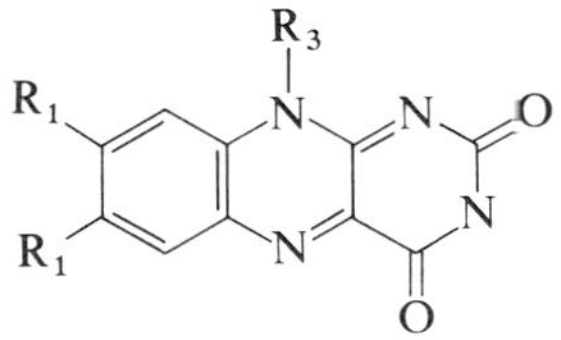 [flavin structure] R_1 = H, R_2 = CH_3, R_3 = C_4H_9 (*1*) R_1 = H, R_2 = $C_{16}H_{33}$, R_3 = C_4H_9 (*2*) R_1 = CH_3, R_2 = CH_3, R_3 = tetra-*o*-acetylribityl (*3*) 2i4-$Cl_2C_6H_3CHO$	*1*, $10^6 k_\Psi = 11.3\ M^{-1}\ sec^{-1}$; *1* + CTAB, $10^6 k_\Psi = 592\ M^{-1}\ sec^{-1}$	Shinkai et al., 1980a

Table 12.6 (*Continued*)

Reaction	Effect of Surfactant	Reference
4-ClC_6H_4CHO	*1*, $10^6 k_\Psi = 0.49\ M^{-1}\ sec^{-1}$; *1* + CTAB, $10^6 k_\Psi = 107\ M^{-1}\ sec^{-1}$; *2* + CTAB, $k_\Psi = 126\ M^{-1}\ sec^{-1}$	
C_6H_5CHO	*1*, $10^6 k_\Psi = 0.10\ M^{-1}\ sec^{-1}$; *1* + CTAB, $10^6 k_\Psi = 25.3\ M^{-1}\ sec^{-1}$; *2* + CTAB, $k_\Psi = 16.8\ M^{-1}\ sec^{-1}$	
C_3H_7CHO	*1*, $10^6 k_\Psi = 0.01\ M^{-1}\ sec^{-1}$; *1* + CTAB, $10^6 k_\Psi = 0.23\ M^{-1}\ sec^{-1}$; *2* + CTAB, $k_\Psi = 0.57\ M^{-1}\ sec^{-1}$	
HCHO	*1*, $10^6 k_\Psi < 0.001\ M^{-1}\ sec^{-1}$; *1* + CTAB, $10^6 k_\Psi < 0.001\ M^{-1}\ sec^{-1}$; *2* + CTAB, $k_\Psi = 0.04\ M^{-1}\ sec^{-1}$	
2,4-$Cl_2C_6H_3COCOOH$	*3*, $10^6 k_\Psi < 0.1\ M^{-1}\ sec^{-1}$; *3* + CTAB, $10^6 k_\Psi < 0.1\ M^{-1}\ sec^{-1}$	
4-$ClC_6H_4COCOOH$	*3*, $10^6\ k_\Psi < 0.1\ M^{-1}\ sec^{-1}$; *3* + CTAB, $10^6\ k_\Psi = 275\ M^{-1}\ sec^{-1}$	
$C_6H_5COCOOH$	*3*, $10^6 k_\Psi < 0.01\ M^{-1}\ sec^{-1}$; *3* + CTAB, $10^6 k_\Psi = 1.4\ M^{-1}\ sec^{-1}$	
$CH_3COCOOH$	*3*, $10^6 k_\Psi < 0.001\ M^{-1}\ sec^{-1}$; *3* + CTAB, $10^6 k_\Psi = 0.05\ M^{-1}\ sec^{-1}$	
Flavin mediated oxidations, 30.0°C		Shinkai et al., 1980b

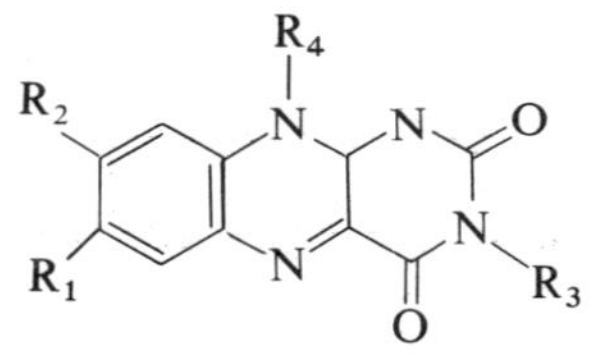

$R_1 = R_2 = H, R_3 = Me, R_4 = Et$ (*1*) $R_1 = R_2 = H, R_3 = C_{16}H_{33}, R_4 = Br$ (*2*) $R_1 = R_2 = R_3 = Me, R_4 = Et$ (*3*) $R_1 = R_3 = Me, R_2 =$ $CH_2ImCH_2CH(COOH)NHCOCH_3, R_4 = Et$ (*4*) $R_1 = R_3 = Me, R_2 =$ $CH_2ImCH_2CH(COOH)NHCOC_{13}H_{37}, R_4 = Et$ (*5*)	
Oxidation of 1-benzyl-1,4-dihydronicotinamide	*1*, $k_0 = 10\ M^{-1}\ sec^{-1}$; *1* + CTAB, $k_\Psi = 11.4\ M^{-1}\ sec^{-1}$; *1* + SDS, $k_\Psi = 6.50\ M^{-1}\ sec^{-1}$; *1* + Brij-35, $k_\Psi = 6.1\ M^{-1}\ sec^{-1}$; *2* + CTAB, $k_\Psi = 9.92\ M^{-1}\ sec^{-1}$; *2* + SDS, $k_\Psi = 1.16\ M^{-1}\ sec^{-1}$; *2* + Brij-35, $k_\Psi = 7.2\ M^{-1}\ sec^{-1}$; *3*, $k_0 = 32.4\ M^{-1}\ sec^{-1}$; *4*, $k_0 = 48.1\ M^{-1}\ sec^{-1}$; *4* + CTAB, $k_\Psi = 73.0\ M^{-1}\ sec^{-1}$; *4* + SDS, $k_\Psi = 20.5\ M^{-1}\ sec^{-1}$; *4* + Brij-35, $k_\Psi = 42.3\ M^{-1}\ sec^{-1}$; *5*, $k_0 = 585\ M^{-1}\ sec^{-1}$; *5* + CTAB, $k_\Psi = 235\ M^{-1}\ sec^{-1}$; *5* + SDS, $k_\Psi = 91.3\ M^{-1}\ sec^{-1}$; *5* + Brij-35, $k_\Psi = 390\ M^{-1}\ sec^{-1}$
Oxidation of nitroethane carbanion	*1*, $k_0 \sim 0$; *1* + CTAB, $k_0 \sim 0$; *2* + CTAB, $k_\Psi = 3.63 \times 10^{-4}\ M^{-1}\ sec^{-1}$; *4* + CTAB, $k_\Psi = 9.9 \times 10^{-3}\ M^{-1}\ sec^{-1}$; *5*, $k_0 = 1.3 \times 10^{-4}\ M^{-1}\ sec^{-1}$; *5* + CTAB, $k_\Psi = 5.71 \times 10^{-1}\ M^{-1}\ sec^{-1}$
Oxidation of thiophenol	*1*, $k_\Psi \sim 0$; *1* + CTAB, $k_\Psi = 3320\ M^{-1}\ sec^{-1}$; *2*, $k_\Psi \sim 0$; *2* + CTAB, $k_\Psi = 9440\ M^{-1}\ sec^{-1}$; *4*, $k_\Psi \sim 0$; *4* + CTAB, $k_\Psi = 4350\ M^{-1}\ sec^{-1}$; *5*, $k_0 \sim 0$; *5* + CTAB, $k_\Psi = 6650\ M^{-1}\ sec^{-1}$

Table 12.6 *(Continued)*

Reaction	Effect of Surfactant	Reference
Flavin mediated oxidations, 30.0°C, pH = 7.52, $[MeFl] = 5.0 \times 10^{-5}\ M$		Shinkai et al., 1980c
R-thiazolium + ArCHO ($ArCOCO_2^-$) $\underset{k_{-1}}{\overset{k_1}{\rightleftharpoons}}$ R-thiazolium-2-C(O⁻)(X)Ar, $X{=}H(CO_2^-)$ $\xrightarrow{k_2}$ R-thiazoline=C(OH)Ar; $\xrightarrow{MeFl}$ R-thiazolium-2-COAr $\xrightarrow{H_2O}$ $ArCO_2H$; $\xrightarrow{ArCHO}$ benzoin		
$R = (CH_2)_{15}CH$		
ArCHO		
$2,4\text{-}Cl_2C_6H_3CHO$	CTAB, $k_\Psi = 114\ M^{-1}\ sec^{-1}$	
$4\text{-}ClC_6H_4CHO$	CTAB, $k_\Psi = 17.4\ M^{-1}\ sec^{-1}$	
C_6H_5CHO	CTAB, $k_\Psi = 3.47\ M^{-1}\ sec^{-1}$; Brij-35, $k_\Psi = 0.068\ M^{-1}\ sec^{-1}$; SDS, $k_\Psi \sim 0$	
HCHO	CTAB, $k_\Psi = 0.020\ M^{-1}\ sec^{-1}$	
C_3H_7CHO	CTAB, $k_\Psi = 0.130\ M^{-1}\ sec^{-1}$	
$C_7H_{15}CHO$	CTAB, $k_\Psi = 0.738\ M^{-1}\ sec^{-1}$	
$ArCOCO_2^-$		
$4\text{-}ClC_6H_4COCO_2H$	CTAB, $k_\Psi = 0.01\ M^{-1}\ sec^{-1}$	
CH_3COCO_2H	CTAB, $k_\Psi = 0.23\ M^{-1}\ sec^{-1}$	

Oxidative cleavage of 1,1-bis(*p*-chlorphenyl)-2,2,2-trichloroethanol to form $CHCl_3$ and 4,4′-dichlorobenzophenone, 30.0°C, pH varied	CTAB, $k_\Psi/k_0 \sim 200$; hexadecyldimethyl(2-hydroxyethyl)ammonium chloride, $k_\Psi/k_0 \sim 345$; SDS, $k_\Psi/k_0 \sim -$	Nome et al., 1980
Diazo coupling of N_2^+, Br^- 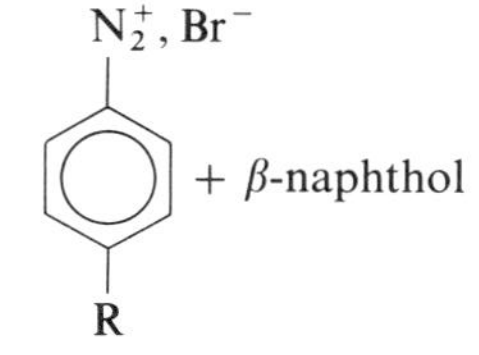 + β-naphthol; $R = nC_{16}H_{33}\overset{+}{N}(CH_3)_2CH_2$ (I); $R = (CH_3)_3\overset{+}{N}CH_2$ (II); 4.0°C, pH = 5.9–7.8	$k_I/k_{II} \sim 164$ (pH = 5.95), 98 (pH = 7.10), 26 (pH = 7.8)	Moss and Rav-Acha, 1980
Dioxygen stability of cobaltous complexes	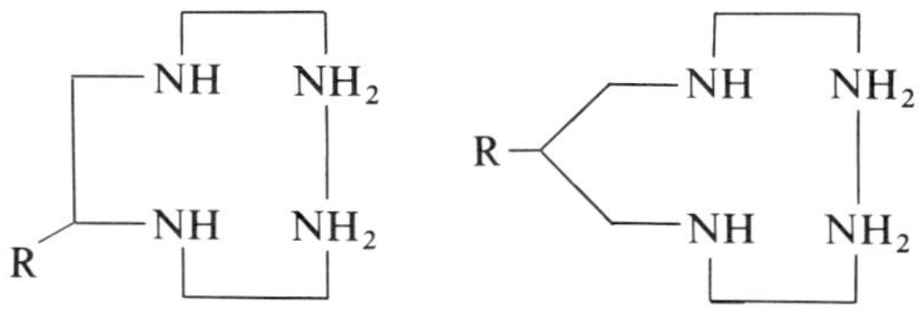 R = 12 (annelides), $k_\Psi/k_0 \sim + +$	Simon et al., 1980

Table 12.7. Energy and Electron Transfer and Charge Separation in Aqueous Micelles

Micellar System	Process	Results	Experimental Techniques	Reference
0.1 *M* SDS	Phenothiazine (PTH) photoionization: $PTH \xrightarrow{h\nu} PTH^T + PTH^{+} + e_{aq}^-$; electron transfer from PTH^T to Eu^{3+} and Cu^{2+}	Photoionization ($\Phi \sim 0.5$) and electron transfer are more efficient but triplet yield is smaller in micelles than in MeOH	Laser flash photolysis	Alkaitis et al., 1975a
0.05 *M* SDS, 5.0×10^{-3} CTAB	Phenothiazine (PTH) photoionization: $PTH \xrightarrow{h\nu} PTH^T + PTH^{+} + e_{aq}^-$; in presence of duroquinone (CQ) in SDS, $DQ^{\cdot-}$ is formed in presence of naphthoquinone sulfonate (NQS) in CTAB, $NQS^{\cdot-}$ and PTH^{+} are formed	Photoionization is rationalized by rapid tunneling of an electron from excited PTH through the double layer into unoccupied electronic redox levels; $DQ^{\cdot-}$ is formed by a two-step process	Laser flash photolysis	Alkaitis et al., 1975b
0.1 *M* SDS, 4.0×10^{-2} *M* DTAC	*N,N,N',N'*-Tetramethylbenzidine (TMB) photolysis: $TMB \rightarrow TMB^T + TMB^{+} + e_{aq}^-$ TMB^T + duroquinone (DQ) $\underset{k_{-1}}{\overset{k_1}{\rightleftharpoons}} DQ^{\cdot-} + TMB^{+}$ $TMB^T + Eu^{3+} \underset{k_{-2}}{\overset{k_2}{\rightleftharpoons}} Eu^{2+} + TMB^{\cdot-}$	Ion:triplet yield ratio is 36 greater in SDS than MeOH; back reaction (k_{-2}) is retarded by SDS; $k_1 = 2.3 \times 10^{10}\ M^{-1}\ sec^{-1}$ (MeOH); $k_{-1} = 2.9 \times 10^{7}\ M^{-1}\ sec^{-1}$ (MeOH); $k_2 = 6.4 \times 10^{10}\ M^{-1}\ sec^{-1}$ (MeOH), $2.3 \times 10^{7}\ M^{-1}\ sec^{-1}$ (SDS); $k_{-2} = 1.4 \times 10^{7}\ M^{-1}\ sec^{-1}$ (MeOH), $8.5 \times 10^{2}\ M^{-1}\ sec^{-1}$ (SDS)	Laser flash photolysis	Alkaitis and Grätzel, 1976
10^{-2} *M* CTAB + 10^{-2} *M* NaBr	Pyrene (P) $\xrightarrow{h\nu} P^T + p^{+} + q$ $e_{aq}^- + N_2O \rightarrow N_2 + \dot{O}H + OH^-$ $\dot{O}H + Br^- \rightarrow Br\cdot$ $Br\cdot + Br^- \rightarrow Br_2^{\cdot-}$ $Br_2^{\cdot-} \rightarrow Br_3^{\cdot-} + Br^-$ $P^T + Br_2^{\cdot-} \xrightarrow{k_1} p^{+} + 2Br^-$ (*1*)	Reaction *1* is of two exponential; fast process, $k_1 = 2 \times 10^{6}\ sec^{-1}$ (independent of CTAB concentration), and a slow process, $k_1 = 1 \times 10^{9}\ M^{-1}\ sec^{-1}$; residence time of $Br_2^{\cdot-}$ in micelle is substantially longer than the diffusion time of the radical between micelles.	Combined laser flash photolysis and pulse radiolysis	Frank et al., 1976
0.1 *M* SDS	Methylene blue (MB) sensitized photooxidation of diphenyl isobenzofuran (DF) $MB \xrightarrow{h\nu} MB^T$ $MB^T + O_2(^3\Sigma) \rightarrow O_2^*(^1\Delta) + MB$ $O_2^*(^1\Delta) + DF \rightarrow DFO_2$	Singlet oxygen, $O_2^*(^1\Delta)$, produced by energy transfer from MB^T in the aqueous phase diffuses into micelles where it reacts with DF	Steady state photolysis	Gorman et al., 1976

10^{-2} *M* SDS	Photoreduction of methylene blue by EDTA	Micelles retard the rate of photoreduction	Steady state photolysis	Usui et al., 1976
DTAC	Duroquinone (DQ) sensitized photooxidation of CO_3^{2-}: $DQ \xrightarrow{h\nu} DQ^T$ $DQ^T + CO_3^{2-} \xrightarrow{1} DQ^{\cdot-} + CO_3^-$ $CO_3^- + DQ \xrightarrow{2} CO_3 + DQ^{\cdot-}$ $CO_3 \xrightarrow{3} CO_2 + \frac{1}{2}O_2$ $CO_2 + OH^- \rightarrow CO_3^{2-} + H_2O$	Rapid electron transfer from CO_3^{2-}, located at the micellar Stern layer, to DQ, residing in the interior (step 1), followed by step 2, with $k_2 = 2 \times 10^9\ M^{-1}\ sec^{-1}$; The net reaction is $2DQ + 2OH^- \rightarrow 2DQ^{\cdot-} + \frac{1}{2}O_2 + H_2O$	Steady state and laser flash photolysis, pulse radiolysis	Scheerer and Grätzel, 1976
SDS, Brij, DTAC	Behavior of pyrene-*N*,*N*-dimethylaniline intramolecular exciplex	Micelle solubilization obviates complicating intermolecular effects	Steady state fluorescence, laser flash photolysis	Masuhara et al., 1977
SDS	Electron transfer from duroquinone triplet to Fe^{2+}	Electron transfer is accelerated	Laser flash photolysis	Scheerer and Grätzel, 1977
SDS	Naphthalene (N) to TbCl energy transfer: $N \xrightarrow{h\nu} {}^1N^* \xrightarrow{isc} {}^3N^*$ ${}^3N^* + {}^3N^* \xrightarrow{k_1}$ quenching ${}^3N^* + Tb^{3+} \xrightarrow{k_2}$ energy transfer	In water there is no energy transfer, k_1 prevails; in SDS micelles, containing less than 1 N per micelle, and high $[Tb^{3+}]$, k_2 predominates	Steady state and nanosecond time resolved fluorescence spectrophotometry	Escabi-Perez et al., 1977
CTACl, SDS, Brij-35	Photosensitized electron transfer from *N*,*N*-dimethylaniline and *N*,*N*-dimethylaniline sulfonate to excited pyrene	Micelles facilitate charge transfer	Steady state fluorescence, laser flash photolysis	Waka et al., 1978
SDS, Triton X-100, Igepal CO-630	Surfactant derivative of tris(2,2′-bipyridyl)ruthenium, $[RuC_{13}(bpy)_3]^{2+}$, sensitized reduction of methylviologen, MV^{2+}, using cystein, cysH, as donor: $[RuC_{13}(bpy)_3]^{2+} \xrightarrow{h\nu} [RuC_{18}(bpy)_3]^{2+*}$ $[RuC_{13}(bpy)_3]^{2+*} + MV^{2+} \rightleftharpoons [RuC_{18}(bpy)_3]^{3+} + MV^{\cdot+}$ $[RuC_{13}(bpy)_3]^{3+} + cysH \rightleftharpoons [RuC_{18}(bpy)_3]^{2+} + cysH^{\cdot+}$ $cysH^{\cdot+} \rightarrow cys\cdot + H^+$ $cys\cdot + cys\cdot \rightarrow$ cys-cys	Micelles solubilize sensitizer; conversion of light to storable ($MV^{\cdot+}$) chemical energy at the expense of the donor is demonstrated	Steady state photolysis	Kalyanasundaram, 1978
SDS	$[RuC_{18}(bpy)_3]^{2+} + X \xrightarrow{k} X = eq, Zn^+, Co^+, Cd^+$	e_{aq}^- reaction is reduced, metal ion reactions are enhanced by SDS micelles; $[Ruc_{18}(bpy)_3]^{2+}$ resides in the micellar environment	Steady state fluorescence, laser flash photolysis, pulse radiolysis	Meisel et al., 1978a
SDS	Methylviologen chloride $(MV^{2+}Cl_2^-) + eq^- \xrightarrow{k} MV^{\cdot+}$	MV^{2+} associates with micellar SDS; $k = 5 \times 10^{10}\ M^{-1}\ sec^{-1}$ (H_2O), $(1\text{–}6)10^9\ M^{-1}\ sec^{-1}$ (various concentrations of SDS	Pulse radiolysis	Rodgers et al., 1978
CTAB	Reduction of manganese *trans*-1,2-diaminocyclohexane-*N*,*N*,*N*′,*N*-tetraacetate (Mncydta) by $Cocydta^{2-}$ and CoEDTA	Micellar CTAB enhances rates of reductions by 160- and 600-fold, respectively	Absorption spectroscopy	Bhalekar and Engberts, 1978

Table 12.7 *(Continued)*

Micellar System	Process	Results	Experimental Techniques	Reference
SDS, CTAB, Igepal	3-Aminoperylene photoionization	e_{aq}^- yield ↑ SDS, ↓ CTAB; e_{aq}^- yield increases linearly with laser intensity in SDS, but in CTAB and Igepal it varies with the square of laser intensity	Laser flash photolysis	Thomas and Piciulo, 1978, 1979
SDS	Pyrene photoionization	Photoionization is monophotonic	Laser flash photolysis	Hall, 1978
SDS	2-Acetonaphthone (AN) sensitized photooxidation of diphenyl isobenzofuran (DF): $AN \xrightarrow{h\nu} AN^T$ $AN^T + O_2(^3\Sigma) \rightarrow O_2^*(^1\Delta) + AN$ $O_2^*(^1\Delta) + DF \rightarrow DFO_2$	Lifetime and reactivity of singlet oxygen, $O_2^*(^1\Delta)$, in SDS are the same as in H_2O	Laser flash photolysis	Groman and Rodgers, 1978
SDS, DTAC	Pyrene (P) sensitized photooxidation of diphenylisobenzofuran (DF): $P \xrightarrow{h\nu} P^T$ $P^T + O_2(^3\Sigma) \rightarrow O_2^*(^1\Delta) + P$ $O_2^*(^1\Delta) + DF \xrightarrow{k} DFO_2$	Quantum yield for singlet oxygen, $O_2^*(^1\Delta)$, and k-values are higher in micelles than in MeOH; empty micelles scavenge $O_2^*(^1\Delta)$	Steady state photolysis	Miyoshi and Tomita, 1978
DTAC	Pyrene (P) sensitized photooxidation of diphenylisobenzofuran (DF) + indole + tryptophan	Indole and tryptophan enhanced the oxygen singlet oxidation of DF	Steady state fluorescence spectroscopy and photolysis	Miyoshi and Tomita, 1979a
Triton X-100	Chlorophyll-*a* (Chl*a*) sensitized photooxidation of imidazole in the presence of *p*-nitrosodimethylaniline (RNO)	Singlet oxygen, $O_2^*(^1\Delta)$, is detected by following the bleaching of KNO; high yield of $O_2^*(^1\Delta)$ is observed	Steady state photolysis	Kraljic et al., 1979
SDS, CTAB, Brij-35, Igepal CO-630, Igepal CO-660	2-Acetonaphthone (AN) and methylene blue (MB) sensitized photooxidation of diphenylisobenzofuran (DF)	AN is in micellar, MB is in aqueous phase; singlet oxygen, $O_2^*(^1\Delta)$, reactivity with DF is insensitive to the site of its production	Laser flash photolysis	Lindig and Rodgers, 1979
DOTAC, CTAB	Thiazine dye (D) sensitized photooxidation of 9,10-dimethylanthracene (DMA): $D \xrightarrow{h\nu} D^T$ $D^T + O_2(^3\Sigma) \rightarrow O_2^*(^1\Delta)$ $DMA + O_2^*(^1\Delta) \rightarrow DMAO_2$	Interaction of the sensitizer with the micelle alters the protonation of the triplet, hence the ability of singlet oxygen, $O_2^*(^1\Delta)$ formation and DMA oxidation	Steady state irradiation, flash photolysis	Bagno et al., 1979
SDS	Electron transfer from photoexcited pyrene, anthracene, benzopyrene, perylene benzanthracene, benzopentaphene, dibenzanthracene, 7,12-dimethylbenzanthracene, 3-methylcholanthrane, to tryptophan	The micelle serves to solubilize the hydrocarbons; no correlation found between the carcinogenic or the photodynamic activity of the hydrocarbons and the electron transfer behavior	Laser flash photolysis	Putna et al., 1979

SDS, CTAB, Brij-35	9,10-methylphenothiazine, MPTH, oxidation	$MPTH^{+}$ residing in the highly polar environment of aqueous SDS is extremely stable	Voltametry	McIntire and Blount, 1979
SDS, CTAB, CTAN, Triton X-100	Electron transfer from oxtacyanomolybolate, hexacyanoferrate and Fe^{2+} to $RuC_{18}(bpy)_3^{3+}$	Micelles enhance electron transfer rates	Stopped-flow absorption spectrophotometry	Pelizzetti and Pramauro, 1979a
SDS, CTAB, CTAN, Triton X-100	Electron transfer from phenothiazine (PH) to Fe(III)	Cationic and nonionic micelles decrease, anionic micelles increase (up to 10^4-fold) the electron transfer rate	Stopped-flow absorption spectrophotometry	Pelizzetti and Pramauro, 1979b
Sodium dodecyl benzoate, polyoxythylene, octylphenyl ether	Zinc(II)tetraphenylporphyrin (AnTPP) photosensitized reduction of 1,2,6-trimethyl-3,5-bisethoxycarbonylpyridinium perchlorate (P): $ZnTPP \xrightarrow{h\nu} ZnTPP^{*}$ $ZnTPP^{*} + P \rightleftharpoons P\cdot + ZnTPP^{+}$ $P\cdot$ + ascorbic acid → 1,2,6-trimethyl-1,2-dihydro-3,5-bisethoxycarbonylpyridine $ZnTPP^{+}$ + ascorbic acid → ZnTPP	Anionic micelles supress the backreaction: $P\cdot + ZnTPP^{+} \rightarrow ZnTPP^{*} + P$	Steady state photolysis and fluorescence spectroscopy	Ogata et al., 1979
CTACl, Triton X-100	Surfactant derivative of tris(2,2'-bipyridyl)ruthenium, $[RuC_{12}(bpy)_3]^{2+}$, sensitized photoreduction of methylviologen, MV^{2+}, dodecyl viologen, $C_{12}V^{2+}$, in the presence of *N,N'*dimethylaniline as donor	Photoreduction occurs at the micelle surface	Flash photolysis	Tsutsui et al., 1979
SDS	Energy transfer from naphthalene, bromonaphthalene, biphenyl, biphenyl, and phenanthrane triplet to Tb^{3+} and Eu^{3+}	Aromatic triplet leaves and reenters the micelle prior to energy transfer; rate constants for energy transfer are in the order of $10^5\ M^{-1}\ sec^{-1}$; kinetics and mechanism are discussed	Fluorescence spectroscopy and flash photolysis	Almgren et al., 1979a
SDS	Electron transfer equilibrium between 9,10-anthraquinone-2-sulfonate, AQS, and duroquinone, DQ: $AQS^{\cdot -} + DQ \underset{k_{-1}}{\overset{k_1}{\rightleftharpoons}} AQS + DQ^{\cdot -}$	Micellar effects are rationalized in terms of AQS, $AQS^{\cdot -}$, and $DQ^{\cdot -}$ partitioning in the aqueous solution and DQ distributed between the micellar and aqueous phases with $K_D/N \sim 150\ M^{-1}$	Pulse radiolysis	Almgren et al., 1979b
SDS, DTAC, BDCAC, SDBS, polyoxyethylene octylphenyl ether	1,2,4,5-tetracyanobenzene–benzyl alcohol charge transfer (CT) complex formation	Micelles affect CT complex formation	Fluorescence and laser flash photolysis	Masuhara et al., 1979

Table 12.7 *(Continued)*

Micellar System	Process	Results	Experimental Techniques	Reference
$Cu(SD)_2$, $Ni(SD)_2$, $Co(SD)_2$	Methyl phenothiazine, MPTH, and *N*,*N*′-dimethyl-5,11-dihydrodroindolo[3,2,-6]carbazole, DI, sensitized reduction of divalent metallic, M^{2+}, micellar counterions: $MPTH + M^{2+} \xrightarrow{h\nu} MPTH^{\ddagger} + M^+$	Electron is transferred from excited donor, located within the micelle, to the counterion extremely rapidly; M^+ escapes into the bulk phase; back reaction is retarded	Steady state and laser flash photolysis, fluorescence spectroscopy	Moroi et al., 1979a
$SDS/Eu(DS)_3$, $Zn(DS)_2/Eu(DS)_3$	Methylphenothiazine, MPTH, sensitized reduction of Eu^{3+} micellar counterion: $MPTH + Eu^{3+} \underset{}{\overset{h\nu}{\rightleftharpoons}} MPTH^{\ddagger} + Eu^{2\ddagger}$	Forward process is extremely rapid; the back reaction consists of intramicellar (fast) and intermicellar (slower) recombinations; a kinetic model is provided	Laser flash photolysis	Moroi et al., 1979b
SDS	Energy transfer between rhodamine 6-G and 3,3′-diethylthiacarbocyanine iodide	Micelles enhance energy transfer	Absorption and fluorescence spectroscopy	Kusumoto and Sato, 1979
Functional crown ether surfactant O O HN NH O O $(CH_2)_{13}CH_3$	Cyanine dye photosensitized reduction of crown-micelle entrapped Ag^+ and Ag^0	Stable Ag^0 formed	Absorption spectroscopy, steady state irradiation, laser flash photolysis	Humphry-Baker et al., 1979
SDS	e_{aq}^-, tryptamine, and indole propionic acid triplet quenching by nitroxide spin labels	Quenching rates are affected by micelles	Laser flash photolysis	Gresset et al., 1979
SDS	Direct photophysical $(O_2^3\Sigma_g^- \rightarrow O_2^*\Delta^1)q + 1\nu$, at high O_2 pressures and methylene blue or rose bengal photosensitized singlet oxygen, $O_2^{*1}\Delta g$, formation detected by its reaction with diphenylisobenzofuran (DF) both in D_2O and H_2O	O_2 partition coefficient between aqueous and micellar phases is determined to be 2.8 ± 0.1; $O_2^{*1}\Delta q$ entry rate into micelle $= 1.0 \times 10^8$ sec^{-1} (at low $[O_2]$), 9×10^9 M^{-1} sec^{-1} (at high $[O_2]$); $O_2^{*1}\Delta q$ exit rate from micelle $= 3.7 \times 10^7$ sec^{-1} (at low $[O_2]$), 3×10^9 M^{-1} sec^{-1} (at high $[O_2]$)	Steady state and laser flash photolysis	Matheson and Massoudi, 1980

CTACl	Zinc porphyrin (ZnP) sensitized reduction of duroquinone (DQ) and ($C_{11}DQ$): structure of 2,3,5-trimethyl-6-undecyl-1,4-benzoquinone (ring substituents H_3C, CH_3, H_3C, $C_{11}H_{23}$; two C=O) $ZnP^T - DQ^T(\text{or } C_{11}DQ^T) \rightarrow ZnP^{\dot{+}} + DQ^{\dot{-}}$ (or $C_{11}DP^{\dot{-}}$)	Both inter- and intramicellar electron transfer occurs; charge separation, particularly for micelle intercalated $C_{11}DQ$, is accomplished by ejection of $ZnP^{\dot{+}}$ from the micelle	Laser flash photolysis	Pileni and Grätzel, 1980a
SDS	Proflavin dissociation PFH_2^+ (proflavin, with ring N^+–H, H_2N and $\overset{+}{N}H_3$ substituents) $\underset{}{\overset{pK_1}{\rightleftharpoons}}$ PFH^+ (ring N^+–H, H_2N and NH_2 substituents) $\overset{pK_2}{\rightleftharpoons}$ PF (ring N^+–H, H_2N and NH_2 substituents) $PFH^+ \xrightarrow{h\nu} {}^1(PFH^+)^* \xrightarrow{isc} {}^3(PFH^+)^*$ $PFH^+ \xrightarrow{h\nu} PFH^{2\dot{+}} + e^-$ $2\,{}^3(PFH^+)^* \rightarrow PFH^{2\dot{+}} + PFH\cdot$ $PFH^{2\dot{+}} \rightarrow PF^{\dot{+}} + H^+$	$pK_2 = 9.5$ (H_2O), 12.5 (SDS); ${}^1(pK_1)^* =$ 1.5 (H_2O), 3.5 (SDS); ${}^1(pK_2)^* = 12.5$ (H_2O), 12.5 (SDS); $3(pK_1)^* = 4$ (SDS) Monophotonic photoionization and T–T annihilation in H_2O; biphotonic photoionization in SDS; H_2 evolution in PF/TEOA/Pt system	Laser flash photolysis, steady state and nanosecond time resolved fluorescence spectroscopy	Pileni and Grätzel, 1980b

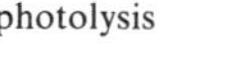

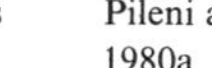

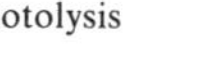
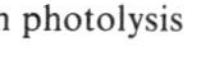
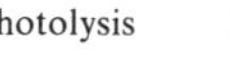
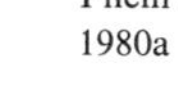
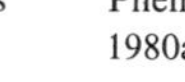
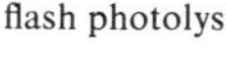
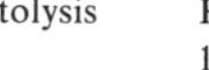

Table 12.7 *(Continued)*

Micellar System	Process	Results	Experimental Techniques	Reference
	Triethanolamine (TEOA) reductively quenches $^3(PFH^+)^*$: $^3(PFH)^* + TEOA \rightleftharpoons PFH\cdot + TEOA^{\dot{+}}$ $PFH\cdot \rightleftharpoons PF^{\dot{-}} + H^+$			
CPC, Brij-35-CPC	Quenching of pyrene fluorescence by hexadecylpyridinium chloride (CPC)	Inter- and intramicellar quenching are observed	Steady state fluorescence	Sapre et al., 1980
SDS	Tris(2,2′-bipyridyl)ruthenium, $[Ru(bpy)_3]^{2+}$ sensitized reduction of alkyl viologens, RMV^{2+}: $[Ru(bpy)_3]^{2+} \xrightarrow{h\nu} [Ru(bpy)_3]^{2+*}$ $[Ru(bpy)_3]^{2+*} \rightleftharpoons RMV^{\dot{+}} + [Ru(bpy)_3]^{3+}$ RMV^2 = dimethyl, dibenzyl, and diheptylviologen	Inter- and intramicellar quenching are observed	Laser flash photolysis	Rodgers and Becker, 1980
SDS, CTAB	Surfactant derivative of tris(2,2′-bipyridyl)ruthenium, $[RuC_{18}(bpy)_3]^{2+}$, sensitized reduction of methyl viologen, MV^{2+}	Electron transfer quenching is more efficient in SDS where it is dynamic and totally intramicellar	Steady state luminescence, laser flash photolysis	Schmehl and Whitten, 1980
Functional micellar $C_{14}MV^{2+}$, CTACl + $C_{14}MV^{2+}$	Tris(2,2′-bipyridyl)ruthenium, $[Ru(bpy)_3]^{2+}$, sensitized reduction of $CH_3-{}^+N\langle C_5H_4\rangle-\langle C_5H_4\rangle N^+-(CH_2)_{13}CH_3, 2Cl^-$ $C_{14}MV^{2+}$: $[Ru(bpy)_3]^{2+} \xrightarrow{h\nu} [Ru(bpy)_3]^{2+*}$ $[Ru(bpy)_3]^{2+*} + C_{14}MV^{2+} \underset{k_{-1}}{\overset{k_1}{\rightleftharpoons}} C_{14}MV^{\dot{+}}$	$k_1 = (8 \pm 0.8)10^8\ M^{-1}\ sec^{-1}$, $[Ru(bpy)_3]^{2+} = 1 \times 10^{-4}\ M$, $[C_{14}MV^{2+}] = 5 \times 10^{-4}\ M$; CTACl micelles do not affect k_1; $k_{-1} = (4 \pm 0.4)10^9\ M^{-1}\ sec^{-1}$ (H_2O), $<10^7\ M^{-1}\ sec^{-1}$ (CTACl); effective charge separation is rationalized by entrapment of $C_{14}MV^{\dot{+}}$ in + charge micelle where k_{-1} is blocked	Laser flash photolysis	Brugger and Grätzel, 1980
CTACl	Surfactant derivative of tris(2,2′-bipyridyl)ruthenium, $[RuC_{12}(bpy)_3]^{2+}$, sensitized reduction of *N*-butylphenothiazine, BPTZ: $[RuC_{12}(bpy)_3]^{2+} \xrightarrow{h\nu} [RuC_{12}(bpy)_3]^{2+*}$ $[RuC_{12}(bpy)_3]^{2+*} + BPTZ \underset{k_{-1}}{\overset{k_1}{\rightleftharpoons}} [RuC_{12}(bpy)_3]^{\dot{+}} + BPTZ^{\dot{+}}$	Increased temperatures decreased k_{-1} in micelles	Flash photolysis, steady state fluorescence	Takayanagi et al., 1980
CTAB, SDS, Triton X-100	Diphenylhexatriene (DPH) cation and anion radical formation, carotene (car) cation and anion radical formation; electron transfer from biphenyl to car and DPH	e^-_{aq} reactivities and electron transfer are affected by micelles	Laser flash photolysis, pulse radiolysis	Almgren and Thomas, 1980

CTAB, SDS	Electron transfer between benzene diols (adrenaline, 3,5-dichlorobenzene-1,2-diol, 2,3-dihydroxybenzoic acid, 3,4-dihydroxybenzoic acid, 3,4-dihydroxybenzonitrile, 4-nitrobenzene-1,2-diol, 5,6-dihydroxybenzene-1,3-disulfonic acid, 2,5-dihydroxyacetophenone) and phenanthroline complexes of Fe^{3+} (tris(1,10-phenanthroline)iron(II), tris(5-methyl-1,10-phenanthroline)iron(II), tris(4,7-dimethyl-1,10-phenanthroline)iron(II), tris(3,4,7,8-tetramethyl-1,10-phenanthroline)iron(II))	Anionic micells enhance, cationic micelles retard reaction rates	Stopped flow absorption spectrophotometry	Pelizzetti et al., 1980
SDS	Energy transfer between rhodamine 6G and 3,3′-diethylthiacarbocyanine iodide	Micelles enhance energy transfer	Picosecond time resolved fluorescence spectroscopy	Sato et al., 1980
SDS	Energy transfer from photoexcited 1,3-dioctylalloxazine (DOA) to proflavine (PF^+) and acriflavin (AF^+) cations	Micelles enhance energy transfer efficiency; DOA is incorporated into micellar interior; AF^+ and PF^+ are electrostatically attracted to micellar surface; micellar effects are rationalized in terms of Poisson's distribution of donors and acceptors	Fluorescence and fluorescence polarization measurements	Matsuo et al., 1980a
Functional micellar 1-ethyl-1′-cetyl-4,4-bipyridinium dibromide ($C_{16}C_2V^{2+}$), CTACl + $C_{16}C_2V^{2+}$	Electron exchange between $C_{16}C_2V^{2+}$ and $C_{16}C_2V^{+\cdot}$ in CTACl + $C_{16}C_2V^{2+}$	$C_{16}C_2V^{+\cdot}$ formed by photosensitization of a long chain $Ru(bpy)_3^{2+}$, incorporated into micelle	Esr line broadening	Takuma et al., 1981
10^{-2} *M* Triton X-100	Photoreduction of methylviologen by zinc tetraphenylporphyrin	Back electron transfer is intercepted by EDTA	Laser flash photolysis	Pradevan and Pileni, 1981
Functional micellar viologens $CnMV^{2+}$ ($n = 12, 14, 16, 18$), CTACl + $CnMV^{2+}$	$Ru(bpy)_3^{2+}$ and zinc tetrakis(*N*-methylpyridyl)porphyrin photosensitized reduction of $CnMV^{2+}$	Back electron transfer is suppressed	Steady state irradiation, laser flash photolysis	Brugger et al., 1981
SDS	Quenching of micelle bound Tb^{3+} luminescence by NO_2^-	Apparent quenching rate is appreciably decreased	Fluorescence quenching, laser flash photolysis	Grieser, 1981

Table 12.8. Miscellaneous Radical Reactions in Aqueous Micelles

Reaction	Effect of Micelles	Reference
Nitroso-*t*-butane + OH (generated from water by pulse radiolysis)	SDS, sodium octyl sulfate: $RCH_2SO_4^- + \dot{O}H \rightarrow R\dot{C}HSO_4^- + H_2O$; $R\dot{C}HSO_4^- + t\text{-Bu—N=O} \rightarrow t\text{-BuN}\dot{O}(CH_2R)SO_4^-$. Reaction occurs in the micellar cage, where collision probability is 2×10^4 greater than that in H_2O	Bakalik and Thomas, 1977
Co-60 γ-ray induced addition of hydrogen sulfite to 1-dodecane	Sodium dodecane sulfonate, no induction, G(-1-dodecene) ~ 860	Miyata et al., 1977
Trans → cis isomerization of [structure: $\overset{+}{N}—C_{18}H_{37}$; $ClC_6H_4SO_3^-$ p]	Substrate + CTAB, $k_\Psi/k_0 \sim 1$	Quina and Whitten, 1977
Disproportionation of pulse radiolytically generated Br_2^-: $Br_2^- + Br_2^- \xrightarrow{k} Br_2 + 2Br^-$	CTABr: $2Br_2^-(H_2O) \xrightarrow{k_1} 2Br^- + Br_2$; $Br_2^-(H_2O) + Br_2^-(M) \xrightarrow{k_2} 2Br^- + Br_2$; $2Br_2^-(M) \xrightarrow{k_3} 2Br^- + Br_2$; $Br_2^-(M) \underset{k_+}{\overset{k_-}{\rightleftharpoons}} Br_2^-(H_2O)$. $k_\Psi/k_0 \sim$ —, kinetic analysis of data lead to Br_2^- residence time in micelles, $1/k- = 1.5 \times 10^{-5}$ sec	Proske and Henglein, 1978
Formation and disappearance radicals formed in pulse irradiated N_2O saturated solutions of 4-(6′-dodecyl)benzenesulfonate (i.e., $\dot{O}H$ + aromatic ring)	< CMC, $2k_{decay} = 4.2 \times 10^8\ M^{-1}\ sec^{-1}$; > CMC $2k_L \uparrow$; kinetic equations are derived	Henglein and Proske, 1978
Radiation initiated peroxidation of linoleic and linolenic acid	Cs^+ and Rb^+ increase rates of peroxidation in micelles	Raleigh and Kremers, 1978
Styrene photopolymerization in *N*-laurylpyridinium azide	Polymerization rate decreases with increasing surfactant concentration; CTACl and hexadecyltrimethylammonium azide do not initiate photopolymerization	Takeishi et al., 1978

Decomposition of 2,3,7,8-tetrachlorodibenzoparadioxin	Hexadecylpyridinium chloride, $k_{\Psi}/k_0 \simeq +$	Botré et al., 1978, 1979
Alkaline photohydrolysis of 3,5-dinitroanisole	*N*-Tetradecyl-*N*,*N*,*N*-trimethylammonium chloride, $\Phi_{\Psi}/\Phi_0 \sim -$; data are analyzed in terms of the ion-exchange model	Bonilha et al., 1979
Photocyclization of *N*-methyldiphenylamine to *N*-methylcarbazole	SDS, CTAB, CTACl, $k_{\Psi}/k_0 \sim 1$	Roessler and Wolff, 1980
Trans → cis isomerization 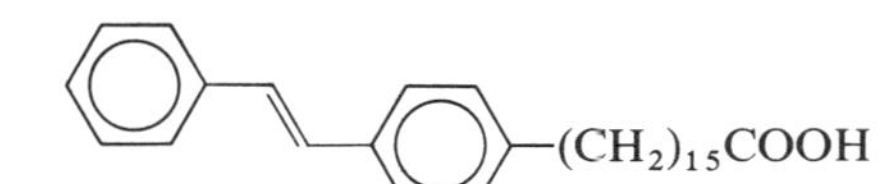	$\Phi_{trans \to cis} = 0.39 \pm 0.02$ (CTAB), 0.50 ± 0.02 (CH_2Cl_2); $\Phi_f = 0.18$ (CTAB), 0.04 (CH_2Cl_2)	Russell et al., 1980
Photodecomposition of 1,3-diphenyl-2-propanone	SDS, amount of escaping benzyl radicals from radical pairs (see equation 11.43) are enhanced by 30% in the presence of a magnetic field of 70 mT	Hayashi et al., 1980
Bilirubin IXa photoisomerization	$t_{1/2}$ isomerization, pH 5–8 = 10 min (CTAB), no detectable isomerization (SDS, Tween-20)	Au and Hutchinson, 1980
Light induced homolysis of R = $C_{12}H_{25}$	$\Phi = 0.20$ (SDS), 0.22 (CTAB), $< 10^{-3}$ (MeOH), 0.23 ($CHCl_3$)	Lerner et al., 1980

Table 12.9. Energy and Electron Transfer and Charge Separation in Reversed Micelles

Micellar System	Process	Results	Experimental Technique	Reference
Dodecylpyridinium chloride in $CHCl_3$	Dimerization of 2,3-dichloro-5,6-dicyano-*p*-benzoquinone	Dimerization occurs in micellar interior	Stopped flow, T-jump	Yamagishi and Watanabe, 1976a
Dodecylpyridinium chloride in $CHCl_3$	Electron transfer between 7,7,8,8-tetracyanoquinonedimethane anion radical (M^+TCNQ^-) and 2,3-dichlo-5,6-dicyano-*p*-benzoquinone (DDQ): $M^+TCNQ^- + DDQ \rightleftharpoons TCNQ + M^+DDQ^-$	Micellar rate enhancement is rationalized by ion-pair interactions	Stopped flow	Yamagishi and Watanabe, 1976b
Aerosol-OT in heptane	Energy and electron transfer from biphenyl anion radical and triplet to pyrene sulfonic acid, Cu^{2+} and H_{30+}	Transfer efficiency depends on microenvironment of the acceptor and size of the water pool	Pulse radiolysis, flash photolysis	Wong and Thomas, 1977
DAP in cyclohexane	Singlet oxygen quenching by 1,3-diphenylisobenzofuran (DF) and NaN_3 using pyrene (P) as sensitizer	${}^3P \xrightarrow{O_2} {}^1O_2 \longrightarrow DF$; ${}^1O_2 \rightarrow$ empty micelle; ${}^1O_2 \rightarrow {}^3O_2$; ${}^1O_2 \rightarrow$ reversed micelle entrapped N_3^-	Steady state photolysis, absorption spectroscopy	Miyoshi and Tomita, 1979b
DAP in cyclohexane	Effects of amines (A) on singlet oxygen quenching by 1,3-diphenylisobenzofuran (DF) using fluorescein (F) as sensitizer	${}^1F \xrightarrow{h\nu} {}^1F^* \longrightarrow {}^3F \longrightarrow \lvert F^- \cdots A^+ \rvert \xrightarrow{DF} DF(ox)$ (I); ${}^3F \xrightarrow{O_2} {}^1O_2$; ${}^1O_2 \xrightarrow{DF} DF(ox)$ (II); ${}^1O_2 \xrightarrow{A} A + {}^3O_2$; ${}^1O_2 \rightarrow {}^3O_2$ A inhibits DF oxidation in reversed micelles; route I is inhibited	Steady state photolysis, absorption spectroscopy	Miyoshi and Tomita, 1979c, 1980b
Dioctyl ester of sulfosuccinic acid in cyclohexane	Singlet → singlet energy transfer from toluene to sodium salicylate ion and from fluoranthane to rose bengal	^{1}Donor* (in hydrocarbon) → ^{1}acceptor* (in surfactant entrapped water pool) by long range transfer mechanism	Laswer flash photolysis	Rodgers and Burrows, 1979

DAP in cyclohexane	Singlet oxygen, produced by fluorescein (F) photosensitization, is quenched by 1,3-diphenylisobenzofuran (DF)	F is in reversed micellar cavities, while DF is in cyclohexane; Φ increases with decreasing size of water pools	Steady state photolysis, absorption and emission spectroscopy	Miyoshi and Tomita, 1980a
DAP in toluene	Tris(2,2′-bipyridine)ruthenium(II), Ru^{2+} (in H_2O), photosensitized electron transfer to 1,1′-dihexadecyl-4,4′-bipyridinium chloride, HV^{2+} (at interface), using EDTA (in H_2O) as donor; photosensitized reduction of 4-dimethylaminoazobenzene dye, mediated by benzylnicotinamide, BNA^+, and $Ru(bpy)_3^{2+}$ using EDTA as donor	oxidation products ← EDTA; Ru^{2+} —$h\nu$→ Ru^{2+}; Ru^{3+}; HV^{2+} → HV^{+}; water; interphase oxidation products ← EDTA; Ru^{2+} —$h\nu$→ Ru^{2+*}; Ru^{3+}; BNA^+; BNA^{red}; dye · H_2; dye; water; interphase; toluene	Steady state photolysis	Willner et al., 1979
DAP in toluene	Tris(2,2′-bipyridine)ruthenium(II), Ru^{2+}, or Zn-tetraphenylporphyrin sulfonate, or Zn-tetramethylpyridinium porphyrin sensitized electron transfer to methylviologen or to propyl viologen sulfonate using diphenylthiol as donor	MV^{2+} —$h\nu$→ Ru^{2+}; MV^{+}; Ru^{3+}; C_6H_5—S—S—C_6H_5; HSC_6H_5; water; interface; toluene Oxidized donor, diphenyldisulfide is extracted into continuous organic phase	Steady state photolysis	Willner et al., 1981
Aerosol-OT in heptane or cyclohexane	Tris(2,2′-bipyridine)ruthenium(II) sensitized reduction of dimethyl-, dibenzyl-, dipheptyl-4,4′-bipyridinium salts and duroquinone	Back reaction is extremely fast in surfactant solubilized water pools	Fluorescence quenching, laser flash photolysis	Rodgers and Becker, 1980
Aerosol-OT in isooctane	Rate constants of excess electron attachment (k) to reversed micelles are determined	With increasing solubilized water concentrations, k increased to $\sim 10^{15}\ M^{-1}\ sec^{-1}$	Conductometric detection of picosecond pulse radiolysis	Bakale et al., 1980

Table 12.10. Miscellaneous Reactions in Reversed Micelles

Reaction	Effects of Micelles	Reference
$RNHCONH_2 \rightleftharpoons RNHCO\overset{+}{N}H_3 \xrightarrow{slow} RN{=}C{=}O + H_2O + NH_4^+ \xrightarrow{fast} RNHCO_2H \xrightarrow[H_2O]{fast} RNH_3^+ + CO_2^+ + H_2O$	DAP (benzene), $\sim k_\Psi/k_0 \simeq 3 \times 10^3$	Mollett and O'Connor, 1976
$R = -C_6H_4NO_2p$ at 51.0°C		
Iodination of RCOR′ at 25.0°C	DAP (hexane), $\sim k_\Psi/k_0 \simeq 10^6$–$10^7$, $\sim k_\Psi/k_{DAP(H_2O)} \sim 10^2$	Seno et al., 1976
R = Me, R′ = Me		
R = Me, R′ = Et		
R = Et, R′ = Et		
R = *i*Pr, R′ = *i*Pr		
R = *i*Bu, R′ = *i*Bu		
R = Me, R′ = Ph		
R = Et, R′ = Ph		
Hydration of CH_3CHO in D_2O at 0°, 25.0°, and 36.0°C	Triton X-100 (CCl_4), $k_\Psi/k_{D_2O} \sim (1.5$–$5.4)10^7$, $k_\Psi^{H+}/k_{D_2O}^{H+} \sim 2.4$–$3.6$	El Seoud, 1976

ESR of $FeCl_3$, $MnCl_2 \cdot 4H_2O$, $CuSO_4 \cdot 5H_2O$, 30.0°C	Dodecylpyridinium chloride ($CHCl_3$), chelatelike complex formation between metal ion and surfactant aggregate	Masui et al., 1977a
$FeCl_3$-dodecylpyridinium chloride formation, 20.0°C	Dodecylpyridinium chloride ($CHCl_3$), complex formation is a function of H_2O is reversed micelles	Masui et al., 1977b
Hydrolysis of *p*-nitrophenyl acetate, 30.0°C	DAP (toluene), 1-propylamine, $k_\Psi/k_0 \sim +$; MeOH, $k_\Psi/k_0 \sim -$; AcOH, $k_\Psi/k_0 \sim -$	Kitahara and Kon-no, 1977
p-Nitrophenyl acetate + dodecylamine, 25.0°C	DAP (benzene), $k_\Psi/k_0 \sim 53$	El Seoud et al., 1977
p-Nitrophenyl acetate + imidazol (or methylimidazole), 25.0°C	DAP (benzene), $k_\Psi/k_0 \sim 284$; addition of water decreases k_Ψ	
p-Nitrophenyl acetate + dodecylamine, 25.0°C	Aerosol-OT (benzene), $k_\Psi/k_0 \sim -$	
p-Nitrophenyl acetate + imidazole (or methylimidazole), 25.0°C	Aerosol-OT (benzene), $k_\Psi/k_0 \sim -$; addition of water increases k_Ψ	
p-Nitrophenyl acetate + imidazole (or 2-methylimidazole or 2-isopropylimidazole or 2-undecylimidazole), 30.0°C	Aerosol-OT (or didocyldimethylammonium chloride or bromide) in CCl_4, $k_\Psi/k_0 \sim +$	Kon-no and Kitahara, 1978; Kon-no et al., 1978
Hydrolysis of methyl and dodecyl-*p*-nitrophenyl carbonates, 25.0°C	DAP (benzene), $k_\Psi/k_0 \sim +$; hydrolysis is proportional to $[DAP]^2$; $[H_2O] < 10^{-1}$, rate independent on $[H_2O]$; $[H_2O] > 10^{-1}$, rate decreases with increasing $[H_2O]$	Kondo et al., 1978
Hydrolysis of sucrose, HCl, 20.0°C	Dodecylbenzene sulfonic acid (benzene), $k_\Psi/k_{H_2O} \sim 400$; increasing H_2O decreases k_Ψ	
$(CH_2Cl)_2C{=}O + H_2O \underset{k_d}{\overset{k_h}{\rightleftharpoons}} (CH_2Cl)_2C(OH)_2$,	Aerosol-OT (CCl_4), $k^h_\Psi/k^h_{H_2O\ diox} \sim 10^2$–$10^3$,	El Seoud et al., 1978
$K_h = k_h/k_d$, 25.0°C	$k^d_\Psi/k^d_{H_2O\ diox} \sim 64$–$508$	El Seoud et al., 1978
$Co(H_2O)X_2 + 2H_2O \overset{K_1}{\rightleftharpoons} Co(H_{20})_4X_2$	DAP ($CHCl_3$), $\ln K_1 = 1.17$ (X = Cl), 1.10 ($X = NO_3$)	Sunamoto and Hamada, 1978
$[Co(X)_4]^{2-} + 2H_2O \overset{K_2}{\rightleftharpoons} [Co(H_2O)_2X_4]^{2-}$	CTAB ($CHCl_3$), $\ln K_2 = 6.64$ ($X = NO_3$), 6.23 (X = Cl)	
Cu^{2+} promoted hydrolysis of norleucine *p*-nitrophenyl ester	Aerosol-OT (CCl_4), $k_\Psi/k_0 \sim +$	Sunamoto et al., 1978

Table 12.10 *(Continued)*

Reaction	Effects of Micelles	Reference
Spontaneous polymerization of alanyl adenylate	Aerosol-OT + hexadecyltrimethylammonium bicarbonate (benzene), polypeptide formation in 95% yield in the 350–3500 MW range	Armstrong et al., 1978
Emulsion polymerization of methyl methyacrylate using ferrocene and CCl_4 as initiator	SDS; $k_\Psi/k_0 \sim +$	Tsunooka and Tanaka, 1978
$RCOOC_6H_4NO_2$-*p* + $R'NH_2$, $RCOOC_6H_4NO_2$-*p* + R″/*m*, 25.0°C R = methyl, butyl, octanoyl, dodecanoyl, hexadecanoyl R′ = butyl, octyl, dodecyl, hexadecyl R″ = methyl, butyl, octyl, dodecyl, hexadecyl	DAP (benzene), $k_\Psi/k_0 \sim +$; increasing $[H_2O]$ decreases k_Ψ; aerosol-OT (benzene), $k_\Psi/k_0 \sim +$; k_Ψ increases with decreasing chain length of R″	El Seoud et al., 1979
Hydrolysis of *p*-nitrophenylacetate, octanoate, and L-alanine *p*-nitrophenyl ester hydrochloride + imidazole and L-histidine, 30.0°C	Sodium octanoate (hexanol), $k_\Psi/k_0 \sim +$; decreasing $[H_2O]$ increases k_Ψ	Fujii et al., 1979a
Metal ion promoted hydrolysis of norleucine *p*-nitrophenyl ester, 25.0°C	DAP ($CHCl_3$), surfactant acts as general acid-base catalyst, metal ion has no additional catalytic effect; AOT ($CHCl_3$), spontaneous hydrolysis is inhibited, Cu^{2+}, Ni^{2+}, Co^{2+}, Zn^{2+}, and Mg promote catalysis	Sunamoto and Hamada, 1979
$Ni^{2+} + Mur^- \rightleftharpoons Ni^{2+}/Mur^-$ (outer-sphere complex) $\rightleftharpoons (NiMur)^+$ (inner sphere complex), 25.0°C	Aerosol-OT (heptane), rapid exchange of reactants between aqueous pools followed by rate limiting loss of a solvated water molecule from the metal-ion in a water pool	Robinson et al., 1979

Hydrolysis of *p*-nitrophenyl acetate	Sodium octanoate (hexanol), $k_\Psi/k_0 \sim +$; pH within water pool is not responsible for enhanced k_Ψ	Fujii et al., 1979a
Dissociation constant of phenol red	Sodium octanoate (hexanol), $pK_a^\Psi = 8.3$, $pK_a^\circ = 7.8$	Fujii et al., 1979b
α-Chymotryps in catalyzed hydrolysis of *N*-acetyl-L-tryptophan methyl ester	Aerosol-OT (heptane), $k_\Psi/k_0 \sim -$ (pH = 7), + (pH > 7)	Menger and Yamada, 1979
Decomposition of *p*-nitrophenyl acetate, 15.0°C, 35.0°C	DAP (benzene, toluene, cyclohexane), $k_\Psi/k_0 \sim +$; mechanism discussed in terms of micellar catalysis and in terms arising from a bimolecular reaction between the ester and the components of DAP	O'Connor and Ramage, 1980a, 1980b
Decomposition of *p*-nitrophenyl acetate, hexanoate, 25.0°C	$CH_3(CH_2)_nNH_3^+\ ^-O_2C(CH_2)_mCH_3$, $n = 3, 5, 7, 9,$ or 11, $m = 1, 2, 6,$ or 7 (benzene), $k_\Psi/k_0 \sim +$; k_Ψ is affected by changes in m and n	O'Connor and Ramage, 1980c
Decomposition of *p*-nitrophenyl acetate, 25.0°C	$CH_3(CH_2)_{11}NH_3^+\ ^-O_2C_6H_4X$-*p*, X = H, Cl, Br, Me, or MeO (benzene), $k_\Psi/k_0 \sim +$; k_Ψ is affected by the basicity of the phenol	O'Connor and Ramage, 1980d
$HClO_4$, $OClO_4$, imidazole (Im), and *N*-methylimidazole (MeIm) catalyzed reversible hydration of 1,3-dichloroacetone	Aerosol-OT (CCl_4), $k_\Psi/k_0 \sim 15$ ($HClO_4$), 209 (Im), 144 (MeIm)	El Seoud and da Silva, 1980
Benzidine rearrangement of hydrazobenzene-3,3′-disulfonic acid	CTAB ($CHCl_3$), $k_\Psi/k_0 \sim +$	Sunamoto and Horiguchi, 1980
Schiff base formation between pyridoxal and amino acids (alanine, arginine, methionine), 25.0°C	Aerosol-OT (CCl_4), $k_\Psi/k_0 \sim 10$–100	Kondo et al., 1980
Aquation of tris(oxalato)chromate(III) anion, 25.0°C	DAP, octylammonium tetradecanoate (benzene), $k_\Psi/k_0 \sim 10^6$–10^8; k_Ψ concentration profile is complex	O'Connor and Ramage, 1980e

Table 12.10 (*Continued*)

Reaction	Effect of Micelles	Reference
Formation of polynuclear cupric halides, 25.0°C	CTABr or CTACl ($CHCl_3$): $[H_2O]/[detergent]$ — species < 1 — $\left[Cu\langle X_2\rangle Cu\right]_n^{2-} \cdot 2R_4\overset{+}{N}$ polynuclear bridged by halogens $\rightleftharpoons$ (H_2O) 2–4 — $[Cu(H_2O)_nX_{m-n}]^x$ monomeric $\rightleftharpoons$ (H_2O) > 4 — $[Cu(H_2O)_6]^{2+} \cdot 2X^{2-}$ monomeric	Sunamoto et al., 1980b

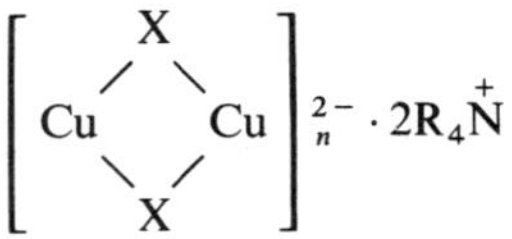

$CoCl_4^{2-} + 4H_2O \rightleftharpoons CoCl_2(H_2O)_4 + 2Cl^-$ + $2H_2O$ $\rightleftharpoons$ $Co(H_2O)_6^{2+} + 2Cl^-$	Dodecylpyridinium chloride ($CHCl_3$), $k_\Psi/k_0 \sim +$; intra- and intermicellar reactions are separated	Yamagishi et al., 1980
Oxidation of acridine orange (AOH) and its N-alkyl derivatives ($AO^+{-}(CH_2)_nH$), $n = 1$–18, by $Ce(SO_4)_2$, 20.0°C	Hexadecylpyridinium chloride ($CHCl_3$), $k_\Psi/k_0 \sim +$; k_Ψ is greatest when $n = 1$	Yamagishi et al., 1981

Table 12.11. Reactivities in Microemulsions

Reaction	Effect of Microemulsion	Experimental Technique	Reference
Complex formation between M^{2+} and the side chains of N^{α}-dodecanoyl-L-lysinol, N^{α}-dodecanoyl-L-glutaminol, and N^{α}-dodecanoyl-L-methioninol	W/o, hexadecyltrimethylammonium perchlorate, 2-propanol, hexane, weak metal–ligand complexes are observed	Potentiometry, absorption, and epr spectroscopy	Smith et al., 1977, 1978
Dissociation of 1-methyl-4-cyanoformyl pyridinium oximate, tetraphenyl porphin dication, and reaction of crystal violet with sodium hydroxide	O/w, Tween, sodium cetyl sulfate, cetylpyridinium bromide, cetyltrimethylammonium bromide, potassium oleate, cyclohexane, benzene, equilibria and rates are altered	Absorption spectroscopy	Mackay et al., 1977
Emulsion polymerization of hydrophilic monomers (*p*-vinylbenzene sulfonate, sodium vinylbenzene sunfonate, 2-sulfoethyl acrylate, acrylic acid, acrylamide, vinylbenzyltrimethylammonium chloride, and 2-aminoethyl methacrylate hydrochloride) using benzoyl or lauroyl peroxides as initiators	W/o, Span 60, xylene, polymerization rates are enhanced	Kinetics, electron microscopy	Vanderhoff et al., 1962
Murexide–metal ion complexation	W/o, aerosol-OT, heptane, rate depends on droplet size and concentration	Stopped-flow spectroscopy	Robinson and Steytler, 1978
Electron transfer from diphenylamine to photoexcited duroquinone in the lipid interior of the microemulsion, electron transfer from photoexcited *N*-methylphenothiazine (in the microemulsion) to methylviologen (on the microemulsion surface)	O/w, sodium hexadecylsulfate potassium oleate, benzene, hexadecane, pentanol, cyclohexanol, charge separation is facilitated	Laser flash photolysis	Kiwi and Grätzel, 1978
Chlorophyll photodegradation	O/w, sodium cetyl sulfate, pentanol, microemulsions stabilize chlorophyll		Jones and MacKay, 1978
Chlorophyll sensitized photoreduction of methyl red and crystal violet	O/w, sodium cetyl sulfate, mineral oil, 1-pentanol, micelles affect rates	Steady state irradiation, absorption spectroscopy	Jones et al., 1979
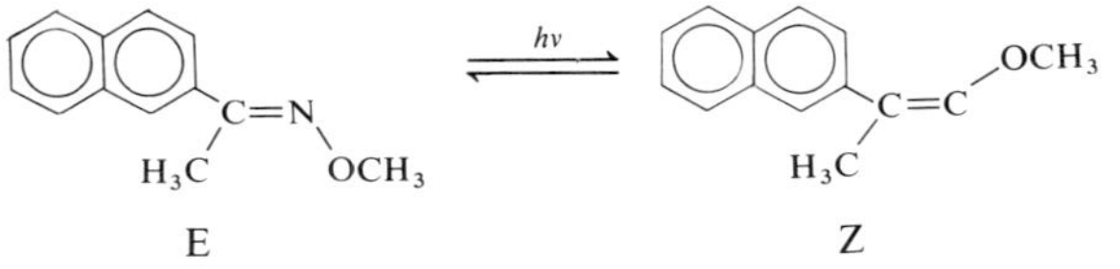 $(H_3C)C{=}N(OCH_3)$ (E) $\underset{}{\overset{h\nu}{\rightleftharpoons}}$ $(H_3C)C{=}C(OCH_3)$ (Z)	O/w and w/o, SDS, benzene, butanol, photoisomerization occurs at interface	Product analysis	Rico et al., 1980
Dissociation of chlorophenol red, 19.0°C, 25.0°C	O/w, CTAB, Brij, sodium cetyl sulfate, alkyl polyethoxyethanols, *pK*'s are altered	Spectroscopy	MacKay et al., 1980

Cu^{2+}(aq) + mesotetraphenyl porphine: ($TPPH_2$) $\rightleftharpoons$ CuTPP + 2HT	W/o, CTAB, Alfonic 1412-60 ethoxylate, sodium hexadecyl sulfate, $k_\Psi/k_0 \sim +$	Spectrophotometry	Keiser et al., 1980
Quenching of pyrene sulfonic acid, pyrene-butyric acid, pyrene dodecanoic acid fluorescence by CH_2I_2, Tl^+, dimethylaniline, and CH_3NO_2, excimer formation and photoionization	O/w, potassium oleate, stearate, hexadecane, hexanol, microemulsions structures are inferred	Fluorescence spectroscopy, laser flash photolysis	Gregoritch and Thomas, 1980
Quenching of pyrene fluorescence by $Ru(bpy)_3^{2+}$ 9-methylanthracene, $TlNO_3$, and O_2, fluorescence spectra of pyrene-3-carboxaldehyde and pyrenebutyric acid, pyrene photoionization	O/w, SDS, dodecane, pentanol, microemulsion structures are inferred	Fluorescence spectroscopy, laser flash photolysis	Almgren et al., 1980
Chlorophyll-*a* sensitized redox processes: Chl-$a \xrightarrow{h\nu}$ Chl-a^T Chl-a^T + methylviologen (MV^{2+}) $\rightarrow MV^{\dot{+}}$ + Chl-$a^{\dot{+}}$ Chl-a^+ + ascorbate ion (ASC^-) $\rightarrow$ ASC + Chl-*a*	O/w, sodium hexadecyl sulfate, hexadecane, 1-pentanol, charge separation, H_2 evolution in the presence of PtO_2	Light scattering, laser flash photolysis	Kiwi and Grätzel, 1980
Tetra-*t*-butyl phthalocyanine (H_2Pc) in hydrocarbon phase, mediated hydrogen atom abstraction from 1-pentanol (RCH_2OH) cosurfactant, in presence of oxygen peroxyradical and singlet oxygen are formed, which react with the 5,5-dimethyl-1-pyrrolinyl-1-oxy (DMPO) spin trap to form a diamagnetic product, $S_2O_8^{2-}$ (in aqueous phase photolysis) to give $SO_4^{\dot{-}}$	O/w, Tween 60, hexadecane, 1-pentanol: $H_2Pc + h\nu \rightarrow H_2Pc'$ $H_2Pc' \rightarrow H_2Pc^3$ $H_2Pc^3 + RCH_2OH \rightarrow R\dot{C}HOH + H_2Pc^{red}$ $H_2Pc^3 + O_2 \rightarrow H_2P_c + O_2'$ $O_2' + DMPO \rightarrow$ diamagnetic products $R\dot{C}HOH + O_2 \rightarrow RC(\dot{O}_2)HOH$ $R\dot{C}H(OH) + DMPO \rightarrow O_2$ adduct $SO_4^{\dot{-}} + DMPO \rightarrow O_2$ adduct $SO_4^{\dot{-}} + OH^- \rightarrow SO_4^{2-} + \dot{O}H$ $SO_4^{\dot{-}} + RCH_2OH \rightarrow R\dot{C}HOH$	Absorption, epr spectroscopy, O_2 monitoring	Harbour and Hair, 1980
Zinc tetraphenylporphyrin (ZnTPP) photosensitized electron transfer to duroquinone (DQ), long chain substituted duroquinone ($C_{11}DQ$), and methylviologen (MV^{2+})	O/w, sodium hexadecyl sulfate, hexadecane, *n*-pentanol: $ZnTPP \rightarrow {}^1ZnTPP^* \rightarrow {}^3ZnTPP^*$ ${}^3ZnTPP^* + DQ \rightleftharpoons DQ^- + ZnTPP^{\dot{+}}$ ${}^3ZnTPP^* + C_{11}DQ \rightleftharpoons C_{11}DQ^{\dot{-}} + ZnTPP^{\dot{+}}$ ${}^3ZnTPP^* + MV^{2+} \rightleftharpoons ZnTPP^{\dot{+}} + MV^{\dot{+}}$ ${}^3ZnTPP^* \rightarrow ZnTPP$ Charge separation is promoted	Laser flash photolysis	Pileni, 1980a, 1980b
Reaction of *p*-nitrophenyl diphenyl, dihexyl, and diethyl phosphate esters with ^-OH and F^-	O/w, hexadecane, Brij-96, CTAB, butanol: the effective nucleophile concentration in the microdroplet is lower than in bulk; intrinsic rate constants are similar in both environments	Absorption spectrophotometric determination of rate constants	Mackay and Hermansky, 1981

Table 12.12. Energy and Electron Transfer and Charge Separation in Monolayers and in Organized Multilayers

System	Process	Results	Experimental Technique	Reference
Long chain aliphatic alcohols in monolayers on surface of aqueous solution containing methylviologen	Electron transfer from alcohols to methylviologen	Primary alcohols are oxidized smoothly and obey 0 order kinetics; behavior of secondary alcohols is more complex	Steady state irradiation, gas chromatography	Brown et al., 1976
Long chain $Ru(bpy)_3^{2+}$ complex and methylviologen organized in multilayers (cadmium arachidate on solid support)	Electron transfer quenching luminescence	Separation of acceptor–donor by two or more layers obviated quenching; when quenching occurred it was followed by rapid back reaction	Fluorescence spectroscopy	Sprintschnik et al., 1976, 1977
Cyanine dye in monolayers	Electron transfer quenching	Quenching occurs over statistical spacings of up to 75 Å	Fluorescence spectroscopy	Seefeld et al., 1977
Metal(Al)-multilayer-Al junctions, prepared by successive deposition of arachidic acid monolayers containing bis-[3-stearylbenzthiazol(2)-trimethincyanineperchlorate] dye I, and 12-(4′-dimethylamino-4-nitroazo-3-benzoyl)-stearic acid (dye II)	Excitation of dye I by light results in electron flow to the π-system acting as the conducting element to dye II	Temperature dependent photocurrents are observed	Capacitance and dissipation factor measurements by a capacitance bridge in the absence and in the presence of illumination with light at 500 nm	Polymeropoulos et al., 1978
(*N,N′*-dioctadecyloxacyanine) and (*N,N′*-dioctadecyl-4,4′-bipyridinium ion) organized in multilayers	Fluorescence quenching by electron transfer	Electron transfer is observed if donor and acceptor are in the same monolayer or at the hydrophilic interface between adjacent monolayers; no fluorescence quenching is observed when donor and acceptor are separated by the long hydrophobic portions of the monolayers or by two fatty acid interlayers	Fluorescence quenching	Möbius, 1978a, 1978b
Organized cadmium arachide multilayers on glass slides containing Zn, Sn, and Pd porphyrins	Photoaddition and reduction with DMA, $SuCl_2$, Et_3N, and ascorbic acid located in aqueous subphase; quenching of Pd porphyrin phosphorescence by amine in multilayers	There is considerable migration of supposedly "anchored" reagents	Steady state photolysis, fluorescence and spectroscopy, product analysis by liquid chromatography	Mercer-Smith and Whitten, 1979

Aluminium-multilayer-mercury junctions, prepared by successive deposition of stearic acid monolayers containing chlorophyll-*a*, quinones, and squalene	Photochemical electron transfer	Chl-*a* triplets eject electrons into the conduction band of the lipid layer; incorporation of electron acceptors with saturated side chains does not improve photoresponse, but ubiquinone and plastoquinone reduced internal resistance and increased efficiency; squalene aids tunnelling of electrons	Photovoltage-photocurrent measurements, fluorescence spectroscopy	Janzen and Bolton, 1979; Janzen, 1979
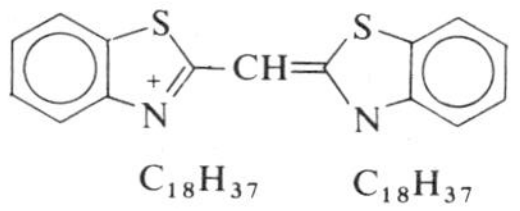 Organized arachidic acid and methylarachide multilayers containing 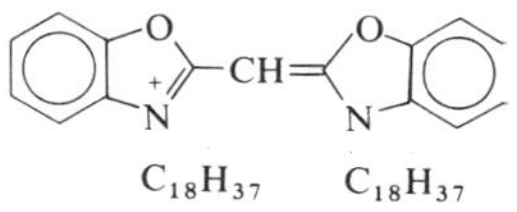 $C_{18}H_{37}$ $C_{18}H_{37}$ donor $C_{18}H_{37}$ $C_{18}H_{37}$ acceptor	Energy transfer from donor to acceptor in multilayers	Energy transfer efficiency depends on organization, mechanisms are proposed	Fluorescence spectroscopy	Möbius and Kuhn, 1979; Kuhn, 1979
Aluminium-multilayer-aluminium (or barium) junctions, prepared by successive deposition of arachidic acid monolayers containing cyanine dyes as electron donors and acceptors	Photocurrents are observed	Vectorial electron transfer through molecular functional unit toward the metal with the smaller work function	Capacitance and dissipation factor measurements by a capacitance bridge in the presence and in the absence of illumination with light at 500 nm	Polymeropoulos et al., 1980
12-(1-Pyrene)dodecanoic acid (PDA)–oleic acid monolayer or air–water interface	Monomer excimer dynamics	Excimer:monomer = mole fraction of PDA:oleic acid	Fluorescence spectroscopy	Loughran et al., 1980

Table 12.13. Miscellaneous Reactions in Monolayers and in Organized Multilayers

Reaction	Effects of Monolayers or Multilayers	Reference
Polymerization of tricosa-10,12-diynoic acid	Substrate can be polymerized to give stable multilayers	Tiecke et al., 1976
Hydrogenation of 17-octadecenoic acid	Oriented monolayers of the substrate on Pt foils; hydrogenation rate is affected by varying the pH of the subphase, incorporating metal ions into the monolayer, varying the rigidity of the film	Richard et al., 1978
Dimerization and isomerization of [structure: 4-styryl-N-octadecylpyridinium, $\overset{+}{N}-C_{18}H_{37}$, X^-] $X = ClC_6H_4SO_3^-p$ $X = Br^-$ $X = BrC_6H_4SO_3^-p$ $X = BF_4^-$	Substrates + arachidic acid, unlike in water, the major pathway is dimerization	Quina and Whitten, 1977
Dimerization of mesoporphyrin IX dioctadecyl ester (*1*), *meso*-tetra(4-carboxyphenyl)porphyrin tetradihydrocholesteryl ester (*2*), *meso*-tetra(*o*-acetylamidophenyl)porphyrin (*3*), *meso*-tetra(*o*-hexylamidophenyl)porphyrin (*4*), and *meso*-tetra(*o*-hexadecylamidophenyl)porphyrin (5)	Monolayers with arachidic acid; complete dimerization for *1*, partial dimerization of *3*, no dimerization for *2* and *4*	Whitten et al., 1978
Incorporation of Cu^{2+} into *1*, *2*, *3*, *4*, and *5*	Monolayers; relative solution rates of metallation at 25.0°C, *3*(110) > *5*(50) > *4*(38) > *1*(10)	
Light induced Co photoejection from Ru(II) and Os(II) porphyrin complexes	Monolayers; +	

Isomerization of thioindigo dyes with alkyl side chains, and *N*-octadecyl-4-stilbazole and *N*-octadecyl-1-(4-pyridyl)-4-(phenyl)-1,3-butadiene	Monolayers, only cis → trans isomerization	
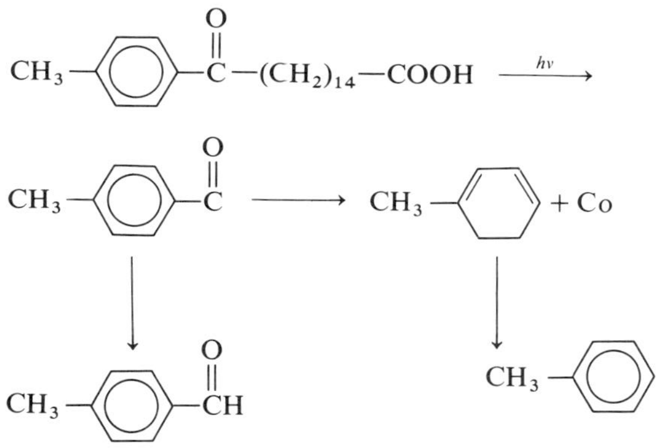 $CH_3{-}C_6H_4{-}C({=}O){-}(CH_2)_{14}{-}COOH \xrightarrow{h\nu}$ $CH_3{-}C_6H_4{-}\dot{C}{=}O \longrightarrow CH_3{-}C_6H_7 + Co$ $CH_3{-}C_6H_4{-}C({=}O)H$; $CH_3{-}C_6H_5$	Monolayers; Φ decreases	
Photooxidation of protoporphyrin IX	Monolayers, $k_\Psi/k_0 \sim ++$	Horsey and Whitten, 1978
Type II photoelimination of 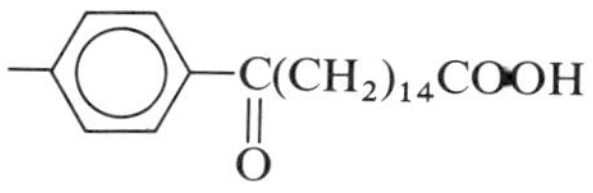 $-C_6H_4{-}C(CH_2)_{14}COOH$ (with $C{=}O$)	Arachidic acid monolayer, elimination is effectively eliminated	Worsham et al., 1978
Hydrolysis of $\{(bpy)_2Ru^{II}[bpy(COOR_1)(COOR_2)]\}^{2+}$ $R_1 = R_2 = C_{18}H_{37}$ $R_1 = R_2 = H$ $R_1 = C_{18}H_{37}, R_2 = H$ $R_1 = C_{18}H_{37}, R_2 = C_2H_5$ $R_1 = R_2 = C_2H_5$	Hydrolysis is retarded in monolayers	Gaines et al., 1978; Valenty, 1979

Table 12.13 (*Continued*)

Reaction	Effects of Monolayers or Multilayers	Reference
Photopolymerization of pentacosa-10,12-diynoic acid in multilayers	Φ depends on conversion; 3,3′-distearyl thiacarbocyanine iodide, incorporated into multilayers, acts as a sensitizer	Fouassier et al., 1979
H_3C CH_3 … N … O … NO_2, N–$C_{18}H_{37}$ spiropyran ⇌ (545 nm / 366 nm) H_3C CH_3 … CH=CH … $\overset{\oplus}{N}$ … O … NO_2, N–$C_{18}H_{33}$ merocyanine	Photoisomerization is slower in monolayers	Polymeropoulos and Möbius, 1979
Reactivity of 16-ferrocenylhexadecanoic acid with I_2, Ag^+, 7,7,8,8-tetracyano-*p*-quinodimethane, tetracyanoethylene	In monolayers I_2 is reactive but other reagents are inactive	Seiders et al., 1979
Chlorophyll-*a* in phosphatidylcholine	Stability (oxidation, pheophytin formation) is enhanced in monolayers and multilayers	Iriyama, 1980

Table 12.14. Energy and Electron Transfer and Charge Separation in BLMs

System	Process	Results	Experimental Technique	Reference
BLMs prepared from dierucoyllecithin, containing chlorophyll-*a* + chlorophyll-*b*; *p*-bis[2(4-methyl-5-phenyloxazolyl)]benzene + chlorophyll-*a*; 1,6-diphenyl-1,3,5-hexatriene + chlorophyll-*b*	Energy transfer	Energy transfer observed when the mean distance is < 100 Å	Fluorescence spectroscopy	Pohl, 1972
BLMs prepared from oxidized cholesterol, chlorophyll, chloroplast, and other membrane extracts; Fe^{2+}/Fe^{3+}, ascorbic acid, and biliproteins are placed in aqueous compartments	Electron transfer across BLM	Photoconductivity, sensitized photoconductivity, and charge separations are observed as functions of the constituents forming BLM (effects of chain lengths, degree of saturation, and headgroup character of a series of synthetic and natural lipids), applied voltage, and temperature; biliproteins enhance the photovoltage and reverse the electron flow when placed on the reducing side; they show no effect if placed on the oxidizing side of a very large redox gradient	EMF measurements in the absence and in the presence of illumination	Berns, 1976
BLMs prepared from glyceromonooleate, containing magnesium octaethylporphyrin; aqueous compartments contain buffer (phosphate or malate), 1.0 *M* KCl, and acceptor (methylviologen)	H_3O^+ and/or ^-OH transport across bilayer is mediated by photogenerated magnesium octaethylporphyrin cation; no photoinitiated fluxes are observed in absence of acceptor in aqueous phase	aqueous phase | BLM | aqueous phase H_2O | $H_2O{-}P^+ \longrightarrow H_2O{-}P^+$ | H_2O HB | ↑ ↓ | B^- H_3O^+ | $HO - P \longleftarrow HO - P$ | ^-OH	Electrical responses of doped BLM is observed following a single ~1 μsec laser flash	Young and Feldberg, 1979
BLMs prepared from oleic acid (R—COOH) containing methylene blue, MB; Becquerel effect studied on Pt electrode coated with this BLM; aqueous compartment(s) contain KCl, $K_4[Fe(CN)_6] - K_3[Fe(CN)_6]$	Photocurrent measurements	$MB + RCO_2^- \rightleftharpoons R{-}CO_2MB$ $R{-}CO_2MB + h\nu \rightarrow R{-}CO_2MB^*$ $R{-}CO_2MB^* + O_2 \rightarrow R{-}CO_2\dot{M}B_{OX}$ $R{-}CO_2MB^* + O_2 \rightarrow R{-}CO_2MB + O_2^*$ $R{-}CO_2MB + O_2^* \rightarrow MB + X$ $R{-}CO_2MB_{OX} \rightarrow MB + R{-}CO_{2OX} \rightarrow X$	Photoresponse measurements in electrical cells	Schreiber and Dupeyrat, 1979

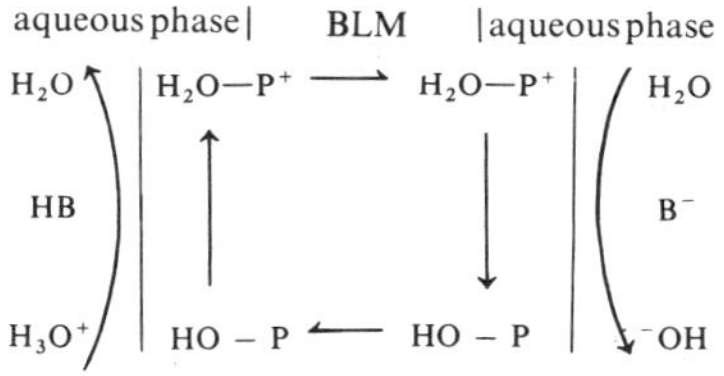

Table 12.14 (*Continued*)

System	Process	Results	Experimental Technique	Reference
BLMs prepared from phosphatidyl ethanolamine and glycerol monodein containing carbocyanine dyes, or chorophyllin	Photocurrent measurements	Performance of aparatus is described	Measurements of light flash induced voltage transients with 10 nsec resolution	Huebner, 1979
BLMs prepared from oxidized cholesterol containing 4-[*p*-(dipentylamino)styryl]-1-methylpyridinium iodide (di-5-ASP)	Electronic shift of the absorption spectrum of di-5-ASP is induced by a transmembrane electric shift	Di-5-ASP is a good "charge-shift" probe	Absorption and fluorescence spectroscopy and polarization, potential measurements	Loew et al., 1979a, 1979b
BLMs prepared from soybean phospholipids containing reaction centers from *Rhodopseudomonas sphaeroides* R-26, aqueous compartments contain cytochrome-c (donor) and ubiquinone (acceptor)	Photocurrent measurements	Photoelectric signal arises from the transfer of electrons from a secondary donor to a secondary acceptor, localized on opposite sides of the BLM; theoretical model developed	Light induced photocurrent measurements	Schönfeld et al., 1979
BLMs prepared from phosphatidylethanolamine, glycerol monoolein, air-oxidized cholesterol, and phosphatidylcholine, containing 3,3′-bis[α-(trimethylammonium)-methyl]azobenzene, bis-Q	cis-bis-Q $\leftrightharpoons$ trans-bis-Q photoisomerization	Voltage transients induced are positive for trans-to-cis and negative for cis-to-trans photoisomerization; equilibrium position of bis-Q changes upon isomerization	Measurements of laser flash induced voltage transients	Duchek and Huebnar, 1979
BLMs prepared from phosphatidylserine, phosphatidylcholine phosphatidylethanolamine, cholesterol containing chlorophyll, chlorophyllin, bacteriorhodopsin; aqueous compartments contain $FeCl_3$, $K_4[^-Fe(CN)_6]$, quinones, NaI, $Na_2S_2O_7$, methyl viologen, flavin mononucleotide	Energy and electron transfer across BLMs	Potentials of photochemical energy conversion are developed	Absorption, fluorescence, spectroscopy, photocurrent measurements	Tien, 1979, 1980

BLMs prepared from phosphatidylcholine, phosphatidylserine, phosphatidylethanol amine containing chlorophyll; aqueous compartments contain KCl, $FeCl_3$, chymotrypsin, bacteriorhodopsin containing liposomes	Photocurrent measurements	Charge separation is accomplished	Photoelectrospectrometry	Lopez and Tien, 1980
BLMs prepared from docosylamine, deposited on an indium-tin oxide electrode containing cyanine dyes	pH dependent sequential electron transfer	BLMs act as pH sensitive antenna, semiconductor, and electron injection and proton switched energy-electron-transfer device	Absorption and emission spectroscopy; photocurrent, capacitance measurements	Fromherz and Arden, 1980
BLMs prepared from egg phosphatidylcholine and cholesterol containing 9-anthroyloxy fatty acids as extrinsic fluorescence probes		Bilayer structure is perturbed	Digital impedance measurements	Ashcroft et al., 1980
BLMs prepared photosynthetic bacterial reaction center counting ubiquinone containing ferrocytochrome C	Flash induced electron transfer	Reaction center is considered to be functionally vectorial and structurally asymmetric	Absorption spectroscopy, flash induced electron transfer measurements	Packham et al., 1980
BLMs prepared from phosphatidylcholine containing octadecylamine and bacteriorhodopsin; aqueous compartments contain KCl or NaOH	pH dependent transient photocurrents	Mechanism for proton transfer is proposed	Capacitance measurements	Seta et al., 1980
BLMs prepared from glycerol monooleate and magnesium ctiochlorin, MgC; aqueous compartments contain ferrocyanide, ferricyanide	Charge induced electron transport	Data consistent with an electron-hopping mechanism in which the transmembrane electron transport occurs between membrane bound donors (MsC) and acceptors (MgC^+)	Conductance measurements	Feldberg et al., 1981

Table 12.15. Energy and Electron Transfer and Charge Separation in Vesicles

Vesicle	Process	Results	Experimental Technique	Reference
Liposomes prepared from egg lecithin and from mono- and digalactosyl diglycerides	Concentration quenching of chlorophyll	Half quenching concentrations range from 1.4×10^{-3} to 7.0×10^{-2}	Absorption spectroscopy, steady state and picosecond time resolved fluorescence spectroscopy	Beddard et al., 1976
Liposomes prepared from dipalmitoyl lecithin by sonication; pyrene or phenothiazine cosonicated	Photoionization	e_q^- and the cation radical are observed; photoionization yield increases markedly above the phase transition temperature	Laser flash photolysis	Barber et al., 1976
Liposomes prepared by shaking egg lecithin + chlorophyll (Chl) + β-carotene + $FeCl_3$ + $FeCl_2$ + KOAc buffer; ascorbic acid (Asc) added outside	Photoinitiated charge transport [diagram: inner circle $FeCl_3$ + $FeCl_3$; Chl; Asc]	Redox gradient created by external addition of ascorbic acid, chlorophyll aggregates are detected	Absorption spectroscopy	Mangel, 1976
Liposomes prepared from phosphatidylcholine + chlorophyll (Chl) + carbonylcyanide-D-trifluoromethyxyphenyl hydrazone by sonication $FeCl_3$ entrapped in inner compartment	Light induced O_2 formation [diagram: inner circle $FeCl_3$; Chl]	$Chl + h\nu \rightarrow Chl^*$ $Chl^* + Fe^{3+} \rightarrow Chl^+ + Fe^{2+}$ $Chl^+ + Fe^{2+} \rightarrow Chl + Fe^{3+}$ $Chl^+ + Chl^* \rightarrow Chl + Chl^+$ $Chl^+ + \frac{1}{2}H_2O \rightarrow Chl + \frac{1}{4}O_2 + H^+$	Light induced O_2 production determined by oxygen electrode	Toyoshima et al., 1977

Liposomes prepared from DL-α-dipalmitoyl phosphatidylcholine, L-α-dimyristoyl phosphatidylcholine, L-α-phosphatidylcholine by EtOH injection of the lipid along with the donor (*p*-terphenyl, TP, diphenyloctatetraene, DPO, chlorophyll-a, Chl-a, β-carotene) and acceptor (DPO, Chl-a)	Singlet-singlet energy transfer for: TP → DPO; DPO → Chl-a; β-carotene → Chl-a 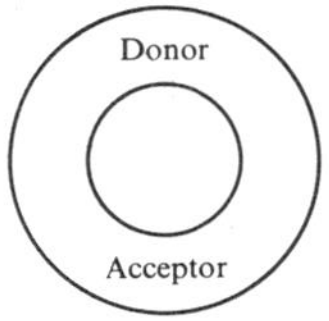 Donor Acceptor	Energy loss due to collision quenching is reduced in liposomes; there is no energy transfer for β-carotene → Chl-a	Absorption and fluorescence spectroscopy	Mehreteab and Strauss, 1978
Liposomes prepared egg yolk phosphatidylcholine by EtOH—DMF injection along with phylloquinone (H_2 carrier), decachloro-*m*-carborane (H^+ carrier), *N*,*N*′-di(1-hexadecyl)-2,2′-bipyridine-4,4′-dicarboxamide)-bis(2,2′-bipyridine)ruthenium (Ru^{2+}); EDTA trapped inside liposomes, hexadecylviologen (HV^{2+}) added outside	Photosensitized electron transport across liposome bilayers 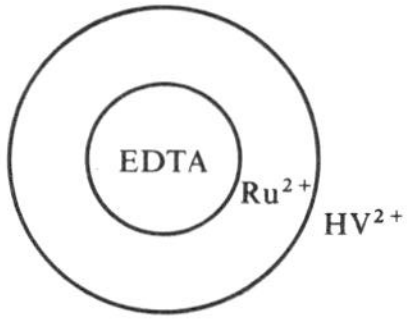 EDTA Ru^{2+} HV^{2+}	Electron is transferred from photosensitized Ru^{2+} to HV^{2+} across the liposome bilayer; EDTA reforms Ru^{2+} and is consumed	Steady state photolysis	Ford et al., 1978, 1979
Liposomes prepared from phosphatidylcholine by ultrasonication along with chlorophyll (Chl); potassium ascorbate (Asc) trapped inside liposomes, $CuCl_2$ added outside	Chlorophyll sensitized photoreduction of Cu^{2+} 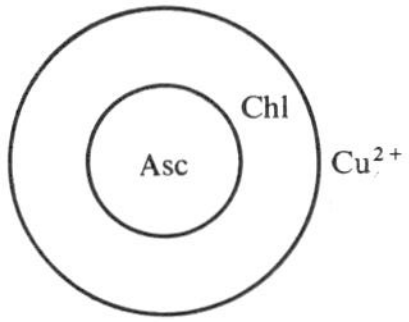 Chl Asc Cu^{2+}	Reduction is coupled through the bilayer	Esr	Kurihara et al., 1979a, 1980

Table 12.15 (*Continued*)

Vesicle	Process	Results	Experimental Technique	Reference
Liposomes prepared from phosphatidylcholine by ultrasonication along with chlorophyll; $K_3Fe(CN)_6$ added outside	Chlorophyll sensitized photoreduction of $Fe(CN)_6^{3-}$	Reduction is coupled through the bilayer; photoreduction rate is enhanced by ion carriers (carbonylcyanide or *m*-chlorophenylhydrazone)	Absorption spectroscopy, pH measurements	Kurihara et al., 1979b
Liposomes prepared lecithin by ultrasonication $Fe(CN)_6^{3-}$ trapped inside liposome, $FeCl_2$ and methylene blue (MB) added to outside	Methylene blue photosensitized reduction of $Fe(CN)_6^{3-}$	$MB_{outside} \xrightarrow{h\nu} MB^*_{outside}$ $MB^*_{outside} + Fe^{2+} \rightarrow LMB_{outside} + Fe^{3+}$ $LMB_{outside} \rightarrow LMB_{inside}$ $LMB_{inside} + Fe(CN)_6^{3-} \rightarrow MB + Fe(CN)_6^{2-}$	Steady state photolysis analysis by absorption spectroscopy	Sudo and Toda, 1979a, 1979b
Dihexadecylphosphate surfactant vesicles prepared by sonication along with pyrene (P)	Photoionization of pyrene in vesicle bilayers: $P \xrightarrow{h\nu} P^{\dot{+}} + P^T + e^-_{aq}$	Electron expelled into bulk solution; anionic vesicle prevents charge recombination	Laser flash photolysis	Escabi-Perez et al., 1979
Dioctadecyldimethylammonium chloride (DODAC) surfactant vesicles prepared by sonication along with 2-hydroxy-1-[ω(1-pyreno)decanoyl]-*sn*-glycero-3-phosphatidylcholine (lysopyrene); trisodium 8-hydroxy-1,3,6-pyrenetrisulfonate (pyranine) added outside	Energy transfer	Energy transfer efficiency in presence of DODAC vesicles is 43%, in their absence is 3%; vesicle structure is substantiated by determining energy transfer subsequent to resonication	Steady state and time resolved fluorescence spectroscopy	Nomura et al., 1980

Liposomes prepared from dipalmitoyl phosphatidylcholine and surfactant vesicles prepared from dioctadecyldimethylammonium chloride by shaking (multicompartment) and by sonication (single compartment) containing alloxazines and isoalloxazines: Alloxazine: $R = C_4H_9$, C_8H_{17}, or $C_{12}H_{25}$ Isoalloxazine: $R_1 = R_2 = C_4H_9$ $R_1 = C_8H_{17}$, $R_2 = C_4H_9$ $R_1 = C_{16}H_{33}$, $R_2 = C_4H_9$	Singlet-singlet energy transfer from alloxazines to isoalloxazines determined at different temperature	Efficiencies of energy transfer and energy loss is greater for multicompartment than single walled vesicles and increase with increasing alkyl chain length of the acceptor	Absorption and fluorescence spectroscopy and fluorescence polarization	Aso et al., 1980
Liposomes prepared from egg yolk lecithin by sonication along with dyes; dyes: neutral red, phenosafranine, indigodisulfonate, thionine, toluylene; donor (D) = sodium ascorbate, acceptor (A) = $K_3Fe(CN)_6$	Electron transfer System A 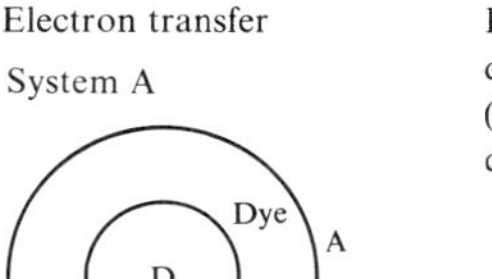 System B 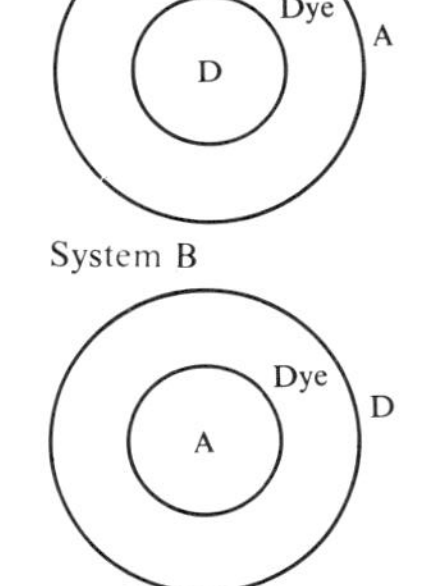	Electron transport is observed for some dye even in the absence of irradiation (it is then controlled by the EM of the dye)	Absorption spectroscopy	Sudo et al., 1980

Table 12.15 (*Continued*)

Vesicle	Process	Results	Experimental Technique	Reference
Liposomes prepared from dipalmitoyl phosphatidyl choline and dimyristoyl phosphatidylcholine; surfactant vesicles prepared from dihexadecyldimethylammonium bromide by sonication along with (*N*,*N*′-didodecyl-2,2′-bipyridine-4,4′-dicarboamide)-bis(2,2′-bipyridine)ruthenium(II)$^{2+}$, $RuC_{12}bpy^{2+}$, and *N*,*N*-dimethylaniline, DMA	Electron transfer from DMA to photosensitized $RuC_{12}bpy^{2+}$ [diagram labels: $RuC_{12}bpy^{2+}$, DMA]	DMA^+ formed is expelled from vesicles, back reaction between DMA^+ and $RuC_{12}bpy^+$ retarded by charge repulsions	Laser flash photolysis	Nagamura et al., 1980
Liposomes prepared from dipalmitoyl-D,L-α-phosphatidylcholine by sonication along with monododecyl substituted zinc *meso*-tetra(4-pyridyl)porphyrin bromide (ZnP) as sensitizer and vitamin K_1 $(VK)_+$ or 1,3-dibutylalloxazine (DBA) or 1,3-didodecylalloxazine (DDA) as electron mediator (EM); EDTA entrapped in interior, disodium 9,10-anthraquinone-2,6-disulfonate (AQDS) added outside	Photosensitized electron transfer [diagram labels: ZnP, EM, EDTA, Pzn, AQDS]	Two-step activation of the sensitizer at the inner and outer surfaces of vesicles account for electron transfer; DBA is the most effective mediator	Steady state photolysis followed by absorption spectroscopy	Matsuo et al., 1980a, 1980b
Liposomes prepared from dimyristoyl phosphatidylcholine surfactant vesicles prepared from dihexadecyldimethylammonium bromide by sonication containing (*N*,*N*′-didodecyl-2,2′-bipyridine-4,4′-dicarboxamide)-bis(2,2′-bipyridine)ruthenium(II)$^{2+}$ ($RuC_{12}bpy^{2+}$) and *N*-butylphenothiazine (BPT)	Electron transfer from BPT to photoexcited $RuC_{12}bpy^{2+}$ [diagram labels: $RuC_{12}bpy^{2+}$, BPT]	Back electron transfer reaction is retarded	Absorption and fluorescence spectroscopy, flash photolysis	Takayanagi et al., 1980

Surfactant vesicles prepared from dioctadecyldimethylammonium chloride by sonication containing *N*-methylphenothiazine (MPTH) or *N*-dodecylphenothiazine (DPTH) (donors) and an amphiphatic ruthenium complex, $RuC_{18}(bpy)_3^{2+}$	Electron transfer from MPTH (or DPTH) to photoexcited $RuC_{18}(bpy)_3^{2+}$ 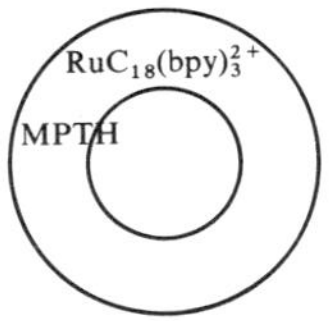 [diagram: $RuC_{18}(bpy)_3^{2+}$, MPTH]	$MPTH^+$ formed is expelled into vesicle interior and exterior; addition of controlled amounts (optimally 1.0×10^{-3} *M*) of NaCl maximizes the amount of $MPTH^{\ddagger}$ expelled into bulk water and prevents back reaction; $DPTH^{\ddagger}$ is not expelled.	Absorption and emission spectroscopy, laser flash photolysis	Infelta et al., 1980
Surfactant vesicles prepared from 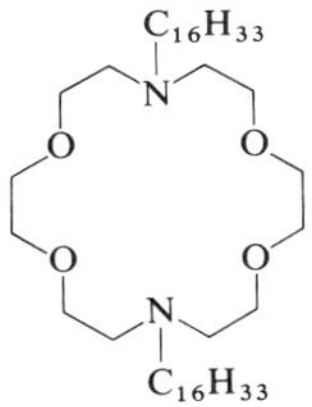 [structure: crown ether with N–$C_{16}H_{33}$ groups] by sonication	Photoinduced electron transfer from 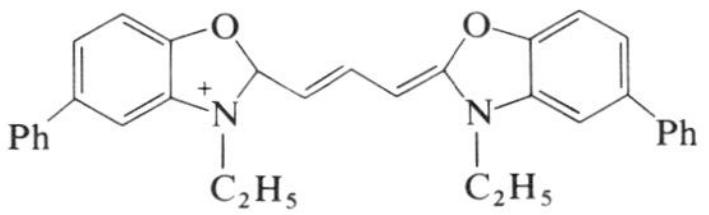 [structure: Ph, N–C_2H_5, O, Ph, N–C_2H_5] and amphiphatic ruthenium complex $RuC_{18}(bpy)_3^{2+}$ to Ag^+ entrapped in crown ether cavity	Ag° formed is stabilized	Absorption and emission spectroscopy, laser flash photolysis, quasi-elastic light scattering	Monserrat et al., 1980

Table 12.15 *(Continued)*

Vesicle	Process	Results	Experimental Technique	Reference
Surfactant vesicles prepared from dioctadecyldimethylammonium chloride by sonication containing zinc porphyrin, zinc tetraphenyl porphyrin (P), and duroquinone(DQ)	Photosensitized electron transfer P DQ	$P + DQ \xrightarrow{h\nu} P^{\ddagger} + DQ^{\bar{\cdot}}$; salts decrease the yield of $P^{\ddagger}$	Laser flash photolysis, quasi-elastic light scattering	Pileni, 1980a
Liposomes prepared from dipalmitoyl phosphatidylcholine by sonication containing the fluorophore (pyrene or pyrenedecanoic acid) and the quencher (*N*,*N*-dimethylaniline, *p*-isopropyl-*N*,*N*-dimethylaniline, *N*,*N*-dicetylaniline, and *p*-*N*,*N*-dimethylanilene sulfonate	Fluorescence quenching	Solubilization sites of probes and quenchers are inferred	Steady state and time resolved fluorescence spectroscopy	Kano et al., 1980a
Liposomes prepared from dipalmitoyl phosphatidylcholine and dimyristoyl phosphatidylcholine by sonication containing pyrene and *N*,*N*-dimethylaniline	Heteroexcimer formation	Quenching kinetics is similar to that in homogeneous solution	Fluorescence spectroscopy and laser flash photolysis	Waka et al., 1980a, 1980b
Liposomes prepared from phosphatidylcholine, chlorophyll-1, and duroquinone	Quenching of chlorophyll-a triplet and electron transfer to quinone	Quenching is due to collisional deexcitation electron transfer to quinone results in long lived transfer products	Laser flash photolysis	Hurley et al., 1980

Surfactant vesicles prepared from dihexadecyl phosphate by sonication; sensitizer ($Ru(bpy)_3^{2+}$), electron donor (EDTA), acceptor (methylviologen, MV^{2+}), and catalyst (PtO_2) are organized differently in the compartments provided by the vesicles	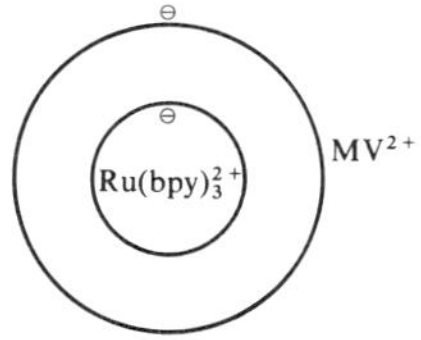 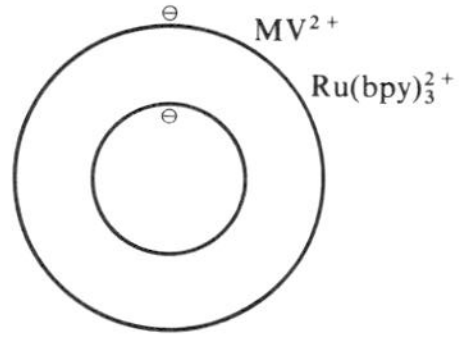 Photosensitized electron transfer, charge separation, and hydrogen production System I 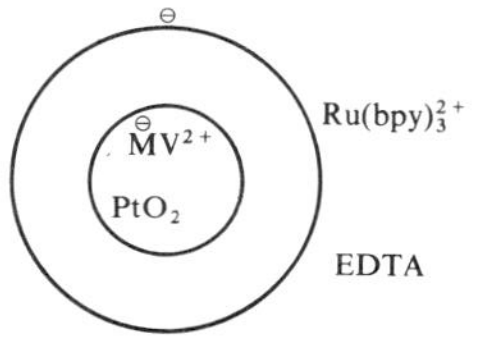 $\ominus$ MV^{2+} PtO_2 $Ru(bpy)_3^{2+}$ EDTA System II $\ominus$ $Ru(bpy)_3^{2+}$ MV^{2+} System III $\ominus$ MV^{2+} $Ru(bpy)_3^{2+}$ System IV $\ominus$ $Ru(bpy)_3^{2+}$ MV^{2+} 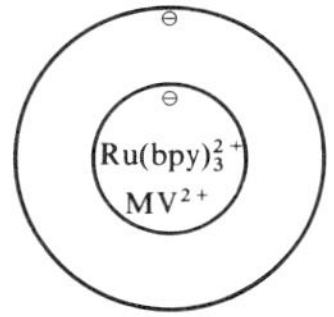	System I is the most efficient; complete neutralization of the inner surface, but only partial neutralization of the outer surface favors electron transfer, charge separation, and hydrogen generation in presence of PtO_2; efficient electron transfer, but equally efficient reaction occurs in Systems III and IV.	Absorption and fluorescence spectroscopy, steady state photolysis, laser flash photolysis	Tunuli and Fendler, 1981

Table 12.15 *(Continued)*

Vesicle	Process	Results	Experimental Technique	Reference
Surfactant vesicles prepared dioctadecyldimethylammonium chloride by sonication, $Ru(bpy)_3^{2+}$ and MV^{2+} added outside; alternatively, long chain $Ru(bpy)_3^{2+}$, $C_{12}Ru(bpy)_3^{2+}$ are anchored onto the vesicle and MV^{2+} and EDTA added outside	Photosensitized electron transfer and charge separation	Effects of potential field on electron transfer and charge separation are quantitatively examined	Fluorescence spectroscopy, laser flash photolysis	Tunuli and Fendler, 1982a, 1982b
Surfactant vesicles prepared from dioctadecyldimethylammonium chloride by sonication containing zinc *meso*-tetra(4-pyridyl)porphyrin ($ZnC_{12}TPyP^+$); MV^{2+} entrapped in inner surface, $Ru(bpy)_3^{2+}$ attached to outer surface	$Ru(bpy)_3^{2+}$ sensitized reduction MV^{2+} to $MV^{\ddagger}$ and charge separation	PtO_2 acts as catalyst to reform MV^{2+}	Laser flash photolysis	Nagamura et al., 1981
Liposomes prepared from egg yolk phosphatidylcholine containing [*N*,*N*′-di(1-hexadecyl)-2,2′-bipyridine-4,4′-dicarboxamide]-bis(2,2′-bipyridine)ruthenium(2+) (Ru^{2+}), valinomycin, and gramicidine as ionophores (ION), EDTA in the inner aqueous compartment, and heptylviologen (C_7V^{2+})	Photosensitized reduction of C_7V^{2+} to $C_7V^{\ddagger}$ and charge separation [diagram labels: EDTA, ION, Ru^{2+}, Ru^{2+}, C_7V^{2+}]	Addition of valinomycin and creation of suitable transmembrane potential enhanced $C_7V^{\ddagger}$ formation	Steady state photolysis	Laane et al., 1981
Functionally polymerized redox active and chemically dissymmetrical surfactant vesicles, with added $Ru(bpy)_3^{2+}$, EDTA, and PtO_2	Photosensitized electron transfer and charge separation	Efficient electron transfer and charge separation	Steady state photolysis, laser flash photolysis	Tundo et al., 1982
Liposomes prepared from dipalmitoyl phosphatidylcholine and dioleylphosphatidylcholine	Chlorophyll-a triplet quenching by quinone	Electron transfer and charge separation	Laser flash photolysis	Hurley et al., 1981

Table 12.16. Miscellaneous Reactions in Vesicles

Reaction	Effect of Vesicle	Reference
Dimerization of *N,N'*-distearylquinocarboxyanine iodide	Phosphatidylcholine liposomes (sonicated); equilibrium constant for dimerization depends on the physical state of the acyl chains in the liposome; $K = 3.2\ M$ (25°C), $1.3 \times 10^3\ M$ (−10°C)	Kurihara et al., 1977
Hydrolysis of *p*-nitrophenyl acetate and nonaoate, 30°C, pH = 8.9; in the presence of *N*-dodecylbenzohydroxamic acid, *N*-dodecyl-4-imidazolecarboxamide, *N*-methylcholohydroxamic acid, *N*-(4-imidazolylmethyl)-cholohydroxamic acid, *N*-[2-(4-imidazolyl)ethyl]cholamide, and cholesteryl 4-imidazolecarboxylate as nucleophiles	Surfactant vesicles prepared from: $[CH_3-(CH_2)_{n-1}]_2\overset{+}{N}(CH_3)_2\ Br^-$; $n = 12$; $2C_{12}\overset{+}{N}2Ci$; $n = 18$: $2C_{18}\overset{+}{N}2Ci$ by sonication; $k_\Psi/k_0 \sim 200–300$ (*p*-nitrophenyl nonaoate), 10 (*p*-nitrophenyl acetate); rate enhancement $2C_{18}\overset{+}{N}2Ci > 2C_{12}\overset{+}{N}2Ci >$ CTAB micelles	Okahata et al., 1979
Decarboxylation of 6-nitrobenzisoxazole-3-carboxylate, 30°C	Surfactant vesicles prepared from: $(C_nH_{2n-1})(C_mH_{2m+1})\overset{+}{N}(CH_3)_2\ Br^-$; $n = m = 12, 14, 16, 18$; $n = 18$, $m = 8, 10, 12, 14$ by sonication, $k_\Psi/k_0 \sim +$; rate enhancement depends on fluidity	Kunitake et al., 1980

Table 12.16 *(Continued)*

Reaction	Effect of Vesicle	Reference
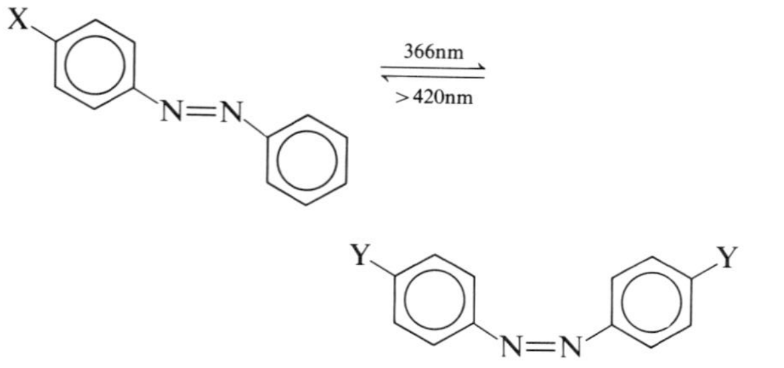 X-C$_6$H$_4$-N=N-C$_6$H$_5$ $\underset{>420nm}{\overset{366nm}{\rightleftharpoons}}$ Y-C$_6$H$_4$-N=N-C$_6$H$_4$-Y; $X = (CH_3)_3\overset{+}{N}CH_2O^-$, Br^-, $Y = CH_3(CH_2)_{11}O^-$ (*1*); $X = (CH_3)_3\overset{+}{N}(CH_2)_4O^-$, Br^-, $Y = CH_3(CH_2)_{11}O^-$ (*2*)	Dipalmitoyl-D,L-α-phosphatidylcholine cosonicated with *1* or *2*; properties (osmotic shrinkage rates and bromothymole blue release) of cis *1* or *2* containing liposomes are different from those containing trans *1* or *2*; light can be used on as "on-off switch" to alter vesicle properties	Kano et al., 1980b
Chlorophyllin photooxidation	Phosphatidylcholine liposome, $k_\Psi/k_0 \sim -$; phosphatidylethanolamine liposome, $k_\Psi/k_0 \sim +$	Stillwell and Karimi, 1980
Photodimerization of parinaric acid	Liposomes prepared from dipalmitoyl and dilauryl phosphatidylcholine; rate is sensitive to thermal phase transition	Morgan et al., 1980
trans ⇆ cis isomerization of the azobenzene unit in $CH_3(CH_2)_{11}-C_6H_4-N{=}N-C_6H_4-O-(CH_2)_3-O-P(=O)(O^-)-O-(CH_2)_2\overset{+}{N}(CH_3)_3$ (*1*)	Properties of disk line aggregates formed from *1* upon sonication are affected by photoisomerization	Okahata et al., 1980

Chlorophyllase catalyzed hydrolysis of chlorophyll	Sonicated egg lecithin, $k_\Psi/k_0 \sim +$	Terpstra, 1980
Hydrogenation of $^-C{=}C^-$ in liposomes using water soluble chlorotris-(sodium diphenylphosphinobenzene-*m*-sulfonate)-rhodium(I)tetrahydrate	Liposomes prepared from purified soya lecithin; catalyst can be easily removed after hydrogenation	Madden et al., 1980
Azobis[(2-*n*-butylcarboxy)propane], R_2N_2, decomposition: $R_2N_2 \rightarrow (2R\cdot)_{cage} + N_2$; $(2R\cdot)_{cage} \rightarrow 2R\cdot$; $(2R\cdot)_{cage} \rightarrow RR$; $R\cdot + O_2 \rightarrow RO_2\cdot$	Multicompartment liposomes, prepared from L-α-dimristoylphosphatidylcholine and L-α-dilauroylphosphatidylcholine; cage escape depressed in liposomes	Winterle and Mill, 1980
trans → cis isomeration of 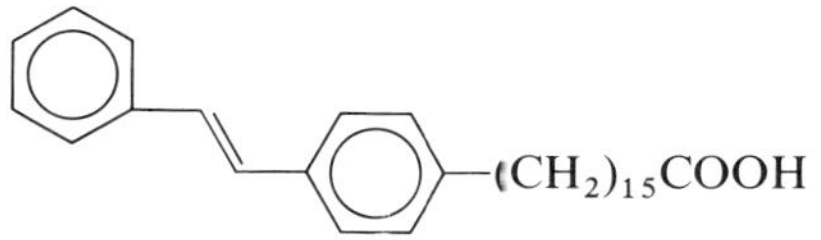 [stilbene structure]–$(CH_2)_{15}COOH$	Liposomes prepared from egg lecithin (*1*), vesicles prepared from didoceyldimethylammonium bromide (*2*), and from mixtures of *1* + *2* and *1* + dicetylphosphate; $\Phi_{trans\ cis}$ smaller below phase transition than above	Russell et al., 1980
$^1O_2$ + tocopherol $\xrightarrow{k_q}$ $^3O_2$ + tocopherol; $^1O_2$ + tocopherol $\xrightarrow{k_R}$ reaction products; 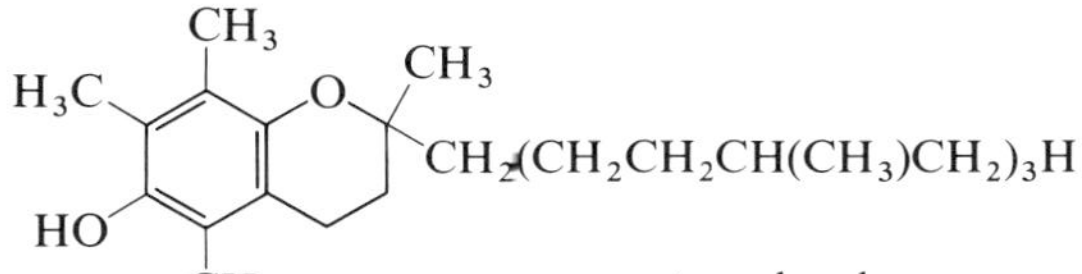 [tocopherol structure: H_3C, CH_3, HO, CH_3, O, CH_3, $CH_2(CH_2CH_2CH(CH_3)CH_2)_3H$] tocopherol	Single- and multicompartment liposomes prepared from phosphatidylcholine, k_q/k_R is a measure of the microenvironment	Fragata and Bellemare, 1980
7-Dehydrocholesterol $\xrightarrow{h\nu}$ previtamin D_3 (precholecalciferol) + vitamin D_3 (cholecalciferol) + tachysterol$_3$ + lumisterel$_3$	Liposomes prepared from dilauryl-L-α-, distearoyl-L-α-, dimyristoyl-L-α-, and dipalmitoyl-L-α-phosphatidylcholine; yield and distribution of photoproducts are altered	Moriarty et al., 1980

Table 12.16 (*Continued*)

Reaction	Effect of Vesicle	Reference
Hydrolysis of L-, and D-*N*-benzyloxycarbonylphenylalanine *p*-nitrophenyl esters, pH = 6.55, *T* varied	Surfactant vesicles prepared from *N*,*N*-didodecyl-N^{α}-[6-(trimethylammonio)-hexanoyl]histidineamine bromide by sonication; $k_{\Psi}/k_{0} \sim +$, $k_{L}/K_{D} \sim 4.4$, enantioselectivity is decreasing with increasing temperature	Murakami et al., 1981a
Solvent dependent charge transfer transition (absorption spectroscopy) of dicyano(1,2,3,7,13,17,18,19-octamethyltetradehydrocorrinato)cobalt(III)	Surfactant vesicles prepared from *N*,*N*-didodecyl-N^{α}-[6-trimethylammoniohexanoyl]-alaninamide bromide by sonication; two separate substrate binding sites, one near the polar head and other in the hydrophobic interior, are identified	Murakami et al., 1981b
Imidazole or hydroxide ion mediated hydrolysis of 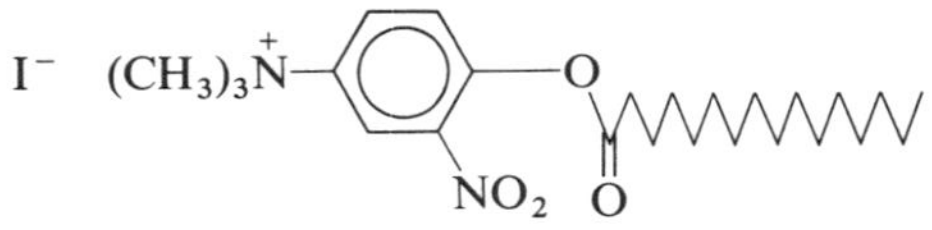 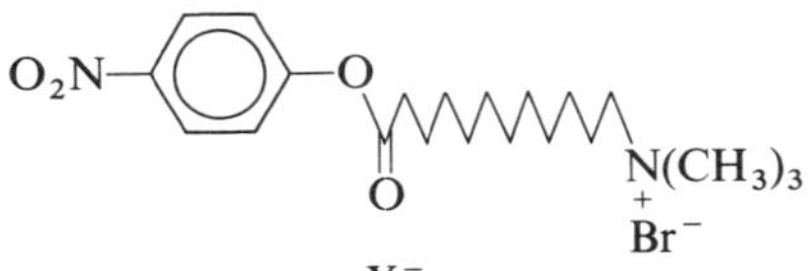 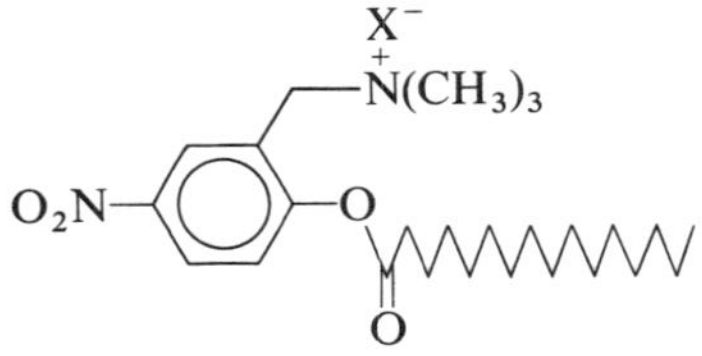 I^{-} $(CH_3)_3\overset{+}{N}$–$C_6H_3(NO_2)$–O–C(=O)–(CH$_2$)$_n$; O_2N–C_6H_4–O–C(=O)–(CH$_2$)$_n$–$\overset{+}{N}(CH_3)_3$ Br^{-}; O_2N–C_6H_3(CH$_2$–$\overset{+}{N}(CH_3)_3$ X^{-})–O–C(=O)–(CH$_2$)$_n$	Liposomes prepared from egg lecithin by sonication; $k_{\Psi}/k_{0} \sim -$, rate retardations, particularly for *3*, is rationalized by assuming substrate intercalation into vesicle such that electrostatic stabilization of transition state is not possible (as it is possible in H_2O)	Fatah and Loew, 1981

Hydrolysis of 5,5′-dithiobis(2-nitrobenzoic acid), function of pH and T	Surfactant vesicles prepared from dioctadecyldimethylammonium chloride by sonication; $k_{\Psi}/k_0 \sim +$, rate enhancement and observed kinetics are accommodated in terms of theories developed for micellar catalysis (see Chapter 11.1)	Fendler and Hinze, 1981
Thiolysis of *p*-nitrophenyl octanoate by *n*-heptylmercaptan, pH = 4–6	Surfactant vesicles prepared from dioctadecyldimethylammonium chloride by sonication, $k_{\Psi}/k_0 \sim +$; rate enhancement and observed kinetics are rationalized in terms of the ion-exchange formation, developed for micellar catalysis (see Chapter 11.1)	Cuccovia et al., 1981
Bilirubin photoisomerization	Surfactant vesicles prepared from dioctadecyldimethylammonium chloride by sonication; bilirubin conformation is affected by vesicles	Tran and Beddard, 1981

Table 12.17. Reactivities in Macrocylic Compounds

Reaction	Effect of Macrocycle	Reference
Charge transfer interaction of 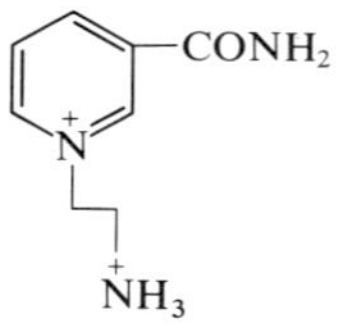 pyridinium ($CONH_2$, $\overset{+}{N}H_3$)	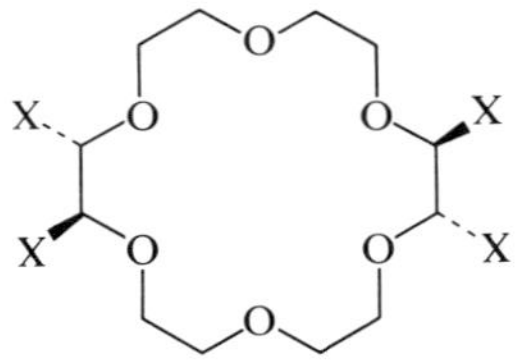 X = L-$CONHCHCH_2$(3-indole) with CO_2Me on the CH; Very stable charge transfer complex formed	Behr et al., 1976
Quenching of naphthalene phosphorescence by metal ions, M^+	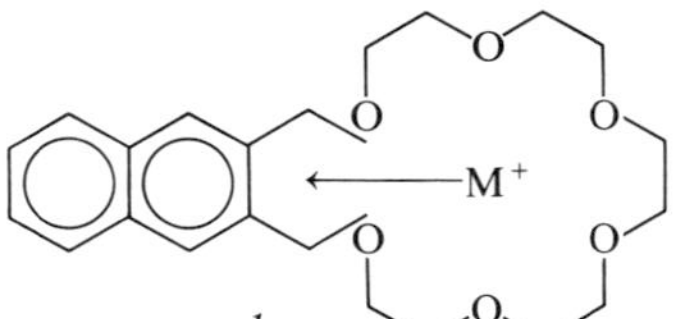 *1*, 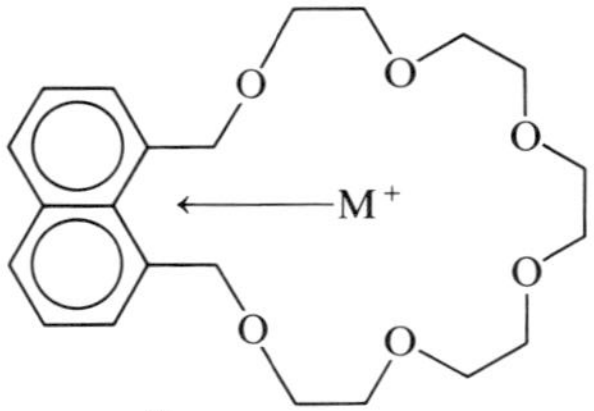 *2*; $M = Na^+, K^+, Rb^+, Cs^+$, Perturbation by metal ions are different for *1* and *2*	Sousa and Larson, 1977

Phase transfer reaction between potassium thiocyanate (in H_2O) and *p*-nitrobenzyl bromide ($CHCl_3$)	18-crown-6, benzo-18-crown-6, dibenzo-18-crown-6, *syn*-bis(methylbenzo)-18-crown-6, anti-bis(methylbenzo)-18-crown-6, bis(acetylbenzo)-18-crown-6, bis[(1-hydroxyethyl)benzo]-18-crown-6, bis(*tert*-butylbenzo)-18-crown-6, bis(*n*-pentylbenzo)-18-crown-6, bis[(1-hydroxyisopentyl)benzo]-18-crown-6, bis(heptanoylbenzo)-18-crown-6, bis(*n*-heptylbenzo)-18-crown-6, bis(*n*-decylbenzo)-18-crown-6, bis(tetradecanoylbenzo)-18-crown-6, bis[(1-hydroxytetradecyl)benzo]-18-crown-6, bis(*n*-tetradecylbenzo)-18-crown-6, dicyclohexano-18-crown-6, bis(methylcyclohexano)-18-crown-6, bis(*n*-decylcyclohexano)-18-crown-6, bis(tetradecanoylbenzo)-18-crown-6, bis[(1-hydroxytetradecyl)benzo]-18-crown-6, bis(*n*-tetradecylbenzo)-18-crown-6, dicyclohexano-18-crown-6, bis(methylcyclohexano)-18-crown-6, bis(*n*-decylcyclohexano)-18-crown-6, crypt 222, decaglyme; catalytic efficiencies are compared, within the series of substituted dibenzo crown ethers, variations in reaction rates correlated only with the electronic effects of substituents; binding ability and ligand lipophilicity, catalytic efficiencies	Stott et al., 1980
Hydrolysis of $RC(=O)OC_6H_4NO_2$-*p*, $R = CH_3, (CH_2)_4CH_3, (CH_2)_8CH_3, (CH_2)_{10}CH_3, (CH_2)_{14}CH_3$, cyclohexyl, CH_2-cyclohexyl, $CH(CH_3)$-cyclohexyl, CH_2-3,5-di-CH_3-cyclohexyl, CH_2-cyclodecyl, $CH_2C_6H_5$, CH_2-1-naphthyl	$(CH_2)_9$ C=O $(CH_2)_{10}$ O=C HN CH_2—CH_2 N NH Hydrolysis rates correlate with substrate partitioning coefficient between octanol and water by: $\log k = 0.45(\pm 0.09)\pi = 0.53(\pm 0.39)$	Hansch, 1978

Table 12.17 *(Continued)*

Reaction	Effect of Macrocycle	Reference
Excited state proton transfer: 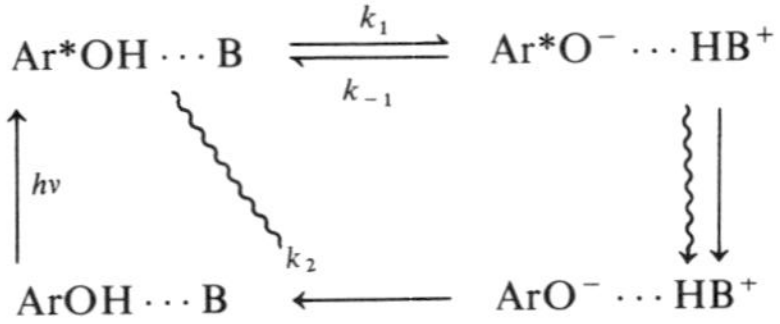 $Ar^{*}OH \cdots B \underset{k_{-1}}{\overset{k_1}{\rightleftharpoons}} Ar^{*}O^{-} \cdots HB^{+}$ $h\nu$ ↑ ; k_2 ; ↓ $ArOH \cdots B \longleftarrow ArO^{-} \cdots HB^{+}$ ArOH = 2-naphthol, 1-naphthol, 4-chloro-1- naphthol	18-crown-6; loss of electronic excitation energy may occur during the proton transfer in the hydrogen bond; efficiencies of radiationless deactivation processes are controlled by the nature of the excited proton donor	Lemmetyinen et al., 1980
Fluorescence quenching (by $TbCl_3$, glycine-L-Trp), and excimer formation	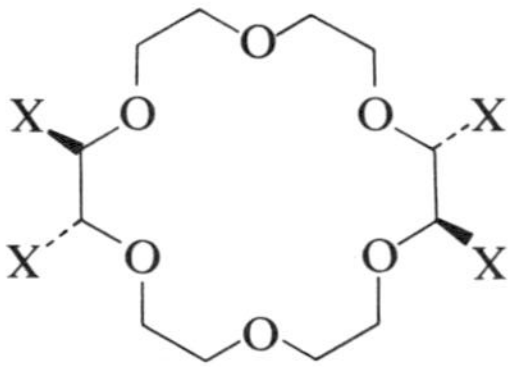 X = D-CONHCH(CO_2Me)CH_2(3-indole) X = L-CONHCH(CO_2Me)CH_2(3-indole) X = CONH-1-pyrene 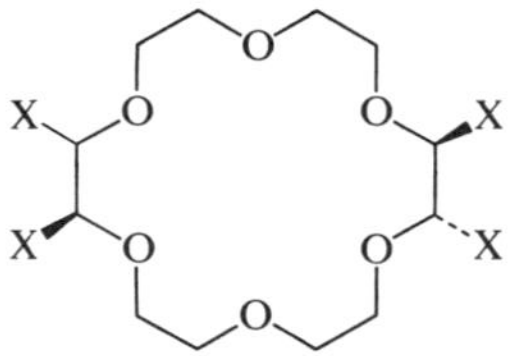	Tundo and Fendler, 1980

	$X = $ D-$CONHCH(CO_2Me)CH_2$(3-indole) $X = $ L-$CONHCH(CO_2Me)CH_2$(3-indole) X = CONH-1-pyrene Chiral recognition in binding; *K*(Gly-L-Trp/D-crown-pyrene)/*K*(Gly-L-Trp/L-crown-pyrene) = 2.4 is realized	
Cis–trans photoisomerization	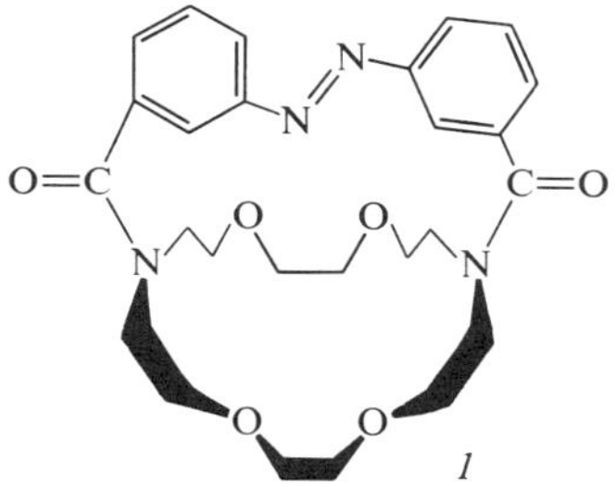 1 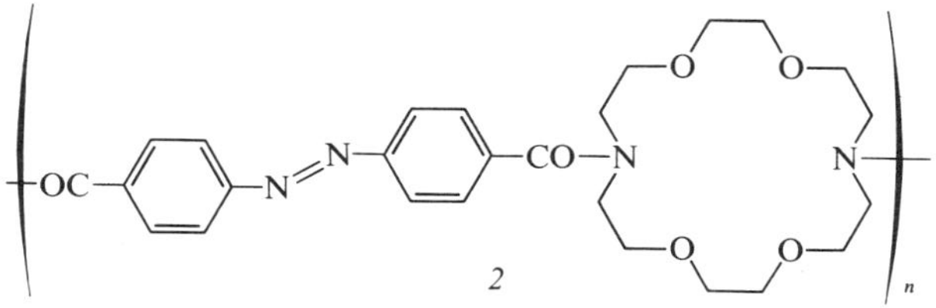 2 *trans-1* binds Li^+ and Na^+ preferentially, whereas *cis-1* favors Rb^+ and K^+; *trans-2* binds Na^+, *cis-2* does not; crown size is photolytically expanded, chemical functions may be controlled by light	Shinkai et al., 1980d
Photosolvolysis of 4′-(1-acetoxy)ethylbenzo-15-crown-5	Complex formation with alkali metals decreases solvolysis rate	Suzuki and Tada, 1980

Table 12.17 *(Continued)*

Reaction	Effect of Macrocycle	Reference
Eu^{2+} luminescence	15-crown-5, 18-crown-6, dicyclohexyl-18-crown-6, dicyclohexyl-24-8; luminescence increases upon complexation	Adachi et al., 1980
$RCH_2CN \xrightarrow[H_2O]{tBuOK} RCOOH$	18-crown-6, $k_{\Psi}/k_0 \sim +$	DiBiase et al., 1980
Cis → trans photoisomerization	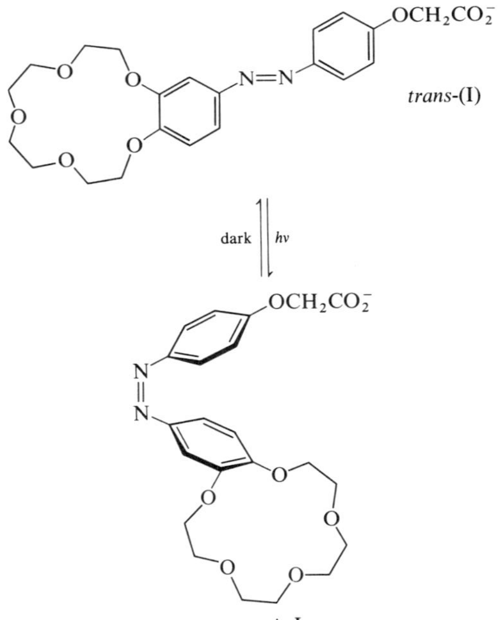 $OCH_2CO_2^-$ N=N O O O O O *trans*-(I) dark $h\nu$ $OCH_2CO_2^-$ N=N O O O O O *cis*-I Photoisomerized *cis*-I extracts K^+ more selectively than *trans*-I, metal ions affect photoisomerization	Shinkai et al., 1980e, 1981a
Anion activation, cation participation	Effects of cryptates are illustrated on examples, principles are discussed	Lehn, 1980

Table 12.18. Reactivities in Cyclodextrins

Reaction	Effect of Cyclodextrin	Reference
Triplet energy transfer between the excited cyclodextrin host and ground state guests: 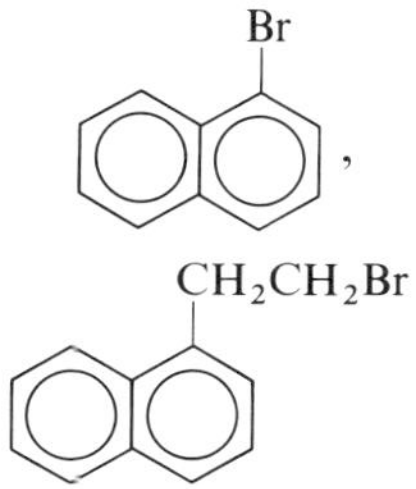 1-bromonaphthalene (Br), 1-naphthyl-CH_2CH_2Br	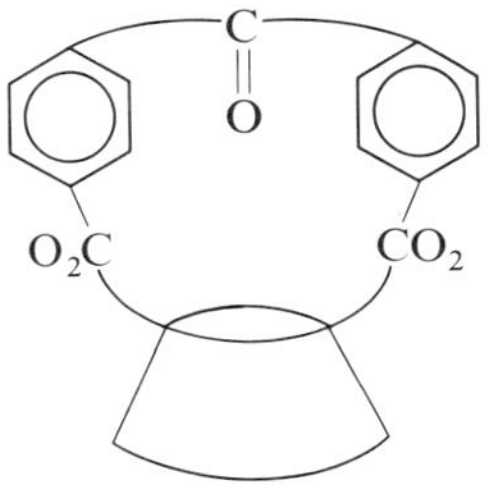 host; $\Phi \approx 50$–60%	Tabushi et al., 1977
Hydrolysis of 2,4-dinitrophenyl sulfate in 0.1 *N* NaOH, 40°C	β-cyclodextrin, $k_{\Psi}/k_0 \sim 1.0$	Sunamoto et al., 1977
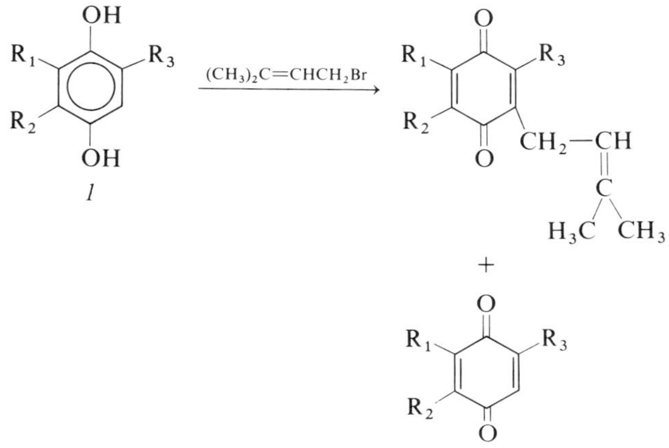 *1* $\xrightarrow{(CH_3)_2C{=}CHCH_2Br}$ *2* + *3*	see yields below	Tabushi et al., 1978

Starting material	Yield, % *2*	Yield, % *3*
1a	7	41
1a + β-cyclodextrin	53	30
2a	18	49
2a + β-cyclodextrin	35	47
3a	11	55
3a + β-cyclodextrin	39	45

a: $R_1 = R_2 = R_3 = CH_3$
b: $R_1 = H, R_2 = R_3 = CH_3$
c: $R_1 = R_2 = OCH_3, R_3 = CH_3$

Table 12.18 (*Continued*)

Reaction	Effect of Cyclodextrin	Reference
Photooxidation of stable carbanions	β-cyclodextrin, quantum yield is affected	Yamada et al., 1978
Cleavage of β-cyclodextrin *trans*-cinnamate by quimclidine and piperidine	$k_{amines}/k_0 \sim 27, 13$	Komiyama and Bender, 1979
2-methyl-1,4-naphthalenediol + RBr $\xrightarrow{O_2}$ 2-methyl-3-R-1,4-naphthoquinone + 2-methyl-1,4-naphthoquinone; R = $CH_2CH{=}CH_2$, $CH_2CH{=}CHCH_3$, $CH_2C(CH_3){=}CH_2$, $CH_2CH{=}C(CH_3)_2$	β-cyclodextrin, $k_\Psi/k_0 \sim +$; one step preparation of vitamin K_1 or K_2 analogs is catalyzed	Tabushi et al., 1979

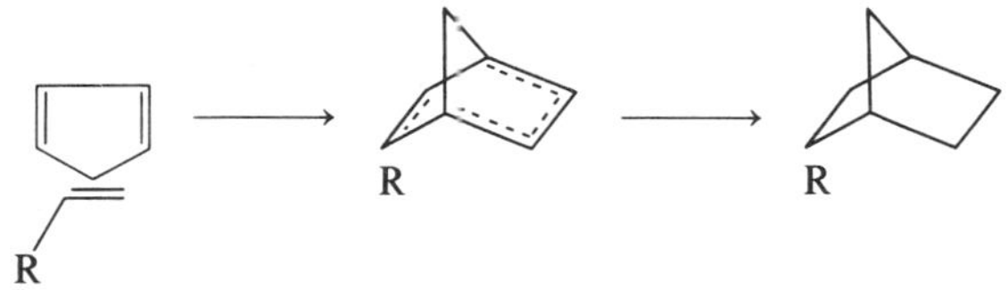

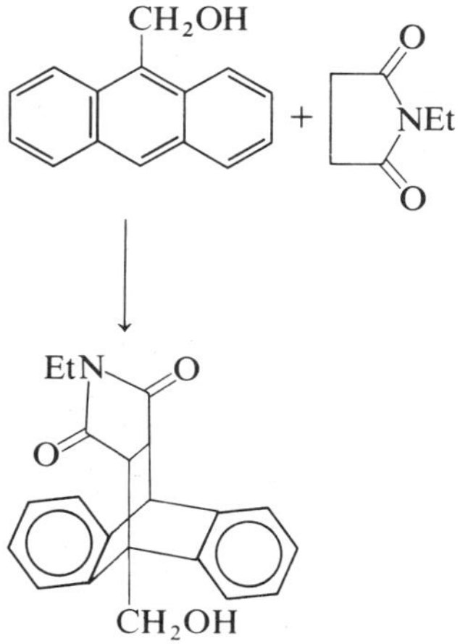

$R = COCH_3, CN$

β-cyclodextrin, $k_\Psi/k_0 \sim +++$

Rideout and Breslow, 1980

Acylation and deacylation of nitrophenyl acetates, adamantyl esters, ferrocenyl esters

Capped β-cyclodextrins:

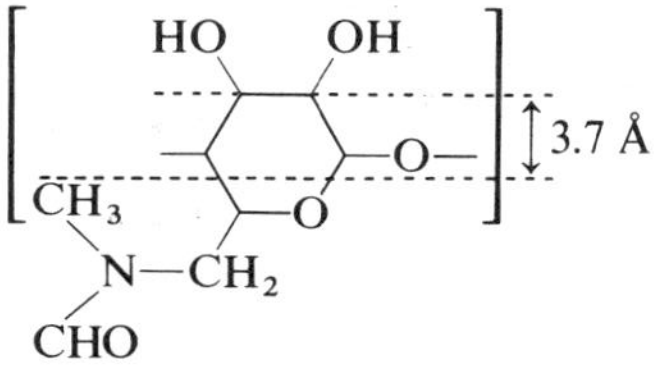

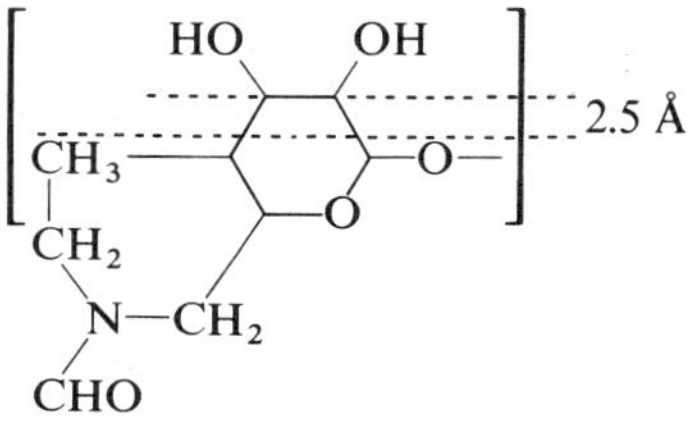

$k_\Psi/k_0 \sim 10^6$–10^7

Breslow et al., 1980

Table 12.18 *(Continued)*

Reaction	Effect of Cyclodextrin	Reference
Hydrolyses of alkyl benzoates	α-cyclodextrin, $k_\Psi/k_0 \sim +$	Komiyama and Hirai, 1980a
Hydrolysis of phenyl and substituted phenylacetates	β-cyclodextrin, $k_\Psi/k_0 \sim +$; rate enhancements rationalized in terms of the time averaged positions of the substrates in the host	Komiyama and Hirai, 1980b, 1980c
Cleavage of *p*- and *m*-nitrophenyl 1-adamantene carboxylates, 16°C	α, β, γ-cyclodextrins, $k_\Psi/k_0 \sim -$	Komiyama and Inoue, 1980a
Hydrolysis of 2,2,2-trifluoroethyl 4-nitrobenzoate, pH = 8.6, 16°C	α-cyclodextrin, $k_\Psi/k_0 \sim +$	Komiyama and Inoue, 1980b
Cleavage of *p*-nitrophenyl 1-adamantane acetate, 16°C	β and γ-cydodextrins, $k_\Psi/k_0 \sim +$	Komiyama and Inoue, 1980c
Cleavage of: $CH_3C(=O)—XC_6H_4NO_2—p$ $p—NO_2C_6H_4C(=O)—X—C_2H_5$ X = S (*1*) X = S (*3*) X = 0 (*2*) X = 0 (*4*) ester + cyclodextrin $\overset{K_d}{\rightleftharpoons}$ complex; ester $\xrightarrow{k_{un}}$ products; complex $\xrightarrow{k_c}$ products	α, β-cyclodextrins, affect rate determining formation of tetrahedral intermediates for *1* and *2* and that for its decomposition for *3* and *4*; α-cyclodextrin, $k_c^\Psi/k_e \sim +$ for *1, 2, 3*; 1 for *4*; β-cyclodextrin, $k_c^\Psi/k_c^e \sim +$ for *1, 2, 3*; − for *4*	Komiyama and Bender, 1980
Ninhydrin reduction	β-cyclodextrin-dihydronicotinamide, $k_\Psi/k_0 \sim +$	Kojima et al., 1980

Methanol, DMSO, DMF, acetone, and acetonitrile (guests)

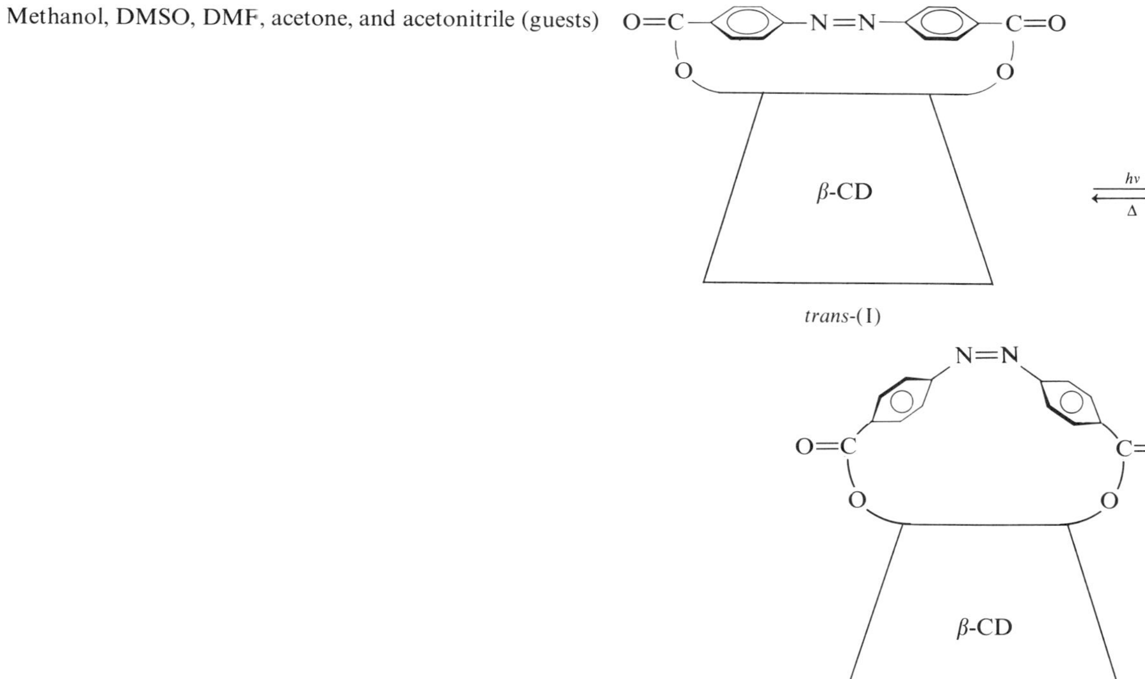

1:2 W host:guest in *cis*-I, nonstoichiometric behavior in *trans*-I

Ueno et al., 1980a

Table 12.18 (*Continued*)

Reaction	Effect of Cyclodextrin	Reference
Sodium α-naphthylacetate fluorescence, 25°C, pH = 8.7 tris buffer	α-cyclodextrin, fluorescence negligible; β-cyclodextrin, increased monomer fluorescence; γ-cyclodextrin, excimer fluorescence observed; two guests are accommodated in γ-cyclodextrin	Ueno et al., 1980b
Tosylation of cyclodextrins	α-, β-cyclodextrin, regiospecific tosylation at C-3 of one glucose unit in cyclodextrin	Onozuka et al., 1980
Decarboxylation of trichloroacetate acid	β-cyclodextrin, $k_{\Psi}/k_0 \sim +$	Motozato et al., 1980
$C_6H_5{-}CH_2SCH_3 + I_2 + H_2O \longrightarrow C_6H_5{-}CH_2S(=O)CH_3 + 2HI$ 25°C, pH = 7.2	β-cyclodextrin phosphate, $k_{\Psi}/k_0 \sim ++$	Eiki and Tagaki, 1980
Fluorescence quenching of pyrenes and naphthalenes by diethylamine and di-n-butylamine in H_2O	β-cyclodextrin, quenching rate increase is rationalized by heteroexcimer formation	Kano et al., 1980c

Acylation of β-cyclodextrin by:

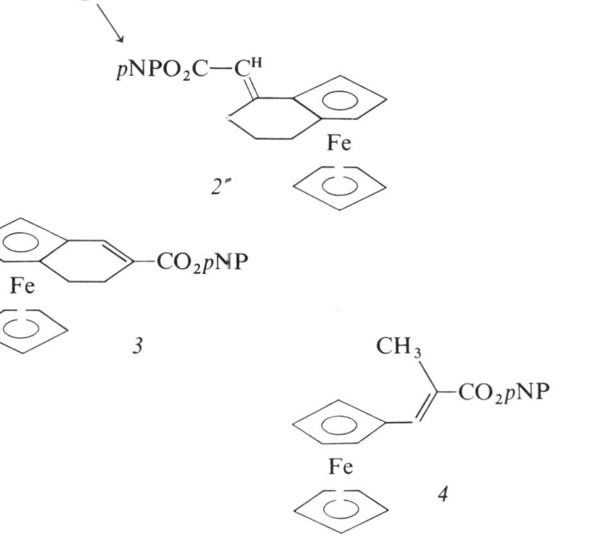

β-cyclodextrin, $k_\Psi/k_0 > 10^6$, twenty-fold enantioselectivity is observed in the rate of acylation (between *2′* and *2″*)

Trainor and Breslow, 1981

Table 12.18 (*Continued*)

Reaction	Effect of Cyclodextrin	Reference
Hydrolysis of *p*-nitrophenyl acetate, 26°C	Azobenzene capped β-cyclodextrin (*1*) hydrolysis is accelerated by light since *cis-1* with deeper cavity is formed: 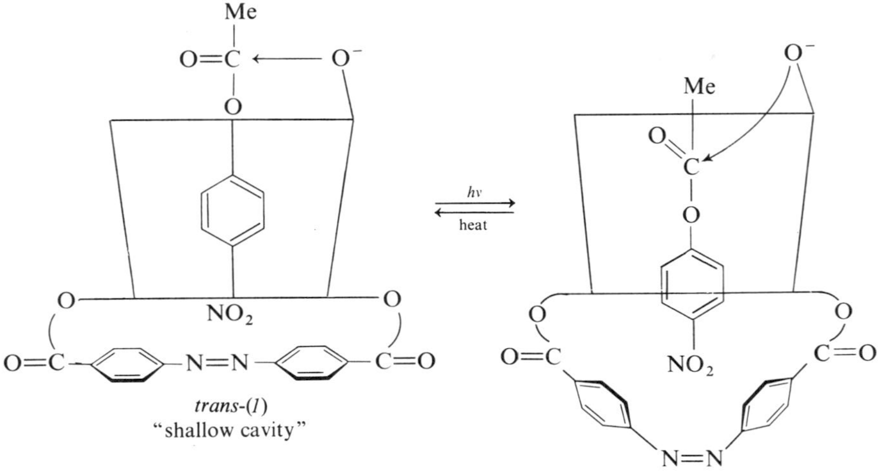 Me, O=C ← O⁻, O, NO_2, O=C, N=N, C=O, *trans-(1)* "shallow cavity" $\underset{heat}{\overset{h\nu}{\rightleftarrows}}$ *cis-(1)* "deep cavity"	Uneo et al., 1981a
Fluorescence of α-naphthyloxyacetic	γ-cyclodextrin, fluorescence intensity slightly enhanced; in presence of cyclohexanol fluorescence intensity markedly increases ∵ cyclohexanol acts as a space regulator narrowing the cavity of γ-cyclodextrin	Ueno et al., 1981b

Table 12.19. Energy and Electron Transfer and Charge Separation in Polyions

System	Process	Results	Experimental Technique	Reference
Polymeric sensitizers: 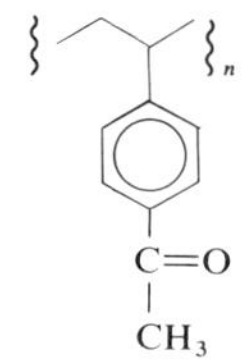 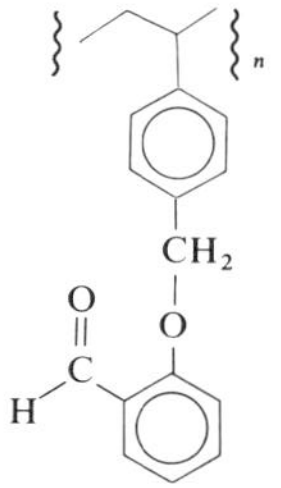 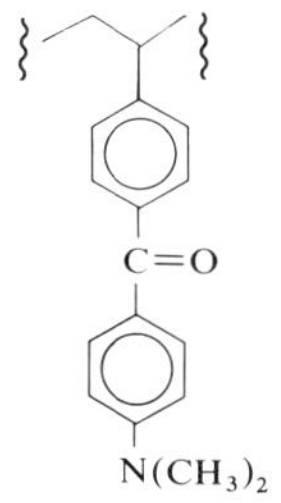	Energy storage reaction: 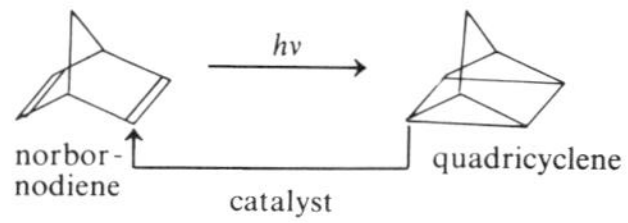 norbornadiene $\xrightarrow{h\nu}$ quadricyclene; quadricyclene $\xrightarrow{\text{catalyst}}$ norbornadiene	$\Phi_{\text{polymeric sensitizers}}$ = 0.3–0.6; $\Phi_{\text{acetophenone}}$ = 0.6	Steady state photolysis, gas chromatographic analysis	Hautala et al., 1977
$C=O$, CH_3, n				
CH_2, O, O, C, H, n				
$C=O$, $N(CH_3)_2$				

Table 12.19 *(Continued)*

System	Process	Results	Experimental Technique	Reference
Poly(vinyl sulfate)	Luminescence quenching of $Ru(bpy)_3^{2+*}$ by Cu^{2+} and Fe^{3+}	Dynamic quenching occurs in the reduced volume of the polymer field	Steady state and time resolved luminescence	Jonah et al., 1979
Poly(vinyl sulfate)	Luminescence quenching of $Ru(bpy)_3^{2+*}$ by O_2, nitrobenzene, $Ru(acac)_3$, Co(acac), Fe(III), ferric nitrilotriacetate	Quenching rate decrease, charge separation is controlled by the potential field of the polyelectrolyte	Luminescence quenching, flash photolysis	Meyerstein et al., 1978; Meisel et al., 1978b
Poly(1-naphthylmethyl methacrylate, poly[2-(1-naphthyl)ethyl methacrylate], copolymerized 9-anthrylmethylmethacrylate	Singlet energy transfer from naphthalene to antracene chromophores in polymers	Energy can be efficiently transferred between like chromophores on a polymer chain; polymer chains can be efficient antennas for collection and transport of photoexcitation energy	Fluorescence decay and lifetime measurements	Holden and Guillet, 1980

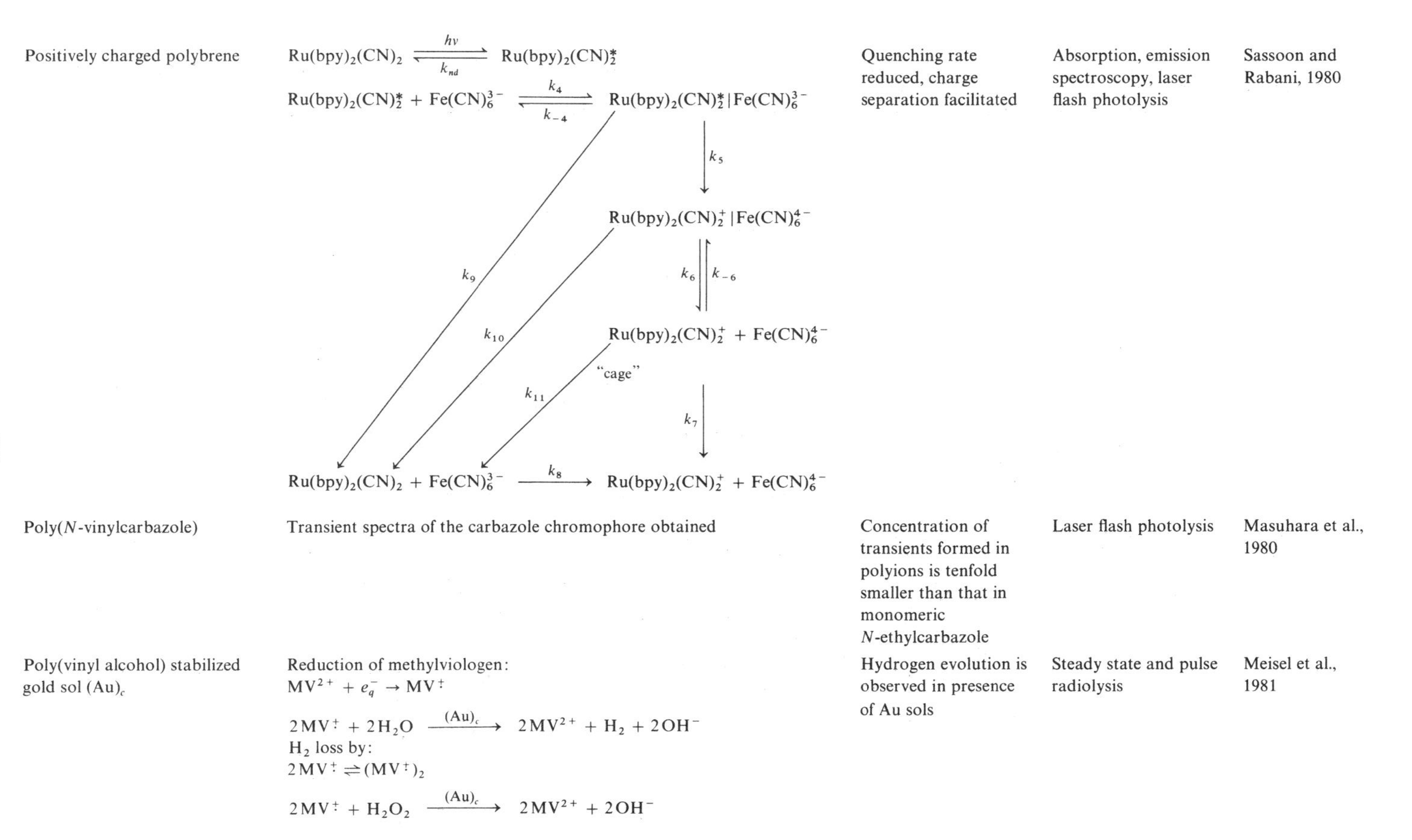

Positively charged polybrene	$Ru(bpy)_2(CN)_2 \underset{k_{nd}}{\overset{h\nu}{\rightleftharpoons}} Ru(bpy)_2(CN)_2^*$ $Ru(bpy)_2(CN)_2^* + Fe(CN)_6^{3-} \underset{k_{-4}}{\overset{k_4}{\rightleftharpoons}} Ru(bpy)_2(CN)_2^* \mid Fe(CN)_6^{3-}$ $\xrightarrow{k_5} Ru(bpy)_2(CN)_2^+ \mid Fe(CN)_6^{4-} \underset{k_{-6}}{\overset{k_6}{\rightleftharpoons}} Ru(bpy)_2(CN)_2^+ + Fe(CN)_6^{4-}$ ("cage") $\xrightarrow{k_7} Ru(bpy)_2(CN)_2^+ + Fe(CN)_6^{4-}$ k_9, k_{10}, k_{11}: $\rightarrow Ru(bpy)_2(CN)_2 + Fe(CN)_6^{3-}$ $Ru(bpy)_2(CN)_2 + Fe(CN)_6^{3-} \xrightarrow{k_8} Ru(bpy)_2(CN)_2^+ + Fe(CN)_6^{4-}$	Quenching rate reduced, charge separation facilitated	Absorption, emission spectroscopy, laser flash photolysis	Sassoon and Rabani, 1980
Poly(*N*-vinylcarbazole)	Transient spectra of the carbazole chromophore obtained	Concentration of transients formed in polyions is tenfold smaller than that in monomeric *N*-ethylcarbazole	Laser flash photolysis	Masuhara et al., 1980
Poly(vinyl alcohol) stabilized gold sol $(Au)_c$	Reduction of methylviologen: $MV^{2+} + e_q^- \rightarrow MV^{\dagger}$ $2MV^{\dagger} + 2H_2O \xrightarrow{(Au)_c} 2MV^{2+} + H_2 + 2OH^-$ H_2 loss by: $2MV^{\dagger} \rightleftharpoons (MV^{\dagger})_2$ $2MV^{\dagger} + H_2O_2 \xrightarrow{(Au)_c} 2MV^{2+} + 2OH^-$	Hydrogen evolution is observed in presence of Au sols	Steady state and pulse radiolysis	Meisel et al., 1981

Table 12.19 *(Continued)*

System	Process	Results	Experimental Technique	Reference
Ruthenium complexes (RuB_3^{2+}) and colloidal platinum stabilized by viologen polymers: poly-*N*-vinylbenzyl-*N'*-*n*-propyl-4,4'-bipyridinium bromide, PV(100) polychloromethylstyrene containing 55% propylviologen units	$RuB_3^{2+} + \underset{(py(100))}{V^{2+}—V^{2+}—V^{2+}} \overset{h\nu}{\rightleftharpoons} {}^*RuB_3^{2+} + V^{2+}—V^{2+}—V^{2+}$ ${}^*RuB_3^{2+} + V^{2+}—V^{2+}—V^{2+} \xrightarrow{k_q} (RuB_3^{3+} \cdots V^{+}—V^{2+}—V^{2+})$ $(RuB_3^{3+} \cdots V^{+}—V^{2+}—V^{2+}) \xrightarrow{k_b} (RuB_3^{2+} \cdots V^{2+}—V^{2+}—V^{2+}) \rightarrow RuB_3^{2+} + V^{2+}—V^{2+}—V^{2+}$ $(RuB_3^{3+} \cdots V^{+}—V^{2+}—V^{2+}) \rightleftharpoons RuB_3^{3+} + V^{+}—V^{2+}—V^{2+}$ $RuB_3^{3+} \xrightarrow{EDTA \rightarrow EDTA_{ox}} RuB_3^{2+}$ $(RuB_3^{3+} \cdots V^{+}—V^{2+}—V^{2+}) \overset{k_m}{\rightleftharpoons} (RuB_3^{3+} \cdots V^{2+}—V^{2+}—V^{2+})$ $(RuB_3^{3+} \cdots V^{2+}—V^{2+}—V^{2+}) \rightleftharpoons RuB_3^{3+} + V^{2+}—V^{2+}—V^{2+}$ $RuB_3^{3+} \xrightarrow{EDTA \rightarrow EDTA_{ox}} RuB_3^{2+}$	In presence of PVP stabilized Pt 10 μmole/hr H_2 evolved	Steady state photolysis, luminescence quenching	Nishijima et al., 1981

Table 12.20. Miscellaneous Reactivities in Polyelectrolytes

Reaction	Effect of Polyelectrolyte	Reference
Alkaline hydrolysis of *p*-nitrophenyl acetate, 4-acetoxy-3-nitrobenzoic acid, 4-acetoxy-3-nitrobenzearsonic acid, 30°C, pH = 8.0	Poly-4-vinyl-*N*-propylpyridinium bromide, poly-4-vinyl-*N*-*n*-butylpyridinium bromide, copolymers of 4-vinyl-*N*-benzylpyridinium chloride and 4-vinyl-*N*-*n*-cetylpyridinium bromide, diethyldiallylammonium chloride–sulphur dioxide copolymer, $k_\Psi/k_0 \sim +$	Kitano et al., 1976
Hydrolysis of *p*-nitrophenyl acetate and other esters, dehydrogenations, esterification, olefin hydration, epoxidation, oxidations, oligomerization, and photosensitization	Vinyl-, imidazole, hydroxamic acid, ethylene imine functionalized polymers, metal containing polymers, $k_\Psi/k_0 \sim +$	Manecke and Storck, 1978
End-to-end cyclization in polystyrene	Kinetics probed by pyrene excimer formation	Winnik et al., 1980
Reaction of flavins with 2-mercaptoethanol, glutathione, thiophenol, and 1,4-butanediol	Polymer bound flavins, $k_\Psi/k_0 \sim$ 30–6000	Shinkai et al., 1980f
6-nitrobenzisoxazole-3-carboxylate (COO^-, O_2N) $\xrightarrow{-CO_2}$ 2-cyano-4-nitrophenolate (CN, O^-, O_2N), 30°C	Water soluble poly(vinylpyridines) quaternized by octyl, dodecyl, octadecyl, and docosyl groups, $k_\Psi/k_0 \sim +$	Shinkai et al., 1981b
$Co(NH_3)_5^{2+} + OH^- \rightarrow$ Alkaline fading of crystal violet	Copolymers of styrene and acrylic acid, $k_\Psi/k_0 \sim -$	Ishiwatari et al., 1981
Hydrolysis of *p*-nitrophenyl acylates	Polyethylene imine, $k_\Psi/k_0 \sim +$; micellar and polyelectrolyte "catalyses" are compared	Klotz et al., 1981

REFERENCES

Adachi, G.-Y., Tomokiyo, K., Sorita, K., and Shiokawa, J. (1980). *J. Chem. Soc. Chem. Comm.*, 914–915. Luminescence of Divalent Europium Complexes with Crown Ethers and Polyethylene Glycols.

Alkaitis, S. A. and Grätzel, M. (1976). *J. Am. Chem. Soc.* **98**, 3549–3554. Laser Photoionization and Light-Initiated Redox Reactions of Tetramethylbenzidine in Organic Solvents and Aqueous Micellar Solution.

Alkaitis, S. A., Beck, G., and Grätzel, M. (1975a). *J. Am. Chem. Soc.* **97**, 5723–5729. Laser Photoionization of Phenothiazine in Alcoholic and Aqueous Micellar Solution. Electron Transfer from Triplet States to Metal Ion Acceptors.

Alkaitis, S. A., Grätzel, M., and Henglein, A. (1975b). *Ber. Bunsenges. Phys. Chem.* **79**, 541–546. Laser Photoionization of Phenothiazine in Micellar Solution. II. Mechanism and Light Induced Redox Reactions with Quinones.

Allen, R. J. and Bunton, C. A. (1976). *Bioinorg. Chem.* **5**, 241–252. Micellar Effects upon the Reaction of Cobalt Complexes with Alkyl Halides.

Almgren, M. and Rydholm, R. (1979). *J. Phys. Chem.* **83**, 360–364. Influence of Counterion Binding on Micellar Reaction Rates. Reaction Between *p*-Nitrophenyl Acetate and Hydroxide Ion in Aqueous Cetyltrimethylammonium Bromide.

Almgren, M. and Thomas, J. K. (1980). *Photochem. Photobiol.* **31**, 329–335. Interfacial Electron Transfer Involving Radical Ions of Carotene and Diphenylhexatriene in Micelles and Vesicles.

Almgren, M., Grieser, F., and Thomas, J. K. (1979a). *J. Am. Chem. Soc.* **101**, 2021–2026. Energy Transfer from Triplet Aromatic Hydrocarbons to Tb^{3+} and Eu^{3+} in Aqueous Micellar Solutions.

Almgren, M., Grieser, F., and Thomas, J. K. (1979b). *J. Phys. Chem.* **83**, 3232–3236. One-Electron Redox Potentials and Rate of Electron Transfer in Aqueous Micellar Solution. Partially Solubilized Quinones.

Almgren, M., Grieser, F., and Thomas, J. K. (1980). *J. Am. Chem. Soc.* **102**, 3188–3193. Photochemical and Photophysical Studies of Organized Assemblies. Interaction of Oils, Long-Chain Alcohols, and Surfactants Forming Microemulsions.

Anoardi, L. and Tomellato, V. (1977). *J. Chem. Soc. Chem. Commun.*, 401–402. Catalysis of Amide Hydrolysis due to Micelles Containing Imidazole and Hydroxy Functional Groups.

Anoardi, L., Buzzacarini, F., Fornasier, R., and Tonellato, V. (1978a). *Tetrahedron Lett.*, 3945–3948. A Powerful Nucleophilic Micellar Reagent. Synthesis and Properties of α-Oximino-Ketone-Functionalized Surfactants.

Anoardi, L., Fornasier, R., Sostero, D., and Tonellato, V. (1978b). *Gazz. Chim. Ital.* **108**, 707–708. Functional Micellar Catalysis. Synthesis and Properties of Thiocholine-Type Surfactants.

Arai, K. and Ogiwara, Y. (1978). *Bull. Chem. Soc. Jpn.* **51**, 182–184. The Hydrolysis of Carbohydrates in the Presence of a Reversed Micelle. II. The Hydrolysis of Sucrose in Benzene.

Armstrong, D. W., Seguin, R., McNeal, C. J., Macfarlane, R. D., and Fendler, J. H. (1978). *J. Am. Chem. Soc.* **100**, 4605–4606. Spontaneous Polypeptide Formation from Amino Acyl Adenylates in Surfactant Aggregates.

Ashcroft, R. G., Thulborn, K. R., Smith, J. R., Coster, H. G. L., and Sawyer, W. H. (1980). *Biochim. Biophys. Acta* **602**, 299–308. Perturbations to Lipid Bilayers by Spectroscopic Probes as Determined by Dielectric Measurements.

Aso, Y., Kano, K., and Matsuo, T. (1980). *Biochim. Biophys. Acta* **599**, 403–416. Energy Transfer in Artificial Membrane Systems. Singlet-Singlet Energy Transfer from Alloxazine to Isoalloxazine in Dipalmitoyl Phosphatidylcholine Liposomes and Dialkylammonium Chloride Vesicles.

Au, Y. N. and Hutchinson, D. W. (1980). *Biochem. J.* **191**, 657–659. The Photoinduced Isomerization of Bilirubin in Cationic Detergent Solutions.

Bagno, O., Soulignac, J. C., and Joussot-Dubien, J. (1979). *Photochem. Photobiol.* **29**, 1079–1081. pH Dependence of Sensitized Photooxidation in Micellar Anionic and Cationic Surfactants, Using Thiazine Dyes.

Bakale, G., Beck, G., and Thomas, J. K. (1980). *J. Phys. Chem.* **85**, 1062–1674. Electron Capture in Water Pools of Reversed Micelles.

Bakalik, D. P. and Thomas, J. K. (1977). *J. Phys. Chem.* **81**, 1905–1908. Micellar Catalysis of Radical Reactions. A Spin Trapping Study.

Barber, D. J. W., Morris, D. A. N., and Thomas, J. K. (1976). *Chem. Phys. Lett.* **37**, 481–484. Laser Induced Photoionization in Lipid Aggregates.

Beckmann, L. S. and Brown, D. G. (1976). *Biochim. Biophys. Acta* **428**, 720–725. The Interaction Between Vitamin B-12 and Micelles in Aqueous Solution.

Beddard, G. S., Carlin, S. E., and Porter, G. (1976). *Chem. Phys. Lett.* **43**, 27–32. Concentration Quenching of Chlorophyll Fluorescence in Bilayer Lipid Vesicles and Liposomes.

Behr, J. P., Lehn, J.-M., and Vierling, P. (1976). *J. Chem. Soc. Chem. Commun.*, 621–623. Stable Ammonium Cryptates of Chiral Macrocyclic Receptor Molecules Bearing Amino-acid Stable Chains.

Berndt, D. C. and Sendelbach, L. E. (1977). *J. Org. Chem.* **42**, 3305–3306. Micellar Catalyzed Reaction of Hydroxamic Acids.

Berndt, D. C., Utrapiromsuk, N., and Jaglan, S. S. (1979). *J. Org. Chem.* **44**, 136–138. Substituent Effects in Micellar Catalysis.

Berns, D. S. (1976). *Photochem. Photobiol.* **24**, 117–139. Photosensitive Bilayer Membranes as Model Systems for Photobiological Processes.

Bhalekar, A. A. and Engberts, J. B. F. N. (1978). *J. Am. Chem. Soc.* **100**, 5914–5920. Electron Transfer Reactions of Transition Metal Aminocarboxylates in the Presence of Micelle-Forming Surfactants. Catalysis by Cetyltrimethylammonium Bromide of the Reduction of $Mn(cydta)^-$ by $Co(edta)^{2-}$ and $Co(cydta)^{2-}$.

Blandamer, M. J. and Reid, D. J. (1975). *J. Chem. Soc. Faraday Trans. I* **71**, 2156–2161. Solvent Effect on the Catalysis by Cetyltrimethylammonium Bromide Micelles of the Reaction Between Hydroxide Ions and 2,4-Dinitrochlorobenzene.

Blandamer, M. J., Burgess, J., and Chambers, J. G. (1976). *Inorg. Chim. Acta* **17**, L37–L39. Effects of Micellar Agents on Peroxodisulfate Oxidations and on the Aquation of Cobalt(III) and Iron(II) Complexes.

Bonilha, J. B. S., Chaimovich, H., Toscano, V. G., and Quina, F. (1979). *J. Phys. Chem.* **83**, 2463–2470. Photophenomena in Surfactant Media. 2. Analysis of the Alkaline Photohydrolysis of 3,5-Dinitroanisole in Aqueous Micellar Solutions of *N*-Tetra-decyl-*N*,*N*,*N*-trimethylammonium Chloride.

Botré, C., Memoli, A., and Alhaique, F. (1978). *Environ. Sci. Technol.* **12**, 335–336. TCDD Solubilization and Photodecomposition in Aqueous Solutions.

Botré, C., Memoli, A., and Alhaique, F. (1979). *Environ. Sci. Technol.* **13**, 228–231. On the Degradation of 2,3,7,8-Tetrachlorodibenzoparadioxin (TCDD) by Means of a New Class of Chloroiodides.

Breslow, R., Czarniecki, M. F., Emert, J., and Hamaguchi, H. (1980). *J. Am. Chem. Soc.* **102**, 762–770. Improved Acylation Rates Within Cyclodextrin Complexes from Flexible Capping of the Cyclodextrin and from Adjustment of the Substrate Geometry.

Brown, J. M., and Darwent, J. R. (1979a). *J. Chem. Soc. Chem. Commun.*, 169–170. Hydrophobic Effects in the Micellar Reactions of Peroxide Nucleophiles.

Brown, J. M. and Darwent, J. R. (1979b). *J. Chem. Soc. Chem. Commun.*, 171–172. Proximity Effects in the Reactions of Surfactant *p*-Nitrophenyl Esters with Peroxide Nucleophiles.

Brown, J. M. and Lynn, J. L., Jr. (1980). *Ber. Bunsenges. Phys. Chem.* **84**, 95–100. Structural and Catalytic Aspects of Functional Micelles. Ester Hydrolysis by Hydroxamic Acids Bound to Cationic Surfactants.

Brown, N. M. D., Cowley, D. J., and Murphy, W. J. (1976). *J. Chem. Soc. Perkin Trans. II*, 1769–1773. The Photooxidation of Aliphatic Alcohols as Monolayers on Aqueous Solutions of 1,1′-Dimethyl-4,4′-bipyridylium (Paraquat) Dichloride.

Brown, J. M., Chaloner, P. A., and Colens, A. (1979). *J. Chem. Soc. Perkin Trans. II*, 71–76. Acyl Transfer Reactions in Functional Micelles Studied by Proton Magnetic Resonance at 270 MHz.

Brown, J. M., Bunton, C. A., Diaz, S., and Ihara, Y. (1980). *J. Org. Chem.* **45**, 4169–4174. Dephosphorylation in Functional Micelles. The Role of the Imidazole Group.

Broxton, T. J., Deady, L. W., and Duddy, N. W. (1978). *Aust. J. Chem.* **31**, 1525–1532. Micellar Catalysis of Aniline Hydrolysis.

Brugger, P.-A. and Grätzel, M. (1980). *J. Am. Chem. Soc.* **102**, 2461–2463. Light-Induced Charge Separation by Functional Micellar Assemblies.

Brugger, P.-A., Infelta, P. P., Braun, A. M., and Grätzel, M. (1981). *J. Am. Chem. Soc.* **103**, 320–326. Photoredox Reactions in Functional Micellar Assemblies. Use of Amphiphatic Redox Relays to Achieve Light Energy Conversion and Charge Separation.

Bunton, C. A. and Diaz, S. (1976). *J. Org. Chem.* **41**, 33–39. Kinetic Solvent Deuterium Isotope Effects on the Micellar-Catalyzed Hydrolysis of Trisubstituted Phosphate Esters.

Bunton, C. A. and Ihara, Y. (1977). *J. Org. Chem.* **42**, 2865–2869. Micellar Effects on Dephosphorylation and Deacylation by Oximate Ions.

Bunton, C. A. and McAneny, M. (1976). *J. Org. Chem.* **41**, 36–39. Micellar Effects on the Hydrolysis of *p*-Nitrobenzoyl Choline and the Related *N*-Hexadecyl Ester.

Bunton, C. A. and McAneny, M. (1977). *J. Org. Chem.* **42**, 475–482. Catalysis of Reactions of *p*-Nitrobenzoyl Phosphate by Functional and Nonfunctional Micelles.

Bunton, C. A. and Paik, C. H. (1976). *J. Org. Chem.* **41**, 40–44. Reactions of Carbocations with a Nucleophilic Surfactant and Related Alkoxide Ions.

Bunton, C. A. and Rubin, R. J. (1976). *J. Am. Chem. Soc.* **98**, 4236–4246. Benzidine Rearrangement in the Presence and Absence of Micelles. Evidence for Rate-Limiting Proton Transfer.

Bunton, C. A. and Sepulveda, L. (1979). *Isr. J. Chem.* **18**, 298–303. Micellar Catalyzed Dephosphorylation. The Role of Hydrophobicity of the Nucleophile.

Bunton, C. A. and Wright, J. L. (1975). *Tetrahedron* **31**, 3013–3017. Micellar Effects upon the Reactions of Amino Acids and Their Derivatives with 2,6-Dinitro-4-trifluoromethylbenzene Sulfonate Ion.

Bunton, C. A., Diaz, S., Romsted, L. S., and Valenzuela, O. (1976a). *J. Org. Chem.* **41**, 3037–3040. The Effect of Substrate Micellization on the Hydrolysis of *n*-Decyl Phosphate.

Bunton, C. A., Ihara, Y., and Wright, J. L. (1976b). *J. Org. Chem.* **41**, 2520–2526. Reactions of Activated Arene Sulfonates with Oxygen and Nitrogen Nucleophiles. Hydroxide Ion and Micellar Catalysis.

Bunton, C. A., Diaz, S., van Fleteren, G. M., and Paik, C. (1978a). *J. Org. Chem.* **43**, 258–261. Effect of Monoalkyl Phosphates upon Micellar-Catalyzed Dephosphorylation and Deacylation.

Bunton, C. A., Savelli, G., and Sepulveda, L. (1978b). *J. Org. Chem.* **43**, 1925–1929. Role of Glucose and Related Compounds in Micellar and Nonmicellar Nucleophilic Reactions.

Bunton, C. A., Carrasco, N., Huang, S. K., Paik, C. H., and Romsted, L. S. (1978c). *J. Am. Chem. Soc.* **100**, 5420–5425. Reagent Distribution and Micellar Catalysis of Carbocation Reactions.

Bunton, C. A., Rivera, F., and Sepulveda, L. (1978d). *J. Org. Chem.* **43**, 1166–1173. Micellar Effects upon Hydrogen Ion and General Acid Catalyzed Hydration of 1,4-Dihydropyridines.

Bunton, C. A., Romsted, L. S., and Savelli, G. (1979a). *J. Am. Chem. Soc.* **101**, 1253–1259. Tests of the Pseudophase Model of Micellar Catalysis: Its Partial Failure.

Bunton, C. A., Cerichelli, G., Ihara, Y., and Sepulveda, L. (1979b). *J. Am. Chem. Soc.* **101**, 2429–2435. Micellar Catalysis and Reactant Incorporation in Dephosphorylation and Nucleophilic Substitution.

Bunton, C. A., Romsted, L. S., and Thamavit, C. (1980). *J. Am. Chem. Soc.* **102**, 3900–3903. The Pseudophase Model of Micellar Catalysis. Addition of Cyanide Ion to *N*-Alkylpyridinium Ions.

Cairns-Smith, A. G. and Rasool, S. (1978). *J. Chem. Soc. Perkin Trans. II*, 1007–1010. Reactions of *p*-Nitrophenyl Dodecanoate in Normal and Modified Cholic Acid Micelles.

Chaimovich, H., Blanco, A., Chayet, L., Costa, L. M., Monteiro, P. M., Bunton, C. A., and Paik, C. (1975). *Tetrahedron* **31**, 1139–1143. Micellar Catalysis of the Reaction of 2,4-Dinitrofluorobenzene with Phenoxide and Thiophenoxide Ions.

Cho, M. J. and Allen, M. A. (1978). *Int. J. Pharm.* **1**, 281–297. Quantitative Assessment of the Negative Catalytic Effects of a Cationic Surfactant Myristyl-γ-picolinium Chloride of the Specific-Acid Catalyzed Epimerization of 15(S)-15-Methyl Prostaglandin $F_{2\alpha}$.

Croce, M. and Okamoto, Y. (1979). *J. Org. Chem.* **44**, 2100–2103. Cationic Micellar Catalysis of the Aqueous Alkaline Hydrolyses of 1,3,5-Triaza-1,3,5-trinitrocyclohexane and 1,3,5,7-Tetraaza-1,3,5,7-tetranitrocyclooctane.

Cuccovia, I. M., Schröter, E. H., Monteiro, P. M., and Chaimovich, H. (1978). *J. Org. Chem.* **43**, 2248–2252. Effect of Hexadecyltrimethylammonium Bromide on the Thiolysis of *p*-Nitrophenyl Acetate.

Cuccovia, I. M., Quina, F. H., and Chaimovich, H. (1981). To be published. A Remarkable Enhancement of the Rate of Ester Hydrolysis by Synthetic Amphiphile Vesicles.

De Albrizzio, J. P. and Cordes, E. H. (1979). *J. Colloid Interface Sci.* **68**, 292–294. Water Activity at the Surface of Ionic Micelles as Measured by the Extent of Hydration of 3-Formul-*N*-tetradecylpyridinium Bromide.

DiBiase, S. A., Wolak, R. P., Jr., Dishong, D. M., and Gokel, G. W. (1980). *J. Org. Chem.* **45**, 3630–3634. Crown Cation Complex Effects. 10. Potassium *t*-Butoxide Mediated Penultimate Oxidative Hydrolysis of Nitriles.

Dieckmann, S., and Frahm, J. (1979). *J. Chem. Soc. Faraday Trans. I* **75**, 2199–2210. Kinetic Investigation and Numerical Analysis of a Micelle-Catalyzed Metal Complex Formation.

Duchek, J. R. and Huebner, J. S. (1979). *Biophys. J.* **27**, 317–321. Voltage Transients from Photoisomerizing Azo Dye in Bilayer Membranes.

Eiki, T. and Tagaki, W. (1980). *Chem. Lett.*, 1063–1066. β-Cyclodextrin Phosphate. A Remarkable Catalyst for the Iodine Oxidation of Benzyl Methyl Sulfide to the Sulfoxide in Water.

Eiki, T., Tomuro, K., Aoshima, S., Suda, M., and Tagaki, W. (1980). *Nippon Kagaku Kaishi*, 461. Micellar Effects of Anionic Surfactant on the Hydrolysis and Aminolysis of *p*-Alkylphenyl Phosphatosulfates.

El Seoud, O. A. (1976). *J. Chem. Soc. Perkin Trans. II*, 1497–1501. Hydration of Acetaldehyde Catalyzed by Micellar Triton X-100 in Carbon Tetrachloride.

El Seoud, O. A. and da Silva, M. J. (1980). *J. Chem. Soc. Perkin Trans. II*, 127–331. Kinetics of the Reversible Hydration of 1,3-Dichloroacetone Catalysed by Aerosol-OT-Solub izediAcids and Bases in Carbon Tetrachloride.

El Seoud, O. A., Martins, A., Barbur, L. P., da Silva, M. J., and Aldrigue, V. (1977). *J. Chem. Soc. Perkin Trans. II*, 1674–1678. Kinetics of the Reaction of *p*-Nitrophenyl Acetate with Amines in the Presence of Dodecylammonium Propionate and Aerosol-OT Aggregates in Benzene.

El Seoud, O. A., da Silva, M. J., and Barbur, L. P. (1978). *J. Chem. Soc. Perkin Trans. II*, 331–335. Kinetics of the Reversible Hydration of 1,3-Dichloroacetone Catalysed by Micellar Aersol-OT in Carbon Tetrachloride.

El Seoud, O. A., Pivêtta, F., El Seoud, M. I., Farah, J. P. S., and Martins, A. (1979). *J. Org. Chem.* **44**, 4832–4836. Kinetics of the Aminolysis and Hydrolysis of *p*-Nitrophenyl Carboxylates in the Presence of Dodecylammonium Propionate and Aerosol-OT Aggregates in Benzene.

Epstein, J., Cannon, P., Jr., Swidler, R., and Baraze, A. (1977). *J. Org. Chem.* **42**, 759–762. Amplification of Cyanide Ion Production by the Micellar Reaction of Keto Oximes with Phosphono- and Phosphorofluoridates.

Epstein, J., Kaminski, J. J., Bodor, N., Enever, R., Sowa, J., and Higuchi, T. (1978). *J. Org. Chem.* **43**, 2816–2921. Micellar Acceleration of Organophosphate Hydrolysis by Hydroximinomethylpyridinium Type Surfactants.

Escabi-Perez, J. R., Nome, F., and Fendler, J. H. (1977). *J. Am. Chem. Soc.* **99**, 7749–7754. Energy Transfer in Micellar Systems. Steady State and Time Resolved Luminescence of Aqueous Micelle Solubilized Naphthalene and Terbium Chloride.

Escabi-Perez, J. R., Romero, A., Lukac, S., and Fendler, J. H. (1979). *J. Am. Chem. Soc.* **101**, 2231–2233. Aspects of Artificial Photosynthesis. Photoionization and Electron Transfer in Dihexadecylphosphate Vesicles.

Fatah, A. A. and Loew, L. M. (1981). *J. Am. Chem. Soc.*, to be published. Inhibition of Intramolecular Electrostatic Catalysis by a Lipid Vesicle Membrane.

Feldberg, S. W., Armen, G. H., Bell, J. A., Chang, C. K., and Wang, C.-B. (1981). *Biophys. J.* **34**, 149–163. Electron Transport Across Glycerol Monooleate Bilayer Lipid Membranes Facilitated by Magnesium Etiochlorin.

Fendler, J. H. and Fendler, E. J. (1975). *Catalysis in Micellar and Macromolecular Systems*, Academic Press, New York.

Fendler, J. H. and Hinze, W. L. (1981). *J. Am. Chem. Soc.*, **103**, 5439–5447. Reactivity Control in Micelles and Surfactant Vesicles. Kinetics and Mechanism of Base Catalyzed Hydrolysis of 5,5′-Dithiobis(2-nitrobenzoic acid) in Water, in Hexadecyltrimethylammonium Bromide Micelles and in Dioctadecyldimethylammonium Chloride Surfactant Vesicles.

Finiels, A. and Geneste, P. (1979). *J. Org. Chem.* **44**, 2036–2038. Micellar Catalysis in the Oximation Reaction of Aliphatic and Cyclic Ketones. Hydrophobic Interactions.

Fischer, M., Knoche, W., Robinson, B. H., MacLagan, J. H., and Wedderburn, J. H. M. (1979). *J. Chem. Soc. Faraday Trans. I* **75**, 119–131. Metal-Ion Complexation in the Presence of Surfactants. Part I. Mechanism of pH-Dependent Reaction between Nickel(II) and Murexide in Aqueous Solution and Application in the Reaction to Study of Micellar Phenomena.

Flamini, V., Linda, P., and Savelli, G. (1975). *J. Chem. Soc. Perkin Trans. II*, 421–423. Micellar Effects on Heteroaromatic Compounds. Part I. Nucleophilic Substitution of 2-Chloroquinoxaline with Hydroxide Ion.

Ford, W. E., Otvos, J. W., and Calvin, M. (1978). *Nature* **274**, 507–508. Photosensitized Electron Transport Across Phospholipid Vesicle Walls.

Ford, W. E., Otvos, J. W., and Calvin, M. (1979). *Proc. Natl. Acad. Sci. USA* **76**, 3590–3593. Photosensitized Electron Transport Across Vesicle Walls: Quantum Yield Dependence on Sensitizer Concentration.

Fornasier, R. and Tonellato, U. (1980). *J. Chem. Soc. Faraday Trans. I* **76**, 1301–1310. Functional Micellar Catalysis. Part 3. Quantitative Analysis of the Catalytic Effects due to Functional Micelles and Comicelles.

Foussier, J. P., Ticke, B., and Wegner, G. (1979). *Isr. J. Chem.* **18**, 227–232. The Photochemistry of the Polymerization of Diacetylenes in Multilayers.

Fragata, M. and Bellemare, F. (1980). *Chem. Phys. Lipids* **27**, 93–99. Model of Singlet Oxygen Scavenging by α-Tocopherol in Biomembranes.

Frank, A. J., Grätzel, M., Henglein, A., and Janata, E. (1976). *Int. J. Chem. Kinet.* **8**, 817–824. Kinetics of the Heterogeneous Electron Transfer Reaction of Triplet Pyrene in Micelles to Br_2^- Radicals in Aqueous Solution.

Fromherz, P. and Arden, W. (1980). *J. Am. Chem. Soc.* **102**, 6211–6218. pH-Modulated Pigment Antenna in Lipid Bilayer on Photosensitized Semiconductor Electrode.

Fujii, H., Kawai, T., and Nishikawa, H. (1979a). *Bull. Chem. Soc. Jpn.* **52**, 1978–1983. Hydrolytic Reactions of Para-Nitrophenyl Esters in Reversed Micellar Systems.

Fujii, H., Kawai, T., and Nishikawa, H. (1979b). *Bull. Chem. Soc. Jpn.* **52**, 2051–2055. Determination of pH in Reversed Micelles.

Funasaki, N. (1977). *J. Colloid Interface Sci.* **62**, 336–343. The Kinetics and Equilibrium of Alkaline Fading of Triphenylmethane Dyes at the Micellar Surface of *N*-Dodecyl-*β*-alanine.

Funasaki, N. (1978). *J. Colloid Interface Sci.* **64**, 461–469. The Effect of a Cationic Surfactant on Basic Hydrolysis of *p*-Nitrophenyl Esters.

Gaines, G. L., Jr., Behnken, P. E., and Valenty, S. J. (1978). *J. Am. Chem. Soc.* **100**, 6549–6559. Monolayer Films of Surfactant Ester Derivatives of Tris(2,2′-bipyridine)ruthenium$(II)^{2+}$.

Gani, V. (1977). *Tetrahedron Lett.*, 2277–2280. Inhibition of Alkaline Hydrolysis of *p*-Nitrophenyl Acetate by Hydroxylated Micelles Derived from a Pyridinium Salt.

Gorman, A. A. and Rodgers, M. A. J. (1978). *Chem. Phys. Lett.* **55**, 52–54. Lifetime and Reactivity of Singlet Oxygen in an Aqueous Micellar System: A Pulsed Nitrogen Laser Study.

Gorman, A. A., Lovering, G., and Rodgers, M. A. J. (1976). *Photochem. Photobiol.* **23**, 399–403. The Photosensitized Formation and Reaction of Singlet Oxygen, $O_2(\Delta)$, in Aqueous Micellar Systems.

Gregoritch, S. J. and Thomas, J. K. (1980). *J. Phys. Chem.* **84**, 1491–1495. Photochemistry in Microemulsions: Photophysical Studies in Oleate/Hexanol/Hexadecane, Oil in Water Microemulsion.

Gresset, C., Daveloose, D., Leterrier, F., Bazin, M., and Santus, R. (1979). *Chem. Phys. Lett.* **64**, 440–441. 265 nm Laser Flash Spectroscopy of Indoles in the Presence of Nitroxide Radicals. A Model for Probing Protein-Membrane Interactions.

Grieser, F. (1981). *J. Phys. Chem.* **85**, 928–932. Nitrite Quenching of Terbium Luminescence in Sodium Dodecyl Sulfate Solutions.

Hall, G. E. (1978). *J. Am. Chem. Soc.* **100**, 8260–8264. Comment on the Communication "Photoionization by Green Light in Micellar Solution."

Hansch, C. (1978). *J. Org. Chem.* **43**, 4889–4850. Hydrophobic Effects in Host-Guest Interactions. Hydrolysis of Nitrophenyl Carboxylates.

Harbour, J. R. and Hair, M. L. (1980). *J. Phys. Chem.* **84**, 1500–1503. Detection of Radicals in Microemulsions Using Spin Trapping Techniques.

Hautala, R. R., Little, J., and Sweet, E. (1977). *Sol. Energy* **19**, 503–508. The Use of Functionalized Polymers as Photosensitizers in an Energy Storage Reaction.

Hayashi, H., Sakaguchi, Y., and Nagakura, S. (1980). *Chem. Lett.*, 1149–1152. Laser-Photolysis Study of the External Magnetic Field Effect upon the Photodecomposition Reaction of 1,3-Diphenyl-2-propanone in a Micelle.

Henglein, A. and Proske, T. (1978). *J. Am. Chem. Soc.* **100**, 3706–3709. Formation and Disappearance of Free Radicals, and the Micellar Equilibrium in the Detergent Sodium 4-(6′-Dodecyl)benzenesulfonate.

Hiramatsu, K. (1977). *Biochim. Biophys. Acta* **490**, 209–215. Cleavage of the S–S Bond in 5,5′-Dithiobis-(2-nitrobenzoic Acid) in the Presence of a Cationic Detergent. An Approach to the Cleavage of the S–S Bond in Bovine Plasma Albumin.

Holden, D. A. and Guillet, J. E. (1980). *Macromolecules* **13**, 289–295. Singlet Electronic Energy Transfer in Polymers Containing Naphthalene and Anthracene Chromophores.

Holzwarth, J., Knoche, W., and Robinson, B. H. (1978). *Ber. Bunsenges. Phys. Chem.* **82**, 1001–1005. "Catalysis" of Metal Complex Formation on Micelle Surfaces. The Reaction Between Divalent Metal Ions and PaDA in the Presence of Sodium Dodecyl Sulphate.

Horsey, B. E. and Whitten, D. G. (1978). *J. Am. Chem. Soc.* **100**, 1293–1295. Environmental Effects on Photochemical Reactions: Contrast in the Photooxidation Behavior of Protoporphyrin IX in Solution, Monolayer Films, Organized Monolayer Assemblies and Micelles.

Huebner, J. S. (1979). *Photochem. Photobiol.* **30**, 233–241. Apparatus for Recording Light Flash Induced Membrane Voltage Transients with Ions Resolution.

Humphry-Baker, R., Grätzel, M., Tundo, P., and Pelizetti, E. (1979). *Angew. Chem. Int. Ed. Eng.* **18**, 630–631. Complexes of Nitrogen-Containing Crown Ether Surfactants with Stable Silver Atoms.

Hurley, J. K., Castelli, F., and Tollin, G. (1980). *Photochem. Photobiol.* **32**, 79–86. Chlorophyll Photochemistry in Condensed Media. II. Triplet State Quenching and Electron Transfer to Quinone in Liposomes.

Hurley, J. K., Castelli, F., and Tollin, G. (1981). *Photochem. Photobiol.*, to be published. Chlorophyll-Quinone Photochemistry in Liposomes: Mechanisms of Radical Formation and Decay.

Ihara, Y. (1978). *J. Chem. Soc. Chem. Commun.*, 984–985. Stereoselective Reaction of *p*-Nitrophenyl *N*-Acyl-phenylalanines with *N*-Acyl-L-histidine in Mixed Micelles.

Ihara, Y. (1980). *J. Chem. Soc. Perkin Trans. II*, 1483–1487. Stereoselective Micellar Catalysis. Reactions of Amino-Acid Ester Derivatives with *N*-Acyl-L-histidine in Micelles.

Ihara, Y., Nango, M., and Kuroki, N. (1980). *J. Org. Chem.* **45**, 5009–5011. Stereoselective Micellar Bifunctional Catalysis.

Infelta, P. P., Grätzel, M., and Fendler, J. H. (1980). *J. Am. Chem. Soc.* **102**, 1479–1483. Aspects of Artificial Photosynthesis. Photosensitized Electron Transfer and Charge Separation in Cationic Surfactant Vesicles.

Iriyama, K. (1980). *J. Membr. Biol.* **52**, 115–120. Preparation of Multilayers Containing Chlorophyll-a and/or Phosphatidylcholine and Chemical Stability of Chlorophyll-a Molecules in the Multilayers.

Ishiwatari, T., Maruno, T., Okubo, M., Okubo, T., and Ise, N. (1981). *J. Phys. Chem.* **85**, 47–50. "Catalytic" Effects of Electrically Charged Polymer Lattices on Interionic Reactions.

Jagt, J. C., and Engberts, J. B. F. N. (1977). *J. Am. Chem. Soc.* **99**, 916–921. Micellar Catalysis of Proton Transfer Reactions. 1. Hydrolysis of Covalent Arylsulfonylmethyl Perchlorates in the Presence of CTABr. Catalysis by Sulfinate, Formate, and Hydroxide Ions, and the Effect of Mechanical Agitation.

James, A. D., and Robinson, B. H. (1978). *J. Chem. Soc. Faraday Trans. I* **74**, 10–12. Micellar Catalysis of Metal-Complex Formation.

Janzen, A. F. and Bolton, J. R. (1979). *J. Am. Chem. Soc.* **101**, 6342–6348. Photochemical Electron Transfer in Monolayer Assemblies. 2. Photoelectric Behavior in Chlorophyll-a Acceptor Systems.

Janzen, A. F., Bolton, J. R., and Stillman, M. J. (1979). *J. Am. Chem. Soc.* **101**, 6337–6341. Photochemical Electron Transfer in Monolayer Assemblies. 1. Spectroscopic Study of Radicals Produced in Chlorophyll-a/Acceptor Systems.

Jonah, C. D., Matheson, M. S., and Meisel, D. (1979). *J. Phys. Chem.* **83**, 257–261. Dynamic Quenching of $Ru(bpy)_3^{2+*}$ Emission in the Potential Field of a Polyelectrolyte.

Jones, C. E. and MacKay, R. A. (1978). *J. Phys. Chem.* **82**, 63–65. Reactions in Microemulsions. 3. Photodegradation of Chlorophyll.

Jones, C. E., Jones, C. A., and Mackay, R. A. (1979). *J. Phys. Chem.* **83**, 805–810. Reaction in Microemulsions. 4. Kinetics of Chlorophyll Sensitized Photoreduction of Methyl Red and Crystal Violet by Ascorbate.

Kalyanasundaram, K. (1978). *J. Chem. Soc. Chem. Commun.*, 628–630. Photoredox Reactions in Micellar Solutions Sensitized by Surfactant Derivative of Tris(2,2′-bipyridyl)ruthenium(II).

Kano, K., Kawazumi, H., Ogawa, T., and Sunamoto, J. (1980a). *Chem. Phys. Lett.* **74**, 511–514. Fluorescence Quenching of Pyrene and Pyrenedecanoic Acid by Various Kinds of *N,N*-Dialkylanilines in Dipalmitoylphosphatidylcholine Liposomes.

Kano, K., Tanaka, Y., Ogawa, T., and Shimomura, M., Okahata, Y., and Kunitake, T. (1980b). *Chem. Lett.*, 421–424. Photoresponsive Membranes. Regulation of Membrane Properties by Photoreversible cis→trans Isomerization of Azobenzenes.

Kano, K., Takenoshita, I., and Ogawa, T. (1980c). *Chem. Lett.*, 1035–1038. Three Component Complexes of Cycloheptaamylose. Fluorescence Quenching of Pyrenes and Naphthalenes in Aqueous Media.

Keiser, B., Holt, S. L., and Barden, R. E. (1980). *J. Colloid Interface Sci.* **73**, 290–293. Copper Incorporation into Tetraphenylphorphine in a Water-in-Oil Microemulsion. A Mechanism for Facilitating the Transport of Copper(II) Across the Interphase.

Kim, O.-K. (1977). *J. Polym. Sci. Lett. Ed.* **15**, 287–294. Effect of Micellar Complexation of Bisulfite on the Polymerization of Methyl Acrylate.

Kim, O.-K. and Griffith, J. R. (1976). *J. Colloid Interface Sci.* **55**, 191–196. Micellar Interaction and Its Catalytic Role in the Polymerization of Acrylamide Catalyzed by Bisulfite.

Kitahara, A. and Kon-no, K. (1977). In *Micellization, Solubilization and Microemulsions* (K. L. Mittal, Ed.), Plenum Press, New York, pp. 675–693. Solubilization and Catalysis of Polar Substances in Nonaqueous Surfactant Solutions.

Kitano, H., Tanaka, M., and Okubo, T. (1976). *J. Chem. Soc. Perkin Trans. II*, 1074–1077. Polyelectrolyte Catalysis of the Alkaline Hydrolysis of Neutral and Anionic Esters.

Kiwi, J. and Grätzel, M. (1978). *J. Am. Chem. Soc.* **100**, 6314–6320. Dynamics of Light-Induced Redox Processes in Microemulsion Systems.

Kiwi, J. and Grätzel, M. (1980). *J. Phys. Chem.* **84**, 1503–1507. Chlorophyll-a Sensitized Redox Processes in Microemulsion Systems.

Klotz, I. M., Drake, E. N., and Sisido, M. (1981). *Bioorg. Chem.* **10**, 63–74. Comparison of Biomimetic Catalytic Properties of Modified Polyethylenimines with Those of Micelles.

Kojima, M., Toda, F., and Hattori, K. (1980). *Tetrahedron Lett.* **21**, 2271–2724. The Cyclodextrin-Nicotinamide Compound as a Dehydrogenase Model Simulating Apoenzyme-Coenzyme-Substrate Ternary Complex System.

Komiyama, M. and Bender, M. L. (1979). *Bioorg. Chem.* **8**, 249–254. Nucleophilic Acceleration of the Cleavage of β-Cyclodextrin *trans*-Cinnamate by Amines.

Komiyama, M. and Bender, M. L. (1980). *Bull. Chem. Soc. Jpn.* **53**, 1073–1076. The Cyclodextrin-Accelerated Cleavage of Thiocarboxylic S-Esters.

Komiyama, M. and Hirai, H. (1980a). *Chem. Lett.*, 1251–1254. General Base Catalyses by α-Cyclodextrin in the Hydrolyses of Alkyl Benzoates.

Komiyama, M. and Hirai, H. (1980b). *Chem. Lett.*, 1467–1470. Time-Averaged Conformations of the Inclusion Complexes of β-Cyclodextrin with *t*-Butylphenols.

Komiyama, M. and Hirai, H. (1980c). *Chem. Lett.*, 1471–1474. Relationship Between the Cyclodextrin Catalyses in the Cleavages of Phenyl Acetates and the Time-Averaged Conformations of the Inclusion Complexes.

Komiyama, M. and Inoue, S. (1980a). *Bull. Chem. Soc. Jpn.* **53**, 2330–2333. Retardation of the Cleavages of Nitrophenyl 1-Adamantenecarboxylates by Cyclodextrins.

Komiyama, M. and Inoue, S. (1980b). *Bull. Chem. Soc. Jpn.* **53**, 3334–3337. α-Cyclodextrin-Catalyzed Hydrolysis of 2,2,2-Trifluoroethyl 4-Nitrobenzoate.

Komiyama, M. and Inoue, S. (1980c). *Bull. Chem. Soc. Jpn.* **53**, 3266–3269. Catalysis of the Hydrolysis of *p*-Nitrophenyl 1-Adamantaneacetate by Cyclodextrins.

Kondo, H., Fujiki, K., and Sunamoto, J. (1978). *J. Org. Chem.* **43**, 3584–3588. Reversed Micellar Catalysis. Catalysis of Dodecylammonium Propionate Reversed Micelles in the Hydrolysis of Alkyl *p*-Nitrophenyl Carbonates.

Kondo, H., Yoshinaga, H., and Sunamoto, J. (1980). *Chem. Lett.*, 973–976. Facile Schiff Base Formation Between Pyridoxal and Amino Acid in Reversed Micelles.

Kon-no, K. and Kitahara, A. (1978). *J. Colloid Interface Sci.* **67**, 477–482. Effect of Ionic Surfactants on Aminolysis of *p*-Nitrophenyl Acetate by Imidazoles in Carbon Tetrachloride.

Kon-no, K., Kitahara, A., and Fujiwara, M. (1978). *Bull. Chem. Soc. Jpn.* **51**, 3165–3169. Carrier Effect of Reversed Micelle on Imidazole Catalyst.

Kraljić, I., Barboy, N., and Leicknam, J.-P. (1979). *Photochem. Photobiol.* **30**, 631–633. Photosensitized Formation of Singlet Oxygen by Chlorophyll a in Neutral Aqueous Micellar Solutions with Triton X-100.

Kuhn, H. (1979). *J. Photochem.* **10**, 111–132. Synthetic Molecular Organizates.

Kunitake, T. and Sakamoto, T. (1979). *Bull Chem. Soc. Jpn.* **52**, 2624–2629. Nucleophilic Ion Pairs. 8. Facile Nucleophilic Cleavage of Dinitrophenyl Sulfate in the Presence of Micellar Zwitterionic Hydroxamates.

Kunitake, T., Okahata, Y., and Sakamoto, T. (1975). *Chem. Lett.*, 459–462. The Hydrolysis of *p*-Nitrophenyl Acetate by a Micellar Bifunctional Catalyst.

Kunitake, T., Okahata, Y., and Sakamoto, T. (1976). *J. Am. Chem. Soc.* **98**, 7799–7806. Multifunctional Hydrolytic Catalyses. 8. Remarkable Acceleration of the Hydrolysis of *p*-Nitrophenyl Acetate by Micellar Bifunctional Catalysts.

Kunitake, T., Okahata, Y., Tanamachi, S., and Ando, R. (1979). *Bull. Chem. Soc. Jpn.* **52**, 1967–1971. Nucleophilic Ion Pairs. 6. Catalytic Hydrolysis of *p*-Nitrophenyl Acetate by Zwitterionic Hydroxamate Nucleophiles in Representative Micellar Systems.

Kunitake, T., Okahata, Y., Ando, R., Shinkai, S., and Hirakawa, S. (1980). *J. Am. Chem. Soc.* **102**, 7877–7881. Decarboxylation of 6-Nitrobenzisoxazole-3-carboxylate Catalyzed by Ammonium Bilayer Membranes. A Comparison of the Catalytic Behavior of Micelles, Bilayer Membranes, and Other Aggregates.

Kurihara, K., Toyoshima, Y., and Sukigara, M. (1977). *J. Phys. Chem.* **81**, 1833–1837. Phase Transition and Dye Aggregation in Phospholipid-Amphiphatic Dye Liposome Bilayers.

Kurihara, K., Sukigara, M., and Toyoshima, Y. (1979a). *Biochim. Biophys. Acta* **547**, 117–126. Photoinduced Charge Separation in Liposomes Containing Chlorophyll a. I. Photoreduction of Copper(II) by Potassium Ascorbate Through Liposome Bilayer Containing Purified Chlorophyll a.

Kurihara, K., Toyoshima, Y., and Sukigara, M. (1979b). *Biochem. Biophys. Res. Commun.* **88**, 320–326. Photoinduced Charge Separation in Liposomes Containing Chlorophyll a. II. The Effect of Ion Transport Across Membrane on the Photoreduction of $Fe(CN)_6^{3-}$.

Kurihara, K., Toyoshima, Y., and Sukigara, M. (1980). *Nippon Kagaku Zasshi*, 1499–1505. Photoredox Reactions in Liposome Systems and Its Application to the Light Energy into Chemical Energy.

Kusumoto, Y. and Sato, H. (1979). *Chem. Phys. Lett.* **68**, 13–16. Energy Transfer Between Rhodamine-6G and 3,3′-Diethylthiacarbocyanine Iodide Enhanced in the Premicellar Region.

Laane, C., Ford, W. E., Otvos, J. W., and Calvin, M. (1981). *Proc. Natl. Acad. Sci. USA* **78**, 2017–2020. Photosensitized Electron Transport Across Lipid Vesicle Walls: Enhancement of Quantum Yield by Ionophores and Transmembrane Potentials.

Lemmetyinen, H., Demyashkevich, A. B., and Kuzmin, M. G. (1980). *Chem. Phys. Lett.* **73**, 98–101. Proton Transfer Photoreactions in Hydrogen Complexes and the Formation of Ion Pairs by Aromatic Hydroxy Compounds in the Presence of Crown Ethers.

Lehn, J.-M. (1980). *Pure Appl. Chem.* **52**, 2303–2319. Cryptate Inclusion Complexes. Effects on Solute-Solute and Solute-Solvent Interactions and on Ionic Reactivity.

Lerner, D. A., Ricchiero, F., and Giannotti, C. (1980). *J. Phys. Chem.* **84**, 3007–3011. Quantum Yields of Photolysis of Amphipathic Alkylcobaloximes in Mixed Micelles. A Cooperative Effect?

Lindog, B. A. and Rodgers, M. A. J. (1979). *J. Phys. Chem.* **83**, 1683–1688. Laser Photolysis Studies of Singlet Molecular Oxygen in Aqueous Micellar Dispersions.

Loew, L. M., Simpson, L., Hassner, A., and Alexanian, V. (1979a). *J. Am. Chem. Soc.* **101**, 5439–5440. An Unexpected Blue Shift Caused by Differential Solvation of a Chromophore Oriented in a Lipid Bilayer.

Loew, L. M., Scully, S., Simpson, L., and Waggoner, A. S. (1979b). *Nature* **281**, 497–499. Evidence for a Charge-Shift Electrochromic Mechanism in a Probe of Membrane Potential.

Lopez, J. R. and Tien, H. T. (1980). *Biochim. Biophys. Acta* **597**, 433–444. Photoelectrospectrometry of Bilayer Lipid Membranes.

Loughran, T., Hatlee, M. D., Patterson, L. K., and Kozak, J. J. (1980). *J. Chem. Phys.* **72**, 5791–5797. Monomer-Excimer Dynamics in Spread Monolayers. I. Lateral Diffusion of Pyrene Dodecanoic Acid at the Air-Water Interface.

McIntire, G. L. and Blount, H. N. (1979). *J. Am. Chem. Soc.* **101**, 7720–7721. Electrochemistry in Ordered Systems. I. Oxidative Electrochemistry of 10-Methylphenothiazine in Anionic, Cationic, and Nonionic Micellar Systems.

Mackay, R. A. and Hermansky, C. (1981). *J. Phys. Chem.* **85**, 739–744. Phosphate Ester-Nucleophile Reactions in Oil-in-Water Microemulsions.

Mackay, R. A., Letts, K., and Jones, C. (1977). In *Micellization, Solubilization, and Microemulsions* (K. L. Mittal, Ed.), Plenum Press, New York, pp. 801–815. Interactions and Reactions in Microemulsions.

Mackay, R. A., Jacobson, K., and Tourian, J. (1980). *J. Colloid Interface Sci.* **76**, 515–524. Measurement of pH and p*K* in o/w Microemulsions.

Madden, T. D., Peel, W. E., Quinn, P. J., and Chapman, D. (1980). *J. Biochem. Biophys. Methods* **2**, 19–27. The Modulation of Membrane Fluidity by Hydrogenation Processes. IV. Homogeneous Catalysis of Liposomes Using a Water Soluble Catalyst.

Manecke, G. and Storck, W. (1978). *Angew. Chem. Int. Ed. Eng.* **17**, 657–670. Polymeric Catalysts.

Mangel, M. (1976). *Biochim. Biophys. Acta* **430**, 459–466. Properties of Liposomes that Contain Chloroplast Pigments: Photosensitivity and Efficiency of Energy Conversion.

Martinek, K., Levashov, A. V., and Berezin, I. V. (1975a). *Tetrahedron Lett.*, 1275–1278. Mechanism of Catalysis by Functional Micelles Containing a Hydroxy Group. Model of Action of Serine Proteinases.

Martinek, K., Osipov, A. P., Yatsimirski, A. K., Dadali, V. A., amd Berezin, I. V. (1975b). *Tetrahedron Lett.*, 1279–1282. Reactivity of Imidazole Derivatives on Their Being Acylated in the Surface Layer of Cationic Micelles.

Martinek, K., Osipov, A. P., Yatsimirskii, A. K., and Berezin, I. V. (1975c). *Tetrahedron* **31**, 709–718. Mechanism of Micellar Effects in Imidazole Catalysis. Acylation of Benzimidazole and Its *N*-Methyl Derivative by *p*-Nitrophenyl Carboxylates.

Maruthamuthu, M. and Usha, G. (1979). *Curr. Sci.* **48**, 1034–1035. Micellar-Catalyzed Cerium(IV)-Acetone Reaction.

Masuhara, H., Kaji, K., and Mataga, N. (1977). *Bull. Chem. Soc. Jpn.* **50**, 2084–2087. Fluorescence and Laser Photolysis Studies on the Intramolecular Exciplex Systems in Micellar Solutions.

Masuhara, H., Tanabe, H., and Mataga, N. (1979). *Chem. Phys. Lett.* **63**, 273–276. Fluorescence and Laser Photolysis Studies on 1,2,4,5-Tetracyanobenzene at Complexes in Micellar Solutions.

Masuhara, H., Ohwada, S., Yamamoto, K., Mataga, N., Itaya, A., Okamoto, K., and Kusabayashi, S. (1980). *Chem. Phys. Lett.* **70**, 276–278. Laser Induced Formation of Transient Polyelectrolyte in Solution.

Masui, T., Watanabe, F., and Yamagishi, A. (1977a). *J. Colloid Interface Sci.* **61**, 388–393. EPF Studies of $FeCl_3$, $MnCl_2 \cdot 4H_2O$ and $CuSO_4 \cdot 5H_2O$ Solubilized by Dodecylpyridinium Chloride in Chloroform.

Masui, T., Watanabe, F., and Yamagishi, A. (1977b). *J. Phys. Chem.* **81**, 494–496. Temperature-Jump Study on the Aquation of the Iron(III) Complex by Dodecylpyridinium Chloride Solubilized Water Pool in Chloroform.

Matheson, I. B. C. and Massoudi, R. (1980). *J. Am. Chem. Soc.* **102**, 1942–1948. Photophysical and Photosensitized Generation of Singlet Molecular Oxygen ($^1\Delta_g$) in Micellar Solutions at Elevated Pressures. Measurement of Singlet Molecular Oxygen Solvent to Micelle Transfer Rates via Both Molecular Diffusion and Energy Transfer.

Matsuo, T., Itoh, K., Takuma, K., Hashimoto, K., and Natamura, T. (1980a). *Chem. Lett.*, 1009–1012. A Concerted Two-Step Activation of Photoinduced Electron Transport Across Lipid Membrane.

Matsuo, T., Aso, Y., and Kano, K. (1980b). *Ber. Bunsenges. Phys. Chem.* **84**, 146–152. Energy Transfer in Micellar Systems. Fluorescence Investigation of Energy Transfer from Dialkylalloxazine to Dye Cations in the Presence of Anionic Micelles.

Mehreteab, A. and Strauss, G. (1978). *Photochem. Photobiol.* **28**, 369–375. Energy Transfer and Energy Losses in Bilayer Membrane Vesicles (Liposomes).

Meisel, D., Matheson, M. S., and Rabani, J. (1978a). *J. Am. Chem. Soc.* **100**, 117–122. Photolytic and Radiolytic Studies of $Ru(bpy)_3^{2+}$ in Micellar Solutions.

Meisel, D., Rabani, J., Meyerstein, D., and Matheson, M. S. (1978b). *J. Phys. Chem.* **82**, 985–990. Concentration Effects on the Quenching of Tris(2,2′-bipyridine)ruthenium(II) Emission in Polyvinylsulfate Solutions.

Meisel, D., Mulac, W. A., and Matheson, M. S. (1981). *J. Phys. Chem.* **85**, 179–187. Catalysis of Methyl Viologen Radical Reactions by Polymer-Stabilized Gold Sols.

Melhado, L. L. and Gutsche, C. D. (1978). *J. Am. Chem. Soc.* **100**, 1850–1856. Association Phenomena. 2. Catalysis of the Decomposition of Acetyl Phosphate by Chelate Micelles and by Amine-Ammonium Micelles.

Menger, F. M. and Yamada, K. (1979). *J. Am. Chem. Soc.* **101**, 6731–6734. Enzyme Catalysis in Water Pools.

Mercer-Smith, J. A. and Whitten, D. G. (1979). *J. Am. Chem. Soc.* **101**, 6620–6625. Photoreactions of Metalloporphyrins in Supported Monolayer Assemblies and at Assembly-Solution Interfaces. Reductive Addition of Palladium Complexes with Surfactants and Water Soluble Dialkylanilines.

Meyer, G. (1979). *J. Org. Chem.* **44**, 3983–3984. Specific Effect of Micellar Microenvironment on an Intramolecular Nucleophilic Anionic Reaction.

Meyer, G. and Viout, P. (1977). *Tetrahedron* **33**, 1959–1961. Specificity of the Catalytic Effects of Hydroxy Cationic Micelles.

Meyerstein, D., Rabani, J., Matheson, M. S., and Meisel, D. (1978). *J. Phys. Chem.* **82**, 1879–1885. Charge Separation in Photoinitiated Electron Transfer Reactions Induced by a Polyelectrolyte.

Minch, M. J., Giaccio, M., and Wolff, R. (1975). *J. Am. Chem. Soc.* **97**, 3766–3772. Effect of Cationic Micelles on the Acidity of Carbon Acids and Phenols. Electronic and ^1H Nuclear Magnetic Resonance Spectral Studies of Nitro Carbanions in Micelles.

Minch, M. J., Chen, S.-S., and Peters, R. (1978). *J. Org. Chem.* **43**, 31–33. Specificity in the Micellar Catalysis of a Hofmann Elimination.

Miyata, T., Sakumoto, A., and Washino, M. (1977). *Bull. Chem. Soc. Jpn.* **50**, 2950–2955. The Kinetics of the Radical Addition of Hydrogensulfite Ions to 1-Dodecene in a Micellar Solution of Sodium 1-Dodecanesulfonate.

Miyoshi, N. and Tomita, G. (1978). *Z. Naturforsch.* **33b**, 622–627. Production and Reaction of Singlet Oxygen in Aqueous Micellar Solutions Using Pyrene as Photosensitizer.

Miyoshi, N. and Tomita, G. (1979a). *Photochem. Photobiol.* **29**, 527–530. Effects of Indole and Tryptophan on Furan Oxidation by Singlet Oxygen in Micellar Solutions.

Miyoshi, N. and Tomita, G. (1979b). *Z. Naturforsch.*.**34b**, 339–343. Quenching of Singlet Oxygen by Sodium Azide in Reversed Micellar Systems.

Miyoshi, N. and Tomita, G. (1979c). *Z. Naturforsch.* **34b**, 1552–1555. Fluorescein-Photosensitized Furan Oxidation in Methanolic and Reversed Micellar Solutions, Part I. Effects of Amines.

Miyoshi, N. and Tomita, G. (1980a). *Z. Naturforsch.* **35b**, 731–735. Singlet Oxygen Production Photosensitized by Fluorescein in Reversed Micellar Solutions.

Miyoshi, N. and Tomita, G. (1980b). *Z. Naturforsch.* **35b**, 107–111. Fluorescein-Photosensitized Furan Oxidation in Methanolic and Reversed Micellar Solutions, Part II. Kinetic Analysis.

Möbius, D. (1978a). In *Topics in Surface Chemistry* (E. Kay and P. S. Bagus, Eds.), Plenum Press, New York, pp. 75–101. Monolayer Assemblies.

Möbius, D. (1978b). *Ber. Bunsenges. Phys. Chem.* **82**, 848–858. Designed Monolayer Assemblies.

Möbius, D. and Kuhn, D. (1979). *Isr. J. Chem.* **18**, 375–384. Monolayer Assemblies of Dyes to Study the Role of Thermal Collisions in Energy Transfer.

Mollett, K. J. and O'Connor, J. (1976). *J. Chem. Soc. Perkin Trans. II*, 369–374. Hydrolysis of Phenylureas. Part III. Micellar Effects on the Solubilization and Decomposition of 4-Methyl- and 4-Nitrophenylurea.

Monserrat, K., Grätzel, M., and Tundo, P. (1980). *J. Am. Chem. Soc.* **102**, 5527–5529. Light-Induced Charge Injection in Functional Crown Ether Vesicles.

Morgan, C. G., Hudson, B., and Wolber, P. K. (1980). *Proc. Natl. Acad. Sci. USA* **77**, 26–30. Photochemical Dimerization of Parinaric Acid in Lipid Bilayers.

Moriarty, R. M., Schwartz, R. N., Lee, C., and Curtis, V. (1980). *J. Am. Chem. Soc.* **102**, 4257–4259. Formation of Vitamin D_3 in Synthetic Lipid Multibilayers. A Model for Epidermal Photosynthesis.

Moroi, Y., Braun, A. M., and Grätzel, M. (1979a). *J. Am. Chem. Soc.* **101**, 567–572. Light-Initiated Electron Transfer in Functional Surfactant Assemblies. 1. Micelles with Transition Metal Counterions.

Moroi, Y., Infelta, P. P., and Grätzel, M. (1979b). *J. Am. Chem. Soc.* **101**, 573–579. Light-Initiated Redox Reactions in Functional Micellar Assemblies. 2. Dynamics in Europium(III) Surfactant Solutions.

Moss, R. A. and Alwis, K. W. (1980). *Tetrahedron Lett.*, 1303–1306. A Surfactant Hydroperoxide.

Moss, R. A. and Rav-Acha, C. (1980). *J. Am. Chem. Soc.* **102**, 5045–5047. Diazo Coupling Reactions with a Functional Micellar Reagent. Circumvention of Hartley's Rules.

Moss, R. A. and Sanders, W. J. (1979). *Tetrahedron Lett.*, **19**, 1669–1670. Micellar Alkylation: A Methylating Surfactant.

Moss, R. A., Nahas, R. C., Ramaswami, S., and Sanders, W. J. (1975). *Tetrahedron Lett.*, 3379–3382. A Comparison of Hydroxyl and Imidazole Functionalized Micellar Catalysis in Ester Hydrolyses.

Moss, R. A., Lukas, T. J., and Nahas, R. C. (1977a). *Tetrahedron Lett.*, 3851–3854. New Amino Acid-Functionalized Surfactants: Preparation and Catalytic Properties.

Moss, R. A., Nahas, R. C., and Ramaswami, S. (1977b). *J. Am. Chem. Soc.* **99**, 627–629. Sequential Bifunctional Micellar Catalysis.

Moss, R. A., Lukas, T. J., and Nahas, R. C. (1978a). *J. Am. Chem. Soc.* **100**, 5920–5927. Preparation and Kinetic Properties of Cysteine Surfactants.

Moss, R. A., Bizzigotti, G. O., Lukas, T. J., and Sanders, W. J. (1978b). *Tetrahedron Lett.*, 3661–3664. A Thiocholine Surfactant: Preparation and Kinetic Properties.

Moss, R. A., Nahas, R. C., and Lukas, T. J. (1978c). *Tetrahedron Lett.*, 507–510. A Cysteine-Functionalized Micellar Catalyst.

Moss, R. A., Bizzogotti, G. O., and Huang, C.-W. (1980). *J. Am. Chem. Soc.* **102**, 754–762. Nucleophilic Esterolytic and Displacement Reactions of a Micellar Thiocholine Surfactant.

Motozato, Y., Furuya, Y., Matsumoto, T., and Nishihara, T. (1980). *Bull. Chem. Soc. Jpn.* **53**, 2578–2581. The Kinetics of the β-Cyclodextrin-Catalyzed Decarboxylation of Trichloroacetic Acid in an Alkaline Solution.

Murakami, Y., Nakano, A., and Matsumoto, K. (1979a). *Bull. Chem. Soc. Jpn.* **52**, 2996–3004. Catalytic Efficiency of Synthetic Micellar Catalysts Bearing a Mercapto Group as the Reaction Center.

Murakami, Y., Nakano, A., Matsumoto, K., and Iwamoto, K. (1979b). *Bull. Chem. Soc. Jpn.* **52**, 3573–3578. Catalytic Efficiency of Cationic Micellar Catalysts Bearing a Mercapto Group as the Reaction Center.

Murakami, Y., Nakano, A., Yoshimatsu, A., and Fukuya, K. (1981a). *J. Am. Chem. Soc.* **103**, 728–730. Functionalized Vesicular Assembly. Enantioselective Catalysis of Ester Hydrolysis.

Murakami, Y., Aoyama, Y., Nakano, A., Tada, T., and Fukuya, K. (1981b). *J. Am. Chem. Soc.*, to be published. Novel Substrate-Binding Property of Synthetic Membrane Vesicles Involving an Amino Acid Residue as a Molecular Component.

Nagamura, T., Takuma, K., Tsutsui, Y., and Matsuo, T. (1980). *Chem. Lett.*, 503–506. Photoinduced Electron Transfer Between Amphipathic Rughtnium(II) Complex and *N,N*-Dimethylaniline in Synthetic Bilayer Membranes and Phospholipid Liposomes.

Nagamura, T., Matsuo, T., and Fendler, J. H. (1981). *Chem. Phys. Lett.*, to be published. Aspects of Artificial Photosynthesis: Functionalized Zinc Porphyrin Sensitized Electron Transfer and Charge Separation in Surfactant Vesicles.

Nishijima, T., Nagamura, T., and Matsuo, T. (1981). *J. Polym. Sci. Polym. Lett.* **19**, 65–73. Hydrogen Generation by Visible Light. Irradiation of Ruthenium Complexes and Colloidal Platinum Stabilized by Viologen Polymers in Aqueous Solutions.

Nome, F., Schwingel, E. W., and Ionescu, L. G. (1980). *J. Org. Chem.* **45**, 705–710. Micellar Effects on the Base-Catalyzed Oxidative Cleavage of a Carbon–Carbon Bond in 1,1-Bis(*p*-chlorophenyl)-2,2,2-trichloroethanol.

Nomura, T., Escabi-Perez, J. R., Sunamoto, J., and Fendler, J. H. (1980). *J. Am. Chem. Soc.* **102**, 1484–1488. Aspects of Artificial Photosynthesis. Energy Transfer in Cationic Surfactant Vesicles.

O'Connor, C. J. and Ramage, R. E. (1980a). *Aust. J. Chem.* **33**, 757–770. The Reactivity of *p*-Nitrophenylesters with Surfactants in Nonaqueous Solvents. Part I. *p*-Nitrophenyl Acetate in Benzene Solutions in Dodecylammonium Propionate.

O'Connor, C. J. and Ramage, R. E. (1980b). *Aust. J. Chem.* **33**, 771–777. The Reactivity of *p*-Nitrophenylesters with Surfactants in Nonaqueous Solvents. Part 2. The Effects of Added Water in Benzene Solutions of Dodecylammonium Propionate, and the Reaction in Toluene and Cyclohexane.

O'Connor, C. J. and Ramage, R. E. (1980c). *Aust. J. Chem.* **33**, 779–784. The Reactivity of *p*-Nitrophenyl Esters with Surfactants in Nonaqueous Solvents. III. The Effect of Altering the Headgroups of Alkylammonium Carboxylates Dissolved in Benzene.

O'Connor, C. J. and Ramage, R. E. (1980d). *Aust. J. Chem.* **33**, 1301–1311. The Reactivity of *p*-Nitrophenyl Esters with Surfactants in Nonaqueous Solvents. IV. *p*-Nitrophenyl Acetate in Benzene Solutions of Dodecylammonium Phenoxides.

O'Connor, C. J. and Ramage, R. E. (1980e). *Aust. J. Chem.* **33**, 695–698. The Reactivity of Tris(oxalato)chromate(III) Anion in Reversed Micellar System.

O'Connor, C. J. and Tan, A.-H. (1980). *Aust. J. Chem.* **33**, 747–755. Micellar Catalysed Hydrolysis of Amides. 4-Nitroacetanilide in Hexadecyltrimethylammonium Bromide.

Ogata, Y., Takagi, K., and Tanabe, Y. (1979). *J. Chem. Soc. Perkin Trans. II*, 1069–1071. Photoinduced Reduction of Pyridinium Ions Catalyzed by Zinc(II) Tetraphenylporphyrin.

Okahata, Y., Ando, R., and Kunitake, T. (1979). *Bull. Chem. Soc. Jpn.* **52**, 3647–3653. Catalytic Hydrolysis of *p*-Nitrophenyl Esters in the Presence of Representative Ammonium Aggregates. Specific Activation of a Cholesteryl Nucleophile Bound to a Dialkylammonium Bilayer Membrane.

Okahata, Y., Ihara, H., Shimomura, M., Tawaki, S., and Kunitake, T. (1980). *Chem. Lett.*, 1169–1172. Formation of Dish-like Aggregates from Single-Chain Phosphocholine Amphiphiles in Water.

Okamoto, Y. and Wang, J. Y. (1977). *J. Org. Chem.* **42**, 1261–1262. Micellar Effects on the Reaction of 2,4,6-Trinitrotoluene with Amines.

Okun, J. D. and Archer, M. C. (1977). *J. Natl. Cancer Inst.* **58**, 409–411. Kinetics of Nitrosamine Formation in the Presence of Micelle-Forming Surfactants.

Ong, J. T. H. and Kostenbauder, H. B. (1976). *J. Pharm. Sci.* **65**, 1713–1718. Effect of Micellization on Rate of Cupric Ion-Promoted Hydrolysis of Dicarboxylic Acid Hemiesters.

Onozuka, S., Kojima, M., Hattori, K., and Toda, F. (1980). *Bull. Chem. Soc. Jpn.* **53**, 3221–3224. The Regiospecific Mono Tosylation of Cyclodextrins.

Packham, N. K., Packham, C., Mueller, P., Tiede, D. M., and Dutton, P. L. (1980). *FEBS Lett.* **110**, 101–106. Reconstitution of Photochemically Active Reaction Centers in Planar Phospholipid Membranes. Light-Induced Electrical Currents Under Voltage-Clamped Conditions.

Pelizzetti, E. and Pramauro, E. (1979a). *Inorg. Chem.* **18**, 882–883. Micellar Effect on Electron Transfer. 1. Electron Transfer of Tris(2,2′-bipyridine)ruthenium in Micellar Solutions.

Pelizzetti, E. and Pramauro, E. (1979b). *Ber. Bunsenges. Phys. Chem.* **83**, 996–1000. Electron Transfer Between Phenothiazine and Iron(III) in Micellar Systems [1,2].

Pelizzetti, E. and Pramauro, E. (1980). *Anal. Chim. Acta* **117**, 403–406. Acid–Base Titrations of Substituted Benzoic Acids in Micellar Systems.

Pelizzetti, E., Pranauro, E., and Croce, D. (1980). *Ber. Bunsenges. Phys. Chem.* **84**, 265–270. Electron Transfer Between Benzenediols and Tris(1,10-phenanthroline)iron(III) in Micellar System 1.

Pileni, M.-P. (1980a). *Chem. Phys. Lett.* **71**, 317–321. Zinc-Porphyrin Sensitized Reduction of Simple and Functional Quinones in Vesicle Systems.

Pileni, M.-P. (1980b). *Chem. Phys. Lett.* **75**, 540–544. Zinc-Porphyrin Sensitized Redox Processes in Microemulsions.

Pileni, M.-P. and Grätzel, M. (1980a). *J. Phys. Chem.* **84**, 1822–1825. Zinc-Porphyrin Sensitized Reduction of Functional Quinones in Micellar Systems.

Pileni, M.-P. and Grätzel, M. (1980b). *J. Phys. Chem.* **84**, 2402–2406. Light-Induced Reactions of Proflavin in Aqueous and Micellar Solutions.

Pillersdorf, A. and Katzhendler, J. (1979a). *J. Org. Chem.* **44**, 934–942. Catalytic Dipolar Micelles. 7. Catalytic Effects of Positively Charged Hydroxylic Micelles on the Hydrolysis of Phenyl and Benzoate Esters.

Pillersdorf, A. and Katzhendler, J. (1979b). *J. Org. Chem.* **44**, 549–554. Dipolar Micelles. 8. Hydrolysis of Substituted Phenyl Esters in a Hydroxamic Acid Surfactant.

Pillersdorf, A. and Katzhendler, J. (1979c). *Isr. J. Chem.* **18**, 330–338. Dipolar Micelles. 9. The Mechanism of Hydrolysis of Cationic Long Chained Benzoate Esters in Choline and Homocholine-Type Micelles.

Pohl, G. W. (1972). *Biochim. Biophys. Acta* **288**, 248–253. Energy Transfer in Black Lipid Membranes.

Politi, M., Cuccovia, I. M., and Chaimovich, H. (1978). *Tetrahedron Lett.* **2**, 115–118. Effect of Hexadecyltrimethylammonium Bromide on the Hydrolysis of *N*-Alkyl-4-cyanopyridinium Ions.

Polymeropoulos, E. E. and Möbius, D. (1979). *Ber. Bunsenges. Phys. Chem.* **83**, 1215–1222. Photochromism in Monolayers.

Polymeropoulos, E. E., Möbius, D., and Kuhn, H. (1978). *J. Chem. Phys.* **68**, 3918–3931. Photoconduction in Monolayer Assemblies with Functional Units of Sensitizing and Conducting Molecular Components.

Polymeropoulos, E. E., Möbius, D., and Kuhn, H. (1980). *Thin Solid Films* **68**, 173–190. Monolayer Assemblies with Functional Units of Sensitizing and Conducting Molecular Components: Photovoltage, Dark Conduction and Photoconduction in Systems with Aluminium and Barium Electrodes.

Pradevan, G. O. and Pileni, M. P. (1981). *J. Chim. Phys.* **78**, 203–205. Reaction Photoredox de la Tetraphenyl Porphyrine de Zinc en Solution dans le Triton.

Proske, T. and Henglein, A. (1978). *Ber. Bunsenges. Phys. Chem.* **82**, 711–713. The Influence of Cationic Micelles on the Dismutation of Br_2^--Radical Anions in Aqueous Solution (A Pulse Radiolysis Study).

Putna, L., Reske, G., and Schmidt, H. (1979). *Photochem. Photobiol.* **30**, 723–725. Flash Photolytic Investigations of the Electron Transfer Between Aromatic Hydrocarbon Radicals and Tryptophan in Micellar Solutions.

Quina, F. H. and Whitten, D. G. (1977). *J. Am. Chem. Soc.* **99**, 877–883. Photochemical Reactions in Organized Monolayer Assemblies. 4. Photodimerization, Photoisomerization, and Excimer Formation with Surfactant Olefins and Dienes in Monolayer Assemblies, Crystals, and Micelles.

Raleigh, J. A. and Kremers, W. (1978). *Int. J. Rad. Biol.* **34**, 439–447. Promotion of Radiation Peroxidation in Models of Lipid Membranes by Caesium and Rubidium Counterions: Micellar Linoleic and Linolenic Acid.

Rav-Acha, C., Chevion, M., Katzhendler, J., and Sarel, S. (1978). *J. Org. Chem.* **43**, 591–595. Dipolar Micelles. 6. Catalytic Effect of Betaine-like Micelles on the Hydrolysis of Substituted Phenyl Esters.

Reinsborough, U. C. and Robinson, B. H. (1979). *J. Chem. Soc. Faraday Trans. I* **75**, 2395–2405. Micellar Catalysis of Metal Complex Formation. Part II. Kinetics of the Reaction between Ni_{aq}^{2+} and Various Neutral Bidentate Ligands in the Presence of Sodium Dodecyl Sulphate Micelles in Aqueous Solution.

Richard, M. A., Deutch, J., and Whitesides, G. M. (1978). *J. Am. Chem. Soc.* **100**, 6613–6625. Hydrogenation of Oriented Monolayers of ω-Unsaturated Fatty Acids Supported on Platinum.

Rico, I., Maurette, M. T., Oliveros, E., Riviere, M., and Latters, A. (1980). *Tetrahedron* **36**, 1779–1783. Reactivite Chemique et Photochimique dans les Milieux Micellaires de les Microemulsions. III. Mise en Evidence de L'Importance des Processus Interfaciaux pour une Reaction de Photoisomerisation.

Rideout, D. C. and Breslow, R. (1980). *J. Am. Chem. Soc.* **102**, 7816–7817. Hydrophobic Acceleration of Diels–Alder Reactions.

Robinson, B. H., and Steytler, D. (1978). *Ber. Bunsenges. Phys. Chem.* **82**, 1012. Ion Reactivity in Micro-emulsion Systems.

Robinson, B. H. and White, N. C. (1978). *J. Chem. Soc. Faraday Trans. I.* **74**, 2625–2636. Kinetics and Mechanism of Fast Metal-Ligand Substitution Processes in Aqueous and Micellar Solutions Studied by Means of a Dye-Laser Photochemical Relaxation Technique.

Robinson, B. H., Steytler, D. C., and Tack, R. D. (1979). *J. Chem. Soc. Faraday Trans. I* **75**, 481–496. Ion Reactivity in Reversed-Micellar Systems. Kinetics of Reactions Between Micelles Containing Hydrated Nickel(II) and Murexide-Containing Micelles in the System Aerosol-OT + Water + Heptane.

Rodgers, M. A. J. and Becker, J. C. (1980). *J. Phys. Chem.* **84**, 2762–2768. Electron-Transfer Quenching of the Luminescent State of the Tris(bipyridyl)ruthenium(II) Complex in Micellar Media.

Rodgers, M. A. J. and Burrows, H. D. (1979). *Chem. Phys. Lett.* **66**, 238–243. Singlet-Singlet Energy Transfer Across Reversed Micellar Phase Boundaries.

Rodgers, M. A. J., Foyt, D. C., and Simek, Z. (1978). *Radiat. Res.* **75**, 296–304. The Effect of Surfactant Micelles on the Reaction of Hydrated Electrons and Dimethyl Viologen.

Roessler, N. and Wolff, T. (1980). *Photochem. Photobiol.* **31**, 547–552. Oxygen Quenching in Micellar Solution and Its Effect on the Photocyclization of *N*-Methyldiphenylamine.

Russell, J. C., Costa, S. B., Seiders, R. P., and Whitten, D. G. (1980). *J. Am. Chem. Soc.* **102**, 5678–5679. Photochemistry of a Surfactant Stilbene in Organized Media: A Probe for Hydrophobic Sites in Micelles and Other Assemblies.

Sapre, A. V., Rao, K. V. S. R., and Rao, K. N. (1980). *J. Phys. Chem.* **84**, 2281–2287. Rapid Excitation Quenching by Host Micelles. Observations from Hexadecylpyridinium Chloride and Brij-35 Systems Containing Pyrene Scintillator.

Sassoon, R. E. and Rabani, J. (1980). *J. Phys. Chem.* **84**, 1319–1325. Charge Separation in the Photoelectron Transfer Reaction Between Dicyanobis(2,2′-bipyridine)ruthenium(II) and Ferricyanide Ion in a Polyelectrolyte Solution.

Sato, H., Kusumoto, Y., Nakashima, N., and Yoshihara, K. (1980). *Chem. Phys. Lett.* **71**, 326–329. Picosecond Study of Energy Transfer Between Rhodamine 6G and 3,3′-Diethylthiacarbocyanine Iodide in the Premicellar Region: Förster Mechanism with Increased Local Concentration.

Scheerer, R. and Grätzel, M. (1976). *Ber. Bunsenges. Phys. Chem.* **80**, 979–982. Photo-induced Oxidation of Carbonate Ions by Duroquinone, a Pathway of Oxygen Evolution from Water by Visible Light.

Scheerer, R. and Grätzel, M. (1977). *J. Am. Chem. Soc.* **99**, 865–871. Laser Photolysis Studies of Duroquinone Triplet State Electron Transfer Reactions.

Schmehl, R. H. and Whitten, D. G. (1980). *J. Am. Chem. Soc.* **102**, 1938–1941. Intramicellar Electron Transfer Quenching of Excited States. Determination of the Binding Constant and Exchange Rates for Dimethyl Viologen.

Schönfeld, M., Montal, M., and Feher, G. (1979). *Proc. Natl. Acad. Sci. USA* **76**, 6351–6355. Functional Reconstitution of Photosynthetic Reaction Centers in Planar Lipid Bilayers.

Schreiber, G. and Dupeyrat, M. (1979). *Bioelectrochem. Bioenerg.* **6**, 427–440. Photosensitive Lipid Membranes. I. The Collodion-Oleic Acid, Methylene Blu Sensitized Membrane in Air Saturated KCl 1×10^{-1} *M* Solution as an Experimental Model for Biological Photosensitization Processes.

Seefeld, K.-P., Möbius, D., and Kuhn, H. (1977). *Helv. Chim. Acta* **60**, 2608–2632. Electron Transfer in Monolayer Assemblies with Incorporated Rutherniurn(II) Complexes.

Seiders, R. P., Brookhare, M., and Whitten, D. G. (1979). *Isr. J. Chem.* **18**, 272–278. Synthesis and Reactivity of a Surfactant Ferrocene. Environmental Effects on Oxidation of Ferrocene in Organized Media.

Seno, M., Araki, K., and Shiraishi, S. (1976). *Bull. Chem. Soc. Jpn.* **49**, 1901–1905. Iodination Reactions of Ketones in the Reversed Micellar Systems of Dodecylammonium Propionate in Hexane.

Seno, M., Kousaka, K., and Kise, H. (1979). *Bull. Chem. Soc. Jpn.* **52**, 2970–2974. Reduction of Methylene Blue with L-Ascorbic Acid or L-Cysteine in Micellar Systems.

Seta, P., D'Epenoux, B., and Gavach, C. (1980). *Bioelectrochem. Bioenerg.* **7**, 539–551. Electrochemical Analysis of Transmembrane Proton Fluxes Photogenerated by Bacteriorhodopsin in BLM.

Shiffman, R., Chevion, M., Katzhendler, J., Rav-Acha, C., and Sarel, S. (1977). *J. Org. Chem.* **42**, 856–863. Catalytic Dipolar Micelles. 3. Substrate and Surfactant Structural Effects in the Hydrolyses of Substituted Phenyl Esters in Presence and in A sence of Dipolar Cationic Micelles: Mechanistic Considerations.

Shinkai, S. and Kunitake, T. (1976a). *Bull. Chem. Soc. Jpn.* **49**, 3219–3223. Micellar Activation of Nucleophilic Reactivity of Coenzyme A and Glutathione Toward *p*-Nitrophenyl Acetate.

Shinkai, S. and Kunitake, T. (1976b). *Chem. Lett.*, 109–112. Nucleophilic Ion Pairs. Facile Cleavage of an Amide Substrate by Hydroxamate Anions.

Shinkai, S. and Kunitake, T. (1976c). *J. Chem. Soc. Perkin Trans. II* 9, 980–985. Nucleophilic Ion Pairs. Part II. Micellar Catalysis of Proton Abstraction by Hydroxamate Anions.

Shinkai, S. and Kunitake, T. (1977). *Bull. Chem. Soc. Jpn.* **50**, 2400–2405. Coenzyme Models. IX. Micellar Catalysis of Isoalloxazine (Flavin) Oxidation of Dithiol.

Shinkai, S., Ando, R., and Kunitake, T. (1975). *Bull. Chem. Soc. Jpn.* **48**, 1914–1917. Coenzyme Models. II. Effect of Micellar Environments on the Acid-Catalyzed Hydration of 1,4-Dihydropyridine Derivatives.

Shinkai, S., Ando, R., and Kunitake, T. (1976a). *Bull. Chem. Soc. Jpn.* **49**, 3652–3655. Coenzyme Models. VII. Micellar Catalysis of Acid-Catalyzed Hydration of 1-Dodecyl-1,4-dihydronicotinamide.

Shinkai, S., Sakuma, Y., and Yoneda, F. (1976b). *J. Chem. Soc. Chem. Commun.*, 986–987. Micellar Catalysis of Oxidation of Nitroethane by Isoalloxazine (Flavin).

Shinkai, S., Ando, R., and Yoneda, F. (1977). *Chem. Lett.*, 147–150. Facile Oxidation of Thiophenol by Isoalloxazine(flavin) Bound to a Cationic Micelle.

Shinkai, S., Ide, T., and Manabe, O. (1978a). *Bull. Chem. Soc. Jpn.* **51**, 3655–3656. Coenzyme Models. XIII. Micellar Catalysis of 1,4-Dihydronicotinamide Reduction of Isoalloxazines and Acridinium Ion.

Shinkai, S., Ide, T., and Manabe, O. (1978b). *Chem. Lett.*, 583–586. Micellar Catalysis of Cyanide Ion-Assisted Isoalloxazine (Flavin Analog) Oxidation of Aldehydes to Carboxylic Acids.

Shinkai, S., Yamashita, T., Kusano, Y., Ide, T., and Manabe, O. (1980a). *J. Am. Chem. Soc.* **102**, 2335–2340. Coenzyme Models. 20. Flavin Trapping of Carbanion Intermediates as Catalyzed by Cyanide Ion and Cationic Micelle.

Shinkai, S., Kusano, Y., and Manabe, O. (1980b). *J. Chem. Soc. Perkin Trans. II*, 1111–1115. Coenzyme Models. Part 24. Micellar Catalysis of Flavin-Mediated Reactions. Influence of the Flavin Structure on Reactivity.

Shinkai, S., Yamashita, T., Kusano, Y., and Manabe, O. (1980c). *J. Org. Chem.* **45**, 4947–4952. Coenzyme Models. 26. Facile Oxidation of Aldehydes and α-Keto Acids by Flavin as Catalyzed by Thiazolium Ion and Cationic Micelle.

Shinkai, S., Nakaji, T., Nishida, Y., Ogawa, T., and Manabe, O. (1980d). *J. Am. Chem. Soc.* **102**, 5860–5865. Photoresponsive Crown Ethers. 1. Cis-Trans Isomerism of Azobenzene as a Tool to Enforce Conformational Changes of Crown Ethers and Polymers.

Shinkai, S., Shigematsu, K., Ogawa, T., Minami, T., and Manabe, O. (1980e). *Tetrahedron Lett.* 4463–4466. Ion Extraction by a Crown Ether with a Photoresponsive Carboxylate Cap.

Shinkai, S., Yamada, S., Ando, R., and Kunitake, T. (1980f). *Bioorg. Chem.* **9**, 238–247. A Flavoenzyme Model: Facile Oxidation of Thiols by a Flavin Immobilized in Cationic Polyelectrolytes.

Shinkai, S., Makaji, T., Ogawa, T., Shigematsu, K., and Manabe, O. (1981a). *J. Am. Chem. Soc.* **103**, 111–115. Photoresponsive Crown Ethers. 2. Photocontrol of Ion Extraction and Ion Transport by a Bis(crown ether) with a Butterfly-like Motion.

Shinkai, S., Hirakawa, S., Shimomura, M., and Kunitake, T. (1981b). *J. Org. Chem.* **46**, 868–872. Decarboxylation of 6-Nitrobenzisoxazole-3-carboxylate Anion Catalyzed by Systematically Quaternized Poly(4-vinylpyridines). "Average Side-Chain Length" as a Useful Index for Relative Hydrophobicity of the Polymer Domain.

Simon, J., Le Moigne, J., Markovitsi, D., and Dayantis, J. (1980). *J. Am. Chem. Soc.* **102**, 7247–7252. Annelides. 3. Complexation of Dioxygen in Organized Cobaltous Complex Assemblies. A New Approach to Kinetic Studies in Micellar Phases.

Smith, G. D., Garrett, B. B., Holt, S. L., and Barden, R. E. (1977). *Inorg. Chem.* **16**, 558–562. Properties of Metal Complexes in the Interphase of an Oil Continuous Microemulsion. 2. Interaction of Copper(II) with the Side Chains of Lysine, Glutamine, and Methionine.

Smith, G. D. Barden, R. E., and Holt, S. L. (1978). *J. Coord. Chem.* **8**, 157–160. Properties of Metal Complexes in the Interphase of an Oil Continuous Microemulsion. 3. Interaction of Copper(II) with the Side Chain of Tryptophan.

Sousa, L. R. and Larson, J. M. (1977). *J. Am. Chem. Soc.* **99**, 307–310. Crown Ether Model Systems for the Study of Photoexcited State Response to Geometrically Oriented Perturbers. The Effect of Alkali Metal Ions on Emission from Naphthalene Derivatives.

Spinelli, D., Noto, R., Consiglio, G., Werber, G., and Buccheri, F. (1978). *J. Org. Chem.* **43**, 4042–4044. Studies on Decarboxylation Reactions. 3. Micellar Catalysis in the Decarboxylation of 5-Amino-1,3,4-thiadiazole-2-carboxylic Acid.

Sprintschnik, G., Sprintschnik, H. W., Kirsch, P. P., and Whitten, D. G. (1976). *J. Am. Chem. Soc.* **98**, 2337–2338. Photochemical Cleavage of Water: A System for Solar Energy Conversion Using Monolayer-Bound Transition Metal Complexes.

Sprintschnik, G., Sprintschnik, H. W., Kirsch, P. P., and Whitten, D. G. (1977). *J. Am. Chem. Soc.* **99**, 4947–4954. Preparation and Photochemical Reactivity of Surfactant Ruthenium(II) Complexes in Monolayer Assemblies of Water-Solid Interfaces.

Srivastava, S. K. and Katiyar, S. S. (1980). *Ber. Bunsenges. Phys. Chem.* **84**, 1214–1219. Micellar Effects upon the Reaction of Setoglaucin Carbocation with Cyanide Ion.

Stillwell, W. and Karimi, S. (1980). *Biochem. Biophys. Res. Commun.* **95**, 1049–1055. Effect of Various Water-Soluble Components of Phospholipids on the Photooxidation of Chlorophyllin.

Stott, P. E., Bradshaw, J. S., and Parish, W. W. (1980). *J. Am. Chem. Soc.* **102**, 4810–4815. Modified Crown Ether Catalysts, 3. Structural Parameters Affecting Phase Transfer Catalysis by Crown Ethers and a Comparison of Effectiveness of Crown Ethers to that of Other Phase Transfer Catalysts.

Sudo, Y. and Toda, F. (1979a). *J. Chem. Soc. Chem. Commun.*, 1044–1045. Photoinduced Reduction of Potassium Hexacyanoferrate(III) by Iron(II) Chloride in the Liposome System.

Sudo, Y. and Toda, F. (1979b). *Nature* **279**, 807. Photoinduced Electron Transport Across Phospholipid Wall of Liposome Using Methylene Blue.

Sudo, Y., Kawashima, T., and Toda, F. (1980). *Chem. Lett.*, 355–358. The Trans-membrane Electron Transport Coupled with Dye Redox Cycle in the Liposome System.

Sunamoto, J. and Hamada, T. (1978). *Bull. Chem. Soc. Jpn.* **51**, 3130–3135. Solvochromism and Thermochromism of Cobalt(II) Complexes Solubilized in Reversed Micelles.

Sunamoto, J. and Hamada, T. (1979). In *Fundamental Research in Homogeneous Catalysis* (M. Tsutsui, Ed.), Vol. 3, Plenum Press, New York, pp. 809–818. Amino Acid Ester Hydrolysis as Promoted by "Naked" Copper(II) Ion and "Free" Water in Reversed Micelles.

Sumamoto, J. and Horiguchi, D. (1980). *Nippon Kagaku Kaishi*, 475–481. Control of the Reaction Path in Reversed Micelles. Acid-Catalyzed and Noncatalyzed Reactions of Hydrazobenzene-3,3′-disulfonic Acid.

Sunamoto, J., Kondo, H., Okamoto, H., and Taira, K. (1977). *Bioorg. Chem.* **6**, 95–102. A Sulfotransferase Model. Covalent Participation of a Macrocyclic Oxime in the Hydrolysis of 2,4-Dinitrophenyl Sulfate.

Sunamoto, J., Kondo, H., and Akimaru, K. (1978). *Chem. Lett.*, 821–824. Enhanced Reactivities of "Naked" Metal Ion and Water Afforded in Reversed Micelles.

Sunamoto, J., Kondo, H., Yanase, F., and Okamoto, H. (1980a). *Bull. Chem. Soc. Jpn.* **53**, 1361–1365. Kinetic Studies of the *N,N*-Type Smiles Rearrangement.

Sunamoto, J., Kondo, H., Hamada, T., Yamamoto, S., Matsuda, Y., and Murakami, Y. (1980b). *Inorg. Chem.* **19**, 3668–3673. Formation of Polynuclear Cupric Halides in Cationic Reversed Micelles.

Suzuki, A., and Tada, M. (1980). *Chem. Lett.*, 515–516. Effect of Guest Cations on the Photosolvolysis of 4′-(1-Acetoxy)ethylbenzo-15-crown-5.

Tabushi, I., Fujita, K., and Yuan, L. C. (1977). *Tetrahedron Lett.*, 2503–2506. Specific Host-Guest Energy Transfer by Use of β-Cyclodextrin Rigidly Capped with a Benzophenone Moiety.

Tabushi, I., Kuroda, Y., Fujita, K., and Kawakubo, H. (1978). *Tetrahedron Lett.*, 2083–2086. Cyclodextrin as a Ligase-oxidase Model. Specific Allylation-oxidation of Hydroquinone Derivatives Included by β-Cyclodextrin.

Tabushi, I., Yamamura, K., Fujita, K., and Kawakubo, H. (1979). *J. Am. Chem. Soc.* **101**, 1019–1026. Specific Inclusion Catalysis by β-Cyclodextrin in the One Step Preparation of Vitamin K_1 or K_2 Analogues.

Tagaki, W., Kobayashi, S., Kurihara, K., Kurashima, A., Yoshida, Y., and Yano, Y. (1976). *J. Chem. Soc. Chem. Commun.*, 843–844. Effect of Cationic Micelles on the ElcB Mechanism of the Hydrolysis of Substituted *p*-Nitrophenyl Acetate.

Tagaki, W., Kobayashi, S., and Fukushima, D. (1977). *J. Chem. Soc. Chem. Commun.*, 29–30. Facile Acyl Transfer from Imidazole to the Hydroxy-Group in a Cationic Micelle in the Hydrolysis of *p*-Nitrophenyl Acetate.

Tagaki, W., Fukushima, D., Eiki, T., and Yano, Y. (1979). *J. Org. Chem.* **44**, 555–563. Functional Micelles. 8. Micellar Catalysis of the Hydrolysis of Substituted Phenyl Carboxylates by Imidazole Containing Cationic Surfactants.

Takayanagi, T., Nagamura, T., and Matsuo, T. (1980). *Ber. Bunsenges. Phys. Chem.* **84**, 1125–1129. Photoinduced Electron Transfer Between Amphipathic Ruthenium(II) Complex and *N*-Butylphenothiazine in Various Microenvironments.

Takeishi, M., Yoshida, H., Niino, S., and Hayama, S. (1978). *Macromol. Chem.* **179**, 1387–1391. Photopolymerization of Styrene in Water in the Presence of *N*-Laurylpyridinium Azide.

Takuma, K., Sakamoto, K., Nagamura, T., and Matsuo, T. (1981). *J. Phys. Chem.* **85**, 619–621. Novel Properties of the Self-Assembling Amphiphatic Viologen System. 1. A Study of Electron-Exchange Reactions in Micellar Systems.

Taniguchi, Y., Inoue, O., and Suzuki, K. (1979). *Bull. Chem. Soc. Jpn.* **52**, 1327–1329. Polymer Effects Under Pressure. III. Hydrolysis of Normal Alkyl Acetate Catalyzed by Dodecyl Hydrogen Sulfate Micelle.

Terpstra, W. (1980). *Biochim. Biophys. Acta* **600**, 36–47. Influence of Lecithin Liposomes on Chlorophyllase Catalyzed Chlorophyll Hydrolysis. Comparison of Intramembraneous and Solubilized Phaeodactylum Chlorophyllase.

Thomas, J. K. and Picuilo, P. (1978). *J. Am. Chem. Soc.* **100**, 3239–3240. Photoionization by Green Light in Micellar Solution.

Thomas, J. K. and Picuilo, P. (1979). *J. Am. Chem. Soc.* **101**, 2502–2503. Photoionization by Green Light.

Tiecke, B., Wegner, G., Naegele, D., and Ringsdorf, H. (1976). *Angew. Chem. Int. Ed. Eng.* **15**, 764–765. Polymerization of Tricosa-10,12-dyonic Acid in Multilayers.

Tien, H. T. (1979). In *Photosynthesis in Relation to Model Systems* (J. Barber, Ed.), Elsevier, Amsterdam, pp. 115–173. Photoeffects in Pigmented Bilayer Lipid Membranes.

Tien, H. T. (1980). *Sep. Sci. Technol.* **15**, 1035–1058. Energy Conversion via Pigmented Bilayer Lipid Membranes.

Tomida, H., Yotsuyanagi, T., and Ikeda, K. (1978). *Chem. Pharm. Bull.* **26**, 148–154. Effect of Micellar Interactions on Base-Catalyzed Hydrolysis of Procaine.

Tonellato, U. (1976). *J. Chem. Soc. Perkin Trans. II*, 771–776. Catalysis of Ester Hydrolysis by Cationic Micelles of Surfactants Containing the Imidazole Ring.

Tonellato, U. (1977). *J. Chem. Soc. Perkin Trans. II*, 821–827. Functional Micellar Catalysis. Part 2. Ester Hydrolysis Promoted by Micelles Containing the Imidazole Ring and the Hydroxy-Group.

Tong, L. K. J. and Glesmann, M. C. (1977). *J. Am. Chem. Soc.* **99**, 6991–6995. Kinetics of Dye Formation by Oxidative Coupling with a Micelle-Forming Coupler.

Toyoshima, Y., Morino, M., Motoki, H., and Sukigara, M. (1977). *Nature* **265**, 187–189. Photo-oxidation of Water in Phospholipid Bilayer Membranes Containing Chlorophyll a.

Trainor, G. L. and Breslow, R. (1981). *J. Am. Chem. Soc.* **103**, 154–158. High Acylation Rates and Enantioselectivity with Cyclodextrin Complexes of Rigid Substrates.

Tran, C. D. and Beddard, G. S. (1981). *Proc. Natl. Acad. Sci. USA*, to be published. Circularly Polarized Luminescence and Picosecond Fluorescence Lifetimes of Bilirubin.

Tsunooka, M. and Tanaka, M. (1978). *J. Polym. Sci. Polym. Lett.* **16**, 119–120. Photochemical Reactions of High Polymers. XVII. Micellar Effect on the Photopolymerization of Methacrylate with Ferrocene-Carbon Tetrachloride.

Tsutsui, Y., Takuma, K., Nishijima, T., and Matsuo, T. (1979). *Chem. Lett.*, 617–620. Photoreduction and Redox Catalysis of an Amphiphathic Derivative of Tris(2,2′-bipyridine)ruthenium(II) at the Micellar Surface.

Tundo, P. and Fendler, J. H. (1980). *J. Am. Chem. Soc.* **102**, 1760–1762. Photophysical Investigations of Chiral Recognition in Crown Ethers.

Tundo, P., Kurihara, K., Prieto, N. E., Kippenberger, D. J., Politi, M. J., and Fendler, J. H. (1982). *J. Am. Chem. Soc.*, to be published. Chemically Dissymmetrical Polymerized Surfactant Vesicles. Syntheses and Utilization in Artificial Photosynthesis.

Tunuli, M. S. and Fendler, J. H. (1981). *J. Am. Chem. Soc.* **103**, 2507–2513. Aspects of Artificial Photosynthesis. Photosensitized Electron Transfer Across Bilayers, Charge Separation and Hydrogen Production in Anionic Surfactant Vesicles.

Tunuli, M. S. and Fendler, J. H. (1982a). To be published, Advances in Chemistry Series, Vol. XX, American Chemical Society, Washington, D.C. Aspects of Artificial Photosynthesis. The Role of Potential Gradients in Promoting Charge Separation in the Presence of Surfactant Vesicles.

Tunuli, M. S. and Fendler, J. H. (1982b). *J. Phys. Chem.*, to be published. Aspects of Artificial Photosynthesis. Field Effects, Kinetics, Energetics and Mechanisms of Photoinduced Charge Separation in the Absence and in the Presence of Highly Charged Surfactant Vesicles.

Uekama, K., Hirayama, F., and Yamasaki, S. (1978). *Bull. Chem. Soc. Jpn.* **51**, 1229–1230. Micellar Effects on Base-Catalyzed Isomerization of Prostaglandin A1 and Prostaglandin A2.

Ueno, A., Saka, R., and Osa, T. (1980a). *Chem. Lett.*, 29–32. Interactions of Organic Solvents with Photoresponsive Capped Cyclodextrin in Aqueous Solution.

Ueno, A., Takahashi, K., and Osa, T. (1980b). *J. Chem. Soc. Chem. Commun.*, 921–922. One Host–Two Guests Complexation Between γ-Cyclodextrin and Sodium α-Naphthylacetate as Shown by Excimer Fluorescence.

Ueno, A., Takahashi, K., and Osa, T. (1981a). *J. Chem. Soc. Chem. Commun.*, 94–95. Photocontrol of Catalytic Activity of Capped Cyclodextrin.

Ueno, A., Takahashi, K., Hino, Y., and Osa, H. (1981b). *J. Chem. Soc. Chem. Commun.*, 194–195. Fluorescence Enhancement of α-Naphthyloxyacetic Acid in the Cavity of γ-Cyclodextrin, Assisted by a Space Regulating Molecule.

Ueoka, R. and Matsumoto, Y. (1980). *Chem. Lett.*, 35–38 (1980). Structural Effect of Mixed Micelles in the Deacylation of *p*-Nitrophenyl Esters by Specific Hydroxamic Acids.

Ueoka, R. and Ohkubo, K. (1978). *Tetrahedron Lett.*, 4131–4134. Catalytically Active Comicelles of Hydroxamic Acids and Surfactants for Selective Hydrolysis of Nonionic and Anionic Esters.

Ueoka, R., Kato, M., and Ohkubo, K. (1977). *Tetrahedron Lett.* **77**, 2163–2166. Hydrophobic Forces in the Stereoselective Hydrolyses of 3-Nitro-4-acyloxybenzoic Acid Substrates Catalyzed by Hydroxamic Acids.

Ueoka, R., Shimamoto, K., Maezato, Y., and Ohkubo, K. (1978). *J. Org. Chem.* **43**, 1815–1817. Hydrophobic Forces in Selective Hydrolyses of Nonionic *p*-Nitrophenyl Ester and Anionic 3-Nitro-4-acyloxybenzoic Acid Substrates by Hydroxamic Acids.

Ueoka, R., Matsuura, H., Nakahata, S., and Ohkubo, K. (1980). *Bull. Chem. Soc. Jpn.* **53**, 347–352. The Micelle Promoted Selective Hydrolysis of Anionic and Nonionic Ester Substrates by Hydroxamic Acids and Histidine Derivatives.

Usui, Y., Kodera, S., and Nishida, Y. (1976). *Chem. Lett.*, 1329–1332. Photoreduction of Methylene Blue Bound to Sodium Dodecyl Sulfate Micelles.

Valenty, S. J. (1979). *J. Am. Chem. Soc.* **101**, 1–8. Chemical Reactions in Monolayer Films. Chromatography, a Multicompartment Trough, and the Hydrolysis of Surfactant Ester Derivatives of Tris(2,2′-bipyridine)ruthenium(II)$^{2+}$.

Van de Langkruis, G. B. and Engberts, J. B. F. N. (1979). *J. Org. Chem.* **44**, 141–143. Micellar Catalysis of Proton-Transfer Reactions. 2. Hydrolysis of Covalent *p*-Tolylsulfonylmethyl Perchlorate Catalyzed by Arenesulfinate Anions in the Presence of CTAB. Irrelevance of the Hydrophobicity of the Arene Moiety of the Sulfinate.

Vanderhoff, J. W., Bradford, E. B., Tarkowski, H. L., Shaffer, J. B., and Wiley, R. M. (1962). In Advances in Chemistry Series, Vol. **34**, American Chemical Society, Washington, D.C., pp. 32–51. *Inverse Emulsion Polymerization.*

Waka, Y., Hamamoto, K., and Mataga, N. (1978). *Chem. Phys. Lett.* **53**, 242–246. Pyrene-*N,N*-Dimethylaniline Heteroexcimer Systems in Aqueous Micellar Solutions.

Waka, Y., Hamamoto, K., and Mataga, N. (1980a). *Photochem. Photobiol.* **32**, 27–35. Heteroexcimer Systems in Aqueous Micellar Solutions.

Waka, Y., Mataga, N., and Tanaka, F. (1980b). *Photochem. Photobiol.* **32**, 335–340. Behavior of Heteroexcimer Systems in Single Bilayer Liposomes.

Wexler, D., Pillersdorf, A., Shiffman, R., Katzhendler, J., and Sarel, S. (1978). *J. Chem. Soc. Perkin Trans. II*, 479–486. Dipolar Micelles. Part 4. Effect of Catalytic Micelles on the Hydrolysis of Neutral and Positively Charged Esters.

Whitten, D. G., Eaker, D. W., Horsey, B. E., Schmell, R. H., and Worsham, P. R. (1978). *Ber. Bunsenges. Phys. Chem.* **82**, 858–867. Photochemical and Thermal Reactions of Porphyrins and Organic Surfactants in Monolayer Assemblies. Modification of Reactivity in Condensed Hydrophobic Microemulsions.

Willner, I., Ford, W. E., Otvos, J. W., and Calvin, M. (1979). *Nature* **280**, 823–824. Photoinduced Electron Transfer Across a Water-Oil Boundary as a Model for Redox Reaction Separation.

Willner, I., Ford, W. E., Otvos, J. W., and Calvin, M. (1981). To be published, Advances in Chemistry Series, Vol. XX, American Chemical Society, Washington, D.C. Endoenergetic Photoinduced Reactions in Water-in-Oil Microemulsions; The Photosensitized Reductions of Viologens by Thiophenol.

Winnik, M. A., Redpath, T., and Richards, D. H. (1980). *Macromolecules* **13**, 328–335. The Dynamics of End-to-End Cyclization in Polystyrene Probed by Pyrene Excimer Formation.

Winterle, J. S. and Mill, T. (1980). *J. Am. Chem. Soc.* **102**, 6336–6338. Free-Radical Dynamics in Organized Lipid Bilayers.

Wong, M. and Thomas, J. K. (1977). In *Micellization, Solubilization and Microemulsions* (K. L. Mittal, Ed.), Plenum Press, New York, pp. 647–664. Some Kinetic Studies in Reversed Micellar System-Aerosol-OT (Diisoctylsulfosuccinate)/H_2O/Heptane Solution.

Worsham, P. R., Eaker, D. W., and Whitten, D. G. (1978). *J. Am. Chem. Soc.* **100**, 7091–7093. Ketone Photoreactivity as a Probe of the Microenvironment: Photochemistry of the Surfactant Ketone 16-Oxo-16-*p*-tolylhexadecanoic Acid in Monolayers, Micelles, and Solution.

Yamada, K., Shigehiro, K., Tomizawa, T., and Iida, H. (1978). *Bull. Chem. Soc. Jpn.* **51**, 3302–3306. The Effect of Surfactants and β-Cyclodextrin on the Photooxidation of Stable Carbanions.

Yamada, K., Shosenji, H., and Ihara, H. (1979a). *Chem. Lett.*, 491–494. The Hydrolysis of *p*-Nitrophenyl of α-Amino Acids by *N*-Lauroyl L or D-Histidine in Cationic Micelles.

Yamada, K., Shosenji, H., Ihara, H., and Otsubo, Y. (1979b). *Tetrahedron Lett.*, 2529–2523. Enantioselectively Catalyzed Hydrolysis of *p*-Nitrophenyl Esters of *N*-Protected L-Amino Acids by *N*-Lauroyl L or D-Histidine in CTABr Micelles.

Yamagishi, A. and Watanabe, F. (1976a). *Chem. Lett.*, 143–146. Dimerization of Anion Radical in Micelles.

Yamagishi, A. and Watanabe, F. (1976b). *J. Colloid Interface Sci.* **59**, 181–183. Surfactant Effects on the Electron-Transfer Reaction between TCNQ Anion Radical and 2,3-Dichloro-5,6-dicyano-*p*-benzoquinone in Chloroform.

Yamagishi, A., Masui, T., and Watanabe, F. (1980). *J. Phys. Chem.* **80**, 34–40. Temperature-Jump Study of Aquation Equilibria of Cobalt(II) Ion Solubilized in Reversed Micelles.

Yamagishi, A., Masui, T., and Watanabe, F. (1981). *J. Phys. Chem.* **85**, 281–285. Selective Activation of Reactant Molecules by Reversed Micelles.

Yano, Y., Yoshida, Y., Kurashima, A., Tamura, Y., and Tagaki, W. (1979). *J. Chem. Soc. Perkin Trans. II*, 1128–1132. Micellar Catalyzed Elimination Reactions of *p*-Substituted Phenethyl Bromides and Related Compounds in Alkaline Solutions.

Young, R. C. and Feldberg, S. W. (1979). *Biophys. J.* **27**, 237–255. Photoinitiated Mediated Transport of H_3O^+ and/or OH^- Across Glycerol Monooleate Bilayers Doped with Magnesium Octaethylporphyrin.

Young, P. R. and Hou, K. C. (1979). *J. Org. Chem.* **44**, 947–950. Entropy-Related Rate Accelerations in the Micelle-Bound Carboxylate-Catalyzed Iodine Oxidation of Diethyl Sulfide.

CHAPTER 13

SOLAR ENERGY CONVERSION IN MEMBRANE MIMETIC SYSTEMS

Photochemical solar energy conversion is a vitally important and extremely active area of research. Key laboratories publish "a paper a week" (Rawls, 1981). The progress is annually reviewed (Archer, 1975, 1976, 1977, 1978, 1979). Photochemical solar energy conversion has been extensively treated in recent books and reviews (Porter and Archer, 1976; Claesson and Engström, 1977; Almgren, 1978; Calvin, 1978; Hautala et al., 1979; Barber, 1979; Gerischer and Katz, 1979; Whitten, 1980; Grätzel, 1980; Kiwi et al., 1981). Only the barest essentials are outlined, therefore, in this chapter. Thermodynamic and kinetic requirements for energy conversion and the principles of artifical photosynthesis are briefly delineated. Exploitation of potential gradients in membrane mimetic systems for enhanced electron transfer and charge separation is illustrated with reference to our own work. Reference should be made to the exhaustive data tabulations on energy and electron transfer and charge separation in aqueous (Table 12.7) and reversed (Table 12.9) micelles, microemulsions (Table 12.11), monolayers (Table 12.12), bilayers (Table 12.14), vesicles (Table 12.15), and polyelectrolytes (Table 12.19).

1. PRINCIPLES OF PHOTOCHEMICAL ENERGY STORAGE AND CONVERSION

The thermodynamics of converting photochemical energy to chemical energy are shown in reaction 13.1:

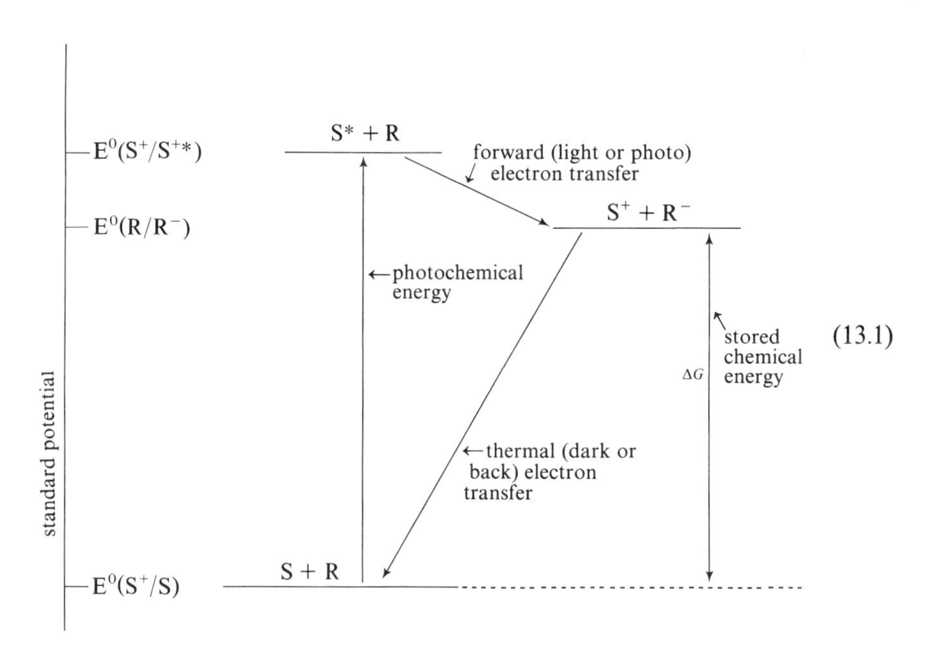

(13.1)

where S is the sensitizer and R is the electron relay. The excited state of the sensitizer is a better electron donor as well as a better electron acceptor than the ground state, S. Absorbtion of light (*hv*) can drive, therefore, a redox reaction nonspontaneously and result in the storage of energy, ΔG, in S^+ and R^-. In reaction 13.1 S* functions as an electron donor. S* may just as well accept an electron from R, which then functions as an electron donor. The point is illustrated by the behavior of the metal-to-ligand charge transfer excited state of the tris(2,2′-bipyridine)ruthenium cation ($Ru(bpy)_3^{2+*}$). This excited state is capable of reducing the methylviologen (paraquat) cation, MV^{2+},

$$Ru(bpy)_3^{2+*} + MV^{2+} \longrightarrow Ru(bpy)_3^{3+} + MV^{\dagger} \qquad (13.2)$$

or oxidizing *N*-methylphenothiazine, MPTH,

$$Ru(pby)_3^{2+*} + MPTH \longrightarrow Ru(bpy)_3^{+} + MPTH^{\dagger} \qquad (13.3)$$

A comparison of the reduction potentials of the redox couples participating in reaction 13.2, $E^0[Ru(bpy)_3^{3+}/Ru(bpy)_3^{2+}] = 1.26$ V and $E^0(MV^{2+}/MV^{\dagger}) = -0.44$ V, reveals this process to be endoergic by 1,7 eV with respect to $Ru(bpy)_3^{2+}$.

This energy provides the driving force for the thermal back electron transfer:

$$Ru(bpy)_3^{3+} + MV^{+\cdot} \longrightarrow Ru(bpy)_3^{2+} + MV^{2+} \quad (13.4)$$

Indeed in homogeneous solution the predominance of reaction 13.4 precludes the storage of photochemical energy in the charge separated species of $Ru(bpy)_3^{3+}$ and/or $MV^{+\cdot}$.

Ideally, the reduced relay and the oxidized sensitizer should be thermodynamically capable of generating hydrogen and oxygen from water:

$$2R^- + 2H_2O \longrightarrow 2R + 2OH^- + H_2\uparrow \quad (13.5)$$

(2 e^- reduction

$$4S^+ + 2H_2O \longrightarrow 4S + 4H^+ + O_2\uparrow \quad (13.6)$$

(4 e^- oxidation)

The thermodynamic requirements for water cleavage (reactions 13.5 and 13.6) are that $E^0(S/S^+) > 1.23$ V and $E^0(R/R^-) < 0$ V. Although a fair number of sensitizers (e.g., ruthenium, porphyrin, and acridin derivatives) and relay compounds (viologen, Eu^{3+}, V^{3+}, and Rh^{3+} derivatives) meet these requirements, the multielectron steps (two for reduction and four for oxidation) demand the use of catalysts that lower the energy barrier and/or alleviate the intermediates ($\cdot$H in reaction 13.5 and $\cdot$OH in reaction 13.6).

Kinetics of photosensitized electron processes are described by the general scheme:

$$S^* + R \underset{k_{-D}}{\overset{k_D}{\rightleftharpoons}} S^*R \underset{k_{-et}}{\overset{k_{et}}{\rightleftharpoons}} S^+|R^- \quad (13.7)$$

$$S^+|R^- \xrightarrow{k_t} S + R \qquad S^+|R^- \xrightarrow{k_s} S^+ + R^-$$

The first step, diffusion of the reaction partners together (governed by k_D), leads to the formation of a precursor complex ($S^*|R$). Electron transfer in this complex (k_{et}) results in the formation of a successor complex ($S^+|R^-$). This species may dissociate into ions S^+ and R^- (k_s) or undergo thermal back reaction to reform S and R (k_t). Applying the steady state approximation to reaction 13.7, the observed second order overall rate constant can be expressed by

$$k_{obs} = \frac{k_D}{1 + k_t/k_s + k_{-D}/k_{et}(1 + k_{et}/k_s + k_t/k_s)} \quad (13.8)$$

The value of k_{obs} for a given S and R interaction can be assessed on the basis of either classical or quantum mechanical theories. An approach for obtaining the calculated rate constant, $k_{\text{obs}}^{\text{calc}}$, is provided by equation 13.9 (Tunuli and Fendler, 1982):

$$k_{\text{obs}}^{\text{calc}} = \frac{k_D}{1 + \dfrac{k_D}{K_A(k_s + k_t)}\left[\exp\left(\Delta G\ddagger/RT\right) + \exp\left(\Delta G/RT\right)\right]} \tag{13.9}$$

where K_A is the association contant ($K_A = k_D/k_{-D}$), obtainable from

$$K_A = \frac{4\pi N a^3}{3000} \exp\left(\mu(a)/kT\right) \tag{13.10}$$

with $\mu(a) = z_1 z_2 d^2/a\varepsilon_s$, where a is the distance of closest approach, ε_s is the static dielectric constant, and z_1 and z_2 are the charges on the reacting species. The diffusion rate constant k_D in equation 13.9 can be calculated from

$$k_D = \frac{4\pi(D_1 + D_2)afN}{1000} \tag{13.11}$$

where the diffusion coefficient $D_i(D_i = D_1$ or $D_2)$ and electrostatic factor are given by

$$D_i = \frac{kT}{6\pi\eta r_i} \tag{13.12}$$

and

$$f = \frac{\mu(a)}{kT}\left[\exp\left(\frac{\mu a}{kT} - 1\right)\right] \tag{13.13}$$

where η is the viscosity of the medium and r_i is the radius of the species.

ΔG has been estimated from Marcus's (1964, 1965) theory by

$$\Delta G\ddagger = \frac{\Delta G}{2} + \left[\left(\frac{\Delta G}{2}\right)^2 + \Delta G\ddagger(0)^2\right]^{1/2} \tag{13.14}$$

$$\Delta G\ddagger(0) = \frac{e^2}{4}\left[\frac{1}{2r_1} + \frac{1}{2r_2} - \frac{1}{a}\right]\left(\frac{1}{\varepsilon_o} - \frac{1}{\varepsilon_s}\right) \tag{13.15}$$

where $\Delta G\ddagger(0)$ is the activation free energy for isoenergetic ($\Delta G = 0$) electron transfer situation and ε_o is the optical dielectric constant of the medium using adjusted parameter as proposed by Rehm and Weller (1970). Reasonable correlations have been obtained between the calculated and experimentally determined rate constants for these electron transfer processes (Bock et al., 1975; Carapelluci and Mauzerall, 1975; Scheerer and Grätzel, 1977; Tunuli and Fendler, 1982).

Efficient photolytic water decomposition in the light of these thermodynamic and kinetic requirements imposes manifold restrictions on the selections of S and R. Absorption of the sensitizer should correspond to the solar spectrum. Its excited state should form with a high quantum yield, be sufficiently long-lived, and undergo efficient forward electron transfer, which results in charge separation. The redox potentials of the primary and, if needed, the secondary donors and acceptors should meet the thermodynamic requirements of reactions 13.5 and 13.6. For viable catalytic systems S, R, and all other donors and acceptors used should remain chemically stable (i.e., have high turnover, be nontoxic, and reasonably priced).

2. ARTIFICIAL PHOTOSYNTHESIS

Much has been learned by imitating natural photosynthesis. In plant photosynthesis light is absorbed by two pigment systems: photosystem I (PSI) and photosystem II (PSII) (Govindjee, 1975). PSI, believed to be on the outer side of the thylakoid membrane, is responsible for reducing carbon dioxide. PSII, considered to be on the inner side of the membrane, is involved in the oxidation of water to molecular oxygen. The transfer of electrons from PSII to PSI is coupled to the production of adenosine 5′-triphosphate ATP (Figure 13.1). The precise, yet incompletely understood, arrangements in the membrane are responsible for the efficient energy deposition and transmission, for prevention of charge recombination, and for creating a proton gradient essential for photophosphorylation (Boyer et al., 1977).

In an artificial system complexities of the thylakoid membrane are eliminated and PSI and PSII are substituted by the relay and sensitizer, which, subsequent to electron transfer, are catalytically regenerated:

O_2 S S^* R H_2
catalyst catalyst (13.16)
H_2O S^+ R^- H_2O

It is only very recently that artificial cyclic water splitting has been realized by using doped TiO_2 (Kiwi et al., 1980; Borgarello et al., 1981). Generally, equation 13.16 is simplified to design half cells where only H_2 or O_2 is produced. Since

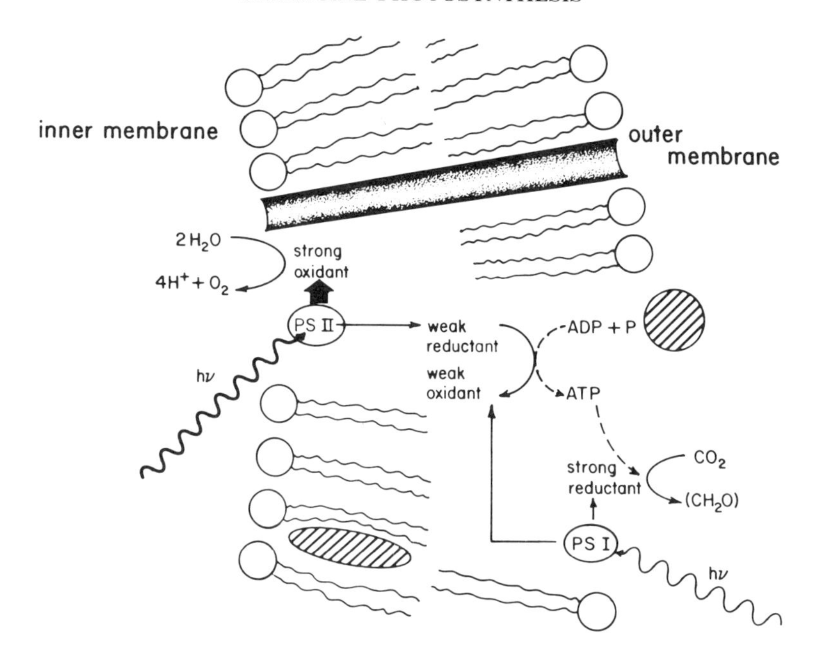

Figure 13.1 Schematic representation of a portion of the thylakoid membrane, thought to be involved in photosynthesis. Photosystem I (PSI), upon light ($h\nu$) absorption, produces a strong reductant and a weak oxidant. Photosystem II (PSII), upon light ($h\nu$) absorption, produces a strong oxidant and a weak reductant. Electron flow from the weak reductant to the weak oxidant is coupled to phosphorylation, which converts adenosine diphosphate (ADP) and inorganic phosphate (P) to adenosine triphosphate (ATP). With the aid of ATP the strong reductant produced by PSI reduces carbon dioxide to carbohydrate. The strong oxidant, produced by PSII, oxidizes water molecules to molecular oxygen and protons, released in the inner membrane. Protons can be transmitted through channels going through the bilayers (indicated by the dotted area). Cholesterol (▨) adds to the rigidity of the membrane and proteins (▨) may be bound to the surface (taken from Fendler, 1980).

these half cells are necessarily noncyclic, their use requires sacrificial electron donors (D) for H_2 production or acceptors (A) for O_2 production:

$$\begin{array}{ccc} D^+ & \curvearrowleft & D \\ S + R & \xrightarrow{h\nu} & S^+ + R^- \\ & PtO_2 & \\ H_2O & \curvearrowright & H_2 \end{array} \tag{13.17}$$

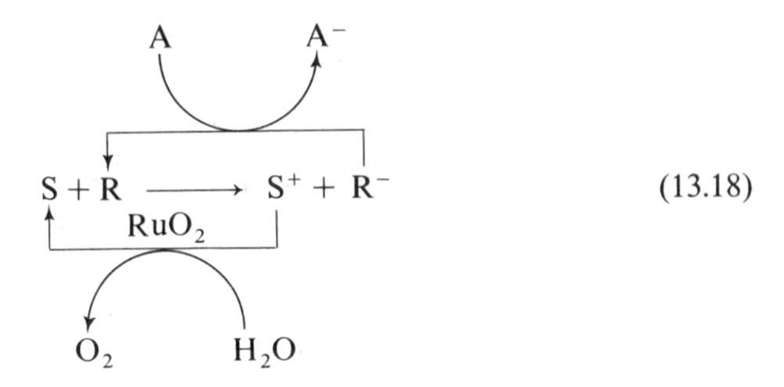

(13.18)

Investigations of these sacrificial half cells have provided valuable insights into the mechanisms of the various pathways indicated in equation 13.7. Membrane mimetic agents have been extensively utilized in these studies (see Tables 12.7, 12.9, 12.11, 12.32, 12.14, 12.15, and 12.19). The role of potential gradients in promoting electron transfer and charge separation is illustrated for surfactant vesicles in the final section of this chapter.

3. EXPLOITATION OF POTENTIALS FOR ENERGY AND ELECTRON TRANSFER AND CHARGE SEPARATION IN SURFACTANT VESICLES

Advantages of surfactant vesicles over other systems are that they are able to organize large numbers of sensitizers, electron donors, and acceptors per aggregate and that they are amenable to electrostatic modification and chemical functionalization. Importantly, unlike natural membranes, which are composed mostly of zwitterionic lipids, surfactant vesicles are highly charged and have high charge densities on their surfaces. These charges create appreciable surface, charge separation, diffusion, and Donnan potentials. Exploitation of the various potentials in photochemical solar energy conversion is illustrated by the following examples (Tunuli and Fendler, 1982).

Enhanced Energy and Electron Transfer on Charged Vesicle Surfaces

Localization of molecules in biological matrices is an essential requirement for many processes. Energy transfer *in vivo* photosynthesis is largely dependent, for example, on the precise location of chlorophyll molecules in the chloroplast (Porter, 1978a, 1978b). An average distance of approximately 15 Å between chlorophylls is considered to be ideal for efficient energy transfer without self-quenching. Ionic surfactant vesicles attract oppositely charged species onto their surfaces. Intravesicular energy and electron transfer readily occur in the potential field of the aggregate at reduced dimensionalities (Adam and Delbrück, 1968; Richter and Eigen, 1974; Eigen, 1974).

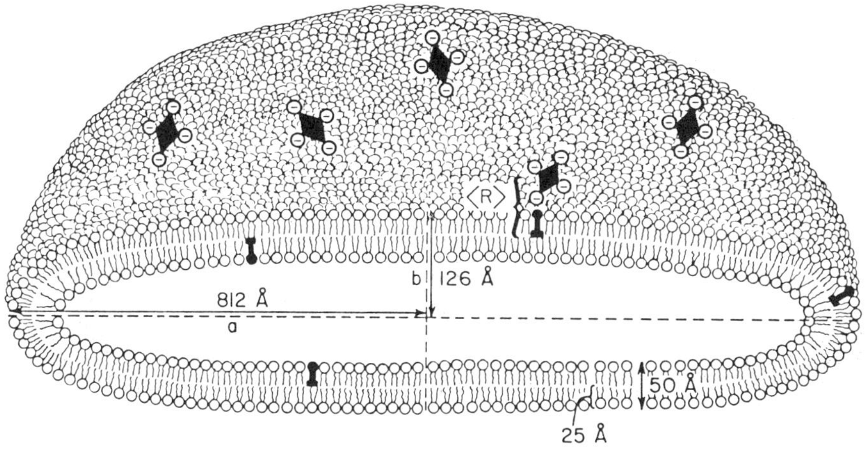

Figure 13.2 Schematic representation of a well-sonicated cationic DODAC surfactant vesicle, based on low-angle laser light scattering and photon correlation spectroscopy. Proposed positions of lysopyrene ( ) and pyranine ( ) and the average distance between them ($\langle R \rangle$) are also indicated (taken from Nomura et al., 1980).

Efficient intramolecular energy transfer has been observed in (DODAC), vesicles (Nomura et al., 1980). The donor, 2-hydroxy-1-[ω-(1-pyrene)decanoly]-*sn*-glycero-3-phosphatidylcholine (lysopyrene), was localized in the hydrophobic bilayer of the vesicles. The acceptor, trisodium 8-hydroxy-1,3,6-pyrenetrisulfonate (pyranine), having four negative charges, was attracted to the outer surface of positively charged DODAC vesicles (Figure 13.2). Depending on the concentration of pyranine, energy transfer efficiencies up to 43% have been observed (Nomura et al., 1980). Conversely, energy transfer efficiencies in the absence of vesicles were less than 3%. The apparent rate constant for energy transfer quenching, $6.2 \times 10^{11}\ M^{-1}\ \text{sec}^{-1}$, is the consequence of an approximate thousandfold increase of acceptor concentration on the vesicle surface.

Efficient photosensitized electron transfer has also been observed from $Ru(bpy)_3^{2+*}$ to methylviologen (reaction 13.2) on the outer and inner surfaces of anionic dihexadecylphosphate (DHP) surfactant vesicles (Systems III and IV, respectively, in Figure 13.3; Tunuli and Fendler, 1981a). The apparent rate constant for reaction 13.2, $(4\text{–}5) \times 10^{11}\ M^{-1}\ \text{sec}^{-1}$, in Systems III and IV is three orders of magnitude greater than that found in water ($2 \times 10^{8}\ M^{-1}\ \text{sec}^{-1}$). Electron transfer is likely to occur by diffusion or hopping on the vesicle surface, Under typical conditions approximately 60 molecules of $Ru(bpy)_3^{2+}$ and 300 molecules of MV^{2+} associate with each DHP vesicle. Taking charge repulsions into consideration, average areas for $Ru(pby)_3^{2+}$ and MV^{2+} molecules are estimated to be 400 $Å^2$ and 200 $Å^2$, respectively. Since the surface area of a DHP vesicle is 1.2×10^{77} $Å^2$ (Herrmann and Fendler, 1979), the maximum area the reactive partners need to cover prior to collision is only 200 $Å^2$. This value is orders of magnitude smaller than 10^5 $Å^2$ estimated for $\bar{d}^2$, the square of the

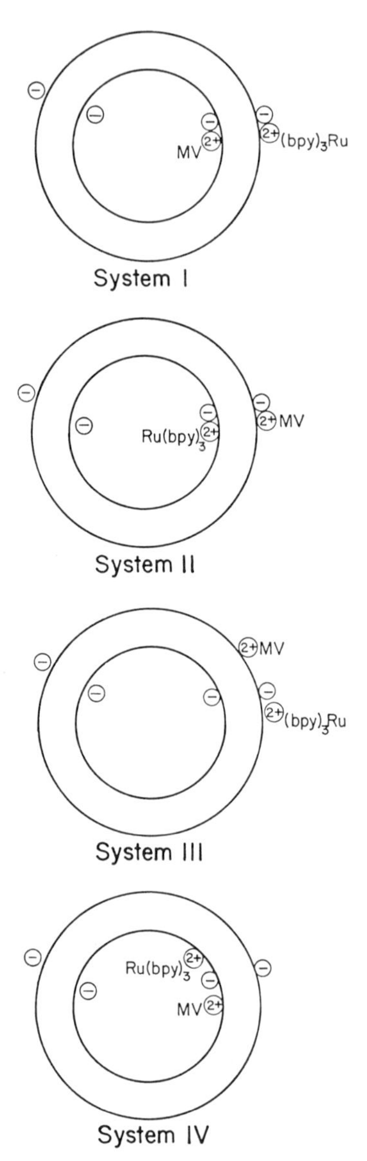

Figure 13.3 Schematics of the different arrangements of the sensitizer, tris(2,2′-bipyridine)ruthenium cation ($Ru(bpy)_3^{2+}$), and acceptor methylviologen (MV^{2+}), on DHP surfactant vesicle surfaces (taken from Tunuli and Fendler, 1981a).

mean diffusive displacement of $Ru(bpy)_3^{2+}$ and MV^{2+} (Rodgers and Becker, 1980). Thus the reactive partners can readily find each other on the surface of the DHP vesicles within their lifetimes. Unfortunately, the close proximity also results in much enhanced back reaction (reaction 13.4). Different organization is needed, therefore, to accomplish the desired efficiency in energy conversion, that is, to enhance the rate of the forward electron transfer (reaction 13.2, for example)

Table 13.1. Calculated and Observed Rate Constants for the Electron Transfer $Ru(bpy)_3^{2+*}$ + L-Cysteine → $Ru(bpy)_3^+$ + L-Cysteine$^+$ in DODAC Vesicles

pH	k_{obs}^{exp}, M^{-1} sec^{-1}	k_{obs}^{cal}, M^{-1} sec^{-1}	Calculated[a]	
			ΔG, kcal mole^{-1}	$\Delta G\ddagger$, kcal mole^{-1}
3.0	3.4×10^7	6.7×10^7	−25.8	1.4
8.4	8.2×10^7	1.3×10^8	−36.4	1.0
11.5	1.2×10^8	1.7×10^8	−44.3	0.8

[a] Calculated by means of equation 13.9, using $K(k_s + k_t) = 6.8 \times 10^8$ M^{-1} sec^{-1} as adjustable parameter.

and at the same time reduce the back reaction (reaction 13.4, for example). Exploitation of potentials to accomplish this goal are illustrated in the following sections.

Influence of Field Effect

Since electron transfer rates are directly related to the field, a judicious manipulation of the distance of a sensitizer and an electron acceptor (or donor) from a highly charged surface across the Stern layer is expected to result in altered efficiencies. This expectation has been realized in achieving effective charge separation under the influence of a positive electric field, generated by DODAC vesicles (Tunuli and Fendler, 1981c). The rate constant for electron transfer from L-cysteine to the excited state of $Ru(bpy)_3^{2+}$,

$$Ru(bpy)_3^{2+*} + \text{L-cysteine} \longrightarrow Ru(bpy)_3^{+} + \text{L-cysteine}^{+} \quad (13.19)$$

has been determined by laser flash photolysis (Tunuli and Fendler, 1981b). Satisfactory agreement has been obtained between the experimentally observed rate constants, k_{obs}^{exp}, and those calculated, k_{obs}^{calc} (equation 13.9), on the basis of the presence of an electric field (Table 13.1).

Considering the assumptions involved, the agreement between k_{obs}^{exp} and k_{obs}^{calc} is quite remarkable.

Effects of Electrolyte Gradients on the Partitioning Between the Inner and the Outer Compartments of Radicals Expelled from Vesicle Bilayers

Exit of photogenerated species from the bilayers of surfactant vesicles can be directed preferentially to the bulk solution (as opposed to the inner compartment of the vesicle) by setting up suitable potentials. Electron transfer from MPTH solubilized within the hydrophobic bilayers of DODAC surfactant vesicles to a long chain derivative of tris(2,2′-bipyridine)ruthenium cation $RuC_{18}(bpy)_3^{2+}$, anchored onto the inner and outer surfaces of DODAC, has been examined (Infelta et al., 1980). Electron transfer resulted in the formation of *N*-methylphenothiazine cation radical, $MPTH^{\ddagger}$:

$$RuC_{18}(bpy)_3^{2+*} + MPTH \longrightarrow RuC_{18}(pby)_3^{\ddagger} + MPTH^{\ddagger} \quad (13.20)$$

The $MPTH^{\ddagger}$ formed can disappear by a geminate-type of back electron transfer at the very site of its creation:

$$RuC_{18}(bpy)_3^{\ddagger} + MPTH^{\ddagger} \xrightarrow{\text{geminate}} RuC_{18}(bpy)_3^{2+} + MPTH \quad (13.21)$$

or, due to the potential gradient, exit into the DODAC entrapped water pool or into the bulk solution. The $MPTH^{\ddagger}$ expelled into the bulk solution is long lived since electrostatic repulsion between this species and the positively charged vesicle surface decreases the probability of back reaction. Preferential expulsion of $MPTH^{\ddagger}$ is accomplished by the addtion of NaCl to the outside of already formed vesicles. This has three important consequences. First, the number of sites where the local electrostatic field prevented the existence of $MPTH^{\ddagger}$ is reduced. Second, a dissymmetry is created between the inner and outer surface potential of the vesicles, which will increase the fraction of $MPTH^{\ddagger}$ exiting into the bulk solution. Third, the reduced net charge on the aggregates increases the rate of back reaction. The amount of $MPTH^{\ddagger}$ produced and the amount expelled into the bulk aqueous solutions were maximized in the presence of 1.0×10^{-3} *M* NaCl. Under this condition there was still a sufficient electrostatic repulsion between $MPTH^{\ddagger}$ and the charged surface of the vesicles to slow down considerably the undesirable charge recombination reactions (Infelta et al., 1980).

The Role of Potential Gradients Across Bilayers to Facilitate Electron Transfer

Creation of appropriate potentials may also assist electron transfer across surfactant vesicle bilayers. Electron transfer from $Ru(bpy)_3^{2+*}$ to MV^{2+} (reaction 13.2) has been examined by placing the sensitizer on the outer surface and the electron acceptor in the inner surface (System I in Figure 13.3) or *vice versa* (System II Figure 13.3) of anionic DHP surfactant vesicles (Tunuli and Fendler, 1981a). System I is much more efficient than System II. In System I all the negative charges on the inner surface of DHP vesicles are neutralized by MV^{2+}, whereas there is only partial neutralization of the outer surface by $Ru(bpty)_3^{2+}$. The gradient, created by the diffusion and Donnan potentials, were though to facilitate the flow of electrons from the outer to the inner surface of the vesicle. More recently, it became apparent that electron transfer occurs on the same surface of DHP vesicles, subsequent to photoinduced transmembrane diffusion of the acceptor (Lee *et al.*, 1982). Addition of EDTA to aqueous solution of System I resulted in the reformation of $Ru(bpy)_3^{2+}$. If PtO_2 is incorporated in the interiors of DHP vesicles of the same system, $MV^{\ddagger}$ is reoxidized with concomitant hydrogen formation

$$2MV^{\ddagger} + H_2O \xrightarrow{PtO_2} H_2 + 2OH^- + 2MV^{2+} \qquad (13.22)$$

Photolysis of this system leads, therefore, to the net consumption of only EDTA at very low stoichiometric $Ru(bpy)_3^{2+}$, MV^{2+}, and PtO_2 concentrations (Tunuli and Fendler, 1981a).

An even more efficient hydrogen generating system has been realized in a chemically dissymmetrical surfactant vesicle prepared from a redox active polymerizable surfactant (Figure 6.27; Tundo et al., 1982). Efficient electron transfer has been observed from photosensitized $Ru(bpy)_3^{2+}$, attached electrostatically to the outer surface of the dissymmetrical vesicles, to the functional viologen moiety localized in the vesicle interiors (Tundo et al., 1982).

REFERENCES

Adam, G. and Delbrück, M. (1968). In *Structural Chemistry and Molecular Biology* (A. Rich and N. Davidson, Eds.), Freeman & Co., San Francisco, pp. 198–215. Reduction of Dimensionality in Biological Diffusion Processes.

Almgren, M. (1978). *Photochem. Photobiol.* **27**, 603–609. Thermodynamic and Kinetic Limitations on the Conversion of Solar Energy into Storable Chemical Free-Energy.

Archer, M. D. (1975). In *Photochemistry, Specialists Periodical Report*, Vol. 6, The Chemical Society, London, pp. 737–764. Photochemical Aspects of Solar Energy Conversion.

Archer, M. D. (1976). In *Photochemistry, Specialists Periodical Report*, Vol. 7, The Chemical Society, London, pp. 559–584. Photochemical Aspects of Solar Energy Conversion.

Archer, M. D. (1977). In *Photochemistry, Specialists Periodical Report*, Vol. 8, The Chemical Society, London, pp. 569–590. Photochemical Aspects of Solar Energy Conversion.

Archer, M. D. (1978). In *Photochemistry, Specialists Periodical Report*, Vol. 9, The Chemical Society, London, pp. 601–622. Photochemical Aspects of Solar Energy Conversion.

Archer, M. D. (1979). In *Photochemistry, Specialists Periodical Report*, Vol. 10, The Chemical Society, London, pp. 611–629. Photochemical Aspects of Solar Energy Conversion.

Barber, J. (1979). *Photosynthesis in Relation to Model Systems*, Elsevier, New York.

Bock, C. R., Meyer, T. J., and Whitten, D. G. (1975). *J. Am. Chem. Soc.* **97**, 2909–2911. Photochemistry of Transition Metal Complexes. The Mechanism and Efficiency of Energy Conversion by Electron-Transfer Quenching.

Borgarello, E., Kiwi, J., Pelizetti, E., Visca, M., and Grätzel, M. (1981). *Nature* **289**, 158–160. Photochemical Cleavage of Water by Photocatalysis.

Boyer, P. D., Chance, B., Ernster, L., Mitchell, P., Racker, E., and Slater, E. C. (1977). *Annu. Rev. Biochem.* **46**, 957–966. Coupling Mechanisms in Capture, Transmission, and Use of Energy.

Calvin, M. (1978). *Acc. Chem. Res.* **11**, 369–374. Simulating Photosynthetic Quantum Conversion.

Carapelluci, P. A. and Mauzerall, D. (1975). *Ann. New York Acad. Sci.* **244**, 214–238. Photosynthesis and Porphyrin Excited State Redox Reactions.

Claesson, S. and Engström, M. (1977). *Solar Energy—Photochemical Conversion and Storage*, National Swedish Board for Energy Source Development, Stockholm.

Eigen, M. (1974). In *Quantum Statistical Mechanics in the Natural Sciences* (B. Kuseneglu, S. L. Mintz, and S. Widmayer, Eds.), Plenum Press, New York, pp. 37–58. Reduction of Dimensionality.

Fendler, J. H. (1980). *J. Phys. Chem.* **84**, 1485–1491. Microemulsions, Micelles and Vesicles as Media for Membrane Mimetic Photochemistry.

Gerischer, H. and Katz, J. J. (1979). *Light Induced Charge Separation in Biology and Chemistry*, Verlag Chemie, New York.

Govindjee (1975). *Bioenergetics of Photosynthesis*, Academic Press, New York.

Grätzel, M. (1980). *Angew. Chem. Int. Ed. Eng.* **84**, 981–991. Photochemical Methods for the Conversion of Light into Chemical Energy.

Hautala, R. R., King, R. B., and Kutal, C. (1979). *Solar Energy, Chemical Conversion and Storage*, The Humana Press, Clifton, N.J.

Herrmann, U. and Fendler, J. H. (1979). *Chem. Phys. Lett.* **64**, 279–274. Low Angle Laser Light Scattering and Photon Correlation Spectroscopy in Surfactant Vesicles.

Infelta, P. P., Grätzel, M., and Fendler, J. H. (1980). *J. Am. Chem. Soc.* **102**, 1479–1483. Aspects of Artificial Photosynthesis. Photosensitized Electron Transfer and Charge Separation in Cationic Surfactant Vesicles.

Kiwi, J., Borgarello, E., Pelizetti, E., Visca, M., and Grätzel, M. (1980). *Angew. Chem. Int. Ed. Eng.* **19**, 646–647. Cylic Water Decomposition by Visible Light: Drastic Increase in the Yield of Hydrogen and Oxygen with Difunctional Redox Catalysis.

Kiwi, J., Kalyanasundaram, K., and Grätzel, M. (1981). *Visible Light Induced Cleavage of Water into Hydrogen and Oxygen in Colloidal Microheterogeneous Systems*, Springer-Verlag, Heidelberg, Germany.

Lee, L. L.-C., Hurst, J. K., Politi, M., Kurihara, K. (1982). *J. Am. Chem. Soc.*, in press. Photoinduced Diffusion of Methyl Viologen Across Anionic Surfactant Vesicle Bilayers.

Marcus, R. A. (1964). *Annu. Rev. Phys. Chem.* **15**, 155–196. Chemical and Electrochemical Electron-Transfer Theory.

Marcus, R. A. (1965). *J. Chem. Phys.* **43**, 679–701. On the Theory of Electron-Transfer Reactions. VI. Unified Treatment for Homogeneous and Electrode Reactions.

Nomura, T., Escabi-Perez, J. R., Sunamoto, J., and Fendler, J. H. (1980). *J. Am. Chem. Soc.* **102**, 1484–1488. Aspects of Artificial Photosynthesis. Energy Transfer in Cationic Surfactant Vesicles.

Porter, G. (1978a). *Proc. Roy. Soc. London Ser. A* **362**, 281–303. In Vitro Models for Photosynthesis.

Porter, G. (1978b). *Pure Appl. Chem.* **50**, 263–271. Pure and Applied Photochemistry.

Porter, G. and Archer, M. (1976). *Interdiscip. Sci. Rev.* **1**, 119–143. In Vitro Photosynthesis.

Rawls, R. (1981). *Chem. Eng. News*, June 15, 26–27. Membrane Models Used to Probe Photosynthesis.

Rehm, D. and Weller, A. (1970). *Isr. J. Chem.* **8**, 259–271. Kinetics of Fluorescence Quenching by Electron and H-Atom Transfer.

Richter, P. H. and Eigen, M. (1974). *Biophys. Chem.* **2**, 255–263. Diffusion Controlled Reaction Rates in Spheroidal Geometry. Application to Repressor-Operator Association and Membrane Bound Enzymes.

Rodgers, M. A. J. and Becker, J. C. (1980). *J. Phys. Chem.* **84**, 2762–2768. Electron-Transfer Quenching of the Luminescent State of the Tris(bipyridyl)ruthenium(II) Complex in Micellar Media.

Scheerer, R. and Grätzel, M. (1977). *J. Am. Chem. Soc.* **99**, 865–871. Laser Photolysis Studies of Duroquinone Triplet State Electron Transfer Reactions.

Tundo, P., Kurihara, K., Prieto, N. E., Kippenberger, D. J., Politi, M., and Fendler, J. H. (1982). J. *Am. Chem. Soc.*, to be published. Chemically Dissymmetrical Polymerized Surfactant Vesicles. Syntheses and Utilization in Artificial Photosynthesis.

Tunuli, M. S. and Fendler, J. H. (1981a). *J. Am. Chem. Soc.* **103**, 2507–2513. Aspects of Artificial Photosynthesis. Photosensitized Electron Transfer Across Bilayers, Charge Separation and Hydrogen Production in Anionic Surfactant Vesicles.

Tunuli, M. S. and Fendler, J. H. (1982). To be published, Advances in Chemistry Services, Vol. XX, American Chemical Society, Washington, D.C. Aspects of Artificial Photosynthesis. The Role of Potential Gradients in Promoting Charge Separation in the Presence of Surfactant Vesicles.

Tunuli, M. S. and Fendler, J. H. (1981b). Unpublished results.

Whitten, D. G. (1980). *Acc. Chem. Res.* **13**, 83–90. Photoinduced Electron Transfer Reactions of Metal Complexes in Solution.

CHAPTER 14

MISCELLANEOUS APPLICATIONS

Membrane mimetic agents are extensively used in a large variety of fields. The subject has blossomed into maturity. The present treatment would be incomplete without drawing attention to the vitally important utilization of organized assemblies in such diverse areas as drug delivery, analytical chemistry, tertiary oil recovery, molecular self-organization, transport, and recognition. Each of these areas could merit a full book-length treatment. Indeed numerous comprehensive books and review articles have recently become available. This chapter consists, therefore, only of an annotated list of references to the available summaries. This should by no means be construed to lessen the significance and potential of applied membrane mimetic chemistry.

1. DRUG ENCAPSULATION

The idea of using a drug carrier is relatively simple. A suitable carrier would efficiently deliver the drug to its target where it would be released, preferentially, in a controllable fashion. This would allow the use of low dosages and hence it would diminish harmful toxic, allergic, and immunological side effects. Subsequent to delivery, the carrier would biodegrade without any deleterious effect. This description is, of course, idealized. In reality, to date, no such perfect system has been developed. Investigations of liposomes and surfactant vesicles as potential drug carriers has been prompted by their biodegradabilities. Primary publications, focused mostly on applications, have appeared in ever increasing number. Table 14.1 presents an annotated guide to the large number of reviews and books that summarize the results and provide insight into developing target directed carrier systems. The utilization of liposomes as target directed drug carriers is hampered by their nonselective distributions. Meaningful design of suitable systems will have to await the elucidation of factors affecting drug entrapment and retainment in vesicles as well as that of their *in vivo* fate at the molecular level (Fendler, 1980b).

Table 14.1. Annotated References to Reviews and Books Related to Drug Carriers in Membrane Mimetic Systems

Annotation	Reference
Effects of additives in liposomes are reviewed	Sadao, 1972
Interactions of antibiotics with liposomes are reviewed (53 refs.)	Bangham *et al.*, 1971
Potential of drug and enzyme entrapment in liposomes and their targeting are reviewed	Gregoriadis, 1973
Enzyme entrapments in liposomes are summarized (22 refs.)	Gregoriadis, 1974
Membrane moderated biocompatible drug delivery systems are reviewed	Michaels, 1974
Potential and types of membrane mediated drug delivery are reviewed (42 refs.)	Richardson, 1975
Uses of liposomes as drug carriers and cell modifiers are reviewed (45 refs.)	Ryman, 1975
Immunogenecity of liposomes is discussed (8 refs.)	Kinsky, 1976
Preparation, properties, and application of liposomes and their interactions with cells are reviewed (225 refs.)	Tyrrell et al., 1976
Applications of liposomes in pharmacology are reviewed (27 refs.)	Tentsova et al., 1976
Use of liposomes as delivery agents for drugs, vaccines, and hormones is summarized (29 refs.)	Colley and Ryman, 1976
Use of liposomes in target directed drug delivery is reviewed (198 refs.)	Gregoriadis, 1976a, 1976b
Bioavailability of drugs and their complexation with membranes are summarized (42 refs.)	Gaber and Rondelet, 1976
Applications of carrier mediated drug transport are discussed (110 refs.)	Gregoriadis, 1977
Drug-membrane-carrier interactions and mechanisms are discussed (37 refs.)	Tritton et al., 1977
Use of liposomes as drug carriers is reviewed (16 refs.)	Wielinga, 1977
Immunological properties of model membranes is reviewed (70 refs.)	Kinsky and Nicolotti, 1977
Use of liposomes as drug carriers is reviewed (97 refs.)	Fendler and Romero, 1977
Use of liposomes for the administration of drugs is reviewed (19 refs.)	Lichtenberg, 1977
Liposomes as carriers are surveyed	Puisieux et al., 1977
Applications of liposomes are surveyed	Marx, 1978

Table 14.1 *(Continued)*

Annotation	Reference
Interactions of liposomes with mammalian cells as related to target directed drug delivery are reviewed (184 refs.)	Pagano and Weinstein, 1978
Drug carriers are surveyed (40 refs.)	Trouet, 1978
Liposomes as protein carriers and their medical applications are surveyed (15 refs.)	Gregoriadis, 1978
Medical applications of liposomes are reviewed (54 refs.)	Bachhawat et al., 1978
Properties and biological effects of liposomes and their uses in pharmacology and toxicology are reviewed (220 refs.)	Kimelberg and Mayhew, 1978
Use of liposomes for target directed drug delivery is reviewed (43 refs.)	Ladygina et al., 1978
Use of liposomes, macromolecules, albumins, microspheres, and nanocapsules as lysosomotropic carriers is reviewed (101 refs.)	Couvreur et al., 1978
Use of membranes for drug administration is reviewed (56 refs.)	Nakano, 1978
Proceedings of the New York Academy of Sciences Symposium on the properties of liposomes and their uses as drug carriers	Papahadjopoulos, 1978
Liposomes as carriers for biologically active molecules are surveyed (75 refs.)	Theoharides, 1978
Use of liposomes as nontoxic and degradable drug carriers is discussed (78 refs.)	Dousset et al., 1979
Liposomes as drug carriers are surveyed (28 refs.)	Giomini et al., 1979
Use of liposomes as carriers for cancer drugs is reviewed (47 refs.)	Yamanaka et al., 1979
Use of liposomes in biology and medicine is surveyed (5 refs.)	Helgeland and Christiansen, 1979
Physical chemical characterization of liposomes as drug carriers is surveyed	Romero-Mella, 1979
Drug-liposome-membrane interactions are reviewed (60 refs.)	Sunamoto and Shiromita, 1979
Potential of liposomes as drug carriers in cancer chemotherapy is reviewed (59 refs.)	Kay and Richardson, 1979
Use of liposome encapsulated drugs to treat leishmaniasis is surveyed (11 refs.)	Alving and Steck, 1979
Use of liposomes as carriers of anti-inflammatory steroids is reviewed (101 refs.)	Knight and Shaw, 1979

Table 14.1 (*Continued*)

Annotation	Reference
Intra-articular liposomal therapy is surveyed (23 refs.)	Thomas et al., 1979
Use of vitamin B_{12} in target directed drug delivery is surveyed (28 refs.)	Fendler, 1979
Drug carriers in biology and medicine are summarized (211 refs.)	Gregoriadis, 1979
Use of liposomes in immunobiology is reviewed (82 refs.)	Uemura, 1979
Immunological applications of liposomes are reviewed (111 refs.)	Yusuda nad Tadakuma, 1979
Use of synthetic surfactant vesicles as drug carriers is surveyed	Fendler, 1980a
Preparation and properties of liposomes and their use as drug carriers are reviewed (295 refs.)	Ryman and Tyrrell, 1980
Use of liposomes as drug carriers is surveyed (77 refs.)	Kellaway, 1980
Physical chemical properties of liposomes and their use as drug carriers are surveyed (15 refs.)	Fuhrhop, 1980
Pharmacological properties of liposomes are reviewed (56 refs.)	Toffano and Bruni, 1980
Liposomes as drug delivery systems are reviewed	Juliano and Derek, 1980
Liposomes as drug carriers are highlighted (5 refs.)	Fifield, 1980
Therapeutic applications of liposomes are surveyed (62 refs.)	Ghosh and Bachhawat, 1980
Tailoring liposomes for drug carrying is highlighted (28 refs.)	Gregoriadis, 1980
Liposomes as carriers are highlighted	Saunders, 1980
Therapeutic potentials of liposomes in myocardial diseased are reviewed (47 refs.)	Wikman-Coffelt and Mason, 1980
Proceedings of a National Symposium on Liposomes and Immunology, March 14–15, 1980, Houston, Texas, 22 contributions	Tom and Six, 1980
Properties of liposomes and their therapeutic uses are treated in 14 contributions	Gregoriadis and Allison, 1980
Use of liposomes as drug carriers is highlighted (10 refs.)	Juliano, 1981
Newer applications of liposomes as carriers for pharmacological agents are reviewed (24 refs.)	Fraley and Papahadjopoulos, 1981

2. ANALYTICAL CHEMISTRY

The ability of membrane mimetic agents to solubilize reactants, to alter their microenvironments, and hence to affect their product distributions, stereochemistries, dissociation constants, redox potentials, and reactivities (see Chapters 9 and 10) have been exploited in a variety of analytical applications. Surfactants have been found to be useful for solubilizing analytical reagents, for enhancing given analytical reactions needed for quantitative determinations, for increasing sensitivities, stabilities, and selectivities. Existing methods have been improved and entirely new ones have been developed for ultraviolet-visible, atomic, and emission spectroscopic and electrochemical determinations. Room temperature observations of phosphorescence and delayed fluorescence in aqueous micellar solutions (Kalyanasundaram et al., 1977; Turro et al., 1978; Humphry-Baker et al., 1978; Turro and Aikawa, 1980; Skrilec and Love, 1980) illustrate nicely the power of the technique. Analytical applications of surfactants have been exhaustively treated by Hinze in his critical review containing 340 references (Hinze, 1979) and in his forthcoming book (Hinze, 1982).

3. OTHER APPLICATIONS

Surfactants play crucial roles in enhanced oil recovery. They reduce oil-water interfacial tensions to extremely low values. This in turn mobilizes oil entrapped by capillary forces. Schumacher's (1978) book should be consulted for technological details and the review by Bansal and Shah (1977) for appropriate colloid chemistry.

Membrane mimetic agents can be used as models for investigating self-assembly and self-replication (Deamer, 1978). A relatively simple model has been proposed for the origin of the genetic code (Nagyvary and Fendler, 1974) and for the accumulation of chemicals from the primordial ocean (Stillwell, 1980). The proposed physical chemical model for the primitive codon assignment (Nagyvary and Fendler, 1974), based on the suggested composition of the primordial (Lasaga et al., 1971), assumed the formation of micelles at the ocean hydrocarbon interface. Micelles provided compartments into which amino acids, nucleotides, and nucleosides may be taken up selectively. The selectivity depends on polar and hydrophobic interactions. Additionally, since micelles are known to catalyse reactions (Fendler and Fendler, 1975), amino acid and nucleotide condensation and polymerization may well be enhanced in this medium. Verification of the proposed hypothesis required, in part, information on the partitioning of amino acids, nucleosides, and nucleotides between bulk solvents and micelles and the demonstration of enhanced polycondensations. Selective partitionings of cationic amino acids (Fendler et al., 1975), nucleotides, and nucleosides (Nagyvary et al., 1976) have been demonstrated in micelles. Similarly, both nucleotides (Armstrong et al., 1977) and amino acids (Armstrong et al., 1978) were shown to undergo efficient polycondensations in micellar environments.

Black lipid membranes (BLMs) have been used for modeling the transport of ions and small molecules across the cell walls (see Chapter 5). This type of studies can be profitably extended to liposomes and surfactant vesicles. Additionally, appropriate systems can be designed for mimicking recognition (Tabushi et al., 1979), receptor-protein interactions (Citri and Schramm, 1980), and information transmission (Hirata and Axelrod, 1980). It has been suggested that phospholipid micelles are responsible for the transport of solutes across the biological membranes, and hence are a central part of energy coupling (Green et al., 1980).

Exploration of these and related, as yet unrecognized, applications of membrane mimetic agents are fully expected in the immediate future. Only lack of imagination will limit this highly multidisciplinary approach.

REFERENCES

Alving, C. R. and Steck, E. A. (1979). *Trends Biochem. Sci.* **4**, N175–177. The Use of Liposome-Encapsulated Drugs in Leishmaniasis.

Armstrong, D. W., Nome, F., Fendler, J. H., and Nagyvary, J. (1977). *J. Mol. Evol.* **9**, 213–224. Novel Prebiotic Systems: Nucleotide Oligomerization in Surfactant Entrapped Water Pools.

Armstrong, D. W., Seguin, R., McNeal, C. J., Macfarlane, R. D., and Fendler, J. H. (1978). *J. Am. Chem. Soc.* **100**, 4605–4606. Spontaneous Polypeptide Formation from Amino Acyl Adenylates in Surfactant Aggregates.

Bachhawat, B. K., Surolia, A., Dorai, T. D., and Poder, S. K. (1978). *Perspect. Incl. Microbiol., Proc. Symp. 1st*, pp. 100–112. Affinity Characteristics of Artificial Cells—Medical Applications.

Bangham, A., Cohen, E., Hill, M., Johnson, S., and Singer, M. (1971). *Mol. Mech. Antibiot. Action Protein Biosynth. Membr., Proc. Symp.* (publ. 1972) (E. Munoz, F. Garcia-Ferrandiz, and D. Vazquez, Eds.), Elsevier, Amsterdam, pp. 694–706. Model Membranes and an Occasional Antibiotic.

Bansal, V. K. and Shah, D. O. (1977). In *Micellization, Solubilization and Microemulsions* (K. L. Mittal, Ed.), Plenum Press, New York, pp. 87–113. Micellar Solutions for Improved Oil Recovery.

Citri, Y. and Schramm, M. (1980). *Nature* **287**, 297–300. Resolution, Reconstitution and Kinetics of the Primary Action of a Hormone Receptor.

Colley, M. C. and Ryman, B. E. (1976). *Trends Biochem. Sci.* **1**, 203–205. The Liposome: From Membrane Model to Therapeutic Agent.

Coster, H. G. L. and Hope, A. B. (1974). *Proc. Aust. Physiol. Pharmacol. Soc.* **5**, 2–9. Membrane Models and the Physical Properties of Biological Membranes.

Couvreur, P., Kante, B., and Roland, M. (1978). *Pharm. Acta Helv.* **53**, 341–347. Perspectives in the Use of the Microdispersed Form as an Intracellular Vehicle.

Deamer, D. W. (1978). *Light Transducing Membranes. Structure, Function, and Evolution*, Academic Press, New York.

Dousset, J. C., Dousset, N., Soula, G., and Douste-Blazy, C. (1979). *Lyon Pharm.* **30**, 83–91. Liposomes: A Therapeutic Future?

Fendler, J. H. (1979). *Vitam. B_{12}, Proc. Eur. Symp. 3rd*, pp. 695–710. Vitamin B_{12} in Membrane Mimetic Agents—Theoretical Considerations and Practical Applications.

Fendler, J. H. (1980a). *Acc. Chem. Res.* **13**, 7–13. Surfactant Vesicles as Membrane Mimetic Agents: Characterization and Utilization.

Fendler, J. H. (1980b). In *Liposomes in Biological Systems* (G. Gregoriadis, Ed.), John Wiley, New York, pp. 87–100. Optimizing Drug Entrapment in Liposomes. Chemical and Biophysical Considerations.

Fendler, J. H. and Fendler, E. J. (1975). *Catalysis in Micellar and Macromolecular Systems*, Academic Press, New York.

Fendler, J. H. and Romero, A. (1977). *Life Sci.* **20**, 1109–1120. Liposomes as Drug Carriers.

Fendler, J. H., Nome, F. J., and Nagyvary, J. (1975). *J. Mol. Evol.* **6**, 215–232. Compartmentalization of Amino Acids in Surfactant Aggregates. Partitioning Between Water and Aqueous Sodium Dodecanoate and Between Hexane and Dodecylammonium Propionate Trapped Water in Hexane.

Fifield, R. (1980). *New Sci.* **88**, 150–153. Liposomes: Bags of Biological Potentials.

Fraley, R. and Papahadjopoulos, D. (1981). *Trends Pharmacol. Sci.*, 77–80. New Generation Liposomes: The Engineering of an Efficient Vehicle for Intracellular Delivery of Nucleic Acids.

Fuhrhop, J. (1980). *Nachr. Chem., Tech. Lab.* **28**, 792–794, 796–797. Synthetic Vesicle with Mono- or Bilayer Membranes.

Gaber, J., and Rondelet, J. (1976). *J. Pharm. Belg.* **31**, 451–466. Bioavailability of Drugs. Complexity of Biological Membranes and Methods of "In Vitro" Stimulation.

Ghosh, P. and Bachhawat, B. K. (1980). *J. Sci. Ind. Res.* **39**, 689–696. Therapeutic Application of Liposomes.

Giomini, M., Giulani, A. M., Gattegno, D., and Conti, F. (1979). *Farmaco, Ed. Prat.* **34**, 3–14. New Type of Carriers for Biologically Active Substances: Liposomes.

Green, D. E., Fry, M., and Blondin, G. A. (1980). *Proc. Natl. Acad. Sci. USA* **77**, 257–261. Phospholipids as the Molecular Instruments of Ion and Solute Transport in Biological Membranes.

Gregoriadis, G. (1973). *New Sci.*, 890–893. Molecular Trojan Horses.

Gregoriadis, G. (1974). In *Insolubilized Enzymes* (M. Salmona, G. Saronio, C. Garatini, and S. Raven, Eds.), New York, pp. 165–177. Enzyme or Drug Entrapment in Liposomes. Possible Biomedical Applications.

Gregoriadis, G. (1976a). *New Engl. J. Med.* **293**, 704–710. The Carrier Potential of Liposomes in Biology and Medicine (Part 1).

Gregoriadis, G. (1976b). *New Engl. J. Med.* **295**, 765–770. The Carrier Potential of Liposomes in Biology and Medicine (Part 2).

Gregoriadis, G. (1977). *Nature* **265**, 407–411. Targeting of Drugs.

Gregoriadis, G. (1978). *Enzyme Eng.* **4**, 187–192. Liposomes as Carriers of Proteins: Possible Medical Applications.

Gregoriadis, G. (1979). In *Drug Carriers in Biology and Medicine*, (G. Gregoriadis, Ed.), Academic Press, London, pp. 287–341. Liposomes.

Gregoriadis, G. (1980). *Nature* **283**, 814. Tailoring Liposome Structure.

Gregoriadis, G. and Allison, A. C. (1980). *Liposomes in Biological Systems*, John Wiley, Chichester.

Helgeland, L. and Christiansen, E. N. (1979). *Kjemi* **39**, 22–24, 27. Liposomes—Important Aids in Biology and Medicine.

Hinze, W. L. (1979). In *Solution Chemistry of Surfactants* (K. L. Mittal, Ed.), Plenum Press, New York, pp. 79–127. Use of Surfactant and Micellar Systems in Analytical Chemistry.

Hinze, W. L. (1982), to be published. Use of Surfactants in Analytical Chemistry.

Hirata, F. and Axelrod, J. (1980). *Science* **209**, 1082–1090. Phospholipid Methylation and Biological Signal Transmission.

Humphry-Baker, R., Moroi, Y., and Grätzel, M. (1978). *Chem. Phys. Lett.* **58**, 207–210. Perturbation Studies of Photophysics of Arenes in Functionalized Micellar Assemblies. Drastic Phosphorescence Enhancements.

Juliano, R. L. (1981). *Trends Pharmacol. Sci.* **2**, 39–42. Liposomes as a Drug Delivery System.

Juliano, R. L. and Derek, L. (1980). In *Drug Delivery Systems* (R. L. Juliano, Ed.), Oxford University Press, New York, pp. 189–236. Liposomes as a Drug Delivery System.

Kalyanasundaram, K., Grieser, F., and Thomas, J. K. (1977). *Chem. Phys. Lett.* **51**, 501–505. Room Temperature Phosphorescence of Aromatic Hydrocarbons in Aqueous Micellar Solutions.

Kay, S. B. and Richardson, U. J. (1979). *Cancer Chemother. Pharmacol.* **3**, 81–85. Potential of Liposomes as Drug-Carrier in Cancer.

Kellaway, W. (1980). *Manuf. Chem. Aerosol News* **51**, 33–34. Liposomes—Model Membranes for Modern Medicine.

Kimelberg, H. K. and Mayhew, E. G. (1978). *CRC Crit. Rev. Toxicol.* **6**, 25–79. Properties and Biological Effects of Liposomes and Their Uses in Pharmacology and Toxicology.

Kinsky, S. C. (1976). *J. Biochem.* **79**, 24P–25P. Immunogenecity of Liposomal Model Membranes Sensitized with *N*-subs. Phosphatidylethanol Amine Derivatives.

Kinsky, S. C. and Nicolotti, R. A. (1977). *Annu. Rev. Biochem.* **46**, 49–67. Immunological Properties of Model Membranes.

Knight, C. G. and Shaw, I. H. (1979). *Front. Biol.* **48** (Lysosomes Appl. Biol. Ther. V6), 575–599. Liposomes as Carriers of Anti-inflammatory Steroids.

Ladygina, G. A., Tentsova, A. I., and Zizina, O. S. (1978). *Farmatsiya* (*Moscow*) **27**, 52–57. Use of Liposomes for the Directed Delivery of Drugs to Organs and Tissues.

Lasaga, A. C., Holland, H. D., and Dwyer, M. J. (1971). *Science* **174**, 53–55. Primordial Oil Slick.

Lichtenberg, D. (1977). *Isr. Pharm. J.* **20**, 176–179. Use of Phospholipid Liposomes for the Administration of Drugs.

Marx, J. L. (1978). *Science* **199**, 1056–1058, 1128. Liposomes: Research Applications Grows.

Michaels, A. S. (1974). *Am. Chem. Soc., Coat. Plast. Chem.* **34**, 559–562. Therapeutic Systems for Controlled Administration of Drugs. New Applications of Membrane Science.

Nagyvary, J. and Fendler, J. H. (1974). *Origins Life* **5**, 357–362. Origin of the Genetic Code: A Physical-Chemical Model of Primitive Codon Assignment.

Nagyvary, J., Bradburg, E., Harvey, J. A., Nome, F., Armstrong, D., and Fendler, J. H. (1976). *Precambrian Res.* **3**, 509–516. Novel Prebiotic Model Systems. Interactions of Nucleosides and Nucleotides with Aqueous Micellar Sodium Dodecanoate.

Nakano, M. (1978). *Maku* **3**, 386–392. Use of Membranes for Drug Administration.

Pagano, R. E. and Weinstein, J. N. (1978). *Annu. Rev. Biophys. Bioeng.* **7**, 435–468. Interactions of Liposomes with Mammalian Cells.

Papahadjopoulos, D. (1978). *Ann. New York Acad. Sci.* **308**, 462 pp. Liposomes and Their Uses in Biology and Medicine.

Puisieux, F., Luong, T. T., and Moufti, A. (1977). *Pharm. Acta Helv.* **52**, 305–318. Liposomes, Possible Carriers of the Active Principles.

Richardson, K. T., Jr., (1975). *Trans. New Orleans Acad. Ophthalmol.* (*Symp. Glaucoma*) **23**, 50–67. Membrane-Controlled Drug Delivery.

Romero-Mella, A. F. (1979). *Diss. Abstr. Int. B* **39**, 4902; dissertation, Texas A & M University, College Station, Tex. Physical Chemical Characterization of Amphiphilic Vesicles and Drug Entrapment Therein.

Ryman, B. E. (1975). *Proc. Int. Congr. Pharmacol. 6th* (*Publ. 1976*) **5**, 91–103. The Use of Liposomes as Carriers of Drugs and Other Cell-Modifying Molecules.

Ryman, B. E., and Tyrrell, D. A. (1980). *Essays Biochem.* **16**, 49–98. Liposomes—Bags of Potential.

Sadao, T. (1972). *Yukagaku* **21**, 298. Liposomes, a Model for Biological Membranes.

Sadao, T. (1973). *Hyomen* **11**, 289, Liposomes.

Saunders, R. (1980). *S. African J. Sci.* **76**, 297–298. Liposomes as Carriers in Biological Systems.

Schumacher, M. M. (1978). *Enhanced Oil Recovery. Secondary and Tertiary Methods.* Noyes Data Corp., Park Ridge, N.J.

Skrilec, M. and Love, L. J. C. (1980). *Anal. Chem.* **52**, 1559–1564. Room Temperature Phosphorescence Characteristics of Substituted Arenes in Thallium Lauryl Sulfate Micelles.

Stillwell, W. (1980). *Origins Life* **10**, 277–292. Facilitated Diffusion as a Method for Selective Accumulation of Materials from the Primordial Oceans by a Lipid Vesicle Protocell.

Sunamoto, J. and Shiromita, M. (1979). *Hyomen* **17**, 357–373. Liposome as Model of Interaction Between Biomembrane and Drug.

Tabushi, I., Kobuke, Y., and Imuta, J. (1979). *Nucleic Acid Res.* **6**, 175–179. Molecular Recognition of Nucelotides by Means of Ionic Interactions in Hydrophobic Media.

Tentsova, A. J., Kovaleva, N. S., Yarova, E. A., and Ivkov, N. N. (1976). *Farmatsiya (Moscow)* **25**, 82–85. Liposomes and Possibilities of Their Use in Pharmacy and Pharmacology.

Theoharides, T. (1978). *Folia Biochim. Biol. Graeca* **14**, 11–21. Liposomes as Carriers of Biological Active Molecules.

Thomas, D., Page, P., and Phillips, N. C. (1979). *Front. Biol.* **48** (Lyosomes Appl. Biol. Ther. V6), 601–624. Intra-articular Liposomal Therapy.

Toffano, G. and Bruni, A. (1980). *Pharmacol. Res. Comm.* **12**, 829–845. Pharmacological Properties of Phospholipid Liposomes.

Tom, B. H. and Six, H. R. (1980). *Liposomes and Immunobiology (Proc. Nat. Symp.)*, Elsevier/North Holland New York.

Tritton, T. R., Murphee, S. A., and Sartorelli, A. C. (1977). *Biochem. Pharmacol.* **26**, 2319–2323. Characterization of Drug-Membrane Interactions Using Liposome Systems.

Trouet, A. (1978). *Eur. J. Cancer* **14**, 105–111. Increased Selectivity of Drugs by Linking to Carriers.

Turro, N. J. and Aikawa, M. (1980). *J. Am. Chem. Soc.* **102**, 4866–4870. Phosphorescence and Delayed Fluorescence in Micellar Solutions.

Turro, N. J., Liu, K. C., Chow, M. F., and Lee, P. (1978). *Photochem. Photobiol.* **27**, 523–529. Convenient and Simple Methods for the Observation of Phosphorescence in Fluid Solutions. Internal Heavy Atom and Micellar Effects.

Tyrrell, D. A., Heath, T. D., Colley, C. M., and Ryman, B. E. (1976). *Biochim. Biophys. Acta* **457**, 259–302. New Aspects of Liposomes.

Uemura, K. (1979). *Sinshu Igaku Zasshi* **27**, 659–671. Lipid Model Membranes (Liposomes) in Immunology.

Wielinga, B. Y. (1977). *Pharm. Weekbl.* **112**, 665–670. Liposomes as Drug Carriers.

Wikman-Coffelt, J. and Mason, D. T. (1980). *Adv. Heart Disease* **3**, 175–191. Liposomal Therapeutic Potential in Myocardial Infarction: Carrier-Transport Concept of Lipid Vesicles for Target Delivery of Drugs and Biological Substances into Diseased Heart Muscle.

Yamanaka, N., Koizumi, K., and Ota, K. (1979). *Gan to Kagaku Ryoho* **6**, 225–230. Cancer Chemotherapy by Considering the Specificity of Cancer Cells and Tissues: Improvement of the Effects of Cancer Chemotherapy Using Liposomes.

Yoshinori, T. and Kurihara, K. (1977). *Yukagaku* **26**, 597–605. Liposome System and Its Application.

Yusuda, T. and Tadakuma, T. (1979). *Taisha* **16** (11-gatsu Rinji Zokango), 1975–1985. Immunological Applications of Liposomal Membranes.

INDEX

The italicized numbers refer to Tables in which the information is contained.